Engineering Materials

This series provides topical information on innovative, structural and functional materials and composites with applications in optical, electrical, mechanical, civil, aeronautical, medical, bio- and nano-engineering. The individual volumes are complete, comprehensive monographs covering the structure, properties, manufacturing process and applications of these materials. This multidisciplinary series is devoted to professionals, students and all those interested in the latest developments in the Materials Science field, that look for a carefully selected collection of high quality review articles on their respective field of expertise.

More information about this series at http://www.springer.com/series/4288

Teik-Cheng Lim

Mechanics of Metamaterials with Negative Parameters

Springer

Teik-Cheng Lim
School of Science and Technology
Singapore University of Social Sciences
Singapore, Singapore

ISSN 1612-1317 ISSN 1868-1212 (electronic)
Engineering Materials
ISBN 978-981-15-6445-1 ISBN 978-981-15-6446-8 (eBook)
https://doi.org/10.1007/978-981-15-6446-8

This Springer imprint is published by the registered company Springer Nature Singapore Pte Ltd.
The registered company address is: 152 Beach Road, #21-01/04 Gateway East, Singapore 189721, Singapore

To Esther Ceng and Isabel Lim

Preface

Most monographs on metamaterials deal with electromagnetic aspects. Very few involve mechanical aspects. Although metamaterials need not necessarily exhibit negative properties, it is those that manifest negative behavior that provide wider options for designing materials that can outperform what can be achieved by conventional materials. This monograph introduces the reader to negative Poisson's ratio (auxetic) materials, negative thermal expansion (NTE) materials, negative compressibility (NC) materials, negative moisture expansion (NME) materials and negative stiffness (NS). In addition, sign-switching materials are also added. These are materials that can toggle between positive and negative characteristics. This monograph is suitable as a textbook for an introductory graduate course in mechanical metamaterials with negative properties, and can be used as a reference by researchers and practicing engineers in academia and industry.

Singapore Teik-Cheng Lim

Contents

Chapter 1
Introduction

Abstract This chapter gives a brief introduction to metamaterials starting from its definition to the generic (including electromagnetic) metamaterials, through mechanical metamaterial, and finally to the mechanics of metamaterials with special emphasis on negative properties.

Keywords Metamaterials · Mechanical metamaterials · Mechanics of metamaterials

1.1 Mechanical Metamaterials

From a combination of the Greek word $\mu\varepsilon\tau\acute{\alpha}$ (meta), which means "beyond," and the Latin word *materia*, which means "matter" or "material," today the term "metamaterials" refers to materials that are microarchitectured such that their effective properties are primarily controlled by the microstructural geometries rather than by the base materials. The term "metamaterials" has been popularly adopted within the broad field of electromagnetic metamaterial, which incorporates photonic, or optical, metamaterials. The repeating patterns of the optical metamaterials are smaller than the wavelengths of the phenomena they affect. There is no lack of resources in the field of electromagnetic metamaterials. Interested readers are referred to the following books in this area as of the year 2019 (Andreone et al. 2011; Borja 2017; Brener et al. 2019; Cai and Shalaev 2010; Caloz and Itoh 2006; Canet-Ferrer 2019; Capolino 2009; Chipouline and Küppers 2018; Choudhury 2017; Choudhury et al. 2016; Cui et al. 2010, 2016; Denz et al. 2010; Diest 2013; Eleftheriades and Balmain 2005; Engheta and Ziolkowski 2006; Hao and Mittra 2009; Hess and Gric 2018; Krowne and Zhang 2007; Lee et al. 2016; Lheurette 2013; Li and Huang 2013; Maier 2011; Maradudin 2011; Marqués et al. 2007; Munk 2009; Nair et al. 2018; Nakano 2016; Noginov and Podolskiy 2011; Pendry 2007; Ramakrishna and Grzegorczyk 2009; Rout and Sonkusale 2017; Sarychev and Shalaev 2007; Shvets and Tsukerman 2011; Smolyaninov 2018a, b; Solymar and Shamonina 2009; Tong 2018; Vanbésien 2012; Werner 2017; Werner and Kwon 2014; Zayats and Maier 2013; Zouhdi et al. 2002; Zouhdi et al. 2009).

T. Lim, *Mechanics of Metamaterials with Negative Parameters*, Engineering Materials,
https://doi.org/10.1007/978-981-15-6446-8_1

Alongside this progress is the development of mechanical metamaterials. As the name suggests, mechanical metamaterials refer to materials whose mechanical properties are due to the designed geometrical properties of the microstructure. The capability to produce microarchitectured materials is further facilitated by 3D printing and other technological advancements in additive manufacturing that are able to design material architecture at the microscale level. Due to the manner by which they are designed, metamaterials can be made to function in certain ways that cannot be achieved by conventional materials (Bertoldi et al. 2017; Che et al. 2017; Christensen et al. 2015; Dudek et al. 2018; Gao et al. 2018; Huang et al. 2019; Jackson et al. 2018; Matlack et al. 2018; Mirzaali et al. 2018a, b; Pan et al. 2019; Powell 2018; Sujardi et al. 2019; Vangelatos et al. 2019; Wu et al. 2019; Yang and Ma 2019; Zadpoor 2016; Zhao et al. 2019).

1.2 Mechanical Metamaterials with Negative Parameters

Although mechanical metamaterials, in general, are artificially microstructured for achieving desired mechanical properties, great interest has especially been placed on those with negative values. This is because metamaterials with negative parameters are not only able to perform in certain ways that cannot be accomplished by conventional materials, but whose counter-intuitive behavior opens up opportunities to design materials and structures with ground-breaking novelties that inspire inventions. Mechanical metamaterials with negative parameters include, but not limited to, the following:

- negative Poisson's ratio (NPR), or auxetic, materials
- negative thermal expansion (NTE) materials
- negative compressibility (NC) materials
- negative stiffness phases or (micro)structures
- negative hygroscopic expansion (NHE) or negative moisture expansion (NME) materials
- negative swelling materials.

In addition, mechanical metamaterials with negative properties have been attained by a combination of multiple negative properties. To date, sufficient progress has been made on doubly negative and triply negative materials. By focusing on the mechanics of metamaterials, it therefore implies that the subject matter is within the domain of mechanical metamaterials, and that the approaches adopted for obtaining the effective negative properties are primarily based on the principles of mechanics.

1.3 Doubly Negative and Triply Negative Metamaterials

As the name suggests, a doubly negative metamaterial is one which manifests two kinds of negative properties. One such category is the combination of auxetic and negative thermal expansion features. This has been demonstrated using the examples of connected triangles (Grima et al. 2007), natrolite (Grima et al. 2012), and interconnected array of rings and sliding rods (Lim 2017). Using a combination of bimaterial strips and rigid beams to form rectangular cells, it been shown that the rectangles transform into shapes that approximate re-entrant geometry, which is associated with negative Poisson's ratio (Li et al. 2016; Lim 2019a), and at the same time the transformation itself exhibits negative thermal expansivity. These microstructures were then extended such that the square cells can be thermally deformed to produce interconnected star network (Lim 2019b), which is known to be auxetic, while the transformation from square cells to star cells demonstrates negative thermal expansivity. In the category for the combination of negative thermal expansion and negative compressibility, this has been proven to exist in methanol monohydrate (Fortes et al. 2011) and a unimode metamaterial (Dudek et al. 2016), while examples of combined auxetic and negative compressibility have been established using 3D cellular structures (Grima et al. 2012) and wine-rack-like carbon allotropes and related poly(phenylacetylene) systems (Degabriele et al. 2019). Systems comprising both auxetic and negative stiffness elements have been developed (Hewage et al. 2016; Dudek et al. 2018), as do those that manifest both negative thermal and negative moisture expansivities; the latter has been termed negative hygrothermal expansion (NHTE) materials (Lim 2019c, d).

Although being less established in comparison with doubly negative metamaterials, triply negative metamaterials offer greater design options for the engineering and materials design practitioners. Some examples of triply negative metamaterials include the combination of negative thermal expansion, negative linear compressibility, and negative Poisson's ratio induced by specific topology in $Zn[Au(CN)_2]_2$ (Wang et al. 2017). A combination of auxetic, negative stiffness, and negative bulk modulus has been established by using instability (Jia and Wang 2019), while a system that exhibits negative thermal expansion, negative compressibility, and negative moisture expansion has been analyzed (Lim 2019e); the latter has been termed negative environmental expansion due to its reversed reaction in response to the trio of environmental fluctuations—temperature, pressure, and moisture.

1.4 International Conferences

Since the year 2004, a series of auxetics conferences has been held almost every year. Initially named the Workshop on Auxetics and Related Systems, the scope has been expanded over the years to include "other materials and models with 'negative' characteristics," with special attention on negative mechanical properties. Although

Table 1.1 List of auxetics conferences arranged by year

Year	Auxetics conference	Location
2004	Workshop on auxetics and related systems	Bedlewo, Poland
2005	2nd workshop on auxetics and related systems	Bedlewo, Poland
2006	International conference and 3rd workshop on auxetics and anomalous systems	Exeter, UK
2007	4th international workshop on auxetics and related systems	Msida, Malta
2008	2nd conference and 5th international workshop on auxetics and related systems	Bristol, UK
2009	6th international workshop on auxetics and related systems	Bolton, UK
2010	3rd international conference and the 7th international workshop on auxetics and related systems	Gozo, Malta
2011	8th workshop on auxetics and related systems	Szczecin, Poland
2012	4th international conference and 9th international workshop on auxetics and related systems	Bolton, UK
2014	5th international conference on auxetics and other materials and models with "negative" characteristics, and the 10th international workshop on auxetics and related systems	Poznan, Poland
2015	6th international conference on auxetics and other materials and models with "negative" characteristics, and the 11th international workshop on auxetics and related systems	Malta
2016	7th international conference on auxetics and other materials and models with "negative" characteristics, and the 12th international workshop on auxetics and related systems	Szymbark, Poland
2017	8th international conference on auxetics and other materials and models with "negative" characteristics, and the 13th international workshop on auxetics and related systems	Heraklion, Crete, Greece
2018	9th international conference on auxetics and other materials and models with "negative" characteristics, and the 14th international workshop on auxetics and related systems	Sheffield, UK
2019	10th international conference on auxetics and other materials and models with "negative" characteristics, and the 15th international workshop on auxetics and related systems	Bedlewo, Poland

a number of similarly named conferences have emerged in recent years, the most authoritative ones are the auxetics conference series. See Table 1.1.

1.5 A Note on the Nomenclature Adopted

The symbols adopted in this book follow their usual meaning in the literature. For example, it is well established that the symbol α has been used for indicating the coefficient of linear expansion (CTE) (e.g., Grima et al. 2007, 2012; Lim 2017;

Boley and Weiner 1997; Nowacki 1987; Hetnarski and Eslami 2009). However, the symbol β has been used in the past to indicate both the material compressibility (e.g., Fortes et al. 2011; Dudek et al. 2016; Grima et al. 2012; Degabriele et al. 2019; Fine and Millero 1973; Baughman et al. 1998; Miller et al. 2015; Lim 2017) as well as the coefficient of moisture expansion (CME) or the coefficient of hygroscopic expansion (CHE) (e.g., Lim 2019c, d; Kaw 2006; Mallick 2008; Gibson 2012; Lim 2018). As such, in this book, different symbols have to be assigned for material compressibility and the CME or CHE. Under some circumstances, it has been found to be convenient if the expansion coefficient is generally indicated by α. As such, the use of superscripts helps to differentiate the type of expansion coefficient. Hence, $\alpha^{(T)}$ refers to the CTE, in which "T" indicates thermal or temperature, while $\alpha^{(P)}$ the compressibility; the letter "P" denotes pressure (e.g., Lim 2019f, 2020). As for CME or CHE, $\alpha^{(C)}$ is adopted wherein the letter "C" corresponds to the moisture or hygroscopic concentration.

References

Andreone A, Cusano A, Cutolo A, Galdi V (eds) (2011) Selected topics in photonic crystals and metamaterials. World Scientific, Singapore

Baughman RH, Stafström S, Cui C, Dantas SO (1998) Materials with negative compressibilities in one of more dimensions. Science 279(5356):1522–1524

Bertoldi K, Vitelli V, Christensen J, van Hecke M (2017) Flexible mechanical metamaterials. Nat Rev Mater 2(11):17066

Boley BA, Weiner JH (1997) Theory of thermal stresses. Dover Publications, New York

Borja AL (ed) (2017) Metamaterials—Devices and applications. InTechOpen, London

Brener I, Liu S, Staude I, Valentine J, Holloway C (2019) Dielectric metamaterials: fundamentals, designs, and applications. Elsevier, Cambridge

Cai W, Shalaev VM (2010) Optical metamaterials: fundamentals and applications. Springer, New York

Caloz C, Itoh T (2006) Electromagnetic metamaterials: transmission line theory and microwave applications. Wiley, Hoboken

Canet-Ferrer J (ed) (2019) Metamaterials and metasurfaces. InTechOpen, London

Capolino F (ed) (2009) Metamaterials handbook: theory and phenomena of metamaterials. CRC Press, Boca Raton

Che K, Yuan C, Wu J, Qi HJ, Meaud J (2017) Three-dimensional-printed multistable mechanical metamaterials with a deterministic deformation sequence. J Appl Mech 84(1):011004

Chipouline A, Küppers F (2018) Optical metamaterials: qualitative models. Springer, Cham

Choudhury B (ed) (2017) Metamaterial inspired electromagnetic applications: role of intelligent systems. Springer, Singapore

Choudhury B, Menon A, Jha RM (2016) Active terahertz metamaterial for biomedical applications. Springer, Singapore

Christensen J, Kadic M, Kraft O, Wegener M (2015) Vibrant times for mechanical metamaterials. MRS Commun 5(3):453–462

Cui TJ, Smith D, Liu R (eds) (2010) Metamaterials: theory, design, and applications. Springer, New York

Cui TJ, Tang WX, Yang XM, Mei ZL, Jiang WX (2016) Metamaterials: beyond crystals, noncrystals, and quasicrystals. CRC Press, Boca Raton

Degabriele EP, Attard D, Grima-Cornish JN, Caruana-Gauci R, Gatt R, Evans KE, Grima JN (2019) On the compressibility properties of the wine-rack-like carbon allotropes and related poly(phenylacetylene) systems. Phys Status Solidi B 256(1):1800572
Denz C, Flach S, Kivshar YS (eds) (2010) Nonlinearities in periodic structures and metamaterials. Springer, Berlin
Diest K (ed) (2013) Numerical methods for metamaterial design. Springer, Dordrecht
Dudek KK, Attard D, Caruana-Gauci R, Wojciechowski KW, Grima JN (2016) Unimode metamaterials exhibiting negative linear compressibility and negative thermal expansion. Smart Mater Struct 25(2):025009
Dudek KK, Wojciechowski KW, Dudek MR, Gatt R, Mizzi L, Grima JN (2018a) Potential of mechanical metamaterials to induce their own global rotational motion. Smart Mater Struct 27(5):055007
Dudek KK, Gatt R, Dudek MR, Grima JN (2018b) Negative and positive stiffness in auxetic magneto-mechanical metamaterials. Proc R Soc A 474(2215):20180003
Eleftheriades GV, Balmain KG (eds) (2005) Negative-refraction metamaterials: fundamental principles and applications. Wiley, Hoboken
Engheta N, Ziolkowski RW (eds) (2006) Metamaterials: physics and engineering explorations. Wiley, Hoboken
Fine RA, Millero FJ (1973) Compressibility of water as a function of temperature and pressure. J Chem Phys 59(10):5529–5536
Fortes AD, Suard E, Knight KS (2011) Negative linear compressibility and massive anisotropic thermal expansion in methanol monohydrate. Science 331(6018):742–746
Gao Z, Liu D, Tomanek D (2018) Two-dimensional mechanical metamaterials with unusual Poisson ratio behavior. Phys Rev Appl 10(6):064039
Gibson RF (2012) Principles of composite material mechanics, 3rd edn. CRC Press, Boca Raton
Grima JN, Farrugia PS, Gatt R, Zammit V (2007) Connected triangles exhibiting negative Poisson's ratios and negative thermal expansion. J Phys Soc Jpn 76(2):025001
Grima JN, Gatt R, Zammit V, Cauchi R, Attard D (2012a) On the negative Poisson's ratio and thermal expansion in natrolite. In: Meunier M (ed) Industrial applications of molecular simulatons. CRC Press, Boca Raton, pp 135–152
Grima JN, Caruana-Gauci R, Attard D, Gatt R (2012b) Three-dimensional cellular structures with negative Poisson's ratio and negative compressibility properties. Proc Royal Soc A 468(2146):3121–3138
Hao Y, Mittra R (2009) FDTD modeling of metamaterials: theory and applications. Artech House, Norwood
Hess O, Gric T (2018) Phenomena of optical metamaterials. Elsevier, Amsterdam
Hetnarski RB, Eslami MR (2009) Thermal stresses—Advanced theory and applications. Springer, New York
Hewage TAM, Alderson KL, Alderson A, Scarpa F (2016) Double-negative mechanical metamaterials displaying simultaneous negative stiffness and negative Poisson's ratio properties. Adv Mater 28(46):10323–10332
Huang Y, Zhang X, Kadic M, Liang G (2019) Stiffer, stronger and sentrosymmetrical class of pentamodal mechanical metamaterials. Materials 12(21):3470
Jackson JA, Messner MC, Dudukovic NA, Smith WL, Bekker L, Moran B, Golobic AM, Pascall AJ, Duoss EB, Loh KJ, Spadaccini CM (2018) Field responsive mechanical metamaterials. Sci Advances 4(12):eaau6419
Jia Z, Wang L (2019) Instability-triggered triply negative mechanical metamaterial. Phys Rev Appl 12(2):024040
Kaw AK (2006) Mechanics of composite materials, 2nd edn. CRC Press, Boca Raton
Krowne CM, Zhang Y (eds) (2007) Physics of negative refraction and negative index materials: optical and electronic aspects and diversified approaches. Springer, Berlin
Lee YP, Rhee JY, Yoo YJ, Kim KW (2016) Metamaterials for perfect absorption. Springer, Singapore
Lheurette É (ed) (2013) Metamaterials and wave control. Wiley, Hoboken

Li J, Huang Y (2013) Time-domain finite element methods for Maxwell's equations in metamaterials. Springer, Berlin
Li D, Ma J, Dong L, Lakes RS (2016) A bi-material structure with Poisson's ratio tunable from positive to negative via temperature control. Mater Lett 181:285–288
Lim TC (2017a) Auxetic and negative thermal expansion structure based on interconnected array of rings and sliding rods. Phys Status Solidi B 254(12):1600775
Lim TC (2017b) 2D structures exhibiting negative area compressibility. Phys Status Solidi B 254(12):1600682
Lim TC (2018) A negative hygroscopic expansion material. Mater Sci Forum 928:277–282
Lim TC (2019a) A class of shape-shifting composite metamaterial honeycomb structures with thermally-adaptive Poisson's ratio signs. Compos Struct 226:111256
Lim TC (2019b) 2D metamaterial with in-plane positive and negative thermal expansion and thermal shearing based on interconnected alternating bimaterials. Mater Res Express 6(11):115804
Lim TC (2019c) A reinforced kite-shaped microstructure with negative linear and area hygrothermal expansions. Key Eng Mater 803:272–277
Lim TC (2019d) A composite metamaterial with sign switchable elastic and hygrothermal properties induced by stress direction and environmental change reversals. Compos Struct 220:185–193
Lim TC (2019e) Negative environmental expansion for interconnected array of rings and sliding rods. Phys Status Solidi B 256(1):1800032
Lim TC (2019f) Composite metamaterial with sign-switchable coefficients of hygroscopic, thermal and pressure expansions. Adv Compos Hybrid Mater 2(4):657–669
Lim TC (2020) Composite metamaterial square grids with sign-flipping expansion coefficients leading to a type of Islamic design. SN Applied Sciences 2(5):918
Maier SA (ed) (2011) World scientific handbook of metamaterials and plasmonics. World Scientific, Singapore
Mallick PK (2008) Fiber-reinforced composites, 3rd edn. CRC Press, Boca Raton
Maradudin AA (ed) (2011) Structured surfaces as optical metamaterials. Cambridge University Press, Cambridge
Marqués R, Martín F, Sorolla M (2007) Metamaterials with negative parameters: theory, design, and microwave applications. Wiley, Hoboken
Matlack KH, Serra-Garcia M, Palermo A, Huber SD, Daraio C (2018) Designing perturbative metamaterials from discrete models. Nat Mater 17(4):323–328
Miller W, Evans KE, Marmier A (2015) Negative linear compressibility in common materials. Appl Phys Lett 106(23):231903
Mirzaali MJ, Caracciolo A, Pahlavani H, Janbaz S, Vergani L, Zadpoor AA (2018a) Multi-material 3D printed mechanical metamaterials: rational design of elastic properties through spatial distribution of hard and soft phases. Appl Phys Lett 113(24):241903
Mirzaali MJ, Janbaz S, Strano M, Vergani L, Zadpoor AA (2018b) Shape-matching soft mechanical metamaterials. Scient Rep 8:965
Munk BA (2009) Metamaterials: critique and alternatives. Wiley, Hoboken
Nair RU, Dutta M, Mohammed Yazeen PS, Venu KS (2018) EM material characterization techniques for metamaterials. Springer, Singapore
Nakano H (2016) Low-profile natural and metamaterial antennas. Wiley, Hoboken
Noginov MA, Podolskiy VA (eds) (2011) Tutorials in metamaterials. CRC Press, Boca Raton
Nowacki W (1987) Thermoelasticity, 2nd edn. Pergamon Press, Oxford
Pan F, Li Y, Li Z, Yang J, Liu B, Chen Y (2019) 3D pixel mechanical metamaterials. Adv Mater 31(25):1900548
Pendry JB (2007) Fundamentals and applications of negative refraction in metamaterials. Princeton University Press, New Jersey
Powell D (2018) Mechanical metamaterials bend the rules of everyday physics. Proc Nat Acad Sci 115(11):2545–2547
Ramakrishna SA, Grzegorczyk TM (2009) Physics and applications of negative refractive index materials. CRC Press, Boca Raton

Rout S, Sonkusale S (2017) Active metamaterials: terahertz modulators and detectors. Springer, Cham
Sarychev AK, Shalaev VM (2007) Electrodynamics of metamaterials. World Scientific, Singapore
Shvets G, Tsukerman I (2011) Plasmonics and plasmonic metamaterials. World Scientific, Singapore
Smolyaninov II (2018a) Hyperbolic metamaterials. IOP Publishing, Bristol
Smolyaninov II (2018b) Metamaterial multiverse. IOP Publishing, Bristol
Solymar L, Shamonina E (2009) Waves in metamaterials. Oxford University Press, Oxford
Sujardi JU, Gao L, Du H, Li X, Xiong X, Fang NX, Lu Y (2019) Mechanical metamaterials and their engineering applications. Adv Eng Mater 21(3):1800864
Tong XC (2018) Functional metamaterials and metadevices. Springer, Cham
Vanbésien O (2012) Artificial materials. Wiley, Hoboken
Vangelatos Z, Komvopoulos K, Grigoropoulos CP (2019) Vacancies for controlling the behavior of microstructured three-dimensional mechanical metamaterials. Math Mech Solids 24(2):511–524
Wang L, Luo H, Deng S, Sun Y, Wang C (2017) Uniaxial negative thermal expansion, negative linear compressibility, and negative Poisson's ratio induced by specific topology in $Zn[Au(CN)_2]_2$. Inorg Chem 56(24):15101–15109
Werner DH (ed) (2017) Broadband metamaterials in electromagnetics: technology and applications. CRC Press, Boca Raton
Werner DH, Kwon DH (eds) (2014) Transformation electromagnetics and metamaterials: fundamental principles and applications. Springer, London
Wu W, Hu W, Qian G, Liao H, Xu X, Berto F (2019) Mechanical design and multifunctional applications of chiral mechanical metamaterials: a review. Mater Des 180:107950
Yang H, Ma L (2019) Multi-stable mechanical metamaterials by elastic buckling instability. J Mater Sci 54(4):3509–3526
Zadpoor AA (2016) Mechanical meta-materials. Mater Horiz 3(5):371–381
Zayats AV, Maier SA (2013) Active plasmonics and tuneable plasmonic metamaterials. Wiley, Hoboken
Zhao Z, Yuan C, Lei M, Yang L, Zhang Q, Chen H, Qi HJ, Fang D (2019) Three-dimensionally printed mechanical metamaterials with thermally tunable auxetic behavior. Phys Rev Appl 11(4):044074
Zouhdi S, Sihvola A, Arsalane M (eds) (2002) Advances in electromagnetics of complex media and metamaterials. Springer, Dordrecht
Zouhdi S, Sihvola A, Vinogradov AP (eds) (2009) Metamaterials and plasmonics: fundamentals, modelling, applications. Springer, Dordrecht

Chapter 2
Auxetic Microstructures

Abstract This chapter gives an overview of auxetic microstructures by considering the various geometries and their intermittent microstructures via gradation. This is followed by some comments on analogies between different auxetic models. Finally, an example is given for the extension of a 2D auxetic model to a 3D one, as well as an example of a rarely discussed auxetic model.

Keywords Auxetic microstructures · Microstructural gradation · Microstructural relationships

2.1 Introduction

Various geometrical microstructures have been conceptualized for elucidating the negative Poisson's ratio observation of naturally occurring and technologically processed auxetic materials. These microstructural geometries include, but are not limited to, the following:

(a) re-entrant models in 2D (Evans et al. 1991; Masters and Evans 1996; Smith et al. 2000; Li et al. 2019; Hou et al. 2019; Zhang et al. 2019; Wang et al. 2020) and 3D (Friis et al. 1988; Choi and Lakes 1995),
(b) double arrowhead in 2D (Larsen et al. 1997) and a related composite (Shilko et al. 2008) as well as in 3D (Lim 2016),
(c) interconnected star models in 2D (Grima et al. 2005a) and its related version in the form of interconnected petals (smoothed stars) (Wang et al. 2017, 2019; Wang and Poh 2018; Kumar et al. 2019), as well as interconnected stars in 3D (Rad et al. 2015)
(d) nodule-fibril model in 2D with rectangular nodules (Alderson and Evans 1995, 1997), hexagonal nodules (Lim and Acharya 2009), circular nodules (Jiang et al. 2016a) and eye-shaped (or oval) nodules (Smardzewski 2019), as well as in 3D with cuboid or tetragonal nodules (tethered-nodule model) (Gaspar et al. 2011) and spherical nodules (Lim 2015a)
(e) rotating unit models such as the rotating squares (Grima and Evans 2000; Grima et al. 2007a, 2008a; Attard et al. 2009a), rotating rectangles (Grima et al. 2004, 2005b, 2011a), rotating triangles (Grima and Evans 2006; Grima et al. 2007b,

T. Lim, *Mechanics of Metamaterials with Negative Parameters*, Engineering Materials,
https://doi.org/10.1007/978-981-15-6446-8_2

2012; Chetcuti et al. 2014; Gao et al. 2018), rotating rhombi and parallelograms (Grima et al. 2008b; Attard and Grima 2008; Attard et al. 2009b), rotating tetrahedrals (Grima et al. 2008c), rotating cubes (Jun et al. 2019), rotating cuboids (Attard and Grima 2012), rotating octahedrons in tetrahedral framework (Tanaka et al. 2019), and hierarchical rotating rigid units (Gatt et al. 2015; Dudek et al. 2017),

(f) slit perforated models (Grima and Gatt 2010; Taylor et al. 2013; Grima et al. 2016; Mizzi et al. 2019),

(g) hard cyclic hexamers (Wojciechowski 1987; Wojciechowski 1989; Wojciechowski and Branka 1989) and other hard cyclic polymers (Wojciechowski and Tretiakov 2000; Wojciechowski et al. 2003),

(h) swastika models (Smith et al. 2000; Gaspar et al. 2005; Jiang and Li 2018; Farrugia et al. 2019a; Bahaloo and Li 2019; Jin et al. 2019) and other missing rib models (Lim et al. 2014),

(i) chiral and anti-chiral models in 2D (Prall and Lakes 1997; Spadoni et al. 2009; Spadoni and Ruzzene 2012; Alderson et al. 2010a, b; Miller et al. 2010; Lorato et al. 2010; Abramovitch et al. 2010; Chen et al. 2013; Pozniak and Wojciechowski 2014; Ki et al. 2019; Niu et al. 2019; Su et al. 2019; Chen et al. 2019a; Zhu et al. 2019) and in 3D (Ha et al. 2016a; b; Fu et al. 2017; 2018; Wu et al. 2018; Farrugia et al. 2019b; Zheng et al. 2019; Reasa and Lakes 2019; Li et al. 2020),

(j) interlocking and sliding models (Ravirala et al. 2007; Hewage et al. 2016; Lim 2019a),

(k) indented sheet models, which include the egg-rack model (Grima et al. 2005c; Li et al. 2017; Smardzewski and Wojciechowski 2019), a related foldable macrostructure (Grima et al. 2011b), folded or origami sheets (Wei et al. 2013; Schenk and Guest 2013; Lv et al. 2014; Eidini and Paulino 2015; Yasuda and Yang 2015; Kamrava et al. 2017; Janbaz et al. 2017), dimpled sheet (Javid et al. 2015), and uneven graphene sheets (Coluci et al. 2008; Grima et al. 2015; Wen et al. 2019),

(l) helical yarns (Miller et al. 2009, 2012; Sloan et al. 2011; Wright et al. 2012), plied yarns (Ge et al. 2016; Ng and Hu 2017) and stitched-through yarns (Lim 2014), auxetic textiles (Ge and Hu 2013; Ge et al. 2013; Wang and Hu 2014a, b; Wang et al. 2014; Glazzard and Breedon 2014) and auxetic tubular braids (Jiang and Hu 2018; Chen et al. 2019b; Jiang et al. 2020),

(m) liquid crystalline polymeric models (He et al. 1998, 2005; Boba et al. 2016),

(n) instability-induced auxetic models (Bertoldi et al. 2008; Shim et al. 2012; Babaee et al. 2013; Javid et al. 2016; Jia and Wang 2019),

(o) ring-rod assembly models in 2D and 3D (Lim 2017a, 2019b),

(p) linkage mechanism models (Broeren et al. 2019; Lim 2019c),

(q) Voronoi structures in 2D (Bouakba et al. 2012; Li et al. 2016) and 3D (Gao et al. 2019),

(r) other models such as the 3D pre-buckled lattices (Albertini et al. 2019), star perforations (Li et al. 2018), and composites with hollow tetrahedral inclusions (Hou and Hu 2015).

Due to the rapid expansion in the area of auxetics, it is inevitable that some models are not listed above while others may be subsumed under the listed models due to certain similar features. Various reviews on auxetic materials have been furnished, some of which are more generic in nature (Lakes 1993; Alderson 1999; Yang et al. 2004; Alderson and Alderson 2007; Liu and Hu 2010; Greaves et al. 2011; Prawoto 2012; Critchley et al. 2013; Saxena et al. 2016; Lakes 2017; Lim 2017b; Kolken and Zadpoor 2017; Ren et al. 2018) while others are scoped toward specific subfields and/or applications (Darja et al. 2013; Jiang et al. 2016b; Park and Kim 2017; Duncan et al. 2018; Kwietniewski and Miedzińska 2019; Pasternak and Dyskin 2019; Surjadi et al. 2019; Wu et al. 2019); in addition, there has been two monographs on auxetics (Lim 2015b; Hu et al. 2019) to date. An overview of auxetic microstructures is best understood by comparing and classifying their various geometrical structures. These have been done in most of the abovementioned reviews, and therefore are not repeated here, except for a few for illustration purposes. The proceeding topics illustrate how each type of auxetic microstructure is related to another by gradually altering some parts of the microstructures, followed by an example where a 2D axuetic microstructure is extended to its 3D version and an example of a relatively overlooked microstructural geometry.

2.2 Gradation Between 2D Auxetic Microstructures

Figure 2.1 demonstrates a few examples on how the prescription of incremental geometrical changes gives rise to a series of auxetic models of the same family. With reference to Fig. 2.1a, we have a re-entrant model on one extreme and a nodule-fibril model with rectangular nodules on the other extreme. It is easy to see that as the width of the rectangular block diminishes, they reduce to vertical ribs in the re-entrant model. While contact region between the fibril and the rectangular blocks is defined by the width of the rectangular blocks (or the area due to the width and the depth of the blocks), the contact region in the case of re-entrant model diminishes to a point contact (or a line defined by the depth of the ribs). For both extremes, the contact regions are fixed. This is not so for the intermediate auxetic models. In the circular (or cylinder) nodule model, there exists contact points (or contact lines) between the fibrils (or sheets) with the circles (or cylinders) when the fibrils or sheets are horizontal. Upon vertical compression, the fibril or sheet slope increases while the contact arc length (or contact curved area) increases. The same is expected in the case of the eye-shaped nodules, but with greater increase in contact arc length (or contact curved area) for initial compression due to the oval-shaped nodules. Suppose the re-entrant model is rotated by 90° as shown in the extreme left of Fig. 2.1b, it can be seen that the re-entrant "properties" are somewhat manifested in the metamaterials investigated by Qin et al. (2018), whereby the middle left and middle right models exhibit mild and large auxeticities, respectively. Recently, Meena and Singamneni (2019) introduced the so-called S-shaped rib that exhibits hollow "X" and hollow

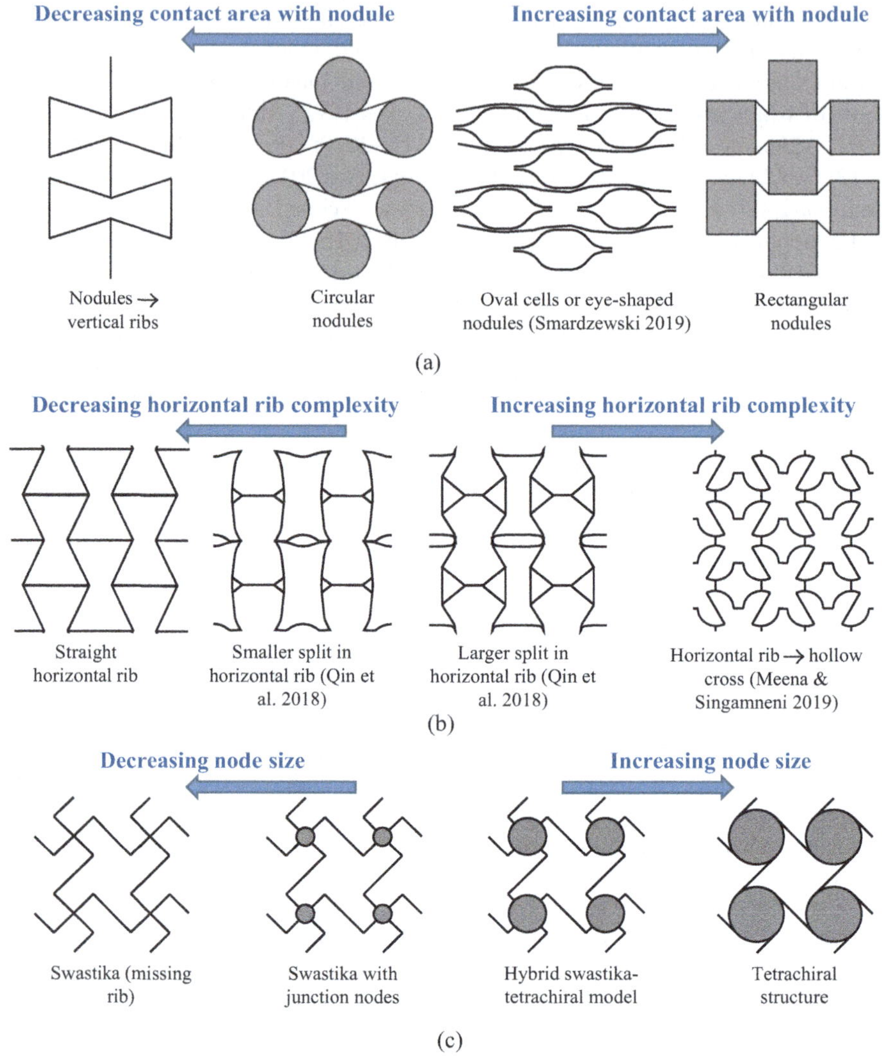

Fig. 2.1 Gradation between the **a** re-entrant and nodule-fibril models, **b** re-entrant and the "S" structure, and **c** swastika and tetrachiral models

"+" cells, as shown on the extreme right of Fig. 2.1b, which correspond to the re-entrant cell and the horizontal ribs, respectively. During the initial stage of vertical compression, the hollow "X" regions collapse while the hollow "+" regions remain comparatively undeformed, thereby corresponding to the collapse of the re-entrant cells at (almost) fixed horizontal rib dimension. The missing rib model shown on the extreme left of Fig. 2.1c was so named due to the removal of ribs from a rhombic mesh. To be more precise, these are herein called the tilted-swastika model to differentiate them from other missing rib, or rib deletion, models. Suppose the "cross" parts of the swastikas are remodeled as circular nodes, one obtains the tetrachiral model depicted on the extreme right of Fig. 2.1c. If the "cross" parts of the swastikas are rigid, they are effectively the circular nodes of the tetrachiral model.

Figure 2.2 shows how a few auxetic models can be related to each other in a loop by gradual alteration to the connecting parts and the corresponding void shapes. Take, for example, the rotating square model displayed in Fig. 2.2a where the squares are connected to one another by pin joints to permit rotation. Suppose

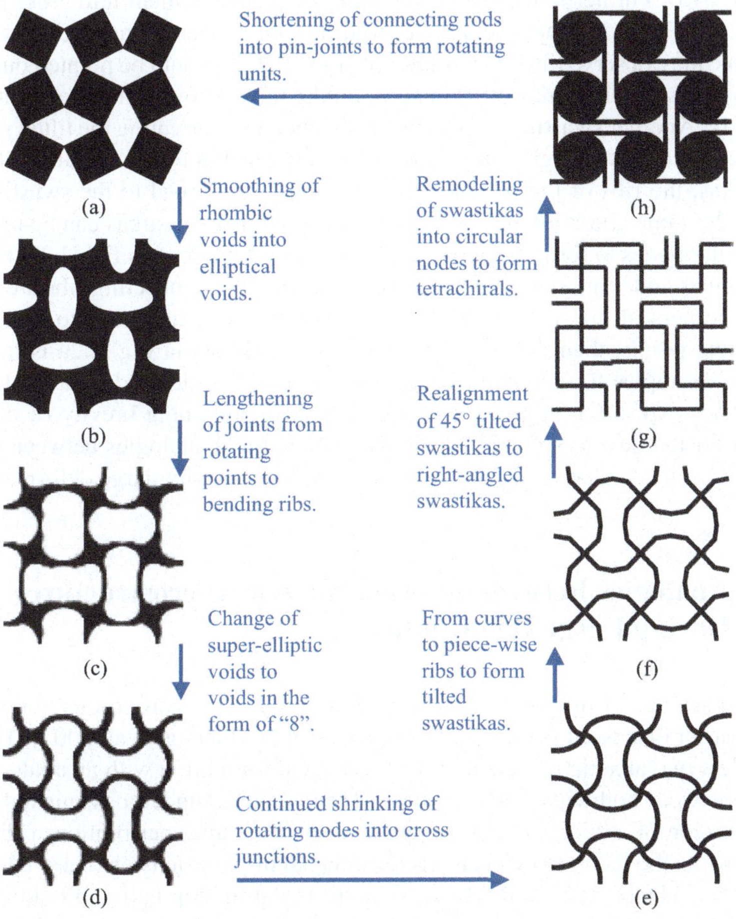

Fig. 2.2 Gradation from the **a** rotating squares model to the **b** perforated sheet model, and a topologically optimized **c** linear and **d** nonlinear designs (Clausen et al. 2015) toward the **e** smoothed curved ribs and its **f** piece-wise version, that exhibits tilted swastikas, as well as the **g** right-angled swastikas, in which the crosses are remodeled into circular nodes to form the **h** anti-tetrachiral model, and finally, the shortening of the connecting robs and the changing of circles to squares lead back to **a**

the squares are distorted such that the rhombic voids transform into ellipses, which leads to the "point" joints being widened, one obtains a pattern that resembles an instability-triggered auxeticity from uniaxial compression of a sheet with square array of circular holes (Bertoldi et al. 2010). Suppose the distortion trend continues, the elliptical voids transform into superelliptical ones, which approximate rectangles with rounded corners, while the joints begin to form connecting ribs, as indicated in Fig. 2.2c. Continuing this trend with decreasing node sizes, we have the voids taking on the shape of "8" or "∞", as shown in Fig. 2.2d. Figure. 2.2c, d resembles the topology optimization study of auxetic microstructures by Clausen et al. (2015). Further reduction of the node size converts them into cross junctions denoted in Fig. 2.2e. Suppose these curved ribs are converted into piece-wise linear ribs, one may observe tilted swastikas. See Fig. 2.2f. It should be pointed out that the alternating swastikas are mirror images, unlike those shown on the extreme left of Fig. 2.1c. Another swastika model can be obtained by reorienting the tilted swastikas to give the connected right-angled swastikas furnished in Fig. 2.2g. Unlike the tilted swastikas, the ribs of the right-angled swastikas are aligned to the swastika array. Using the same alteration in Fig. 2.1c, the right-angled swastikas can be remodeled as circular nodes to form the anti-tetrachiral model furnished in Fig. 2.2h. Finally, if the circular nodes are replaced by squares such that the connecting ribs are attached to the corners of the squares, and if the connecting ribs are made to shorten until becoming points, then the rotating square model shown in Fig. 2.2a is recovered. Further examples for geometry-based gradation of auxetic models are aplenty, but no further exposition is given in the interest of maintaining brevity of this book, except for the next topic which deals with geometrical analogies between different auxetic models undergoing microstructural evolution during large deformation.

2.3 Analogies Between Evolved Auxetic Microstructures During Large Deformation

The occurrence of microstructural evolution in auxetic microporous polymers has been earlier discussed pertaining to its processing (Alderson et al. 2007). This topic considers two auxetic models undergoing large deformation with an analogy established between both sets of deformations. The first model undergoes uniaxial in-plane compression of a thick sheet with square array of circular perforations (Shim et al. 2013) while the second model entails the uniaxial in-plane tension of a slit perforated thick sheet (Dubrovski et al. 2019), as indicated at the top half and bottom half of Fig. 2.3, respectively. For visual aid, a 2-by-2 cluster of rotating units is identified from each model at the original state and indicated as "start." The rotating units are assigned different colors for ease of tracking from the original states, or the "start" positions, to the "end" positions. It is clear that the original state of the former is geometrically analogous to the final fully opened state of the latter, as shown on the left side of Fig. 2.3. Likewise, the final collapsed state of the former is geometrically

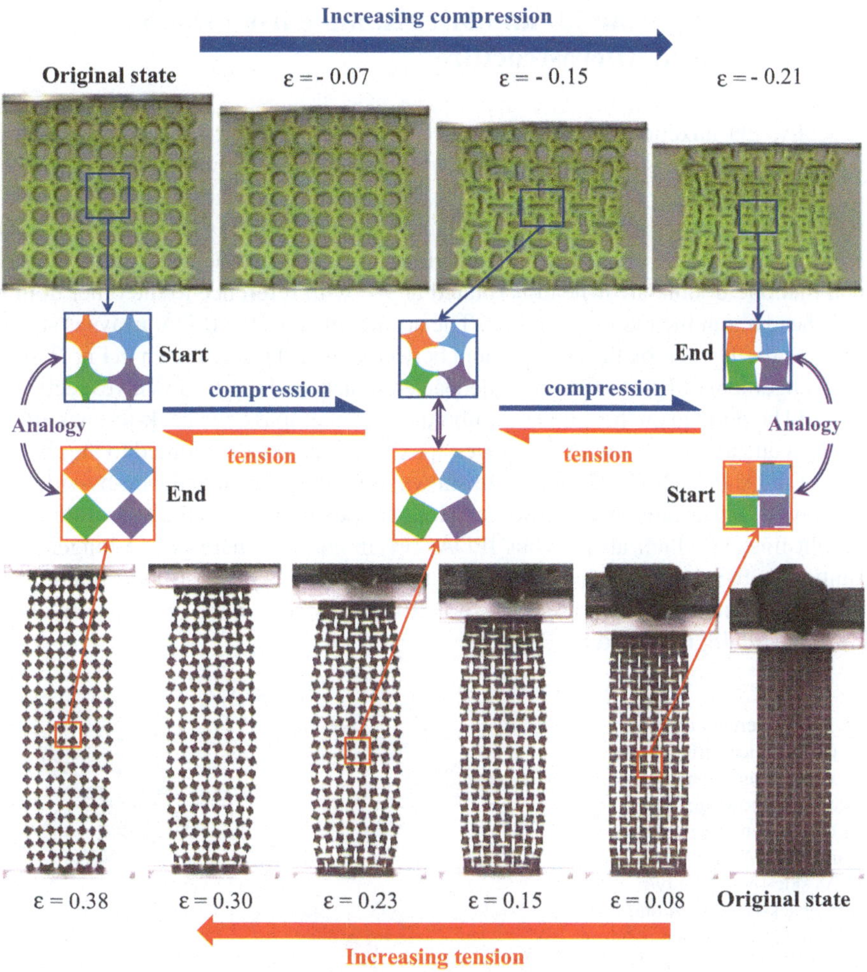

Fig. 2.3 Deformation-triggered microstructural evolution, as exemplified by the compression of sheets with square array of circular perforation (Adapted from Shim et al. (2013) with permission from The Royal Society of Chemistry) (top) and tension of slit perforated sheet (Dubrovski et al. 2019) (bottom)

analogous to the original closed state of the latter, as observed on the right side of Fig. 2.3. The intermediate states, i.e., the intermediate compression of the former and the intermediate tension of the latter reveal voids of elliptical and rhombic shapes, respectively. In other words, the gradation of one microstructural model during a large deformation can be matched against that of another microstructural model under opposing load directions.

2.4 Extending from 2D and 3D: Example from Double Arrowhead Microstructure

Typically, 2D auxetic microstructures are extended to their 3D versions. In this section, an example is made for the case where the double arrowhead microstructure is extended to its 3D version (Lim 2016). The Intersecting Pair of Double Arrowheads (IPDA) for generating this 3D auxetic microstructure is shown in Fig. 2.4a, showing two double arrowheads intersecting one another at their axis of symmetry, such that one double arrowhead is rotated at 90° with reference to the other double arrowhead about the intersection line. The arrangement of the IPDA arrays is shown in Fig. 2.4b–d, whereby the arrangement between two IPDAs is shown in Fig. 2.4b–d for arrangement of IPDAs side-by-side, back-to-back, and top-to-down, respectively. It should be pointed out here that the cuboids that encompass two back-to-back IPDA overlaps one another because the longer apex from one IPDA is joined to the shorter apex of the shorter IPDA. Figure 2.4e shows four IPDAs arranged side-by-side and top-down. This arrangement shows four longer apex pointing to the left while at the straight line, in which all the four IPDAs are in contact, there is one longer apex pointing to the right. In other words, an arrangement of 3 by 3 IPDAs pointing in one direction inherently introduces 2 by 2 IPDAs pointing in the opposite direction, as schematically indicated by Fig. 2.4f.

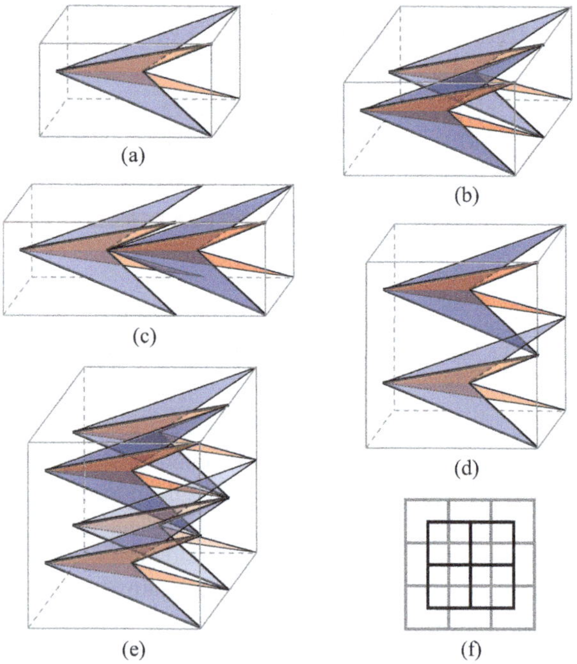

Fig. 2.4 Schematics for the 3D auxetic microstructure of the intersecting double arrowhead, showing: **a** a single representative volume element (IPDA), **b** two IPDAs side-by-side, **c** two IPDAs back-to-back, **d** one IPDA on top of another, **e** four IPDAs (2 by 2) with apex pointing in one direction resulting in an apex pointing in the opposite direction, and **f** nine (3 by 3) IPDAs denoted in gray that inherently introduces four (2 by 2) IPDAs denoted in black

The Cartesian coordinate system selected for the IPDA is such that the straight line joining the longer apex and shorter apex, which is also the intersection line between the two double arrowheads, forms the x-axis, while the other two orthogonal axes parallel to the sides of the IPDA are the y'-axis and z'-axis (see Fig. 2.5a). In this analysis, a rotated axes of y-axis and z-axis, as shown in Fig. 2.5b, are used because one of the double arrowheads falls on the x-y plane, while the intersecting double arrowhead lies on the x-z plane. It should be pointed out that if the geometry of both intersecting double arrowheads is identical, then not only is $\varepsilon_y = \varepsilon_z$, it must be noted that additionally we have $\varepsilon_y = \varepsilon_{y'}$ and $\varepsilon_y = \varepsilon_{z'}$. This is because the original dimensions measured along the y'-axis and z'-axis are greater than those along the y'-axis and z'-axis by a factor of $\sqrt{2}$, and that the displacements measured along the y'-axis and z'-axis are also greater than those along the y'-axis and z'-axis by a factor of $\sqrt{2}$. Hence, the Poisson's ratios for both the x, y', and z' and the x, y, and z coordinate systems are similar.

Perusal to Fig. 2.5c shows the major and minor linkages are of lengths l_1 and l_2, respectively, subtending at angles θ_1 and θ_2 from the x-axis, such that $l_1 > l_2 > 0$ and $0 < \theta_1 < \theta_2 < \pi/2$. Since point A is the point at which an IPDA is joined to neighboring IPDAs along the side-to-side and top-to-down directions, we let the lateral dimension be represented by y_0, which is the perpendicular distance of point A from the x-axis. On the other hand, the neighboring IPDA joined from back-to-back is at point B, i.e., point O of the IPDA in Fig. 2.5c is joined to point B of another IPDA on its left while point B of the IPDA in Fig. 2.5c is joined to point O of another

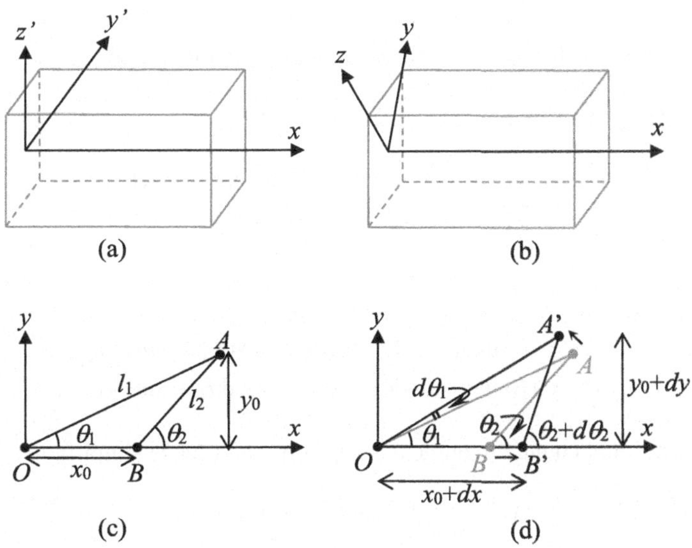

Fig. 2.5 **a** A Cartesian coordinate system (x, y', and z') for the IPDA, **b** a rotated Cartesian coordinate system (x, y, and z) for analysis, **c** a quarter of the IPDA (i.e., half of an arrowhead) before displacement, and **d** a quarter of the IPDA after displacement within the same x-y plane with the original location indicated in gray

IPDA on its right. Hence, the displacement of point B in the x-direction causes a similar displacement to the neighboring IPDA on the right. As such we let the axial dimension be represented by x_0, which is the distance of B from the origin. From Fig. 2.5c,

$$x_0 = \overline{OB} = l_1 \cos\theta_1 - l_2 \cos\theta_2 \tag{2.4.1}$$

and

$$y_0 = \left(\overline{OA}\right)_y = l_1 \sin\theta_1 = l_2 \sin\theta_2 \tag{2.4.2}$$

from which we have the relation

$$\frac{l_1}{l_2} = \frac{\sin\theta_2}{\sin\theta_1} \tag{2.4.3}$$

Arising from displacement, point A moves to point A' at constant linkage length of l_1 such that the angle θ_1 changes by $\mathrm{d}\theta_1$; point B slides along the x-axis to B' at constant linkage length of l_2 such that the angle θ_2 changes by $\mathrm{d}\theta_2$. With reference to Fig. 2.5d,

$$x_0 + \mathrm{d}x = \overline{OB'} = l_1 \cos(\theta_1 + \mathrm{d}\theta_1) - l_2 \cos(\theta_2 + \mathrm{d}\theta_2) \tag{2.4.4}$$

and

$$y_0 + \mathrm{d}y = \left(\overline{OA'}\right)_y = l_1 \sin(\theta_1 + \mathrm{d}\theta_1) = l_2 \sin(\theta_2 + \mathrm{d}\theta_2) \tag{2.4.5}$$

thereby giving displacements

$$\begin{aligned} \mathrm{d}x = {} & l_1(\cos\theta_1 \cos \mathrm{d}\theta_1 - \sin\theta_1 \sin \mathrm{d}\theta_1 - \cos\theta_1) \\ & - l_2(\cos\theta_2 \cos \mathrm{d}\theta_2 - \sin\theta_2 \sin \mathrm{d}\theta_2 - \cos\theta_2) \end{aligned} \tag{2.4.6}$$

and

$$\begin{aligned} \mathrm{d}y & = l_1(\sin\theta_1 \cos \mathrm{d}\theta_1 + \cos\theta_1 \sin \mathrm{d}\theta_1 - \sin\theta_1) \\ & = l_2(\sin\theta_2 \cos \mathrm{d}\theta_2 + \cos\theta_2 \sin \mathrm{d}\theta_2 - \sin\theta_2) \end{aligned} \tag{2.4.7}$$

For infinitesimal change in angle, Eqs. (2.4.6) and (2.4.7) simplify to

$$\mathrm{d}x = -l_1 \mathrm{d}\theta_1 \sin\theta_1 + l_2 \mathrm{d}\theta_2 \sin\theta_2 \tag{2.4.8}$$

and

$$\mathrm{d}y = l_1 \mathrm{d}\theta_1 \cos\theta_1 = l_2 \mathrm{d}\theta_2 \cos\theta_2 \tag{2.4.9}$$

respectively. These displacement components give the strain components

$$\varepsilon_x = \frac{\mathrm{d}x}{x_0} = -\frac{l_1 \mathrm{d}\theta_1 \sin\theta_1 - l_2 \mathrm{d}\theta_2 \sin\theta_2}{l_1 \cos\theta_1 - l_2 \cos\theta_2} \tag{2.4.10}$$

and

$$\varepsilon_y = \frac{\mathrm{d}y}{y_0} = \frac{\mathrm{d}\theta_1}{\tan\theta_1} = \frac{\mathrm{d}\theta_2}{\tan\theta_2} \tag{2.4.11}$$

Using the relation described by Eq. (2.4.3), we have the strain ratio

$$\frac{\varepsilon_x}{\varepsilon_y} = \tan\theta_1 \tan\theta_2 \tag{2.4.12}$$

Recalling the definition of Poison's ratio being $\upsilon_{ij} = -\varepsilon_j/\varepsilon_i$, we obtain

$$\upsilon_{yx} = \frac{1}{\upsilon_{xy}} = -\tan\theta_1 \tan\theta_2 \tag{2.4.13}$$

Since the half angles of the double arrowheads comply with $0 < \theta_1 < \theta_2 < \pi/2$, it follows that both υ_{yx} and υ_{xy} are negative. If the double arrowhead in the *x-z* plane is identical to that of the *x-y* plane, then $\upsilon_{yx} = \upsilon_{zx}$ and $\upsilon_{xy} = \upsilon_{xz}$.

Suppose the geometry of the linkages lying on the *x-z* plane differs from those on the *x-y* plane, as shown in Fig. 2.6, in which the major and minor linkage lengths L_1 and L_2, and their subtending half angles ϕ_1 and ϕ_2 on the *x-z* plane correspond to l_1, l_2, θ_1 and θ_2 respectively on the *x-y* plane, then

$$\upsilon_{zx} = \frac{1}{\upsilon_{xz}} = -\tan\phi_1 \tan\phi_2 \tag{2.4.14}$$

by comparison, in which a relation similar to Eq. (2.4.3)

$$\frac{L_1}{L_2} = \frac{\sin\phi_2}{\sin\phi_1} \tag{2.4.15}$$

holds. While both Equations (2.4.3) and (2.4.15) describe geometrical relationship within its own plane, the relation between both planes can be established by recognizing common axial length

$$\begin{aligned} l_1 \cos\theta_1 &= L_1 \cos\phi_1 \\ l_2 \cos\theta_2 &= L_2 \cos\phi_2 \end{aligned} \tag{2.4.16}$$

and common displacement along the *x*-axis

$$l_1 \mathrm{d}\theta_1 \sin\theta_1 - l_2 \mathrm{d}\theta_2 \sin\theta_2 = L_1 \mathrm{d}\phi_1 \sin\phi_1 - L_2 \mathrm{d}\phi_2 \sin\phi_2 \tag{2.4.17}$$

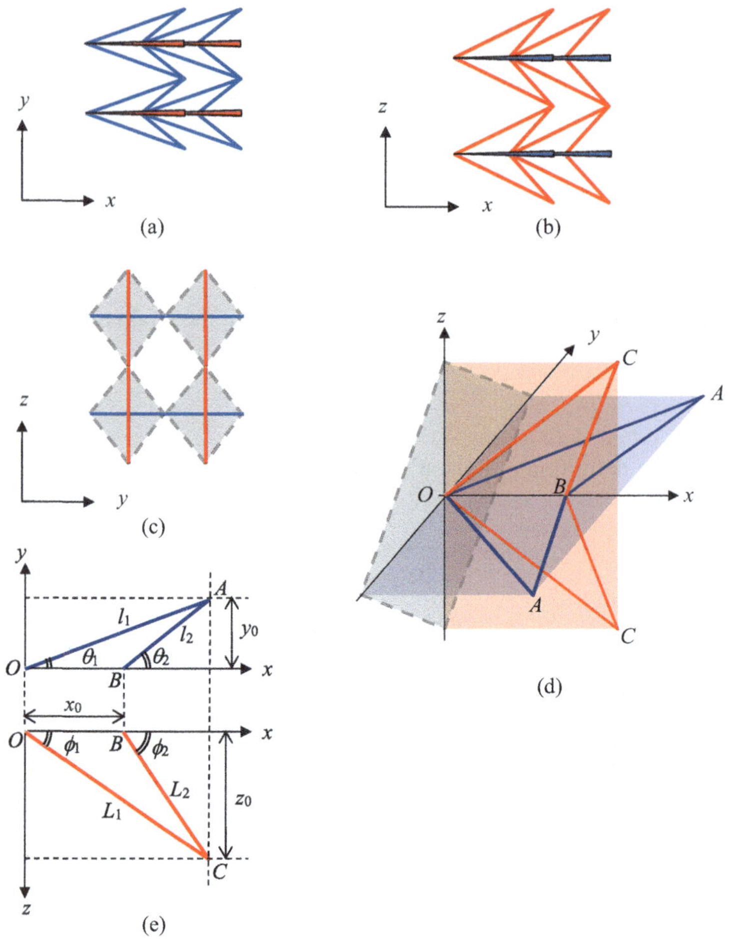

Fig. 2.6 3D representation of intersecting double arrowheads, showing: **a** 2 × 2 × 2 IPDAs projected on *x*-*y* plane, **b** 2 × 2 × 2 IPDAs projected on x-z plane, **c** 2 × 2 × 2 IPDAs projected on *y*-*z* plane, **d** isometric view of a single IPDA, and **e** comparison of geometries for each of the half intersecting double arrowheads

Since auxeticity is achievable in two perpendicular planes *x*-*y* and *x*-*z*, it follows that auxeticity is also achievable in the third orthogonal plane *y*-*z*. The geometrical properties of the linkages in one plane (l_1, l_2, θ_1, and θ_2) can be related to the geometrical properties in the other plane (L_1, L_2, ϕ_1, and ϕ_2) by taking the ratio of l_1 to L_2 from Eqs. (2.4.3) and (2.4.15) to give

$$\frac{l_1}{L_2} = \frac{l_2}{L_1} \frac{\sin\theta_2 \sin\phi_2}{\sin\theta_1 \sin\phi_1} \tag{2.4.18}$$

The variables l_2 and L_1 in Eq. (2.4.18) can be substituted by L_2 and l_1, respectively, by using Eq. (2.4.16) to give

$$\frac{l_1}{L_2} = \sqrt{\frac{\tan\theta_2 \sin 2\phi_2}{\tan\phi_1 \sin 2\theta_1}} \tag{2.4.19}$$

A 2D planar auxeticity is graphically represented in Fig. 2.7 based on Eq. (2.4.13), which is in reference to the *x*-*y* plane. A family of υ_{yx} and υ_{xy} versus θ_2 were plotted for various θ_1. The trend shown in Fig. 2.7 is valid for the x-z plane, i.e., $\upsilon_{yx} = \upsilon_{zx}$ and $\upsilon_{xy} = \upsilon_{xz}$, if ϕ_1 and ϕ_2 are identical to θ_1 and θ_2 respectively, thereby leading to transversely isotropic 3D auxeticity with the *y*-axis and *z*-axis forming the plane of isotropy.

An overview on the auxeticity of the proposed structure is displayed in Fig. 2.8 in terms of υ_{yx} contour plot as a function of θ_1 and θ_2 with the condition $\theta_1 < \theta_2$ as defined by the structural geometry set out in Fig. 2.5c and the condition $\theta_2 < \pi/2$ to confine the plots within the auxetic region. The former condition requires that the contour plot be made undefined for $\theta_1 > \theta_2$ in Fig. 2.8, i.e., only the portion of the contour plots below the dashed diagonal lines are shown. As expected from

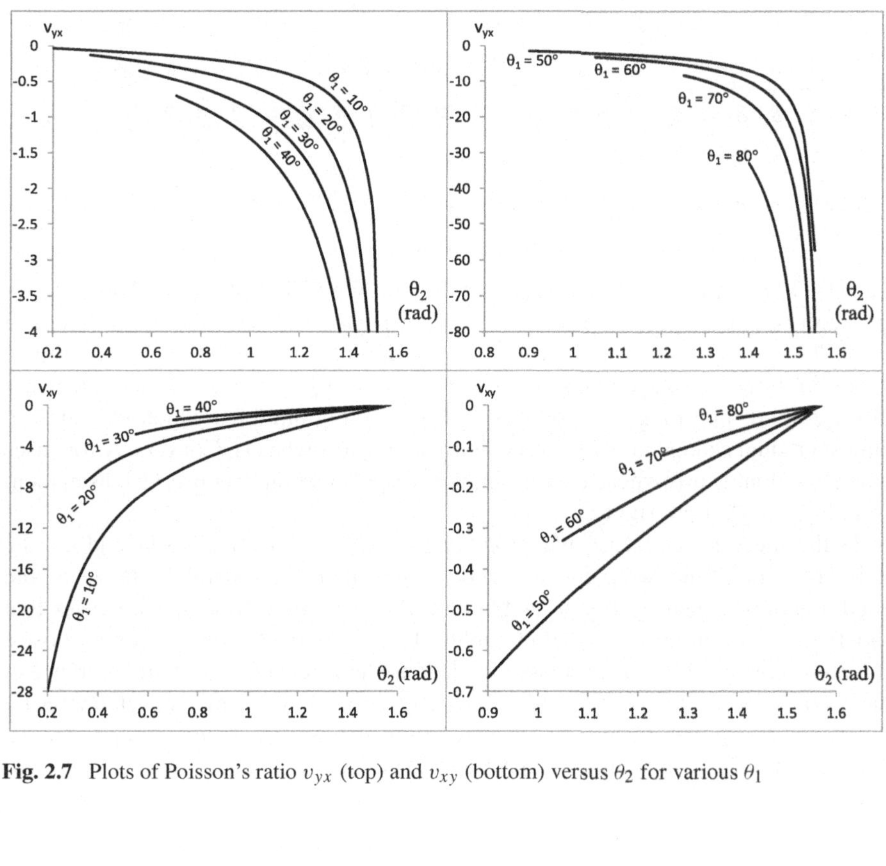

Fig. 2.7 Plots of Poisson's ratio υ_{yx} (top) and υ_{xy} (bottom) versus θ_2 for various θ_1

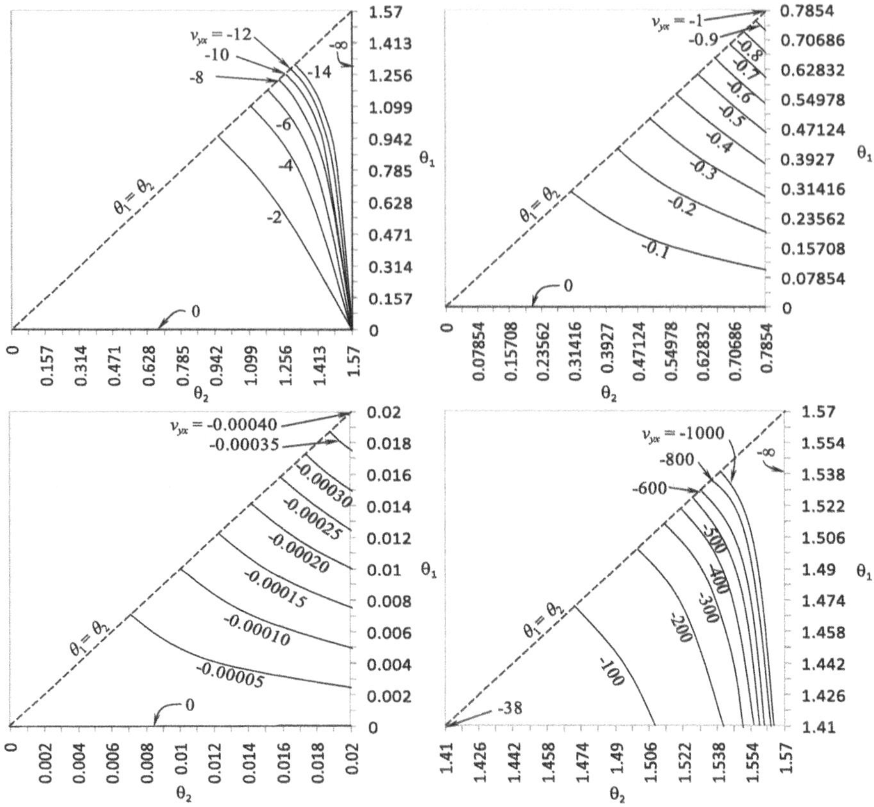

Fig. 2.8 Contour plots of υ_{yx} as function of θ_1 and θ_2

Eq. (2.4.13), Figure 2.8 (top left) shows the limiting values of $\upsilon_{yx} = 0$ and $\upsilon_{yx} \to -\infty$ corresponding to $\theta_1 = 0$ and $\theta_2 = \pi/2$, respectively. Due to the sharp change in υ_{yx} as $\theta \to \pi/2$, a narrower range of υ_{yx} contour plot is shown in Fig. 2.8 (top right); of special interest is the condition $\upsilon_{yx} = -1$ at $\theta_1 = \theta_2 = \pi/4$. A further narrowing of the contour plot range is displayed in Fig. 2.8 (bottom left), which indicates an almost gradual change in υ_{yx} for very sharp double arrowheads. Conversely for cases where the double arrowheads are blunt, very large auxeticity is observed, as evident from Fig. 2.8 (bottom right).

In the more generic case where ϕ_1 and ϕ_2 are not identical to θ_1 and θ_2, the Poisson's ratio of the two planes υ_{yx} and υ_{zx} are calculated separately; however, the conditions laid down by Eqs. (2.4.16) and (2.4.17) apply. Nine special cases that satisfy these conditions, as listed in Table 2.1, are selected for illustration purposes. It is convenient to express the subtending half angles of the linkages with reference to the x-axis in terms of their tangents. A full description of the geometry is attainable by two broad approaches. One approach is by listing all four half angles or their tangents. Another approach is by listing both the intraplane linkage length ratio (i.e., l_1/l_2 and

Table 2.1 Summary of special case Poisson's ratios where the geometry of the mutually intersecting double arrowheads are different

Cases	$\tan\begin{bmatrix}\theta_1 & \theta_2 \\ \phi_1 & \phi_2\end{bmatrix}$	$\frac{l_1}{l_2}$	$\frac{L_1}{L_2}$	$\frac{l_1}{L_2}$	Poisson's ratios	Schematics showing relative linkage lengths
1	$\frac{1}{2}\begin{bmatrix}1 & 2 \\ 2 & 4\end{bmatrix}$	$\sqrt{\frac{5}{2}}$	$\sqrt{\frac{8}{5}}$	1	$\upsilon_{xy} = \upsilon_{zx} = \frac{1}{\upsilon_{yx}} = \frac{1}{\upsilon_{xz}} = -2$	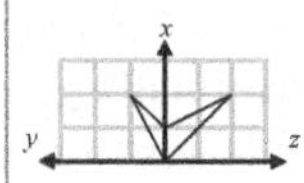
2	$\frac{1}{3}\begin{bmatrix}1 & 3 \\ 3 & 9\end{bmatrix}$	$\sqrt{5}$	$\sqrt{\frac{9}{5}}$	1	$\upsilon_{xy} = \upsilon_{zx} = \frac{1}{\upsilon_{yx}} = \frac{1}{\upsilon_{xz}} = -3$	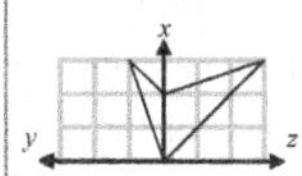
3	$\frac{1}{6}\begin{bmatrix}4 & 6 \\ 6 & 9\end{bmatrix}$	$\sqrt{\frac{13}{8}}$	$\sqrt{\frac{18}{13}}$	1	$\upsilon_{xy} = \upsilon_{zx} = \frac{1}{\upsilon_{yx}} = \frac{1}{\upsilon_{xz}} = -\frac{3}{2}$	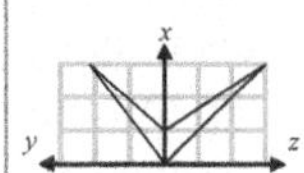
4	$\frac{1}{3}\begin{bmatrix}2 & 6 \\ 3 & 9\end{bmatrix}$	$\sqrt{\frac{13}{5}}$	$\sqrt{\frac{9}{5}}$	$\sqrt{\frac{13}{10}}$	$\upsilon_{xy} = \frac{1}{\upsilon_{yx}} = -\frac{3}{4}$ $\upsilon_{zx} = \frac{1}{\upsilon_{xz}} = -3$	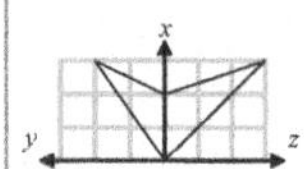
5	$\frac{1}{6}\begin{bmatrix}2 & 3 \\ 6 & 9\end{bmatrix}$	$\sqrt{2}$	$\sqrt{\frac{18}{13}}$	$\sqrt{\frac{10}{13}}$	$\upsilon_{xy} = \frac{1}{\upsilon_{yx}} = -6$ $\upsilon_{zx} = \frac{1}{\upsilon_{xz}} = -\frac{3}{2}$	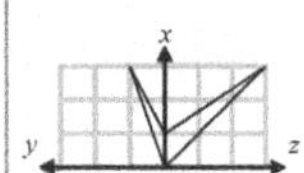
6	$\frac{1}{2}\begin{bmatrix}2 & 4 \\ 3 & 6\end{bmatrix}$	$\sqrt{\frac{8}{5}}$	$\sqrt{\frac{13}{10}}$	$\sqrt{\frac{4}{5}}$	$\upsilon_{xy} = \frac{1}{\upsilon_{yx}} = -\frac{1}{2}$ $\upsilon_{zx} = \frac{1}{\upsilon_{xz}} = -\frac{9}{2}$	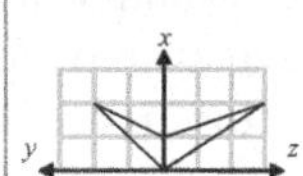
7	$\frac{1}{2}\begin{bmatrix}1 & 2 \\ 3 & 6\end{bmatrix}$	$\sqrt{\frac{5}{2}}$	$\sqrt{\frac{13}{10}}$	$\sqrt{\frac{1}{2}}$	$\upsilon_{xy} = \frac{1}{\upsilon_{yx}} = -2$ $\upsilon_{zx} = \frac{1}{\upsilon_{xz}} = -\frac{9}{2}$	
8	$\frac{1}{3}\begin{bmatrix}1 & 3 \\ 2 & 6\end{bmatrix}$	$\sqrt{5}$	$\sqrt{\frac{13}{5}}$	$\sqrt{\frac{13}{10}}$	$\upsilon_{xy} = \frac{1}{\upsilon_{yx}} = -3$ $\upsilon_{zx} = \frac{1}{\upsilon_{xz}} = -\frac{4}{3}$	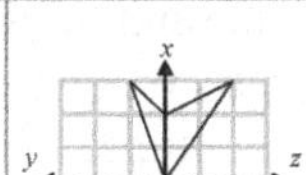
9	$\frac{1}{6}\begin{bmatrix}2 & 3 \\ 4 & 6\end{bmatrix}$	$\sqrt{2}$	$\sqrt{\frac{13}{8}}$	$\sqrt{\frac{5}{4}}$	$\upsilon_{xy} = \frac{1}{\upsilon_{yx}} = -6$ $\upsilon_{zx} = \frac{1}{\upsilon_{xz}} = -\frac{2}{3}$	

L_1/L_2) and one interplane linkage ratio (l_1/L_1, l_2/L_2, l_2/L_1 or l_1/L_2). Both the half angle tangents and the linkage length ratios, as well as graphical representations, of the nine special cases are furnished in Table 2.1. The calculated Poisson's ratio implies that a change in the linkage length ratios significantly alters the Poisson's ratio, thereby showing that the auxeticity of such a system is not only achievable in 3D, but also can be effectively tailor-made.

With the possibility of attaining tailor-made auxeticity, it is now suggested that a functionally graded plate that exhibits a change in Poisson's ratio from negative to positive is possible, akin to the case of functionally graded (FG) beam consisting of honeycomb structures in which the re-entrant angles change values from the top surface to the bottom surface (Lim 2002b). Unlike the previous work, the use of intersecting double arrowhead is easier to achieve due to its simpler geometrical description and therefore can be extended from a FG beam undergoing single curvature bending to FG plates undergoing double curvature bending. A 2D representation of a functionally graded bulk solid changing from auxetic IPDAs of intersecting double arrowheads to conventional IPDAs of intersecting rhombi is shown in Fig. 2.9a. The semi-auxetic (or partial auxetic) material as shown in Fig. 2.9b, i.e., $\theta_1 < 90° < \theta_2$ with $\phi_1 < \phi_2 < 90°$ or alternatively $\theta_1 < \theta_2 < 90°$ with $\phi_1 < 90° < \phi_2$, is analogous to that based on combined re-entrant and hexagonal microstructures in different planes (Lim 2004); it further allows the possibility of producing an "opposing" functionally graded material that exhibits change from auxetic to conventional from top to bottom in one plane but conventional to auxetic from top to bottom in another plane, as shown in Fig. 2.9c.

2.5 Ring-Rod Assembly as a Unique Example of Auxetic Microstructure

In this example, a 2D analysis of single ring structures is firstly presented, followed by a 3D analysis of double-ring structures. Thereafter, some special cases are considered, while discussion on its NTE properties is given in Sect. 11.11. A 2D auxetic behavior can be attained based on the ring and a pair of sliding rods. Each rod is attached to the inner surface of the ring and protrudes from a small hole diametrically opposite, as shown in Fig. 2.10a. It is herein assumed that the ring is thin and flexible, while the rod is rigid. The dimensions of such a ring with a pair of sliding rods are determined from the ends of the protruding rods and the outer surface of the ring perpendicular to the rods. When the rods are compressed as shown in Fig. 2.10b, the ring transforms into an ellipse with the major axis coinciding with rod alignment while the minor axis is formed in the transverse direction, thereby leading to overall contraction of the ring-rod structure along the rod direction and in their transverse directions. When the rods are pulled as indicated in Fig. 2.10c, the ring becomes narrow in the direction of the sliding rods, thereby leading to ring elongation in the direction perpendicular to the rods. This gives a 2D auxetic behavior.

To prevent each ring from rotating about the axis perpendicular to the sliding rods, the sliding rods are arranged in an alternate manner to cancel any rotating effects. Figure 2.11a shows the alternating arrangement of the sliding rods in connecting neighboring rings in the direction of the sliding rods. The arrangement of the sliding rods is also alternated, as illustrated in Fig. 2.11b, in the direction perpendicular to the sliding rods. Therefore, when the alternating arrangement of the sliding rods is implemented in both directions, we have a rectangular array of ring and sliding rod shown in Fig. 2.11c. In what follows, we herein consider the small deformation

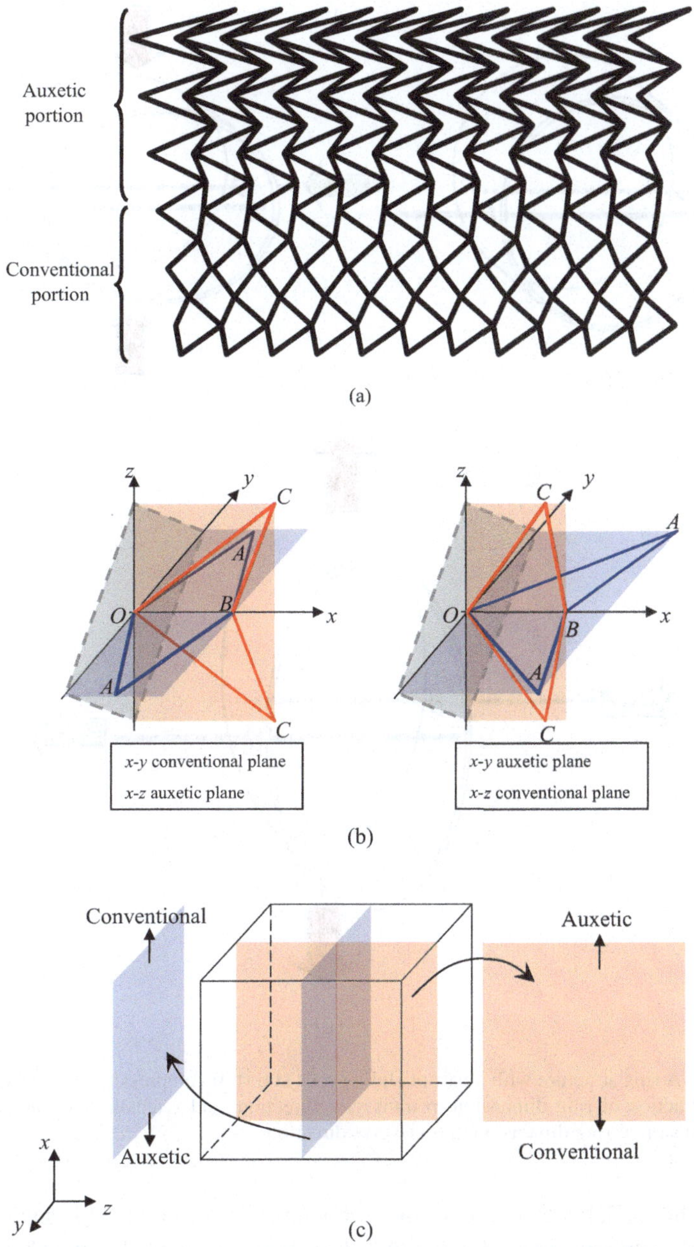

Fig. 2.9 **a** A 2D representation of a functionally graded bulk solid changing from auxetic IPDAs as indicated by intersecting double arrowhead from the top to conventional IPDAs as shown by the intersecting rhombus toward the bottom, **b** schematics of semi-auxetic (or partial auxetic) structure with the *x*-*y* and *x*-*z* axes forming the conventional and auxetic planes, respectively (left), and auxetic and conventional planes, respectively (right), and **c** the arrangement of functionally graded material with decreasing and increasing auxeticity on *x*-*y* and *x*-*z* planes, respectively, with increasing *x*

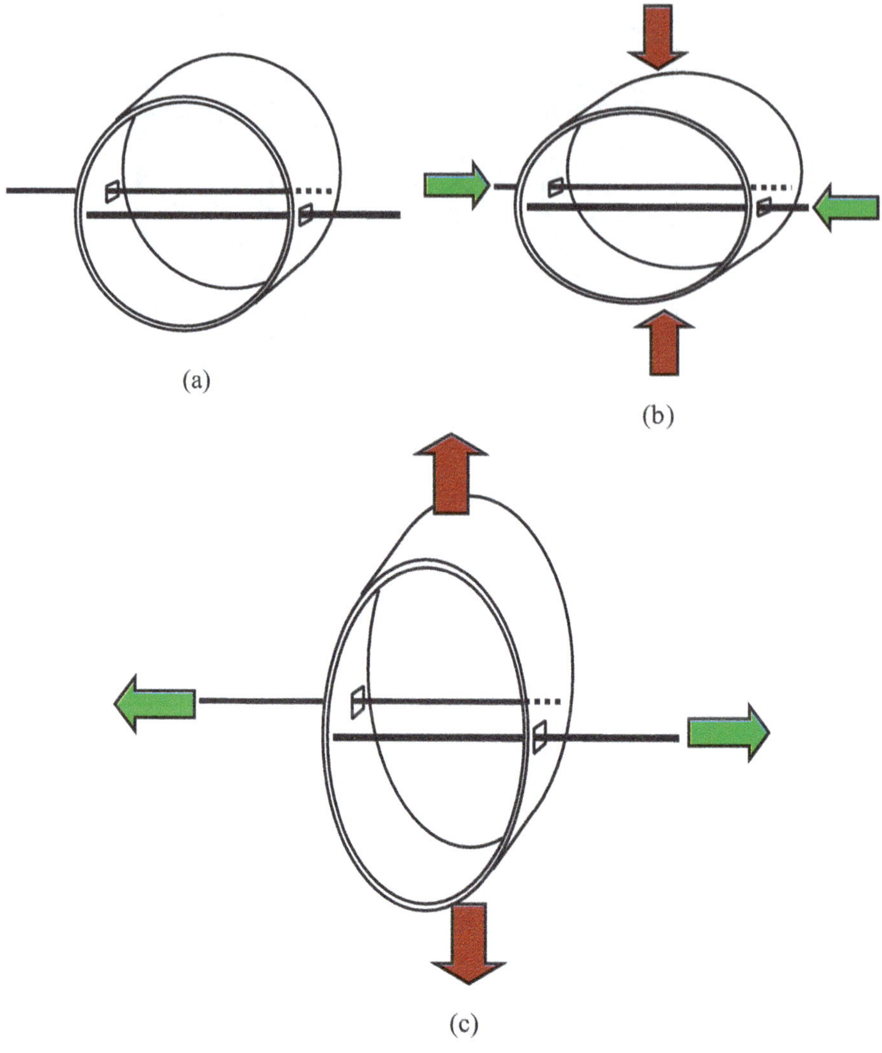

Fig. 2.10 **a** A ring structure with a pair of sliding rods at rest, **b** compressive load on sliding rods causes contraction of ring dimension in transverse direction, and **c** tensile load on sliding rods causes expansion of ring dimension in transverse direction

nature of the model development. Suppose a thin ring of radius R is subjected to a tensile or compressive load of magnitude P on two opposite sides, the change in the ring dimension along the load line is (Timoshenko 1948)

$$\delta_P = \pm\left(\frac{\pi}{4} - \frac{2}{\pi}\right)\frac{PR^3}{EI} \tag{2.5.1}$$

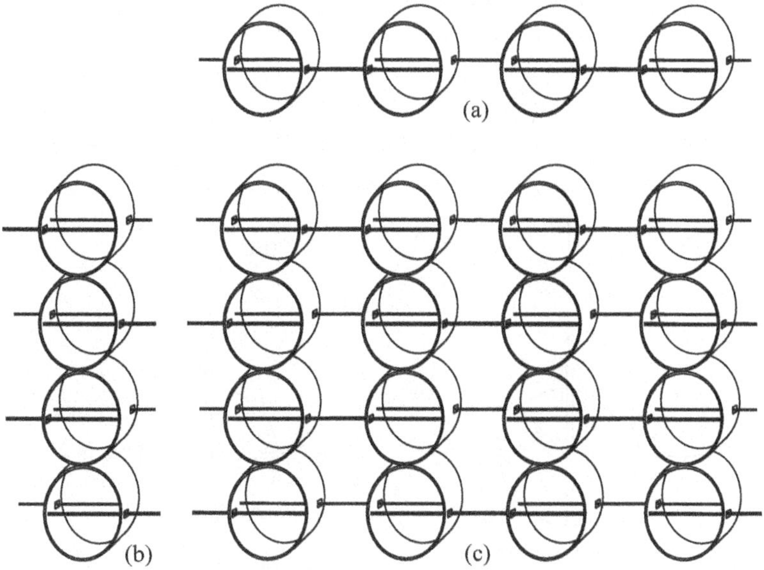

Fig. 2.11 Alternate arrangement to prevent turning of rings about any axis: **a** a series of four connected rings parallel to the sliding rods, **b** a series of four connected rings perpendicular to the sliding rods, and **c** an array of 4 by 4 rings

while the change in the ring dimension perpendicular or transverse to the load line is

$$\delta_T = \mp\left(\frac{2}{\pi} - \frac{1}{2}\right)\frac{PR^3}{EI} \tag{2.5.2}$$

where I is the second moment area of the ring's cross section about its neutral axis, and E is Young's modulus of the ring material. The upper and lower signs on the RHS of Eqs. (2.5.1) and (2.5.2) refer to the application of tensile and compressive loads, respectively, on the sliding rods.

Poisson's ratio, being defined as

$$v_{xz} = -\frac{\varepsilon_z}{\varepsilon_x} \tag{2.5.3}$$

can be used for the ring structure shown in Fig. 2.12 whereby a tensile load of P is applied on A and B to give

$$v_{xz} = -\frac{\Delta\overline{KL}}{\overline{KL}} / \frac{\Delta\overline{AB}}{\overline{AB}} = -\frac{\Delta\overline{KL}}{\Delta\overline{AB}} \times \frac{\overline{AB}}{\overline{KL}} \tag{2.5.4}$$

Suppose we let the portion of rods measured radially outward from the thin ring be l_1 and l_3 along the x and z axes, respectively, then we have the lengths of short

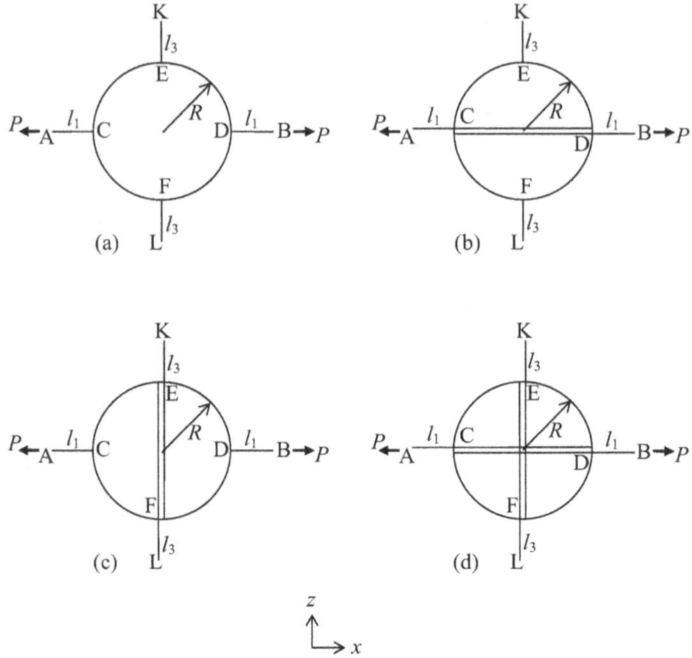

Fig. 2.12 Single ring structure exhibiting **a** conventional behavior, **b** auxetic behavior, **c** auxetic behavior, and **d** conventional behavior

rods that are fixed to the outer surface of the ring as l_1 and l_3 along the x and z axes, respectively, while the lengths of the sliding rods that are fixed to the inner surface of the rings are $(2R + l_1)$ and $(2R + l_3)$ along the x and z axes, respectively. This gives

$$v_{xz} = -\frac{\Delta \overline{KL}}{\Delta \overline{AB}}\left(\frac{R + l_1}{R + l_3}\right) \tag{2.5.5}$$

where $|\Delta \overline{AB}| = |\Delta \overline{CD}|$ and $|\Delta \overline{KL}| = |\Delta \overline{EF}|$ or

$$|\Delta \overline{AB}| = |\delta_P| = \frac{\pi^2 - 8}{4\pi}\frac{PR^3}{EI} \tag{2.5.6}$$

and

$$|\Delta \overline{KL}| = |\delta_T| = \frac{8 - 2\pi}{4\pi}\frac{PR^3}{EI} \tag{2.5.7}$$

The denominators of Eqs. (2.5.6) and (2.5.7) are made common to facilitate convenient calculations of Poisson's ratio in Table 2.2, whereby appropriate signs are

Table 2.2 Summary of effective Poisson's ratio based on $\Delta\overline{AB} > 0$ for the ring structure

Schematics. See Fig. 2.12 for details	Effective Poisson's ratio based on $\Delta\overline{AB} = \frac{\pi^2-8}{4\pi}\frac{PR^3}{EI}$	Remarks
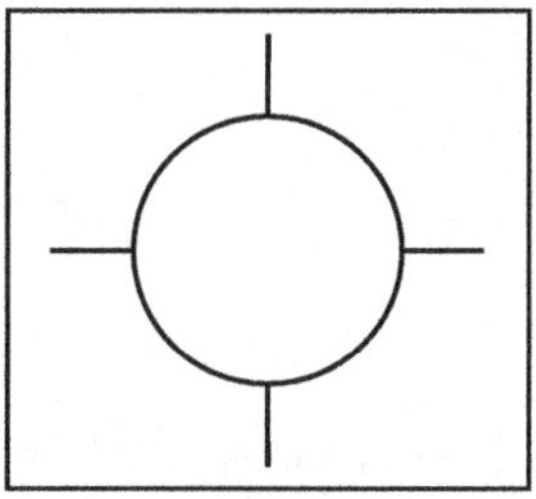	$\left\lvert\Delta\overline{KL}\right\rvert = -\frac{8-2\pi}{4\pi}\frac{PR^3}{EI}$ $\therefore \upsilon_{xz} = +\frac{8-2\pi}{\pi^2-8}\left(\frac{R+l_1}{R+l_3}\right)$	Conventional
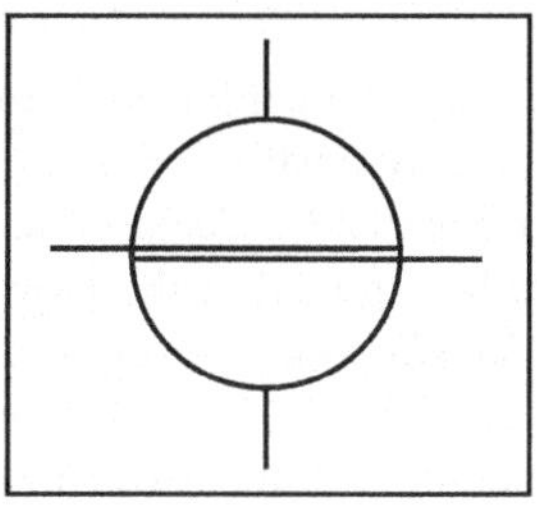	$\left\lvert\Delta\overline{KL}\right\rvert = +\frac{8-2\pi}{4\pi}\frac{PR^3}{EI}$ $\therefore \upsilon_{xz} = -\frac{8-2\pi}{\pi^2-8}\left(\frac{R+l_1}{R+l_3}\right)$	Auxetic
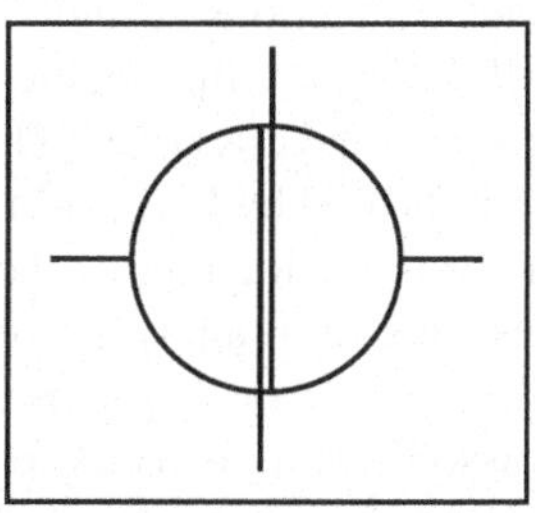	$\left\lvert\Delta\overline{KL}\right\rvert = +\frac{8-2\pi}{4\pi}\frac{PR^3}{EI}$ $\therefore \upsilon_{xz} = -\frac{8-2\pi}{\pi^2-8}\left(\frac{R+l_1}{R+l_3}\right)$	Auxetic
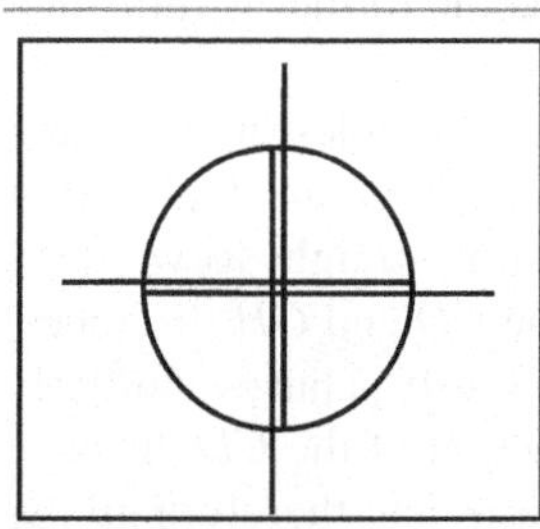	$\left\lvert\Delta\overline{KL}\right\rvert = -\frac{8-2\pi}{4\pi}\frac{PR^3}{EI}$ $\therefore \upsilon_{xz} = +\frac{8-2\pi}{\pi^2-8}\left(\frac{R+l_1}{R+l_3}\right)$	Conventional

incorporated into $\Delta\overline{KL}$ based on $\Delta\overline{AB} > 0$. For clarity, details shown in Fig. 2.12 are removed from Table 2.2. Calculations of υ_{xz} are summarized in Table 2.2 based on Eqs. (2.5.5)–(2.5.7). It is clear that of the four single ring structures displayed in Fig. 2.12, two of them—Fig. 2.12b, c—are geometrically equivalent, but differ only in the loading direction; loading is applied on the sliding rods and fixed rods,

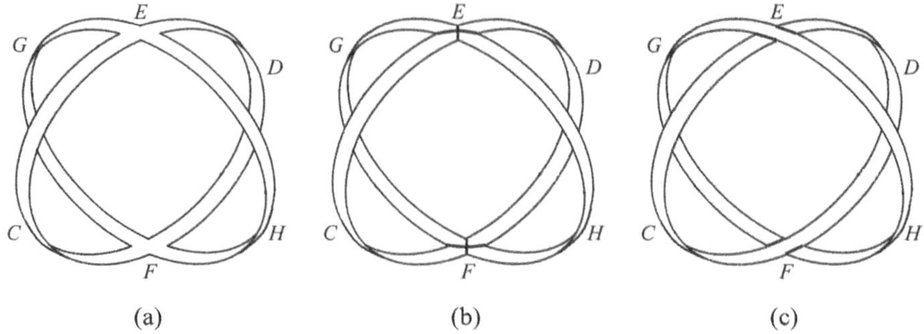

Fig. 2.13 **a** A double-ring fabricated from **b** four half rings, and **c** two incomplete rings

respectively. Hence v_{zx} for the single ring structure in Fig. 2.12b can be inferred from v_{xz} of Fig. 2.12c and vice versa. For square array, i.e., $l_1 = l_3$, we have $v_{zx} = v_{xz}$.

A 3D analysis is made herein for a double-ring shown in Fig. 2.13a. The double-ring is essentially a pair of rings of equal thickness and radius, joined at the pole. There are two possible methods to produce the double-ring. The first method requires four half rings, referred to as ECF, EDF, EGF, and EHF in Fig. 2.13b, such that perfect bonds are made at the poles E and F. The second method requires two rings as indicated by ECFDE and FGEHF in Fig. 2.13c and cut at locations E and F, respectively, to allow attachment with one another. Regardless of how the double-ring is made, the following analysis on the effective Poisson's ratio and effective CTE of the double-ring is based on perfect and seamless bonding indicated by Fig. 2.13a. Two sets of analyses are presented, i.e., Poisson's ratio analysis for loading of the double-ring along the pole direction (i.e., z-axis), and Poisson's ratio analysis for loading of the double-ring along the equator direction (i.e., either x-axis or y-axis). The effective CTE analysis is furnished in Sect. 11.11. Poisson's ratio analysis for pole-loading is simpler than that for equator-loading; hence, the former is furnished before that latter.

To qualitatively illustrate the auxeticity of the pole-loaded double-ring with and without the sliding rods, Figure 2.14 shows four sample examples of this double-ring being stretched in the z-axis. The first example, Fig. 2.14a, is straight forward, as stretching of the double-ring at points E and F contracts the CD and GH distances, thereby exhibiting conventional behavior. The double-ring also behaves similarly for the structure shown in Fig. 2.14b. However, the contraction of the CD distance pushes out the sliding rods such that the AB distance increases and therefore giving auxetic behavior in the x-z plane. Likewise the contraction of CD and GH, due to the stretching apart of E and F indicated by Fig. 2.14c, leads to elongation of the AB and IJ distances. For this structure, the pair of sliding rods AD and CB is placed slightly above the equator while the pair of sliding rods GJ and IH is placed slightly below the equatorial plane. The arrangement results in auxeticity being demonstrated in both the x-z and y-z planes, i.e., fully auxetic behavior. However, the trend reverses when a pair of sliding rods is implemented along the z-axis direction. Perusal to

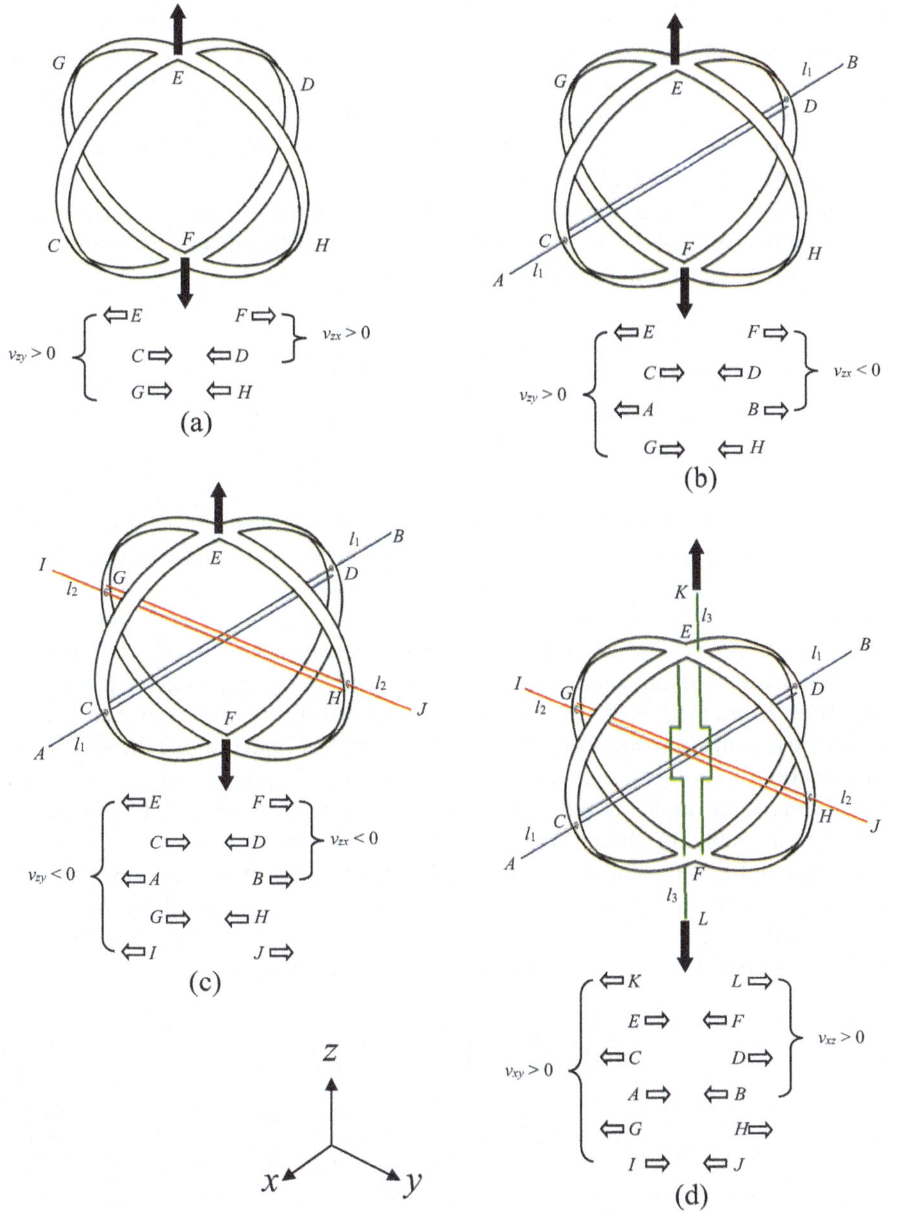

Fig. 2.14 Double-rings exhibiting conventional behavior for **a** and **d**, partial auxeticity for **b**, and full auxeticity for **c**, based on loading in *z*-axis (pole-loading)

Fig. 2.14d shows that the application of tension on *K* and *L* causes contraction to the *EF* distance, thereby extending the *CD* and *GH* distances, which in turn gives rise to the shortening of the *AB* and *IJ* distances, respectively—therefore the structure is conventional. Due to geometrical reason, the middle portion of the pairs of sliding rods *EL* and *KF* are made to avoid the other two pairs of the sliding rods in the vicinity of the double-ring's center.

For a load applied at the pole of a double-ring, the actual load on each ring is halved, and as a consequence, both the change in the pole dimension as well as the changes in the equatorial dimensions are halved, i.e.

$$\Delta\overline{EF} = \pm\left(\frac{\pi}{8} - \frac{1}{\pi}\right)\frac{PR^3}{EI} \tag{2.5.8}$$

$$\Delta\overline{CD} = \Delta\overline{GH} = \mp\left(\frac{1}{\pi} - \frac{1}{4}\right)\frac{PR^3}{EI} \tag{2.5.9}$$

where the upper and lower signs correspond to the application of tensile and compressive loads, respectively, on points *E* and *F*. In the following calculations, we consider eight combinations of double-ring with pairs of fixed short rods and/or pairs of long sliding rods. The fixed short rods are attached to the outer surface of the ring and are of lengths l_1, l_2 and l_3 aligned in the *x*, *y* and *z* axes, respectively. The long sliding rods are attached to the inner surface of the ring and are of lengths $2R + l_1$, $2R + l_2$ and $2R + l_3$ oriented in the *x*, *y* and *z* axes, respectively, so that the partial lengths measured radially outward from the ring surface are l_1, l_2 and l_3 along the three axes. The locations of points *A* to *L*, the Cartesian axes, and the pole-loading direction for the schematics in Table 2.3 follow the convention described in Fig. 2.14, but are not shown in the table for clarity. In the calculation of Poisson's ratio, a tensile load *P* is applied on the ends of either the pair of fixed rods or the pair of sliding rods, i.e., *K* and *L*, so that the strain in the loading direction (*z*-axis) is positive. We further recognize that the elongation, or shortening magnitudes, of *KL*, *AB*, and *IJ* must be equal to those of *EF*, *CD*, and *GH* on the basis of rigid rods, i.e.

$$\left|\Delta\overline{KL}\right| = \left(\frac{\pi}{8} - \frac{1}{\pi}\right)\frac{PR^3}{EI} \tag{2.5.10}$$

$$\left|\Delta\overline{AB}\right| = \left|\Delta\overline{IJ}\right| = \left(\frac{1}{\pi} - \frac{1}{4}\right)\frac{PR^3}{EI} \tag{2.5.11}$$

For pole-loading, Poisson's ratio

$$v_{zx} = -\frac{\varepsilon_x}{\varepsilon_z} \tag{2.5.12}$$

and

Table 2.3 Summary of effective Poisson's ratio and effective CTE for the double-ring structure

Schematics	Effective υ based on z-axis loading with $\Delta\overline{KL} = \left(\frac{\pi}{8} - \frac{1}{\pi}\right)\frac{PR^3}{EI}$	Effective υ based on x-axis loading with $\Delta\overline{AB} = \{\ldots\}PR^3/EI$. See Eqs. (2.5.30), (2.5.31), (2.5.32) for $\{\ldots\}$, $(\ldots)$ and $[\ldots]$, respectively
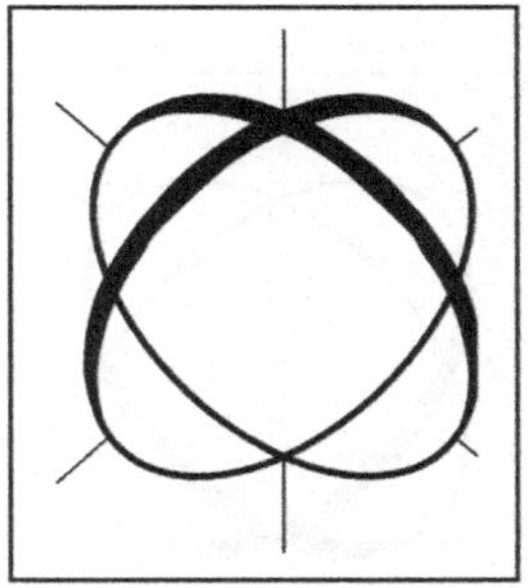	$\begin{Bmatrix} \Delta(\overline{AB}) \\ \Delta(\overline{IJ}) \end{Bmatrix} = \left(\frac{1}{\pi} - \frac{1}{4}\right)\frac{PR^3}{EI}\begin{Bmatrix} -1 \\ -1 \end{Bmatrix}$ $\therefore \begin{matrix} \upsilon_{zx} = +\frac{8-2\pi}{\pi^2-8}\left(\frac{R+l_3}{R+l_1}\right) \\ \upsilon_{zy} = +\frac{8-2\pi}{\pi^2-8}\left(\frac{R+l_3}{R+l_2}\right) \end{matrix}$ Conventional	$\Delta\overline{KL} = -(\ldots)PR^3/EI$ $\Delta\overline{IJ} = +[\ldots]PR^3/EI$ $\therefore \begin{matrix} \upsilon_{xz} = +\left[\frac{\pi^2-8}{4-\pi} - \frac{8-2\pi}{\pi^2-8}\right]^{-1}\frac{R+l_1}{R+l_3} \\ \upsilon_{xy} = -\left[\frac{1}{2}\left(\frac{\pi^2-8}{4-\pi}\right)^2 - 1\right]^{-1}\frac{R+l_1}{R+l_2} \end{matrix}$ Partially auxetic
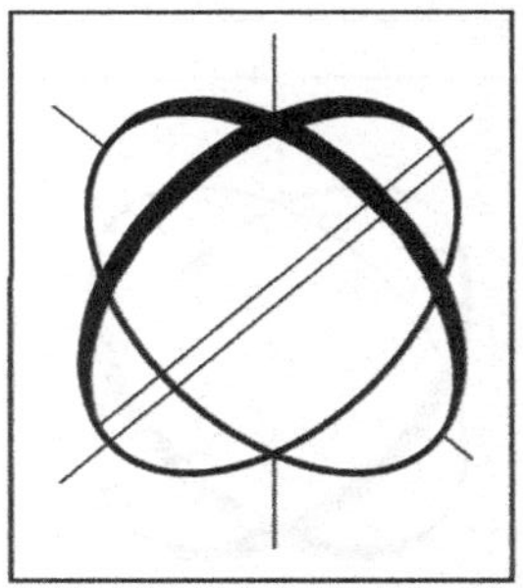	$\begin{Bmatrix} \Delta(\overline{AB}) \\ \Delta(\overline{IJ}) \end{Bmatrix} = \left(\frac{1}{\pi} - \frac{1}{4}\right)\frac{PR^3}{EI}\begin{Bmatrix} +1 \\ -1 \end{Bmatrix}$ $\therefore \begin{matrix} \upsilon_{zx} = -\frac{8-2\pi}{\pi^2-8}\left(\frac{R+l_3}{R+l_1}\right) \\ \upsilon_{zy} = +\frac{8-2\pi}{\pi^2-8}\left(\frac{R+l_3}{R+l_2}\right) \end{matrix}$ Partially auxetic	$\Delta\overline{KL} = +(\ldots)PR^3/EI$ $\Delta\overline{IJ} = -[\ldots]PR^3/EI$ $\therefore \begin{matrix} \upsilon_{xz} = -\left[\frac{\pi^2-8}{4-\pi} - \frac{8-2\pi}{\pi^2-8}\right]^{-1}\frac{R+l_1}{R+l_3} \\ \upsilon_{xy} = +\left[\frac{1}{2}\left(\frac{\pi^2-8}{4-\pi}\right)^2 - 1\right]^{-1}\frac{R+l_1}{R+l_2} \end{matrix}$ Partially auxetic

(continued)

Table 2.3 (continued)

Schematics	Effective υ based on z-axis loading with $\Delta\overline{KL} = \left(\frac{\pi}{8} - \frac{1}{\pi}\right)\frac{PR^3}{EI}$	Effective υ based on x-axis loading with $\Delta\overline{AB} = \{\ldots\}PR^3/EI$. See Eqs. (2.5.30), (2.5.31), (2.5.32) for $\{\ldots\}$, $(\ldots)$ and $[\ldots]$, respectively
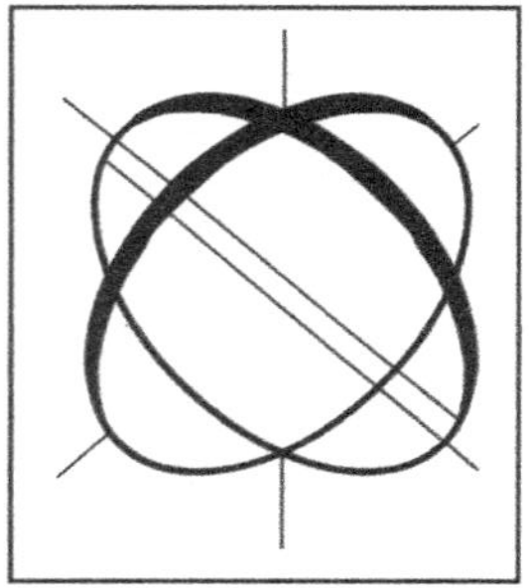	$\begin{Bmatrix} \Delta(\overline{AB}) \\ \Delta(\overline{IJ}) \end{Bmatrix} = \left(\frac{1}{\pi} - \frac{1}{4}\right)\frac{PR^3}{EI}\begin{Bmatrix} -1 \\ +1 \end{Bmatrix}$ $\therefore \begin{matrix} \upsilon_{zx} = +\frac{8-2\pi}{\pi^2-8}\left(\frac{R+l_3}{R+l_1}\right) \\ \upsilon_{zy} = -\frac{8-2\pi}{\pi^2-8}\left(\frac{R+l_3}{R+l_2}\right) \end{matrix}$ Partially auxetic	$\Delta\overline{KL} = -(\ldots)PR^3/EI$ $\Delta\overline{IJ} = -[\ldots]PR^3/EI$ $\therefore \begin{matrix} \upsilon_{xz} = +\left[\frac{\pi^2-8}{4-\pi} - \frac{8-2\pi}{\pi^2-8}\right]^{-1}\frac{R+l_1}{R+l_3} \\ \upsilon_{xy} = +\left[\frac{1}{2}\left(\frac{\pi^2-8}{4-\pi}\right)^2 - 1\right]^{-1}\frac{R+l_1}{R+l_2} \end{matrix}$ Conventional
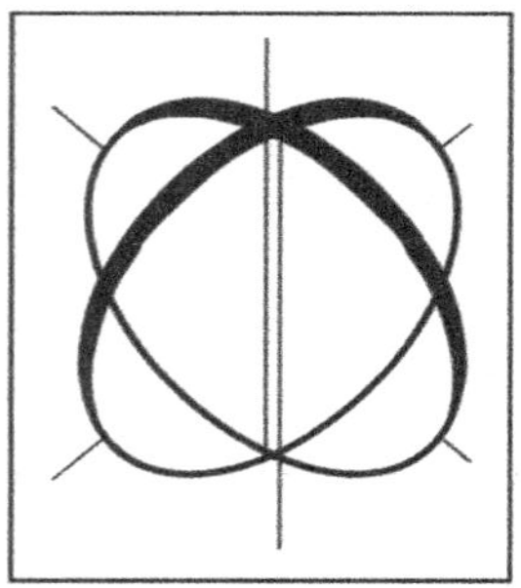	$\begin{Bmatrix} \Delta(\overline{AB}) \\ \Delta(\overline{IJ}) \end{Bmatrix} = \left(\frac{1}{\pi} - \frac{1}{4}\right)\frac{PR^3}{EI}\begin{Bmatrix} +1 \\ +1 \end{Bmatrix}$ $\therefore \begin{matrix} \upsilon_{zx} = -\frac{8-2\pi}{\pi^2-8}\left(\frac{R+l_3}{R+l_1}\right) \\ \upsilon_{zy} = -\frac{8-2\pi}{\pi^2-8}\left(\frac{R+l_3}{R+l_2}\right) \end{matrix}$ Fully auxetic	$\Delta\overline{KL} = +(\ldots)PR^3/EI$ $\Delta\overline{IJ} = +[\ldots]PR^3/EI$ $\therefore \begin{matrix} \upsilon_{xz} = -\left[\frac{\pi^2-8}{4-\pi} - \frac{8-2\pi}{\pi^2-8}\right]^{-1}\frac{R+l_1}{R+l_3} \\ \upsilon_{xy} = -\left[\frac{1}{2}\left(\frac{\pi^2-8}{4-\pi}\right)^2 - 1\right]^{-1}\frac{R+l_1}{R+l_2} \end{matrix}$ Fully auxetic

(continued)

Table 2.3 (continued)

Schematics	Effective v based on z-axis loading with $\Delta\overline{KL} = \left(\frac{\pi}{8} - \frac{1}{\pi}\right)\frac{PR^3}{EI}$	Effective v based on x-axis loading with $\Delta\overline{AB} = \{\ldots\}PR^3/EI$. See Eqs. (2.5.30), (2.5.31), (2.5.32) for $\{\ldots\}$, $(\ldots)$ and $[\ldots]$, respectively
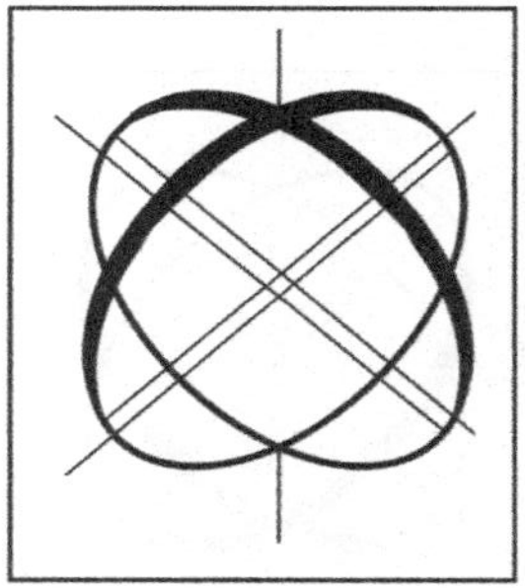	$\begin{Bmatrix} \Delta(\overline{AB}) \\ \Delta(\overline{IJ}) \end{Bmatrix} = \left(\frac{1}{\pi} - \frac{1}{4}\right)\frac{PR^3}{EI}\begin{Bmatrix} +1 \\ +1 \end{Bmatrix}$ $\therefore \begin{matrix} v_{zx} = -\frac{8-2\pi}{\pi^2-8}\left(\frac{R+l_3}{R+l_1}\right) \\ v_{zy} = -\frac{8-2\pi}{\pi^2-8}\left(\frac{R+l_3}{R+l_2}\right) \end{matrix}$ Fully auxetic	$\Delta\overline{KL} = +(\ldots)PR^3/EI$ $\Delta\overline{IJ} = +[\ldots]PR^3/EI$ $\therefore \begin{matrix} v_{xz} = -\left[\frac{\pi^2-8}{4-\pi} - \frac{8-2\pi}{\pi^2-8}\right]^{-1}\frac{R+l_1}{R+l_3} \\ v_{xy} = -\left[\frac{1}{2}\left(\frac{\pi^2-8}{4-\pi}\right)^2 - 1\right]^{-1}\frac{R+l_1}{R+l_2} \end{matrix}$ Fully auxetic
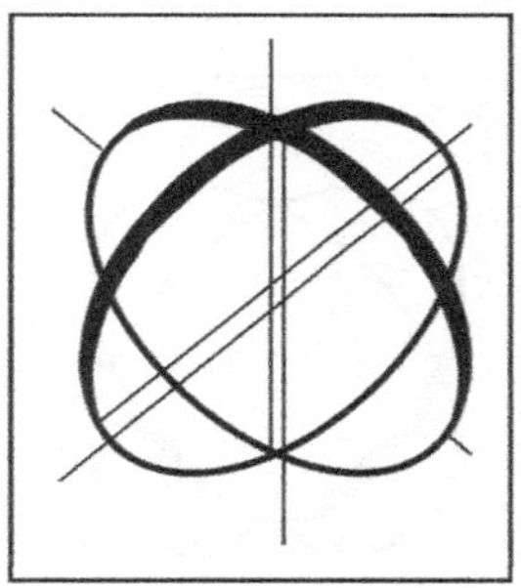	$\begin{Bmatrix} \Delta(\overline{AB}) \\ \Delta(\overline{IJ}) \end{Bmatrix} = \left(\frac{1}{\pi} - \frac{1}{4}\right)\frac{PR^3}{EI}\begin{Bmatrix} -1 \\ +1 \end{Bmatrix}$ $\therefore \begin{matrix} v_{zx} = +\frac{8-2\pi}{\pi^2-8}\left(\frac{R+l_3}{R+l_1}\right) \\ v_{zy} = -\frac{8-2\pi}{\pi^2-8}\left(\frac{R+l_3}{R+l_2}\right) \end{matrix}$ Partially auxetic	$\Delta\overline{KL} = -(\ldots)PR^3/EI$ $\Delta\overline{IJ} = -[\ldots]PR^3/EI$ $\therefore \begin{matrix} v_{xz} = +\left[\frac{\pi^2-8}{4-\pi} - \frac{8-2\pi}{\pi^2-8}\right]^{-1}\frac{R+l_1}{R+l_3} \\ v_{xy} = +\left[\frac{1}{2}\left(\frac{\pi^2-8}{4-\pi}\right)^2 - 1\right]^{-1}\frac{R+l_1}{R+l_2} \end{matrix}$ Conventional

(continued)

Table 2.3 (continued)

Schematics	Effective υ based on z-axis loading with $\Delta\overline{KL} = \left(\frac{\pi}{8} - \frac{1}{\pi}\right)\frac{PR^3}{EI}$	Effective υ based on x-axis loading with $\Delta\overline{AB} = \{\ldots\}PR^3/EI$. See Eqs. (2.5.30), (2.5.31), (2.5.32) for $\{\ldots\}$, $(\ldots)$ and $[\ldots]$, respectively
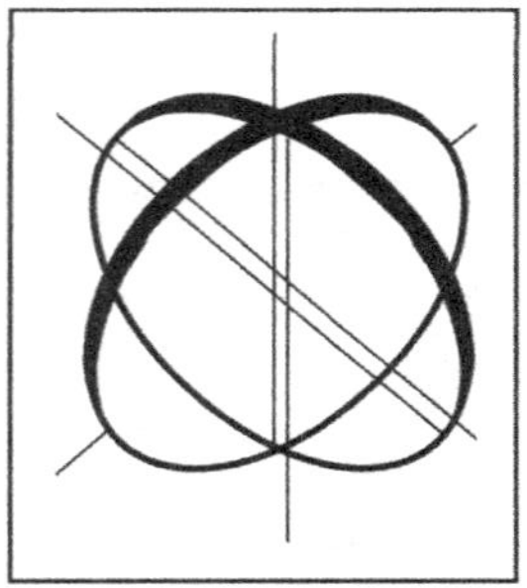	$\begin{Bmatrix} \Delta(\overline{AB}) \\ \Delta(\overline{IJ}) \end{Bmatrix} = \left(\frac{1}{\pi} - \frac{1}{4}\right)\frac{PR^3}{EI}\begin{Bmatrix} +1 \\ -1 \end{Bmatrix}$ $\therefore \begin{matrix} \upsilon_{zx} = -\frac{8-2\pi}{\pi^2-8}\left(\frac{R+l_3}{R+l_1}\right) \\ \upsilon_{zy} = +\frac{8-2\pi}{\pi^2-8}\left(\frac{R+l_3}{R+l_2}\right) \end{matrix}$ Partially auxetic	$\Delta\overline{KL} = +(\ldots)PR^3/EI$ $\Delta\overline{IJ} = -[\ldots]PR^3/EI$ $\therefore \begin{matrix} \upsilon_{xz} = -\left[\frac{\pi^2-8}{4-\pi} - \frac{8-2\pi}{\pi^2-8}\right]^{-1}\frac{R+l_1}{R+l_3} \\ \upsilon_{xy} = +\left[\frac{1}{2}\left(\frac{\pi^2-8}{4-\pi}\right)^2 - 1\right]^{-1}\frac{R+l_1}{R+l_2} \end{matrix}$ Partially auxetic
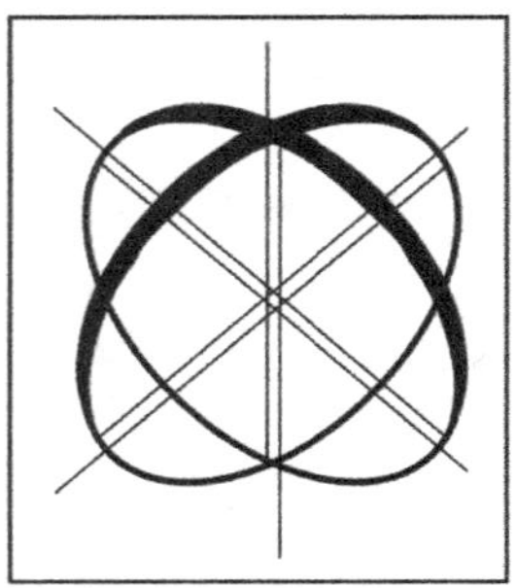	$\begin{Bmatrix} \Delta(\overline{AB}) \\ \Delta(\overline{IJ}) \end{Bmatrix} = \left(\frac{1}{\pi} - \frac{1}{4}\right)\frac{PR^3}{EI}\begin{Bmatrix} -1 \\ -1 \end{Bmatrix}$ $\therefore \begin{matrix} \upsilon_{zx} = +\frac{8-2\pi}{\pi^2-8}\left(\frac{R+l_3}{R+l_1}\right) \\ \upsilon_{zy} = +\frac{8-2\pi}{\pi^2-8}\left(\frac{R+l_3}{R+l_2}\right) \end{matrix}$ Conventional	$\Delta\overline{KL} = -(\ldots)PR^3/EI$ $\Delta\overline{IJ} = +[\ldots]PR^3/EI$ $\therefore \begin{matrix} \upsilon_{xz} = +\left[\frac{\pi^2-8}{4-\pi} - \frac{8-2\pi}{\pi^2-8}\right]^{-1}\frac{R+l_1}{R+l_3} \\ \upsilon_{xy} = -\left[\frac{1}{2}\left(\frac{\pi^2-8}{4-\pi}\right)^2 - 1\right]^{-1}\frac{R+l_1}{R+l_2} \end{matrix}$ Partially auxetic

$$v_{zy} = -\frac{\varepsilon_y}{\varepsilon_z} \quad (2.5.13)$$

can be written for the double-ring structure in Fig. 2.14 as

$$v_{zx} = -\frac{\Delta\overline{AB}}{\overline{AB}} \bigg/ \frac{\Delta\overline{KL}}{\overline{KL}} = -\frac{\Delta\overline{AB}}{\Delta\overline{KL}} \times \frac{\overline{KL}}{\overline{AB}} \quad (2.5.14)$$

and

$$v_{zy} = -\frac{\Delta\overline{IJ}}{\overline{IJ}} \bigg/ \frac{\Delta\overline{KL}}{\overline{KL}} = -\frac{\Delta\overline{IJ}}{\Delta\overline{KL}} \times \frac{\overline{KL}}{\overline{IJ}} \quad (2.5.15)$$

respectively. In terms of double-ring geometry, Poisson's ratios are written as

$$v_{zx} = -\frac{\Delta\overline{AB}}{\Delta\overline{KL}}\left(\frac{R+l_3}{R+l_1}\right) \quad (2.5.16)$$

and

$$v_{zy} = -\frac{\Delta\overline{IJ}}{\Delta\overline{KL}}\left(\frac{R+l_3}{R+l_2}\right) \quad (2.5.17)$$

Specific calculations of v_{zx} and v_{zy} from Eqs. (2.5.16) and (2.5.17) are furnished in Table 2.3 with incorporation of appropriate signs for Eqs. (2.5.10) and (2.5.11).

To qualitatively illustrate the auxeticity of the equator-loaded double-ring with and without the sliding rods, Figure 2.15 shows four sample examples of this double-ring being stretched in the x-axis. Figure 2.15a shows a double-ring without sliding rods being pulled at points C and D such that the EF distance decreases while the GH distance increases, hence producing auxeticity in the x-y plane while the x-z plane remains conventional. In the case of a pair of sliding rods aligned in the x-direction, as shown in Fig. 2.15b, the application of stretching force on points A and B brings C and D closer together and, consequently, EF and GH distances increases and decreases, respectively. This gives auxetic and conventional behavior in the x-z and x-y planes, respectively. Figure 2.15a, b is therefore partially auxetic structures. Figure 2.15c gives an example of two pairs of sliding rods. The deformation of the double-ring in Fig. 2.15c is similar to that in Fig. 2.15b; however, the motion of G and H toward each other translates to increasing distance of I and J. The increasing distance of EF and IJ arising from stretching of AB therefore exhibits auxeticity in both x-y and x-z planes and hence full auxeticity. The final example of three pairs of sliding rods illustrated in Fig. 2.15d gives the same ring deformation as those in Fig. 2.15b, c. However, the elongation of the EF distance leads to contraction of the KL distance; the resulting auxetic and conventional characteristics in the x-y and x-z

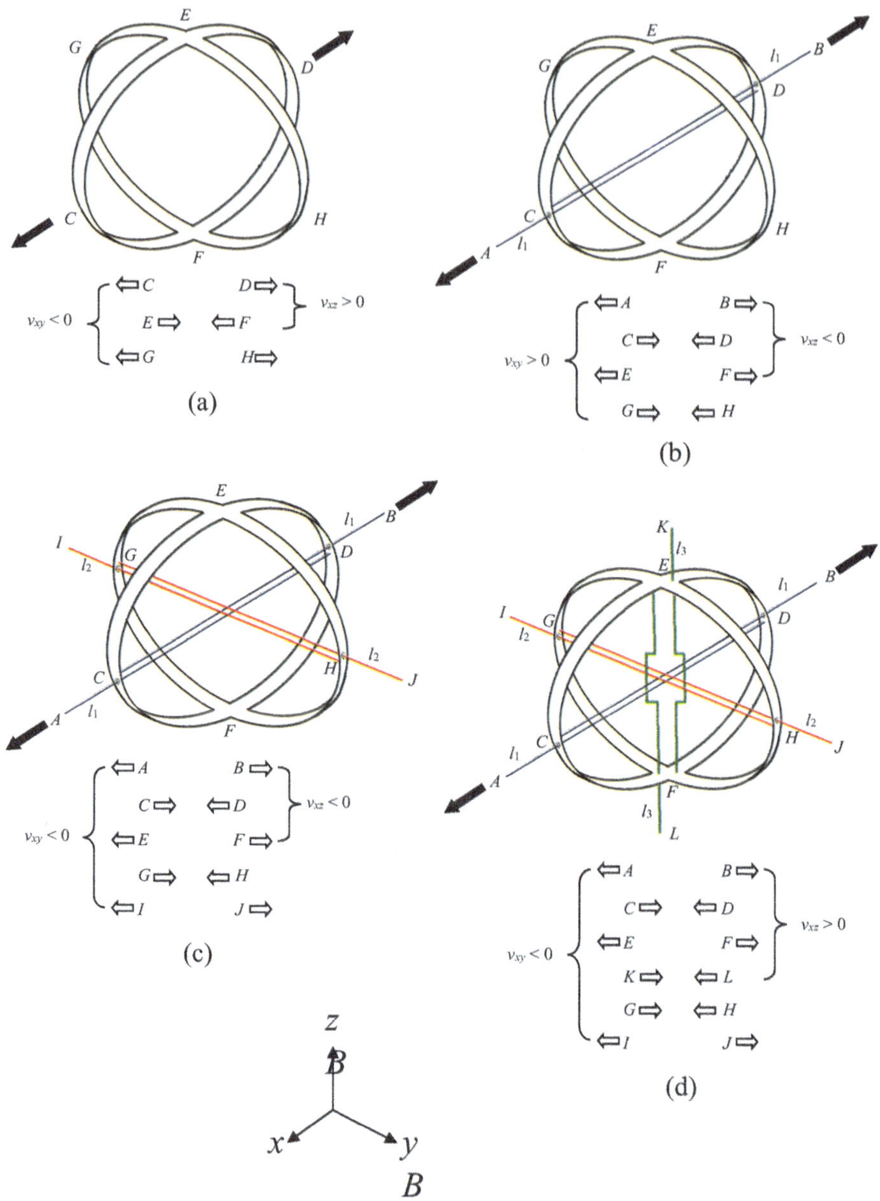

Fig. 2.15 Double-rings exhibiting auxeticity on **a** *x*-*y* plane, **b** *x*-*z* plane, **c** both *x*-*y* and *x*-*z* planes, and **d** *x*-*y* plane, based on loading in *x*-axis (equator-loading)

planes, respectively, puts this double-ring and triple pair sliding rod structure under the category of partially auxetic structure.

In the 3D analysis of Poisson's ratio for equator-loading, a two-stage calculation is imposed. The first stage assumes that the double-ring consists of two unbonded rings, in which one of them is loaded at two opposite points indicated by C and D in Fig. 2.15a. The arising gap between both unbonded rings at E and F is then eliminated in the second stage calculation by applying an internal load between both rings at E and F, thereby leading to changes in the GH distance as well as a second change in the CD distance.

In the first stage calculation, a load P is applied on C and D such that the CD distance changes by

$$\Delta\left(\overline{CD}\right)_1 = \pm\delta_{P1} = \pm\left(\frac{\pi}{4} - \frac{2}{\pi}\right)\frac{PR^3}{EI} \tag{2.5.18}$$

where subscript "#1" refers to the first stage calculation, while the EF distance changes by

$$\Delta\left(\overline{EF}\right)_1^{\#1} = \mp\delta_{T1} = \mp\left(\frac{2}{\pi} - \frac{1}{2}\right)\frac{PR^3}{EI} \tag{2.5.19}$$

where superscript "#1" refers to the first ring indicated by CEDF. The upper and lower signs correspond to the application of tensile and compressive loads, respectively.

By geometrical compatibility, the EF distance for both rings must be made equal. Hence, an internal force Q is applied to E and F on both rings to bridge the gap. Let $\delta_{Q2}^{\#1}$ and $\delta_{Q2}^{\#2}$ be the change for the EF distance in the first ring (CEDF) and second ring (GEHF), respectively, due solely to the application of Q in the second stage calculation such that

$$\left|\delta_{Q2}^{\#1}\right| + \left|\delta_{Q2}^{\#2}\right| = \left|\delta_{T1}\right| \tag{2.5.20}$$

where subscript "2" refers to the second stage calculation while superscript "#2" refers to the second ring indicated by GEHF. Let Q be the internal force on the first ring to change the EF distance by $\delta_{Q2}^{\#1}$,

$$\Delta\left(\overline{EF}\right)_2^{\#1} = \pm\delta_{Q2}^{\#1} = \pm\left(\frac{\pi}{4} - \frac{2}{\pi}\right)\frac{QR^3}{EI} \tag{2.5.21}$$

while the resulting change in the CD distance is

$$\Delta\left(\overline{CD}\right)_2 = \mp\delta_{P2} = \mp\left(\frac{2}{\pi} - \frac{1}{2}\right)\frac{QR^3}{EI} \tag{2.5.22}$$

Recognizing that Q is also an opposite internal force on E and F of the second ring, we have the change in EF distance for the second ring

$$\Delta\left(\overline{EF}\right)_2^{\#2} = \mp\delta_{Q2}^{\#2} = \mp\left(\frac{\pi}{4} - \frac{2}{\pi}\right)\frac{QR^3}{EI} \tag{2.5.23}$$

and the corresponding change in GH distance

$$\Delta\overline{GH} = \pm\delta_{W2} = \pm\left(\frac{2}{\pi} - \frac{1}{2}\right)\frac{QR^3}{EI} \tag{2.5.24}$$

Since $\left|\delta_{Q2}^{\#1}\right| = \left|\delta_{Q2}^{\#2}\right|$ as evidenced from Eqs. (2.5.21) and (2.5.23), it is possible to express the internal force Q in terms of the applied load P from Eq. (2.5.20)

$$\left|\delta_{Q2}^{\#1}\right| = \left|\delta_{Q2}^{\#2}\right| = \frac{1}{2}\left|\delta_{T1}\right| \tag{2.5.25}$$

to give

$$Q = \frac{4-\pi}{\pi^2 - 8}P \tag{2.5.26}$$

Expressing all the deformations in terms of applied load P, it is possible to obtain the overall change in dimension. The overall change to the CD distance is summed from Eqs. (2.5.18) and (2.5.22) from both calculation stages to give

$$\Delta\overline{CD} = \Delta\left(\overline{CD}\right)_1 + \Delta\left(\overline{CD}\right)_2 = \pm\left\{\frac{\pi^2 - 8}{4\pi} \mp \frac{(4-\pi)^2}{2\pi\left(\pi^2 - 8\right)}\right\}\frac{PR^3}{EI} \tag{2.5.27}$$

The overall change to the EF distance can be obtained by two ways. The first way is to sum up the change of EF distance on the first ring from both stages of calculation using Eqs. (2.5.19) and (2.5.21). The second, and more convenient, way is to recognize that the second stage calculation for the change in EF distance for the second ring, i.e., Equation (2.5.23), is the overall change in the EF distance itself. Hence

$$\Delta\overline{EF} = \Delta\left(\overline{EF}\right)_1^{\#1} + \Delta\left(\overline{EF}\right)_2^{\#1} = \Delta\left(\overline{EF}\right)_2^{\#2} = \mp\left(\frac{1}{\pi} - \frac{1}{4}\right)\frac{PR^3}{EI} \tag{2.5.28}$$

Substituting Eq. (2.5.26) into Eq. (2.5.24), we have the change in GH distance

$$\Delta\overline{GH} = \pm\left[\frac{(4-\pi)^2}{2\pi\left(\pi^2 - 8\right)}\right]\frac{PR^3}{EI} \tag{2.5.29}$$

Note that the expressions contained within the parentheses {...}, (...) and [...] in Eqs. (2.5.27), (2.5.28) and (2.5.29), respectively, are positive values; the extension and contraction are taken care by the upper and lower signs, which correspond to the application of tensile and compressive loads, respectively. In the following calculations, we consider again eight combinations of double-ring with pairs of short fixed rods and/or pairs of long sliding rods. The fixed short rods (l_1, l_2, l_3) and long sliding rods $(2R + l_1, 2R + l_2, 2R + l_3)$ are oriented along the x, y, z axes. The locations of points A to L, the Cartesian axes, and the equator-loading direction for the schematics follow the convention described in Fig. 2.15 and are not shown in Table 2.3 for clarity. Similarly in the calculation of Poisson's ratio, a tensile load P is applied on the ends of either the pair of fixed rods or the pair of sliding rods, i.e., A and B, so that the strain in the loading direction (x-axis) is positive. We further recognize that the elongation magnitudes of AB, KL, and IJ must be equal to the elongation magnitudes of CD, EF, and GH on the basis of rigid rods, i.e.

$$\left|\Delta\overline{AB}\right| = \left\{\frac{\pi^2 - 8}{4\pi} \mp \frac{(4-\pi)^2}{2\pi\left(\pi^2 - 8\right)}\right\}\frac{PR^3}{EI} \tag{2.5.30}$$

$$\left|\Delta\overline{KL}\right| = \left(\frac{1}{\pi} - \frac{1}{4}\right)\frac{PR^3}{EI} \tag{2.5.31}$$

$$\left|\Delta\overline{IJ}\right| = \left[\frac{(4-\pi)^2}{2\pi\left(\pi^2 - 8\right)}\right]\frac{PR^3}{EI} \tag{2.5.32}$$

Hence, Poisson's ratio in the x-z plane $v_{xz} = -\varepsilon_z/\varepsilon_x$ is described by Eq. (2.5.4) while Poisson's ratio in the x-y plane $v_{xy} = -\varepsilon_y/\varepsilon_x$ is written as

$$v_{xy} = -\frac{\Delta\overline{IJ}}{\overline{IJ}}\bigg/\frac{\Delta\overline{AB}}{\overline{AB}} = -\frac{\Delta\overline{IJ}}{\Delta\overline{AB}} \times \frac{\overline{AB}}{\overline{IJ}} \tag{2.5.33}$$

In terms of double-ring geometry, Poisson's ratio v_{xz} is described by Eq. (2.5.5), whereby $\Delta\overline{KL}$ for double-ring is half that of a single ring while $\Delta\overline{AB}$ for double-ring is different from that of a single ring. The corresponding Poisson's ratio in the x-y plane is

$$v_{xy} = -\frac{\Delta\overline{IJ}}{\Delta\overline{AB}}\left(\frac{R + l_1}{R + l_2}\right) \tag{2.5.34}$$

when expressed in terms of double-ring geometry. Poisson's ratio for equator-loaded double-ring with pairs of sliding rods and/or fixed rods can be calculated from Eqs. (2.5.5) and (2.5.34) using appropriate incorporation of signs for Eqs. (2.5.30) to (2.5.32); a summary of Poisson's ratio calculations based on equator-loading for the eight specific cases of double-ring is furnished Table 2.3.

Poisson's ratio υ_{zx} and υ_{zy} expressions using Eqs. (2.5.16) and (2.5.17) for pole-loading as well as υ_{xz} and υ_{xy} expressions using Eqs. (2.5.5) and (2.5.34) for equator-loading are summarized in Table 2.3. No expressions were made for equator-loading in the y-axis as υ_{yz} and υ_{yx} can be inferred from υ_{xz} and υ_{xy}. Appropriate signs have been incorporated into Eqs. (2.5.10) and (2.5.11) as well as for Eqs. (2.5.30) to (2.5.32) for pole-loading and equator-loading cases, respectively. The incorporation of signs adheres to the following rules: (a) the extension (or contraction) of CD distance leads to the contraction (or extension) and extension (or contraction) of EF and GH, respectively, and (b) the ends of the sliding rods possess opposite motion from the point of the double-rings where the sliding rods pass through. Here, we introduce the terms "conventional" for loading which leads to positive Poisson's ratio in two planes, "partially auxetic" for loading that gives positive Poisson's ratio in one plane and negative Poisson's ratio in another plane, and "fully auxetic" behavior for loading which gives rise to negative Poisson's ratio in two planes. We can see that when the double-ring structure is fully auxetic under pole-loading, it is also fully auxetic under equator-loading. To a certain extent, the double-ring is partially auxetic under pole-loading if it is also partially auxetic under equator-loading. However, it is of interest to note that, to a large extent, there is a combination of conventional and partially auxetic behavior depending on the loading direction.

The 2D and 3D analyses of ring and double-ring structures, respectively, with combinations of short fixed rods and/or long sliding rods provide Poisson's ratio values that are strongly determined by the ratio of the ring radius and the rod lengths. For example, the 2D analyses summarized in Table 2.2 show that Poisson's ratio magnitudes for all the four combinations are

$$|\upsilon_{xz}| = \frac{8-2\pi}{\pi^2-8}\left(\frac{R+l_1}{R+l_3}\right) = 0.918\frac{R+l_1}{R+l_3} \tag{2.5.35}$$

Likewise, strong influence from ring radius and rod length on the Poisson's ratio is observed for the 3D analysis. Table 2.3 indicates that Poisson's ratio magnitudes for all the eight combinations are

$$|\upsilon_{zx}| = \frac{8-2\pi}{\pi^2-8}\left(\frac{R+l_3}{R+l_1}\right) = 0.918\frac{R+l_3}{R+l_1} \tag{2.5.36}$$

and

$$|\upsilon_{zy}| = \frac{8-2\pi}{\pi^2-8}\left(\frac{R+l_3}{R+l_2}\right) = 0.918\frac{R+l_3}{R+l_2} \tag{2.5.37}$$

for pole-loading (z-axis), while

$$|\upsilon_{xz}| = \left[\frac{\pi^2-8}{4-\pi} - \frac{8-2\pi}{\pi^2-8}\right]^{-1}\frac{R+l_1}{R+l_3} = 0.794\frac{R+l_1}{R+l_3} \tag{2.5.38}$$

and

$$\left|v_{xy}\right| = \left[\frac{1}{2}\left(\frac{\pi^2 - 8}{4 - \pi}\right)^2 - 1\right]^{-1} \frac{R + l_1}{R + l_2} = 0.729\frac{R + l_1}{R + l_2} \tag{2.5.39}$$

for equator-loading (x-axis).

We shall now consider a few special cases. For the first special case of $l_1 = l_2 = l_3$, i.e., the short fixed rods are of equal length l and the long sliding rods are of equal length $2R + l$ such that the single ring and the double-ring structures are arranged in square and cube arrays, respectively, then

$$|v_{xz}| = \frac{8 - 2\pi}{\pi^2 - 8} = 0.918 \tag{2.5.40}$$

for the single ring structure, while

$$|v_{zx}| = \left|v_{zy}\right| = \frac{8 - 2\pi}{\pi^2 - 8} = 0.918 \tag{2.5.41}$$

for the double-ring structure under pole-loading (z-axis), and

$$|v_{xz}| = \left[\frac{\pi^2 - 8}{4 - \pi} - \frac{8 - 2\pi}{\pi^2 - 8}\right]^{-1} = 0.794 \tag{2.5.42}$$

and

$$\left|v_{xy}\right| = \left[\frac{1}{2}\left(\frac{\pi^2 - 8}{4 - \pi}\right)^2 - 1\right]^{-1} = 0.729 \tag{2.5.43}$$

for the double-ring structure under equator-loading (x-axis).

The second special case is where Poisson's ratio magnitude is set at 1. This is attainable for the single ring structure, i.e., $|v_{xz}| = 1$, if

$$\frac{R + l_1}{R + l_3} = \frac{\pi^2 - 8}{8 - 2\pi} = 1.089 \tag{2.5.44}$$

and $|v_{zx}| = 1$ or $\left|v_{zy}\right| = 1$ for the double-ring structure under pole-loading if

$$\frac{R + l_3}{R + l_1} = \frac{\pi^2 - 8}{8 - 2\pi} = 1.089 \tag{2.5.45}$$

or

$$\frac{R + l_3}{R + l_2} = \frac{\pi^2 - 8}{8 - 2\pi} = 1.089 \tag{2.5.46}$$

respectively. For equator-loading of the double-ring, $|\upsilon_{xz}| = 1$ or $|\upsilon_{xy}| = 1$ are obtained if

$$\frac{R + l_1}{R + l_3} = \frac{\pi^2 - 8}{4 - \pi} - \frac{8 - 2\pi}{\pi^2 - 8} = 1.260 \tag{2.5.47}$$

or

$$\frac{R + l_1}{R + l_2} = \frac{1}{2}\left(\frac{\pi^2 - 8}{4 - \pi}\right)^2 - 1 = 1.372 \tag{2.5.48}$$

are satisfied, respectively.

The third special case is made for extreme positive and negative Poisson's ratio. This can be achieved when a pair of rod lengths along one axis is made very large in comparison to the summation of rod lengths and radius in another axis. Specifically,

$$\begin{aligned} |\upsilon_{xz}| \to \infty \Leftrightarrow l_1 \gg R + l_3 \\ |\upsilon_{zx}| \to \infty \Leftrightarrow l_3 \gg R + l_1 \end{aligned} \tag{2.5.49}$$

for the single and double-ring structures, and

$$\begin{aligned} |\upsilon_{xy}| \to \infty \Leftrightarrow l_1 \gg R + l_2 \\ |\upsilon_{zy}| \to \infty \Leftrightarrow l_3 \gg R + l_2 \end{aligned} \tag{2.5.50}$$

for the double-ring structures.

2.6 Concluding Remarks

Thus far the examples of geometrical gradation have been taken from 2D microstructures due to their simplicity and for aiding understanding. Obviously, gradation of 3D auxetic microstructures is more varied due to greater number of possible combinations. This includes 3D partially auxetic models that exhibit auxetic behavior in one plane but non-auxetic behavior in another plane (Lim 2002a, 2007; Wojciechowski 2005; Tretiakov and Wojciechowski 2014; Piglowski et al. 2016; Tretiakov et al. 2018). Similar to previous reviews on auxetic systems, the analogies discussed in Sects. 2.2 and 2.3 cluster auxetic models according to their geometrical similarities. A less intuitive but more meaningful approach is discussed in the next chapter.

References

Abramovitch H, Burgard M, Edery-Azulay L, Evans KE, Hoffmeister M, Miller W, Scarpa F, Smith CW, Tee KF (2010) Smart tetrachiral and hexachiral honeycomb: Sensing and impact detection. Compos Sci Technol 70(7):1072–1079

Albertini F, Dirrenberger J, Molotnikov A, Sollogoub C (2019) Computational investigation of the effective mechanical behavior for 3D pre-buckled auxetic lattices. J Appl Mech 86(11):111003

Alderson A (1999) A triumph of lateral thought. Chem Ind 10:384–391

Alderson A, Alderson KL (2007) Auxetic materials. IMechE J Aerosp Eng 221(4):565–575

Alderson A, Evans KE (1995) Microstructural modelling of auxetic microporous polymers. J Mater Sci 30(13):3319–3332

Alderson A, Evans KE (1997) Modelling concurrent deformation mechanisms in auxetic microporous polymers. J Mater Sci 32(11):2797–2809

Alderson KL, Webber RS, Evans KE (2007) Microstructural evolution in the processing of auxetic microporous polymers. Phys Status Solidi B 244(3):828–841

Alderson A, Alderson KL, Chirima G, Ravirala N, Zied KM (2010a) The in-plane linear elastic constants and out-of-plane bending of 3-coordinated ligament and cylinder-ligament honeycombs. Compos Sci Technol 70(7):1034–1041

Alderson A, Alderson KL, Attard D, Evans KE, Gatt R, Grima JN, Miller W, Ravirala N, Smith CW, Zied K (2010b) Elastic constants of 3-, 4- and 6-connected chiral and anti-chiral honeycombs subject to uniaxial in-plane loading. Compos Sci Technol 70(7):1042–1048

Attard D, Grima JN (2008) Auxetic behavior from rotating rhombi. Phys Status Solidi B 245(11):2395–2404

Attard D, Grima JN (2012) A three-dimensional rotating rigid units network exhibiting negative Poisson's ratios. Phys Status Solidi B 249(7):1330–1338

Attard D, Manicaro E, Grima JN (2009a) On the properties of auxetic rotating stretching squares. Phys Status Solidi B 246(9):2045–2054

Attard D, Manicaro E, Grima JN (2009b) On rotating parallelograms and their potential for exhibiting auxetic behavior. Phys Status Solidi B 246(9):2033–2044

Babaee S, Shim J, Weaver JC, Chen ER, Patel N, Bertoldi K (2013) 3D soft metamaterials with negative Poisson's ratio. Adv Mater 25(36):5044–5049

Bahaloo H, Li Y (2019) Micropolar modeling of auxetic chiral lattices with tunable internal rotation. J Appl Mech 86(4):041002

Bertoldi K, Boyce MC, Deschanel S, Prange SM, Mullin T (2008) Mechanics of deformation-triggered pattern transformations and superelastic behavior in periodic elastomeric structures. J Mech Phys Solids 56(8):2642–2668

Bertoldi K, Reis PM, Wilshaw S, Mullin T (2010) Negative Poisson's ratio behavior induced by an elastic instability. Adv Mater 22(3):361–366

Boba K, Bianchi M, McCombe G, Gatt R, Griffin AC, Richardson RM, Scarpa F, Hamerton I, Grima JN (2016) Blocked shape memory effect in negative Poisson's ratio polymer metamaterials. ACS Appl Mater Interfaces 8(31):20319–20328

Bouakba M, Bezazi A, Scarpa F (2012) FE analysis of the in-plane mechanical properties of a novel Voronoi-type lattice with positive and negative Poisson's ratio configurations. Int J Solids Struct 49:2450–2459

Broeren FGJ, Herder JL, van der Wijk V (2019) On the synthesis of periodic linkages with a specific constant Poisson's ratio. In: Uhl T (ed) Proceedings of the 15th IFToMM World Congress on Mechanism and Machine Science. Advances in mechanism and machine science, vol n 73. Springer Nature, Cham, pp 249–257

Chen YJ, Scarpa F, Liu YJ, Leng JS (2013) Elasticity of antitetrachiral anisotropic lattices. Int J Solids Struct 50(6):996–1004

Chen Z, Liu H, Li W, Mo J, Wang M, Zhang Y, Li J, Jiang Q, Yang W, Tang C (2019a) Chiral metallic glass nanolattices with combined lower density and improved auxeticity. Phys Chem Chem Phys 21(37):20588–20594

Chen Y, Jiang N, Hu H (2019b) Mechanical modeling of an auxetic tubular braided structure: Experimental and numerical analyses. Int J Mech Sci 160:182–191

Chetcuti E, Ellul B, Manicaro E, Brincat JP, Attard D, Gatt R, Grima JN (2014) Modeling auxetic foams through semi-rigid rotating triangles. Phys Status Solidi B 251(2):297–306

Choi JB, Lakes RS (1995) Nonlinear analysis of the Poisson's ratio of negative Poisson's ratio foams. J Compos Mater 29(1):113–128

Clausen A, Wang F, Jensen JS, Sigmund O, Lewis JA (2015) Topology optimized architectures with programmable Poisson's ratio over large deformations. Adv Mater 27(37):5523–5527

Coluci VR, Hall LJ, Kozlov ME, Zhang M, Dantas SO, Galvao DS, Baughman RH (2008) Modeling the auxetic transition for carbon nanotube sheets. Phys Rev B 78(11):115408

Critchley R, Corni I, Wharton JA, Walsh FC, Wood RJK, Stokes KR (2013) A review of the manufacture, mechanical properties and potential applications of auxetic foams. Phys Status Solidi B 250(10):1963–1982

Darja R, Tatjana R, Alenka PC (2013) Auxetic textiles. Acta Chim Slov 60(4):715–723

Dubrovski PD, Noval N, Borovinsek M, Vesenjak M, Ren Z (2019) In-plane behavior of auxetic non-woven fabric based on rotating square unit geometry under tensile load. Polymers 11(6):1040

Dudek KK, Gatt R, Mizzi L, Dudek MR, Attard D, Evans KE, Grima JN (2017) On the dynamics and control of mechanical properties of hierarchical rotating rigid unit auxetics. Sci Rep 7:46529

Duncan O, Shepherd T, Moroney C, Foster L, Venkatraman PD, Winwood K, Allen T, Alderson A (2018) Review of auxetic materials for sports applications: Expanding options in comfort and protection. Appl Sci 8(6):941

Eidini M, Paulino GH (2015) Unraveling metamaterial properties in zigzag-base folded sheets. Sci Adv 1(8):e1500224

Evans KE, Nkansah MA, Hutchinson IJ, Rogers SC (1991) Molecular network design. Nature 353(6340):124

Farrugia PS, Gatt R, Lonardelli EZ, Grima JN, Evans KE (2019a) Different deformation mechanisms leading to auxetic behavior exhibited by missing rib square grid structures. Phys Status Solidi B 256(1):1800186

Farrugia PS, Gatt R, Grima JN (2019b) A novel three-dimensional anti-tetrachiral honeycomb. Phys Status Solidi B 256(1):1800473

Friis EA, Lakes RS, Park JB (1988) Negative Poisson's ratio polymeric and metallic foams. J Mater Sci 23(12):4406–4414

Fu MH, Zheng BB, Li WH (2017) A novel chiral three-dimensional material with negative Poisson's ratio and the equivalent elastic parameters. Compos Struct 176:442–448

Fu M, Liu F, Hu L (2018) A novel category of 3D chiral material with negative Poisson's ratio. Compos Sci Technol 160:111–118

Gao Z, Liu D, Tomanek D (2018) Two-dimensional mechanical metamaterials with unusual Poisson ratio behavior. Phys Rev Appl 10(6):064039

Gao R, Li D, Dong L, Wang X (2019) Numerical analysis of the mechanical properties of 3D random Voronoi structures with negative Poisson's ratio. Phys Status Solidi B 256(7):1800539

Gaspar N, Ren XJ, Smith CW, Grima JN, Evans KE (2005) Novel honeycombs with auxetic behavior. Acta Mater 53(8):2439–2445

Gaspar N, Smith CW, Alderson A, Grima JN, Evans KE (2011) A generalised three-dimensional tethered-nodule model for auxetic materials. J Mater Sci 46(2):372–384

Gatt R, Mizzi L, Azzopardi JI, Azzopardi KM, Attard D, Casha A, Briffa J, Grima JN (2015) Hierarchical auxetic mechanical metamaterials. Sci Rep 5:8395

Ge Z, Hu H (2013) Innovative three-dimensional fabric structure with negative Poisson's ratio for composite reinforcement. Text Res J 83(5):543–550

Ge Z, Hu H, Liu Y (2013) A finite element analysis of a 3D auxetic textile structure for composite reinforcement. Smart Mater Struct 22(8):084005

Ge ZY, Hu H, Liu SR (2016) A novel plied yarn structure with negative Poisson's ratio. J Text Inst 107(5):578–588

Glazzard M, Breedon P (2014) Weft-knitted auxetic textile design. Phys Status Solidi B 251(2):267–272

Greaves GN, Greer AL, Lakes RS, Rouxel T (2011) Poisson's ratio and modern materials. Nat Mater 10(11):823–837

Grima JN, Evans KE (2000) Auxetic behavior from rotating squares. J Mater Sci Lett 19(17):1563–1565

Grima JN, Evans KE (2006) Auxetic behavior from rotating triangles. J Mater Sci 41(10):3193–3196

Grima JN, Gatt R (2010) Perforated sheets exhibiting negative Poisson's ratio. Adv Eng Mater 12(6):460–464

Grima JN, Alderson A, Evans KE (2004) Negative Poisson's ratios from rotating rectangles. Computat Methods Sci Technolo 10(2):137–145

Grima JN, Gatt R, Alderson A, Evans KE (2005a) On the potential of connected stars as auxetic systems. Mol Simul 31(13):925–935

Grima JN, Gatt R, Alderson A, Evans KE (2005b) On the auxetic properties of 'rotating rectangles' with different connectivity. J Phys Soc Jpn 74(10):2866–2867

Grima JN, Williams JJ, Evans KE (2005c) Networked calix[4]arene polymers with unusual mechanical properties. Chem Commun 32:4065–4067

Grima JN, Zammit V, Gatt R, Alderson A, Evans KE (2007a) Auxetic behavior from rotating semi-rigid units. Phys Status Solidi B 244(3):866–882

Grima JN, Farrugia PS, Gatt R, Zammit V (2007b) Connected triangles exhibiting negative Poisson's ratio and negative thermal expansion. J Phys Soc Jpn 76(2):025001

Grima JN, Farrugia PS, Caruana C, Gatt R, Attard D (2008a) Auxetic behavior from stretching connected squares. J Mater Sci 43(17):5962–5971

Grima JN, Farrugia PS, Gatt R, Attard D (2008b) On the auxetic properties of rotating rhombi and parallelograms: a preliminary investigation. Phys Status Solidi B 245(3):521–529

Grima JN, Zammit V, Gatt R, Attard D, Caruana C, Bray TGC (2008c) On the role of rotating tetrahedra for generating auxetic behavior in NAT and related systems. J Non-Cryst Solids 354(35–39):4214–4220

Grima JN, Manicaro E, Attard D (2011a) Auxetic behavior from connected different-sized squares and rectangles. Proc R Soc A 467(2126):439–458

Grima JN, Ravirala N, Galea R, Ellul B, Attard D, Gatt R, Alderson A, Rasburn J, Evans KE (2011b) Modelling and testing of a foldable macrostructure exhibiting auxetic behavior. Phys Status Solidi B 248(1):117–122

Grima JN, Chetcuti E, Manicaro E, Attard D, Camilleri M, Gatt R, Evans KE (2012) On the auxetic properties of generic rotating rigid triangles. Proc R Soc A 468(2139):810–830

Grima JN, Winczewski S, Mizzi L, Grech MC, Cauchi R, Gatt R, Attard D, Wojciechowski KW, Rybicki J (2015) Tailoring graphene to achieve negative Poisson's ratio properties. Adv Mater 27(8):1455–1459

Grima JN, Mizzi L, Azzopardi KM, Gatt R (2016) Auxetic perforated mechanical metamaterials with randomly oriented cuts. Adv Mater 28(2):385–389

Ha CS, Plesha ME, Lakes RS (2016a) Chiral three-dimensional lattices with tunable Poisson's ratio. Smart Mater Struct 25(5):054005

Ha CS, Plesha ME, Lakes RS (2016b) Chiral three-dimensional isotropic lattices with negative Poisson's ratio. Phys Status Solidi B 253(7):1243–1251

He CB, Liu PW, Griffin AC (1998) Toward negative Poisson ratio polymers through molecular design. Macromolecules 31(9):3145–3147

He CB, Liu PW, McMullan PJ, Griffin AC (2005) Toward molecular auxetics: Main chain liquid crystalline polymers consisting of laterally attached para-quaterphenyls. Phys Status Solidi B 242(3):576–584

Hewage TAM, Alderson KL, Alderson A, Scarpa F (2016) Double-negative mechanical metamaterials displaying simultaneous negative stiffness and negative Poisson's ratio properties. Adv Mater 28(46):10323–10332

Hou X, Hu H (2015) A novel 3D composite structure with tunable Poisson's ratio and stiffness. Phys Status Solidi B 252(7):1565–1574
Hou J, Li D, Dong L (2019) Study on band-gap behaviors of 2D hierarchical re-entrant lattice structures. Phys Status Solidi B 256(5):1800693
Hu H, Zhang M, Liu Y (2019) Auxetic textiles. Woodhead, Duxford
Janbaz S, Noordzij N, Widyaratih DS, Hagen CW, Fratila-Apachitei, Zadpoor AA (2017) Origami lattices with free-form surface ornaments. Sci Adv 3(11): eaao1595
Javid F, Smith-Roberge E, Innes MC, Shanian A, Weaver JC, Bertoldi K (2015) Dimpled elastic sheets: a new class of non-porous negative Poisson's ratio materials. Sci Rep 5:18373
Javid F, Liu J, Shim J, Weaver JC, Shanian A, Bertoldi K (2016) Mechanics of instability-induced pattern transformations in elastomeric porous cylinders. J Mech Phys Solids 96:1–17
Jia Z, Wang L (2019) Instability-triggered triply negative mechanical metamaterial. Phys Rev Appl 12(2):024040
Jiang N, Hu H (2018) A study of tubular braided structure with negative Poisson's ratio behavior. Textile Res J 88(24):2810–2824
Jiang Y, Li Y (2018) Novel 3D-printed hybrid auxetic mechanical metamaterial with mhirality-induced sequential cell opening mechanisms. Avd Eng Mater 20(2):1700744
Jiang L, Gu B, Hu H (2016a) Auxetic composite made with multilayer orthogonal structural reinforcement. Compos Struct 135:23–29
Jiang JW, Kim SY, Park HS (2016b) Auxetic nanomaterials: Recent progress and future development. Appl Phys Rev 3:041101
Jiang N, Chen Y, Hu H (2020) Theoretical modeling on the deformation Behavior of auxetic tubular braid made from modified circular braiding technique. Phys Status Solidi B (accepted 1900173)
Jin S, Korkolis YP, Li Y (2019) Shear resistance of an auxetic chiral mechanical metamaterial. Int J Solids Struct 174–175:28–37
Jun E, Lee IS, Kim D, Lee H, Jang WD, Lah MS, Min SK, Choe W (2019) Metal-organic framework based on hinged cube tessellation as transformable mechanical metamaterial. Sci Adv 5(5):eaav4119
Kamrava S, Mousanezhad D, Ebrahimi H, Ghosh R, Vaziri A (2017) Origami-based cellular metamaterial with auxetic, bistable, and self-locking properties. Scient Rep 7:46046
Ki M, Lu X, Zhu X, Su X, Wu T (2019) Research on in-plane quasi-static mechanical properties of gradient tetra-chiral hyper-structures. Adv Eng Mater 21(3):1801038
Kolken HMA, Zadpoor AA (2017) Auxetic mechanical metamaterials. RSC Adv 7(9):5111–5129
Kumar D, Wang ZP, Poh LH, Quek ST (2019) Isogeometric shape optimization of smoothed petal auxetics with prescribed nonlinear deformation. Comput Meth Appl Mech Eng 356:16–43
Kwietniewski M, Miedzińska D (2019) Review of elastomeric materials for application to composites reinforced by auxetics fabrics. Proc Struct Integrity 17:154–161
Lakes R (1993) Advances in negative Poisson's ratio materials. Adv Mater 5(4):293–296
Lakes RS (2017) Negative-Poisson's-ratio materials: auxetic solids. Ann Rev Mater Res 47:63–81
Larsen UD, Sigmund O, Bouwstra S (1997) Design and fabrication of compliant mechanisms and material structures with negative Poisson's ratio. J Microelectromech Syst 6(2):99–106
Li D, Dong L, Yin J, Lakes RS (2016) Negative Poisson's ratio in 2D Voronoi cellular solids by biaxial compression: a numerical study. J Mater Sci 51(14):7029–7037
Li T, Hu X, Chen Y, Wang L (2017) Harnessing out-of-plane deformation to design 3D architected lattice metamaterials with tunable Poisson's ratio. Sci Rep 7:8949
Li D, Yin J, Dong L (2018) Numerical analysis of a two-dimensional open cell topology with tunable Poisson's ratio from positive to negative. Phys Status Solidi RRL 12(3):1700374
Li C, Shen HS, Wang H (2019) Nonlinear bending of sandwich beams with functionally graded negative Poisson's ratio honeycomb core. Compos Struct 212:317–325
Li J, Ha CS, Lakes RS (2020) Observation of squeeze-twist coupling in a chiral 3D isotropic lattice. Phys Status Solidi B (accepted 1900140)
Lim TC (2002a) Material structure for attaining pure Poisson-shearing. J Mater Sci Lett 21(20):1595–1597

Lim TC (2002b) Functionally graded beam for attaining Poisson-curving. J Mater Sci Lett 21(24):1899–1901
Lim TC (2004) Elastic properties of Poisson-shear material. J Mater Sci 39(15):4965–4969
Lim TC (2007) Kinematical studies on rotation-based semi-auxetics. J Mater Sci 42(18):7690–7695
Lim TC (2014) Semi-auxetic yarns. Phys Status Solidi B 251(2):273–280
Lim TC (2015a) Effect of nodule shape for modeling of auxetic microporous polymers. MATEC Web of Conference 34:01002
Lim TC (2015b) Auxetic materials and structures. Springer, Singapore
Lim TC (2016) A 3D auxetic material based on intersecting double arrowheads. Phys Status Solidi B 253(7):1252–1260
Lim TC (2017a) Auxetic and negative thermal expansion structure based on interconnected array of rings and sliding rods. Phys Status Solidi B 254(12):1600775
Lim TC (2017b) Analogies across auxetic models based on deformation mechanism. Phys Status Solidi RRL 11(6):1600440
Lim TC (2019a) An anisotropic auxetic 2D metamaterial based on sliding microstructural mechanism. Materials 12(3):429
Lim TC (2019b) Negative environmental expansion for interconnected array of rings and sliding rods. Phys Status Solidi B 256(1):1800032
Lim TC (2019c) Metamaterials with Poisson's ratio sign toggling by means of microstructural duality. SN Appl Sci 1(2):176
Lim TC, Acharya R (2009) An hexagonal array of fourfold interconnected hexagonal nodules for modeling auxetic microporous polymers: a comparison of 2D and 3D models. J Mater Sci 44(16):4491–4494
Lim TC, Alderson A, Alderson KL (2014) Experimental studies on the impact properties of auxetic materials. Phys Status Solidi B 251(2):307–313
Liu Y, Hu H (2010) A review on auxetic structures and polymeric materials. Sci Res Essays 5(10):1052–1063
Lorato A, Innocenti P, Scarpa F, Alderson A, Alderson KL, Zied KM, Ravirala N, Miller W, Smith CW, Evans KE (2010) The transverse elastic properties of chiral honeycombs. Compos Sci Technol 70(7):1057–1063
Lv C, Krishnaraju D, Konjevod G, Yu H, Jiang H (2014) Origami based mechanical metamaterials. Sci Rep 4:5979
Masters IG, Evans KE (1996) Models for the elastic deformation of honeycombs. Compos Struct 35(4):403–422
Meena K, Singamneni S (2019) A new auxetic structure with significantly reduced stress concentration effects. Mater Des 173:107779
Miller W, Hook PB, Smith CW, Wang X, Evans KE (2009) The manufacture and characterisation of a novel, low modulus, negative Poisson's ratio composite. Compos Sci Technol 69(5):651–655
Miller W, Smith CW, Scarpa F, Evans KE (2010) Flatwise buckling optimization of hexachiral and tetrachiral honeycombs. Compos Sci Technol 70(7):1049–1056
Miller W, Ren Z, Smith CW, Evans KE (2012) A negative Poisson's ratio carbon fibre composite using a negative Poisson's ratio yarn reinforcement. Compos Sci Technol 72(7):761–766
Mizzi L, Grima JN, Gatt R, Attard D (2019) Analysis of the deformation behavior and mechanical properties of slit-perforated auxetic metamaterials. Phys Status Solidi B 256(1):1800153
Ng WS, Hu H (2017) Tensile and deformation behavior of auxetic plied yarns. Phys Status Solidi B 254(12):1600790
Niu Y, Ge J, Liang J, Liao H, Wu W (2019) Effects of disordered circular nodes dispersion and missing ligaments on the mechanical properties of chiral structures. Phys Status Solidi B 256(10):1800586
Park HS, Kim SY (2017) A perspective on auxetic nanomaterials. Nano Converg 4:10
Pasternak E, Dyskin AV (2019) Architectured materials with inclusions having negative Poisson's ratio or negative stiffness. In: Estrin Y, Bréchet Y, Dunlop J, Fratzl P (eds) Architectured materials in nature and engineering. Springer series in materials science, vol 282. Springer Nature, Cham

Piglowski PM, Wojciechowski KW, Tretiakoc KV (2016) Partial auxeticity induced by nanoslits in the Yukawa crystal. Phys Status Solidi RRL 10(7):566–569

Pozniak AA, Wojciechowski KW (2014) Poisson's ratio of rectangular anti-chiral structures with size dispersion of circular nodes. Phys Status Solidi B 251(2):367–374

Prall D, Lakes RS (1997) Properties of a chiral honeycomb with a Poisson's ratio of −1. Int J Mech Sci 39(3):305–314

Prawoto Y (2012) Seeing auxetic materials from the mechanics point of view: A structural review on the negative Poisson's ratio. Comput Mater Sci 58:140–153

Qin H, Yang D, Ren C (2018) Design method of lightweight metamaterials with arbitrary Poisson's ratio. Materials 11(9):1574

Rad MS, Ahmad Z, Alias A (2015) Computational approach in formulating mechanical characteristics of 3D star honeycomb auxetic structure. Adv Mater Sci Eng 2015:650769

Ravirala N, Alderson A, Alderson KL (2007) Interlocking hexagon model for auxetic behavior. J Mater Sci 42(17):7433–7445

Reasa DR, Lakes RS (2019) Cosserat effects in achiral and chiral cubic lattices. J Appl Mech 86(11):111009

Ren X, Das R, Tran P, Ngo T, Xie YM (2018) Auxetic metamaterials and structures: a review. Smart Mater Struct 27(2):023001

Saxena KK, Das R, Calius EP (2016) Three decades of auxetics research—Materials with negative Poisson's ratio: a review. Adv Eng Mater 18(11):1847–1870

Schenk M, Guest SD (2013) Geometry of Miura-folded metamaterials. Proc Natl Acad Sci 110(9):3276–3281

Shilko SV, Petrokovets EM, Pleskachevsky YM (2008) Peculiarities of friction in auxetic composites. Phys Status Solidi B 245(3):591–597

Shim J, Perdigou C, Chen ER, Bertoldi K, Reis PM (2012) Buckling-induced encapsulation of structured elastic shells under pressure. Proc Natl Acad Sci 109(16):5978–5983

Shim J, Shan S, Kosmrlj A, Kang SH, Chen ER, Weaver JC. Bertoldi K (2013) Harnessing instabilities for design of soft reconfigurable auxetic/chiral materials. Soft Matter 9(34):8198–8202

Sloan MR, Wright JR, Evans KE (2011) The helical auxetic yarn—a novel structure for composites and textiles; geometry, manufacture and mechanical properties. Mech Mater 43(9):476–486

Smardzewski J (2019) Experimental and numerical analysis of wooden sandwich panels with an auxetic core and oval cells. Mater Des 183:108159

Smardzewski J, Wojciechowski KW (2019) Response of wood-based sandwich beams with three-dimensional lattice core. Compos Struct 216:340–349

Smith CW, Grima JN, Evans KE (2000) A novel mechanism for generating auxetic behavior in reticulated foams: missing rib foam model. Acta Mater 48(17):4349–4356

Spadoni A, Ruzzene M (2012) Elasto-static micro polar behavior of a chiral auxetic lattice. J Mech Phys Solids 60(1):156–171

Spadoni A, Ruzzene M, Gonella S, Scarpa F (2009) Phononic properties of hexagonal chiral lattices. Wave Motion 46(7):435–450

Su X, Zhu D, Zheng C, Tomovic MM (2019) Frequency response characteristics of finite periodic chiral structures with three ligaments. J Aerosp Eng 233(12):4623–4634

Surjadi JU, Gao L, Du H, Li X, Xiong X, Fang NX, Lu Y (2019) Mechanical metamaterials and their engineering applications. Adv Eng Mater 21(3):1800864

Tanaka H, Suga K, Shibutani Y (2019) Primitive rotation mechanism of periodic stellated octahedron units with shearing edges. Int J Solids Struct (in press)

Taylor M, Francesconi L, Gerendas M, Shanian A, Carson C, Bertoldi K (2013) Low porosity metallic periodic structures with negative Poisson's ratio. Adv Mater 26(15):2365–2370

Timoshenko S (1948) Strength of materials, Part 2 (2nd edition, 9th printing). Van Nostrand, New York

Tretiakov KV, Wojciechowski KW (2014) Partially auxetic behavior in fcc crystals of hard-core repulsive Yukawa particles. Phys Status Solidi B 251(2):383–387

Tretiakov KV, Piglowski PM, Narojczyk JW, Bilski M, Wojciechowski KW (2018) High partial auxeticity induced by nanochannels in [111]-direction in a simple model with Yukawa interactions. Materials 11(12):2550

Wang Z, Hu H (2014a) 3D auxetic warp-knitted spacer fabrics. Phys Status Solidi B 251(2):281–288

Wang Z, Hu H (2014b) Auxetic materials and their potential applications in textiles. Text Res J 84(15):1600–1611

Wang ZP, Poh LH (2018) Optimal form and size characterization of planar isotropic petal-shaped auxetics with tunable effective properties using IGA. Compos Struct 201:486–502

Wang Z, Hu H, Xiao X (2014) Deformation behaviors of three-dimensional auxetic spacer fabrics. Text Res J 84(13):1361–1372

Wang ZP, Poh LH, Dirrenberger J, Zhu Y, Forest S (2017) Isogeometric shape optimization of smoothed petal auxetic structures via computational periodic homogenization. Comput Meth Appl Mech Eng 323:250–271

Wang ZP, Poh LH, Zhu Y, Dirrenberger J, Forest S (2019) Systematic design of tetra-petals auxetic structures with stiffness constraint. Mater Des 170:107669

Wang YC, Lai HW, Ren XJ (2020) Enhanced auxetic and viscoelastic properties of filled reentrant honeycomb. Phys Status Solidi B (accepted 1900184)

Wei ZY, Guo ZV, Dudte L, Liang HY, Mahadevan L (2013) Geometric mechanics of periodic pleated origami. Phys Rev Lett 110(21):215501

Wen Y, Gao E, Hu Z, Xu T, Lu H, Xu Z, Li C (2019) Chemically modified graphene films with tunable negative Poisson's ratios. Nat Commun 10:2446

Wojciechowski KW (1987) Constant thermodynamic tension Monte-Carlo studies of elastic properties of a two-dimensional system of hard cyclic hexamers. Mol Phys 61(5):1247–1258

Wojciechowski KW (1989) Two-dimensional isotropic system with a negative Poisson ratio. Phys Lett A 137(1&2):60–64

Wojciechowski KW (2005) Poisson's ratio of anisotropic systems. Comput Meth Sci Technol 11(1):73–79

Wojciechowski KW, Branka AC (1989) Negative Poisson ratio in a two-dimensional "isotropic" solid. Phys Rev A 40(12):7222–7225

Wojciechowski KW, Tretiakov KV (2000) Elastic constants of dense crystalline phases of twodimensional hard cyclic heptamers. Comput Meth Sci Technol 6(1):101–119

Wojciechowski KW, Tretiakov KV, Kowalik M (2003) Elastic properties of dense solid phases of hard cyclic pentamers and heptamers in two dimensions. Phys Rev E 67(3):036121

Wright JR, Burns MK, James E, Sloan MR, Evans KE (2012) On the design and characterisation of low-stiffness auxetic yarns and fabrics. Text Res J 82(7):645–654

Wu W, Qi D, Liao H, Qian G, Geng L, Niu Y, Liang J (2018) Deformation mechanism of innovative 3D chiral metamaterials. Sci Rep 8:12575

Wu W, Hu W, Qian G, Liao H, Xu X, Berto F (2019) Mechanical design and multifunctional applications of chiral mechanical metamaterials: A review. Mater Des 180:107950

Yang W, Li ZM, Shi W, Xie BH, Yang MB (2004) Review on auxetic materials. J Mater Sci 39(10):3269–3279

Yasuda H, Yang J (2015) Reentrant origami-based metamaterials with negative Poisson's ratio and bistability. Phys Rev Lett 114(18):185502

Zhang J, Zhu X, Yang X, Zhang W (2019) Transient nonlinear responses of an auxetic honeycomb sandwich plate under impact loads. Int J Impact Eng 134:103383

Zheng BB, Zhong RC, Chen X, Fu MH, Hu LL (2019) A novel metamaterial with tension-torsion coupling effect. Mater Des 171:107700

Zhu Y, Zeng Z, Wang ZP, Poh LH, Shao YB (2019) Hierarchical hexachiral auxetics for large elasto-plastic deformation. Mater Res Express 6(8):085701

Chapter 3
Analogies Across Auxetic Models

Abstract This chapter views 2D auxetic models, including 3D deformation models with 2D auxetic behavior, from mechanism perspective instead of geometrical perspective. On this basis, auxetic models across different geometrical groups can be regrouped into clusters that exhibit analogy in deformation mechanism. Factors that are taken into consideration include the identification of corresponding rotation and non-rotation units, as well as linkages/joints between rotation and non-rotation units and non-linkages/non-joints across various auxetic models. As a result, five clusters of auxetic models have been identified, in which auxetic models within each cluster are analogous to each other. The identified clusters are those that exhibit: (1) double periodicity in the rotation direction of their rotating units, (2) synchronized rotation direction of their rotation units, (3) single periodicity in the rotation direction of their rotating units, (4) random rotation of their rotation units, and (5) non-rotation of units. Results from this analogy identification place auxetic models in a systematic representation and will enrich the future development of auxetic models, particularly, those that do not fall within these five clusters.

Keywords Analogies · Auxetics · Geometrical models · Mechanism

3.1 Introduction

This chapter lays out the analogies across different auxetic models; in particular, the attempt herein places less emphasis on the geometrical properties of the auxetic models, but places greater emphasis on the deformation mechanism as a means to establish analogies (Lim 2017a) based on deformation mechanisms. The scope of this chapter is confined to 2D geometry, and where 3D deformations are concerned, consideration is made only for those that exhibit auxetic properties only on one plane so as to focus on the fundamental principles involved.

T. Lim, *Mechanics of Metamaterials with Negative Parameters*, Engineering Materials,
https://doi.org/10.1007/978-981-15-6446-8_3

3.2 Analogies Across Rotating Square, Perforated Sheet, Anti-tetrachiral, Missing Rib, and Re-entrant Models

An obvious comparison between two types of auxetic models is furnished in Fig. 3.1. Here, the rotating squares model by Grima and Evans (2000) compares well with the perforated sheet by Taylor et al. (2013), in which the presence of long and narrow perforation is analogous to a closed-up rotating square model (Fig. 3.1, top).

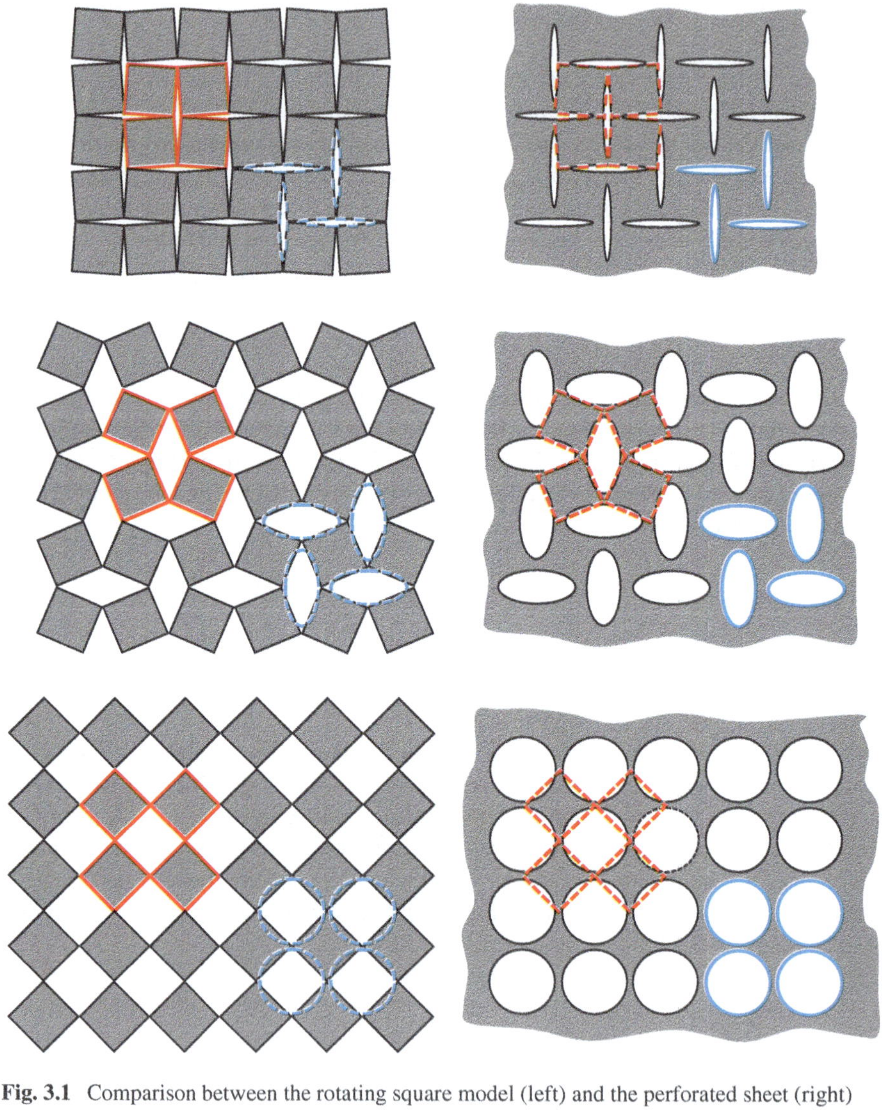

Fig. 3.1 Comparison between the rotating square model (left) and the perforated sheet (right)

On the other extreme, the fully opened rotating square model is analogous to a sheet with circular cut-outs in square array (Fig. 3.1, bottom). The intermediate case of rotating squares in the partially opened position corresponds to the sheet with perforated ellipses (Fig. 3.1, middle). Inasmuch as a uniaxial compression of the fully opened rotating squares (Fig. 3.1, bottom left) leads to the fully closed rotating squares (Fig. 3.1, top left) through the partially opened rotating squares (Fig. 3.1, middle left), a uniaxial compression on the perforated sheet of circular holes (Fig. 3.1, bottom right) results in closing of the holes (Fig. 3.1, top right) via the intermediary state of elliptical perforated holes (Fig. 3.1, middle right). This case of buckling is further discussed in Sects. 3.9 and 3.10. To aid comparison, four neighboring squares indicated in red are transposed on the perforated sheets while four neighboring perforations denoted in blue are transposed on the rotating squares model.

In spite of the rotating nature in a number of auxetic models, one may identify and isolate certain boundaries in which there is no overall rotation. Reference to Fig. 3.2 shows some typical auxetic models, i.e., the rotating squares model by Grima and Evans (2000), the anti-tetrachiral model by Chen et al. (2013), the rotating rigid cross model by Grima et al. (2007), and the re-entrant model by Gibson and Ashby (1988),

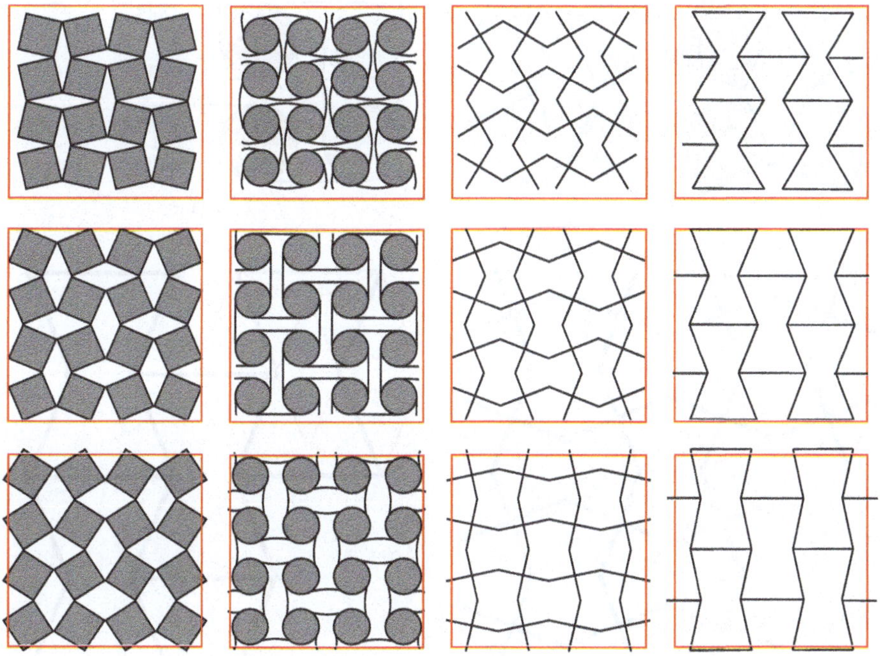

Fig. 3.2 Identification of the overall non-rotating group of the rotating square model (first column), anti-tetrachiral model (second column), a type of missing rib model (third column), and re-entrant model (last column), at various stages of deformation from top to bottom

in which clusters of 2-by-2 rotating elements are grouped within red boundaries that do not encounter overall rotation but exhibit expansion and contraction.

Detailed comparison of these four auxetic models can be made in conjunction with Fig. 3.3. The four rotating squares on the top left of Fig. 3.3a are transposed on the top left of Fig. 3.3b–d, while the four rotating cylinders at the bottom right of Fig. 3.3b are similarly transposed on the bottom right of Fig. 3.3a, c, and d. The rotating hinges for the missing rib model on the top right of Fig. 3.3c is likewise colored on the top right of Fig. 3.3a, b, and d. It can be noticed that the point hinges in Fig. 3.3a, c correspond to linkages that connect the rotating cylinders of Fig. 3.3b, and that the two vertically aligned hinges in Fig. 3.3a, c each correspond to a pair of hinges at the ends of the horizontal linkages of the re-entrant structure. Finally, the rotating linkages of the

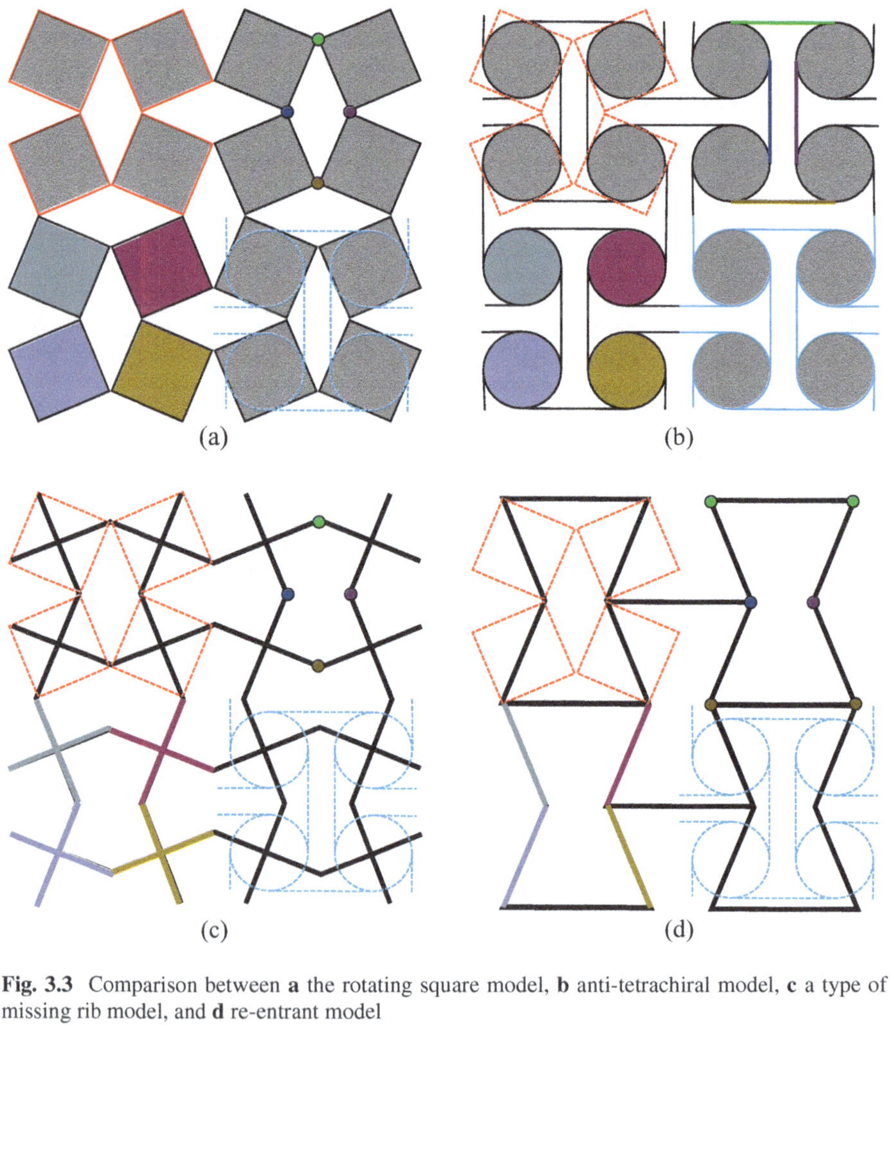

Fig. 3.3 Comparison between **a** the rotating square model, **b** anti-tetrachiral model, **c** a type of missing rib model, and **d** re-entrant model

re-entrant structure indicated at the bottom left of Fig. 3.3d is correspondingly given the same color on the bottom left of Fig. 3.3a–c. The similarities between Fig. 3.3a, c have been pointed out by Grima et al. (2007), in which the ends of the rotating crosses are essentially the hinges for the rotating squares. The rotating crosses in Fig. 3.3c are herein called a type of missing rib model as it is formed from a square grid with alternating vertical and horizontal ribs removed.

Arising from this comparison, an analogy can be established on the auxetic mechanism of the five auxetic models discussed so far. This is illustrated in Fig. 3.4 for the expansion and contraction of the perforated sheet, rotating square, anti-tetrachiral, missing rib, and re-entrant models.

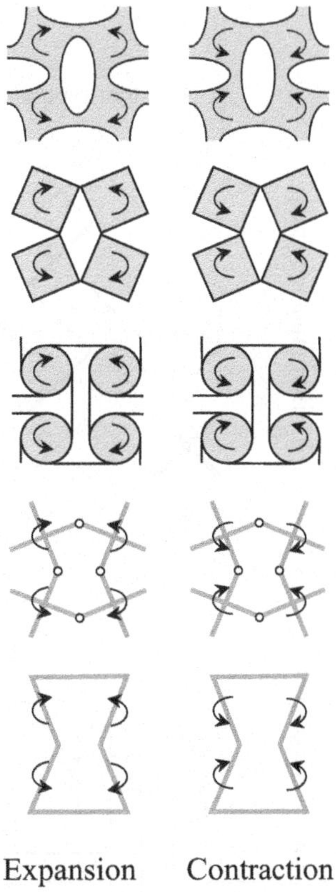

Fig. 3.4 Mechanism similarities for the expansion (left) and contraction (right) of (from top) perforated sheet, rotating square model, anti-tetrachiral model, a type of missing rib model, and re-entrant model

3.3 Analogy Between the Missing Rib Models with the Tetrachiral and Anti-tetrachiral Models

A more established type of missing rib model, as depicted in Fig. 3.5a, was proposed by Smith et al. (2000) and has been investigated by Gaspar et al. (2005) and later by Jiang and Li (2017). This is herein termed the tilted-swastika missing rib model as the swastika ribs are tilted at 45° from the square array alignment of the swastikas. This model is analogous to the tetrachiral model reviewed by Alderson et al (2010) and displayed in Fig. 3.5b. The swastika on the top right of Fig. 3.5a, being colored in red, is transposed onto the top right corner of Fig. 3.5b for comparison against the tetrachiral model. Conversely, the cylinder at the top left of Fig. 3.5b, being colored in blue, is transposed on the top left corner of Fig. 3.5a. Such transposes visually reveal

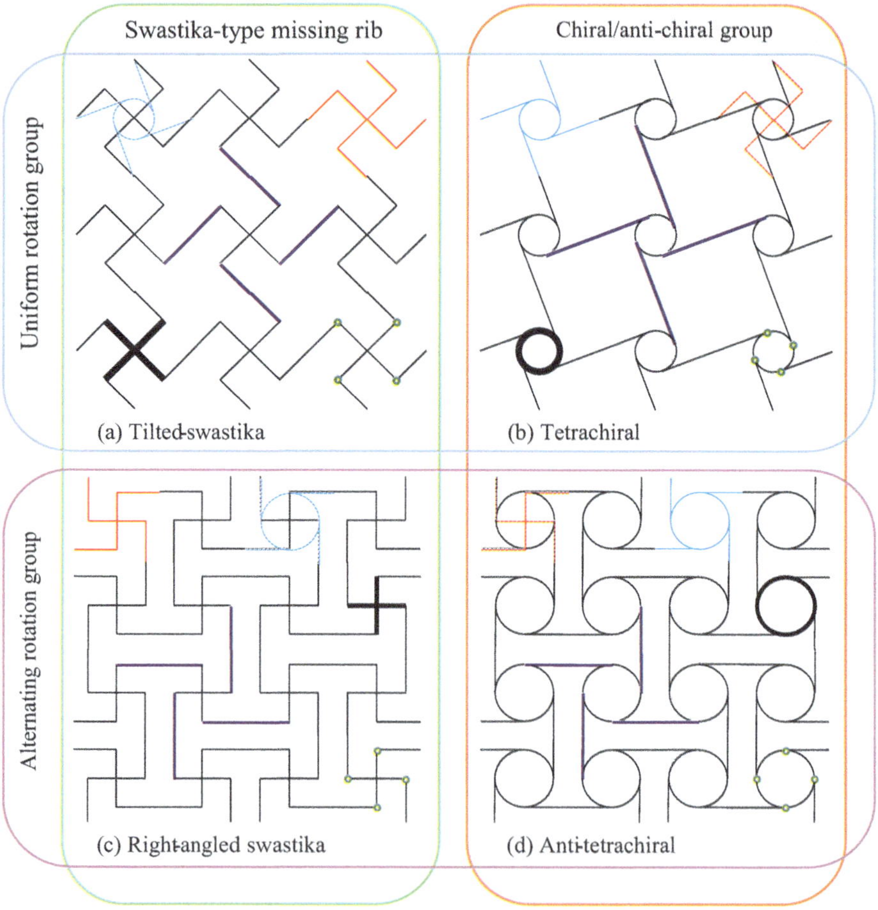

Fig. 3.5 Analogy between swastika-type missing rib model and the (anti)tetrachiral model

the analogy between the tilted-swastika model and the tetrachiral model. The "rigid" parts that rotate are bolded on the bottom left corners of Fig. 3.5a, b. Likewise, the corresponding hinges that permit rotation mechanism are pointed out in green at the bottom right corners of Fig. 3.5a, b. Finally, the corresponding connecting parts for both models are indicated by purple at the middle portion of Fig. 3.5a, b. A similar comparison can be made to reveal the analogy between the right-angled swastika model by Gaspar et al. (2005) and the anti-tetrachiral model (Alderson et al. 2010, Chen et al. 2013) furnished in Fig. 3.5c, d, respectively.

This one-to-one correspondence between both groups of models is colored accordingly for ease of comparison. Such comparison reveals that while the tilted-swastika and the right-angled swastika models belong to the same group by virtue of geometry, and likewise the tetrachiral and anti-tetrachiral models are correctly grouped together again by virtue of their geometrical properties, perspective from mechanism viewpoint suggests otherwise. Reference to Fig. 3.5a, b shows that the rotating units from both the tilted-swastika and the tetrachiral models rotate synchronously, i.e., every rotating unit rotates in the one direction during contraction and reverses during expansion. On the other hand, perusal to Fig. 3.5c, d indicates that every unit rotates in the opposite direction in comparison with its immediate neighbor along both the horizontal and vertical directions, i.e., during contraction half of the units rotate in the opposite direction from the other half and the rotational direction of all these units reverse during expansion.

3.4 Analogies Among the Egg-Rack, Dimpled Sheet, and a Class of Graphene Sheet

The egg-rack model by Grima et al. (2005a) is shown in Fig. 3.6 (top), the dimpled sheet by Javid et al. (2015) is displayed in Fig. 3.6 (middle) and the uneven graphene sheet by Grima et al. (2015) furnished in Fig. 3.6 (bottom) would not be typically clustered together, and for good reason—their geometrical difference is obvious. Their mechanisms, however, exhibit commonalities and therefore these models are analogous to one another. For convenience, we consider one unit of the egg-rack model displayed in Fig. 3.7 (top), in which application of tensile load at the ends of the blue linkage in the x-direction would lead to a similar deformation on the red linkage due to the requirement for the ends of both linkages to be constrained on the same plane indicated by gray. Likewise, the application of x-direction load on the dome rim that coincides with the blue meridian (Fig. 3.7, middle) would unbend the red meridian so as to retain the dome rim on a plane. The imposition of flat base for the egg-rack and dimpled sheet is a result of alternating out-of-plane bulge direction, which provides cancelation effect to the out-of-plane displacement of the unit cell boundary. Hence, auxetic property is observed in the x-y plane but conventional properties are exhibited on the x-z and y-z planes, and this would, therefore, place both the egg-rack and dimpled sheet models as plane auxetics or 2D auxetics. A

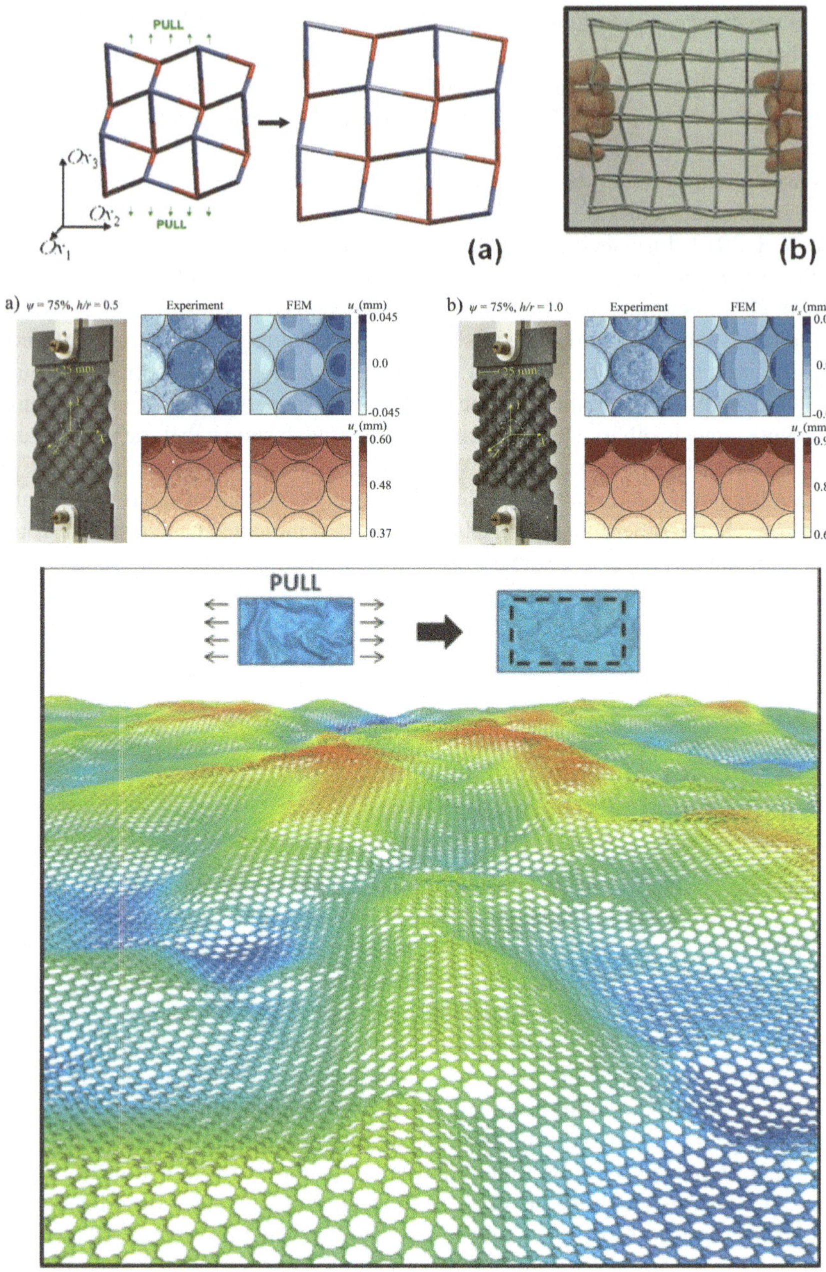

Fig. 3.6 Grouping of the egg-rack model (Top, reproduced from Grima et al. (2005a) with permission of The Royal Society of Chemistry); dimpled sheet (middle, attributed to Javid et al. (2015)); and uneven graphene sheet (bottom, reproduced from Grima et al. (2015) with permission from John Wiley & Sons) into a category

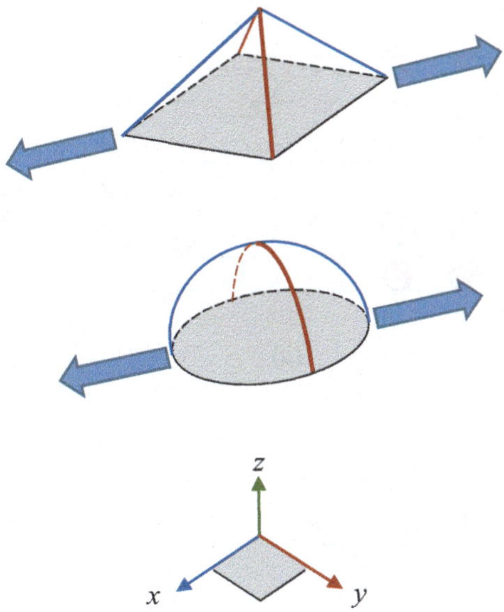

Fig. 3.7 Common mechanism for the deformation of egg-rack model, dimpled sheet, and uneven graphene sheet

similar mechanism can be seen for the case of the uneven graphene sheet, although the bulges deform to different extent due to their irregularities.

3.5 Analogy Between the Metatetrachiral Units and Double Arrowheads

A lesser known cousin to the tetrachiral and anti-tetrachiral models is the metatetrahiral model proposed by Grima et al. (2008). With reference to Fig. 3.8a, the cylinders along the odd columns rotate in the opposite direction to those in the even columns. This mechanism is analogous to the double arrowhead model proposed by Larsen et al. (1997) as shown in Fig. 3.8b. To aid comparison, a 2-by-2 group of cylinders on the top left of Fig. 3.8a is colored in red and transposed onto the top left of Fig. 3.8b, while the double arrowhead at the top right corner of Fig. 3.8b is colored blue and transposed on the top right of Fig. 3.8a. The corresponding rotating parts of both models are similarly colored at the bottom left of Fig. 3.8a, b. Finally, the connecting parts for the metatetrachiral and the double arrowhead model are identified and correspondingly colored as connecting linkages and connecting hinges at the bottom of Fig. 3.8a, b, respectively. Analogies between these two auxetic models at different stages of deformations are depicted in Fig. 3.8c.

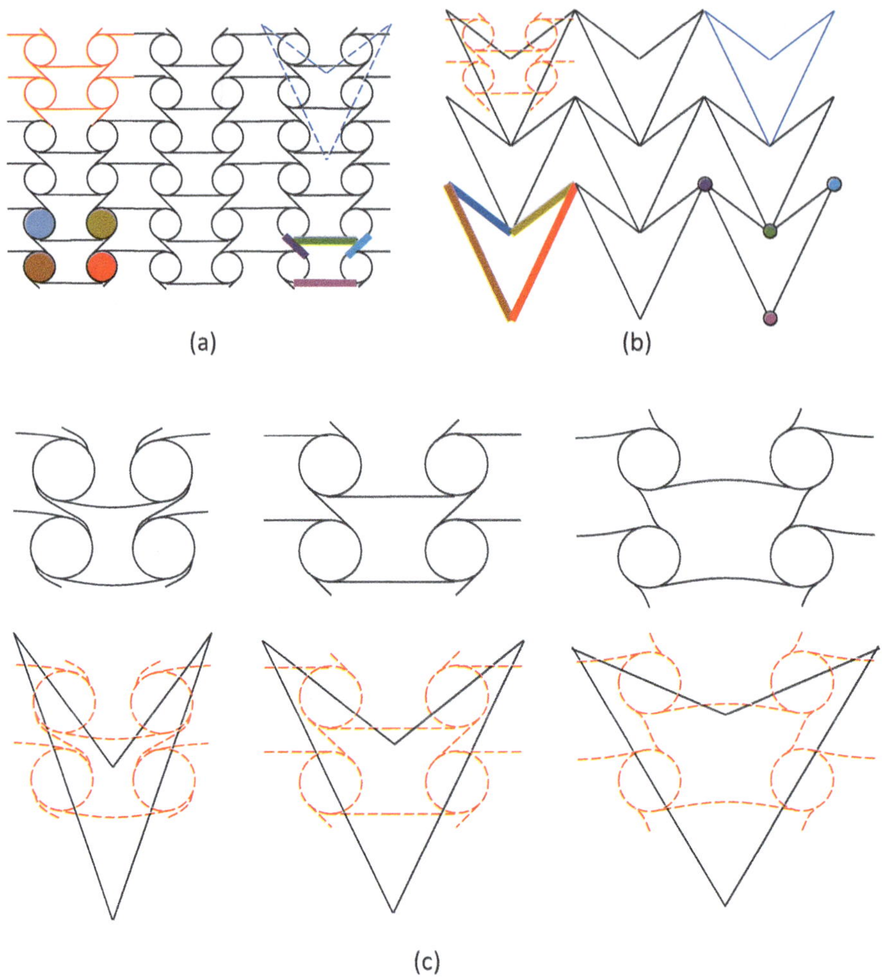

Fig. 3.8 Analogy between **a** metatetrachiral model and **b** double arrowhead model, with their **c** deformation shapes (note the chirals' circular centers coincide with the double arrowhead's linkage mid-spans)

3.6 Analogy Between the Star-Honeycomb Structures and Anti-chiral Models

In this section, an analogy is established between the star-connected models and a type of anti-chiral models. We firstly examine the relationship between the 4-pointed, square array star-connected model that was proposed by Theocaris et al. (1997) and the anti-tetra-octachiral model illustrated in Fig. 3.9a, b, respectively. The red-colored star and its connecting linkages at the top left of Fig. 3.9a are transposed onto the top left of Fig. 3.9b, while the blue-colored anti-tetra-octachiral unit at the

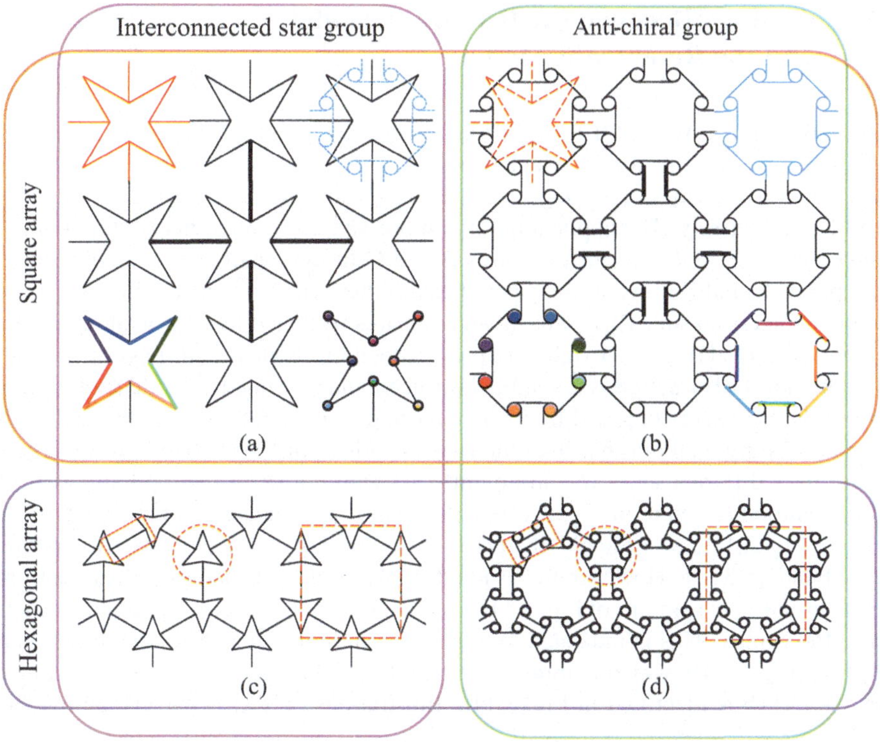

Fig. 3.9 Analogy between **a** 4-pointed, square array star-connected model, **b** anti-octa-tetrachiral model, **c** 3-pointed, hexagonal array star-connected model, and **d** anti-deca-hexa-tetrachiral model

top right of Fig. 3.9b is transposed on the top right of Fig. 3.9a for a one-to-one comparison. Having established this connection, the rotating parts of both models are correspondingly colored at the bottom left of Fig. 3.9a, b. The parts that function as connectors to the rotating units are indicated at the bottom right of Fig. 3.9a, b. For completeness' sake, the parts that connect one unit cell to its four closest neighbors are indicated by bold black lines at the middle of Fig. 3.9a, b. A similar analogy can be established between the 3-pointed, hexagonal array of star-connected model as one of the star-connected models studied by Grima et al. (2005b) and the anti-tetra-hexa-dodecachiral model displayed in Fig. 3.9c, d, respectively. It can be observed that the hexagon, the star and the linkages in Fig. 3.9c are analogous to the anti-dodecachiral, anti-hexachiral, and the anti-tetrachiral units, respectively, as shown in Fig. 3.9d.

3.7 Analogy Between the Ring-Rod Model and the Re-entrant Model

Recently, Lim (2017b, 2019) introduced a ring-rod structure (Fig. 3.10, left) that exhibits auxetic behavior, in which a pair of rods are attached at points B and C, such that the rod passes through an opening on the ring and ending at points A and D, respectively. The application of compressive force at points A and D in the direction of the rod axis pushes points B and C further apart and as a consequent the points E and F move closer toward each other. The motion reverses when the applied load is tensile. This mechanism compares well, and finds its analogy to, the re-entrant model (Fig. 3.10, right). To prevent rotation about the axis passing through EF, the ring-rod structure is assembled in such a manner that the tendency to rotate about the EF axis is opposed by its neighboring ring, as indicated in Fig. 3.11a. It can now be pointed out that both the ring as well as the space between four rings constitutes different set of re-entrant units. As with Fig. 3.10, the top left corner of the re-entrant structure in Fig. 3.11b, colored in red, is transposed onto the top left ring in Fig. 3.11a; accordingly, the points A to F in Fig. 3.11a correspond to points A to F in Fig. 3.11b. However, this is not the only re-entrant analogy—the re-entrant units in alternate columns of Fig. 3.11b is equivalent to the space formed at the center of 2-by-2 rings. This is indicated by the re-entrant unit colored in blue, on the top right of Fig. 3.11b, which is transposed onto Fig. 3.11a. As such the points K to P in Fig. 3.11b is identified in Fig. 3.11a. Illustrations on these two auxetic models

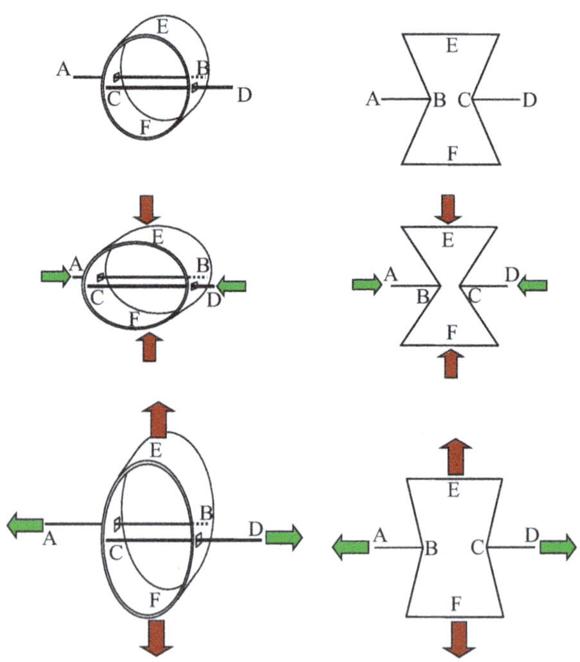

Fig. 3.10 Analogies between a single ring-rod assembly (left) and a single re-entrant structure (right)

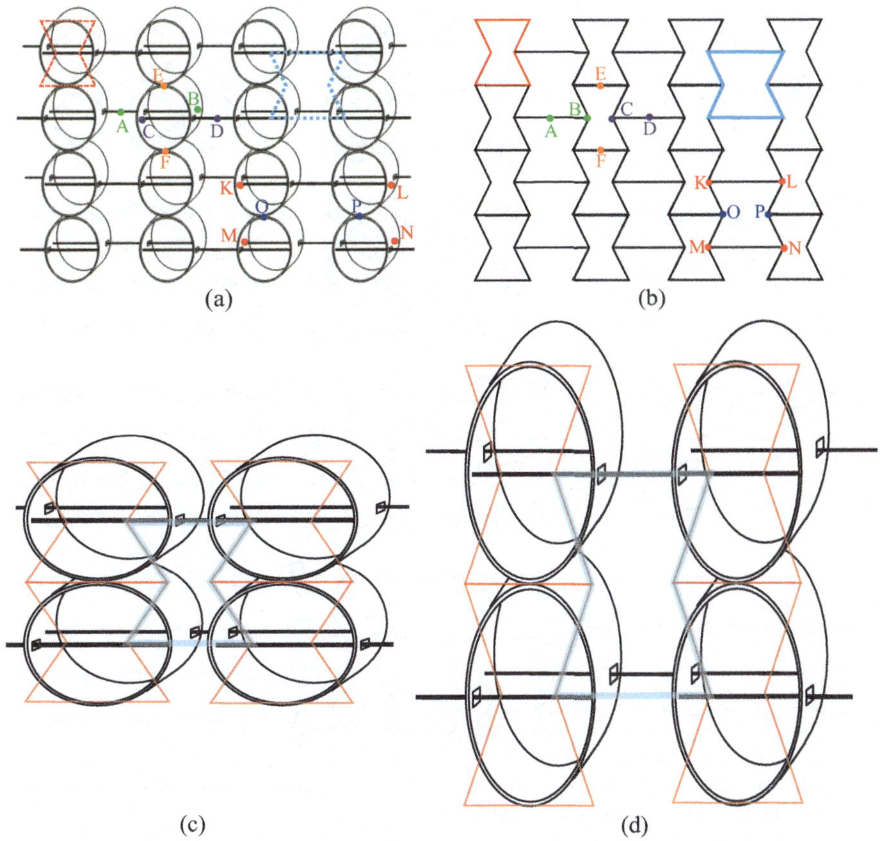

Fig. 3.11 Additional analogies between **a** an array of ring-rod assembly, **b** its relation with the re-entrant structure, as well as their comparison at **c** compressive, and **d** tensile deformation states

at compressive and tensile deformations are furnished in Fig. 3.11c, d, respectively, with the re-entrant models transposed on the ring-rod structures for indicating their analogy at deformed states.

3.8 Analogies Between the Semi-auxetic Yarn Model, the Liquid Crystalline Polymer Chain and the Rotating Square Model

We now shift our attention to the case of semi-auxetic yarn introduced by Lim (2014), which can be categorized as a chain-type auxetic system. The "main" chain is the

Fig. 3.12 Analogies across the semi-auxetic yarn (left), idealized liquid crystalline polymer (center), and the rotating square model (right)

thin inextensible cord that is stitched through a thick elastic cord in a zig-zag triangular pattern, as shown in Fig. 3.12 (top left). When stretched, the inextensible cord straightens, during which the thick elastic cord is simultaneously bent and stretched, thereby pushing other semi-auxetic yarns away if the arrangement of the semi-auxetic yarns are 180° out of phase from their immediate neighbor. The thin inextensible cord and the thick elastic cord for the semi-auxetic yarn find their counterparts with the main polymeric chain and the rigid groups, respectively, of the idealized liquid crystalline polymer (LCP) proposed by He et al. (1998, 2005), as illustrated in Fig. 3.12 (center column). In the latter, stretching of the LCP straightens the main polymer chain as with the straightening of the thin inextensible cord while the rotation of the rigid groups toward alignment in the lateral direction is similar to the lateral in-plane bulging of the thick elastic cord. This lateral alignment of rigid groups pushes the rigid groups from neighboring chain, thereby increasing the lateral distance between the main polymer chains. It is also interesting to note that these two chain-type auxetic systems can be related to the rotating square model as shown in Fig. 3.12 (right). To facilitate comparison, a zig-zag pattern is drawn across every row of the rotating square model. This zig-zag pattern represents the main chain while the two remaining corners are analogous to the thick elastic cord and the rigid groups in the semi-auxetic yarn and LCP, respectively.

3.9 Analogies Between the Buckling-Based Auxeticity and Related Models

Perusal to Fig. 3.13 (top) shows one of the buckling modes encountered by a square honeycomb or square grid, in which the buckled mode is analogous to other auxetic models displayed in Fig. 3.13 (bottom). To facilitate comparison, the ribs of the buckled mode are colored correspondingly with rotating units of other auxetic models. A similar pattern of buckled deformation is observed when the square holes shown at the top left corner of Fig. 3.13 are replaced with circular holes. The cross formed by four holes act as a rotating unit similar to that of the rotating squares model, as observed by Bertoldi et al. (2010) (Fig. 3.14). During the course of rotation, the circular holes transform into elliptical holes and eventually take on the shape of number "8" or symbol "∞", as shown in Fig. 3.14c; this approximates the hole shape arising from buckling of square grid (Fig. 3.13, top right). Therefore, in addition to analogies between buckling-based auxetic models with other auxetic models,

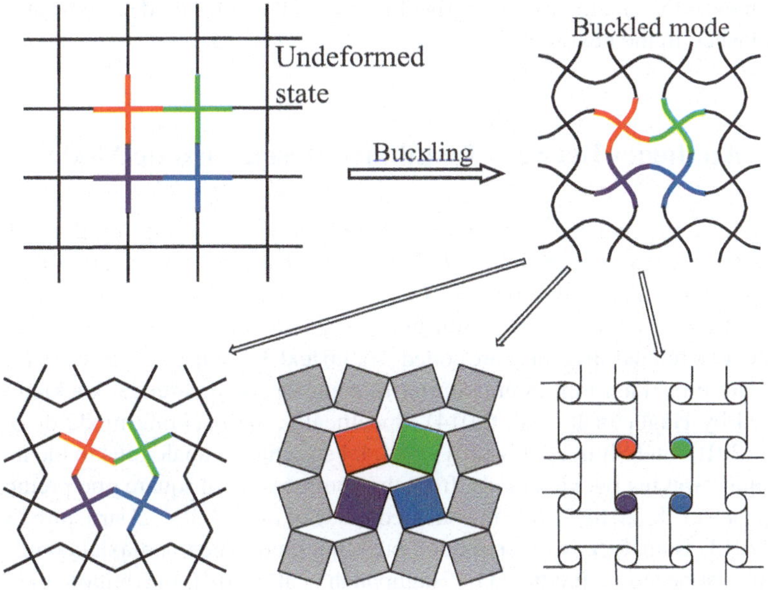

Fig. 3.13 Analogies between the buckled mode of a square grid (top) with other auxetic models (bottom)

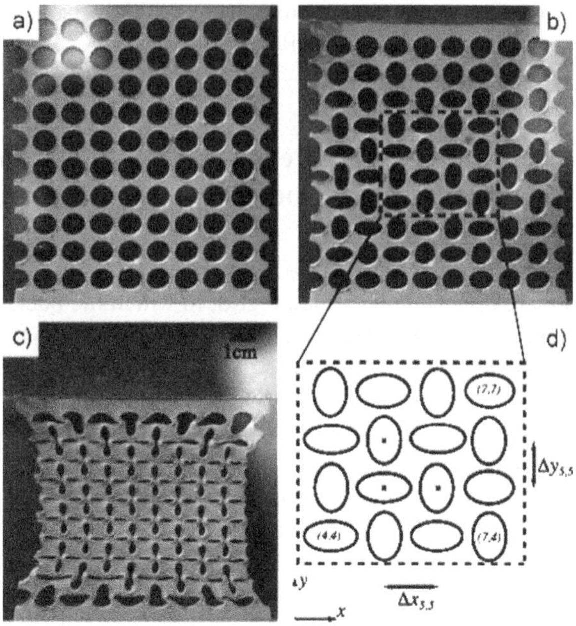

Fig. 3.14 Buckling of perforated sheet with square array of circular holes by Bertoldi et al. (2010) Reproduced with permission from John Wiley & Sons

one can establish analogies among buckling-based auxetic models. Further details are furnished in the next section.

3.10 Analogies Between Instability-Based Auxetic Models

Arising from the pioneering work of instability-based auxeticity by Bertoldi et al. (2010), the same group (Shim et al. 2013) extended their investigation to include hexagonal array and related patterns of circular perforations, with the square array being coded as a 4.4.4.4 structure; the hexagonal structure is coded 3.3.3.3.3.3 while the other two related structures are coded 3.6.3.6 and 3.4.6.4, as furnished in Fig. 3.15 (top). These buckled modes are similar to some of the theoretical buckled modes modeled by Haghpanah et al. (2014). Specifically, the buckling mode of Bertoldi et al.'s (2010) and Shim et al.'s (2013) 4.4.4.4 structure is analogous to the mode II, or the non-swaying buckled mode, from the compression of square grid predicted by Haghpanah et al. (2014), while the buckled mode of the 3.3.3.3.3.3 structure by Shim et al. (2013) is analogous to the mode I buckling mode due to uniaxial compression of regular honeycomb predicted by Haghpanah et al. (2014). In addition, the 3.6.3.6 structure by Shim et al. (2013) buckles in a manner analogous to the mode III, also termed as the flower-like or chiral, buckling of regular honeycomb modeled by Haghpanah et al. (2014). Comparison between these three buckling modes by Shim et al.

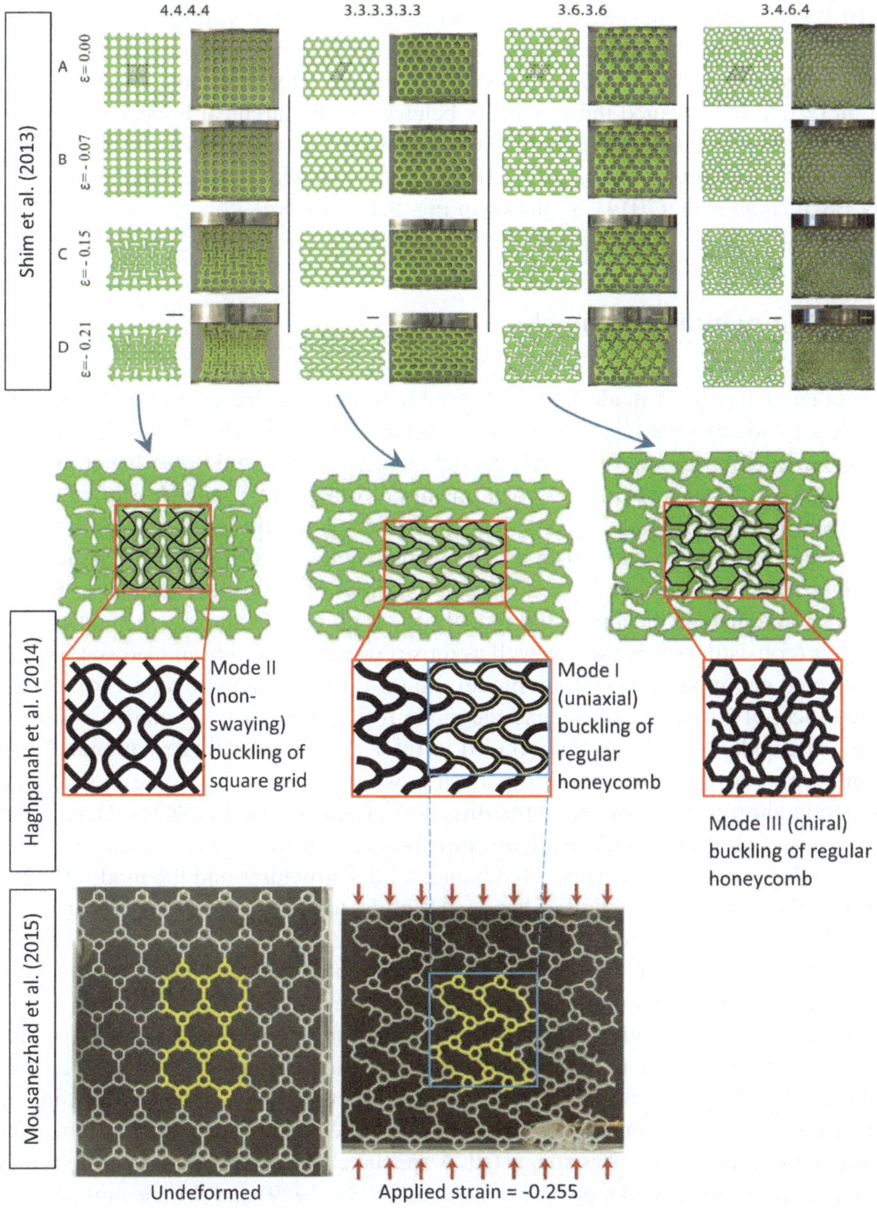

Fig. 3.15 Analogy between buckling of perforated sheet by experiment (top) Reproduced from Shim et al. (2013) with permission of The Royal Society of Chemistry and buckling of honeycombs by analysis (middle) adapted from Haghpanah et al. (2014), with experimental verification (bottom) adapted from Mousanezhad et al. (2015)

(2013) and Haghpanah et al. (2014) are furnished in Fig. 3.15 (middle). The reader's attention is brought to the buckling of the 3.4.6.4 structure by Shim et al. (2013), which resembles that of 3.6.3.6 to a certain extent. More recently, Mousanezhad et al. (2015) investigated the instability behavior of hierarchical honeycomb structures, one of which has been experimentally shown to be analogous to the buckling of 3.3.3.3.3.3 structure by Shim et al. (2013) and mode I buckling of regular honeycomb by Haghpanah et al. (2014), as shown in Fig. 3.15 (bottom).

3.11 Concluding Remarks

An attempt has been made herein to establish analogies across different auxetic models by identifying their deformation patterns. Interestingly, the results suggest that the existing groups of auxetic models can be re-classified into fewer classes due to overlapping similarities in patterns of deformation mechanism from a larger number of geometry-based auxetic models. Majority of auxetic models can be classified under a generic mechanism group in which neighboring units alternate in their rotation direction. Auxetic models that fall within this category are analogous to one another, and include the anti-chiral models, the rotating unit models, the missing rib models (consisting of crosses, as well as those consisting of right-angled swastikas), connected star models, re-entrant model, ring-rod model, semi-auxetic yarn, perforated sheet of 4.4.4.4 structure, and the mode II (i.e., non-swaying mode) of square grid. The code $+ - +-$ is introduced to refer to the alternating rotational directions. Opposed to this mechanism pattern is where all the units rotate in the same direction and reverse only when the direction of applied load changes. Hence, the code $+ + ++$ is proposed for all the chiral models, the missing rib (tilted swastika) model, the perforated sheets of 3.6.3.6 and 3.4.6.4 structures and the mode III (i.e., flower-like or chiral) buckling of hexagonal grids, to reflect their synchronous rotational direction and hence their analogous characteristics. An intermediate case is where the units rotate in the same direction when one considers all the rotating units along one axis but the units rotate in alternating direction when one observes all the rotating units along another axis. Auxetic models under this category are, therefore, analogous to one another, and include the metatetrachiral model, the double arrowhead model, the perforated sheet of 3.3.3.3.3.3 structure, and the mode I buckling of hexagonal grid due to uniaxial compression. To reflect this intermediate mechanism pattern, the code $+ + --$ is suggested. A special class of 2D auxetics can be found from the pioneering works of Wojciechowski (1987, 1989) and Wojciechowski and Branka (1989) which involves rigid hexamers. Auxetic behavior near a phase transition has also been observed in computer simulations of hard cyclic hexamer models by Tretiakov and Wojciechowski (2005). For this case, there is no linking fibrils that mechanically connect neighboring hexamers so, during the course of simulation, the hexamers rotate in a manner that is relatively random in comparison to all the auxetic models discussed in preceding sections. As such the code $\pm\pm\pm\pm$ would intuitively describe the randomness of the rotational direction, including the fact that the rotational direction for each hexamer can change during the simulation. Such randomness is also seen for the generic case of LCP, which are therefore grouped under $\pm\pm\pm\pm$; however, the LCP discussed in Sect. 3.8 is highly idealized for the sake of comparing against the semi-auxetic yarn and rotating squares and so the idealized LCP is classified under $+-+-$. Finally, the code 0 0 0 0 is introduced for models that do not exhibit rotational units. These include the interlocking building block for the Magnox nuclear reactor (Alderson 1999, Muto et al. 1963), as well as the interlocking hexagon model pioneered by Ravirala et al. (2007) and a recent related work extended by Hewage et al. (2016). The egg-rack model and the dimpled sheet involve 3D deformation but their auxeticity is confined to their in-planes, i.e., 2D auxeticity; although rotation exists, observation of their mechanism from the direction normal to their auxetic planes reveal non-rotation—therefore, these models are analogous to the interlocking models and are included in the 0 0 0 0 classification. Finally, while the graphene sheet essentially falls under the 0 0 0 0 broad category, the unevenness of the graphene sheet bulge permits observation of rotation from the direction normal to the graphene sheet, and so to a lesser extent the uneven graphene sheet is associated with the $\pm\pm\pm\pm$ group.

Figure 3.16 summarizes the entire chapter into a table consisting of horizontal grouping by geometry and vertical grouping by analogy of rotational mechanism pattern. Essentially, what appears to be geometrically similar may not necessarily possess similarities in deformation mechanism, and that what appears to be different auxetic geometries could possess similar deformation mechanism, and that it is the deformation mechanism across the different auxetic models that constitute analogy.

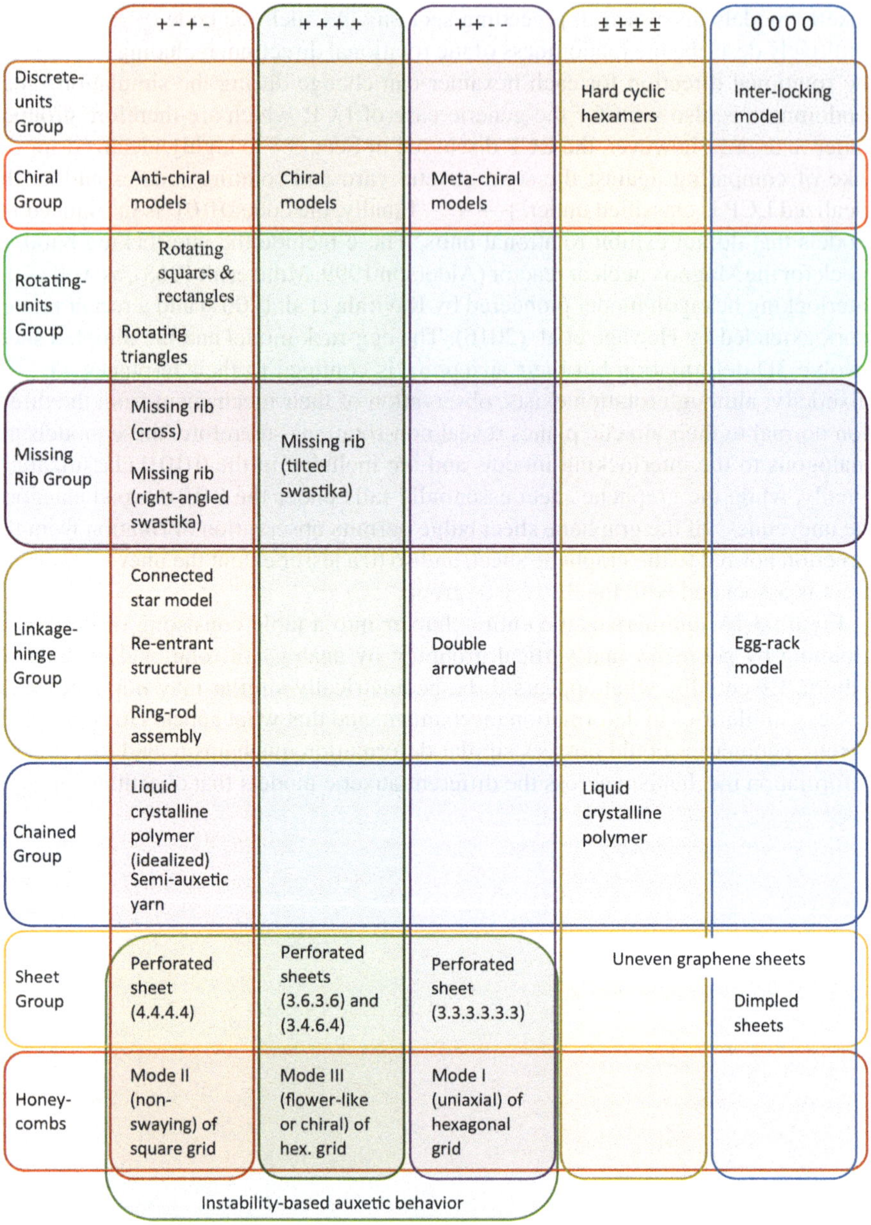

Fig. 3.16 Euler diagram for connecting 2D auxetics by geometrical similarities (rows) and mechanism analogies (columns)

References

Alderson A (1999) A triumph of lateral thought. Chem Ind 10:384–391
Alderson A, Alderson KL, Chirima G, Ravirala N, Zied KM (2010) The in-plane linear elastic constants and out-of-plane bending of 3-coordinated ligament and cylinder-ligament honeycombs. Compos Sci Technol 70(7):1034–1041
Bertoldi K, Reis PM, Wilshaw S, Mullin T (2010) Negative Poisson's ratio behavior induced by an elastic instability. Adv Mater 22(3):361–366
Chen YJ, Scarpa F, Liu YJ, Leng JS (2013) Elasticity of antitetrachiral anisotropic lattices. Int J Solids Struct 50(6):996–1004
Gaspar N, Ren XJ, Smith CW, Grima JN, Evans KE (2005) Novel honeycombs with auxetic behavior. Acta Mater 53(8):2439–2445
Gibson LJ, Ashby MF (1988) Cellular solids: structure & properties. Pergamon Press, Oxford
Grima JN, Evans KE (2000) Auxetic behavior from rotating squares. J Mater Sci Lett 19(17):1563–1565
Grima JN, Williams JJ, Evans KE (2005a) Networked calix[4]arene polymers with unusual mechanical properties. Chem Commun 32:4065–4067
Grima JN, Gatt R, Alderson A, Evans KE (2005b) On the potential of connected stars as auxetic systems. Mol Simul 31(13):925–935
Grima JN, Zammit V, Gatt R, Alderson A, Evans KE (2007) Auxetic behavior from rotating semi-rigid units. Phys Status Solidi B 244(3):866–882
Grima JN, Gatt R, Farrugia PS (2008) On the properties of auxetic meta-tetrachiral structures. Phys Status Solidi B 245(3):511–520
Grima JN, Winczewski S, Mizzi L, Grech MC, Cauchi R, Gatt R, Attard D, Wojciechowski KW, Rybicki J (2015) Tailoring graphene to achieve negative Poisson's ratio properties. Adv Mater 27(8):1455–1459
Haghpanah B, Papadopoulos J, Mousanezhad D, Nayeb-Hashemi H, Vaziri A (2014) Buckling of regular, chiral and hierarchical honeycombs under a general macroscopic stress state. Proc Royal Soc A 470(2167):20130856
He CB, Liu PW, Griffin AC (1998) Toward negative Poisson ratio polymers through molecular design. Macromol 31(9):3145–3147
He CB, Liu PW, McMullan PJ, Griffin AC (2005) Toward molecular auxetics: main chain liquid crystalline polymers consisting of laterally attached para-quaterphenyls. Phys Status Solidi B 242(3):576–584
Hewage TAM, Alderson KL, Alderson A, Scarpa F (2016) Double-negative mechanical metamaterials displaying simultaneous negative stiffness and negative Poisson's ratio properties. Adv Mater 28(46):10323–10332
Javid F, Smith-Roberge E, Innes MC, Shanian A, Weaver JC, Bertoldi K (2015) Dimpled elastic sheets: a new class of non-porous negative Poisson's ratio materials. Scient Rep 5:18373
Jiang Y, Li Y (2017) 3D printed chiral cellular solids with amplified auxetic effects due to elevated internal rotation. Adv Eng Mater 19(2):1600609
Larsen UD, Sigmund O, Bouwstra S (1997) Design and fabrication of compliant mechanisms and material structures with negative Poisson's ratio. J Microelectromech Syst 6(2):99–106
Lim TC (2014) Semi-auxetic yarns. Phys Status Solidi B 251(2):273–280
Lim TC (2017a) Analogies across auxetic models based on deformation mechanism. Phys Status Solidi RRL 11(6):1600440
Lim TC (2017b) Auxetic and negative thermal expansion structure based on interconnected array of rings and sliding rods. Phys Status Solidi B 254(12):1600775
Lim TC (2019) Negative environmental expansion for interconnected array of rings and sliding rods. Phys Status Solidi B 256(1):1800032
Mousanezhad D, Babaee S, Ebrahimi H, Ghosh R, Hamouda AS, Bertoldi K, Vaziri A (2015) Hierarchical honeycomb auxetic metamaterials. Scient Rep 5:18306

Muto K, Bailey RW, Mitchell KJ (1963) Special requirements for the design of nuclear power stations to withstand earthquakes. Proc Inst Mech Eng 177(1):155–203
Ravirala N, Alderson A, Alderson KL (2007) Interlocking hexagon model for auxetic behavior. J Mater Sci 42(17):7433–7445
Shim J, Shan S, Kosmrlj A, Kang SH, Chen ER, Weaver JC. Bertoldi K (2013) Harnessing instabilities for design of soft reconfigurable auxetic/chiral materials. Soft Matter 9(34):8198–8202
Smith CW, Grima JN, Evans KE (2000) A novel mechanism for generating auxetic behavior in reticulated foams: missing rib foam model. Acta Mater 48(17):4349–4356
Taylor M, Francesconi L, Gerendas M, Shanian A, Carson C, Bertoldi K (2013) Low porosity metallic periodic structures with negative Poisson's ratio. Adv Mater 26(15):2365–2370
Theocaris PS, Stavroulakis GE, Panagiotopoulos (1997) Negative Poisson's ratios in composites with star-shaped inclusions: a numerical homogenization approach. Arch Appl Mech 67(4):274–286
Tretiakov KV, Wojciechowski KW (2005) Monte Carlo simulation of two-dimensional hard body systems with extreme values of the Poisson's ratio. Phys Status Solidi B 242(3):730–741
Wojciechowski KW (1987) Constant thermodynamic tension Monte-Carlo studies of elastic properties of a two-dimensional system of hard cyclic hexamers. Mol Phys 61(5):1247–1258
Wojciechowski KW (1989) Two-dimensional isotropic system with a negative Poisson ratio. Phys Lett A 137(1&2):60–64
Wojciechowski KW, Branka AC (1989) Negative Poisson ratio in a two-dimensional "isotropic" solid. Phys Rev A 40(12):7222–7225

Chapter 4
Thin Auxetic Plates

Abstract This chapter discusses some non-circular and non-rectangular thin auxetic plates—such as equilateral triangular plates, elliptical plates, sectorial plates and rhombic plates—with special emphasis on their performances such as bending stress minimization or in terms of deflection minimization.

Keywords Elliptical plates · Equilateral triangular plates · Rhombic plates · Sectorial plates

4.1 Fundamentals

The analysis on the effects of auxeticity on thin plates (Lim 2014a) has been investigated for constrained plates (Strek et al. 2008; Pozniak et al. 2010), circular and rectangular plates (Lim 2013a, b), nanoscale plates (Ho et al. 2014), and for auxetic plates resting on auxetic foundations (Lim 2014b). These have been summarized in Chap. 8 of Lim (2015).

For a plate of Young's modulus E, Poisson's ratio v and thickness h lying on the x, y plane of a Cartesian coordinate system, it is well known that the bending moment relation

$$M_y = -D\left(v\frac{\partial^2 w}{\partial x^2} + \frac{\partial^2 w}{\partial y^2}\right) \tag{4.1.1}$$

and

$$M_x = -D\left(\frac{\partial^2 w}{\partial x^2} + v\frac{\partial^2 w}{\partial y^2}\right) \tag{4.1.2}$$

where

$$D = \frac{Eh^3}{12\left(1 - v^2\right)} \tag{4.1.3}$$

simplifies to

T. Lim, *Mechanics of Metamaterials with Negative Parameters*, Engineering Materials,
https://doi.org/10.1007/978-981-15-6446-8_4

$$M_x + M_y = 0 \Leftrightarrow v = -1 \tag{4.1.4}$$

Likewise, the bending moments under the polar coordinate system

$$M_r = -D\left(\frac{\partial^2 w}{\partial r^2} + \frac{v}{r}\frac{\partial w}{\partial r}\right) \tag{4.1.5}$$

and

$$M_\theta = -D\left(v\frac{\partial^2 w}{\partial r^2} + \frac{1}{r}\frac{\partial w}{\partial r}\right) \tag{4.1.6}$$

obviously gives

$$M_r + M_\theta = 0 \Leftrightarrow v = -1 \tag{4.1.7}$$

Other unique, and even surprising, properties can be found in the auxetic region for other values of Poisson's ratio. More details on the effect of auxeticity on thin plates are illustrated in the following examples in Sects. 4.2–4.5 for equilateral triangular plates, elliptical plates, sectorial plates, and rhombic plates.

4.2 Example: Equilateral Triangular Auxetic Plates

We consider herein three special cases on the basis of loading distribution on a simply supported equilateral triangular plate. A "central" special case is considered in the form of uniformly loaded plate, such that two extreme cases can be identified. In one extreme case, the transverse uniform load is redistributed toward the plate boundary. Since neither deflection nor bending stresses take place when the load is directly on top of the simple supports, we consider line load uniformly distributed near to the simple supports; in practice, this loading condition translates into uniformly distributed bending moment applied at the plate boundary. In the other extreme case, the transverse uniform load is redistributed toward the plate center to form a central point load.

For a simply supported equilateral triangular plate of sides $2l/\sqrt{3}$ as shown in Fig. 4.1, its bending moment distributions are

$$\begin{aligned} M_x &= \frac{M_0}{2}\left[1 + v - 3(1 - v)\frac{x}{l}\right] \\ M_y &= \frac{M_0}{2}\left[1 + v + 3(1 - v)\frac{x}{l}\right] \end{aligned} \tag{4.2.1}$$

in the presence of uniformly distributed edge moment M_0,

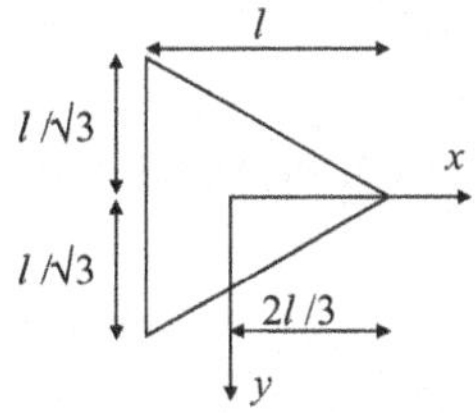

Fig. 4.1 Geometrical description of a simply supported equilateral triangular plate

$$
\begin{aligned}
M_x &= \frac{q_0 l^2}{16}\left[(5-v)\left(\frac{x}{l}\right)^3 - (3+v)\left(\frac{x}{l}\right)^2 - 3(1+3v)\left(\frac{x}{l}\right)\left(\frac{y}{l}\right)^2 \right. \\
&\quad \left. -\frac{2}{3}(1-v)\frac{x}{l} - (1+3v)\left(\frac{y}{l}\right)^2 + \frac{8}{27}(1+v)\right] \\
M_y &= \frac{q_0 l^2}{16}\left[(1-5v)\left(\frac{x}{l}\right)^3 - (1+3v)\left(\frac{x}{l}\right)^2 - 3(3+v)\left(\frac{x}{l}\right)\left(\frac{y}{l}\right)^2 \right. \\
&\quad \left. +\frac{2}{3}(1-v)\frac{x}{l} - (3+v)\left(\frac{y}{l}\right)^2 + \frac{8}{27}(1+v)\right]
\end{aligned}
\tag{4.2.2}
$$

arising from uniformly distributed load q_0, and

$$
\begin{aligned}
M_x &= \frac{P}{4\pi}\left[(1+v)\left(\ln\frac{l\sqrt{3}}{\pi c} - 0.379\right) - \frac{1-v}{2}\right] \\
M_y &= \frac{P}{4\pi}\left[(1+v)\left(\ln\frac{l\sqrt{3}}{\pi c} - 0.379\right) + \frac{1-v}{2}\right]
\end{aligned}
\tag{4.2.3}
$$

at a small distance c from a point load P applied at plate center. Since the bending stress is related to the bending moment as

$$\sigma_{\max} = \frac{6M}{h^2} \tag{4.2.4}$$

the following dimensionless bending stresses are introduced

$$
\begin{aligned}
\sigma^* &= \frac{\sigma_{\max} h^2}{3M_0} \\
\sigma^{**} &= \frac{8\sigma_{\max} h^2}{3q_0 l^2} \\
\sigma^{***} &= \frac{2\pi\sigma_{\max} h^2}{3P}
\end{aligned}
\tag{4.2.5}
$$

so that

$$\sigma_x^* = 1 + v - 3(1-v)\frac{x}{l}$$

$$\sigma_y^* = 1 + \upsilon + 3(1 - \upsilon)\frac{x}{l} \tag{4.2.6}$$

for uniform edge moment M_0,

$$\begin{aligned}
\sigma_x^{**} &= (5 - \upsilon)\left(\frac{x}{l}\right)^3 - (3 + \upsilon)\left(\frac{x}{l}\right)^2 - 3(1 + 3\upsilon)\left(\frac{x}{l}\right)\left(\frac{y}{l}\right)^2 \\
&\quad - \frac{2}{3}(1 - \upsilon)\frac{x}{l} - (1 + 3\upsilon)\left(\frac{y}{l}\right)^2 + \frac{8}{27}(1 + \upsilon) \\
\sigma_y^{**} &= (1 - 5\upsilon)\left(\frac{x}{l}\right)^3 - (1 + 3\upsilon)\left(\frac{x}{l}\right)^2 - 3(3 + \upsilon)\left(\frac{x}{l}\right)\left(\frac{y}{l}\right)^2 \\
&\quad + \frac{2}{3}(1 - \upsilon)\frac{x}{l} - (3 + \upsilon)\left(\frac{y}{l}\right)^2 + \frac{8}{27}(1 + \upsilon)
\end{aligned} \tag{4.2.7}$$

for uniform load q_0, and

$$\begin{aligned}
\sigma_x^{***} &= (1 + \upsilon)\left(\ln\frac{l\sqrt{3}}{\pi c} - 0.379\right) - \frac{1 - \upsilon}{2} \\
\sigma_y^{***} &= (1 + \upsilon)\left(\ln\frac{l\sqrt{3}}{\pi c} - 0.379\right) + \frac{1 - \upsilon}{2}
\end{aligned} \tag{4.2.8}$$

at a small distance c from the central point load P. It is not surprising that, in general, we have (Lim 2016a)

$$\frac{\sigma_x^*}{\sigma_y^*} = \frac{\sigma_x^{**}}{\sigma_y^{**}} = \frac{\sigma_x^{***}}{\sigma_y^{***}} = -1 \tag{4.2.9}$$

when $\upsilon = -1$; this is a consequence from Eq. (4.1.4). In addition, it is worthy to note that

$$\sigma_x^* = \sigma_y^* = \sigma_x^{**} = \sigma_y^{**} = 0 \tag{4.2.10}$$

at the plate center when $\upsilon = -1$.

Equation (4.2.6) is plotted in Fig. 4.2 showing that auxeticity has adverse effect on triangular plates with uniformly distributed bending moment applied at the edge. Specifically, the dimensionless bending moment σ_x^* plotted in Fig. 4.2a is within its lower range for $-1/4 < \upsilon < 1/2$, while the range $-1 \le \upsilon \le -1/4$ leads to higher range of σ_x^* at the plate apex. In the case of dimensionless bending moment furnished in Fig. 4.2b, σ_y^* is maximum at the apex. The magnitude of σ_x^* is confined within $|\sigma_x^*| \le 2$ for $-1/3 \le \upsilon \le 1/2$ while the maximum value of σ_y^* is minimized to $\sigma_y^* = 2.5$ when $\upsilon = 1/2$. Therefore, the bending stress in a simply supported equilateral triangular plate, under uniform applied bending at the sides, is minimized if the plate material is incompressible. When the same bending load, M_0, is applied on a simply supported circular plate or a square plate, the bending moment is uniformly

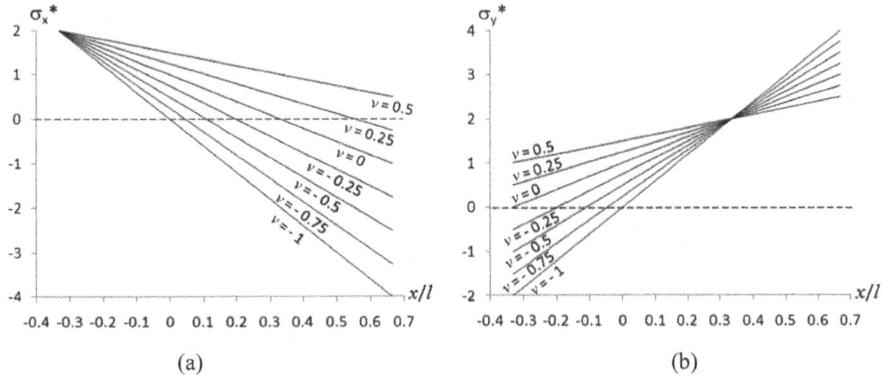

Fig. 4.2 Dimensionless bending stress distributions of **a** σ_x^* and **b** σ_x^* versus x/l arising from uniformly applied bending moment along the plate edge

distributed throughout the plate as M_0 (Timoshenko and Woinowsky-Krieger 1959), i.e., independent from the Poisson's ratio of the plate material—this is unlike the case of triangular plates.

The effect of auxeticity on uniformly loaded triangular plate is shown in Fig. 4.3, in which the dimensionless bending stresses are evaluated along the x-axis. Substituting $y = 0$ into Eq. (4.2.7), plots of σ_x^{**} and σ_y^{**} against x/l are furnished in Fig. 4.3a, b, respectively, for the entire range of Poisson's ratio. Reference to Fig. 4.3a shows that the maximum σ_x^{**} magnitudes for $v = -1, -3/4, -1/2$ take place between the plate center and the plate corners, while the maximum σ_x^{**} magnitudes for $v = -1/4, 0, 1/4, 1/2$ occur between the plate center and the middle of the plate sides. Furthermore, the maximum σ_x^{**} magnitude decreases from $v = -1$ to $v = -1/2$ and from $v = 1/2$ to $v = -1/4$, thereby implying that there exists a minimized maximum σ_x^{**} magnitude between $v = -1/2$ and $v = -1/4$, such that at the optimal Poisson's ratio there exists a minimum and a maximum σ_x^{**} of equal magnitude, i.e., this magnitude is the least of all the maximum σ_x^{**} magnitudes. The exact value of this optimal Poisson's ratio is $v = -1/3$, such that a maximum and a minimum $\sigma_x^{**} = \pm 0.256$ take place at $x/l = -2/15$ and $x/l = 7/15$, respectively. Similarly, in Fig. 4.3b, it can be seen that the maximum σ_y^{**} decreases from $v = 1/2$ to $v = 0$ and also from $v = -1$ to $v = -1/4$. The exact value for the optimal Poisson's ratio that gives the least maximum stress of $\sigma_y^{**} = 0.384$ is $v = -1/15$. Using the Poisson's ratio of $v = -1/15$, we obtain the maximum stress of $\sigma_x^{**} = 0.313 < 0.384$. As such, it is the stress in the y-direction, rather than the x-direction, along the x-axis that determines the greatest extent of stress. Hence, the Poisson's ratio of $v = -1/15$ is selected as the optimal value that minimizes the maximum stress. For comparison with its conventional counterpart and further extent of auxeticity, plots of σ_x^{**} versus x/l for $-1 \le v \le 1/3$ at Poisson's ratio increments of $2/3$ and σ_y^{**} versus x/l for $-1/5 \le v \le 1/5$ at Poisson's ratio increments of $2/15$ are shown in Fig. 4.3c, d, respectively. In comparison to plates of other shapes with similar boundary condition and load pattern, the optimal Poisson's ratio for minimizing bending stresses in

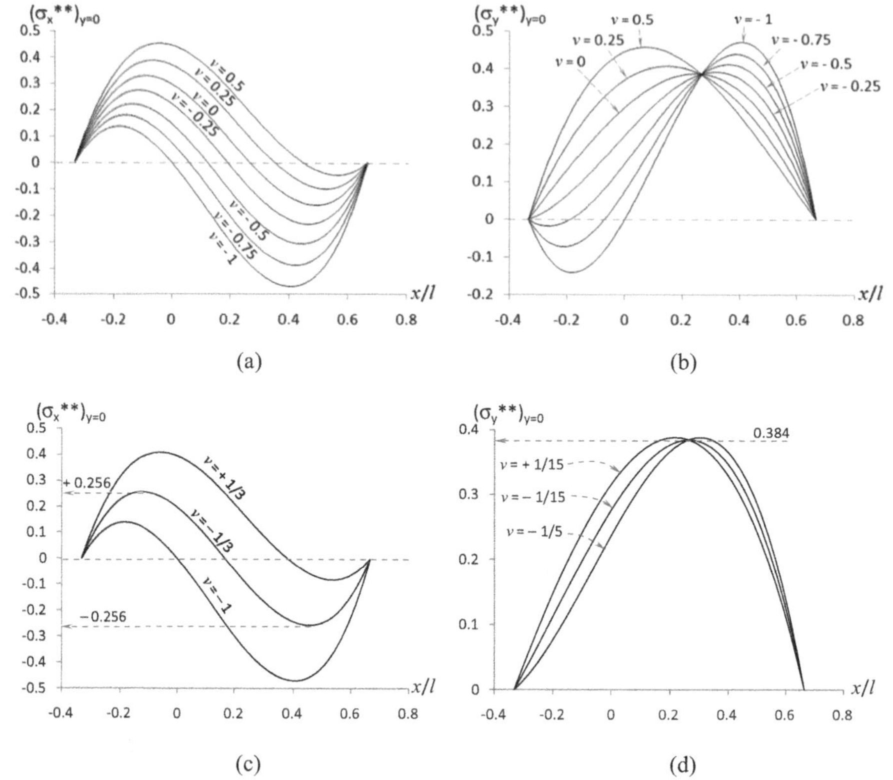

Fig. 4.3 Dimensionless bending stress distributions of **a** σ_x^{**} and **b** σ_y^{**} versus x/l along $y = 0$ arising from uniform load for the entire range of Poisson's ratio, as well as the dimensionless bending stress distribution **c** at the optimal σ_x^{**} at $\upsilon = -1/3$ with σ_x^{**} at $\upsilon = -1/3 \pm 2/3$ for comparison, and **d** at the optimal σ_y^{**} at $\upsilon = -1/15$ with σ_y^{**} at $\upsilon = -1/15 \pm 2/15$ for comparison

circular and square plates are $\upsilon = -1/3$ and $\upsilon = 0.115$, respectively (Lim 2013a, b).

Dimensionless bending stresses under a central point load are plotted only for $0 \le c/l \le 0.1$ based on Eq. (4.2.8) because it is known that bending stresses under point load is maximum at the loading point, as displayed in Fig. 4.4. It is therefore of special interest to note that the bending stresses are constant within $0 \le c/l \le 0.1$ when $\upsilon = -1$, thereby implying that when $\upsilon = -1$, the bending stresses do not approach infinity as $c/l \to 0$. Figure 4.4a suggests that the magnitude of the dimensionless bending stress σ_x^{***} is effectively reduced at plate center ($c/l = 0$) when $\upsilon = -1$, at $0 < c/l < 0.05$ when $\upsilon = -3/4$, and at $0.05 \le c/l \le 0.1$ when $\upsilon = -1/2$, while reference to Fig. 4.4b indicates that the dimensionless bending stress σ_y^{***} is minimized throughout the plate when $\upsilon = -1$. Since both σ_x^{***} and σ_y^{***} tend to infinity at load point $c/l = 0$ for $-1 < \upsilon \le 1/2$, it follows that bending

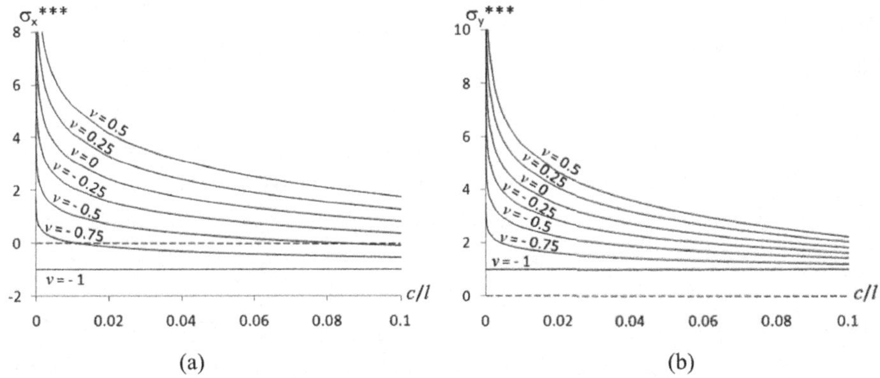

Fig. 4.4 Dimensionless bending stress distributions of **a** σ_x^{***} and **b** σ_y^{***} versus c/l arising from central point load

stress is minimized only when $v = -1$, thereby agreeing with the cases of circular and square plates under central point load.

Results suggest that for simply supported triangular plates, the use of

1. conventional materials are advisable under applied bending at plate boundary
2. mildly auxetic materials are advisable under uniformly distributed load throughout plate and
3. highly auxetic materials are advisable under concentrated load at plate center.

An extremely auxetic isotropic plate ($v = -1$) is consistently the best for limiting the maximum bending stress of a central point load in circular, square, and triangular plates. In the case of uniformly distributed load, mildly auxetic materials are recommended for circular and triangular plates, while mildly conventional materials are suggested for square plates (Lim 2016a).

4.3 Example: Elliptical Auxetic Plates

Consider an elliptical plate of Young's modulus E and thickness h with a and b being the radius of an elliptical plate along the major (x) and minor (y) axes, respectively, as illustrated in Fig. 4.5. It has been shown that under uniform load q, the maximum bending stress occurs at the ends of the shorter principal axis if the plate edge is fully clamped (Timoshenko and Woinowsky-Krieger 1959), and that this bending moment is independent from the Poisson's ratio of the plate material. On the other hand, the bending moment is greatest at the plate center if the plate edge is simply supported (Timoshenko and Woinowsky-Krieger 1959), and that this bending moment is dependent on the Poisson's ratio of the plate material. Due to the complexity of the analytical model for simply supported elliptical plate, the deflection and bending moments at the plate center are typically furnished by the formulas

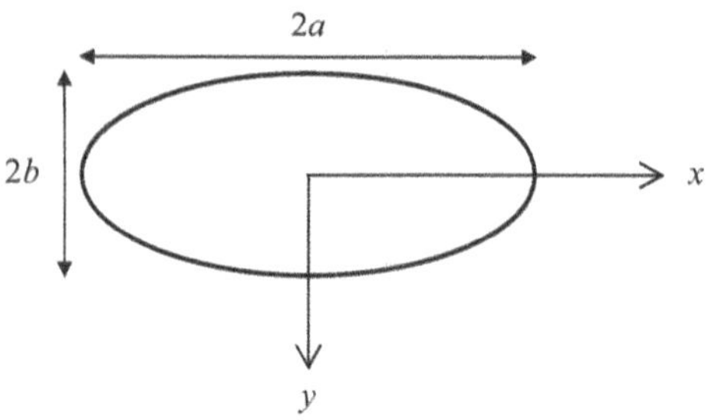

Fig. 4.5 Geometrical description of an elliptical plate

$$(w)_{x=y=0} = \alpha \frac{qb^4}{Eh^3} \tag{4.3.1}$$

$$(M_x)_{x=y=0} = \beta qb^2 \tag{4.3.2}$$

$$(M_y)_{x=y=0} = \beta_1 qb^2 \tag{4.3.3}$$

where the numerical values of α, β, and β_1 have only been made available for $v = 0.3$ (Timoshenko and Woinowsky-Krieger 1959; Ventsel and Krauthammer 2001). In order to conveniently observe the effect of Poisson's ratio for the entire range of isotropic solids, $-1 \le v \le 1/2$, a simplified model based on the Kirchhoff–Love plate theory for an elliptical plate with $a \ne b$ is introduced herein in the form

$$w = w_0\left(1 - \frac{x^2}{a^2} - \frac{y^2}{b^2}\right)\left(C - \frac{x^2}{a^2} - \frac{y^2}{b^2}\right) \tag{4.3.4}$$

where C is determined from boundary condition. This form of expression is selected because its deflection profiles along the principal axes are similar to the deflection profile of a simply supported circular plate, while at the same time catering for the major and minor axes of the elliptical plates. It can be seen that Eq. (4.3.4) fulfills the boundary condition $w = 0$, and that $C = 1$ applies for fully clamped edge. Another boundary condition is that the bending moment perpendicular to the tangent of the simply supported elliptical plate is zero. Hence, applying $(M_y)_{x=0,y=\pm b} = 0$ and $(M_x)_{x=\pm a,y=0} = 0$ for the bending moment equations furnished in Eqs. (4.1.1) and (4.1.2) based on the deflection profile in Eq. (4.3.4)

gives two solutions for C, i.e.,

$$C_1 = \frac{5a^2 + vb^2}{a^2 + vb^2} \tag{4.3.5a}$$

and

$$C_2 = \frac{5b^2 + va^2}{b^2 + va^2} \tag{4.3.5b}$$

respectively. The choice of either C_1 and/or C_2 can be elucidated after the expression of w_0 is established (Lim 2016b). To obtain w_0, there is a need to solve

$$\Delta\Delta w = \frac{q}{D} \tag{4.3.6}$$

where

$$\Delta = \nabla^2 = \frac{\partial^2}{\partial x^2} + \frac{\partial^2}{\partial y^2} \tag{4.3.7}$$

Substituting

$$\begin{Bmatrix} \frac{\partial^4 w}{\partial x^4} \\ \frac{\partial^4 w}{\partial x^2 \partial y^2} \\ \frac{\partial^4 w}{\partial y^4} \end{Bmatrix} = 8w_0 \begin{Bmatrix} \frac{3}{a^4} \\ \frac{1}{a^2 b^2} \\ \frac{3}{b^4} \end{Bmatrix} \tag{4.3.8}$$

into Eq. (4.3.6) leads to

$$w_0 = \frac{q}{D}\left(\frac{24}{a^4} + \frac{16}{a^2 b^2} + \frac{24}{b^4}\right)^{-1} \tag{4.3.9}$$

Hence, Eq. (4.3.4) can be written as

$$w = \frac{q}{D}\left(\frac{24}{a^4} + \frac{16}{a^2 b^2} + \frac{24}{b^4}\right)^{-1}\left(1 - \frac{x^2}{a^2} - \frac{y^2}{b^2}\right)\left(C - \frac{x^2}{a^2} - \frac{y^2}{b^2}\right) \tag{4.3.10}$$

such that when $b = a$ and $x^2 + y^2 = r^2$, it reduces to the familiar plate deflection profile of a simply supported circular plate under uniform load

$$w = \frac{q}{64D}(a^2 - r^2)\left(\frac{5+v}{1+v}a^2 - r^2\right) \tag{4.3.11}$$

where r is the radial distance from the center of the circular plate. In addition, the substitution of $C = 1$ into Eq. (4.3.10) converts it into the deflection profile of a fully clamped elliptical plate (Timoshenko and Woinowsky-Krieger 1959; Ventsel and Krauthammer 2001),

$$w = \frac{q}{D}\left(\frac{24}{a^4} + \frac{16}{a^2 b^2} + \frac{24}{b^4}\right)^{-1}\left(1 - \frac{x^2}{a^2} - \frac{y^2}{b^2}\right)^2 \tag{4.3.12}$$

It is clear that Eqs. (4.3.5a) and (4.3.5b) are equal only for circular plates, but differ for elliptical plates. As a guiding principle on the selection of C, we note that the deflection of simply supported plates exceeds those that are fully clamped, so it is a requirement that the deflection stated in Eq. (4.3.4) must exceed that for fully clamped elliptical plates. Since the expression of w_0 is common for both simply supported and fully clamped elliptical plates, it follows that the deflection of the simply supported elliptical plate can only exceed that of fully clamped plate if $C > 1$. Reference to Eqs. (4.3.5a) and (4.3.5b) reveals that in the case of negative Poisson's ratio, $C_1 < 1$ for $0 < a/b < 1$ and $C_2 < 1$ for $a/b > 1$. A graphical view is shown in Fig. 4.6. As such, both C_1 and C_2 are applicable with their validity range being $a/b \geq 1$ and $a/b \leq 1$, respectively. In the subsequent formulation, we adopt C_1 to facilitate comparison with the exact numerical values of α, β, and β_1, which are based on $a/b \geq 1$ (Timoshenko and Woinowsky-Krieger 1959; Ventsel and Krauthammer 2001).

Based on Eqs. (4.3.5a) and (4.3.10), we have the deflection at the plate center,

$$(w)_{x=y=0} = \frac{qb^4}{Eh^3} \frac{\left(1 - v^2\right)\left(5\frac{a^2}{b^2} + v\right)}{\left(2\frac{b^4}{a^4} + \frac{4}{3}\frac{b^2}{a^2} + 2\right)\left(\frac{a^2}{b^2} + v\right)} \tag{4.3.13}$$

which, upon comparison with Eq. (4.3.1), gives

$$\alpha = \frac{\left(1 - v^2\right)\left(5\frac{a^2}{b^2} + v\right)}{\left(2\frac{b^4}{a^4} + \frac{4}{3}\frac{b^2}{a^2} + 2\right)\left(\frac{a^2}{b^2} + v\right)} \tag{4.3.14}$$

Based on Eqs. (4.3.5a) and (4.3.10) again, the bending moments can be expressed as

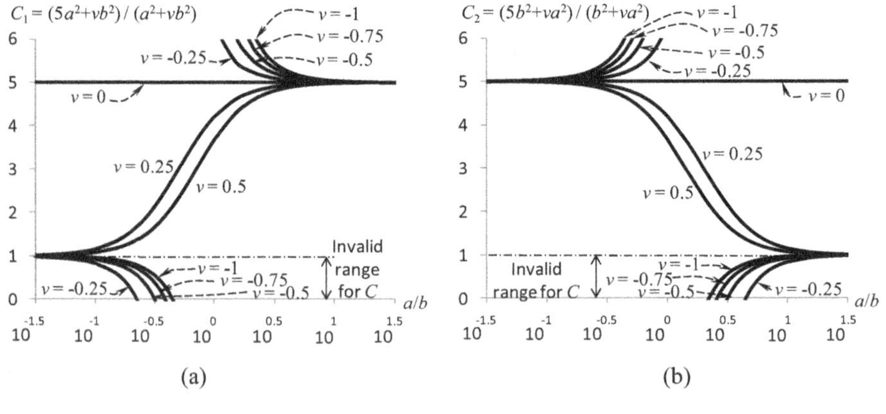

Fig. 4.6 Plots of **a** C_1 and **b** C_2 for $-1 \leq v \leq 1/2$ and $10^{-1.5} \leq a/b \leq 10^{1.5}$

$$\begin{Bmatrix} M_x \\ M_y \end{Bmatrix} = -D \begin{bmatrix} 1 & v \\ v & 1 \end{bmatrix} \begin{Bmatrix} \frac{\partial^2 w}{\partial x^2} \\ \frac{\partial^2 w}{\partial y^2} \end{Bmatrix} \tag{4.3.15}$$

where

$$\begin{aligned} \frac{\partial^2 w}{\partial x^2} &= \frac{q}{D}\left(\frac{24}{a^4}+\frac{16}{a^2b^2}+\frac{24}{b^4}\right)^{-1}\left[-\frac{2}{a^2}\left(1+\frac{5a^2+vb^2}{a^2+vb^2}\right)+\frac{4y^2}{a^2b^2}+\frac{12x^2}{a^4}\right] \\ \frac{\partial^2 w}{\partial y^2} &= \frac{q}{D}\left(\frac{24}{a^4}+\frac{16}{a^2b^2}+\frac{24}{b^4}\right)^{-1}\left[-\frac{2}{b^2}\left(1+\frac{5a^2+vb^2}{a^2+vb^2}\right)+\frac{4x^2}{a^2b^2}+\frac{12y^2}{b^4}\right] \end{aligned} \tag{4.3.16a}$$

At the plate center, Eq. (4.3.16a) simplifies to

$$\begin{aligned} \left(\frac{\partial^2 w}{\partial x^2}\right)_{x=y=0} &= -\frac{2q}{a^2D}\left(1+\frac{5a^2+vb^2}{a^2+vb^2}\right)\left(\frac{24}{a^4}+\frac{16}{a^2b^2}+\frac{24}{b^4}\right)^{-1} \\ \left(\frac{\partial^2 w}{\partial y^2}\right)_{x=y=0} &= -\frac{2q}{b^2D}\left(1+\frac{5a^2+vb^2}{a^2+vb^2}\right)\left(\frac{24}{a^4}+\frac{16}{a^2b^2}+\frac{24}{b^4}\right)^{-1} \end{aligned} \tag{4.3.16b}$$

such that the bending moment $(M_x)_{x=y=0}$ at the plate center can be expressed as Eq. (4.3.2) with

$$\beta = \frac{\left(\frac{b^2}{a^2}+v\right)\left(3\frac{a^2}{b^2}+v\right)}{\left(6\frac{b^4}{a^4}+4\frac{b^2}{a^2}+6\right)\left(\frac{a^2}{b^2}+v\right)} \tag{4.3.17}$$

while the bending moment $\left(M_y\right)_{x=y=0}$ at the plate center can be expressed as Eq. (4.3.3) with

$$\beta_1 = \frac{3+v\frac{b^2}{a^2}}{6\frac{b^4}{a^4}+4\frac{b^2}{a^2}+6} \tag{4.3.18}$$

It can be observed that

$$(\beta)_{\frac{a}{b}\to 1} = (\beta_1)_{\frac{a}{b}\to 1} = \frac{3+v}{16} \tag{4.3.19}$$

and

$$(\beta)_{\frac{a}{b}\to\infty} = \frac{v}{2},\ (\beta_1)_{\frac{a}{b}\to\infty} = \frac{1}{2} \tag{4.3.20}$$

Having obtained the analytical forms of β and β_1, it is now of interest to understand how the bending stresses can be minimized. As such, we herein define the optimal

Poisson's ratio as the value of Poisson's ratio that gives the minimum magnitude to the maximum bending stress. Since $\sigma_{max} = \pm 6M/h^2$, it follows that minimization of the maximum bending stress can be performed by minimizing the bending moment. The bending moment $(M_x)_{x=y=0}$ can be minimized by setting $(M_x)_{x=y=0} = 0$ (i.e., by substituting $\beta = 0$) into Eq. (4.3.17) to give

$$v = -\left(\frac{3a^2}{2b^2} + \frac{b^2}{2a^2}\right) \pm \sqrt{\left(\frac{3a^2}{2b^2} + \frac{b^2}{2a^2}\right)^2 - 3} \tag{4.3.21}$$

Since the range of the upper and lower solutions is $-1 \le v_{upp} < 0$ and $-\infty < v_{low} \le -3$, respectively, for $a/b \ge 1$, the upper solution of Eq. (4.3.21) prescribes the Poisson's ratio that gives $(M_x)_{x=y=0} = 0$ while the lower solution of Eq. (4.3.21) has no physical meaning. This is because the Poisson's ratio range for the upper and lower solutions falls inside and outside the range of Poisson's ratio of isotropic solids, respectively. With reference to the numerator of Eq. (4.3.18), i.e., $3 + v(b/a)^2$, the bending moment $(M_y)_{x=y=0}$ can be minimized by using auxetic materials of $v = -1$ and circular plates (i.e., $a = b$). In order to minimize the effective maximum bending stress, we recall the von Mises stress

$$\sigma_{eff} = \sqrt{\sigma_x^2 + \sigma_y^2 + \sigma_z^2 - \sigma_y\sigma_z - \sigma_z\sigma_x - \sigma_x\sigma_y + 3\left(\tau_{yz}^2 + \tau_{zx}^2 + \tau_{xy}^2\right)} \tag{4.3.22}$$

At the top and bottom surfaces where the bending stresses are maximum, the following shearing stresses are zero, $\tau_{yz} = \tau_{zx} = 0$. Considering an element at the plate center aligned to the principal axes, the in-plane shear stress is zero, $\tau_{xy} = 0$. The out-of-plane stresses at the upper and lower surfaces are $\sigma_z = -q$ and $\sigma_z = 0$, respectively; the former is negligible in comparison to the bending stresses. Hence, the effective stress at the plate center reduces to

$$\sigma_{eff} = \sqrt{\sigma_x^2 + \sigma_y^2 - \sigma_x\sigma_y} \tag{4.3.23}$$

Since $\sigma_{max} = 6M/h^2$, substitution of

$$\begin{Bmatrix} \sigma_x \\ \sigma_y \end{Bmatrix} = \frac{6}{h^2}\begin{Bmatrix} M_x \\ M_y \end{Bmatrix} = \frac{6qb^2}{h^2}\begin{Bmatrix} \beta \\ \beta_1 \end{Bmatrix} \tag{4.3.24}$$

into Eq. (4.3.23) gives $\sigma_{eff} = 6M_{eff}/h^2$ where the effective maximum bending moment is expressed as

$$M_{eff} = \beta_2 q b^2 \tag{4.3.25}$$

in which

$$\beta_2 = \sqrt{\beta^2 + \beta_1^2 - \beta\beta_1} \tag{4.3.26}$$

is the dimensionless parameter for the effective bending moment. Equation (4.3.25) is expressed in a form similar to Eqs. (4.3.2) and (4.3.3) to facilitate comparison.

Substituting $v = 0.3$ into Eqs. (4.3.14), (4.3.17) and (4.3.18) allows a comparison to be made between the exact model (Timoshenko and Woinowsky-Krieger 1959; Ventsel and Krauthammer 2001) and the simplified model developed herein; the latter is listed in Table 4.1 under the columns indicated by "Approx". A graphical comparison is plotted in Fig. 4.7 revealing that the simplified model gives a more severe, and hence more conservative, prediction for a wide range of a/b ratio at $v = 0.3$. Equation (4.3.14) gives an overestimation of α, while Eqs. (4.3.17) and (4.3.18) provide reasonable correlation to the exact values of β and β_1, respectively; the latter models are used for evaluating the bending stresses at the elliptical plate center in the major axis direction, $(M_x)_{x=y=0}$, and minor axis direction, $\left(M_y\right)_{x=y=0}$, in terms of their dimensionless parameters.

Having established the validity of β and β_1 for $v = 0.3$ by comparison against the exact numerical values, it is now of interest to make use of Eqs. (4.3.17) and (4.3.18) to evaluate the effect of negative Poisson's ratio on the bending stresses at the plate center. The dimensionless parameter β, as shown in Fig. 4.8a, suggests that the corresponding bending stress $(\sigma_x)_{x=y=0}$ is minimized (a) by using auxetic materials for elliptical plates that are almost circular $a/b \approx 1$, and (b) by using zero Poisson's ratio materials for elliptical plates that are very long or very narrow $a/b \to \infty$. The Poisson's ratio for maintaining $(\sigma_x)_{x=y=0} = 0$ is plotted in Fig. 4.8b using the upper solution of Eq. (4.3.21), suggesting that stresses is minimized by using mildly auxetic and highly auxetic materials for elliptical plates of high ($a/b \gg 1$) and low ($a/b \approx 1$) aspect ratios, respectively.

The family of curves in Fig. 4.8a can be explained for both $a/b \approx 1$ and $a/b \gg 1$. In the case of $a/b \approx 1$, the boundary condition dictates that both M_x and M_y possess equal sign, although the bending into a synclastic shape is more easily attained when the plate possesses negative Poisson's ratio. For this reason, the value of β

Table 4.1 A comparison between the exact and the approximate α, β, and β_1 for various plate aspect ratio at $v = 0.3$

$\frac{a}{b}$	α (Exact)	α (Approx)	β (Exact)	β (Approx)	β_1 (Exact)	β_1 (Approx)
1	0.7	0.695625	0.206	0.20625	0.206	0.20625
1.1	0.83	0.856504	0.215	0.218724	0.235	0.242313
1.2	0.96	1.008221	0.219	0.226232	0.261	0.274891
1.3	1.07	1.146751	0.223	0.229879	0.282	0.303556
1.4	1.17	1.270527	0.223	0.230719	0.303	0.328353
1.5	1.26	1.379599	0.222	0.229631	0.321	0.349587
2	1.58	1.747544	0.21	0.213323	0.379	0.416949
3	1.88	2.039995	0.188	0.185136	0.433	0.465341
4	2.02	2.143804	0.184	0.171223	0.465	0.481196
5	2.1	2.191473	0.17	0.16402	0.48	0.4882

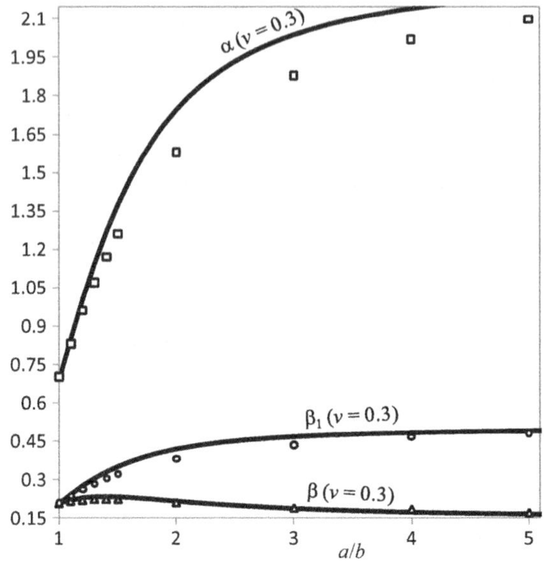

Fig. 4.7 A comparison between the exact (data points) and the approximate (smooth curves) of α, β, and β_1 for various plate aspect ratio (a/b) at $v = 0.3$

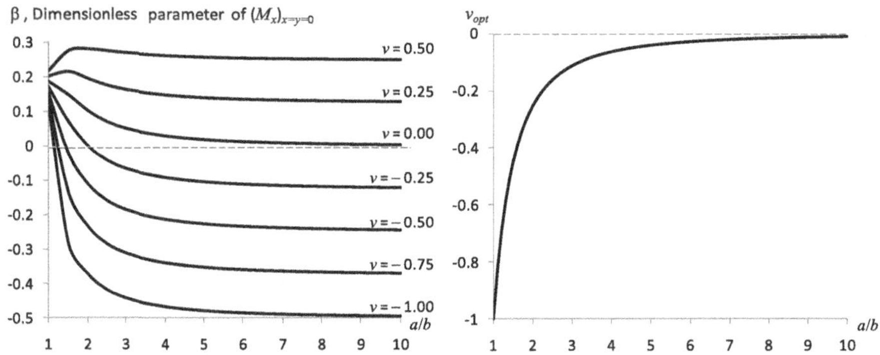

(a) (b)

Fig. 4.8 Bending stress M_x at plate center: **a** effect of plate aspect ratio and Poisson's ratio, and **b** optimal Poisson's ratio for minimizing M_x

decreases slightly as the Poisson's ratio becomes more negative. The case of very long or very narrow elliptical plate $(a/b \gg 1)$ is worth elucidating, in which the value of β converges as $a/b \to \infty$ as mathematically described in Eq. (4.3.20) and plotted in Fig. 4.8a. As the aspect ratio becomes very large, the elliptical plate can be viewed as a very long rectangular plate that is simply supported along $y = \pm b$. Under transverse load, the plate exhibits positive curvature along the y-axis with a corresponding positive M_y. If the plate is conventional, it will exhibit negative curvature along the x-axis, with a positive M_x required to transform the anti-clastic shell into a cylindrical shell. Conversely, if the plate is auxetic, it will exhibit positive curvature along the x-axis, with a negative M_x required to transform the synclastic curve into a cylindrical shell. For this reason, positive and negative M_x are obtained for conventional and auxetic elliptical plates with large aspect ratio as evidenced in Fig. 4.8a. Likewise, if the plate material possesses $v = 0$, the long elliptical plate will transform into a cylindrical-like shell, and therefore $M_x = 0$ as shown in Fig. 4.8a as $a/b \to \infty$.

The other dimensionless parameter β_1, as illustrated in Fig. 4.9, not only confirms that the bending stress $(\sigma_y)_{x=y=0}$ can be minimized by using auxetic materials of $v = -1$ and circular plates (i.e., $a = b$), but clearly shows that the plate aspect ratio and the material's Poisson's ratio play primary and secondary roles, respectively, in influencing $(\sigma_y)_{x=y=0}$. It is also of interest to observe that the bending moments M_x and M_y at the plate center tend to $v/2$ and $1/2$, respectively, as the plate aspect ratio goes to infinity, as proven in Eq. (4.3.20). By way of analogy, this is a direct result of a uniformly loaded infinitely long rectangular plate ($a/b \to \infty$) that is simply supported at $y = \pm b/2$, such that $\partial^2 w/\partial x^2 = 0$.

A graphical description for the von Mises stress at the plate center is furnished as a contour plot in Fig. 4.10, in terms of its dimensionless parameter expressed in Eq. (4.3.26). It can be seen that for $a/b = 1$ plate aspect ratio, the maximum effective stress is minimized if the plate material is chosen such that its Poisson's ratio is approximately $v = -3/4$. As the plate aspect ratio a/b increases, the Poisson's ratio for minimizing the maximum effective stress also increases. Essentially the maximum effective stress is minimized with the use of auxetic and conventional materials for elliptical plates of low and high aspect ratios, respectively.

Table 4.2 puts in perspective the current deflection model for simply supported elliptical plates in relation with the well-known fully clamped elliptical plates as well as with circular plates of both boundary conditions, by using Eq. (4.3.2) such that w_0 and C are reduced for the latter cases. Emphasis is given for the case where

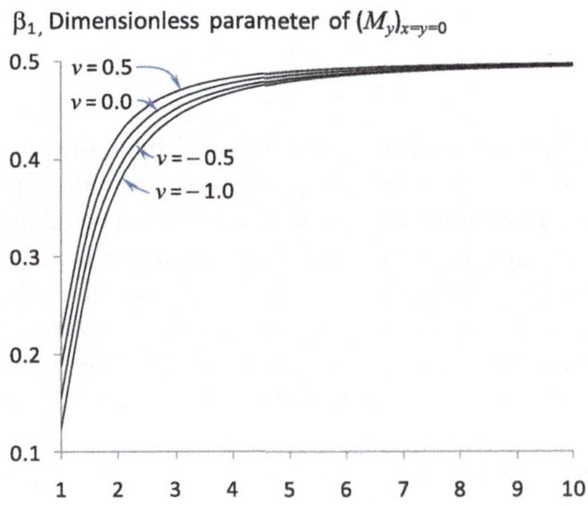

Fig. 4.9 Bending stress M_y at plate center

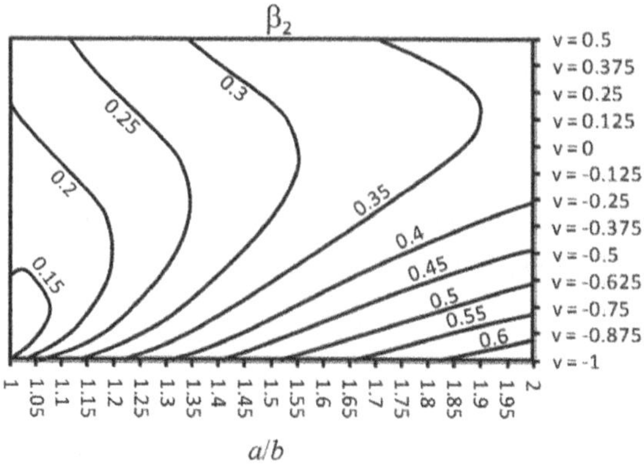

Fig. 4.10 Contour plot of β_2 for indicating the influence of elliptical plate aspect ratio and Poisson's ratio on the effective maximum bending stress

Table 4.2 Reduction of the simply supported elliptical plate deflection model, Eq. (4.3.2), to the deflections of fully clamped elliptical plates, as well as with the deflections of simply supported and fully clamped circular plates

Edge condition	Elliptical plate	Circular plate	Effect of auxeticity
Simply supported	$w_0 = \frac{q}{D}\left(\frac{24}{a^4} + \frac{16}{a^2b^2} + \frac{24}{b^4}\right)^{-1}$ $C = \begin{cases} \frac{5a^2+\upsilon b^2}{a^2+\upsilon b^2}; & a \geq b \\ \frac{5b^2+\upsilon a^2}{b^2+\upsilon a^2}; & a \leq b \end{cases}$	$w_0 = \frac{q}{D}\left(\frac{64}{a^4}\right)^{-1}$ $C = \frac{5+\upsilon}{1+\upsilon}$	Auxetic, $C > 5$ Conventional, $3\frac{2}{3} \leq C \leq 5$ For both auxetic and conventional, $\lim_{a/b\to\infty} C = 5$
Fully clamped	$w_0 = \frac{q}{D}\left(\frac{24}{a^4} + \frac{16}{a^2b^2} + \frac{24}{b^4}\right)^{-1}$ $C = 1$	$w_0 = \frac{q}{D}\left(\frac{64}{a^4}\right)^{-1}$ $C = 1$	No effect

the plate material is auxetic in comparison to conventional ones, in which the coefficient $C = 5$ demarcates simply supported elliptical (including circular) auxetic plates from conventional ones, while no such demarcation exists when the plates are fully clamped. Further details on circular auxetic plates have been discussed earlier (Lim 2013a). Reference to Table 4.2 also shows that a geometrically long or narrow elliptical plate ($a/b \to \infty$) of any Poisson's ratio has mechanical equivalence to a circular plate $a/b = 1$ of zero Poisson's ratio, as both possess $C = 5$.

In conclusion, calculated results for the dimensionless bending stresses at the center of elliptical plates of aspect ratio 1–5 and $\upsilon = 0.3$ reveal reasonably good approximation in comparison to results from exact solution. When extended over the entire range of Poisson's ratio for isotropic solids, generated results suggest that the use of auxetic materials is useful for reducing the maximum bending stresses for elliptical plates of aspect ratio close to 1, while conventional materials are more advantageous for elliptical plates of higher aspect ratio (Lim 2016b).

4.4 Example: Sectorial Auxetic Plates

This section identifies a type of sectorial plate that encounters severe deflection and investigates the effect of material auxeticity on limiting the extent of deflection. Thereafter, this section proposes a new equation, in the form of semi-empirical model, for convenient and reliable quantification for the maximum deflection of the sectorial plate under consideration. Let S, C, and F indicate the simply supported, clamped, and free boundary conditions, respectively. We consider a broad category of sectorial plate possessing radius R and angle α in which both straight sides are simply supported (SS), with the curved side being simply supported (SSS), clamped (SSC) or free (SSF). See Fig. 4.11.

The deflection of such a plate—of thickness h, Young's modulus E and Poisson's ratio ν—due to uniform load q_0 is described as by Mansfield (1989), based on Kirchhoff–Love kinematic assumption, as

$$w = \frac{1}{D}\sum_{m=1}^{\infty}\left\{\frac{q_m r^4}{(16-\mu^2)(4-\mu^2)} + A_m r^{\mu} + B_m r^{\mu+2}\right\}\sin(\mu\theta), \quad m = 1, 2, 3, 4, \ldots. \tag{4.4.1}$$

$$q_m = \frac{2q_0}{m\pi}\left[1-(-1)^m\right] = \frac{4q_0}{m\pi}\begin{Bmatrix} 1 \\ 0 \end{Bmatrix}, \quad \begin{matrix} m = 1, 3, 5, \ldots \\ m = 2, 4, 6, \ldots \end{matrix} \tag{4.4.2}$$

$$\mu = \frac{m\pi}{\alpha} \tag{4.4.3}$$

and D is the plate flexural rigidity described in Eq. (4.1.3). For the exact solution, α is in radians. Among the three types of sectorial plates mentioned above, the SSC sectorial plate experiences the least deflection while the SSF plate encounters the

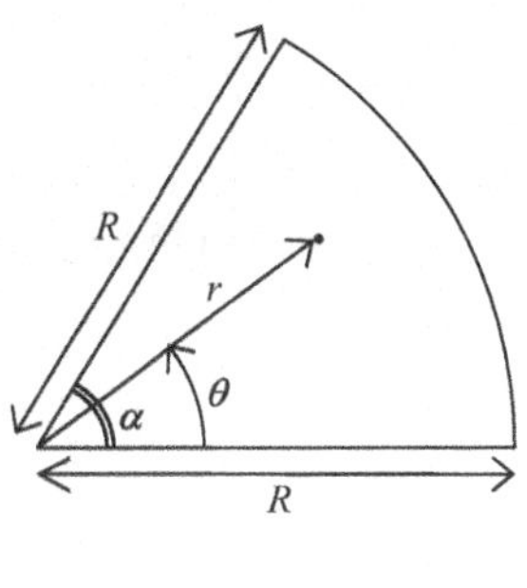

Fig. 4.11 Schematic view of a sectorial plate

most severe deflection—hence the latter class of sectorial plate is considered herein. For such a plate, the coefficients in Eq. (4.4.1) has been obtained by Wang et al. (2000) as

$$A_m = \frac{\frac{4q_0}{m\pi}R^{4-\mu}\left\{(\mu-4)\left[8+\mu(5+v)+v\mu^2\right]\left[\mu(1-v)+2(1+v)\right] + 2\mu(3+v)\left[4(3+v)-v\mu^2\right]\right\}}{2\mu^2\left(16-\mu^2\right)\left(4-\mu^2\right)(1-\mu)(1-v)(3+v)} \tag{4.4.4}$$

$$B_m = -\frac{\frac{4q_0}{m\pi}R^{2-\mu}\left[8+5\mu+v\mu(1+\mu)\right]}{2\mu\left(4-\mu^2\right)(4+\mu)(1+\mu)(3+v)} \tag{4.4.5}$$

where $m = 1, 3, 5, 7, \ldots$. Considering symmetry about $\theta = \alpha/2$ and the free boundary condition at the curved side $r = R$, we recognize that the maximum deflection takes place at $\theta = \alpha/2$ and $r = R$, i.e.,

$$w_{\max} = w_{(r,\theta)=(R,\alpha/2)} \tag{4.4.6}$$

It is further convenient to introduce a dimensionless form for the deflection as

$$w^* = \frac{wD}{q_0 R^4} \tag{4.4.7}$$

Equation (4.4.1) converges rapidly because the use of the first three summative terms ($m = 1, 2, 3$), i.e.,

$$w^*_{\max} = \frac{4}{\pi}\left\{\frac{1}{\left(16-\frac{\pi^2}{\alpha^2}\right)\left(4-\frac{\pi^2}{\alpha^2}\right)} - \frac{8+5\frac{\pi}{\alpha}+v\frac{\pi}{\alpha}\left(1+\frac{\pi}{\alpha}\right)}{2\frac{\pi}{\alpha}\left(4-\frac{\pi^2}{\alpha^2}\right)\left(4+\frac{\pi}{\alpha}\right)\left(1+\frac{\pi}{\alpha}\right)(3+v)} + \frac{\left(\frac{\pi}{\alpha}-4\right)\left[8+\frac{\pi}{\alpha}(5+v)+v\frac{\pi^2}{\alpha^2}\right]\left[\frac{\pi}{\alpha}(1-v)+2(1+v)\right] + 2\frac{\pi}{\alpha}(3+v)\left[4(3+v)-v\frac{\pi^2}{\alpha^2}\right]}{2\frac{\pi^2}{\alpha^2}\left(16-\frac{\pi^2}{\alpha^2}\right)\left(4-\frac{\pi^2}{\alpha^2}\right)\left(1-\frac{\pi}{\alpha}\right)(1-v)(3+v)}\right\}$$

$$-\frac{4}{3\pi}\left\{\frac{1}{\left(16-9\frac{\pi^2}{\alpha^2}\right)\left(4-9\frac{\pi^2}{\alpha^2}\right)} - \frac{8+15\frac{\pi}{\alpha}+3v\frac{\pi}{\alpha}\left(1+3\frac{\pi}{\alpha}\right)}{6\frac{\pi}{\alpha}\left(4-9\frac{\pi^2}{\alpha^2}\right)\left(4+3\frac{\pi}{\alpha}\right)\left(1+3\frac{\pi}{\alpha}\right)(3+v)}\right.$$

$$
+\frac{\left(3\frac{\pi}{\alpha}-4\right)\left[8+3\frac{\pi}{\alpha}(5+v)+9v\frac{\pi^2}{\alpha^2}\right]\left[3\frac{\pi}{\alpha}(1-v)+2(1+v)\right]+6\frac{\pi}{\alpha}(3+v)\left[4(3+v)-9v\frac{\pi^2}{\alpha^2}\right]}{18\frac{\pi^2}{\alpha^2}\left(16-9\frac{\pi^2}{\alpha^2}\right)\left(4-9\frac{\pi^2}{\alpha^2}\right)\left(1-3\frac{\pi}{\alpha}\right)(1-v)(3+v)}\Bigg\}
$$

$$
+\frac{4}{5\pi}\left\{\frac{1}{\left(16-25\frac{\pi^2}{\alpha^2}\right)\left(4-25\frac{\pi^2}{\alpha^2}\right)}-\frac{8+25\frac{\pi}{\alpha}+5v\frac{\pi}{\alpha}\left(1+5\frac{\pi}{\alpha}\right)}{10\frac{\pi}{\alpha}\left(4-25\frac{\pi^2}{\alpha^2}\right)\left(4+5\frac{\pi}{\alpha}\right)\left(1+5\frac{\pi}{\alpha}\right)(3+v)}\right.
$$

$$
\left.+\frac{\left(5\frac{\pi}{\alpha}-4\right)\left[8+5\frac{\pi}{\alpha}(5+v)+25v\frac{\pi^2}{\alpha^2}\right]\left[5\frac{\pi}{\alpha}(1-v)+2(1+v)\right]+10\frac{\pi}{\alpha}(3+v)\left[4(3+v)-25v\frac{\pi^2}{\alpha^2}\right]}{50\frac{\pi^2}{\alpha^2}\left(16-25\frac{\pi^2}{\alpha^2}\right)\left(4-25\frac{\pi^2}{\alpha^2}\right)\left(1-5\frac{\pi}{\alpha}\right)(1-v)(3+v)}\right\} \tag{4.4.8}
$$

is sufficient to give accurate dimensionless maximum deflection.

Reference to Table 4.3 shows that for $v = 0.3$ and $\alpha = \pi/2$, the use of Eq. (4.4.1) with $m = 1, 2, 3$—i.e., Eq. (4.4.8)—would give sufficient accuracy for comparison with results given by Wang et al. (2000) and Timoshenko and Woinowsky-Krieger (1959). In addition, the extension of the dimensionless maximum deflection for the entire range of Poisson's ratio in the case of isotropic solids ($-1 \le v \le 0.5$) shows that the Poisson's ratio plays an important role in influencing the plate deflection. Specifically, the deflection reduces as the Poisson's ratio of the plate material becomes more negative. Hence, it is of interest to understand the extent of deflection reduction arising from material auxeticity. In addition, there is a need to establish the maximum α value such that the plate does not fall off from the supports arising from the effective load line occurring beyond the dotted line shown in Fig. 4.12.

The maximum α under uniform load can be established by recognizing that the sectorial plate will topple from the simply supported straight sides if the load on the triangle area is less than the load outside the triangular area. In the case of uniform load, the maximum α is defined as the plate angle in which the triangle area is equal to half the entire plate area, hence

$$
\frac{1}{2}R^2\sin\alpha = \frac{1}{2}\left(\pi R^2\frac{\alpha}{2\pi}\right) \tag{4.4.9}
$$

Table 4.3 Dimensionless exact maximum deflection computation results for conventional and auxetic SSF sectorial plates with $\alpha = \pi/2$

ν	Timoshenko and Woinowsky-Krieger (1959) for $\upsilon = 0.3$	Wang et al. (2000) for $\upsilon = 0.3$	$w^*_{\max}$ for $-1 \leq \upsilon \leq 0.5$ using $m = 1, 2, 3$	Remarks
−1			0.030775	Auxetic region
−0.9			0.031297	
−0.8			0.031983	
−0.7			0.032846	
−0.6			0.033907	
−0.5			0.035194	
−0.4			0.036746	
−0.3			0.038612	
−0.2			0.040865	
−0.1			0.043599	
0			0.046952	Conventional region
0.1			0.051123	
0.2			0.056409	
0.3	0.0633	0.06328	0.063283	
0.4			0.072529	
0.5			0.085563	

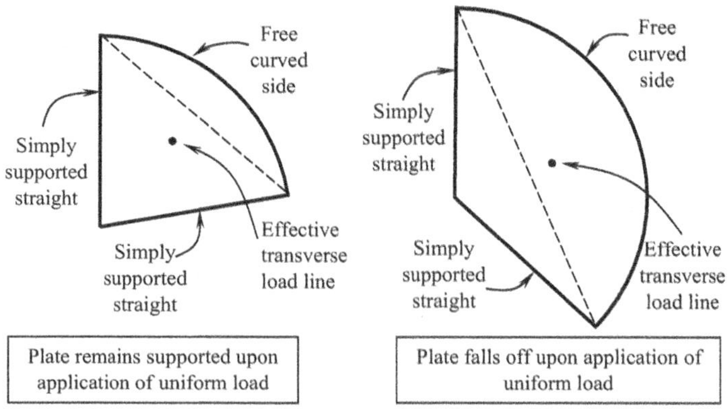

Fig. 4.12 Schematics for showing the effective transverse line load being inside (left) and outside (right) the triangle formed by the simple supports

or $\sin\alpha = \alpha/2$, which gives $\alpha_{\max} = 1.8955$ rad or $\alpha_{\max} = 108.6°$. In the proceeding analysis, the maximum deflection is calculated for $\alpha = 15°$ to $\alpha = 105°$ with an increment of 15°.

Reference to Eq. (4.4.8) shows that, even by considering only the first three terms ($m = 1, 3, 5$), the exact solution for the maximum deflection is cumbersome. Great simplification can be attained by obtaining the dimensionless maximum deflection of the SSF sectorial plate in the form

$$w^*_{\max} = \sum_n a_n v^n \quad (4.4.10)$$

where

$$a_n = a_n(\alpha) \quad (4.4.11)$$

or

$$w^*_{\max} = \sum_n b_n \alpha^n \quad (4.4.12)$$

where

$$b_n = b_n(v) \quad (4.4.13)$$

by semi-empirical modeling (Lim 2017). Both sets of semi-empirical models can be obtained by performing surface fitting on the basis of least square method.

Table 4.4 lists the dimensionless maximum deflection of SSF sectorial plates from $\alpha = 15°$ to $\alpha = 105°$ at the typical Poisson's ratio of $v = 0.3$ using Eq. (4.4.8). To exhibit the negativity effect from the Poisson's ratio, the corresponding dimensionless maximum deflection at $v = -0.3$ was computed, so as to facilitate comparison

Table 4.4 Reduction of maximum SSF sectorial plate deflection as the plate material's Poisson's ratio changes from $v = 0.3$ to $v = -0.3$

Plate angle	$w^*_{\max}$ at typical Poisson's ratio of $v = 0.3$	$w^*_{\max}$ at a negative Poisson's ratio of $v = -0.3$	Change in $w^*_{\max}$ as v reduces from 0.3 to −0.3
$\alpha = 15°$	0.00005916	0.00004438	−24.99%
$\alpha = 30°$	0.0008383	0.0005945	−29.09%
$\alpha = 45°$	0.003943	0.002669	−32.31%
$\alpha = 60°$	0.012041	0.007834	−34.94%
$\alpha = 75°$	0.029421	0.018498	−37.13%
$\alpha = 90°$	0.063283	0.038612	−38.98%
$\alpha = 105°$	0.126602	0.075222	−40.58%

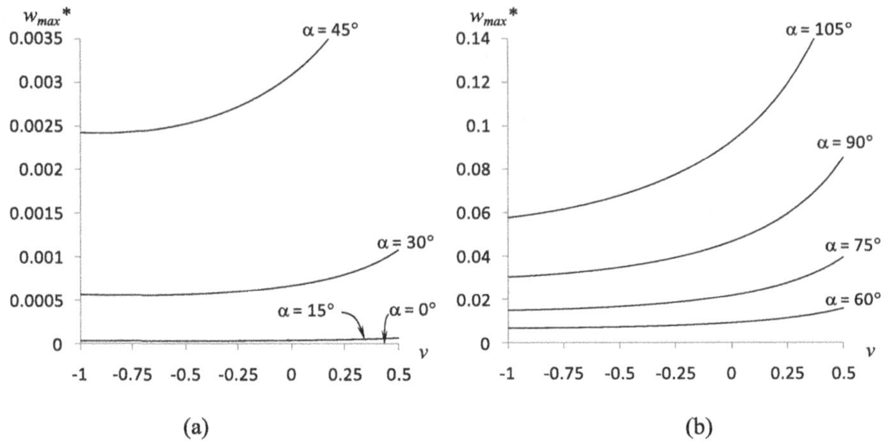

Fig. 4.13 Dimensionless maximum deflection of an SSF sectorial plate versus its Poisson's ratio for **a** $\alpha = 0°, 15°, 30°$ and $45°$, and **b** $\alpha = 60°, 75°, 90°$ and $105°$

between a typical conventional material and an auxetic material with equal Poisson's ratio magnitude. Due to the singularities encountered for $\alpha = \pi/4$ and $\alpha = \pi/2$, computations of $w^*_{\max}$ for these plate angles were made at $\mu \pm 1 \times 10^{-10}$ so that the values of $w^*_{\max}$ based on the upper and lower μ do not differ up to five significant figures, and both the $w^*_{\max}$ results from the upper and lower μ are averaged to give the final $w^*_{\max}$ up to five significant figures for $\alpha = \pi/4$ and $\alpha = \pi/2$.

In assessing the effect of material auxeticity, it is insufficient to merely compare the maximum deflection of an SSF sectorial plate based on a positive value and a negative value of Poisson's ratio. Since the Poisson's ratio of isotropic solids range from $v = -1$ to $v = 1/2$, a more comprehensive comparison between conventional and auxetic SSF sectorial plates can be made by observing the maximum deflection for the entire range of the plate's Poisson's ratio, as plotted in Fig. 4.13, based on Eq. (4.4.8).

The reduction in deflection with decreasing plate angle is trivial. It can be seen that the deflection decreases as the Poisson's ratio of the plate material becomes more negative. This observation is attributed to the anti-clastic shape formed at the vicinity of the plate rim because the deformation of a plate into an anti-clastic shell is facilitated by highly positive Poisson's ratio. To visually indicate the effect of material auxeticity on the plate deformation, Fig. 4.14 demarcates the auxetic region ($-1 \le v < 0$) from the conventional region ($0 \le v \le 1/2$) in the plots of dimensionless maximum deflection versus the sectorial plate angle.

To obtain a highly accurate surface fit, the semi-empirical equation is confined within $30° \le \alpha \le 90°$, to give

$$w^*_{\max} = \sum_{n=0}^{4} a_n v^n \tag{4.4.14}$$

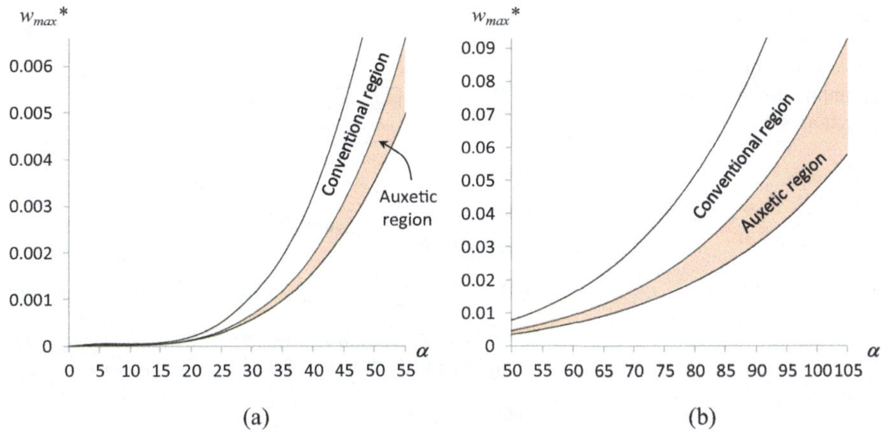

Fig. 4.14 Dimensionless maximum deflection regions of conventional and auxetic SSF sectorial plates versus its angle for **a** narrow-angled sectorial plate ($0° \le \alpha \le 55°$), and **b** wide-angled sectorial plate ($50° \le \alpha \le 105°$)

where

$$\begin{bmatrix} a_0 \\ a_1 \\ a_2 \\ a_3 \\ a_4 \end{bmatrix} = 10^{-6} \begin{bmatrix} 4799.05 & 432.36 & 14.177 & 0.19442 & 0.0016412 \\ 4789.45 & 433.72 & 14.474 & 0.21529 & 0.0016883 \\ 5771.05 & 521.52 & 17.298 & 0.25085 & 0.0020080 \\ 6245.65 & 564.85 & 18.771 & 0.27425 & 0.0021845 \\ 3238.35 & 284.77 & 9.2693 & 0.13301 & 0.0010450 \end{bmatrix} \begin{bmatrix} +\alpha^0 \\ -\alpha^1 \\ +\alpha^2 \\ -\alpha^3 \\ +\alpha^4 \end{bmatrix} \tag{4.4.15}$$

or

$$w^*_{\max} = \sum_{n=0}^{4} b_n \alpha^n \tag{4.4.16}$$

where

$$\begin{bmatrix} +b_0 \\ -b_1 \\ +b_2 \\ -b_3 \\ +b_4 \end{bmatrix} = 10^{-6} \begin{bmatrix} 4799.05 & 4789.45 & 5771.05 & 6245.65 & 3238.35 \\ 432.36 & 433.72 & 521.52 & 564.85 & 284.77 \\ 14.177 & 14.474 & 17.298 & 18.771 & 9.2693 \\ 0.19442 & 0.21529 & 0.25085 & 0.27425 & 0.13301 \\ 0.0016412 & 0.0016883 & 0.0020080 & 0.0021845 & 0.0010450 \end{bmatrix} \begin{bmatrix} \upsilon^0 \\ \upsilon^1 \\ \upsilon^2 \\ \upsilon^3 \\ \upsilon^4 \end{bmatrix} \tag{4.4.17}$$

For obvious reason, the square matrices in Eqs. (4.4.15) and (4.4.17) are transposition of each other. For the convenience of the designer, the sectorial plate angle α in

Fig. 4.15 Comparison between exact (circles) and semi-empirical (curves) maximum dimensionless deflection of an SSF plate

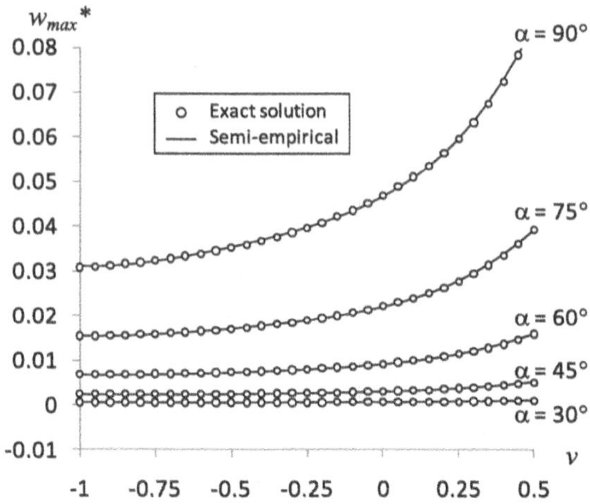

the semi-empirical models of Eqs. (4.4.15) and (4.4.16) is in terms of degrees instead of radians. The validity of the semi-empirical model described by Eqs. (4.4.14) and (4.4.15) or Eqs. (4.4.16) and (4.4.17) is confirmed by making comparison against the exact solution given in Eq. (4.4.8). The use of Eqs. (4.4.14) and (4.4.15) would be more expedient for plotting $w^*_{\max}$ against α for different curves denoting different v, while Eqs. (4.4.16) and (4.4.17) are more convenient for plotting $w^*_{\max}$ versus v for different curves indicating different α. Based on Eqs. (4.4.16) and (4.4.17), graphical plots in Fig. 4.15 for $\alpha = 30°, 45°, 60°, 75°$, and $90°$ exhibit very good agreement between the exact solution and the semi-empirical model. The accuracy of the semi-empirical models with reference to the analytical model can be described by the coefficient of determination at 99.996%.

As the auxetic and conventional regions are defined by $-1 \leq v < 0$ and $0 \leq v \leq 1/2$, respectively, the use of the limiting Poisson's ratio in Eq. (4.4.14) allows representation of

$$\sum_{n=0}^{4} (-1)^n a_n \leq w^*_{\max} < a_0 \tag{4.4.18}$$

for auxetic range and

$$a_0 \leq w^*_{\max} \leq \sum_{n=0}^{4} \frac{a_n}{2^n} \tag{4.4.19}$$

for conventional range, whereby the coefficients a_n for $n = 0, 1, 2, 3, 4$ are defined by Eq. (4.4.15). Alternatively the implementation of the limiting Poisson's ratio in Eq. (4.4.17) permits representation of

$$\begin{bmatrix} 2773.35 \\ 240.08 \\ 7.4983 \\ 0.08874 \\ 0.0008214 \end{bmatrix}^{\mathrm{T}} \begin{bmatrix} +\alpha^0 \\ -\alpha^1 \\ +\alpha^2 \\ -\alpha^3 \\ +\alpha^4 \end{bmatrix} \times 10^{-6} \le w^*_{\max} < \begin{bmatrix} 4799.05 \\ 432.36 \\ 14.177 \\ 0.19442 \\ 0.0016412 \end{bmatrix}^{\mathrm{T}} \begin{bmatrix} +\alpha^0 \\ -\alpha^1 \\ +\alpha^2 \\ -\alpha^3 \\ +\alpha^4 \end{bmatrix} \times 10^{-6} \tag{4.4.20}$$

for auxetic range and

$$\begin{bmatrix} 4799.05 \\ 432.36 \\ 14.177 \\ 0.19442 \\ 0.0016412 \end{bmatrix}^{\mathrm{T}} \begin{bmatrix} +\alpha^0 \\ -\alpha^1 \\ +\alpha^2 \\ -\alpha^3 \\ +\alpha^4 \end{bmatrix} \times 10^{-6} \times 10^{-6} \le w^*_{\max} \le \begin{bmatrix} 9619.64 \\ 868.004 \\ 28.6643 \\ 0.40737 \\ 0.0033257 \end{bmatrix}^{\mathrm{T}} \begin{bmatrix} +\alpha^0 \\ -\alpha^1 \\ +\alpha^2 \\ -\alpha^3 \\ +\alpha^4 \end{bmatrix} \times 10^{-6} \tag{4.4.21}$$

for conventional range, based on the semi-empirical model described in Eq. (4.4.16), whereby the superscript ^{T} indicates matrix transposition. A graphical representation for the conventional and auxetic windows is furnished in Fig. 4.16.

Although the Poisson's ratio range in the auxetic region is twice that in the conventional region, the auxetic window for $w^*_{\max}$ versus α is narrower than that in the conventional region. This is attributed to the diminishing reduction in deflection as the plate material's Poisson's ratio becomes more negative, as shown in Fig. 4.15, i.e.,

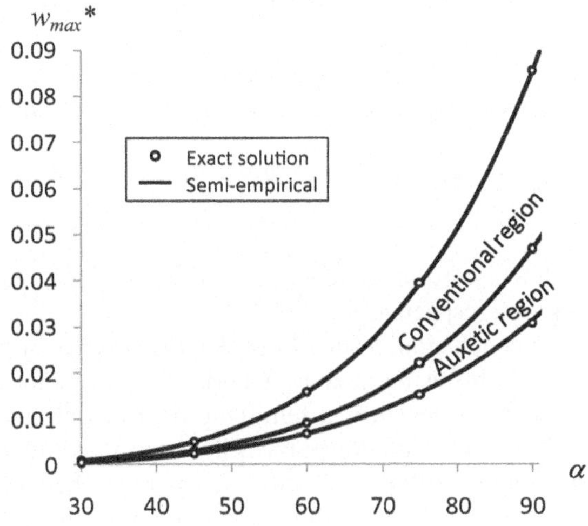

Fig. 4.16 Exact (circles) and semi-empirical (curves) of maximum dimensionless deflection of SSF plates for demarcation of conventional and auxetic regions

$$v_1 < v_2 \Leftrightarrow \left(\frac{\partial w^*_{\max}}{\partial v}\right)_{v_1} < \left(\frac{\partial w^*_{\max}}{\partial v}\right)_{v_2} \tag{4.4.22}$$

To appreciate the convenience of the semi-empirical model, we note that the boundary that demarcates the conventional and auxetic regions, being defined as $v = 0$, is written in the exact solution form

$$\begin{aligned}
w^*_{\max} = \frac{4}{\pi}\Bigg\{ & \frac{1}{\left(16-\frac{\pi^2}{\alpha^2}\right)\left(4-\frac{\pi^2}{\alpha^2}\right)} - \frac{\frac{4}{3}+\frac{5}{6}\frac{\pi}{\alpha}}{\frac{\pi}{\alpha}\left(4-\frac{\pi^2}{\alpha^2}\right)\left(4+\frac{\pi}{\alpha}\right)\left(1+\frac{\pi}{\alpha}\right)} \\
& + \frac{\left(\frac{\pi}{\alpha}-4\right)\left[\frac{4}{3}+\frac{5}{6}\frac{\pi}{\alpha}\right]\left[\frac{\pi}{\alpha}+2\right]+12\frac{\pi}{\alpha}}{\frac{\pi^2}{\alpha^2}\left(16-\frac{\pi^2}{\alpha^2}\right)\left(4-\frac{\pi^2}{\alpha^2}\right)\left(1-\frac{\pi}{\alpha}\right)}\Bigg\} \\
- \frac{4}{3\pi}\Bigg\{ & \frac{1}{\left(16-9\frac{\pi^2}{\alpha^2}\right)\left(4-9\frac{\pi^2}{\alpha^2}\right)} - \frac{\frac{4}{9}+\frac{5}{6}\frac{\pi}{\alpha}}{\frac{\pi}{\alpha}\left(4-9\frac{\pi^2}{\alpha^2}\right)\left(4+3\frac{\pi}{\alpha}\right)\left(1+3\frac{\pi}{\alpha}\right)} \\
& + \frac{\left(\frac{\pi}{\alpha}-\frac{4}{3}\right)\left[\frac{4}{9}+\frac{5}{6}\frac{\pi}{\alpha}\right]\left[3\frac{\pi}{\alpha}+2\right]+4\frac{\pi}{\alpha}}{\frac{\pi^2}{\alpha^2}\left(16-9\frac{\pi^2}{\alpha^2}\right)\left(4-9\frac{\pi^2}{\alpha^2}\right)\left(1-3\frac{\pi}{\alpha}\right)}\Bigg\} \\
+ \frac{4}{5\pi}\Bigg\{ & \frac{1}{\left(16-25\frac{\pi^2}{\alpha^2}\right)\left(4-25\frac{\pi^2}{\alpha^2}\right)} - \frac{\frac{4}{15}+\frac{5}{6}\frac{\pi}{\alpha}}{\frac{\pi}{\alpha}\left(4-25\frac{\pi^2}{\alpha^2}\right)\left(4+5\frac{\pi}{\alpha}\right)\left(1+5\frac{\pi}{\alpha}\right)} \\
& + \frac{\left(\frac{\pi}{\alpha}-\frac{4}{5}\right)\left[\frac{4}{15}+\frac{5}{6}\frac{\pi}{\alpha}\right]\left[5\frac{\pi}{\alpha}+2\right]+\frac{12}{5}\frac{\pi}{\alpha}}{\frac{\pi^2}{\alpha^2}\left(16-25\frac{\pi^2}{\alpha^2}\right)\left(4-25\frac{\pi^2}{\alpha^2}\right)\left(1-5\frac{\pi}{\alpha}\right)}\Bigg\}
\end{aligned} \tag{4.4.23}$$

based on the first three terms $m = 1, 2, 3$ remains cumbersome while the corresponding semi-empirical model is greatly simplified to (Lim 2017)

$$w^*_{\max} = \left(4799.05 - 432.36\alpha + 14.177\alpha^2 - 0.19442\alpha^3 + 0.0016412\alpha^4\right) \times 10^{-6} \tag{4.4.24}$$

with the sectorial plate angle α in Eqs. (4.4.23) and (4.4.24) in units of radians and degrees, respectively.

In conclusion, results show that the severity of plate deflection can be effectively reduced by using auxetic material for the sectorial plate. For convenience to the mechanical designer, a semi-empirical model has been developed herein to allow convenient calculation of the SSF plate's maximum deflection as a function of sectorial angle and Poisson's ratio. Comparison between the semi-empirical model and the exact solution exhibits very good agreement. It is therefore suggested that the deflection of an SSF sectorial plate can be reduced by replacing conventional materials with auxetic materials, and that the proposed semi-empirical model is a useful design equation for providing simple and yet accurate indication of the maximum allowable deflection.

4.5 Example: Rhombic Auxetic Plates

This section develops a set of design equations that is simple to execute and sufficiently accurate for a class of rhombic plates, which are simply supported and uniformly loaded, with special emphasis on the use of auxetic materials.

With reference to Fig. 4.17 for a simply supported rhombic plate of thickness h, sides a, and complementary angles of 2α and $(\pi - 2\alpha)$ under uniformly distributed load of q_0, the deflection of such plate within the context of Kirchhoff plate theory

$$w = w_0 + w_1 \tag{4.5.1}$$

at any point (r, θ) is given by the summation of a particular solution w_0

$$w_0 = \frac{q_0}{D}\frac{r^4}{64}\left(1 - \frac{4\cos 2\theta}{3\cos 2\alpha} + \frac{\cos 4\theta}{3\cos 4\alpha}\right) \tag{4.5.2}$$

which is applicable for all $0 < \alpha < \pi/2$ except $\alpha = \pi/8, \pi/4, 3\pi/8$, whereby the plate's flexural rigidity D is defined in Eq. (4.1.3). Hereafter, we consider the approaches by Morley (1962) and Warren (1964).

In the first approach, the complementary function w_1 is (Morley 1962)

$$w_1 = \frac{q_0}{D}\sum_{m=1}^{\infty}\left(a_m + b_m r^2\right) r^{\lambda_m + 1}\cos(\lambda_m + 1)\theta \tag{4.5.3}$$

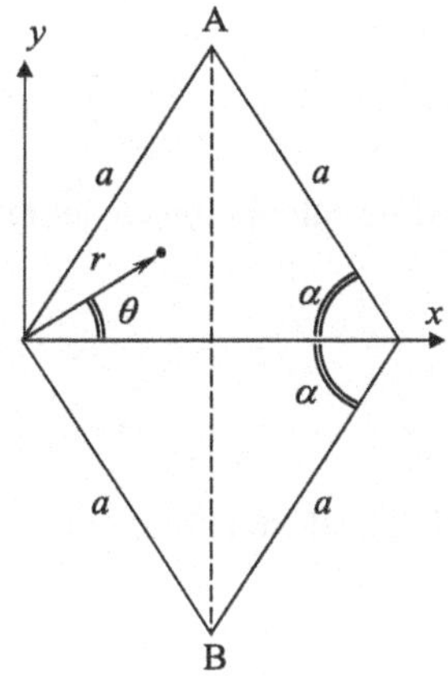

Fig. 4.17 Schematic diagram of a rhombic plate

where

$$\lambda_m + 1 = \frac{(2m-1)\pi}{2\alpha} \tag{4.5.4}$$

The calculations leading to Morley's (1962) solutions of coefficients b_m require

$$\delta \int_{-a \sin\alpha}^{a \sin\alpha} \left[\frac{\partial}{\partial x} \nabla^2 w \right]^2_{x=a \cos\alpha} \mathrm{d}y = 0 \tag{4.5.5}$$

that leads to a system of M equations

$$\begin{bmatrix} A_{11} & A_{12} & \cdots & A_{1M} \\ A_{21} & A_{22} & & \\ \vdots & & \ddots & \\ A_{M1} & & & A_{MM} \end{bmatrix} \begin{Bmatrix} b_1 \\ b_2 \\ \vdots \\ b_M \end{Bmatrix} = \begin{Bmatrix} A_1 \\ A_2 \\ \vdots \\ A_M \end{Bmatrix} \tag{4.5.6}$$

where

$$A_{mn} = A_{nm} = 16(\lambda_m + 2)(\lambda_m + 1)(\lambda_n + 2)(\lambda_n + 1) \int_{-a \sin\alpha}^{a \sin\alpha} \left[r^{\lambda_m + \lambda_n} \cos \lambda_m \theta \ \cos \lambda_n \theta \right]_{x=a \cos\alpha} \mathrm{d}y \tag{4.5.7}$$

and

$$\begin{aligned} \frac{A_{mn}}{4(\lambda_m + 2)(\lambda_m + 1)} &= \frac{1}{2}\left(\frac{1}{\cos 2\alpha} - 1 \right) \int_{-a \sin\alpha}^{a \sin\alpha} \left[r^{\lambda_m + 1} \cos \lambda_m \theta \ \cos\theta \right]_{x=a \cos\alpha} \mathrm{d}y \\ &= \frac{(-1)^{m+1}(1 - \cos 2\alpha) a^{\lambda_m + 2}}{\lambda_m + 1} \end{aligned} \tag{4.5.8}$$

while calculations leading to the solutions of coefficients a_m requires

$$\delta \int_{-a \sin\alpha}^{a \sin\alpha} \left[\frac{\partial w}{\partial x} \right]^2_{x=a \cos\alpha} \mathrm{d}y = 0 \tag{4.5.9}$$

that leads to another system of M equations

$$\begin{bmatrix} A'_{11} & A'_{12} & \cdots & A'_{1M} \\ A'_{21} & A'_{22} & & \\ \vdots & & \ddots & \\ A'_{M1} & & & A'_{MM} \end{bmatrix} \begin{Bmatrix} a_1 \\ a_2 \\ \vdots \\ a_M \end{Bmatrix} = \begin{Bmatrix} A'_1 \\ A'_2 \\ \vdots \\ A'_M \end{Bmatrix} \tag{4.5.10}$$

where

$$A'_{mn} = A'_{nm} = \frac{A_{mn}}{16(\lambda_m + 2)(\lambda_n + 2)} \tag{4.5.11}$$

and

$$A'_m = (\lambda_m + 1) \int_{-a \sin\alpha}^{a \sin\alpha} \left[\chi r^{\lambda_m} \cos\lambda_m \theta\right]_{x=a \cos\alpha} \mathrm{d}y \tag{4.5.12}$$

in which A_{mn} is given by Eq. (4.5.7) while

$$\chi = \frac{r^3}{16}\left[\left(1 - \frac{1}{\cos 2\alpha}\right)\cos\theta + \frac{1}{3}\left(\frac{1}{\cos 4\alpha} - \frac{1}{\cos 2\alpha}\right)\cos 3\theta\right] + \sum_{m=1}^{M} b_m\{(\lambda_m + 2)\cos\lambda_m\theta + \cos(\lambda_m + 2)\theta\}r^{\lambda_m + 2} \tag{4.5.13}$$

In the second approach, the complementary function w_1 is (Warren 1964)

$$w_1 = \frac{q_0}{D}\sum_{n=1}^{\infty}(a_n + b_n r^2)r^{\lambda_n}\cos\lambda_n\theta \tag{4.5.14}$$

where

$$\lambda_n = \frac{(2n-1)\pi}{2\alpha} \tag{4.5.15}$$

The coefficients a_n and b_n are chosen ideally so as to satisfy symmetry boundary conditions along the diagonal AB (see Fig. 4.17). These conditions require

$$\left.\frac{\partial}{\partial x}\nabla^2 w\right|_{\mathrm{AB}} = 0 \tag{4.5.16}$$

and

$$\left.\frac{\partial w}{\partial x}\right|_{\mathrm{AB}} = 0 \tag{4.5.17}$$

which give

$$\left\{a\cos\alpha\left(1-\frac{1}{\cos 2\alpha}\right)+8\sum_{n=1}^{\infty}\lambda_n(\lambda_n+1)b_n r^{\lambda_n-1}\cos(\lambda_n-1)\theta\right\}_{\mathrm{AB}}=0 \tag{4.5.18}$$

and

$$\begin{aligned}\Bigg\{&\frac{r^3}{16}\left[\left(1-\frac{1}{\cos 2\alpha}\right)\cos\theta+\frac{1}{3}\left(\frac{1}{\cos 4\alpha}-\frac{1}{\cos 2\alpha}\right)\cos 3\theta\right]\\&+\sum_{n=1}^{\infty}\lambda_n a_n r^{\lambda_n-1}\cos(\lambda_n-1)\theta\\&+\sum_{n=1}^{\infty}b_n r^{\lambda_n+1}[(\lambda_n+1)\cos(\lambda_n-1)\theta+\cos(\lambda_n+1)\theta]\Bigg\}_{\mathrm{AB}}=0\end{aligned} \tag{4.5.19}$$

Warren (1964) appreciated the fact that the coefficients b_m and a_m are chosen by Morley (1962) so as to minimize the integral square error over AB. This leads to numerous integrals which are tedious to evaluate, and then finally to the solution of M by M simultaneous linear equations for determining these coefficients. Warren (1964) proposed the method of "point matching" that requires satisfying Eqs. (4.5.18) and (4.5.19) exactly at M discrete points along AB, which leads directly to two sets of M by M simultaneous linear equations for determining the said coefficients. Both ways are nevertheless tedious to the designer, and a simplified model would be of great practical importance. To provide simplified design equations for thin rhombic plates, results from Morley (1962) are used for generating semi-empirical models that are very convenient and yet sufficiently accurate.

Perusal to the expressions for the rhombic plate deflection shows that it is a function of load q_0, location (r, θ), half angle of obtuse corner α as well as the Young's modulus, plate thickness, and Poisson's ratio. If one were to express the plate deflection in the form of $D/\left(q_0 a^4\right)$, then this dimensionless plate deflection is independent from the material properties (E, v) of the plate. Based on Morley's (1962) numerical results for the dimensionless deflection as well as the bending moments at the center of the plate for $v = 0.3$ and $45° \leq \alpha \leq 75°$, a curve-fitting exercise yields the following semi-empirical models for the plate center (Lim 2018)

$$\frac{w_{\max}D}{q_0 a^4}=\left\{\begin{matrix}1.907 & 3.578 & 2.069 & 3.397\end{matrix}\right\}\left\{\begin{matrix}+10^{-7}\alpha^3\\-10^{-5}\alpha^2\\+10^{-3}\alpha^1\\-10^{-2}\alpha^0\end{matrix}\right\} \tag{4.5.20}$$

$$\frac{D}{q_0 a^2}\begin{Bmatrix} \frac{\partial^2 w_{\max}}{\partial x^2} \\ \frac{\partial^2 w_{\max}}{\partial y^2} \end{Bmatrix} = \begin{bmatrix} 3.768 & 3.861 & 6.039 \\ 0 & -1.071 & -8.604 \end{bmatrix} \begin{Bmatrix} +10^{-5}\alpha^2 \\ -10^{-3}\alpha^1 \\ +10^{-2}\alpha^0 \end{Bmatrix} \quad (4.5.21)$$

where α in the above semi-empirical models is expressed in degrees. Since the rhombic plate deflection functions are independent from the plate's Poisson's ratio when non-dimensionalized, it follows that this independence holds for Eqs. (4.5.20) and (4.5.21). Nevertheless, the Poisson's ratio plays its role in the bending moment, as described in Eq. (4.5.22).

Verification of the semi-empirical maximum deflection model described in Eq. (4.5.20) is made by plotting it in comparison to the discrete data points by Morley, as shown in Fig. 4.18 (left). Apart from the maximum deflection, another plate response that is of interest to the designer is the maximum stresses. Since the bending stress is related to the bending moment, i.e., $\sigma_{\max} = 6M_{\max}/h^2$, it suffices to evaluate the maximum bending moments. Substitution of Eq. (4.5.21) and $v = 0.3$ into

$$\begin{Bmatrix} M_{x,\max} \\ M_{y,\max} \end{Bmatrix} = -D\begin{bmatrix} 1 & -v \\ -v & 1 \end{bmatrix} \begin{Bmatrix} \frac{\partial^2 w_{\max}}{\partial x^2} \\ \frac{\partial^2 w_{\max}}{\partial y^2} \end{Bmatrix} \quad (4.5.22)$$

allows the semi-empirical bending moments to be compared with the exact results, as plotted in Fig. 4.18 (right), in which $M_{\max} = M_{x,\max} > M_{y,\max}$ for $\alpha > 45°$. Having established their validity, the semi-empirical models are used for plotting the loci of dimensionless maximum bending moments in the conventional $0 \leq v \leq 1/2$, moderately auxetic $-1/2 \leq v < 0$ and highly auxetic $-1 \leq v < -1/2$ regions.

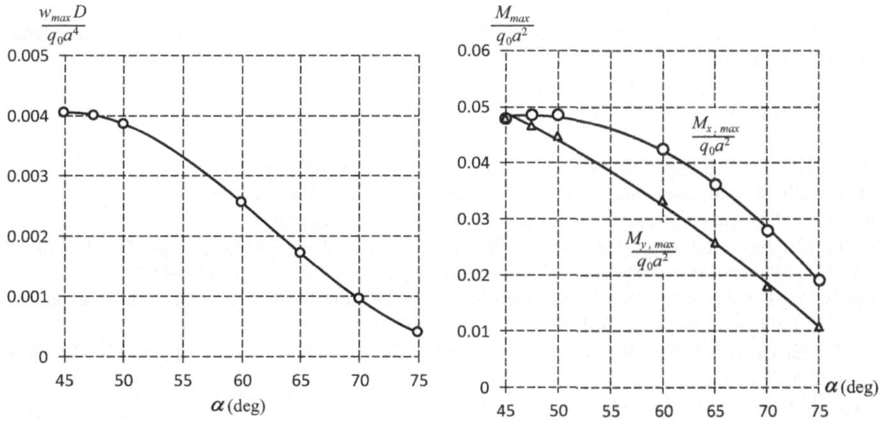

Fig. 4.18 Comparison between the semi-empirical (smooth curves) and Morley's (1962) exact numerical results (data points) for the dimensionless maximum deflection (left) and the dimensionless maximum bending moments along the x- and y-axes (right) for $v = 0.3$ and $45° \leq \alpha \leq 75°$

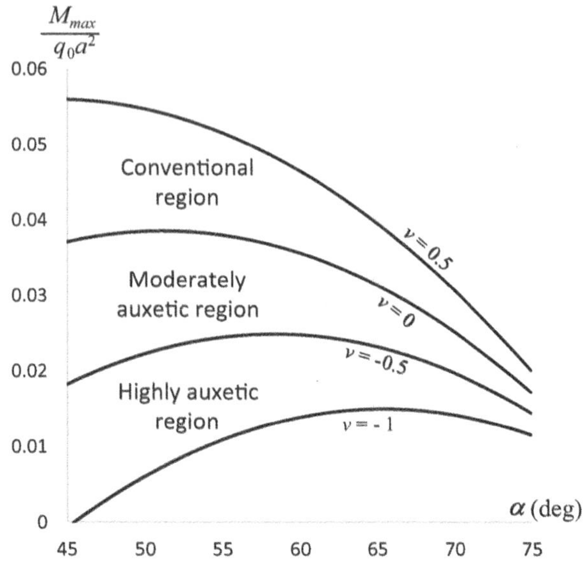

Fig. 4.19 Loci of dimensionless maximum bending stress for conventional, moderately auxetic, and highly auxetic regions

Reference to Fig. 4.19 shows that, with all other parameters fixed, the maximum bending moment—and hence the maximum bending stress—is reduced by the extent of material auxeticity (Lim 2018).

References

Ho DT, Park SD, Kwon SY, Park K, Kim SY (2014) Negative Poisson's ratio in metal nanoplates. Nat Commun 5:3255

Lim TC (2013a) Circular auxetic plates. J Mech 29(1):121–133

Lim TC (2013b) Optimal Poisson's ratio for laterally loaded rectangular plates. J Mater Des Appl 227(2):111–123

Lim TC (2014a) Flexural rigidity of thin auxetic plates. Int J Appl Mech 6(2):1450012

Lim TC (2014b) Auxetic plates on auxetic foundation. Adv Mater Res 974:398–401

Lim TC (2015) Auxetic materials and structures. Springer, Singapore

Lim TC (2016a) Bending stresses in triangular auxetic plates. J Eng Mater Technol 138(1):014501

Lim TC (2016b) Simply-supported elliptical auxetic plates. J Mech 32(4):413–419

Lim TC (2017) An accurate design equation for the maximum deflection in a class of auxetic sectorial plates. Phys Status Solidi B 254(12):1600784

Lim TC (2018) Simplified design equations for a class of rhombic auxetic plates. MATEC Web Conf 206:01009

Mansfield EH (1989) The bending and stretching of plates, 2nd edn. Cambridge University Press, Cambridge

Morley LSD (1962) Bending of a simply supported rhombic plate under uniform normal load. Q J Mech Appl Math 15(4):413–426

Pozniak AA, Kaminski H, Kedziora P, Maruszewski B, Strek T, Wojciechowski KW (2010) Anomalous deformation of constrained auxetic square. Rev Adv Mater Sci 23(2):169–174

Strek T, Maruszewski B, Narojczyk Wojciechowski KW (2008) Finite element analysis of auxetic plate deformation. J Non-Cryst Solids 354(35–39):4475–4480
Timoshenko SP, Woinowsky-Krieger S (1959) Theory of plates and shells, 2nd edn. McGraw-Hill, New York
Ventsel E, Krauthammer T (2001) Thin plates and shells. Marcel Dekker, New York
Wang CM, Reddy JN, Lee KH (2000) Shear deformable beams and plates: relationships with classical solutions. Elsevier, New York
Warren WE (1964) Bending of rhombic plates. AIAA J 2(1):166–168

Chapter 5
Thick Auxetic Plates

Abstract This chapter considers shear deformation in transversely loaded thick plates—such as rectangular plates, equilateral triangular plates, isosceles right triangular plates, hexagonal plates, and regular polygonal plates—with special emphasis on the effect of auxeticity on their performances and the use of Reddy plate theory to extract the shear correction factors of Mindlin plates. Thereafter a comparison is made between the shear deformation in thick plates and buckling of thick columns to establish an analogy between them. Finally, a discussion is made on vibration of thick auxetic plates.

Keywords Mindlin plates · Reddy plates · Shear deformation · Thick plates

5.1 Preamble

The use of Kirchhoff plate theory, or the classical plate theory (CPT), is justifiable for thin plates of thickness-to-length ratio of 1/20 (Yuan and Miller 1992; Reddy 2006). Beyond this limit, shear deformable theories are required to take into consideration the transverse shear deformation. Following the work of Reissner (1945), the first-order shear deformation theory (FSDT) by Mindlin (1951) assumes constant shear strain and shear stress through the plate thickness. Hence, a shear correction factor, κ, is required to correct the difference between the actual transverse shear force distributions and those calculated using the kinematic relationships of the FSDT. However, the Reddy (1984) plate theory, which falls under the third-order shear deformation theory (TSDT), accounts for the variation of transverse shear strain and shear stress through the plate thickness and therefore needs no shear correction. Although the TSDT is more rigorous, the FSDT is nevertheless attractive due to its relative simplicity; accurate prediction can still be achieved using FSDT with judicious selection of shear correction factor. The analysis of shear deformation in auxetic plates within the framework of FSDT has been done for plates for circular (Lim 2013) and rectangular (Lim 2014a) shapes, and are summarized in Chap. 15 of Lim (2015) alongside shear deformation of auxetic beams and columns. This chapter explores the use of Reddy plate theory for extracting the shear correction factors for

T. Lim, *Mechanics of Metamaterials with Negative Parameters*, Engineering Materials,
https://doi.org/10.1007/978-981-15-6446-8_5

Mindlin plates of various shapes, with special emphasis on the effect of material auxeticity.

Consider a flat plate lying on the x-y plane with application of transverse load parallel to the z-axis within a 3D Cartesian coordinate system. The displacement field of a plate according to the Kirchhoff plate theory is written as

$$\begin{aligned} u_x(x, y, z) &= -z\frac{\partial w_0}{\partial x} \\ u_y(x, y, z) &= -z\frac{\partial w_0}{\partial y} \\ w(x, y, z) &= w_0(x, y) \end{aligned} \tag{5.1.1}$$

where (u_x, u_y, w) denotes the displacement components parallel to the x-axis, y-axis, and z-axis, respectively, while w_0 refers to the transverse deflection of a plate point on the mid-plane $z = 0$. Due to the Kirchhoff plate assumption that deflection is attained purely be bending such that straight lines normal to the mid-surface remain straight and normal to the mid-surface after deflection, the rotations are $-\partial w/\partial x$ about the in-plane y-axis (Fig. 5.1, top). The incorporation of shear deformation but with a simplifying assumption of uniform shear strain across the plate thickness, which results in a constant rotation ϕ_x about the y-axis, would therefore require a shear correction factor for the Mindlin plate (Fig. 5.1, middle). This is reflected in the displacement field of a flat plate according to the Mindlin plate theory

$$\begin{aligned} u_x(x, y, z) &= z\phi_x(x, y) \\ u_y(x, y, z) &= z\phi_y(x, y) \\ w(x, y, z) &= w_0(x, y) \end{aligned} \tag{5.1.2}$$

No shear correction is required for the Reddy plate (Fig. 5.1, bottom) due to its allowance for variation in ϕ_x across the plate thickness

$$\begin{aligned} u_x(x, y, z) &= z\phi_x(x, y) - \frac{4z^3}{3h^2}\left(\phi_x + \frac{\partial w_0}{\partial x}\right) \\ u_y(x, y, z) &= z\phi_y(x, y) - \frac{4z^3}{3h^2}\left(\phi_y + \frac{\partial w_0}{\partial y}\right) \\ w(x, y, z) &= w_0(x, y) \end{aligned} \tag{5.1.3}$$

which qualifies it to be suitable for extracting the shear correction factor of Mindlin plates. In addition to the emphasis on material auxeticity, the subsequent sections in this chapter develop models that are tractable. In the proceeding sections, w_0 is denoted as w for brevity.

5.2 Fundamentals

This section establishes, in a generic manner, the fundamentals for extracting the Mindlin plate's shear correction factor using Reddy plate theory for transversely

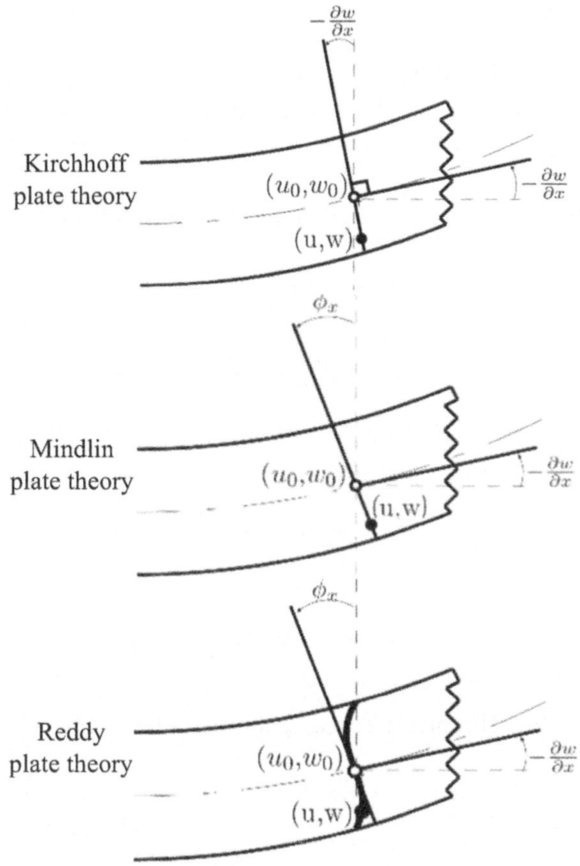

Fig. 5.1 Schematics of displacements based on Kirchhoff (top), Mindlin (middle), and Reddy (bottom) plate theories

loaded simply supported polygonal plates, including irregular ones. For the purpose of the present analysis, the superscripts $(\cdot)^K$, $(\cdot)^M$, and $(\cdot)^R$ refer to the Kirchhoff, Mindlin, and Reddy plates, respectively. For a simply supported polygonal plate of shear modulus G, Young's modulus E, Poisson's ratio v, and thickness h, the relationship between the Mindlin plate deflection, w^M, and the Kirchhoff plate deflection, w^K, has been given by Wang and Alwis (1995) as

$$w^M = w^K + \frac{\overline{M^K}}{\kappa G h} \tag{5.2.1}$$

where the Marcus moment of the Kirchhoff plate

$$\overline{M^K} = \frac{M_x^K + M_y^K}{1 - v} \tag{5.2.2}$$

is related to the Kirchhoff plate deflection as

$$\overline{M^K} = -D\Delta w^K \tag{5.2.3}$$

in which the plate flexural rigidity D has been described by Eq. (4.1.3) while Δ is defined in Eq. (4.3.7). The deflection relationship between Kirchhoff and Reddy plate theories has been given by Reddy and Wang (1998) as

$$w^R = w^K + \frac{1}{Gh}\left(\alpha C_1 D\Delta w^R + C_4\overline{M^K}\right) \tag{5.2.4}$$

or

$$\Delta w^R - \lambda^2 w^R = -\lambda^2\left(w^K + \frac{C_4}{Gh}\overline{M^K}\right) \tag{5.2.5}$$

where $C_4 = 17/14$,

$$280C_1 = 3h^2 = \frac{4}{\alpha} \tag{5.2.6}$$

and

$$\lambda^2 = \frac{Gh}{\alpha C_1 D} = \frac{70Gh}{D} \tag{5.2.7}$$

From the plate flexural rigidity described in Eq. (4.1.3) and the moduli relationship

$$G = \frac{E}{2(1+v)} \tag{5.2.8}$$

we have

$$\frac{D}{G} = \frac{h^3}{6(1-v)} \tag{5.2.9}$$

The relation furnished in Eq. (5.2.9) is useful for recasting Eq. (5.2.1) in a more tractable form

$$w^M = w^K - \frac{h^2}{6\kappa(1-v)}\Delta w^K \tag{5.2.10}$$

Likewise, the use of Eqs. (5.2.6) or (5.2.7) with Eq. (5.2.9) allows Eq. (5.2.4) to be expressed in a more tractable form

$$w^R - \frac{h^2}{420(1-v)}\Delta w^R = w^K - \frac{17h^2}{84(1-v)}\Delta w^K \tag{5.2.11}$$

The forms of CPT-FSDT and CPT-TSDT relationships furnished in Eqs. (5.2.10) and (5.2.11) are used for forming the FSDT-TSDT relationship to extract the shear correction factor of Mindlin plates within the framework of Reddy plate theory. Two approaches are hereafter furnished: the direct approach and the iteration approach.

In the direct approach, we let $w^R = w^M$ (and hence $\Delta w^R = \Delta w^M$), so that Eq. (5.2.11) becomes

$$w^M - \frac{h^2}{420(1-\upsilon)}\Delta w^M = w^K - \frac{17h^2}{84(1-\upsilon)}\Delta w^K \tag{5.2.12}$$

or, with reference to Eq. (5.2.10),

$$\left[w^K - \frac{h^2}{6\kappa(1-\upsilon)}\Delta w^K\right] - \frac{h^2}{420(1-\upsilon)}\left[\Delta w^K - \frac{h^2}{6\kappa(1-\upsilon)}\Delta^2 w^K\right] = w^K - \frac{17h^2}{84(1-\upsilon)}\Delta w^K \tag{5.2.13}$$

Setting $\Delta w^R = 0$ in Eq. (5.2.11) commensurates with imposing zero for the second parenthesis [...] in Eq. (5.2.13), which would lead to $\kappa = 14/17$. If we do not let the entire second parenthesis in Eq. (5.2.13) be zero, but set only $\Delta^2 w^K = 0$, then $\kappa = 5/6$. If we do not neglect any term, then

$$\kappa = \frac{5}{6}\left(1 - \frac{h^2}{420(1-\upsilon)}\frac{\Delta^2 w^K}{\Delta w^K}\right) \tag{5.2.14}$$

Since $\Delta w^K < 0 < \Delta^2 w^K$ for simply supported polygonal plates, the ratio $\Delta^2 w^K / \Delta w^K$ is negative. Therefore, Eq. (5.2.14) can be written as (Lim 2020)

$$\kappa = \frac{5}{6}\left(1 + \frac{h^2}{420(1-\upsilon)}\frac{\Delta^2 w^K}{\left|\Delta w^K\right|}\right) \tag{5.2.15}$$

Knowing the signs of Δw^K and $\Delta^2 w^K$, it can be seen that $\kappa = 5/6$ is a lower bound. Examples of specific shear correction factors obtained for rectangular plates in Sect. 5.3 and isosceles right triangular plates in Sect. 5.5 are subsets of the more generalized Eq. (5.2.15).

In the iteration approach, we begin by setting $\Delta w^R = 0$ initially for Eq. (5.2.11) to give

$$w^R = w^K - \frac{17h^2}{84(1-\upsilon)}\Delta w^K \tag{5.2.16}$$

and hence

$$\Delta w^R = \Delta w^K - \frac{17h^2}{84(1-v)}\Delta^2 w^K \tag{5.2.17}$$

Substituting Eq. (5.2.17) into Eq. (5.2.11) gives

$$w^R = w^K - \frac{h^2}{5(1-v)}\Delta w^K - \frac{17h^4}{84(420)(1-v)^2}\Delta^2 w^K \tag{5.2.18}$$

which leads to

$$\Delta w^R = \Delta w^K - \frac{h^2}{5(1-v)}\Delta^2 w^K - \frac{17h^4}{84(420)(1-v)^2}\Delta^3 w^K \tag{5.2.19}$$

Substituting Eq. (5.2.19) into Eq. (5.2.11) and proceeding similarly, an infinite series expansion

$$\begin{aligned} w^R = w^K &- \frac{h^2 \Delta w^K}{5(1-v)} - \frac{h^4 \Delta^2 w^K}{5(420)(1-v)^2} \\ &- \frac{h^6 \Delta^3 w^K}{5(420)^2(1-v)^3} - \frac{h^8 \Delta^4 w^K}{5(420)^3(1-v)^4} - \cdots \end{aligned} \tag{5.2.20}$$

is obtained, which can be expressed in the following contracted form

$$w^R = w^K - \sum_{n=1}^{\infty} \frac{h^{2n} \Delta^n w^K}{5(420)^{n-1}(1-v)^n} \tag{5.2.21}$$

Recall now that Eqs. (5.2.16), (5.2.18), and (5.2.20) are the series expansions based on one, two, and infinite iterations, respectively. If we let $w^R = w^M$ based on Eqs. (5.2.10) and (5.2.16), the latter being of one iteration, then $\kappa = 14/17$. Suppose we let $w^R = w^M$ based on Eqs. (5.2.10) and (5.2.18), the latter being of two iterations, then $\kappa = 5/6$ if the highest order term is neglected. Now if we let $w^R = w^M$ based on Eqs. (5.2.10) and (5.2.20), the latter being of infinite iterations, then

$$\kappa = \frac{5}{6}\left(1 + \frac{h^2}{420(1-v)}\frac{\Delta^2 w^K}{\Delta w^K} + \frac{h^4}{420^2(1-v)^2}\frac{\Delta^3 w^K}{\Delta w^K} + \frac{h^6}{420^3(1-v)^3}\frac{\Delta^4 w^K}{\Delta w^K} + \cdots\right)^{-1} \tag{5.2.22}$$

or

$$\kappa = \frac{5}{6}\left(1 + \sum_{n=1}^{\infty} \frac{h^{2n}}{420^n (1-v)^n}\frac{\Delta^{n+1} w^K}{\Delta w^K}\right)^{-1} \tag{5.2.23}$$

Truncating the higher-order terms in Eq. (5.2.22), or neglecting terms of $n \geq 2$ in Eq. (5.2.23), we have

$$\kappa = \frac{5}{6}\left(1 + \frac{h^2}{420(1-\upsilon)}\frac{\Delta^2 w^K}{\Delta w^K}\right)^{-1} \tag{5.2.24}$$

which is related to Eq. (5.2.14) by virtue of Maclaurin expansion followed by neglect of higher orders. Considering again that $\Delta w^K < 0 < \Delta^2 w^K$ for simple supports, Eq. (5.2.24) can be expressed as (Lim 2020)

$$\kappa = \frac{5}{6}\left(1 - \frac{h^2}{420(1-\upsilon)}\frac{\Delta^2 w^K}{\left|\Delta w^K\right|}\right)^{-1} \tag{5.2.25}$$

Examples of specific shear correction factors obtained for equilateral triangular plates in Sect. 5.4, hexagonal plates in Sect. 5.6, and the regular polygonal plates in Sect. 5.7 are subsets of the more generalized Eq. (5.2.25).

5.3 Example: Rectangular Auxetic Plates

Figure 5.2 illustrates a simply supported thick rectangular plate of sides a and b, measured along the x and y axes, respectively, while the thickness h is measured along the z-axis. Equating the Mindlin and Reddy plate deflections gives the following

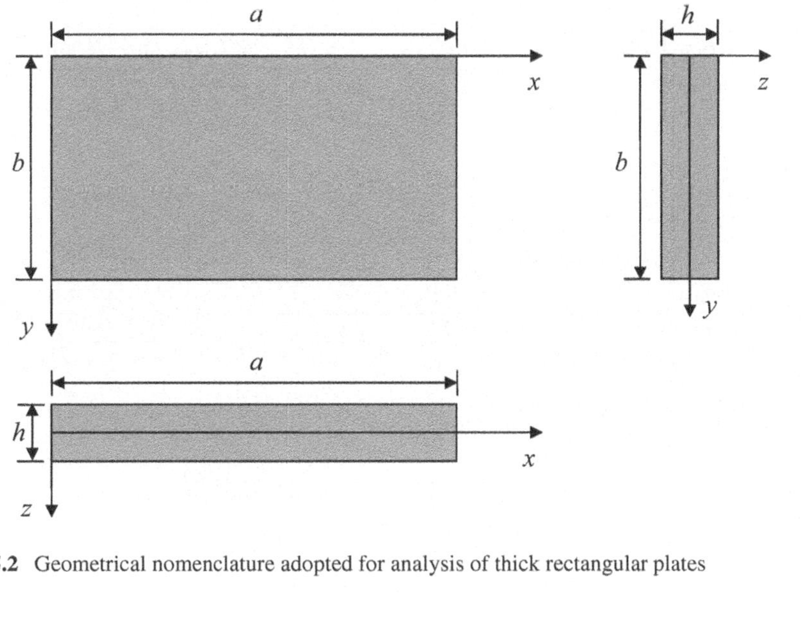

Fig. 5.2 Geometrical nomenclature adopted for analysis of thick rectangular plates

general relationship under the same boundary condition,

$$\frac{1}{\kappa} = \frac{17}{14} - \frac{1}{70}\frac{\Delta w^R}{\Delta w^K} \tag{5.3.1}$$

Perusal to Eq. (5.3.1) suggests that a meaningful exact shear correction factor can be obtained if both the Reddy plate and Kirchhoff plate deflections are known. Neglecting the higher-order term in Eq. (5.3.1) gives a shear correction factor of $\kappa = 14/17$. We shall now consider two types of loads: uniform load and sinusoidal load.

As the Kirchhoff plate deflection for a simply supported rectangular plate under uniform load $q = q_0$ is

$$w^K = \frac{16q_0}{\pi^6 D}\sum_{m=1}^{\infty}\sum_{n=1}^{\infty}\frac{\sin\frac{m\pi x}{a}\sin\frac{n\pi y}{b}}{mn\left(\frac{m^2}{a^2}+\frac{n^2}{b^2}\right)^2} \tag{5.3.2a}$$

with $m, n = 1, 3, 5, \ldots$, we adopt a similar deflection profile for the Reddy plate

$$w^R = A^R\sum_{m=1}^{\infty}\sum_{n=1}^{\infty}\frac{\sin\frac{m\pi x}{a}\sin\frac{n\pi y}{b}}{mn\left(\frac{m^2}{a^2}+\frac{n^2}{b^2}\right)^2} \tag{5.3.2b}$$

where A^R is the amplitude term of the Reddy plate. Substituting the deflection profiles of Kirchhoff and Reddy plates into the relationships described by Eqs. (5.2.10) and (5.2.11) leads to

$$w^M = \frac{16q_0}{\pi^6 D}\left\{\sum_{m=1}^{\infty}\sum_{n=1}^{\infty}\frac{\sin\frac{m\pi x}{a}\sin\frac{n\pi y}{b}}{mn\left(\frac{m^2}{a^2}+\frac{n^2}{b^2}\right)^2} + \frac{\pi^2 h^2}{6\kappa(1-\upsilon)}\sum_{m=1}^{\infty}\sum_{n=1}^{\infty}\frac{\sin\frac{m\pi x}{a}\sin\frac{n\pi y}{b}}{mn\left(\frac{m^2}{a^2}+\frac{n^2}{b^2}\right)}\right\} \tag{5.3.3}$$

and

$$A^R = \frac{16q_0}{\pi^6 D}\frac{\sum_{m=1}^{\infty}\sum_{n=1}^{\infty}\frac{\sin\frac{m\pi x}{a}\sin\frac{n\pi y}{b}}{mn\left(\frac{m^2}{a^2}+\frac{n^2}{b^2}\right)^2} + \frac{17\pi^2 h^2}{84(1-\upsilon)}\sum_{m=1}^{\infty}\sum_{n=1}^{\infty}\frac{\sin\frac{m\pi x}{a}\sin\frac{n\pi y}{b}}{mn\left(\frac{m^2}{a^2}+\frac{n^2}{b^2}\right)}}{\sum_{m=1}^{\infty}\sum_{n=1}^{\infty}\frac{\sin\frac{m\pi x}{a}\sin\frac{n\pi y}{b}}{mn\left(\frac{m^2}{a^2}+\frac{n^2}{b^2}\right)^2} + \frac{\pi^2 h^2}{420(1-\upsilon)}\sum_{m=1}^{\infty}\sum_{n=1}^{\infty}\frac{\sin\frac{m\pi x}{a}\sin\frac{n\pi y}{b}}{mn\left(\frac{m^2}{a^2}+\frac{n^2}{b^2}\right)}} \tag{5.3.4}$$

respectively. Introducing the function

$$f(a,b,x,y) = \frac{\sum_{m=1}^{\infty}\sum_{n=1}^{\infty}\frac{\sin\frac{m\pi x}{a}\sin\frac{n\pi y}{b}}{mn\left(\frac{b}{a}m^2+\frac{a}{b}n^2\right)}}{\sum_{m=1}^{\infty}\sum_{n=1}^{\infty}\frac{\sin\frac{m\pi x}{a}\sin\frac{n\pi y}{b}}{mn\left(\frac{b}{a}m^2+\frac{a}{b}n^2\right)^2}} \tag{5.3.5}$$

allows Eq. (5.3.4) to be contracted as

$$A^R = \frac{16q_0}{\pi^6 D} \frac{1 + \frac{17\pi^2 h^2}{84(1-\upsilon)ab} f(a, b, x, y)}{1 + \frac{\pi^2 h^2}{420(1-\upsilon)ab} f(a, b, x, y)} \tag{5.3.6}$$

The terms a/b in Eq. (5.3.5) and $h/\sqrt{ab}$ in Eq. (5.3.6) indicate the plate aspect ratio and its relative thickness, respectively; for a square plate, these reduce to $a/b = 1$ and h/a. Using Eq. (5.3.6) and equating the Mindlin and Reddy plate deflections gives

$$\begin{aligned} 1 &+ \frac{\pi^2 h^2}{420(1-\upsilon)ab} f(a, b, x, y) + \frac{\pi^2 h^2}{6\kappa(1-\upsilon)ab} f(a, b, x, y) \\ &+ \frac{\pi^4 h^4}{2520\kappa(1-\upsilon)^2 a^2 b^2} (f(a, b, x, y))^2 \\ &= 1 + \frac{17\pi^2 h^2}{84(1-\upsilon)ab} f(a, b, x, y) \end{aligned} \tag{5.3.7}$$

Hence, the shear correction factor $\kappa = 14/17$ is obtained from Eq. (5.3.7) if the terms containing $\pi^2 h^2/420$ and $\pi^4 h^4/2520$ are neglected, and that the usual shear correction factor of $\kappa = 5/6$ is obtained from Eq. (5.3.7) if only the highest order term is neglected. Taking into account the highest order term, we have the exact shear correction factor

$$\kappa = \frac{5}{6}\left[1 + \frac{\pi^2}{420(1-\upsilon)} \frac{h^2}{ab} f(a, b, x, y)\right] \tag{5.3.8}$$

Since the maximum deflection takes place at the plate center, it is practical to consider the shear correction factor therein (Lim 2016a)

$$\kappa = \frac{5}{6}\left[1 + \frac{\pi^2}{420(1-\upsilon)} \frac{h^2}{ab} f\left(a, b, \frac{a}{2}, \frac{b}{2}\right)\right] \tag{5.3.9}$$

where the function described by Eq. (5.3.5) becomes

$$f\left(a, b, \frac{a}{2}, \frac{b}{2}\right) = \frac{\sum_{m=1}^{\infty} \sum_{n=1}^{\infty} \frac{(-1)^{\frac{m+n}{2}-1}}{mn\left(\frac{b}{a}m^2 + \frac{a}{b}n^2\right)}}{\sum_{m=1}^{\infty} \sum_{n=1}^{\infty} \frac{(-1)^{\frac{m+n}{2}-1}}{mn\left(\frac{b}{a}m^2 + \frac{a}{b}n^2\right)^2}} = f\left(\frac{a}{b}\right) \tag{5.3.10}$$

at the plate center. For an extremely long and narrow plate, Eq. (5.3.10) reduces to

$$\lim_{\frac{a}{b}\to\infty} f\left(\frac{a}{b}\right) = \frac{a}{b} \tag{5.3.11}$$

Table 5.1 Computed results of Eq. (5.3.10)

Plate aspect ratio, *a*/*b*	Numerator $\sum_{m=1}^{\infty}\sum_{n=1}^{\infty}\frac{(-1)^{\frac{m+n}{2}-1}}{mn\left(\frac{b}{a}m^2+\frac{a}{b}n^2\right)}$	Denominator $\sum_{m=1}^{\infty}\sum_{n=1}^{\infty}\frac{(-1)^{\frac{m+n}{2}-1}}{mn\left(\frac{b}{a}m^2+\frac{a}{b}n^2\right)^2}$	Ratio $f\left(a,b,\frac{a}{2},\frac{b}{2}\right)$
1.0	0.44895	0.24409	1.839281
1.2	0.44059	0.23578	1.868649
1.5	0.40575	0.20627	1.967082
2.0	0.34749	0.15215	2.283865
3.0	0.24622	0.08167	3.014816
5.0	0.15405	0.031175	4.941460
7.5	0.10389	0.013907	7.470339
10.0	0.07859	0.007824	10.04473

It can be seen that both the numerator and denominator of Eq. (5.3.10) are dependent on the plate aspect ratio a/b. Setting $a \geq b$ for the uniformly loaded plate, Table 5.1 lists the denominator and numerator of Eq. (5.3.10) by performing double series summation. The summation was performed up to $m = n = 41$ in order to obtain sufficient numerical accuracy.

A simple curve-fit based on Table 5.1 gives the function

$$f\left(a,b,\frac{a}{2},\frac{b}{2}\right) = f\left(\frac{a}{b}\right) = -0.0062\left(\frac{a}{b}\right)^3 + 0.1281\left(\frac{a}{b}\right)^2 + 0.1956\left(\frac{a}{b}\right) + 1.4598 \tag{5.3.12}$$

for $1 \leq a/b \leq 10$ with a statistical accuracy of $R^2 = 0.9998$.

Suppose the load distribution takes a sinusoidal form

$$q = q_0 \sin\frac{m\pi x}{a}\sin\frac{n\pi y}{b} \tag{5.3.13}$$

instead of being uniformly distributed—whereby m and n quantify the load waviness along the x and y axes, respectively—then Eqs. (5.3.5) and (5.3.8) reduce to

$$f(a,b,x,y) = \frac{b}{a}m^2 + \frac{a}{b}n^2 \tag{5.3.14}$$

and

$$\kappa = \frac{5}{6}\left[1 + \frac{\pi^2}{420(1-v)}\frac{h^2}{ab}\left(\frac{b}{a}m^2 + \frac{a}{b}n^2\right)\right] \tag{5.3.15}$$

respectively. Although Eq. (5.3.15) can be written in a simpler way as

Table 5.2 Shear correction factor expressions for special cases of rectangular plates under sinusoidal loads

	Simple sinusoidal load distribution ($m = n = 1$)	General sinusoidal load distributions ($m, n \geq 1$)
Square plates $a = b$	$\kappa = \frac{5}{6}\left[1 + \frac{\pi^2}{210(1-\upsilon)}\frac{h^2}{a^2}\right]$	$\kappa = \frac{5}{6}\left[1 + \frac{\pi^2}{420(1-\upsilon)}\frac{h^2}{a^2}(m^2 + n^2)\right]$
Rectangular plates $a \neq b$	$\kappa = \frac{5}{6}\left[1 + \frac{\pi^2}{420(1-\upsilon)}\frac{h^2}{ab}\left(\frac{b}{a} + \frac{a}{b}\right)\right]$	$\kappa = \frac{5}{6}\left[1 + \frac{\pi^2}{420(1-\upsilon)}\frac{h^2}{ab}\left(\frac{b}{a}m^2 + \frac{a}{b}n^2\right)\right]$

$$\kappa = \frac{5}{6}\left[1 + \frac{\pi^2 h^2}{420(1-\upsilon)}\left(\frac{m^2}{a^2} + \frac{n^2}{b^2}\right)\right] \tag{5.3.16}$$

the former is instructive for showing the effect of relative plate thickness $h/\sqrt{ab}$ and aspect ratio a/b, in addition to the Poisson's ratio. Unlike the previous subsection on uniform load, this subsection on sinusoidal load allows one to observe the interlacing effect of load waviness pattern and plate aspect ratio on the shear correction factor.

For the special case of square plate, perusal to Table 5.2 shows that load waviness increases the shear correction factor. Reference to the same table also shows that waviness is strongly influenced by the aspect ratio of the plate; the effect of load waviness along the longer side diminishes as the plate becomes long or very narrow, i.e.,

$$\begin{aligned} \kappa &= \frac{5}{6}\left[1 + \frac{n^2\pi^2}{420(1-\upsilon)}\frac{h^2}{b^2}\right] \quad ; \quad a \gg b \\ \kappa &= \frac{5}{6}\left[1 + \frac{m^2\pi^2}{420(1-\upsilon)}\frac{h^2}{a^2}\right] \quad ; \quad a \ll b \end{aligned} \tag{5.3.17}$$

and consequently the relative thickness is governed by the ratio of the plate thickness to its shorter side.

In determining the range of relative thickness that is applicable for the shear deformation theories, one may classify the plate thickness into four categories: (i) $a/h > 100$, (ii) $20 < a/h < 100$, (iii) $3 < a/h < 20$, and (iv) $a/h < 3$. This implies that one may then adopt the membrane theory for $h/a < 0.01$, CPT for $h/a < 0.05$, shear deformation theories for $h/a < 0.3333$, and elasticity theory for $h/a > 0.3333$. It therefore follows that the TSDT-based shear correction factor for FSDT problems is therefore applicable for relative thickness range of $h/a < 0.3333$. As such the following results were computed for relative thickness up to 0.2 since shear deformation theories are not applicable for relative thickness of 1/3 and above. As with CPT and FSDT, the TSDT is applicable for auxetic materials since the development of these theories is not confined to cases where the Poisson's ratio is positive.

Figure 5.3a, b shows the effect of relative plate thickness and plate aspect ratio on the shear correction factor of a uniformly loaded plate for the entire range of Poisson's ratio. Specifically, the shear correction factor increases when (i) the plate

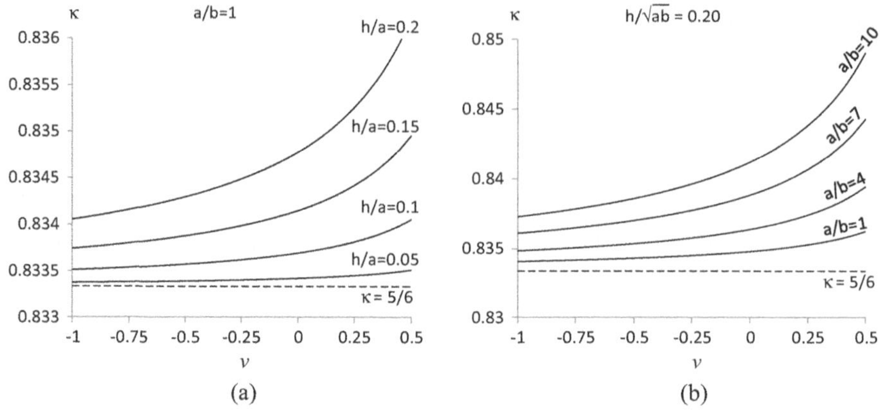

Fig. 5.3 Shear correction factor versus Poisson's ratio of a simply supported rectangular plate under uniform load with **a** variation in relative thickness for a square plate and **b** variation in aspect ratio for a thick plate

becomes thicker, (ii) the plate becomes longer or narrower, and (iii) the Poisson's ratio of the plate material is greater. The curves of the shear correction factors are plotted for the Poisson's ratio of the range $-1 \leq v \leq 1/2$. The dashed lines in this and subsequent figures indicate the lower bound for the shear correction factor, i.e., $\kappa = 5/6$, for comparison. The influence of the plate geometry, in terms of the in-plane aspect ratio and the relative thickness, on the shear correction factor is plotted in Fig. 5.4 for $v = 0.3$.

In the case of sinusoidal load, there is a qualitatively comparable trend in the effect of plate geometry (aspect ratio and relative thickness) and Poisson's ratio on

Fig. 5.4 Influence of aspect ratio and relative thickness on the shear correction factor of a plate with $v = 0.3$

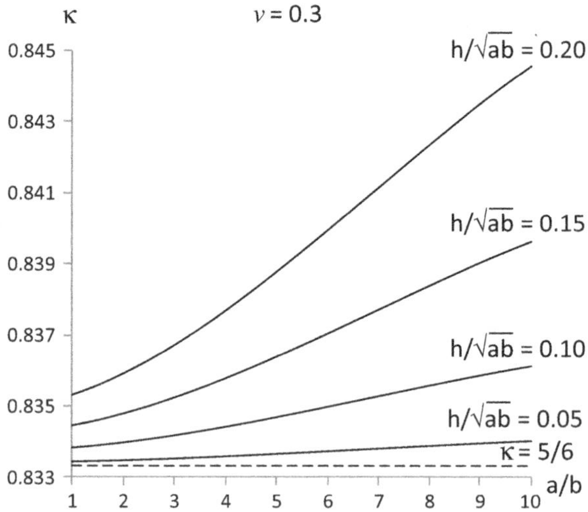

uniform load. In addition, the waviness of the transverse static load increases the shear correction factor, as evidenced in Fig. 5.5 for square plates.

In the special case of square plates, the shear correction factor is unchanged when the load waviness changes direction. For example, the shear correction factor for $(m, n) = (3, 1)$ is similar to that for $(m, n) = (1, 3)$; likewise the shear correction factor for $(m, n) = (5, 1)$ is similar to that for $(m, n) = (1, 5)$. This observation, however, does not hold for rectangular plates. Perusal to Eqs. (5.3.15) or (5.3.17) shows that for very long or very narrow plates, the load waviness measured along the shorter side has a greater influence than that along the longer side, as shown in Fig. 5.6.

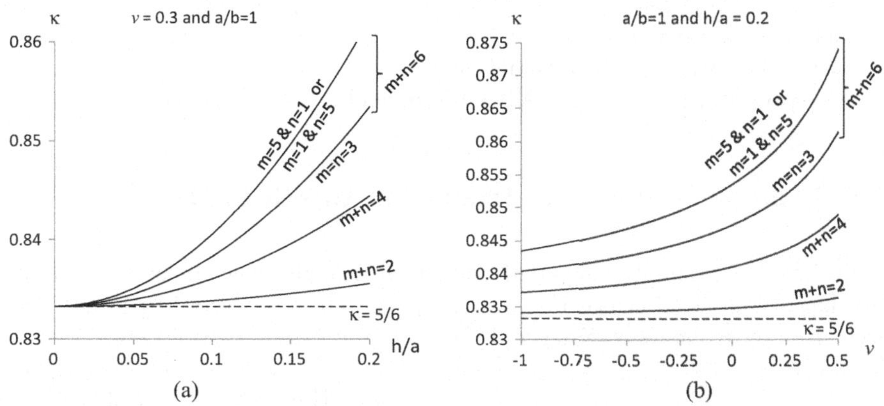

Fig. 5.5 Effect of load waviness on the shear correction factor of a sinusoidally loaded square plate with varying **a** relative thickness and **b** Poisson's ratio

Fig. 5.6 Asymmetric effect of load waviness on shear correction factor of a rectangular plate with Poisson's ratio 0.3 and relative thickness 0.2

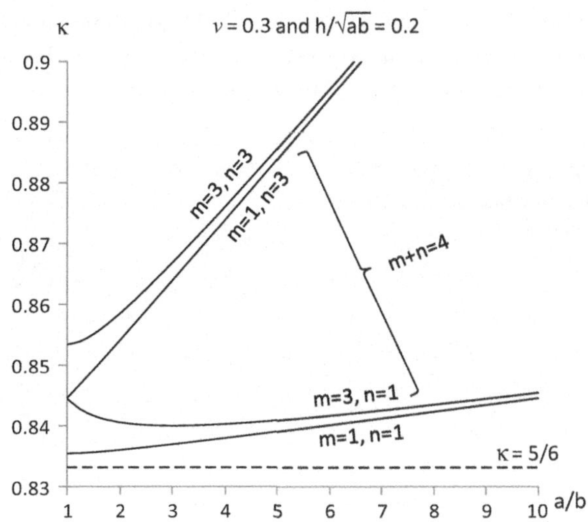

In summary, the results obtained herein for uniform and sinusoidal loads show that the exact shear correction factor is higher than the commonly used shear correction factor of 5/6 under the following conditions:

1. very thick plates
2. very long or narrow plates
3. plates made from large Poisson's ratio (especially incompressible) materials and
4. highly patterned loading pattern, or sinusoidal load with high load waviness.

However, the use of the lower-bound shear correction factor of $\kappa = 5/6$ is valid under the following conditions:

1. marginally thick plates
2. square or almost square plates
3. plates made from auxetic materials and
4. less wavy load pattern, especially uniform loads.

5.4 Example: Equilateral Triangular Auxetic Plates

In the case of a simply supported equilateral triangle plate under a uniformly distributed load as shown in Fig. 5.7, the plate deflection according to CPT has been given by Woinowsky–Krieger (1933) as

$$w^K = \frac{q}{64LD}\left[x^3 - 3xy^2 - L\left(x^2 + y^2\right) + \frac{4}{27}L^3\right]\left[\frac{4}{9}L^2 - \left(x^2 + y^2\right)\right] \quad (5.4.1)$$

Notwithstanding the availability of the CPT model, it can be easily seen that the TSDT-CPT relation given in Eq. (5.2.11) is not readily applicable due to the need for solving for both w^R and Δw^R . Although one may set $\Delta w^R = 0$ in Eq. (5.2.11) in order to conveniently solve for w^R directly, this simplification could possibly affect its accuracy—as will be shown later—thereby resulting in a lower accuracy

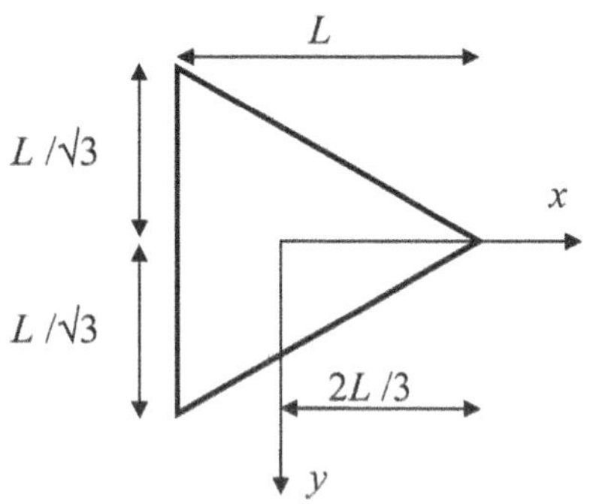

Fig. 5.7 Geometrical description of an equilateral triangular plate

in comparison to the FSDT model. Suppose we set $\Delta w^R = 0$ in Eq. (5.2.11), then substitution of Eq. (5.4.1) and

$$\Delta w^K = \frac{q}{64LD}(-16)\left[x^3 - 3xy^2 - L(x^2 + y^2) + \frac{4}{27}L^3\right] \tag{5.4.2}$$

into Eq. (5.2.11) gives the simplified TSDT deflection as

$$w^R \sim \frac{q}{64LD}\left[x^3 - 3xy^2 - L(x^2 + y^2) + \frac{4}{27}L^3\right]\left[\frac{4}{9}L^2 - (x^2 + y^2) + \frac{68h^2}{21(1-\upsilon)}\right] \tag{5.4.3}$$

In order to directly solve Eq. (5.2.11) for the considered plate, the following solution is proposed herein in a similar form

$$w^R = \frac{q}{64LD}\left\{\left[x^3 - 3xy^2 - L(x^2 + y^2) + \frac{4}{27}L^3\right]\left[\frac{4}{9}L^2 - (x^2 + y^2) + \frac{C_1 h^2}{1-\upsilon}\right] - \frac{C_2 L h^4}{(1-\upsilon)^2}\right\} \tag{5.4.4}$$

where C_1 and C_2 are dimensionless constants. Substituting Eqs. (5.4.1), (5.4.2), (5.4.4), and

$$\Delta w^R = \frac{q}{64LD}(-16)\left[x^3 - 3xy^2 - L(x^2 + y^2) + \frac{4}{27}L^3 + \frac{C_1 L h^2}{4(1-\upsilon)}\right] \tag{5.4.5}$$

into Eq. (5.2.11) leads to

$$\left(C_1 - \frac{16}{5}\right)\left[x^3 - 3xy^2 - L(x^2 + y^2) + \frac{4}{27}L^3\right] = \left(C_2 - \frac{C_1}{105}\right)\frac{Lh^2}{1-\upsilon} \tag{5.4.6}$$

The above equation is satisfied if the coefficients are chosen as

$$C_1 = \frac{16}{5}\ , \quad C_2 = \frac{16}{525} \tag{5.4.7}$$

The exact plate deflection by TSDT as described by Eq. (5.4.4) with Eq. (5.4.7) can also be arrived at by an indirect route, which also serves to confirm the choice of Eqs. (5.4.4) and (5.4.7), by means of iteration in analytical form instead of numerical form. Based on the first approximation, which is the simplified deflection of Eq. (5.4.3), substitution of

$$\Delta w^R \sim \frac{q}{64LD}(-16)\left[x^3 - 3xy^2 - L(x^2 + y^2) + \frac{4}{27}L^3 + \frac{17Lh^2}{21(1-\upsilon)}\right] \tag{5.4.8}$$

together with Eqs. (5.4.1) and (5.4.2) into Eq. (5.2.11) gives the second approximation

$$w^R \approx \frac{q}{64LD}\left\{\left[x^3 - 3xy^2 - L\left(x^2 + y^2\right) + \frac{4}{27}L^3\right]\right.$$
$$\left.\left[\frac{4}{9}L^2 - \left(x^2 + y^2\right) + \frac{16h^2}{5(1-\upsilon)}\right] - \frac{68Lh^4}{2205(1-\upsilon)^2}\right\} \tag{5.4.9}$$

From Eq. (5.4.9), substitution of

$$\Delta w^R \approx \frac{q}{64LD}(-16)\left[x^3 - 3xy^2 - L\left(x^2 + y^2\right) + \frac{4}{27}L^3 + \frac{4Lh^2}{5(1-\upsilon)}\right] \tag{5.4.10}$$

as well as Eqs. (5.4.1) and (5.4.2) into Eq. (5.2.11) gives the third "approximation," which is the converged solution

$$w^R = \frac{q}{64LD}\left\{\left[x^3 - 3xy^2 - L\left(x^2 + y^2\right) + \frac{4}{27}L^3\right]\right.$$
$$\left.\left[\frac{4}{9}L^2 - \left(x^2 + y^2\right) + \frac{16h^2}{5(1-\upsilon)}\right] - \frac{16Lh^4}{525(1-\upsilon)^2}\right\} \tag{5.4.11}$$

Equation (5.4.11) is said to have converged, and is therefore no longer an approximation, for two reasons: (a) proceeding similarly will no longer change the form, and that (b) it is exactly the same as Eq. (5.4.4) with the constants described by Eq. (5.4.7).

Having obtained the plate deflection by exact TSDT, it is now of interest to refine the shear correction factor of Mindlin plates. Substituting Eqs. (5.4.1) and (5.4.2) into Eq. (5.2.1) gives (Wang et al. 2000a; Reddy 2006)

$$w^M = \frac{q}{64LD}\left[x^3 - 3xy^2 - L\left(x^2 + y^2\right) + \frac{4}{27}L^3\right]\left[\frac{4}{9}L^2 - \left(x^2 + y^2\right) + \frac{16D}{\kappa Gh}\right] \tag{5.4.12}$$

Substitution of Eq. (5.2.9) into Eq. (5.4.12) yields

$$w^M = \frac{q}{64LD}\left[x^3 - 3xy^2 - L\left(x^2 + y^2\right) + \frac{4}{27}L^3\right]\left[\frac{4}{9}L^2 - \left(x^2 + y^2\right) + \frac{8h^2}{3\kappa(1-\upsilon)}\right] \tag{5.4.13}$$

Comparing terms between the simplified TSDT model, Eq. (5.4.3) and the FSDT model, Eq. (5.4.13) the shear correction factor is $\kappa = 14/17$. This suggests that, by matching the FSDT and the simplified TSDT, the shear correction factor is a constant. Likewise, the typically adopted shear correction factor $\kappa = 5/6$ is also a constant. When the exact TSDT model in Eq. (5.4.11) is compared against the FSDT model in Eq. (5.4.13), the shear correction factor is

$$\kappa = \frac{5}{6 - \frac{6(h/L)^2}{105(1-\upsilon)}\left[\left(\frac{x}{L}\right)^3 - 3\frac{x}{L}\left(\frac{y}{L}\right)^2 - \left(\frac{x^2}{L^2} + \frac{y^2}{L^2}\right) + \frac{4}{27}\right]^{-1}} \tag{5.4.14}$$

Equation (5.4.14) suggests that, by matching the FSDT and exact TSDT, the "exact" shear correction factor is a function of plate relative thickness (h/L), the Poisson's ratio v, and the plate's in-plane coordinates (x, y). Since one is normally concerned with the maximum deflection, then in the case of uniform load the maximum deflection for the plate under consideration takes place at the center $x = y = 0$ to give (Lim 2016b)

$$\kappa = \frac{5}{6}\left[1 - \frac{9}{140(1-\upsilon)}\left(\frac{h}{L}\right)^2\right]^{-1} \tag{5.4.15}$$

The customary shear correction factor of $\kappa = 5/6$ is therefore valid when the plate's relative thickness (h/L) is small or when the Poisson's ratio is negative.

To compare the present exact TSDT for triangular plates with the FSDT and the simplified TSDT, a dimensionless maximum deflection is introduced as

$$w^*_{\max} = \frac{972D}{qL^4} w_{x=y=0} \tag{5.4.16}$$

so that $w^*_{\max} = 1$ for CPT, while Eqs. (5.4.3), (5.4.4), and (5.4.13) become

$$w^*_{\max} = 1 + \frac{51}{7(1-\upsilon)}\left(\frac{h}{L}\right)^2 \tag{5.4.17}$$

for simplified TSDT,

$$w^*_{\max} = 1 + \frac{36}{5(1-\upsilon)}\left(\frac{h}{L}\right)^2 - \frac{81}{175(1-\upsilon)^2}\left(\frac{h}{L}\right)^4 \tag{5.4.18}$$

for exact TSDT based on Eq. (5.4.11), and

$$w^*_{\max} = 1 + \frac{36}{5(1-\upsilon)}\left(\frac{h}{L}\right)^2 \tag{5.4.19}$$

for FSDT whereupon $\kappa = 5/6$ had been substituted in Eq. (5.4.13).

A family of curves based on Eqs. (5.4.17) and (5.4.18) is plotted in Fig. 5.8, showing the overestimation of the simplified TSDT in relation to the exact TSDT, especially for large relative thickness and large Poisson's ratio. To put into perspective the maximum dimensionless deflection based on both the simplified and the exact TSDT with reference to that of FSDT, calculated dimensionless maximum deflections

Fig. 5.8 Combined effect of Poisson's ratio and relative thickness on the maximum deflection of uniformly loaded simply supported equilateral triangular plate based on exact (continuous curves) and simplified (dashed curves) TSDT

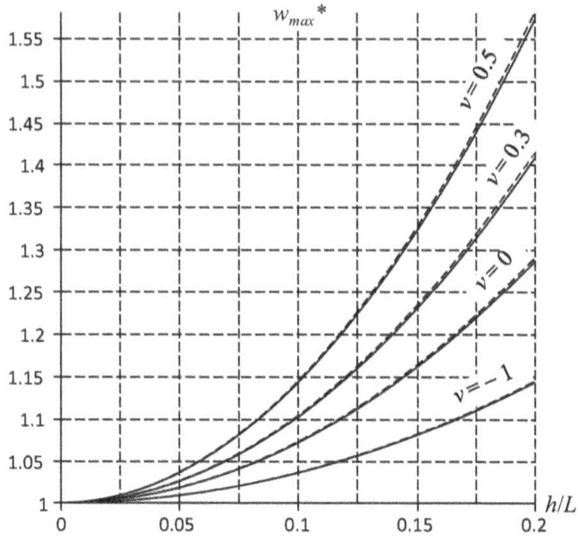

are consolidated in Fig. 5.9 based on the range of relative thickness $0 \leq h/L \leq 0.2$ and the range of Poisson's ratio $-1 \leq \upsilon \leq 1/2$, which applies for isotropic solids.

Figure 5.10 illustrates the effect of Poisson's ratio and relative thickness on the shear correction factor at the plate center, indicating the departure of the shear correction factor from 5/6 as the Poisson's ratio and the relative thickness increase. For very thin plate, such as $h/L \leq 0.01$, the equivalent shear correction factor based on

Fig. 5.9 Dimensionless maximum deflections by the simplified and the exact TSDT with reference to the dimensionless maximum deflection by FSDT

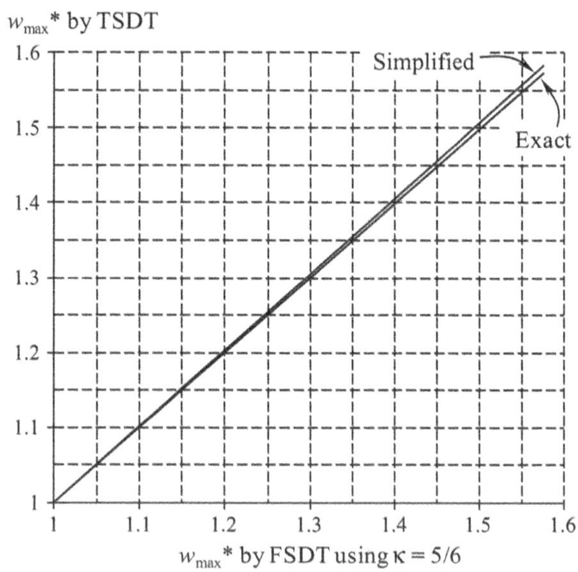

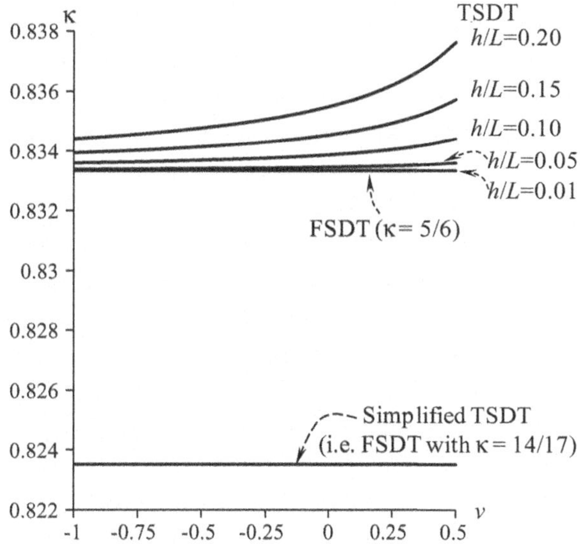

Fig. 5.10 Combined effect of Poisson's ratio and relative plate thickness on the shear correction factor of Mindlin plate based on TSDT (exact and simplified) and FSDT

the exact TSDT is very close to that by FSDT. It is of interest to note that the shear correction factor for the simplified TSDT is furthest from those of the exact TSDT, thereby suggesting that the FSDT is more accurate than the simplified TSDT and that a higher-order analysis need not necessarily give better prediction when certain assumptions are used prematurely during analysis.

Now it can also be shown that the shear correction factor varies with the in-plane coordinates of the triangular plate. Table 5.3 lists the exact shear correction factor using Eq. (5.4.14) for $v = 0.3$ and $h/L = 0.2$. Since the plate boundary encounters no deflection, no calculation is made on the shear correction factor at the boundary. Reference to Table 5.3 reveals that the shear correction factor is lowest at the plate center but greatest near the sides and corners. These results would be of special interests to precision engineers and design engineers, although for different reasons. The analytical model for the shear correction factor, as a function of relative plate thickness and Poisson's ratio, would be of interest to precision engineers to attain TSDT accuracy by using FSDT model with the variable shear correction factor. On the other hand, the analytical model is useful as a basis for design engineers to justify the use of FSDT with constant shear correction factor even for very thick plates so long as the Poisson's ratio is sufficiently negative to ensure that the numerical value of the variable shear correction factor is insignificantly higher than 5/6.

In summary, the validity of the proposed exact form was verified by using the simplified TSDT as the starting equation, such that convergence is rapidly achieved by only two steps of analytical iterations. Using the plate deflection from the exact TSDT as reference, the FSDT (using $\kappa = 5/6$) gives a better accuracy than the simplified TSDT, where the equivalent shear correction factor is $\kappa = 14/17$ for the latter. Using the exact TSDT, comparison of terms with the FSDT gives rise to a

Table 5.3 Distribution of equivalent shear correction factor for Mindlin plates with $h/L = 0.2$ and $\upsilon = 0.3$ using the exact TSDT

		x/L												
		−4/12	−3/12	−2/12	−1/12	0/12	1/12	2/12	3/12	4/12	5/12	6/12	7/12	8/12
y/L	−0.577350	NA												
	−0.494872	NA	0.8883											
	−0.412393	NA	0.8502	0.8484	0.8695									
	−0.329914	NA	0.8441	0.8408	0.8411	0.8450	0.9208							
	−0.247436	NA	0.8417	0.8387	0.8381	0.8386	0.8403	0.8472						
	−0.164957	NA	0.8406	0.8378	0.8371	0.8371	0.8376	0.8388	0.8419	0.8571				
	−0.082479	NA	0.8400	0.8374	0.8367	0.8366	0.8368	0.8373	0.8384	0.8409	0.8480	0.9144		
	0	NA	0.8399	0.8373	0.8366	0.8364	0.8365	0.8370	0.8378	0.8395	0.8431	0.8534	0.9112	NA
	0.082479	NA	0.8400	0.8374	0.8367	0.8366	0.8368	0.8373	0.8384	0.8409	0.8480	0.9144		
	0.164957	NA	0.8406	0.8378	0.8371	0.8371	0.8376	0.8388	0.8419	0.8571				
	0.247436	NA	0.8417	0.8387	0.8381	0.8386	0.8403	0.8472						
	0.329914	NA	0.8441	0.8408	0.8411	0.8450	0.9208							
	0.412393	NA	0.8502	0.8484	0.8695									
	0.494872	NA	0.8883											
	0.577350	NA												

shear correction factor that is a function of Poisson's ratio, relative thickness and in-plane location on the plate. It was also established that the use of the typical shear correction factor $\kappa = 5/6$ for Mindlin plate is suitable when the plate is moderately thick and when the Poisson's ratio is small or negative, while the use of the refined shear correction factor for Mindlin plate, on the basis of the exact TSDT, is advised for very thick plates with large Poisson's ratio.

5.5 Example: Isosceles Right Triangular Auxetic Plates

Figure 5.11 shows an isosceles right triangular plate of sides a and hypotenuse side length of $\sqrt{2}a$. Under uniform load q and simply supported boundary condition, the CPT gives the Kirchhoff plate deflection as

$$w^K = \frac{16qa^4}{\pi^6 D}\left(\sum_{m=1,3,\ldots}^{\infty}\sum_{n=2,4,\ldots}^{\infty}\frac{n\sin\frac{m\pi x}{a}\sin\frac{n\pi y}{a}}{m\left(n^2-m^2\right)\left(m^2+n^2\right)^2} + \sum_{m=2,4,\ldots}^{\infty}\sum_{n=1,3,\ldots}^{\infty}\frac{m\sin\frac{m\pi x}{a}\sin\frac{n\pi y}{a}}{n\left(m^2-n^2\right)\left(m^2+n^2\right)^2}\right) \tag{5.5.1}$$

where D, E, h, and v refer to the flexural rigidity, Young's modulus, thickness, and Poisson's ratio of the plate, respectively.

Since the Mindlin plate deflection according to the FSDT is related to the Kirchhoff plate deflection as described by Eq. (5.2.10), substitution of Eq. (5.5.1) and

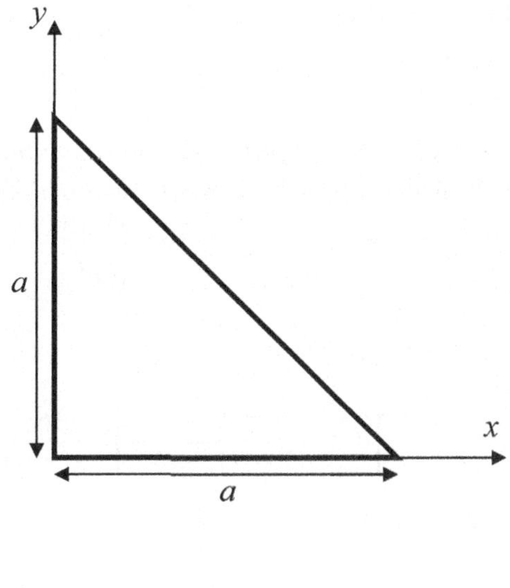

Fig. 5.11 Schematic of an isosceles right triangular plate

$$\Delta w^K = -\frac{16qa^2}{\pi^4 D} \left(\sum_{m=1,3,\ldots}^{\infty} \sum_{n=2,4,\ldots}^{\infty} \frac{n \sin \frac{m\pi x}{a} \sin \frac{n\pi y}{a}}{m(n^2 - m^2)(m^2 + n^2)} + \sum_{m=2,4,\ldots}^{\infty} \sum_{n=1,3,\ldots}^{\infty} \frac{m \sin \frac{m\pi x}{a} \sin \frac{n\pi y}{a}}{n(m^2 - n^2)(m^2 + n^2)} \right) \tag{5.5.2}$$

into Eq. (5.2.10) gives

$$w^M = \frac{16qa^4}{\pi^6 D} \left\{ \phi + \frac{\pi^2 h^2}{6\kappa a^2(1 - v)} \psi \right\} \tag{5.5.3}$$

where

$$\phi = \sum_{m=1,3,\ldots}^{\infty} \sum_{n=2,4,\ldots}^{\infty} \frac{n \sin \frac{m\pi x}{a} \sin \frac{n\pi y}{a}}{m(n^2 - m^2)(m^2 + n^2)^2} + \sum_{m=2,4,\ldots}^{\infty} \sum_{n=1,3,\ldots}^{\infty} \frac{m \sin \frac{m\pi x}{a} \sin \frac{n\pi y}{a}}{n(m^2 - n^2)(m^2 + n^2)^2} \tag{5.5.4}$$

and

$$\psi = \sum_{m=1,3,\ldots}^{\infty} \sum_{n=2,4,\ldots}^{\infty} \frac{n \sin \frac{m\pi x}{a} \sin \frac{n\pi y}{a}}{m(n^2 - m^2)(m^2 + n^2)} + \sum_{m=2,4,\ldots}^{\infty} \sum_{n=1,3,\ldots}^{\infty} \frac{m \sin \frac{m\pi x}{a} \sin \frac{n\pi y}{a}}{n(m^2 - n^2)(m^2 + n^2)} \tag{5.5.5}$$

are introduced for simplicity. Let the Reddy plate deflection according to the TSDT be expressed as

$$w^R = A\phi \tag{5.5.6}$$

Since the Reddy plate deflection according to the TSDT is related to the Kirchhoff plate deflection as described by Eq. (5.2.11), substitution of Eq. (5.5.6) and

$$\Delta w^R = -A\frac{\pi^2}{a^2}\psi \tag{5.5.7}$$

into Eq. (5.2.11) leads to

$$A\left\{ \phi + \frac{\pi^2 h^2}{420a^2(1 - v)} \psi \right\} = \frac{16qa^4}{\pi^6 D} \left\{ \phi + \frac{17\pi^2 h^2}{84a^2(1 - v)} \psi \right\} \tag{5.5.8}$$

so that Eq. (5.5.6) becomes

$$w^R = \frac{16qa^4\phi}{\pi^6 D}\frac{1+\frac{17\pi^2h^2}{84a^2(1-\upsilon)}\frac{\psi}{\phi}}{1+\frac{\pi^2h^2}{420a^2(1-\upsilon)}\frac{\psi}{\phi}} \tag{5.5.9}$$

To obtain the shear correction factor of the Mindlin plate based on the Reddy plate deflection, we let $w^R = w^M$. As a result,

$$\frac{1}{6\kappa} + \frac{\pi^2h^2}{2520\kappa a^2(1-\upsilon)}\frac{\psi}{\phi} = \frac{1}{5} \tag{5.5.10}$$

The usually assumed shear correction factor of $\kappa = 5/6$ in the FSDT is therefore recovered from Eq. (5.5.10) if the highest order term is neglected. Had the term containing Δw^R in Eq. (5.2.11) been neglected, one would obtain a simplified TSDT which gives the approximate Reddy plate deflection

$$w^R \approx w^K - \frac{17h^2}{84(1-\upsilon)}\Delta w^K = \frac{16qa^4}{\pi^6 D}\left\{\phi + \frac{17\pi^2h^2}{84a^2(1-\upsilon)}\psi\right\} \tag{5.5.11}$$

Comparing this with the Mindlin plate deflection described in Eq. (5.5.3) would give $\kappa = 14/17$. Suppose the highest order term in Eq. (5.5.10) is taken into account, we have the exact shear correction factor

$$\kappa = \frac{5}{6}\left(1 + \frac{\pi^2h^2}{420a^2(1-\upsilon)}\frac{\psi}{\phi}\right) \tag{5.5.12}$$

where ϕ and ψ are described by Eqs. (5.5.4) and (5.5.5), respectively. It is of practical importance to evaluate the shear correction factor where the deflection is significant. Therefore, substituting $x = y = a/3$ into Eq. (5.5.12) for the plate centroid, we have

$$\kappa = \frac{5}{6}\left(1 + \frac{\pi^2h^2}{420a^2(1-\upsilon)}\frac{\sum_{m=1,3,\ldots}^{\infty}\sum_{n=2,4,\ldots}^{\infty}\frac{n\sin\frac{m\pi}{3}\sin\frac{n\pi}{3}}{m(n^2-m^2)(m^2+n^2)}}{\sum_{m=1,3,\ldots}^{\infty}\sum_{n=2,4,\ldots}^{\infty}\frac{n\sin\frac{m\pi}{3}\sin\frac{n\pi}{3}}{m(n^2-m^2)(m^2+n^2)^2}}\right) \tag{5.5.13}$$

The shear correction factor at the plate centroid can be obtained by solving the two double summation series in Eq. (5.5.13). The double summation series are evaluated as

$$\sum_{m=1,3,\ldots}^{\infty}\sum_{n=2,4,\ldots}^{\infty}\frac{n\sin\frac{m\pi}{3}\sin\frac{n\pi}{3}}{m(n^2-m^2)(m^2+n^2)} = 0.08817 \tag{5.5.14}$$

and

$$\sum_{m=1,3,\ldots}^{\infty}\sum_{n=2,4,\ldots}^{\infty}\frac{n\sin\frac{m\pi}{3}\sin\frac{n\pi}{3}}{m(n^2-m^2)(m^2+n^2)^2} = 0.019303 \tag{5.5.15}$$

using (m, n) values up to (59, 60) and (29, 30), respectively. The (m, n) values for evaluating Eq. (5.5.14) are higher than those of Eq. (5.5.15) due to a more rapid convergence for the latter, as shown in Fig. 5.12. Substituting Eqs. (5.5.14) and (5.5.15) into Eq. (5.5.13) gives (Lim 2016c)

$$\kappa = \frac{5}{6}\left\{1 + \frac{0.10734}{1-v}\left(\frac{h}{a}\right)^2\right\} \tag{5.5.16}$$

To compare the shear corrections factors $\kappa = 5/6$, $\kappa = 14/17$ and Eq. (5.5.16), curves of these shear correction factors are plotted in Fig. 5.13 using the commonly adopted Poisson's ratio of $v = 0.3$. The shear correction factor of $\kappa = 14/17$ is the least accurate, as truncation of an entire term containing Δw^R in Eq. (5.2.11) was performed prematurely. The shear correction factor of $\kappa = 5/6$ is more accurate, as truncation of the term containing ψ/ϕ is performed at the final stage described by Eq. (5.5.12). Similar to $\kappa = 14/17$, the more accurate shear correction factor of $\kappa = 5/6$ underestimates the actual shear correction factor. Unlike $\kappa = 14/17$, the shear correction factor $\kappa = 5/6$ coincides with the exact shear correction factor as the plate thickness diminishes. In practical applications, the actual shear correction factor exceeds the commonly adopted shear correction factor of $\kappa = 5/6$ due to the relative thickness h/a and the Poisson's ratio v of the plate.

Having established the validity of the obtained refined shear correction factor by comparison against commonly used shear correction factors, it is now of interest to observe the effect from relative thickness and Poisson's ratio of the plate. A graphical representation on the combined effect of relative plate thickness and Poisson's ratio of the plate material is illustrated in Fig. 5.14a, with the case of $\kappa = 5/6$ being incorporated for comparison purposes. When plotted differently, one can observe the conventional and auxetic regions, defined by $-1 \leq v < 0$ and $0 \leq v \leq 1/2$, respectively, as furnished in Fig. 5.14b.

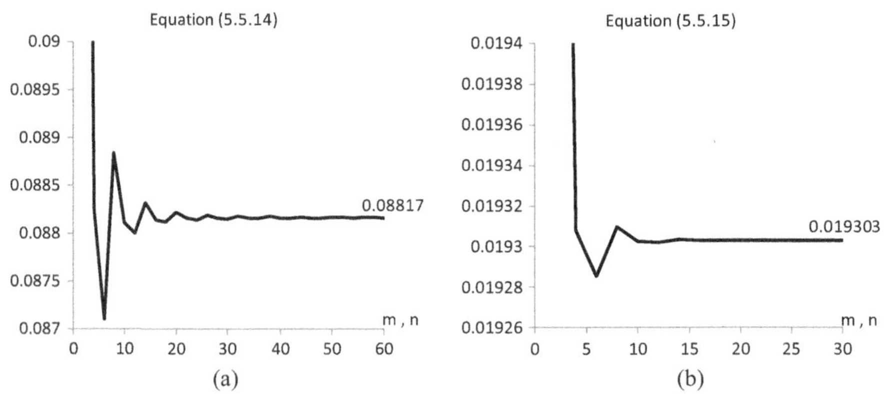

Fig. 5.12 Plots of **a** Eq. (5.5.14) up to $m = 59$, $n = 60$ and **b** Eq. (5.5.15) up to $m = 29$, $n = 30$

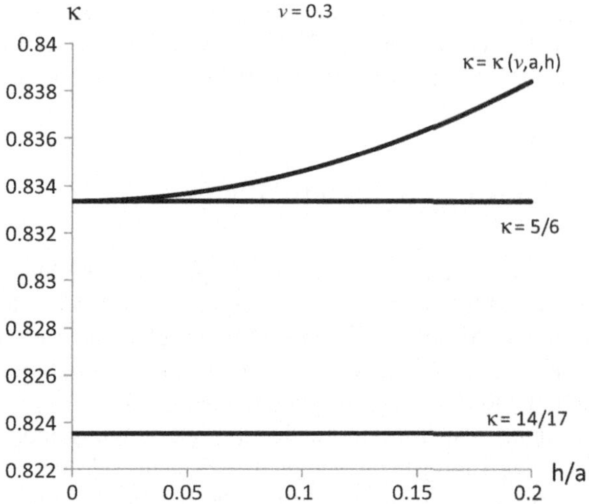

Fig. 5.13 Graphical comparison of Eq. (5.5.16), $\kappa = 5/6$, and $\kappa = 14/17$

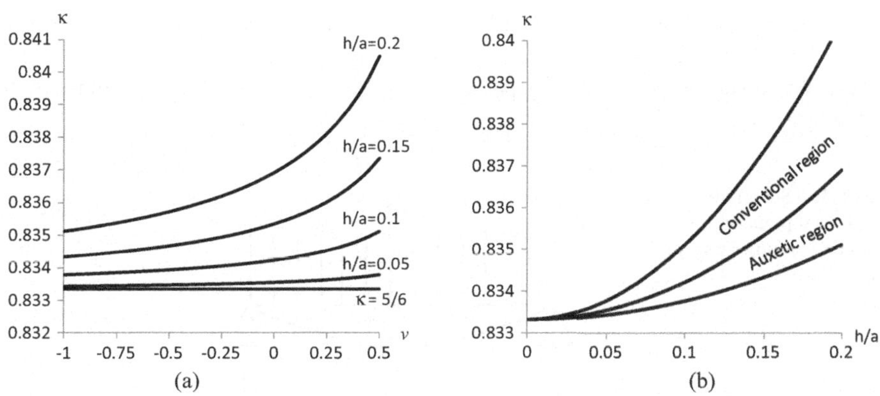

Fig. 5.14 A family of refined shear correction factor curves, **a** plotted against the Poisson's ratio for various relative plate thicknesses, and **b** plotted against the relative plate thickness with clear demarcation of conventional and auxetic regions

For plates that are very thin or whose Poisson's ratio is very negative the shear correction factor approaches 5/6, which is the commonly adopted shear correction factor; for very thick plates or for plates with very large Poisson's ratio, the use of the variable shear correction factor is more accurate. The developed refined shear correction factor also provides an avenue for the use of the constant shear correction factor of 5/6 to be justified. By way of example, the term $(h/a)^2/(1 - v)$ in Eq. (5.5.16) reveals that the shear correction factor of a conventional plate of lesser thickness $h/a = 0.1$ and $v = 0.5$ is equal to the shear correction factor of an auxetic plate of greater thickness $h/a = 0.2$ and $\upsilon = -1$. In other words, a geometrically thick plate is a mechanically thin plate if the plate's Poisson's ratio is sufficiently negative to suppress the shear deformation in favor of bending deformation.

5.6 Example: Hexagonal Auxetic Plates

Figure 5.15 shows schematics of a hexagonal plate of inradius a with simply supported boundary condition and uniform load q. With reference to Fig. 5.15, the deflections according to Mindlin w^M and Reddy w^R plate theories are greater than the deflection of Kirchhoff w^K plate theory due to the incorporation of transverse shear deformation in addition to deflection due to bending.

By method of point matching, the deflection and bending moment results for regular polygonal plates have been obtained with clamped edge (Conway 1960) and simply supported edge (Conway 1961). The bending moments at the center of a simply supported regular hexagonal plate under uniform load, q, have been established by Conway (1961) as

$$(M_x)_{x=y=0} = \left(M_y\right)_{x=y=0} = 0.17578qa^2 \tag{5.6.1}$$

where the distance between two parallel sides is $2a$, as well as the maximum deflection

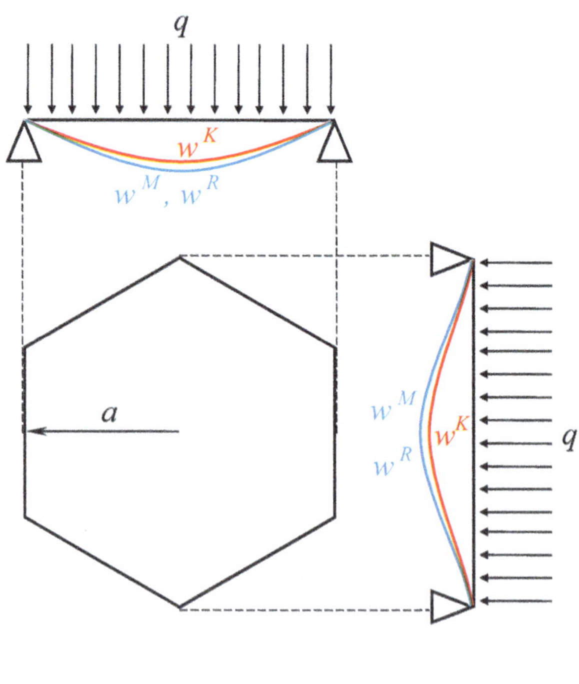

Fig. 5.15 Schematics of hexagonal plate with boundary conditions and load

$$w_{x=y=0} = 0.054788\frac{qa^4}{D} \tag{5.6.2}$$

for Poisson's ratio of $v = 0.3$, where D is defined in Eq. (4.1.3) for a plate of Young's modulus E and thickness h. A very convenient deflection model has been proposed by Lim (2017a) for a hexagonal Kirchhoff plate in the following form

$$w^K = 0.054788\frac{qa^4}{D}\cos\left(\frac{\pi x}{2a}\right)\cos\left(\frac{\pi y}{2a}\right) \tag{5.6.3}$$

whereby substitution of

$$\frac{\partial^2 w^K}{\partial x^2} = \frac{\partial^2 w^K}{\partial y^2} = -\frac{\pi^2}{4a^2}0.054788\frac{qa^4}{D}\cos\left(\frac{\pi x}{2a}\right)\cos\left(\frac{\pi y}{2a}\right) \tag{5.6.4}$$

into Eqs. (4.1.1) and (4.1.2) gives

$$M_x^K = M_y^K = 0.135184(1+v)qa^2\cos\left(\frac{\pi x}{2a}\right)\cos\left(\frac{\pi y}{2a}\right) \tag{5.6.5}$$

The maximum deflection of $w^K_{x=y=0} = 0.054788qa^4/D$ according to Eq. (5.6.3) is exactly the same as that obtained by Conway (1961). Using $v = 0.3$, we have $\left(M_x^K\right)_{x=y=0} = \left(M_y^K\right)_{x=y=0} = 0.17574qa^2$, which has an error of only -0.02% compared with Conway's (1961) result for simply supported hexagonal plate. The proposed form of deflection function is selected because it facilitates convenient execution of FSDT and TSDT for thick plates. The use of circular plates for approximating the hexagonal plate is less appropriate. This is because if we approximate the hexagonal plate by a circular plate, the deflection and bending moment at the plate center are $w_{\max} = 0.063702qa^4/D$ and $M_{\max} = 0.20625qa^2$, respectively, for $v = 0.3$. Even if the circular plate deflection is modified by a factor of 0.86 so as to exactly match its maximum deflection with that by Conway (1961), the corresponding maximum bending moment is $0.17739qa^2$, which deviates by 0.9%. See Table 5.4.

Table 5.4 Deflections and moments at center of a uniformly loaded, simply supported hexagonal plate with $v = 0.3$

	Deflection at plate center, with % error w.r.t. Conway (1961)	Moment at plate center, with % error w.r.t. Conway (1961)
Conway (1961)	$0.054788qa^4/D$	$0.17578qa^2$
Equations (5.6.3) and (5.6.5)	$0.054788qa^4/D$ (error = 0%)	$0.17574qa^2$ (error = − 0.02%)
Circ. plate approx.	$0.063702qa^4/D$ (error = 16%)	$0.20625qa^2$ (error = 17%)
Modified circ. plate	$0.054788qa^4/D$ (error = 0%)	$0.17739qa^2$ (error = 0.9%)

Having established the validity of Eq. (5.6.3) for quantifying both the deflection and bending moment at the plate center vis-à-vis the results by Conway (1961), we shall now proceed to investigate the deflection of thick hexagonal plates considering shear deformation for both conventional and auxetic plate material. In determining the shear deformation characteristics on hexagonal plates, recall that the Mindlin plate deflection w^M is related to the Kirchhoff plate deflection w^K for simply supported edges as furnished in Eq. (5.2.1) while the Reddy plate deflection w^R is related to the Kirchhoff plate deflection as described by Eq. (5.2.4). To pave a way for expressing the FSDT-CPT and TSDT-CPT relationship in a tractable manner, recall that Eqs. (5.2.1) and (5.2.4) can be expressed as Eqs. (5.2.10) and (5.2.11), respectively. Substituting Eq. (5.6.3) into Eq. (5.2.10) gives the hexagonal Mindlin plate deflection

$$w^M = \left[1 + \frac{\pi^2 h^2/a^2}{12\kappa(1-v)}\right] 0.054788 \frac{qa^4}{D} \cos\left(\frac{\pi x}{2a}\right)\cos\left(\frac{\pi y}{2a}\right) \tag{5.6.6}$$

The hexagonal Reddy plate deflection can be obtained by iteration from Eq. (5.2.11), wherein the terms w^K and Δw^K are from Eq. (5.6.6). For the first iteration, we set $\Delta w^R = 0$ to Eq. (5.2.11) to get

$$w^R = \left[1 + \frac{17\pi^2 h^2/a^2}{168(1-v)}\right] 0.054788 \frac{qa^4}{D} \cos\left(\frac{\pi x}{2a}\right)\cos\left(\frac{\pi y}{2a}\right) \tag{5.6.7}$$

In the second iteration, we obtain from Eq. (5.6.7)

$$\Delta w^R = -\frac{\pi^2}{2a^2}\left[1 + \frac{17\pi^2 h^2/a^2}{168(1-v)}\right] 0.054788 \frac{qa^4}{D} \cos\left(\frac{\pi x}{2a}\right)\cos\left(\frac{\pi y}{2a}\right) \tag{5.6.8}$$

for substitution into Eq. (5.2.11) to give

$$w^R = \left[1 + \frac{\pi^2 h^2/a^2}{10(1-v)} - \frac{17\pi^4 h^4/a^4}{141120(1-v)^2}\right] 0.054788 \frac{qa^4}{D} \cos\left(\frac{\pi x}{2a}\right)\cos\left(\frac{\pi y}{2a}\right) \tag{5.6.9}$$

Similarly for the third iteration we obtain from Eq. (5.6.9)

$$\Delta w^R = -\frac{\pi^2}{2a^2}\left[1 + \frac{\pi^2 h^2/a^2}{10(1-v)} - \frac{17\pi^4 h^4/a^4}{141,120(1-v)^2}\right]$$
$$0.054788 \frac{qa^4}{D} \cos\left(\frac{\pi x}{2a}\right)\cos\left(\frac{\pi y}{2a}\right) \tag{5.6.10}$$

which, upon substitution into Eq. (5.2.11), leads to

$$w^R = \left[1 + \frac{\pi^2 h^2/a^2}{10(1-v)} - \frac{\pi^4 h^4/a^4}{8400(1-v)^2} + \frac{17\pi^6 h^6/a^6}{118,540,800(1-v)^3}\right]$$
$$0.054788\frac{qa^4}{D}\cos\left(\frac{\pi x}{2a}\right)\cos\left(\frac{\pi y}{2a}\right) \quad (5.6.11)$$

Likewise for the fourth iteration, we have

$$\Delta w^R = -\frac{\pi^2}{2a^2}\left[1 + \frac{\pi^2 h^2/a^2}{10(1-v)} - \frac{\pi^4 h^4/a^4}{8400(1-v)^2} + \frac{17\pi^6 h^6/a^6}{118,540,800(1-v)^3}\right]$$
$$0.054788\frac{qa^4}{D}\cos\left(\frac{\pi x}{2a}\right)\cos\left(\frac{\pi y}{2a}\right) \quad (5.6.12)$$

from Eq. (5.6.11) for substitution into Eq. (5.2.11) to yield

$$w^R = \left[1 + \frac{\pi^2 h^2/a^2}{10(1-v)} - \frac{\pi^4 h^4/a^4}{8400(1-v)^2} + \frac{\pi^6 h^6/a^6}{7,056,000(1-v)^3} - \frac{17\pi^8 h^8/a^8}{99,574,272,000(1-v)^4}\right]$$
$$0.054788\frac{qa^4}{D}\cos\left(\frac{\pi x}{2a}\right)\cos\left(\frac{\pi y}{2a}\right) \quad (5.6.13)$$

Performing similarly for the fifth iteration, we obtain from Eq. (5.6.13) the following

$$\Delta w^R = -\frac{\pi^2}{2a^2}\left[1 + \frac{\pi^2 h^2/a^2}{10(1-v)} - \frac{\pi^4 h^4/a^4}{8400(1-v)^2} + \frac{\pi^6 h^6/a^6}{7,056,000(1-v)^3}\right.$$
$$\left. - \frac{17\pi^8 h^8/a^8}{99,574,272,000(1-v)^4}\right] 0.054788\frac{qa^4}{D}\cos\left(\frac{\pi x}{2a}\right)\cos\left(\frac{\pi y}{2a}\right) \quad (5.6.14)$$

and substituting it into Eq. (5.2.11) gives

$$w^R$$
$$= \left[1 + \frac{\pi^2 h^2/a^2}{10(1-v)} - \frac{\pi^4 h^4/a^4}{8400(1-v)^2} + \frac{\pi^6 h^6/a^6}{7,056,000(1-v)^3}\right.$$
$$\left. - \frac{\pi^8 h^8/a^8}{5,927,040,000(1-v)^4} + \frac{17\pi^{10} h^{10}/a^{10}}{83,642,388,480,000(1-v)^5}\right]$$
$$0.054788\frac{qa^4}{D}\cos\left(\frac{\pi x}{2a}\right)\cos\left(\frac{\pi y}{2a}\right) \quad (5.6.15)$$

The terms of w^R contained within [] parenthesis of Eqs. (5.6.7), (5.6.9), (5.6.11), (5.6.13), and (5.6.15) are summarized in Table 5.5.

It can now be observed from these five iterations that a total of $n + 1$ terms are generated within the [...] parenthesis of the w^R expression at the nth iteration, with the

Table 5.5 Terms of w^R contained within [] parenthesis based on n number of iterations

n	Terms of w^R contained within the [] parenthesis.
1	$1+\frac{17\pi^2h^2/a^2}{168(1-v)}$
2	$1+\frac{\pi^2h^2/a^2}{10(1-v)}-\frac{17\pi^4h^4/a^4}{141{,}120(1-v)^2}$
3	$1+\frac{\pi^2h^2/a^2}{10(1-v)}-\frac{\pi^4h^4/a^4}{8400(1-v)^2}+\frac{17\pi^6h^6/a^6}{118{,}540{,}800(1-v)^3}$
4	$1+\frac{\pi^2h^2/a^2}{10(1-v)}-\frac{\pi^4h^4/a^4}{8400(1-v)^2}+\frac{\pi^6h^6/a^6}{7{,}056{,}000(1-v)^3}-\frac{17\pi^8h^8/a^8}{99{,}574{,}272{,}000(1-v)^4}$
5	$1+\frac{\pi^2h^2/a^2}{10(1-v)}-\frac{\pi^4h^4/a^4}{8400(1-v)^2}+\frac{\pi^6h^6/a^6}{7{,}056{,}000(1-v)^3}-\frac{\pi^8h^8/a^8}{5{,}927{,}040{,}000(1-v)^4}+\frac{17\pi^{10}h^{10}/a^{10}}{83{,}642{,}388{,}480{,}000(1-v)^5}$

first term being unity. The first n terms are exact because they no longer change with increasing iteration, while the $(n+1)$th term rapidly diminishes with increasing number of iterations. With reference to Table 5.5 for n number of iterations, the general expression of w^R can be written as

$$w^R=\left[1-\frac{1}{10}\sum_{m=1}^{n-1}(-1)^m\frac{\pi^{2m}h^{2m}/a^{2m}}{840^{m-1}(1-v)^m}-(-1)^n\frac{17\pi^{2n}h^{2n}/a^{2n}}{168(840)^{n-1}(1-v)^n}\right]w^K \tag{5.6.16}$$

The shear correction factor can be extracted by comparing terms of w^M and w^R within the [...] parenthesis and, in the case of the first iteration, we have $\kappa=14/17$. Likewise, comparing terms based on the second and third iterations suggests that

$$\kappa=\frac{5}{6}\left(1-\frac{17\pi^2h^2/a^2}{14112(1-v)}\right)^{-1} \tag{5.6.17}$$

and

$$\kappa=\frac{5}{6}\left(1-\frac{\pi^2h^2/a^2}{840(1-v)}+\frac{17\pi^4h^4/a^4}{11854080(1-v)^2}\right)^{-1} \tag{5.6.18}$$

respectively. Performing similarly, we have

$$\kappa=\frac{5}{6}\left(1-\frac{\pi^2h^2/a^2}{840(1-v)}+\frac{\pi^4h^4/a^4}{705,600(1-v)^2}-\frac{17\pi^6h^6/a^6}{9,957,427,200(1-v)^3}\right)^{-1} \tag{5.6.19}$$

and

$$\kappa\frac{5}{6}\left(1-\frac{\pi^2h^2/a^2}{840(1-v)}+\frac{\pi^4h^4/a^4}{705,600(1-v)^2}\right.$$

$$- \frac{\pi^6 h^6 / a^6}{592,704,000(1-\upsilon)^3} + \frac{17\pi^8 h^8 / a^8}{8,364,238,848,000(1-\upsilon)^4} \Bigg)^{-1} \quad (5.6.20)$$

based on the fourth and fifth iterations, respectively. In other words, the general shear correction factor up to the nth iteration is

$$\kappa = \frac{5}{6} \left(\sum_{m=0}^{n-2} (-1)^m \frac{\pi^{2m} h^{2m} / a^{2m}}{840^m (1-\upsilon)^m} + (-1)^{n-1} \frac{17\pi^{2(n-1)} h^{2(n-1)} / a^{2(n-1)}}{16.8(840)^{n-1} (1-\upsilon)^{n-1}} \right)^{-1} \quad (5.6.21)$$

for $n \geq 2$. Since neither Eqs. (5.6.16) nor (5.6.21) are exact except when $n \to \infty$, therefore exact expressions for them are best written in the forms

$$w^R = \left[1 - \frac{1}{10} \sum_{m=1}^{\infty} (-1)^m \frac{\pi^{2m} h^{2m} / a^{2m}}{840^{m-1} (1-\upsilon)^m} \right] w^K \quad (5.6.22)$$

and

$$\kappa = \frac{5}{6} \left(\sum_{m=0}^{\infty} (-1)^m \frac{\pi^{2m} h^{2m} / a^{2m}}{840^m (1-\upsilon)^m} \right)^{-1} \quad (5.6.23)$$

respectively.

To assess the number of iterations required for providing accurate central deflection and shear correction factor, the ratio of w^R / w^K and κ, in Eqs. (5.6.22) and (5.6.23), respectively, have been computed using Poisson's ratio of ± 0.3 for various h/a ratio up to $m = 5$ number of terms. With reference to Table 5.6, the w^R / w^K ratio is accurate up to six decimal points with $m = 2$. For the case of the shear correction factor, accuracy of six decimal points is achieved with $m = 1$, as furnished in Table 5.7.

Table 5.6 w^R / w^K ratio of hexagonal plates at $\upsilon = \pm 0.3$ for various h/a ratio

m	$\upsilon = 0.3$				$\upsilon = -0.3$			
	$h/a =$ 0.20	$h/a =$ 0.15	$h/a =$ 0.10	$h/a =$ 0.05	$h/a =$ 0.20	$h/a =$ 0.15	$h/a =$ 0.10	$h/a =$ 0.05
0	1.000000	1.000000	1.000000	1.000000	1.000000	1.000000	1.000000	1.000000
1	1.056398	1.031724	1.014099	1.003525	1.030368	1.017082	1.007592	1.001898
2	1.056360	1.031712	1.014097	1.003525	1.030357	1.017079	1.007591	1.001898
3	1.056360	1.031712	1.014097	1.003525	1.030357	1.017079	1.007591	1.001898
4	1.056360	1.031712	1.014097	1.003525	1.030357	1.017079	1.007591	1.001898
5	1.056360	1.031712	1.014097	1.003525	1.030357	1.017079	1.007591	1.001898

Table 5.7 Shear correction factor of hexagonal plates at $\upsilon = \pm 0.3$ for various h/a ratio

	$\upsilon = 0.3$				$\upsilon = -0.3$			
m	$h/a =$ 0.20	$h/a =$ 0.15	$h/a =$ 0.10	$h/a =$ 0.05	$h/a =$ 0.20	$h/a =$ 0.15	$h/a =$ 0.10	$h/a =$ 0.05
0	0.833333	0.833333	0.833333	0.833333	0.833333	0.833333	0.833333	0.833333
1	0.833893	0.833648	0.833473	0.833368	0.833635	0.833503	0.833409	0.833352
2	0.833893	0.833648	0.833473	0.833368	0.833635	0.833503	0.833409	0.833352
3	0.833893	0.833648	0.833473	0.833368	0.833635	0.833503	0.833409	0.833352
4	0.833893	0.833648	0.833473	0.833368	0.833635	0.833503	0.833409	0.833352
5	0.833893	0.833648	0.833473	0.833368	0.833635	0.833503	0.833409	0.833352

The rapid convergence therefore suggests that the plate deflection and shear correction factor can be simplified as (Lim 2017a)

$$w^R = \left[1 + \frac{\pi^2}{10(1-\upsilon)} \left(\frac{h}{a} \right)^2 - \frac{\pi^4}{8400(1-\upsilon)^2} \left(\frac{h}{a} \right)^4 \right] w^K \tag{5.6.24}$$

and

$$\kappa = \frac{5}{6} \left(1 - \frac{\pi^2}{840(1-\upsilon)} \left(\frac{h}{a} \right)^2 \right)^{-1} \tag{5.6.25}$$

respectively, while maintaining very good accuracy. These are used for plotting Fig. 5.16.

The conventional and auxetic regions, separated by $\upsilon = 0$, can be compared in Fig. 5.16 (top) by means of the w^R/w^K versus h/a plane (left) and the κ versus h/a plane (right). In spite of the wider range of Poisson's ratio in the auxetic region $-1 \leq \upsilon < 0$ in comparison to the conventional region $0 \leq \upsilon \leq 1/2$, these plots reveal that the auxetic regions within the context of w^R/w^K versus h/a and κ versus h/a planes are narrower than the conventional region. This narrower region is attributed to a more gradual change in w^R/w^K and κ at a more negative Poisson's ratio, as shown in Fig. 5.16 (bottom). Reference to Eqs. (5.6.24) and (5.6.25) also suggests that the extent of shear deformation vis-à-vis the bending deformation is reduced not only by the decreasing plate thickness according to current solid mechanics practice, but also by using auxetic materials for the plates. In the case of hexagonal plates considered herein, the ratio w^R/w^K and shear correction factor are maintained with increasing plate thickness if the plate Poisson's ratio decreases sufficiently.

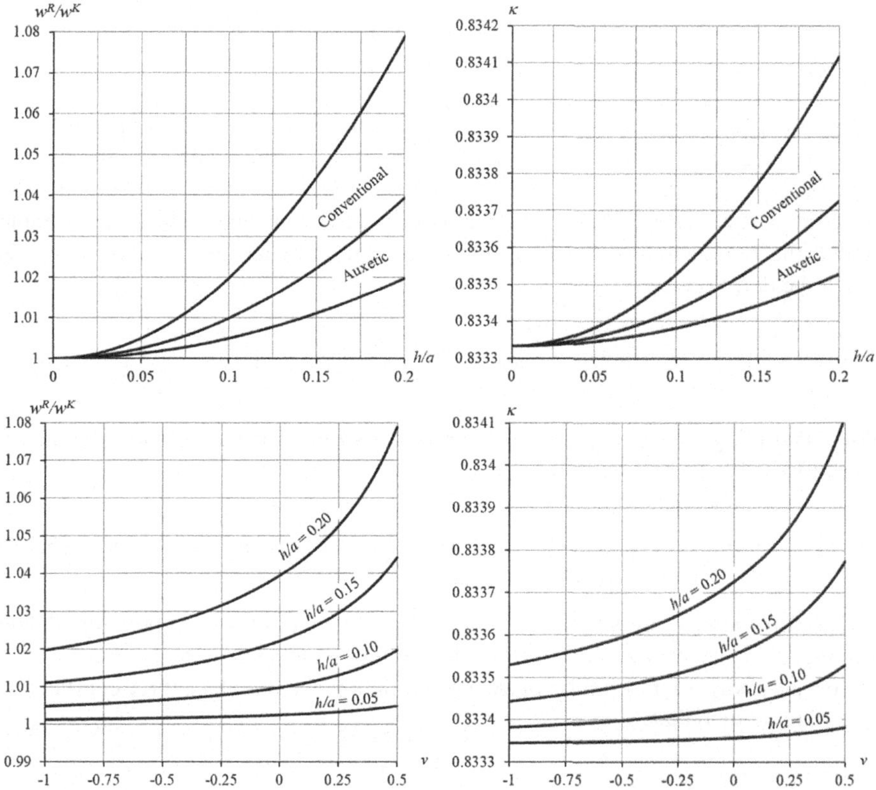

Fig. 5.16 Curves of Reddy-to-Kirchhoff deflection ratio (left column) and shear correction factor (right column) plotted against the relative plate thickness (top row) and Poisson's ratio of plate material (bottom row)

5.7 Example: Regular Polygonal Auxetic Plates

The purposes of this section are twofold. The first objective is to develop a set of shear correction factors using known CPT models for regular polygonal plates (i.e., of sides $n = 3,\ 4,\ 5,\ 6,\ 7,\ 8,\ 9,\ 10,\ 15$), and FSDT-TSDT relationship. The second objective is to develop a semi-empirical model for the shear correction factors so as to facilitate the shear correction factor prediction of other regular polygonal plates (e.g., $n = 11,\ 12,\ 13$, etc.) and to show how the shear correction factor varies with the number of plate sides.

A convenient CPT model is developed as follows. For a simply supported regular polygonal plate of n sides with thickness h, Young's modulus E, and Poisson's ratio ν, the Kirchhoff deflection model under uniform load q has been given by Leissa et al. (1965) as

$$w^K = \sum_{m=0}^{M} \left(A_m r^m + B_m r^{m+2} \right) \cos m\theta + \frac{qr^4}{64D} \quad (5.7.1)$$

where A_m and B_m are determined from boundary conditions, the upper limit M is determined from by the accuracy of solution desired from the point-matching process, D is the plate flexural rigidity, and θ is the angle measured from an axis bisecting a side. To facilitate the development of a convenient shear deformable model, the regular polygonal plate model is approximated by circular plates, in which the deformation is

$$w^K = \frac{q\left(a^2 - r^2\right)}{64D} \left(\frac{5+v}{1+v} a^2 - r^2 \right) \quad (5.7.2)$$

based on solving the circular plate differential equation using the following boundary conditions

$$\left(\frac{dw^K}{dr} \right)_{r=0} = (M_r)_{r=a} = w^K_{r=a} = 0 \quad (5.7.3)$$

The dimension a refers to the radius of the inscribed circle on the polygonal plate (see Fig. 5.17), as well as the radius of the circular plate as $n \to \infty$.

To pave a way for the circular plate approximation, we introduce three coefficients c_0, c_2 and c_4 such that

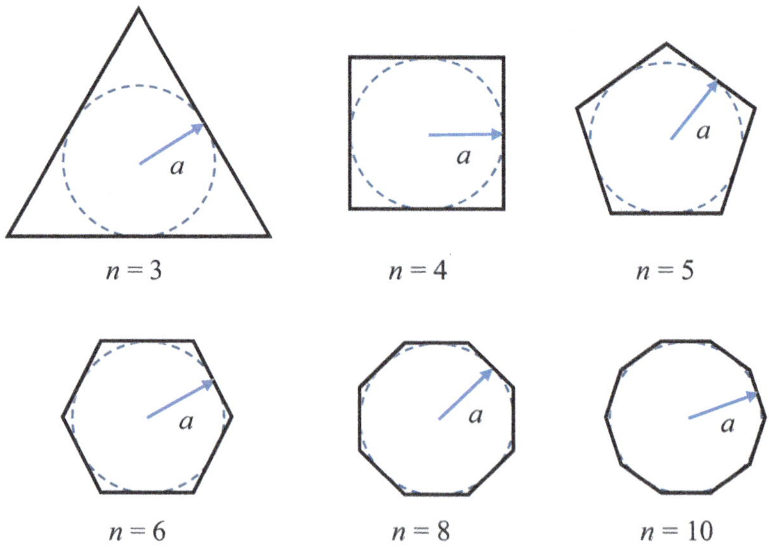

Fig. 5.17 A sample of regular polygonal plates, illustrating the definition of a

$$w^K = \frac{q}{64D}\left(c_0\frac{5+\upsilon}{1+\upsilon}a^4 - 2c_2\frac{3+\upsilon}{1+\upsilon}a^2r^2 + c_4r^4\right) \tag{5.7.4}$$

It can be seen that for circular plates (i.e., $n \rightarrow \infty$), Eq. (5.7.2) is recovered by substituting $c_0 = c_2 = c_4 = 1$ into Eq. (5.7.4). For regular polygonal plates, these coefficients need to be calculated from the boundary conditions

$$w^K_{r=0}\ ,\ \ (M_r)_{r=0}\ ,\ \ \left(\frac{dw^K}{dr}\right)_{(r,\theta)=(a,0)} \tag{5.7.5}$$

furnished by Leissa et al. (1965). See Table 5.8. To view the deflection profiles of the various regular polygonal plates based on the circular plate approximation, a family of dimensionless deflection profiles is plotted in Fig. 5.18.

To obtain the CPT-TSDT relationship, we begin with Eq. (5.2.11) where

$$\Delta w = \frac{\partial^2 w}{\partial r^2} + \frac{1}{r}\frac{\partial w}{\partial r} \tag{5.7.6}$$

To establish the TSDT deflection model, we begin by setting $\Delta w^R = 0$ in Eq. (5.2.11) to give

$$\begin{aligned} \frac{64D}{qa^4}w^R \sim & c_0\frac{5+\upsilon}{1+\upsilon} - 2c_2\frac{3+\upsilon}{1+\upsilon}\frac{r^2}{a^2} + c_4\frac{r^4}{a^4} \\ & + \frac{17}{84}\frac{(h/a)^2}{1-\upsilon}\left[8c_2\frac{3+\upsilon}{1+\upsilon} - 16c_4\frac{r^2}{a^2}\right] \end{aligned} \tag{5.7.7}$$

Table 5.8 Dimensionless central deflections and moments, as well as dimensionless side slopes by Leissa et al. (1965), with corresponding calculated coefficients c_0, c_2 and c_4 based on Eqs. (5.7.4) and (5.7.5)

n	$\frac{D}{qa^4}w^K_{r=0}$	$\frac{(M_r)_{r=0}}{qa^2}$	$\frac{D}{qa^3}\left(\frac{dw^K}{dr}\right)_{(r,\theta)=(a,0)}$	c_0	c_2	c_4
3	0.0833	0.217	−0.1406	1.307653	1.052121	0.421169
4	0.0650	0.192	−0.1078	1.020377	0.930909	0.638277
5	0.0581	0.181	−0.0942	0.912060	0.877576	0.720492
6	0.0548	0.176	−0.0871	0.860257	0.853333	0.772554
7	0.0532	0.174	−0.0833	0.835140	0.843636	0.808738
8	0.0525	0.173	−0.0813	0.824151	0.838788	0.828431
9	0.0523	0.173	−0.0803	0.821011	0.838788	0.844431
10	0.0523	0.174	−0.0798	0.821011	0.843636	0.864738
15	0.0537	0.179	−0.0809	0.842989	0.867879	0.908677
∞	0.0637	0.206	−0.0962	0.999970	0.998788	0.996185

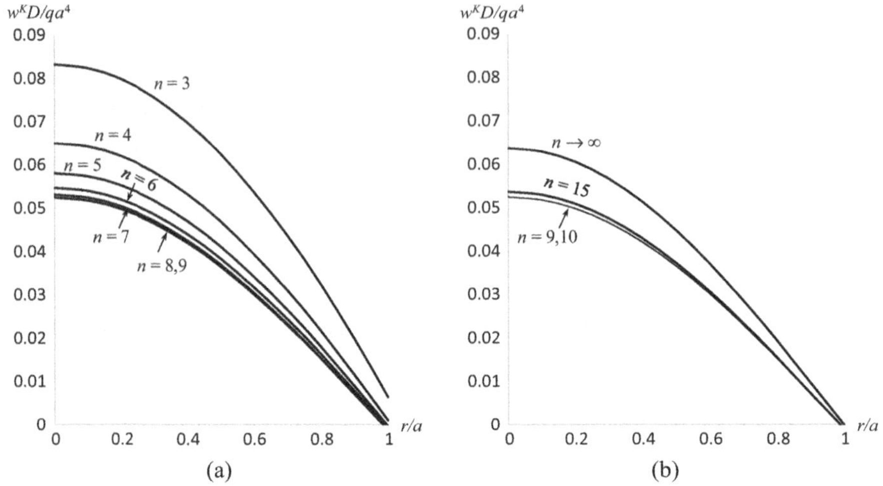

Fig. 5.18 Dimensionless deflection profiles of regular polygonal plates of n sides: **a** $3 \le n \le 9$, and **b** $9 \le n \le \infty$

From this we introduce two coefficients B_1 and B_2 into Eq. (5.7.7) so as to give

$$\frac{64D}{qa^4} w^R = c_0 \frac{5+v}{1+v} - 2c_2 \frac{3+v}{1+v} \frac{r^2}{a^2} + c_4 \frac{r^4}{a^4} + B_1 \frac{(h/a)^2}{1-v} \left[8c_2 \frac{3+v}{1+v} - 16c_4 \frac{r^2}{a^2} - B_2 \frac{(h/a)^2}{1-v} \right] \tag{5.7.8}$$

It can be seen that Eq. (5.7.8) reduces to Eq. (5.7.7) when $B_1 = 17/84$ and $B_2 = 0$. To obtain the exact values of B_1 and B_2 within the framework of the CPT-TSDT relationship, Eqs. (5.7.4) and (5.7.8) are substituted into Eq. (5.2.11) to yield

$$\frac{64D}{qa^4} \left\{ w^K - \frac{17h^2}{84(1-v)} \Delta w^K \right\} = c_0 \frac{5+v}{1+v} - 2c_2 \frac{3+v}{1+v} \frac{r^2}{a^2} + c_4 \frac{r^4}{a^4} + \frac{34}{21} \frac{(h/a)^2}{1-v} \left[c_2 \frac{3+v}{1+v} - 2c_4 \frac{r^2}{a^2} \right] \tag{5.7.9}$$

and

$$\frac{64D}{qa^4} \left\{ w^R - \frac{h^2}{420(1-v)} \Delta w^R \right\} = c_0 \frac{5+v}{1+v} - 2c_2 \frac{3+v}{1+v} \frac{r^2}{a^2}$$

$$+ c_4\frac{r^4}{a^4} + \frac{(h/a)^2}{1-v}$$
$$\left[c_2\frac{3+v}{1+v}\left(8B_1 + \frac{2}{105}\right) - 2c_4\frac{r^2}{a^2}\left(8B_1 + \frac{2}{105}\right)\right.$$
$$\left. - \frac{(h/a)^2}{1-v}B_1\left(B_2 - \frac{16c_4}{105}\right)\right] \tag{5.7.10}$$

on the RHS and LHS of Eq. (5.2.11), respectively. Comparison of terms between Eqs. (5.7.9) and (5.7.10) leads to

$$B_1 = \frac{1}{5}\ ,\quad B_2 = \frac{16}{105}c_4 \tag{5.7.11}$$

Therefore, the TSDT deflection model becomes

$$\frac{64D}{qa^4}w^R = c_0\frac{5+v}{1+v} - 2c_2\frac{3+v}{1+v}\frac{r^2}{a^2} + c_4\frac{r^4}{a^4}$$
$$+ \frac{(h/a)^2}{5(1-v)}\left[8c_2\frac{3+v}{1+v} - 16c_4\frac{r^2}{a^2} - \frac{16c_4}{105}\frac{(h/a)^2}{1-v}\right] \tag{5.7.12}$$

This TSDT deflection model can alternatively be obtained by analytical iteration. From Eq. (5.7.7) we obtain

$$\frac{64D}{qa^4}\Delta w^R \approx -8c_2\frac{3+v}{1+v}\frac{1}{a^2} + 16c_4\frac{r^2}{a^4} - \frac{17}{84}\frac{(h/a)^2}{1-v}\frac{64c_4}{a^2} \tag{5.7.13}$$

which, upon substitution into Eq. (5.2.11), gives rise to

$$\frac{64D}{qa^4}w^R \approx c_0\frac{5+v}{1+v} - 2c_2\frac{3+v}{1+v}\frac{r^2}{a^2} + c_4\frac{r^4}{a^4}$$
$$+ \frac{(h/a)^2}{420(1-v)}\left[672c_2\frac{3+v}{1+v} - 1344c_4\frac{r^2}{a^2} - \frac{17}{84}\frac{(h/a)^2}{1-v}64c_4\right] \tag{5.7.14}$$

From Eq. (5.7.14) we have

$$\frac{64D}{qa^4}\Delta w^R = -8c_2\frac{3+v}{1+v}\frac{1}{a^2} + 16c_4\frac{r^2}{a^4} - \frac{(h/a)^2}{5(1-v)}\frac{64c_4}{a^2} \tag{5.7.15}$$

whereupon substitution into Eq. (5.2.11) yields Eq. (5.7.12). The latter is said to have converged since one obtains Eq. (5.7.15) by calculating Δw^R from Eq. (5.7.12).

To obtain the refined shear correction factor, there is a need to establish the CPT-FSDT relationship. Substituting Eq. (5.7.4) into Eq. (5.2.10) gives

$$\frac{64D}{qa^4}w^M = c_0\frac{5+v}{1+v} - 2c_2\frac{3+v}{1+v}\frac{r^2}{a^2} + c_4\frac{r^4}{a^4} + \frac{(h/a)^2}{6\kappa(1-v)}\left[8c_2\frac{3+v}{1+v} - 16c_4\frac{r^2}{a^2}\right] \quad (5.7.16)$$

which, upon comparing with Eq. (5.7.12), allows the shear correction factor to be extracted as

$$\kappa = \frac{5}{6}\frac{c_2\frac{3+v}{1+v} - 2c_4\frac{r^2}{a^2}}{c_2\frac{3+v}{1+v} - 2c_4\frac{r^2}{a^2} - \frac{2c_4}{105}\frac{(h/a)^2}{1-v}} \quad (5.7.17)$$

The customary adoption of $\kappa = 5/6$ has its merit, for it can be seen that

$$\kappa \approx \frac{5}{6} \leftrightarrow h \ll a \quad (5.7.18)$$

and

$$\lim_{v\to-1} = \frac{5}{6} \quad (5.7.19)$$

From a practical viewpoint, one is interested in the maximum deflection, which takes place at the plate center, wherein the shear correction factor is (Lim 2017b)

$$\kappa = \frac{5}{6}\left[1 - \frac{2c_4}{105c_2}\frac{1+v}{(1-v)(3+v)}\left(\frac{h}{a}\right)^2\right]^{-1} \quad (5.7.20)$$

Figure 5.19 shows the contour plots of shear correction factors at the center of the regular polygonal plates with increments of 5 sides. The Poisson's ratio range adopted for illustration is $-1 \le v \le 1/2$, as this is the range for isotropic solids. In addition to agreeing with previously established finding that geometrically thick plates are mechanically thin plates if the Poisson's ratio are sufficiently negative, Fig. 5.19 also suggests that the shear correction factor increases with increasing polygon sides. However, the increase in the shear correction factor diminishes when the polygon sides are large. Specifically, the shear correction factor increases more significantly from $n = 5$ to $n = 10$, but the increase is negligible from $n = 10$ to $n = 15$. It is therefore sufficient to use the shear correction factor of $n = 10$ for any regular polygonal plates of $n > 10$ if accuracy is not important.

Since most solids possess Poisson's ratio of about 0.3, a contour plot of the shear correction factor at the plate center is plotted in Fig. 5.20 at this value of Poisson's ratio with respect to h/a and n. Reference to Fig. 5.20 indicates that the plate thickness exerts a greater influence on the shear correction factor than the plate shape. An advantage of using the regular polygonal plate vis-à-vis separate plate geometries is that the former permits one to observe the variation of the shear correction factor

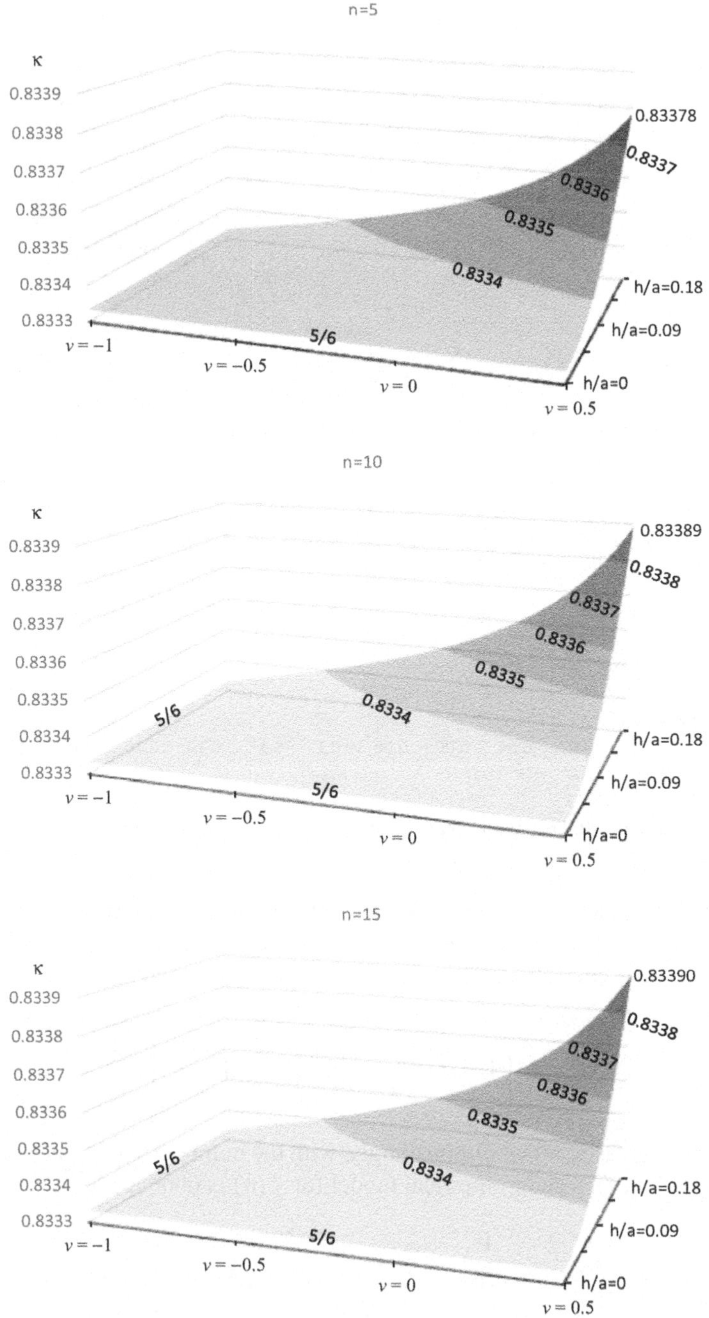

Fig. 5.19 Contour plots of shear correction factor, as functions of Poisson's ratio and plate relative thickness, for pentagonal (top), decagonal (middle), and pentadecagonal (bottom) plates

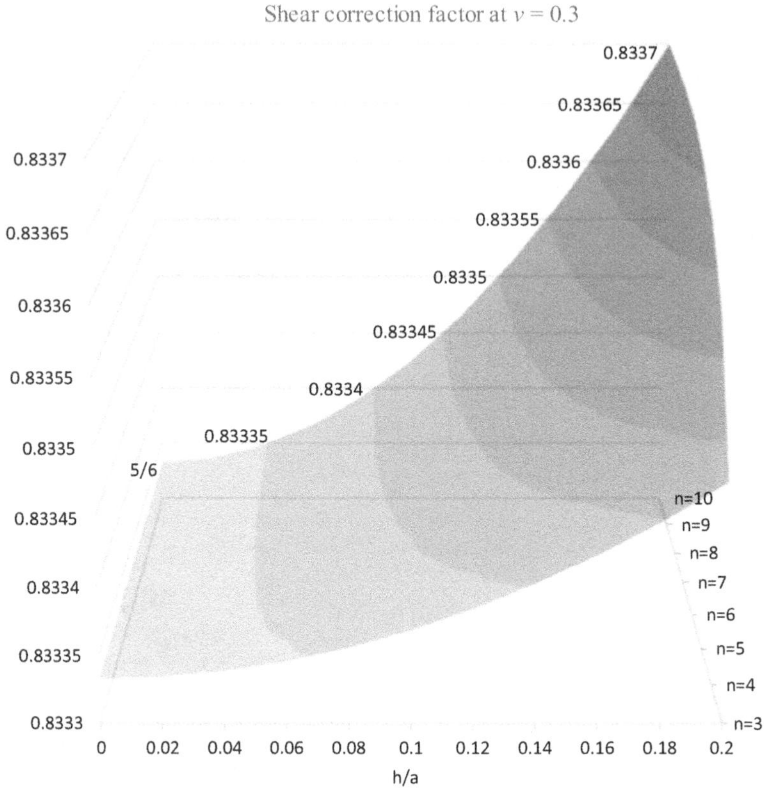

Fig. 5.20 Combined effect of plate shape and thickness on the shear correction factor at $v = 0.3$

with increasing plate sides, thereby allowing one to predict values of κ for cases that have not been considered. With reference to the middle column of Table 5.9, one can see that the shear correction factor can be generally written as

$$\kappa = \frac{5}{6}\left[1 - f(n)\frac{1+v}{(1-v)(3+v)}\left(\frac{h}{a}\right)^2\right]^{-1} \tag{5.7.21}$$

where $f(n)$ increases, albeit diminishingly, with the number of polygon sides. Using the analytical $f(n)$, a semi-empirical model for $f(n)$ is obtained as

$$\frac{1}{f(n)} = 49.4 + \frac{257.572}{(n-1.3)^{2.239}} \tag{5.7.22}$$

with an accuracy indicated by $R^2 = 0.99728$. See Fig. 5.21 for the comparison between the analytical and semi-empirical $f(n)$ values. The semi-empirical model is useful since no thin plate data was furnished by Leissa et al. (1965)

Table 5.9 List of shear correction factors by the analytical method, and those by semi-empirical modeling

Regular polygonal plates	Analytical shear correction factor, κ	Semi-empirical shear correction factor, κ
Triangle ($n = 3$)	$\frac{5}{6}\left[1 - 0.007625\frac{1+v}{(1-v)(3+v)}\left(\frac{h}{a}\right)^2\right]^{-1}$	$\frac{5}{6}\left[1 - 0.007818\frac{1+v}{(1-v)(3+v)}\left(\frac{h}{a}\right)^2\right]^{-1}$
Square ($n = 4$)	$\frac{5}{6}\left[1 - 0.01360\frac{1+v}{(1-v)(3+v)}\left(\frac{h}{a}\right)^2\right]^{-1}$	$\frac{5}{6}\left[1 - 0.012942\frac{1+v}{(1-v)(3+v)}\left(\frac{h}{a}\right)^2\right]^{-1}$
Pentagon ($n = 5$)	$\frac{5}{6}\left[1 - 0.015638\frac{1+v}{(1-v)(3+v)}\left(\frac{h}{a}\right)^2\right]^{-1}$	$\frac{5}{6}\left[1 - 0.015832\frac{1+v}{(1-v)(3+v)}\left(\frac{h}{a}\right)^2\right]^{-1}$
Hexagon ($n = 6$)	$\frac{5}{6}\left[1 - 0.017245\frac{1+v}{(1-v)(3+v)}\left(\frac{h}{a}\right)^2\right]^{-1}$	$\frac{5}{6}\left[1 - 0.017405\frac{1+v}{(1-v)(3+v)}\left(\frac{h}{a}\right)^2\right]^{-1}$
Heptagon ($n = 7$)	$\frac{5}{6}\left[1 - 0.018260\frac{1+v}{(1-v)(3+v)}\left(\frac{h}{a}\right)^2\right]^{-1}$	$\frac{5}{6}\left[1 - 0.018305\frac{1+v}{(1-v)(3+v)}\left(\frac{h}{a}\right)^2\right]^{-1}$
Octagon ($n = 8$)	$\frac{5}{6}\left[1 - 0.018812\frac{1+v}{(1-v)(3+v)}\left(\frac{h}{a}\right)^2\right]^{-1}$	$\frac{5}{6}\left[1 - 0.018853\frac{1+v}{(1-v)(3+v)}\left(\frac{h}{a}\right)^2\right]^{-1}$
Nonagon ($n = 9$)	$\frac{5}{6}\left[1 - 0.019176\frac{1+v}{(1-v)(3+v)}\left(\frac{h}{a}\right)^2\right]^{-1}$	$\frac{5}{6}\left[1 - 0.019206\frac{1+v}{(1-v)(3+v)}\left(\frac{h}{a}\right)^2\right]^{-1}$
Decagon ($n = 10$)	$\frac{5}{6}\left[1 - 0.019524\frac{1+v}{(1-v)(3+v)}\left(\frac{h}{a}\right)^2\right]^{-1}$	$\frac{5}{6}\left[1 - 0.019444\frac{1+v}{(1-v)(3+v)}\left(\frac{h}{a}\right)^2\right]^{-1}$
Hendecagon ($n = 11$)	Not available	$\frac{5}{6}\left[1 - 0.019612\frac{1+v}{(1-v)(3+v)}\left(\frac{h}{a}\right)^2\right]^{-1}$
Dodecagon ($n = 12$)	Not available	$\frac{5}{6}\left[1 - 0.019733\frac{1+v}{(1-v)(3+v)}\left(\frac{h}{a}\right)^2\right]^{-1}$
Tridecagon ($n = 13$)	Not available	$\frac{5}{6}\left[1 - 0.019823\frac{1+v}{(1-v)(3+v)}\left(\frac{h}{a}\right)^2\right]^{-1}$
Tetradecagon ($n = 14$)	Not available	$\frac{5}{6}\left[1 - 0.019893\frac{1+v}{(1-v)(3+v)}\left(\frac{h}{a}\right)^2\right]^{-1}$
Pentadecagon ($n = 15$)	$\frac{5}{6}\left[1 - 0.019943\frac{1+v}{(1-v)(3+v)}\left(\frac{h}{a}\right)^2\right]^{-1}$	$\frac{5}{6}\left[1 - 0.019946\frac{1+v}{(1-v)(3+v)}\left(\frac{h}{a}\right)^2\right]^{-1}$

for $n = 11, 12, 13, 14$, and so the shear correction factors for thick plates of $n = 11, 12, 13, 14$ are not readily available. Comparison between the analytical and semi-empirical models for $f(n)$ in Table 5.9 shows that the semi-empirical model is accurate, the error magnitude being less than 0.25% for $n = 7, 8, 9, 10, 15$, thereby providing confidence for the empirically obtained $f(n)$, and hence κ, for $n = 11, 12, 13, 14$.

By establishing the CPT-TSDT and CPT-FSDT relationships, the subsequently obtained FSDT-TSDT relationship allows the extraction of shear correction factors that reveals its gradual change with increasing number of plate sides. As such, the FSDT description of thick pentagonal, hexagonal, heptagonal, octagonal, and other polygonal plate of high number of sides can be easily computed with the availability of shear correction factors for thick polygonal plates with $n = 3$ to $n = 10$ and $n = 15$ sides. Computed results show that the shear correction factor increases with

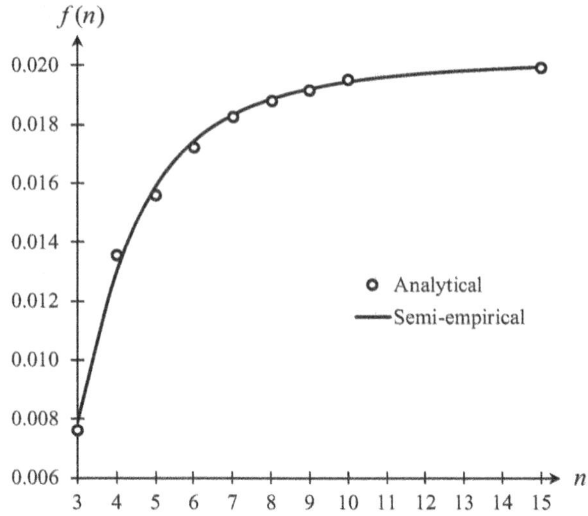

Fig. 5.21 Comparison between the analytical and semi-empirical $f(n)$ values

the number of plate edges, but the rate of increase then tapers off and asymptotically approaches the factor for a circular plate. While a linear interpolation would give a reasonable approximation of the shear correction factor for $n = 11, 12, 13, 14$, a better shear correction factor is obtained from Eqs. (5.7.21) and (5.7.22) as they cater for nonlinear increment with the number of plate sides.

5.8 Example: Buckling of Thick Auxetic Plates

The buckling load of simply supported circular and polygonal Mindlin plates has been given by Wang (1995a, b) as

$$N^M = \frac{N^K}{1 + \frac{N^K}{\kappa Gh}} \tag{5.8.1}$$

where N^K is the buckling load of the Kirchhoff plate, while the corresponding solution for the Reddy plates (Wang and Reddy 1997) is

$$N^R = \frac{N^K\left(1 + \frac{1}{70}\frac{N^K}{Gh}\right)}{1 + \frac{17}{14}\frac{N^K}{Gh}} \tag{5.8.2}$$

Although the extraction of shear correction factor for the case of buckling is straightforward, i.e., $\kappa = 5/6$, an attempt is made herein to show that the alternate shear correction factor of $\kappa = 14/17$ is similarly inferior, and that a generalized refined shear correction factor for buckling is analogous to that of plate deflection. As with Sect. 5.2, two approaches are furnished herein, namely the direct comparison and the series expansion approaches.

In the direct comparison approach, we let $N^M = N^R$, so that

$$1 + \frac{N^K}{\kappa G h} = \frac{1 + \frac{17}{14}\frac{N^K}{Gh}}{1 + \frac{1}{70}\frac{N^K}{Gh}} \tag{5.8.3}$$

or

$$1 + \frac{N^K}{\kappa G h} + \left[\frac{1}{70}\frac{N^K}{Gh} + \frac{1}{70\kappa}\left(\frac{N^K}{Gh}\right)^2 \right] = 1 + \frac{17}{14}\frac{N^K}{Gh} \tag{5.8.4}$$

The neglect of the $N^K/(70Gh)$ term in the denominator of Eq. (5.8.3) is tantamount to imposing zero for the square parenthesis [...] in the LHS of Eq. (5.8.4), which would lead to $\kappa = 14/17$. If we do not let the entire parenthesis in Eq. (5.8.4) be zero, but set only its last term as zero, i.e., $\left(N^K/Gh\right)^2 = 0$, then $\kappa = 5/6$. If we do not neglect any term, then (Lim 2020)

$$\kappa = \frac{5}{6}\left(1 + \frac{1}{70}\frac{N^K}{Gh}\right) \tag{5.8.5}$$

This solution is analogous to Eq. (5.2.15), and it also shows that $\kappa = 5/6$ is a lower bound.

In the series expansion approach, we apply the Maclaurin series expansion on the term $\left(1 + N^K/70Gh\right)$ in the numerator of Eq. (5.8.2) to give

$$N^R = \frac{N^K}{\left(1 + \frac{17}{14}\frac{N^K}{Gh}\right)\left[1 - \frac{N^K}{70Gh} + \left(\frac{N^K}{70Gh}\right)^2 - \left(\frac{N^K}{70Gh}\right)^3 + \cdots\right]} \tag{5.8.6}$$

or, in compact form,

$$N^R = \frac{N^K}{\left(1 + \frac{17}{14}\frac{N^K}{Gh}\right)\sum_{n=0}^{\infty}\left(-\frac{N^K}{70Gh}\right)^n} \tag{5.8.7}$$

Suppose we let $N^M = N^R$ based on Eqs. (5.8.1) and (5.8.6) then $\kappa = 14/17$ is recovered if the series summation in the latter is neglected. Equation (5.8.6) can also be expressed as

$$N^R = \frac{N^K}{1 + \frac{6}{5}\frac{N^K}{Gh}\left[1 - \frac{N^K}{70Gh} + \left(\frac{N^K}{70Gh}\right)^2 - \left(\frac{N^K}{70Gh}\right)^3 + \cdots\right]} \tag{5.8.8}$$

or, more conveniently,

$$N^R = \frac{N^K}{1 + \frac{6}{5}\frac{N^K}{Gh}\sum_{n=0}^{\infty}\left(-\frac{N^K}{70Gh}\right)^n} \tag{5.8.9}$$

so that comparison with Eq. (5.8.1) leads to

$$\kappa = \frac{5}{6}\left[1 - \frac{N^K}{70Gh} + \left(\frac{N^K}{70Gh}\right)^2 - \left(\frac{N^K}{70Gh}\right)^3 + \cdots\right]^{-1} = \frac{5}{6}\left(1 + \frac{N^K}{70Gh}\right) \tag{5.8.10}$$

The shear correction factor of $\kappa = 5/6$ is recovered if the term $N^K/(70Gh)$ is neglected in Eq. (5.8.10). For the sake of comparison, the higher-order terms in the infinite series can be truncated to give (Lim 2020)

$$\kappa = \frac{5}{6}\left(1 - \frac{N^K}{70Gh}\right)^{-1} \tag{5.8.11}$$

thereby revealing its analogy with Eq. (5.2.25).

5.9 Analogy Between Plates Under Transverse and Buckling Loads

Having obtained the refined generalized shear correction factors for transverse deflection and buckling load, one may now consolidate the results. Equations (5.2.15) and (5.2.25) can be combined as

$$\kappa = \frac{5}{6}\left(1 \pm \frac{h^2}{420(1-v)}\frac{\Delta^2 w^K}{|\Delta w^K|}\right)^{\pm 1} \tag{5.9.1}$$

in which the upper and lower signs correspond to the direct approach and the iteration approach, respectively. Similarly, Eqs. (5.8.5) and (5.8.11) can be consolidated as

$$\kappa = \frac{5}{6}\left(1 \pm \frac{N^K}{70Gh}\right)^{\pm 1} \tag{5.9.2}$$

whereby the upper and lower signs correspond to the direct comparison and the series expansion comparison, respectively. These results again suggest that the shear correction factor of $\kappa = 14/17$ is the least accurate, while the customary $\kappa = 5/6$ is the best shear correction factor in the form of a constant; the latter forming the better of the two constant lower bounds. The shear correction factors of Eqs. (5.9.1) and (5.9.2), with the positive signs, are exact within the framework of Reddy plate theory; the same set of shear correction factors, with the negative signs, forms a very tight upper bound. Table 5.10 summarizes the exact and other shear correction factors discussed herein.

To visually observe these generic correction factors, a family of κ curves

$$\kappa = \frac{5}{6}(1 \pm \Psi)^{\pm 1} \tag{5.9.3}$$

are plotted in Fig. 5.22, where Ψ denotes the generic correction term

Table 5.10 Summary of shear correction factors of Mindlin plate, within the framework of matching against Reddy plate

Category	Transverse deflection	Buckling load	Remarks
Variable κ	$\kappa = \frac{5}{6}\left(1 + \frac{h^2}{420(1-\upsilon)}\frac{\Delta^2 w^K}{\lvert\Delta w^K\rvert}\right)^{+1}$	$\kappa = \frac{5}{6}\left(1 + \frac{N^K}{70Gh}\right)^{+1}$	Exact, within Reddy plate framework
Variable κ	$\kappa = \frac{5}{6}\left(1 - \frac{h^2}{420(1-\upsilon)}\frac{\Delta^2 w^K}{\lvert\Delta w^K\rvert}\right)^{-1}$	$\kappa = \frac{5}{6}\left(1 - \frac{N^K}{70Gh}\right)^{-1}$	Tight upper bound, and very accurate
Constant κ	$\kappa = \frac{5}{6}$	$\kappa = \frac{5}{6}$	Tight lower bound, and accurate
Constant κ	$\kappa = \frac{14}{17}$	$\kappa = \frac{14}{17}$	A lower bound, and acceptable

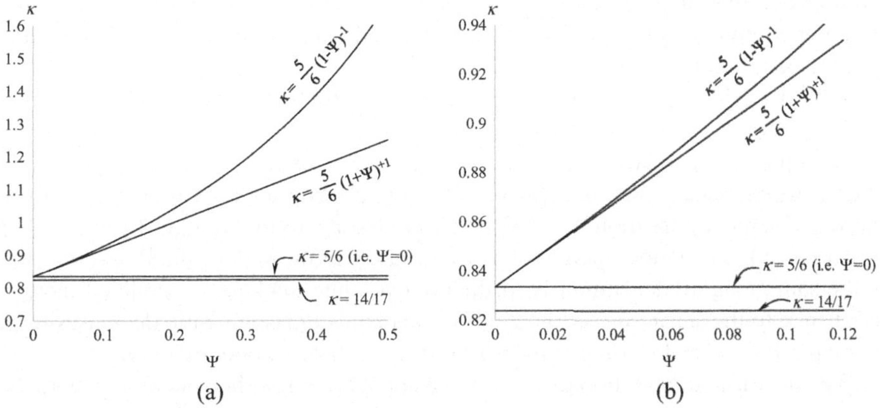

Fig. 5.22 Visual observation of the four shear correction factors discussed: **a** small but significant Ψ and **b** very small and insignificant Ψ

$$\Psi = \begin{cases} \frac{h^2}{420(1-v)} \frac{\Delta^2 w^K}{|\Delta w^K|} \\ \frac{N^K}{70Gh} \end{cases} \tag{5.9.4}$$

such that the exact and the upper bound correspond to the + and − signs, respectively, in Eq. (5.9.3). The tight lower bound $\kappa = 5/6$ is represented by $\Psi = 0$, while the other lower bound $\kappa = 14/17$ refers to $\Psi = -1/85$ and $\Psi = -1/84$ corresponding to the + and - signs, respectively, in Eq. (5.9.3).

Where the attainment of simple but nevertheless accurate shear deformation solution is concerned, the use of

$$\kappa = \frac{5}{6}(1 + \Psi)^{+1} \tag{5.9.5}$$

as shear correction factor for the simpler Mindlin plate would give the Reddy plate solution exactly, and that a tight bound can be attained as

$$\frac{5}{6} < \kappa < \frac{5}{6}(1 - \Psi)^{-1} \tag{5.9.6}$$

for design purposes. The simple change in the signs between Eq. (5.9.5) and the upper bound of Eq. (5.9.6), established herein as the exact and tight upper bound, respectively, suggests that previous work on obtaining refined shear correction factors can be easily extended on the basis of the present generalized analysis. While the choice of obtaining the exact or upper-bound shear correction factors for previous sections was based on convenience for each plate—the plate deflection description being unique for each plate shape—the generalized exact and upper-bound shear correction relationships developed in this section allow very convenient conversion from one to the other. A summary of shear correction factor conversion from the tight upper bound to the exact model, within Reddy plate framework, and vice versa is furnished in Table 5.11.

Table 5.11 shows that the shear correction factors for transversely loaded plates are conserved for varying plate thickness so long as the ratio $(h/a)^2/(1-v)$ is preserved. For example, if the relative plate thickness increases by a factor of $\sqrt{2}$, there is no change to the shear correction factor if the plate's Poisson's ratio decreases from incompressibility condition $v = 1/2$ to zero Poisson's ratio $v = 0$, or from $v = 1/3$ to $v = -1/3$. The latter change refers to a change in the sign of the Poisson's ratio at the same magnitude. Similarly, if the relative plate thickness is doubled there is no change to the shear correction factor if the plate's Poisson's ratio decreases from the upper limit $v = 1/2$ to the lower limit $v = -1$ for isotropic material.

We now turn our attention to an example application for plate buckling where the use of $\kappa = 14/17$ is critical compared to $\kappa = 5/6$. Since $14/17 < 5/6$, it follows that the substitution of these shear correction factors into the RHS of Eq. (5.8.1) leads to

Table 5.11 Conversion between the tight upper bound and the exact shear correction factors for uniformly loaded simply supported polygonal plates of specific shapes on the basis of the generalized irregular polygonal plates

Plate shapes	Schematics	Tight upper-bound shear correction factor	Exact shear correction factor	Lower bound
Rectangular plates of aspect ratio $a/b \geq 1$	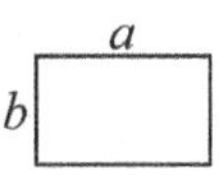	Inferring from Lim (2016a) and this section, $\frac{5}{6}\left[1-\frac{\pi^2}{420(1-\upsilon)}\frac{h^2}{ab}f\left(\frac{a}{b}\right)\right]^{-1}$	From Lim (2016a), $\frac{5}{6}\left[1+\frac{\pi^2}{420(1-\upsilon)}\frac{h^2}{ab}f\left(\frac{a}{b}\right)\right]$	$\frac{5}{6}$
Equilateral triangular plates	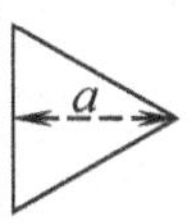	From Lim (2016b), $\frac{5}{6}\left[1-\frac{9}{140(1-\upsilon)}\left(\frac{h}{L}\right)^2\right]^{-1}$	Inferring from Lim (2016b) and this section, $\frac{5}{6}\left[1+\frac{9}{140(1-\upsilon)}\left(\frac{h}{L}\right)^2\right]$	$\frac{5}{6}$
Isosceles right triangular plates	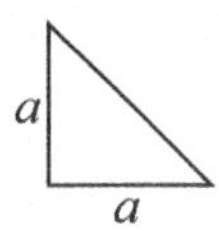	Inferring from Lim (2016c) and this section, $\frac{5}{6}\left[1-\frac{0.10734}{1-\upsilon}\left(\frac{h}{a}\right)^2\right]^{-1}$	From Lim (2016c), $\frac{5}{6}\left[1+\frac{0.10734}{1-\upsilon}\left(\frac{h}{a}\right)^2\right]$	$\frac{5}{6}$
Hexagonal plates	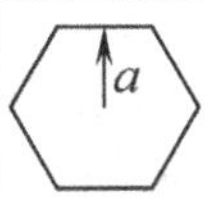	From Lim (2017a), $\frac{5}{6}\left[1-\frac{\pi^2}{840(1-\upsilon)}\left(\frac{h}{a}\right)^2\right]^{-1}$	Inferring from Lim (2017a) and this section, $\frac{5}{6}\left[1+\frac{\pi^2}{840(1-\upsilon)}\left(\frac{h}{a}\right)^2\right]$	$\frac{5}{6}$
Regular polygonal plates of n sides	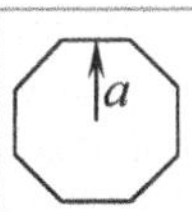	From Lim (2017b), $\frac{5}{6}\left[1-\frac{f(n)(1+\upsilon)}{(1-\upsilon)(3+\upsilon)}\left(\frac{h}{a}\right)^2\right]^{-1}$	Inferring from Lim (2017b) and this section, $\frac{5}{6}\left[1+\frac{f(n)(1+\upsilon)}{(1-\upsilon)(3+\upsilon)}\left(\frac{h}{a}\right)^2\right]$	$\frac{5}{6}$
Irregular polygonal plates	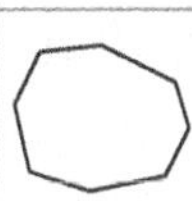	From Lim (2020), $\frac{5}{6}\left[1-\frac{h^2}{420(1-\upsilon)}\frac{\Delta^2 w^K}{\lceil \Delta w^K \rceil}\right]^{-1}$	From Lim (2020), $\frac{5}{6}\left[1+\frac{h^2}{420(1-\upsilon)}\frac{\Delta^2 w^K}{\lceil \Delta w^K \rceil}\right]$	$\frac{5}{6}$

$$\frac{N^K}{1+\frac{17}{14}\frac{N^K}{Gh}} < \frac{N^K}{1+\frac{6}{5}\frac{N^K}{Gh}} \tag{5.9.7}$$

or

$N^M_{=14/17} < N^M_{=5/6}$. Suppose the critical buckling load, N_{cr}, falls between these two bounds, i.e., $N^M_{=14/17} < N_{cr} < N^M_{=5/6}$, then the use of $\kappa = 14/17$ would predict that the critical buckling load has not being attained, while the use of the more accurate $\kappa = 5/6$ reveals that the plate would undergo buckling failure. As such, if $N^M_{=14/17} < N_{cr} < N^M_{=5/6}$, then the use of $\kappa = 14/17$ would be catastrophic while the use of $\kappa = 5/6$ is more conservative.

It can be concluded that the shear correction factor of $\kappa = 14/17$ is consistently shown to be the least accurate for all plate shapes but reasonably valid, while the customary $\kappa = 5/6$ is the best correction factor in the form of a constant. The twin counterpart of the exact shear correction factor, obtained by replacing the + signs with − signs, forms the tight upper bound. While the exact shear correction factor permits the Mindlin plate solution to exactly match the Reddy plate solution, the tight bounds are useful for precision design purposes. The availability of shear correction factors in terms of lower bounds, upper bounds, and exact forms, avails to the engineer more options in the design of thick plates, and to also permit very convenient conversion between the tight upper bound and the exact shear correction factors when one of them is known.

5.10 Vibration of Thick Auxetic Plates

So far previous sections deal with static transverse loads and in-plane buckling loads. From the instability and vibration study of thin auxetic plates (Lim 2014b), analyses have been extended on the instability (Lim 2014c) and vibration (Lim 2014d) characteristics of thick auxetic plates.

The shear correction factor that is commonly used for conventional plates (i.e., positive Poisson's ratio), $\kappa = 5/6$, is obtained by matching the Mindlin constitutive shear forces against those proposed by Reissner (1947). However, this constant shear correction factor does not allow for the effect of Poisson's ratio. When dealing with auxetic solids it is obviously important to allow for the effect of Poisson's ratio because the range of Poisson's ratio for isotropic auxetic solids is twice the Poisson's ratio range for conventional solids. A variable shear correction factor for isotropic plates has been obtained by equating the angular frequency of the first antisymmetric mode of thickness-shear vibration according to the exact 3D theory to the corresponding frequency according to Mindlin's (1951) theory to give (Liew et al. 1998)

$$\kappa^3 - 8\kappa^2 + 8\left(\frac{2-v}{1-v}\right)\kappa - \frac{8}{1-v} = 0 \tag{5.10.1}$$

This variable shear correction factor is selected because the exact 3D theory adopted for extracting Eq. (5.10.1) is not limited to positive Poisson's ratio; the formulation implicitly caters for Poisson's ratio of either signs. It will later be shown that the use of Eq. (5.10.1) instead of $\kappa = 5/6$ gives a better agreement with Reddy's third-order shear deformation theory and has the same effect as the inclusion of rotary inertia. Recall that the circular natural frequency for polygonal Mindlin plates ω_M, without considering rotary inertia, is related to that of Kirchhoff plates ω_K as (Wang et al. 2000a)

$$\omega_M^2 = \frac{\omega_K^2}{1 + \frac{\omega_K h^2}{6(1-v)\kappa}\sqrt{\frac{\rho h}{D}}} \tag{5.10.2}$$

where h, v, ρ, and D are the thickness, Poisson's ratio, density, and flexural rigidity of the plate, respectively. Introducing the dimensionless natural frequency in terms of the natural frequency, density, thickness, and flexural rigidity of the plate

$$\omega^* = \omega\sqrt{\frac{\rho h^5}{D}} \tag{5.10.3}$$

gives

$$\omega_M^* = \frac{\omega_K^*}{\sqrt{1 + \frac{\omega_K^*}{6(1-\upsilon)\kappa}}} \tag{5.10.4}$$

By adopting the choice of dimensionless frequency shown in Eq. (5.10.3), the curves of ω_M^* versus ω_K^* are independent of other plate properties. Considering rotary inertia, the natural frequency of a polygonal Mindlin plate is related to that of Kirchhoff plate as (Wang et al. 2000a)

$$\omega_M^2 = \frac{6\kappa G}{\rho h^2}\left\{1 + \frac{h^2\omega_K}{12}\sqrt{\frac{\rho h}{D}}\left(1 + \frac{2}{\kappa(1-\upsilon)}\right) - \sqrt{\left[1 + \frac{h^2\omega_K}{12}\sqrt{\frac{\rho h}{D}}\left(1 + \frac{2}{\kappa(1-\upsilon)}\right)\right]^2 - \frac{\rho h^2\omega_K^2}{3\kappa G}}\right\} \tag{5.10.5}$$

where G is the shear modulus of the plate material. Using Eq. (5.2.9) such that G is expressed in terms of D, v and h, Eq. (5.10.5) can be written as

$$\omega_M^* = 6\sqrt{\kappa(1-\upsilon)\left\{1 + \frac{\omega_K^*}{12}\left(1 + \frac{2}{\kappa(1-\upsilon)}\right) - \sqrt{\left[1 + \frac{\omega_K^*}{12}\left(1 + \frac{2}{\kappa(1-\upsilon)}\right)\right]^2 - \frac{(\omega_K^*)^2}{18\kappa(1-\upsilon)}}\right\}} \tag{5.10.6}$$

Table 5.12 shows a list of variable κ calculated from Eq. (5.10.1). When the constant κ is compared with the variable κ, the percentage error falls within 10% when the plate's Poisson's ratio is positive but exceeds 10% when the plate's Poisson's ratio is negative. Hence, the variable κ, instead of the constant κ, is to be used when dealing with auxetic plates.

Figure 5.23 shows a family of ω_M^* versus ω_K^* curves for the entire range of Poisson's ratio within the framework of isotropic solids. When κ is assumed constant throughout the entire range of Poisson's ratio of the plate material and the rotary inertia is neglected, as furnished in Fig. 5.23 (top left), the dimensionless natural frequency natural for Mindlin plates approximates that of Kirchhoff plates as the Poisson's ratio of the plate material becomes more negative. This result is consistent to earlier sections, in which the maximum lateral deflection of a uniformly loaded

Table 5.12 Comparison between constant and variable shear correction factors for a plate

Region	Poisson's ratio, υ	Variable shear correction factor, $\kappa = \kappa(\upsilon)$	Constant shear correction factor, $\kappa = 5/6$	Percentage error of $\kappa = 5/6$ with reference to $\kappa = \kappa(\upsilon)$	Remarks
Auxetic region	−1	0.474572	5/6	75.59682	Percentage error exceeds 10%
	−0.9	0.496042	5/6	67.99653	
	−0.8	0.519175	5/6	60.51107	
	−0.7	0.544078	5/6	53.16431	
	−0.6	0.570826	5/6	45.98728	
	−0.5	0.599446	5/6	39.01725	
	−0.4	0.629884	5/6	32.29949	
	−0.3	0.661966	5/6	25.88763	
	−0.2	0.695367	5/6	19.84079	
	−0.1	0.729580	5/6	14.22097	
Conventional region	0	0.763932	5/6	9.084753	Percentage error within 10%
	0.1	0.797638	5/6	4.475129	
	0.2	0.829914	5/6	0.412011	
	0.3	0.860094	5/6	−3.11137	
	0.4	0.887732	5/6	−6.12783	
	0.5	0.912622	5/6	−8.68801	

simply supported Mindlin plate approaches that of a Kirchhoff plate as $\upsilon \to -1$. However, reference to Fig. 5.23 (top right) shows that the natural frequency of the same Mindlin plate generally decreases when rotary inertia is taken into consideration. A similar effect is observed when the variable κ is adopted without considering rotary inertia, as evidenced in Fig. 5.23 (bottom left). When both the variable κ and rotary inertia are taken into account, the natural frequency of the Mindlin plate is further reduced, as plotted in Fig. 5.23 (bottom right). In all the four sets of combinations shown in Fig. 5.23, the natural frequency of Mindlin plates increases at decreasing rate as the plate's Poisson's ratio becomes more negative. This is especially observed when the variable κ is used, in which the ω_M^* versus ω_K^* curves for $-1 \leq \upsilon \leq -0.25$ are very close, i.e., the Poisson's ratio has insignificant incremental effect on the incremental ω_M^*/ω_K^* ratio when the Poisson's ratio is sufficiently negative.

In order to compare how the ω_M^* to ω_K^* relationship varies according to the choice of shear correction factor and the inclusion or exclusion of rotary inertia, a family of ω_M^* versus ω_K^* curves were plotted under the four different combinations of assumptions at fixed Poisson's ratio of the plate. When the plate is incompressible, i.e., the plate's Poisson's ratio is set at the upper limit, all the four curves fall within very close proximity, as shown in Fig. 5.24 (top left). At $\upsilon = 0$, the four curves are slightly

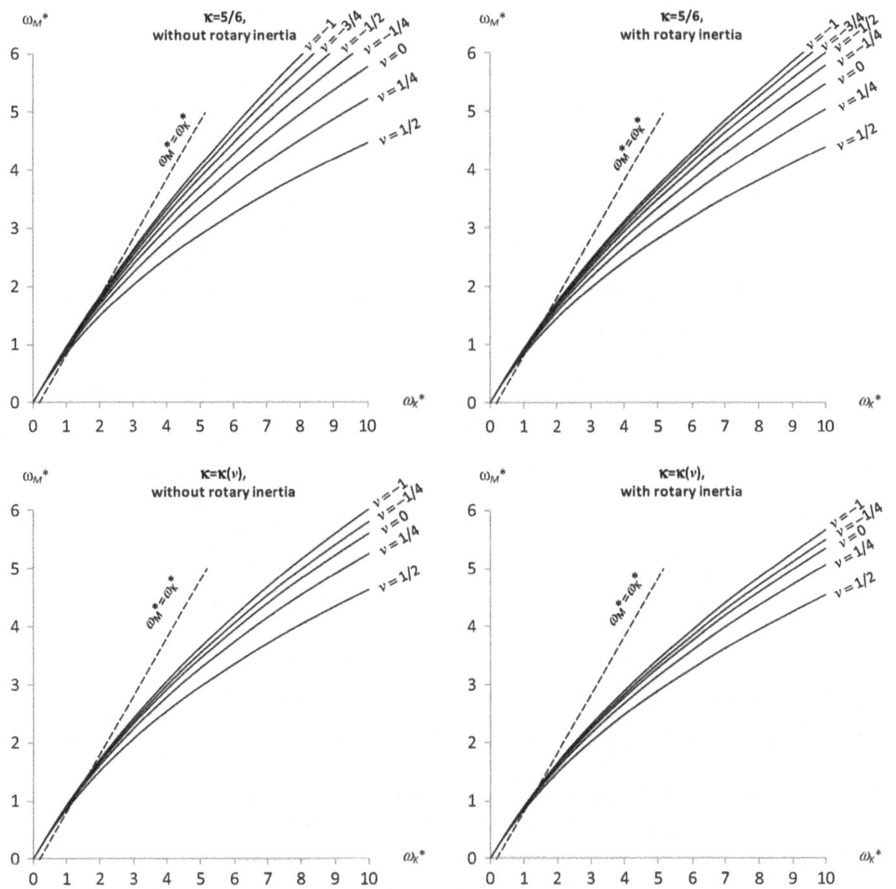

Fig. 5.23 Plots of ω_M^* versus ω_K^* for $-1 \leq \upsilon \leq 1/2$ based on constant shear correction factor (top), variable shear correction factor (bottom), without rotary inertia (left), and with rotary inertia (right)

further apart but are nevertheless moderately close to one another, as depicted in Fig. 5.24 (top right). Figure 5.24 (bottom left), however, exhibits a significant difference to the ω_M^* to ω_K^* relationship at $\upsilon = -1/2$, whereby the natural frequency of Mindlin plates with both simplifying assumptions gives the highest value while the natural frequency of Mindlin plates with variable shear correction factor and rotary inertia gives the lowest value. Interestingly, the ω_M^* to ω_K^* relationships on the assumption of either constant κ or no rotary inertia are almost equal. As the Poisson's ratio becomes more negative, further differentiation is exhibited. As shown in Fig. 5.24 (bottom right) for $\upsilon = -1$, the ω_M^* to ω_K^* relationship is primarily influenced by the choice of shear correction factor and secondarily influenced by the rotary inertia.

In order to clearly observe how ω_M^* relates to ω_K^* for the entire range of Poisson's ratio, a set of ω_M^*/ω_K^* curves was plotted against the plate's Poisson's ratio, as

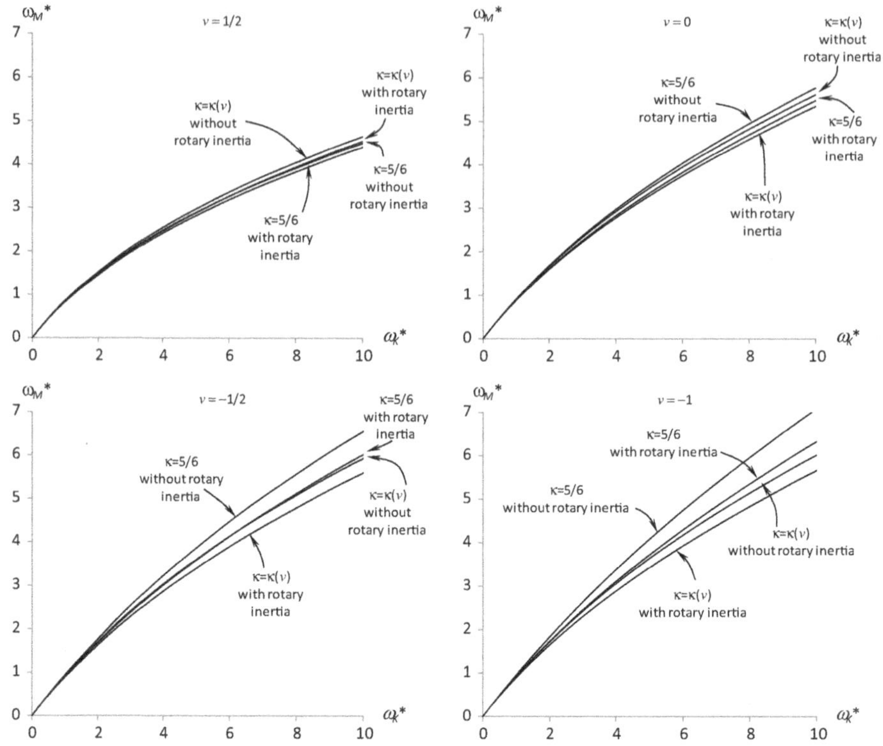

Fig. 5.24 Plots of ω_M^* versus ω_K^* for different combinations of assumptions at $\upsilon = 1/2$ (top left), $\upsilon = 0$ (top right), $\upsilon = -1/2$ (bottom left), and $\upsilon = -1$ (bottom right)

furnished in Fig. 5.25. At very low natural frequencies, such as $\omega_K^* = 1$ shown in Fig. 5.25 (top left), the Mindlin plate's natural frequencies at the conventional region $0 \le \upsilon \le 1/2$ is primarily controlled by rotary inertia. At the lower limit of the plate's Poisson's ratio ($\upsilon = -1$), the Mindlin plate's natural frequency based on the assumption of no rotary inertia at variable shear correction factor is almost equal to that based on the assumption of constant shear correction factor with rotary inertia. At very high natural frequencies, such as $\omega_K^* = 100$ depicted in Fig. 5.25 (bottom right), the Mindlin plate's natural frequency is common at $\upsilon = 0.2$. This implies that at very high natural frequencies, the Mindlin plate's natural frequency can be sufficiently accurate without taking into account the variation of κ with Poisson's ratio and without considering rotary inertia only if the plate's Poisson's ratio is positive. However, if the plate is auxetic, then the Mindlin plate's natural frequency is primarily influenced by κ, whereby the Mindlin natural frequencies considering constant κ are higher than those at variable κ. Figure 5.25 (top right) and (bottom left) show the transition of the ω_M^*/ω_K^* curves from low to high frequencies.

Finally, it is of interest to evaluate the effectiveness of the variable shear correction factor described by Eq. (5.10.1) against the constant shear correction factor, by means

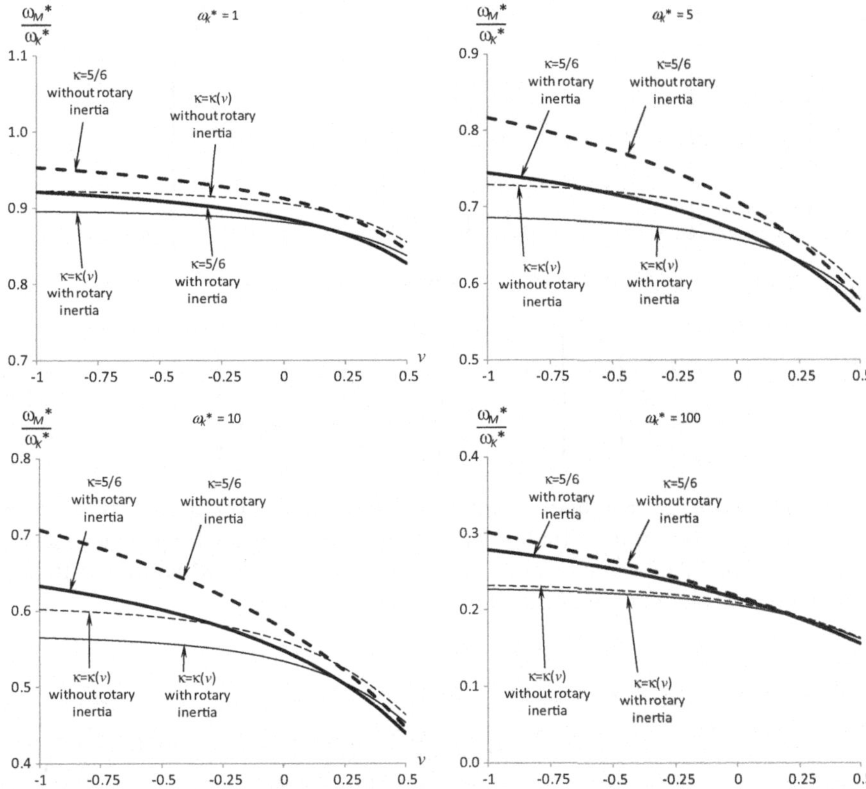

Fig. 5.25 Plots of ω_M^*/ω_K^* versus plate Poisson's ratio for $\omega_K^* = 1$ (top left), $\omega_K^* = 5$ (top right), $\omega_K^* = 10$ (bottom left), and $\omega_K^* = 100$ (bottom right)

of comparison with Reddy's third-order shear deformation theory. Limitations on the present analysis when the plate is very thick or when it is vibrating at very high frequency can be concurrently evaluated using the dimensionless frequency described by Eq. (5.10.3), in which $\omega^* \propto \omega\sqrt{h^5/D}$ or $\omega^* \propto \omega h/\sqrt{E}$. Using the Reddy–Kirchhoff relations (Wang et al. 2000b)

$$\omega_K\sqrt{\frac{\rho h}{D}} = -2\sqrt{\Phi}\cos\left(\frac{\theta + 2\pi}{3}\right) + \frac{a_2}{3a_1} \tag{5.10.7}$$

where

$$\begin{Bmatrix} \Phi \\ \Psi \\ \theta \end{Bmatrix} = \begin{Bmatrix} -\frac{a_3}{3a_1} + \left(\frac{a_2}{3a_1}\right)^2 \\ \frac{a_2 a_3}{6a_1^2} - \frac{a_4}{2a_1} - \left(\frac{a_2}{3a_1}\right)^3 \\ \cos^{-1}\left(\frac{\Psi}{\Phi^{3/2}}\right) \end{Bmatrix} \tag{5.10.8}$$

and

$$\begin{Bmatrix} a_1 \\ a_2 \\ a_3 \\ a_4 \end{Bmatrix} = \begin{Bmatrix} \frac{4D^2}{525K}\left(1+\frac{4\rho h^3\omega_R^2}{315KJ}\right) \\ -\frac{D}{5}\left[1-\frac{4}{K}\left(\frac{\rho h^3\omega_R^2}{252}-\frac{8Gh}{15}\right)+\frac{16L}{21KJ}-\frac{\rho h^3\omega_R^2}{315KJ}\right] \\ -\frac{4D}{5K}\rho h\omega_R^2+\frac{16D}{105KJ}\rho h\omega_R^2\left(1-\frac{\rho h^3\omega_R^2}{15K}\right)-\frac{L}{J} \\ \frac{\rho h\omega_R^2}{J}\left(1-\frac{\rho h^3\omega_R^2}{15K}\right) \end{Bmatrix}. \tag{5.10.9}$$

with

$$\begin{Bmatrix} K \\ J \\ L \end{Bmatrix} = \begin{Bmatrix} \frac{4\rho h^3\omega_R^2}{315}+\frac{8Gh}{15} \\ 1-\frac{4\rho h^3\omega_R^2}{315K} \\ \frac{\rho h^3\omega_R^2}{60}-\frac{\rho h^3\omega_R^2}{15K}\left(\frac{\rho h^3\omega_R^2}{252}-\frac{8Gh}{15}\right) \end{Bmatrix} \tag{5.10.10}$$

as well as the Mindlin–Kirchhoff relations given in Eq. (5.10.4) and (5.10.6), a set of relationship between the Mindlin and Reddy dimensionless frequencies can be obtained. To facilitate a graphical representation between the Mindlin and Reddy plates, it is useful to firstly make the following conversions

$$\frac{8Gh}{15} = \frac{16D}{5h^2}(1-\upsilon) \tag{5.10.11}$$

$$\rho h^3\omega_R^2 = \frac{D}{h^2}\frac{\rho h^5\omega_R^2}{D} = \frac{D}{h^2}\left(\omega_R^*\right)^2 \tag{5.10.12}$$

$$\rho h\omega_R^2 = \frac{D}{h^4}\frac{\rho h^5\omega_R^2}{D} = \frac{D}{h^4}\left(\omega_R^*\right)^2 \tag{5.10.13}$$

so that Eqs. (5.10.9) and (5.10.10) are commonly expressed in terms of plate flexural rigidity and frequency of Reddy plates. Figure 5.26 shows the dimensionless frequencies between the Mindlin and Reddy plates for $\upsilon = -1, -1/2, 0, 1/2$ based on constant (top) and variable (bottom) shear correction factors, as well as without (left) and with (right) rotary inertia.

Perusal to Fig. 5.26 (top left) shows that the Mindlin plate approximates the Reddy plate well for $\upsilon \approx 0$, with the conventional and auxetic Mindlin plate generally overestimating and underestimating the frequency in comparison to the Reddy plate. When only the rotary inertia is taken into account, as shown in Fig. 5.26 (top right), there is good agreement between the Mindlin and Reddy plates for auxetic plates but not for conventional plates. A similar trend is observed when only the variable shear correction factor is incorporated, as shown in Fig. 5.26 (bottom left). However, when both the variable shear correction factor and rotary inertia are taken into account, as shown in Fig. 5.26 (bottom right), the improvement to the Reddy–Mindlin relation is slightly less satisfactory in comparison to the case when only one of them is considered—this could possibly be a result of overcorrection. Nevertheless, it

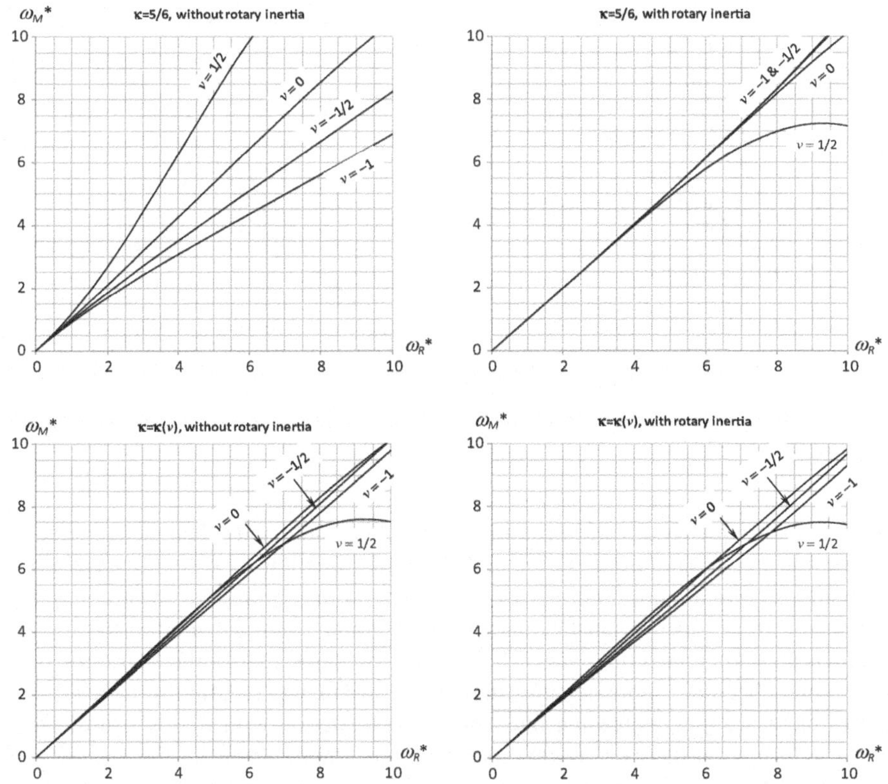

Fig. 5.26 Plots of dimensionless frequencies of Mindlin plates considering constant κ (top), variable κ (bottom), without rotary inertia (left), and with rotary inertia (right) versus the dimensionless frequencies of Reddy plates

is interesting to point out that in the auxetic region the Mindlin and Reddy vibrational frequency relationship is almost linear. Figure 5.27 represents a conventional plate (left) and an auxetic plate (right) so as to allow comparison to be made under Poisson's ratio of equal magnitude $|v| = 1/2$. In the case of incompressible solids, unsatisfactory correlation between the Mindlin and Reddy vibrational frequencies is found when constant shear correction factor is used, especially so when the rotary inertia is neglected. Better correlation is found when the variable shear correction factor is used, although there is no appreciable difference whether rotary inertia is used or not. In the case where $v = -1/2$, good correlations between the Mindlin and Reddy vibrational frequencies are found when variable shear correction factor is used.

It can be concluded that the use of Mindlin plate theory with simplifications (i.e., constant shear correction factor and no rotary inertia) is sufficient to approximate the case Mindlin plates without simplifications (i.e., variable shear correction factor with rotary inertia) when the plate's Poisson's ratio is positive. When the plate is auxetic,

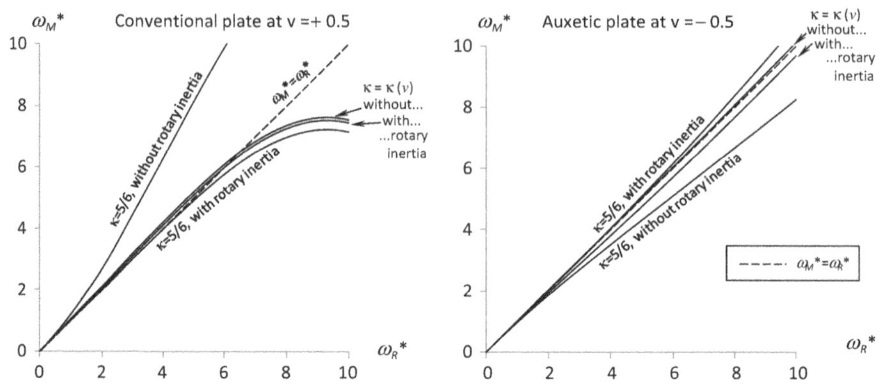

Fig. 5.27 Dimensionless vibrational frequency of Mindlin plates ω_M^* in comparison to Reddy plates ω_R^* at $\upsilon = \pm 1/2$

the natural frequency for the Mindlin plate is overestimated when at least one of the simplifying assumptions is used and the overestimation is especially large when both simplifying assumptions are used. As such, both the variable shear correction factor and rotary inertia must be taken into account when calculating the natural frequency of thick auxetic plates. The results also show that as the Poisson's ratio of the plate becomes more negative, the natural frequency of Mindlin-to-Kirchhoff plates increases at a decreasing rate. Finally, comparison with Reddy's third-order shear deformation theory confirms that sufficient improvement to the accuracy of vibration frequencies is found when either the rotary inertia or the variable shear correction factor is used.

References

Conway HD (1960) The approximate analysis of a certain boundary-value problems. J Appl Mech 27(2):275–277

Conway HD (1961) The bending, buckling, and flexural vibration of a simply supported polygonal plates by point-matching. J Appl Mech 28(2):288–291

Leissa AW, Lo CC, Niedenfuhr FW (1965) Uniformly loaded plates of regular polygonal shape. AIAA J 3(3):566–567

Liew KM, Wang CM, Xiang Y, Kitipornchai S (1998) Vibration of Mindlin plates. Elsevier, Oxford

Lim TC (2013) Shear deformation in thick auxetic plates. Smart Mater Struct 22(8):084001

Lim TC (2014a) Shear deformation in rectangular auxetic plates. J Eng Mater Technol 136(3):031007

Lim TC (2014b) Buckling and vibration of circular auxetic plates. J Eng Mater Technol 136(2):021007

Lim TC (2014c) Elastic stability of thick auxetic plates. Smart Mater Struct 23(4):045004

Lim TC (2014d) Vibration of thick auxetic plates. Mech Res Commun 61:60–66

Lim TC (2015) Auxetic materials and structures. Springer, Singapore

Lim TC (2016a) Improved shear correction factors for deflection of simply supported very thick rectangular auxetic plates. Int J Mech Mater Eng 11(1):13

Lim TC (2016b) Higher-order shear deformation of very thick simply supported equilateral triangular plates under uniform load. Mech Based Des Struct Mach 44(4):514–522

Lim TC (2016c) Refined shear correction factor for very thick simply supported and uniformly loaded isosceles right triangular auxetic plates. Smart Mater Struct 25(5):054001

Lim TC (2017a) Shear deformation in a class of thick hexagonal plates. Phys Status Solidi B 254(12):1700014

Lim TC (2017b) Refined shear correction of polygonal plates with static loads. Eng Comput Mech 170(4):167–173

Lim TC (2020) Extraction of Mindlin plates' shear correction factors from Reddy plate theory. Eng Comput Mech 173(1):37–44

Mindlin RD (1951) Influence of rotatory inertia and shear on flexural motions of isotropic, elastic plates. J Appl Mech 18:31–38

Reddy JN (1984) A simple higher-order theory for laminated composite plates. J Appl Mech 51(4):745–752

Reddy JN, Wang CM (1998) Deflection relationships between classical and third-order plate theories. Acta Mech 130(3–4):199–208

Reddy JN (2006) Theory and analysis of elastic plates. CRC Press, Boca Raton

Reissner E (1945) The effect of transverse shear deformation on the bending of elastic plates. J Appl Mech 12:69–77

Reissner E (1947) On bending of elastic plates. Quart Appl Math 5(1):55–68

Wang CM, Alwis WAM (1995) Simply supported polygonal Mindlin plate deflections using Kirchhoff plates. J Eng Mech 121(12):1383–1385

Wang CM (1995a) Allowance for prebuckling deformations in buckling load relationship between Mindlin and Kirchhoff simply supported plates of general polygonal shape. Eng Struct 17(6):413–418

Wang CM (1995b) Buckling of polygonal and circular sandwich plates. AIAA J 33(5):962–964

Wang CM, Reddy JN (1997) Buckling load relationship between Reddy and Kirchhoff plates of polygonal shape with simply supported edges. Mech Res Commun 24(1):103–108

Wang CM, Reddy JN, Lee KH (2000a) Shear deformable beams and plates: relationships with classical solutions. Elsevier, Oxford

Wang CM, Kitipornchai S, Reddy JN (2000b) Relationship between vibration frequencies of Reddy and Kirchhoff polygonal plates with simply supported edges. J Vib Acoust 122(1):77–81

Yuan FG, Miller RE (1992) Improved rectangular element for shear deformable plates. J Eng Mech 118(2):312–328

Woinowsky-Krieger S (1933) Der Spannungszustand in dicken elastischen Platten. Ingenieur-Archiv 4(4):305–331

Chapter 6
Longitudinal Elastic Waves in Auxetic Solids

Abstract This chapter discusses the longitudinal wave speed in prismatic rods, plates, and bulk solids made from isotropic auxetic materials—as well as their intermediate structural elements—with special emphasis on the changes in cross-sectional area and density while retaining the traditional strength-of-materials flavor in order to keep the practical solutions tractable.

Keywords Cross-sectional area change · Density change · Longitudinal waves

6.1 Preamble

Classical, or elementary, models of wave speed do not take into account the strain, the change in material density and the variation of cross-sectional area perpendicular to the wave direction. Take, for example, the longitudinal wave speed in prismatic rods are illustrated in Fig. 6.1.

The elementary equation of motion for longitudinal wave in a prismatic rod is long well established and can be set up by applying Newton's second law of motion to the elemental volume, shaded in Fig. 6.1, to give

$$F_{x+\mathrm{d}x} - F_x = \mathrm{d}m \frac{\partial^2 u_x}{\partial t^2} \tag{6.1.1}$$

where the elemental mass $\mathrm{d}m = \rho_0 \mathrm{d}V$ and elemental volume $\mathrm{d}V = A_0 \mathrm{d}x$ are substituted so that (Timoshenko and Goodier 1951)

$$A_0(\sigma_x + \mathrm{d}\sigma_x) - A_0\sigma_x = \rho_0 A_0 \mathrm{d}x \frac{\partial^2 u_x}{\partial t^2} \tag{6.1.2}$$

whereby both A_0, the cross-sectional area, and ρ, the rod density, are assumed constant. This simplification allows Eq. (6.1.1) to be reduced to

$$\mathrm{d}\sigma_x = \rho_0 \mathrm{d}x \frac{\partial^2 u_x}{\partial t^2} \tag{6.1.3}$$

T. Lim, *Mechanics of Metamaterials with Negative Parameters*, Engineering Materials,
https://doi.org/10.1007/978-981-15-6446-8_6

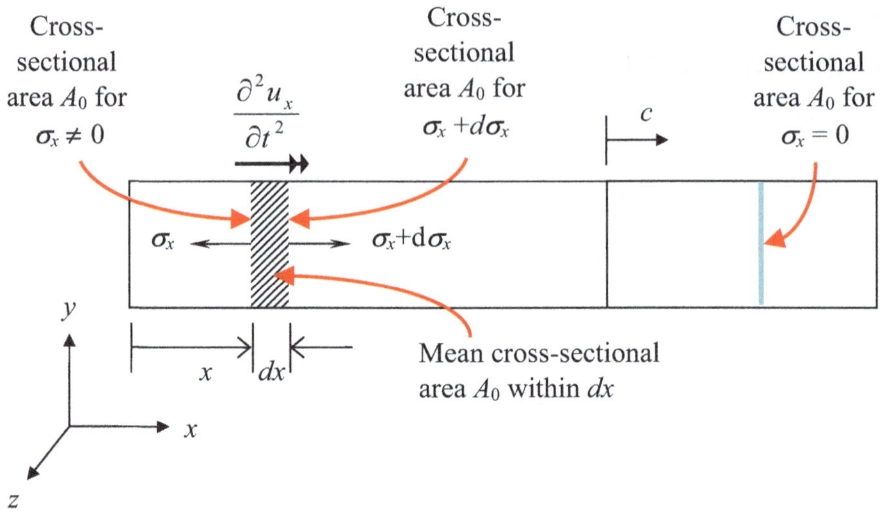

Fig. 6.1 Schematic for analysis of longitudinal wave motion through a prismatic rod assuming no changes to the cross-sectional geometry

Let E be the Young's modulus of the rod material, then the substitution of

$$d\sigma_x = E d\varepsilon_x = E\frac{\partial \varepsilon_x}{\partial x}dx = E\frac{\partial^2 u_x}{\partial x^2}dx \tag{6.1.4}$$

into Eq. (6.1.3) leads to

$$\frac{\partial^2 u_x}{\partial x^2} = \frac{\rho_0}{E}\frac{\partial^2 u_x}{\partial t^2} \tag{6.1.5}$$

Since the longitudinal wave velocity, c, is defined as (Kolsky 1963; Graff 1975)

$$c^2 = \frac{\left(\frac{\partial^2 u_x}{\partial t^2}\right)}{\left(\frac{\partial^2 u_x}{\partial x^2}\right)} \tag{6.1.6}$$

the substitution of Eq. (6.1.5) into Eq. (6.1.6) gives the classical elementary longitudinal wave speed in a prismatic rod as

$$c = \sqrt{\frac{E}{\rho_0}} \tag{6.1.7}$$

The assumption of constant density is exactly valid only when $v = 1/2$, i.e., the absence of volumetric change with mass preservation ensures constancy of density. On the other hand, the assumption of constant cross-sectional area is exactly valid only when $v = 0$. The simultaneous assumptions of constant cross-sectional area and constant density is, therefore, theoretically inconsistent and is only justified as a simplified or design equation for isotropic solids whose Poisson's ratio falls within $0 < v < 1/2$. Due to their negativity of Poisson's ratio, auxetic solids expand laterally when stretched axially and contract laterally when compressed axially; these triaxial expansion and contraction responses to uniaxial tension and compression, respectively, give rise to a large change in the density of auxetic solids in comparison with solids with positive Poisson's ratio. See Fig. 6.2.

This justifies why the change in density must be taken into consideration for wave motion through auxetic solids. When the density changes is accounted for, the change in cross-sectional area is to be considered not only for the sake of consistency, but also because in the case of isotropic solids the range of Poisson's ratio is larger in auxetic materials ($-1 \leq v < 0$) than in conventional ones ($0 \leq v \leq 1/2$). In other words, the change in lateral strain can be as much as the applied strain for auxetic solids while the magnitude change in lateral strain can be at most only half of the applied strain for conventional material. It should also be clarified that the actual problem is that of 3D mechanics; although the stress state is 1D, the strain state is

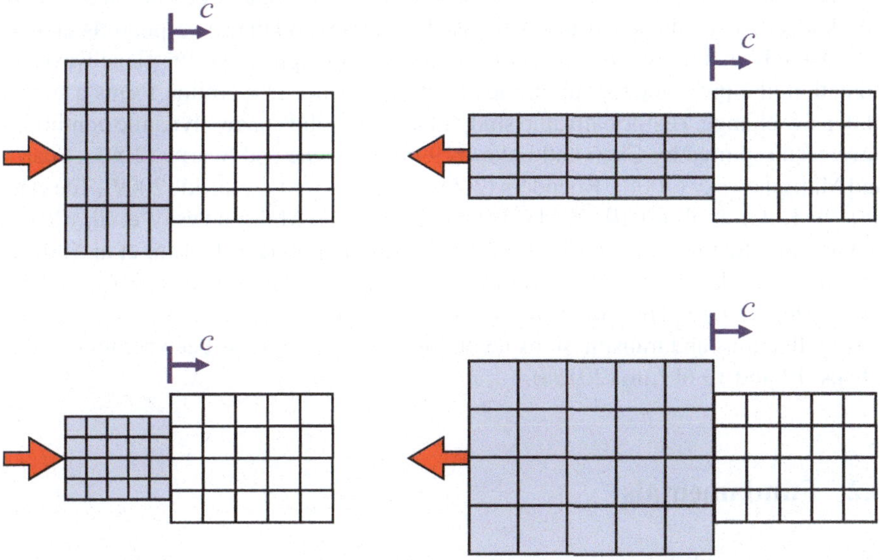

Fig. 6.2 Wave speed, c, as a result of prescribed compressive (left) and tensile (right) strains—indicated by arrows—on rods with Poisson's ratio of positive (top) and negative (bottom) values. Note the larger change in volume, and hence larger change in density, for solids with negative Poisson's ratio

3D. However, the classical elementary treatment is that of 1D mechanics on the basis of 1D stress state with 1D strain state assumption.

The classical approach for obtaining the longitudinal wave speed in prismatic rods has also been extended to those in infinite flat plates and in plane waves of dilatation; the former is equivalent to plates with prescribed zero strain in the width direction while the latter applied for bulk solids where strains orthogonal to the direction of wave propagation is prescribed zero. Note that Eq. (6.1.7) is expressed in terms of Young's modulus and the original, or unstressed, density of the rod material. In the case of longitudinal wave in infinite plates,

$$c = \sqrt{\frac{E}{\rho_0\left(1 - v^2\right)}} \tag{6.1.8}$$

and in the case of plane waves of dilatation

$$c = \sqrt{\frac{E(1 - v)}{\rho_0(1 + v)(1 - 2v)}} \tag{6.1.9}$$

their classical wave speeds additionally include the Poisson's ratio of the material (Kolsky 1963; Graff 1975). Although the Poisson's ratio of the materials is taken into account, it remains that the wave speed in the flat plate does not account for the change in plate thickness under stressed condition and the corresponding change in density. Even though cross-sectional change is not applicable for plane waves of dilatation, the prescribed strain in the direction of wave preparation forms a source of density change. Historically, the study of auxetic solids under dynamic conditions have been reported by Chen and Lakes (1996), Ruzzene and Scarpa (2003), Scarpa and Malischewsky (2008), Koenders (2009), Trzupek and Zielinski (2009), Tee et al. (2010), Kolat et al. (2010, 2011), Hou et al. (2011), Malischewsky et al. (2012), Bianchi and Scarpa (2013), Lim (2013), Goldstein et al. (2014), Lim et al. (2014a, b), Sobieszczyk et al. (2015), Boldrin et al. (2016), Reda et al. (2016), and He and Huang (2018). The effect of auxeticity on various types of waves as well as wave reflections and transmissions involving auxetic solids has been summarized in Chaps. 12 and 13 of Lim (2015a).

6.2 Fundamentals

This section establishes a generic approach for obtaining the longitudinal waves with cross-section and density corrections. Throughout the entire chapter, we set up solids such that the wave propagates parallel to the x-axis of a Cartesian coordinate system, with the y–z plane defining the cross-section of the solids. For the section on wave motion through a cylindrical rod parallel to its axis, it is obviously expedient to use the cylindrical coordinate system. We further confine ourselves to cases where the solids are prismatic such that their cross-sections retain their respective geometry throughout the x-axis in their original unstressed state so that any change to the cross-section arises from the presence of stresses.

The equation of motion for longitudinal wave based on the indicated elemental volume, allowing for variable cross-sectional area and variable density, is written as $A_{x+\mathrm{d}x}\sigma_{x+\mathrm{d}x} - A_x\sigma_x = \mathrm{d}m\left(\partial^2 u_x/\partial t^2\right)$ based on Eq. (6.1.1), whereby $A_x = A$ is the cross-sectional area corresponding to stress σ_x at location x, while $A_{x+\mathrm{d}x} = A + \mathrm{d}A$ is the cross-sectional area corresponding to stress $\sigma_x + \mathrm{d}\sigma_x$ at location $x + \mathrm{d}x$. Note that A_0 is the cross-sectional area corresponding to the original, unstressed, solid, which is not applicable to the elemental volume under consideration. As such, the equation of motion can be expressed as

$$(A + \mathrm{d}A)(\sigma_x + \mathrm{d}\sigma_x) - A\sigma_x = \rho \mathrm{d}V \frac{\partial^2 u_x}{\partial t^2} \tag{6.2.1}$$

where the elemental mass, again, is the product of density ρ and elemental volume $\mathrm{d}V = \overline{A}\mathrm{d}x$. Alternatively, Eq. (6.2.1) can be expressed as

$$\sigma_x \frac{\mathrm{d}A}{A_0} + \frac{A}{A_0}\mathrm{d}\sigma_x + \mathrm{d}\sigma_x \frac{\mathrm{d}A}{A_0} = \rho \overline{A}\mathrm{d}x \frac{\partial^2 u_x}{\partial t^2} \tag{6.2.2}$$

in which the mean cross-sectional area is confined within the elemental volume of length $\mathrm{d}x$ and is assumed to be a simple average of the cross-sectional areas at x and $x + \mathrm{d}x$, i.e.,

$$\overline{A} = \frac{A_x + A_{x+\mathrm{d}x}}{2} = \frac{A + (A + \mathrm{d}A)}{2} = A + \frac{1}{2}\mathrm{d}A \tag{6.2.3}$$

while the altered density ρ can be expressed in term of the original density ρ_0 on the basis of mass conservation $m = m_0$, i.e.,

$$\rho V = \rho_0 V_0 \tag{6.2.4}$$

and that the choice of Eq. (6.2.2) over Eq. (6.2.1) is for the sake of convenience; the ratio of cross-sectional area in the stressed part over that in the unstressed part can be expressed in terms of strain.

Consider an elemental cuboid of unstressed original dimensions x_0, y_0, and z_0 measured along the x-, y- and z-axes, respectively, which undergoes the following deformation

$$x = x_0 + \mathrm{d}x$$
$$y = y_0 + \mathrm{d}y$$

$$z = z_0 + \mathrm{d}z \tag{6.2.5}$$

This gives the original cross-sectional area

$$A_0 = y_0 z_0 \tag{6.2.6}$$

while the cross-sectional area at a distance x from the reference plane is

$$A = (y_0 + \mathrm{d}y)(z_0 + \mathrm{d}z) = y_0 z_0 + z_0 \mathrm{d}y + y_0 \mathrm{d}z + \mathrm{d}y\mathrm{d}z \tag{6.2.7}$$

or, in terms of strain,

$$\frac{A}{A_0} = 1 + \frac{\mathrm{d}y}{y_0} + \frac{\mathrm{d}z}{z_0} + \frac{\mathrm{d}y}{y_0}\frac{\mathrm{d}z}{z_0} = 1 + \varepsilon_y + \varepsilon_z + \varepsilon_y \varepsilon_z \tag{6.2.8}$$

From Eq. (6.2.4), we have

$$\rho xyz = \rho_0 x_0 y_0 z_0 \tag{6.2.9}$$

Dividing Eq. (6.2.9) by $V_0 = x_0 y_0 z_0$ we obtain

$$\rho\left(1 + \frac{\mathrm{d}x}{x_0}\right)\left(1 + \frac{\mathrm{d}y}{y_0}\right)\left(1 + \frac{\mathrm{d}z}{z_0}\right) = \rho_0 \tag{6.2.10}$$

or

$$\rho = \frac{\rho_0}{(1 + \varepsilon_x)\left(1 + \varepsilon_y\right)(1 + \varepsilon_z)} \tag{6.2.11}$$

within the context of Cartesian coordinate system. In the case of wave motion through a cylindrical solid, it is expedient to use the cylindrical coordinate system, in which the original radius r_0 changes to

$$r = r_0 + \mathrm{d}r \tag{6.2.12}$$

This gives the original cross-sectional area

$$A_0 = \pi r_0^2 \tag{6.2.13}$$

while the cross-sectional area at a distance x is

$$A = \pi (r_0 + \mathrm{d}r)^2 = \pi \left(r_0^2 + 2r_0 \mathrm{d}r + (\mathrm{d}r)^2\right) \tag{6.2.14}$$

or, in terms of strain

$$\frac{A}{A_0} = 1 + 2\frac{\mathrm{d}r}{r_0} + \left(\frac{\mathrm{d}r}{r_0}\right)^2 = 1 + 2\varepsilon_\mathrm{r} + \varepsilon_\mathrm{r}^2 \quad (6.2.15)$$

From Eq. (6.2.4), we have

$$\rho x \pi r^2 = \rho_0 x_0 \pi r_0^2 \quad (6.2.16)$$

Dividing Eq. (6.2.16) by $V_0 = x_0 \pi r_0^2$, we obtain

$$\rho\left(1 + \frac{\mathrm{d}x}{x_0}\right)\left(1 + \frac{\mathrm{d}r}{r_0}\right)^2 = \rho_0 \quad (6.2.17)$$

or

$$\rho = \frac{\rho_0}{(1 + \varepsilon_x)(1 + \varepsilon_\mathrm{r})^2} \quad (6.2.18)$$

within the context of cylindrical coordinate system. Although a generic solution can be obtained by proceeding, it is more practical to state the specific structural elements and any imposed boundary condition in subsequent formulation. It is sufficient at this juncture to identify some relations that will be repeatedly used in subsequent sections and, due to their generality, are produced here the constitutive relationships for isotropic solids

$$\begin{Bmatrix} \varepsilon_x \\ \varepsilon_y \\ \varepsilon_z \end{Bmatrix} = \frac{1}{E}\begin{bmatrix} 1 & -v & -v \\ -v & 1 & -v \\ -v & -v & 1 \end{bmatrix}\begin{Bmatrix} \sigma_x \\ \sigma_y \\ \sigma_z \end{Bmatrix} \quad (6.2.19)$$

and

$$\begin{Bmatrix} \sigma_x \\ \sigma_y \\ \sigma_z \end{Bmatrix} = \frac{E}{(1 + v)(1 - 2v)}\begin{bmatrix} 1 - v & v & v \\ v & 1 - v & v \\ v & v & 1 - v \end{bmatrix}\begin{Bmatrix} \varepsilon_x \\ \varepsilon_y \\ \varepsilon_z \end{Bmatrix} \quad (6.2.20)$$

as well as the following relationship between strain and displacement along the direction of wave propagation

$$\mathrm{d}\varepsilon_x = \frac{\partial \varepsilon_x}{\partial x}\mathrm{d}x = \frac{\partial^2 u_x}{\partial x^2}\mathrm{d}x \quad (6.2.21)$$

The constitutive relationships are useful in subsequent sections involving rods $(\sigma_y = \sigma_z = 0)$, plates $(\varepsilon_y = \sigma_z = 0)$, bulk $(\varepsilon_y = \varepsilon_z = 0)$ in Sects. 6.3–6.5. The more complicating conditions in Sects. 6.6–6.8 require the use of boundary parameters for quantifying elastic restraints on rods acting along the radial direction b_r, on plates acting along the width direction b_w, and on slabs acting along the thickness direction b_t such that the lateral strains can be written as functions of the boundary parameters and longitudinal strains. Equation (6.2.21) is useful for converting $d\varepsilon_x$, arising from $d\sigma_x$, into the form of $\left(\partial^2 u_x / \partial x^2\right) dx$ so as to execute Eq. (6.1.6). The approaches are more instructive by considering examples of individual solid geometries and their complicating boundary conditions in the subsequent sections.

6.3 Example: Longitudinal Waves in Prismatic Auxetic Rods

Here, we consider the problem as being 1D stress state with 3D strain state, hence, effectively a 3D mechanics problem, albeit maintaining a classical approach. Adopting Fig. 6.3, the equation of motion for longitudinal wave in a prismatic rod, allowing for variable cross-sectional area A and variable density ρ, is written as Eq. (6.2.2). Substituting $\varepsilon_y = \varepsilon_z = -v\varepsilon_x$ for a laterally unconstrained isotropic rod into Eq. (6.2.8) gives

$$\frac{A}{A_0} = 1 - 2v\varepsilon_x + v^2\varepsilon_x^2 \tag{6.3.1}$$

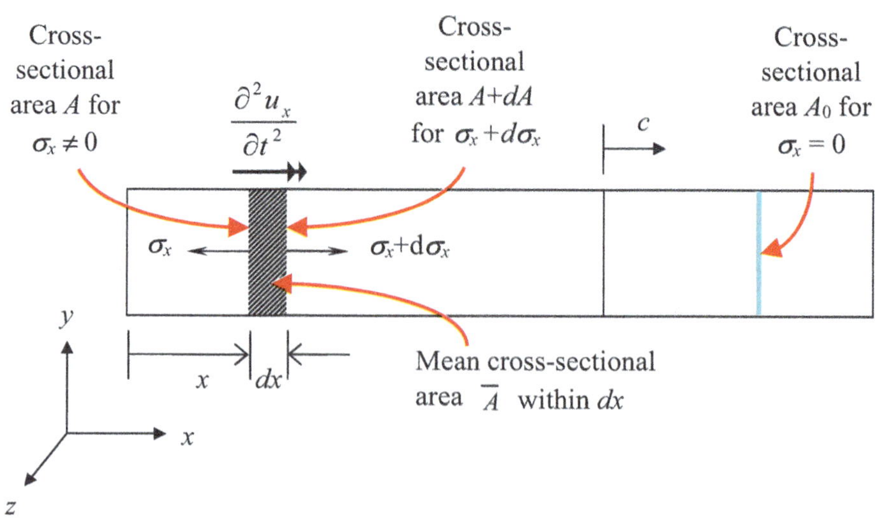

Fig. 6.3 Schematic adopted for analysis of longitudinal wave motion with changes to the cross-sectional area. Note that $A \neq \overline{A} \neq A + dA \neq A_0$, unlike Fig. 6.1

Writing the change in the stresses, strains and cross-sectional areas from x to $x + \mathrm{d}x$ as

$$\left\{\begin{matrix}\sigma_x \\ \varepsilon_x \\ A\end{matrix}\right\}_x \rightarrow \left\{\begin{matrix}\sigma_x + \mathrm{d}\sigma_x \\ \varepsilon_x + \mathrm{d}\varepsilon_x \\ A + \mathrm{d}A\end{matrix}\right\}_{x+\mathrm{d}x} \tag{6.3.2}$$

we have from Eq. (6.3.1)

$$\frac{A + \mathrm{d}A}{A_0} = 1 - 2v(\varepsilon_x + \mathrm{d}\varepsilon_x) + v^2(\varepsilon_x + \mathrm{d}\varepsilon_x)^2 \tag{6.3.3}$$

or its expanded form

$$\frac{A + \mathrm{d}A}{A_0} = 1 - 2v\left(\varepsilon_x + \frac{\partial \varepsilon_x}{\partial x}\mathrm{d}x\right) + v^2\left[\varepsilon_x^2 + 2\varepsilon_x \frac{\partial \varepsilon_x}{\partial x}\mathrm{d}x + \left(\frac{\partial \varepsilon_x}{\partial x}\mathrm{d}x\right)^2\right] \tag{6.3.4}$$

The incremental change in the cross-sectional area, dA, along the infinitesimal thickness dx can then be obtained by subtracting Eq. (6.3.1) from Eq. (6.3.4) to give

$$\frac{\mathrm{d}A}{A_0} = -2v\frac{\partial \varepsilon_x}{\partial x}\mathrm{d}x + 2v^2\varepsilon_x\frac{\partial \varepsilon_x}{\partial x}\mathrm{d}x + v^2\left(\frac{\partial \varepsilon_x}{\partial x}\mathrm{d}x\right)^2 \tag{6.3.5}$$

Substituting Eqs. (6.3.1) and (6.3.5) into Eq. (6.2.3) allows the mean cross-sectional area within dx to be written in terms of longitudinal strain as

$$\frac{\overline{A}}{A_0} = 1 - 2v\varepsilon_x + v^2\varepsilon_x^2 - v\frac{\partial \varepsilon_x}{\partial x}\mathrm{d}x + v^2\varepsilon_x\frac{\partial \varepsilon_x}{\partial x}\mathrm{d}x + \frac{v^2}{2}\left(\frac{\partial \varepsilon_x}{\partial x}\mathrm{d}x\right)^2 \tag{6.3.6}$$

Substituting Eqs. (6.3.1), (6.3.5), (6.3.6), and

$$\left\{\begin{matrix}\sigma_x \\ \mathrm{d}\sigma_x\end{matrix}\right\} = E\left\{\begin{matrix}\varepsilon_x \\ \frac{\partial \varepsilon_x}{\partial x}\mathrm{d}x\end{matrix}\right\} \tag{6.3.7}$$

into Eq. (6.2.2) leads to

$$\begin{aligned}&(1 - 2v\varepsilon_x + v^2\varepsilon_x^2)E\frac{\partial \varepsilon_x}{\partial x}\mathrm{d}x + E\varepsilon_x\left[-2v\frac{\partial \varepsilon_x}{\partial x}\mathrm{d}x + 2v^2\varepsilon_x\frac{\partial \varepsilon_x}{\partial x}\mathrm{d}x + v^2\left(\frac{\partial \varepsilon_x}{\partial x}\mathrm{d}x\right)^2\right] \\ &+ \left[-2v\frac{\partial \varepsilon_x}{\partial x}\mathrm{d}x + 2v^2\varepsilon_x\frac{\partial \varepsilon_x}{\partial x}\mathrm{d}x + v^2\left(\frac{\partial \varepsilon_x}{\partial x}\mathrm{d}x\right)^2\right]E\frac{\partial \varepsilon_x}{\partial x}\mathrm{d}x\end{aligned}$$

$$= \rho\left[1 - 2\upsilon\varepsilon_x + \upsilon^2\varepsilon_x^2 - \upsilon\frac{\partial\varepsilon_x}{\partial x}\mathrm{d}x + \upsilon^2\varepsilon_x\frac{\partial\varepsilon_x}{\partial x}\mathrm{d}x + \frac{\upsilon^2}{2}\left(\frac{\partial\varepsilon_x}{\partial x}\mathrm{d}x\right)^2\right]\mathrm{d}x\frac{\partial^2 u_x}{\partial t^2} \tag{6.3.8}$$

For isotropic rod that is laterally unconstrained, the substitution of $\varepsilon_y = \varepsilon_z = -\upsilon\varepsilon_x$ into Eq. (6.2.11) gives

$$\rho = \frac{\rho_0}{(1+\varepsilon_x)(1-\upsilon\varepsilon_x)^2} \tag{6.3.9}$$

Substituting Eq. (6.3.9) into Eq. (6.3.8) gives

$$\begin{aligned}&(1 - 2\upsilon\varepsilon_x + \upsilon^2\varepsilon_x^2)E\frac{\partial\varepsilon_x}{\partial x}\mathrm{d}x + E\varepsilon_x\left[-2\upsilon\frac{\partial\varepsilon_x}{\partial x}\mathrm{d}x + 2\upsilon^2\varepsilon_x\frac{\partial\varepsilon_x}{\partial x}\mathrm{d}x + \upsilon^2\left(\frac{\partial\varepsilon_x}{\partial x}\mathrm{d}x\right)^2\right]\\&+\left[-2\upsilon\frac{\partial\varepsilon_x}{\partial x}\mathrm{d}x + 2\upsilon^2\varepsilon_x\frac{\partial\varepsilon_x}{\partial x}\mathrm{d}x + \upsilon^2\left(\frac{\partial\varepsilon_x}{\partial x}\mathrm{d}x\right)^2\right]E\frac{\partial\varepsilon_x}{\partial x}\mathrm{d}x\\&= \frac{\rho_0}{(1+\varepsilon_x)(1-\upsilon\varepsilon_x)^2}\left[1 - 2\upsilon\varepsilon_x + \upsilon^2\varepsilon_x^2 - \upsilon\frac{\partial\varepsilon_x}{\partial x}\mathrm{d}x + \upsilon^2\varepsilon_x\frac{\partial\varepsilon_x}{\partial x}\mathrm{d}x\right.\\&\left.+\frac{\upsilon^2}{2}\left(\frac{\partial\varepsilon_x}{\partial x}\mathrm{d}x\right)^2\right]\mathrm{d}x\frac{\partial^2 u_x}{\partial t^2}\end{aligned} \tag{6.3.10}$$

Dividing with $E\mathrm{d}x$, and substituting $\varepsilon_x = \frac{\partial u_x}{\mathrm{d}x}$, we have the following differential equation

$$\begin{aligned}&\left[1 - 2\upsilon\frac{\partial u_x}{\partial x} + \upsilon^2\left(\frac{\partial u_x}{\partial x}\right)^2\right]\frac{\partial^2 u_x}{\partial x^2} + \frac{\partial u_x}{\partial x}\left[-2\upsilon\frac{\partial^2 u_x}{\partial x^2} + 2\upsilon^2\frac{\partial u_x}{\partial x}\frac{\partial^2 u_x}{\partial x^2} + \upsilon^2\left(\frac{\partial^2 u_x}{\partial x^2}\right)^2\mathrm{d}x\right]\\&+\left[-2\upsilon\frac{\partial^2 u_x}{\partial x^2} + 2\upsilon^2\frac{\partial u_x}{\partial x}\frac{\partial^2 u_x}{\partial x^2} + \upsilon^2\left(\frac{\partial^2 u_x}{\partial x^2}\right)^2\mathrm{d}x\right]\frac{\partial^2 u_x}{\partial x^2}\mathrm{d}x\\&= \frac{\rho_0}{E(1+\varepsilon_x)(1-\upsilon\varepsilon_x)^2}\left[1 - 2\upsilon\frac{\partial u_x}{\partial x} + \upsilon^2\left(\frac{\partial u_x}{\partial x}\right)^2 - \upsilon\frac{\partial^2 u_x}{\partial x^2}\mathrm{d}x\right.\\&\left.+\upsilon^2\frac{\partial u_x}{\partial x}\frac{\partial^2 u_x}{\partial x^2}\mathrm{d}x + \frac{\upsilon^2}{2}\left(\frac{\partial^2 u_x}{\partial x^2}\mathrm{d}x\right)^2\right]\frac{\partial^2 u_x}{\partial t^2}\end{aligned} \tag{6.3.11}$$

for describing the longitudinal wave motion in prismatic bars with lateral inertia. At zero Poisson's ratio, Eq. (6.3.11) reduces to

$$\frac{\partial^2 u_x}{\partial x^2} = \frac{\rho_0}{E(1+\varepsilon_x)}\frac{\partial^2 u_x}{\partial t^2} \tag{6.3.12}$$

which further simplifies to Eq. (6.1.5) for negligible strain. Neglecting terms with $\mathrm{d}x$ and $(\mathrm{d}x)^2$, Eq. (6.3.11) becomes

$$\left[1-4v\frac{\partial u_x}{\partial x}+3v^2\left(\frac{\partial u_x}{\partial x}\right)^2\right]\frac{\partial^2 u_x}{\partial x^2}=\frac{\rho_0}{E(1+\varepsilon_x)(1-v\varepsilon_x)^2}\left[1-2v\frac{\partial u_x}{\partial x}+v^2\left(\frac{\partial u_x}{\partial x}\right)^2\right]\frac{\partial^2 u_x}{\partial t^2} \tag{6.3.13}$$

Substituting $\partial u_x/\partial x=\varepsilon_x$ into the square parentheses of Eq. (6.3.13), the phase velocity with lateral inertia and its corresponding density correction can be expressed in terms of strain as (Lim 2015b)

$$c=\sqrt{\left(\frac{\partial^2 u_x}{\partial t^2}\right)/\left(\frac{\partial^2 u_x}{\partial x^2}\right)}=\sqrt{(1+\varepsilon_x)(1-v\varepsilon_x)(1-3v\varepsilon_x)\frac{E}{\rho_0}} \tag{6.3.14}$$

on the basis of Eq. (6.1.6). Here, we observe three special cases in which two of the most obvious are at the limiting Poisson's ratio in the auxetic range, i.e.,

$$c=\begin{cases}(1+\varepsilon_x)\frac{E}{\rho_0}; & v=0\\ (1+\varepsilon_x)\sqrt{(1+3\varepsilon_x)\frac{E}{\rho_0}}; & v=-1\end{cases} \tag{6.3.15}$$

Another special case takes place under Cauchy relation to give the following approximation

$$c=\sqrt{\left(1-\frac{13}{16}\varepsilon_x^2+\frac{3}{16}\varepsilon_x^3\right)\frac{E}{\rho_0}}\approx\sqrt{\frac{E}{\rho_0}};\quad v=\frac{1}{4} \tag{6.3.16}$$

To provide a visual clarity on the interlacing effect of auxeticity and strain on the longitudinal wave velocity, a family of $c\sqrt{\rho_0/E}$ versus v curves are shown in Fig. 6.4 for $\varepsilon_x=-5,\ -2.5,\ 0,\ 2.5,\ 5\%$. These results suggest that the use of elementary wave velocity, as described by Eq. (6.1.7), is valid only when the rod is made from isotropic and conventional materials. However, if the rod is made from isotropic auxetic material, both the Poisson's ratio and strain must be taken in account in order to incorporate their influence on the wave velocity.

To shed insight on the effects from density and cross-section corrections, it is instructive to express from Eq. (6.3.13) the wave speed

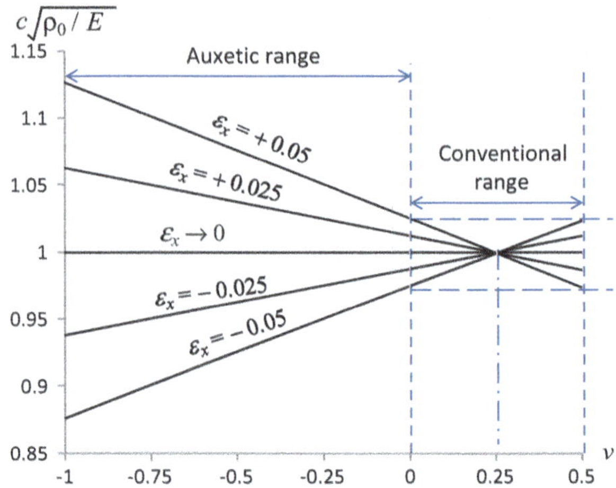

Fig. 6.4 Effect of constant axial strain and rod auxeticity on the velocity of longitudinal waves

$$c = \sqrt{\left(\frac{\partial^2 u_x}{\partial t^2}\right) / \left(\frac{\partial^2 u_x}{\partial x^2}\right)} = \sqrt{\frac{\left[(1+\varepsilon_x)(1-v\varepsilon_x)^2\right]}{\rho_0} E \left[\frac{1-4v\varepsilon_x+3v^2\varepsilon_x^2}{1-2v\varepsilon_x+v^2\varepsilon_x^2}\right]}$$
$$= \sqrt{\frac{E}{\rho_0}\left[(1+\varepsilon_x)(1-v\varepsilon_x)^2\right]\left[\frac{1-3v\varepsilon_x}{1-v\varepsilon_x}\right]} = \sqrt{\frac{E}{\rho_0} f(v,\varepsilon_x) g(v,\varepsilon_x)} \quad (6.3.17)$$

where the terms in the first square parenthesis

$$f(v,\varepsilon_x) = (1+\varepsilon_x)(1-v\varepsilon_x)^2 \quad (6.3.18)$$

reflects the density correction originating from Eq. (6.3.9), while the terms in the second parenthesis

$$g(v,\varepsilon_x) = \frac{1-4v\varepsilon_x+3v^2\varepsilon_x^2}{1-2v\varepsilon_x+v^2\varepsilon_x^2} = \frac{1-3v\varepsilon_x}{1-v\varepsilon_x} \quad (6.3.19)$$

involve the cross-sectional corrections, in which the numerator $1-4v\varepsilon_x+3v^2\varepsilon_x^2$ and denominator $1-2v\varepsilon_x+v^2\varepsilon_x^2$ originate from the LHS and RHS of Eq. (6.3.13), respectively. Had there been no correction then $f(v,\varepsilon_x) = g(v,\varepsilon_x) = 1$ to give the horizontal straight line of $c\sqrt{\rho_0/E}$ versus v as shown in Fig. 6.5 (top left). Suppose only density correction is made, i.e., no cross-sectional correction $g(v,\varepsilon_x) = 1$, then one would obtain Fig. 6.5 (top right), in which the dimensionless wave speed is common only for the condition of incompressibility $v = 1/2$. If only the cross-sectional area correction is implemented, i.e., no density correction $f(v,\varepsilon_x) = 1$

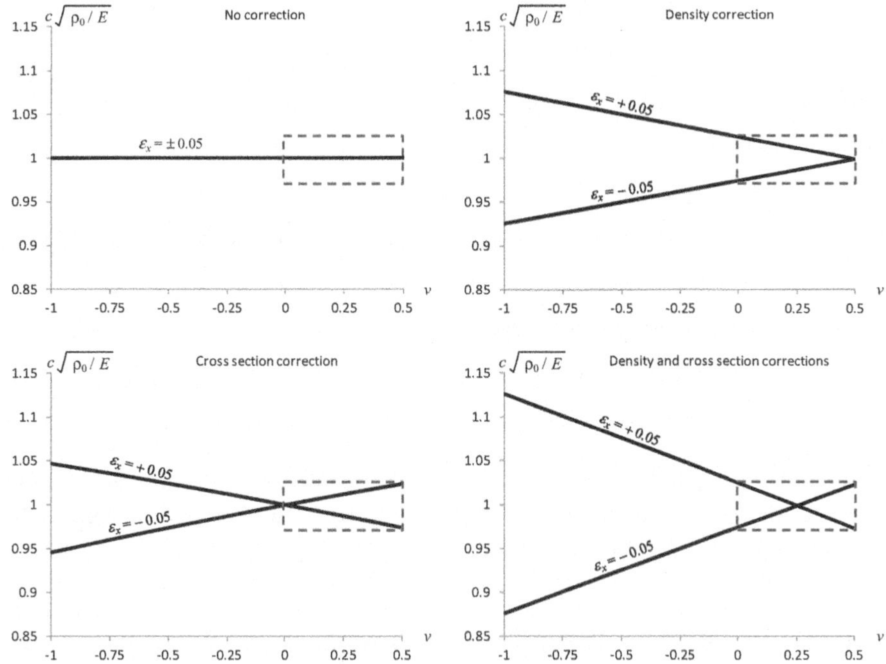

Fig. 6.5 Dimensionless longitudinal wave speed in a prismatic rod based on classical elementary model (top left), density corrected model (top right), cross section-corrected model (bottom left), and combined density and cross section-corrected model (bottom right)

then one would obtain Fig. 6.5 (bottom left), whereby the wave speed is common when there is no change to the cross-sectional area $v = 0$. When both corrections are made, we obtain the plots in Fig. 6.5 (bottom right), which clearly shows that the wave speed's deviation becomes larger when the Poisson's ratio of the rod material is negative. The dashed rectangular box is added for visual aid.

Having shown that the plots of dimensionless wave velocity versus rod Poisson's ratio being almost linear, it is desirable to introduce a simplified model for ease of application by design and other practice-oriented engineers. Expanding the terms inside the square root of Eq. (6.3.14) gives

$$c\sqrt{\frac{\rho_0}{E}} = \sqrt{1 + (1 - 4v)\varepsilon_x + (3v - 4)v\varepsilon_x^2 + 3v^2\varepsilon_x^3} \tag{6.3.20}$$

Using Taylor series expansion

$$\sqrt{1 + \psi} = \sum_{n=0}^{\infty} \frac{(-1)^n (2n)!}{4^n (n!)^2 (1 - 2n)} \psi^n \tag{6.3.21}$$

where

$$\psi = (1 - 4v)\varepsilon_x + (3v - 4)v\varepsilon_x^2 + 3v^2\varepsilon_x^3 \tag{6.3.22}$$

we have a series

$$\sqrt{1+\psi} = 1 + \frac{1}{2}\psi - \frac{1}{8}\psi^2 + \frac{1}{16}\psi^3 - \frac{5}{128}\psi^4 + \frac{7}{256}\psi^5 - \frac{21}{1024}\psi^6 + \cdots \tag{6.3.23}$$

Neglecting higher orders of ε_x leads to

$$c\sqrt{\frac{\rho_0}{E}} \approx 1 + \frac{1-4v}{2}\varepsilon_x \tag{6.3.24}$$

A family of longitudinal wave velocity curves by this design equation in comparison with the theoretical approach described Eq. (6.3.14) is plotted in Fig. 6.6. It can be seen that the design equation gives a very good approximation, especially for $|\varepsilon_x| \leq 5\%$ and even at a considerably large strain magnitude of $|\varepsilon_x| = 10\%$, with only a slight under estimation and overestimation for $v \leq -0.2$ and $v > -0.1$, respectively. The illustration using large strain demonstrates the accuracy of the design model well beyond actual strain encountered in practice.

The validity range of the elementary model in terms of percentage error with reference to the corrected model is listed in Table 6.1. Suppose a strict criterion is imposed such that only a 1% error is allowed, then the range of longitudinal strain

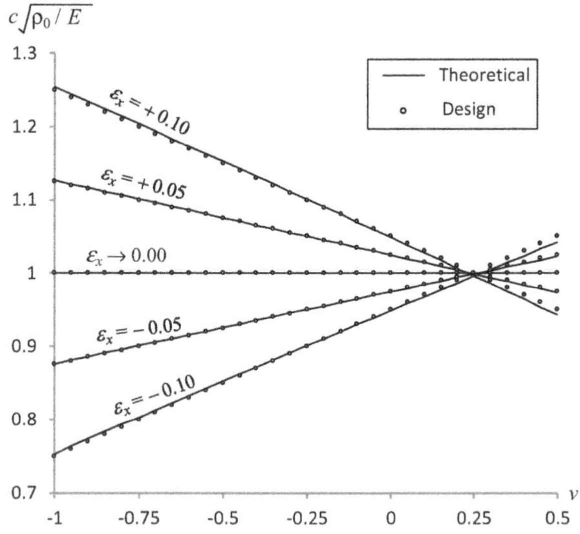

Fig. 6.6 Comparison between the theoretical curves and the proposed design equation

Table 6.1 Validity range of wave velocity by the elementary model as a percentage error with respect to the corrected model

Poisson's ratio of rod	Error within ±1%	Error within ±5%
$v = -1$	$-0.4\% < \varepsilon_x < 0.4\%$	$-2.01\% < \varepsilon_x < 1.99\%$
$v = -3/4$	$-0.5\% < \varepsilon_x < 0.5\%$	$-2.51\% < \varepsilon_x < 2.49\%$
$v = -1/2$	$-0.67\% < \varepsilon_x < 0.67\%$	$-3.35\% < \varepsilon_x < 3.32\%$
$v = -1/4$	$-1\% < \varepsilon_x < 1\%$	$-5.02\% < \varepsilon_x < 4.98\%$

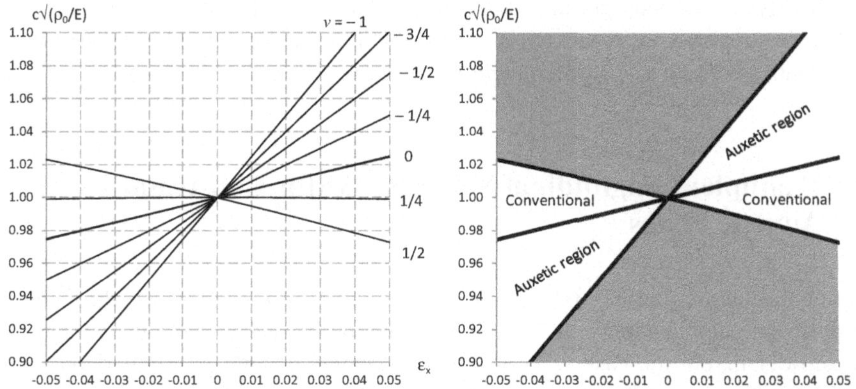

Fig. 6.7 Plots of dimensionless longitudinal wave speed in a prismatic rod versus longitudinal strain for various Poisson's ratio (left) and loci of conventional and auxetic regions (right)

in which the elementary model is valid is very narrow. Outside this validity range, the corrected model is to be used for accuracy.

As an alternative to Fig. 6.4, one may plot the dimensionless wave speed against the longitudinal wave strain for various Poisson's ratio of the rod material, as shown in Fig. 6.7 (left). If the curves are plotted for Poisson's ratio of $v = -1, 0, 1/2$, then the conventional and auxetic regions are demarcated as in Fig. 6.7 (right), whereby the theoretically impossible regions are shaded. These results suggest that the use of elementary wave velocity, as described by Eq. (6.1.7), is valid only when the isotropic rod is made from conventional materials, and especially so when $v = 1/4$. For this reason, the usual derivation of longitudinal wave velocity in prismatic bars neglects any changes to the cross-sectional area as well as any changes to the volume, and hence density, of the bar. However, deviation from the elementary model becomes large in the auxetic region, especially as $v \to -1$; hence, the use of the current wave velocity model provides a more accurate prediction. Plotted results also suggest that auxetic rods can be beneficial if there is a need to control the speed of wave propagation in rods, i.e., the use of highly auxetic rods can slow down or speed up the propagation of compressive and tensile stress waves, respectively.

In summary, the elementary model for describing the longitudinal wave velocity in a prismatic bar is in terms of Young's modulus and density. Density correction is taken into consideration due to the larger change of density in auxetic solids, while lateral inertia is accounted for because the range of Poisson's ratio is larger in the auxetic range than in the conventional range for isotropic solids. When both the density correction and lateral inertia are incorporated, the wave velocity is also a function of the Poisson's ratio and axial strain. Results show that the use of the wave velocity by the elementary model suffices if the rod is made from isotropic conventional materials, but the proposed model must be used if the rod is made from isotropic auxetic materials. Plotted results also suggest that auxetic rods can be beneficial if there is a need to control the speed of wave propagation in rods, i.e., the use of highly auxetic rods can slow down or speed up the propagation of compressive and tensile stress waves, respectively. A proposed design equation has been shown to be reliable even up to a strain magnitude of 10%.

6.4 Example: Longitudinal Waves in Width-Constrained Auxetic Plates

Figure 6.8 illustrates a longitudinal wave propagating in the x-direction through an isotropic plate that lies in the x, y plane, with the thickness being in the z-direction. We consider herein the case whereby the wavelength is long in comparison with the plate thickness. Furthermore, an attempt is made to present a nonlinear model, though the effects of finite wave amplitude are neglected in the interest of allowing the formulation to be more tractable. It should now be pointed out that for a longitudinal wave to propagate through a plate that is constrained from motion in the width direction, the physical problem is that of 2D stress and 2D strain, but 3D mechanics. The reason is that the stresses are in the x–y plane while the strains are in the x–z plane, thereby rendering the stress and strain mechanics effectively in all directions; this differs from the purely 2D approach in the classical elementary treatment.

To apply Eq. (6.2.2) for solving this plate problem, there is a need to obtain a set of stress-strain relationship for the given problem of constrained width, i.e., the plate is arrested from motion in the y-direction, but the plate material is allowed to displace in the wave propagation direction (or x-direction) and in the thickness

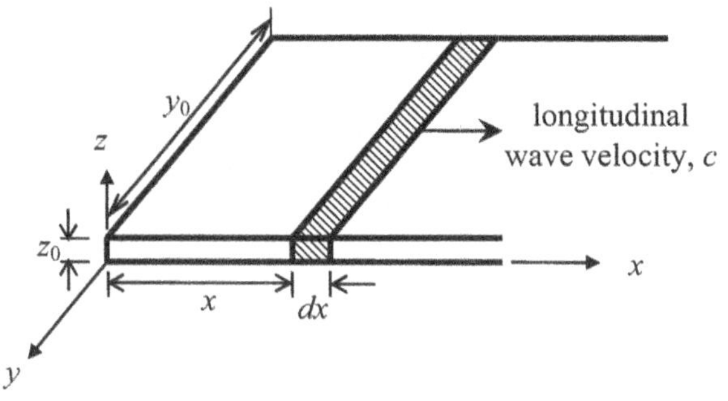

Fig. 6.8 Schematic diagram for a longitudinal wave in a plate

direction (or z-direction). Arising from this constraint, it follows that every point on the plate is constrained from moving in the y-direction. This constraint is inherent for the case of an "infinite" plate, in which the infinite width renders the strain to be zero in the width direction. Without this width constraint, an isotropic plate of finite width will exhibit equal strain in the thickness and width directions, thereby leading to a longitudinal wave velocity similar to that in prismatic bars. Naturally, the stress is assumed zero in the out-of-plane direction as the upper and lower surfaces of the plate are traction free. Applying the restriction of strain in the width direction of the plate in the constitutive relation of Eq. (6.2.19) with

$$\varepsilon_y = \sigma_z = 0 \tag{6.4.1}$$

gives the following relations

$$\sigma_y = v\sigma_x \tag{6.4.2}$$

$$\varepsilon_x = \left(1 - v^2\right)\frac{\sigma_x}{E} \tag{6.4.3}$$

$$\varepsilon_z = -v(1 + v)\frac{\sigma_x}{E} \tag{6.4.4}$$

With reference to Eqs. (6.4.3) and (6.4.4), it is useful to write the following formulae in differential forms

$$\begin{aligned} \sigma_x &= \frac{E}{1 - v^2}\varepsilon_x \\ \mathrm{d}\sigma_x &= \frac{E}{1 - v^2}\mathrm{d}\varepsilon_x \end{aligned} \tag{6.4.5}$$

and

$$\begin{aligned} \varepsilon_z &= -\frac{v}{1 - v}\varepsilon_x \\ \mathrm{d}\varepsilon_z &= -\frac{v}{1 - v}\mathrm{d}\varepsilon_x \end{aligned} \tag{6.4.6}$$

For small strain, we write

$$\varepsilon_z = \frac{\mathrm{d}A}{A_0} \tag{6.4.7}$$

since $\varepsilon_y = 0$ as per definition of infinitesimal strain. In the case of nonlinear analysis, we write the infinitesimal strain

$$d\varepsilon_z = \frac{dA}{A} \tag{6.4.8}$$

Since $\varepsilon_y = 0$, Eq. (6.2.11) becomes

$$\rho = \frac{\rho_0}{(1 + \varepsilon_x)(1 + \varepsilon_z)} \tag{6.4.9}$$

Substituting Eqs. (6.4.5) and (6.4.9) into Eq. (6.2.2) gives

$$\begin{aligned} &\frac{E}{1 - \upsilon^2}\varepsilon_x \frac{dA}{A} + \frac{E}{1 - \upsilon^2} d\varepsilon_x + \frac{E}{1 - \upsilon^2} d\varepsilon_x \frac{dA}{A} \\ &= \frac{\rho_0}{(1 + \varepsilon_x)(1 + \varepsilon_z)} \left(1 + \frac{1}{2}\frac{dA}{A}\right) dx \frac{\partial^2 u_x}{\partial t^2} \end{aligned} \tag{6.4.10}$$

To obtain the wave velocity defined in Eq. (6.1.6), there is a need to express every term on the LHS of Eq. (6.4.10) in terms of $\partial^2 u_x/\partial x^2$, and the term dx on the RHS of Eq. (6.4.10) can be eliminated by writing every term on the LHS of Eq. (6.4.10) in terms of dx. It must be noted that every term on the LHS should have only one $\partial^2 u_x/\partial x^2$ and one dx, while the excess $\partial^2 u_x/\partial x^2$ are to be retained in terms of ε_x. With reference to Eqs. (6.4.6) and (6.4.8), Eq. (6.4.10) becomes

$$\begin{aligned} &\frac{E}{1 - \upsilon^2}\left[1 - \frac{\upsilon}{1 - \upsilon}(\varepsilon_x + d\varepsilon_x)\right] d\varepsilon_x \\ &= \frac{\rho_0}{(1 + \varepsilon_x)\left(1 - \frac{\upsilon}{1-\upsilon}\varepsilon_x\right)}\left[1 - \frac{1}{2}\left(\frac{\upsilon}{1 - \upsilon}\right) d\varepsilon_x\right] dx \frac{\partial^2 u_x}{\partial t^2} \end{aligned} \tag{6.4.11}$$

upon rearranging terms. Substituting Eq. (6.2.21) into the LHS of Eq. (6.4.11) while retaining the terms $(\varepsilon_x + d\varepsilon_x)$ as they are, we obtain the longitudinal wave velocity

$$\begin{aligned} c &= \sqrt{\left(\frac{\partial^2 u_x}{\partial t^2}\right) / \left(\frac{\partial^2 u_x}{\partial x^2}\right)} \\ &= \sqrt{\frac{E}{\rho_0 (1 - \upsilon^2)} \frac{(1 + \varepsilon_x)\left(1 - \frac{\upsilon}{1-\upsilon}\varepsilon_x\right)\left[1 - \frac{\upsilon}{1-\upsilon}(\varepsilon_x + d\varepsilon_x)\right]}{1 - \frac{1}{2}\left(\frac{\upsilon}{1-\upsilon}\right) d\varepsilon_x}} \end{aligned} \tag{6.4.12}$$

Neglecting the terms containing $d\varepsilon_x$, Eq. (6.4.12) reduces to (Lim 2016a)

$$c = \left(1 - \frac{\upsilon}{1 - \upsilon}\varepsilon_x\right)\sqrt{\frac{E(1 + \varepsilon_x)}{\rho_0 (1 - \upsilon^2)}} \tag{6.4.13}$$

Suppose the strain is neglected (i.e., $\varepsilon_x = 0$), Eq. (6.4.13) greatly simplifies to the classical wave velocity of longitudinal waves in plates as described by Eq. (6.1.8).

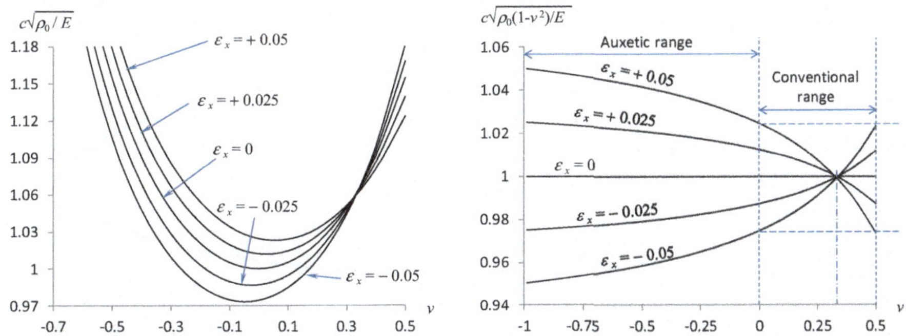

Fig. 6.9 Combined effect of Poisson's ratio and longitudinal strain on wave velocity, normalized against Young's modulus and density (left), and the classical longitudinal wave velocity in plates (right)

Figure 6.9 shows the combined effect of Poisson's ratio and longitudinal strain on the wave velocity in width-constrained plates using Eq. (6.4.13). As a way of non-dimensionalization, the wave velocity plotted in Fig. 6.9 (left) is normalized against the original density ρ_0 and the Young's modulus E of the plate material so as to clearly observe the effect from Poisson's ratio. To reflect the deviation of the wave velocity from that of the classical elementary plate model, the wave velocities in Fig. 6.9 (right) are normalized against Eq. (6.1.8).

Figure 6.9 (right) reveals that the deviation is negligible in the conventional range, i.e., the wave velocity error magnitudes are about 1.25% and 2.5% for $\varepsilon_x = \pm 2.5\%$ and $\varepsilon_x = \pm 5\%$, respectively, when $0 \le v \le 1/2$. Furthermore, Eq. (6.4.13) approximates the classical model when $0.3 < v < 0.4$. Since the Poisson's ratio of most materials is about 0.3, it follows that the use of the classical model is quite accurate for most materials. However, the wave velocity deviates to a larger extent in the auxetic region. This is especially so when $v = -1$, which results in error magnitudes of 2.5% and 5% for $\varepsilon_x = 2.5\%$ and $\varepsilon_x = 5\%$, respectively.

To shed insight on the result, it is useful to rewrite Eq. (6.4.13) in the form

$$c = \sqrt{\frac{E}{\rho_0 \left(1 - v^2\right)} f(v, \varepsilon_x) g(v, \varepsilon_x)} \tag{6.4.14}$$

where

$$f(v, \varepsilon_x) = (1 + \varepsilon_x)\left(1 - \frac{v}{1 - v}\varepsilon_x\right) \tag{6.4.15}$$

is the density correction function, while

$$g(v, \varepsilon_x) = 1 - \frac{v}{1 - v}(\varepsilon_x + \mathrm{d}\varepsilon_x) \approx 1 - \frac{v}{1 - v}\varepsilon_x \tag{6.4.16}$$

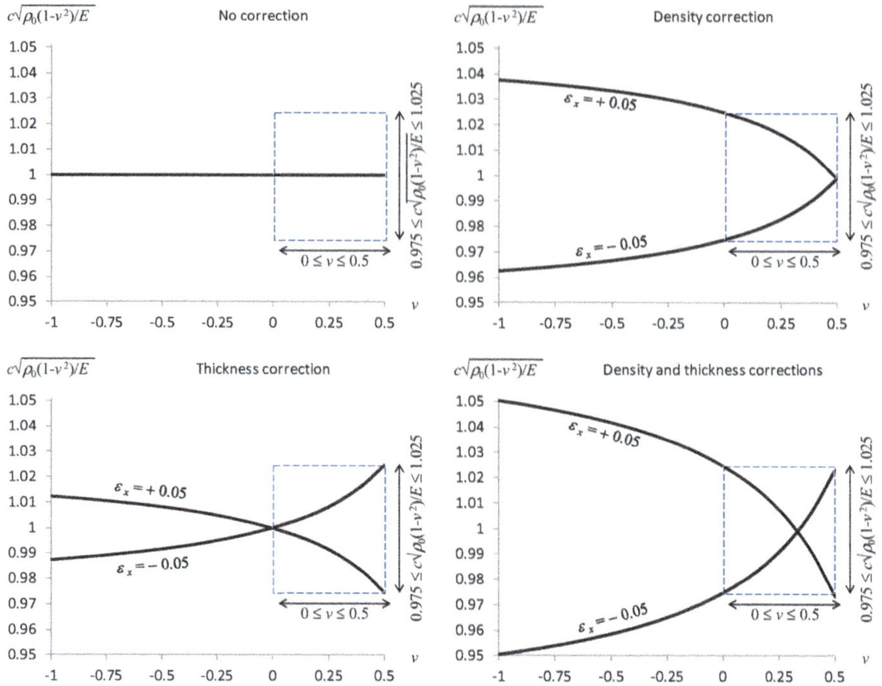

Fig. 6.10 Dimensionless longitudinal wave speed in width-constrained flat plates based on classical elementary model (top left), density corrected model (top right), thickness-corrected model (bottom left), and combined density and thickness-corrected model (bottom right)

is the cross-section or thickness correction function. Using Eqs. (6.4.14)–(6.4.16), Fig. 6.10 shows the plots of dimensionless longitudinal wave velocity in a width-constrained plate considering the classical model, i.e., $f(v, \varepsilon_x) = g(v, \varepsilon_x) = 1$ (top left), the density corrected model $g(v, \varepsilon_x) = 1$ (top right), the thickness-corrected model $f(v, \varepsilon_x) = 1$ (bottom left), and the combined thickness and density corrected model (bottom right), at $\varepsilon_x = \pm 5\%$.

To aid comparison, the range of the normalized wave velocity and Poisson's ratio are being standardized as $0.95 \le c\sqrt{\rho_0(1 - v^2)/E} \le 1.05$ and $-1 \le v \le 1/2$. Although obvious, Fig. 6.10 (top left) is shown for completeness' sake. When only the density correction is taken into consideration, the wave velocity deviates when $v \ne 1/2$, as shown in Fig. 6.10 (top right). This is because $v = 1/2$ is synonymous with incompressibility. When only the thickness correction is applied, the wave velocity deviates when $v \ne 0$, as shown in Fig. 6.10 (bottom left). This is because there is no change in the cross-sectional area when $v = 0$. When both the thickness and the density corrections are taken into account, the wave velocity deviates as shown in Fig. 6.10 (bottom right), i.e., deviation is more significant in the auxetic range than in conventional range, and that the deviation is the most at $v = -1$ and the least at about $v = 1/3$. It can be seen from Fig. 6.10 (bottom right) that for $\varepsilon_x = \pm 5\%$, the range of the normalized wave velocity within the conventional region can be approximated by $0.975 \leq c\sqrt{\rho_0(1-v^2)/E} \leq 1.025$. For comparison with the classical model and the inconsistent models, a dashed rectangle defining $0.975 \leq c\sqrt{\rho_0(1-v^2)/E} \leq 1.025$ and $0 \leq v \leq 1/2$ are superposed in all graphs of Fig. 6.10.

Figure 6.11 (left) shows a family of $c\sqrt{\rho_0(1-v^2)/E}$ curves plotted against the longitudinal strain ε_x. It is interesting to point out that when $v = 1/3$, the curves of $c\sqrt{\rho_0(1-v^2)/E}$ versus ε_x suggest that the wave velocity's deviation from the classical model is negligible. For this reason, the use of Eq. (6.1.8) for modeling the wave velocity of plates is acceptable since the Poisson's ratio of most materials is about $v \sim 0.3$. However, when the Poisson's ratio deviates from 1/3, and especially when the Poisson's ratio is negative, the corrected model is required. The loci of conventional and auxetic regions are furnished in Fig. 6.11 (right), showing that the auxetic region is narrower than the conventional region even though the Poisson's ratio of the auxetic range ($-1 \leq v < 0$) is larger than that for the conventional range ($0 \leq v \leq 1/2$).

In summary, changes to the plate density and thickness have opposing effect on the wave velocity in the conventional region, but they have additive effect in the auxetic region. When both the effects of dimensional (thickness) and density variations are taken into account in the wave model and normalized against the longitudinal wave velocity according to the classical model for plates, it was found that the developed wave velocity model herein coincides with that of the classical model when the Poisson's ratio is about 1/3. Within the range of strain demonstrated the normalized wave velocity exhibits an almost linear correlation with the longitudinal strain, while

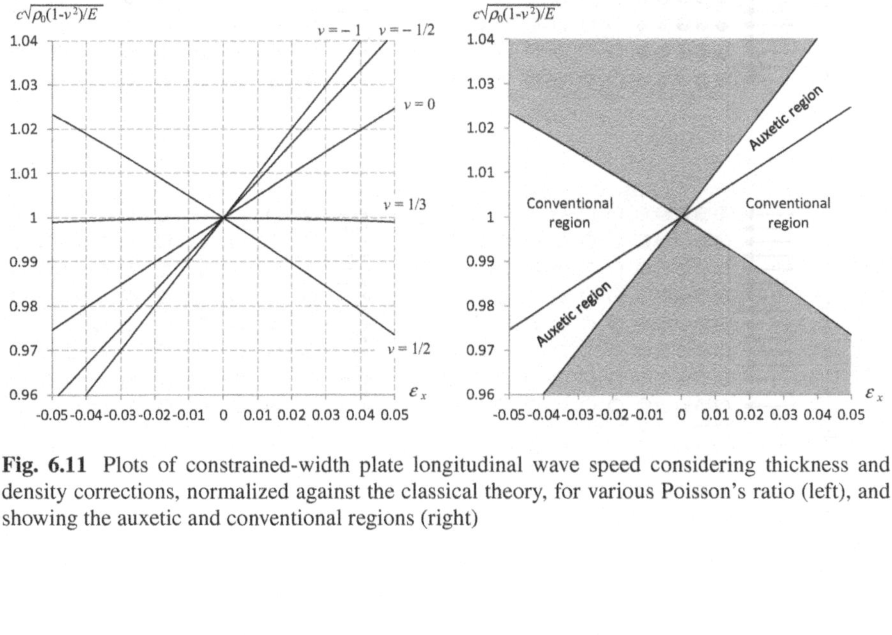

Fig. 6.11 Plots of constrained-width plate longitudinal wave speed considering thickness and density corrections, normalized against the classical theory, for various Poisson's ratio (left), and showing the auxetic and conventional regions (right)

auxeticity suppresses and enhances the wave velocities under compressive and tensile loads, respectively. The results derived from this study suggest that the developed wave velocity model is more accurate than the classical model when dealing with large strains and auxetic materials, and that auxetic materials can be taken advantage of in order to implement passive control of elastic wave speed in plates.

6.5 Example: Longitudinal Waves in Infinite Bulk Auxetic Solids

The opposing nature of the transverse strain of unbounded auxetic solids in comparison with conventional solids translates into opposite transverse stresses in the case of bounded auxetic solids vis-à-vis conventional ones. An example of this case is that of plane waves of dilatation in infinite solids, whereby the application of compressive or tensile impact leads to tensile or compressive lateral stresses, respectively, as opposed to compressive and tensile lateral stresses, respectively, behind the wave front in satisfying zero transverse strain. See Fig. 6.12. Unlike other works on plane waves of dilatation, which considers fluctuating density, the present analysis is confined to

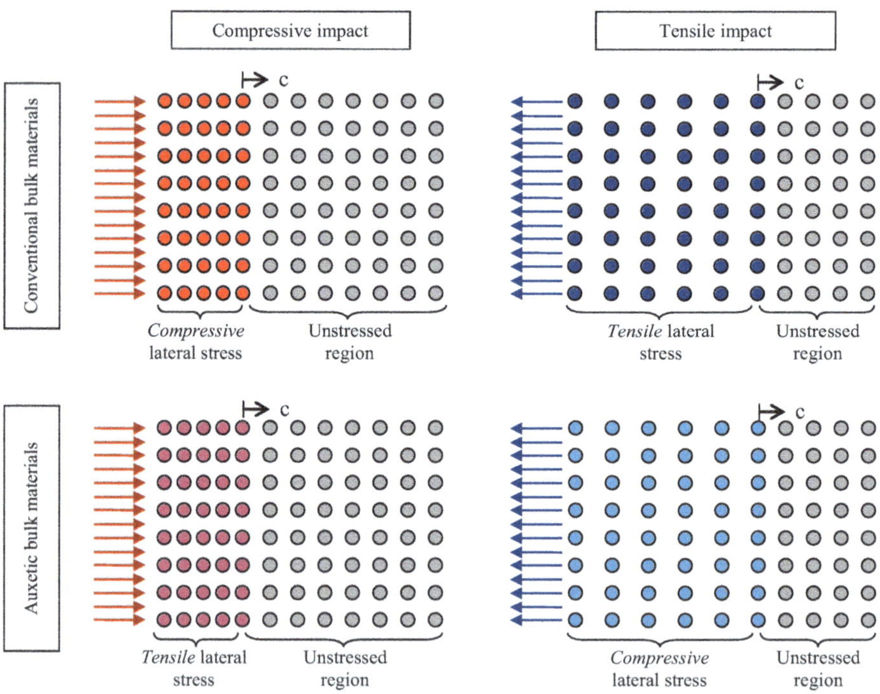

Fig. 6.12 Comparison of lateral stresses for plane waves of dilatation in conventional and auxetic bulk materials

the case where the load at the impacted end is sustained in such a manner so as to maintain a constant strain behind the wave front.

In the previous analyses of wave speed in auxetic rods and plates, *both* the dimensional *and* density changes are incorporated. For example, the analysis of longitudinal wave speed in prismatic auxetic rods takes into consideration the lateral strain while the analysis of longitudinal wave speed in width-constrained plates takes into consideration the thickness strain in addition to accounting for the density variation in both analyses. The analysis of plane waves of dilatation is comparatively straight forward as no dimensional correction is required—only density correction is needed. Nevertheless, analysis on plane waves of dilatation in auxetic solids is justified at this juncture to complement the previous analyses of longitudinal waves in rods and plates.

The classical model for the plane wave of dilatation's speed has been given in Eq. (6.1.9). In the stressed part of the solid, the strain alters the density. Hence,

$$c = \sqrt{\frac{E(1-v)}{\rho(1+v)(1-2v)}} \tag{6.5.1}$$

where ρ is the variable density in the stressed portion of the solid. Although the use of the variable ρ is more accurate, it is the original density ρ_0 data that is easily available for predictive modeling; hence, there is a need to express ρ in terms of ρ_0. Since the boundary condition of the RVE must satisfy $\varepsilon_y = \varepsilon_z = 0$ for plane waves of dilatation propagating in the x-direction, Eq. (6.2.11) simplifies to

$$\rho = \frac{\rho_0}{1+\varepsilon_x} \tag{6.5.2}$$

Using this variable density for Eq. (6.5.1) leads to (Lim 2016b)

$$c = \sqrt{\frac{E(1+\varepsilon_x)(1-v)}{\rho_0(1+v)(1-2v)}} \tag{6.5.3}$$

A possible dimensionless speed for plane waves of dilatation can be introduced in the form

$$c\sqrt{\frac{\rho_0}{E}} = \sqrt{\left(\frac{1+\varepsilon_x}{1+v}\right)\left(\frac{1-v}{1-2v}\right)} \tag{6.5.4}$$

so as to facilitate observation of Poisson's ratio as the sole mechanical property on the wave speed. As shown in Fig. 6.13 (left), the Poisson's ratio and the strain play primary and secondary roles, respectively, on the dimensionless wave speed. Due to the small gap between the curves, a close-up view of the dimensionless wave speed is furnished in Fig. 6.13 (right) for clarity. It is evident that while the dimensionless

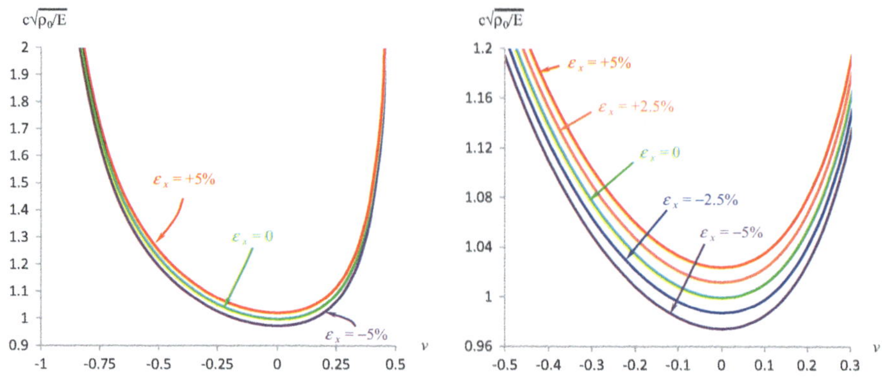

Fig. 6.13 Curves of plane waves of dilatation's dimensionless speed plotted against Poisson's ratio of bulk materials for the entire range of Poisson's ratio of isotropic solids (left) and close-up view (right)

wave speed increases with the magnitude of the Poisson's ratio, there is better control of wave speed in the auxetic region as the change in the dimensionless wave speed in that region is more gradual than in the conventional region.

It should be noted that the dimensionless wave speed described in Eq. (6.5.4) is essentially a normalization against the classical speed of longitudinal waves in prismatic rods, i.e., $c = \sqrt{E/\rho_0}$ and was used in the analysis of longitudinal wave speed in prismatic rods with lateral and density corrections in Sect. 6.3. On the other hand, the longitudinal wave speed in width-constrained plates, considering thickness and density corrections, was normalized against the classical speed of longitudinal waves in plates, i.e., $c = \sqrt{E/\left[\rho_0\left(1 - \upsilon^2\right)\right]}$, in Sect. 6.4. Going by the same line of thought, it is possible to normalize the presently proposed speed of plane waves of dilatation, considering density correction, against the classical speed of plane waves of dilatation as described by Eq. (6.1.9). Hence, we have the normalized speed of plane waves of dilatation as

$$c\sqrt{\frac{\rho_0(1+\upsilon)(1-2\upsilon)}{E(1-\upsilon)}} = \sqrt{1+\varepsilon_x} \tag{6.5.5}$$

The plot of Eq. (6.5.5) against the longitudinal strain is shown in Fig. 6.14. The strain magnitude being small, it can be said that the normalized speed of plane waves of dilatation is approximately linear, i.e.,

$$c\sqrt{\frac{\rho_0(1+\upsilon)(1-2\upsilon)}{E(1-\upsilon)}} \approx 1 + \frac{\varepsilon_x}{2} \tag{6.5.6}$$

It can be seen that Eq. (6.5.3) reduces to the classical elementary wave speed in prismatic rods Eq. (6.1.7) if the strain is chosen such that

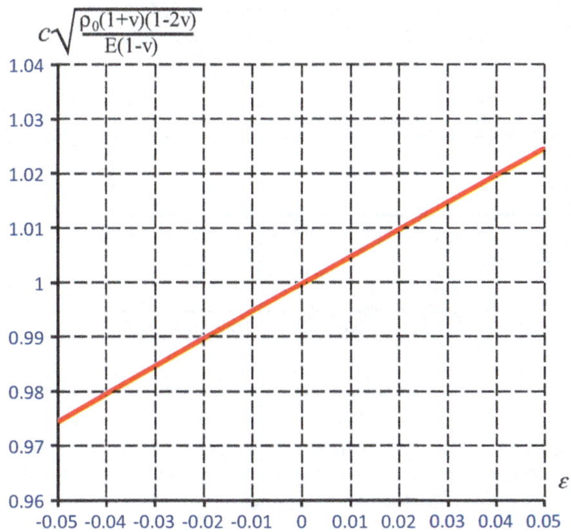

Fig. 6.14 Plane wave of dilatation's speed, normalized against its classical model

$$\varepsilon_x = -\frac{2v^2}{1-v} \tag{6.5.7}$$

In summary, the results reveal that, other than Young's modulus and original density, the Poisson's ratio of the bulk material plays a major role while the longitudinal strain plays a minor role at influencing the wave speed. These results further suggest that the propagation speed of plane waves of dilatation can be better controlled when the Poisson's ratio is negative as the change in the dimensionless wave speed is more gradual in comparison to solids with positive Poisson's ratio.

6.6 Example: Longitudinal Waves in Cylindrical Auxetic Rods with Surface Elastic Restraint in Radial Direction

Consider a cylindrical rod with its lateral surface being elastically restrained in the radial direction but the entire rod cross-section is free to move along the direction of the wave propagation, as indicated in Fig. 6.15, then the magnitudes of the lateral strains would reduce correspondingly.

One can then obtain two extremes of the wave speed. For the first extreme, whereby the elastic constraint diminishes, the wave speed as described in Eq. (6.3.14)—with density and cross-section corrections—is recovered. For the other extreme, whereby the elastic constraint in the radial direction is infinite, the radial strain diminishes. Therefore, the wave speed for the latter is that for plane waves of dilatation described by Eq. (6.5.3), with density correction.

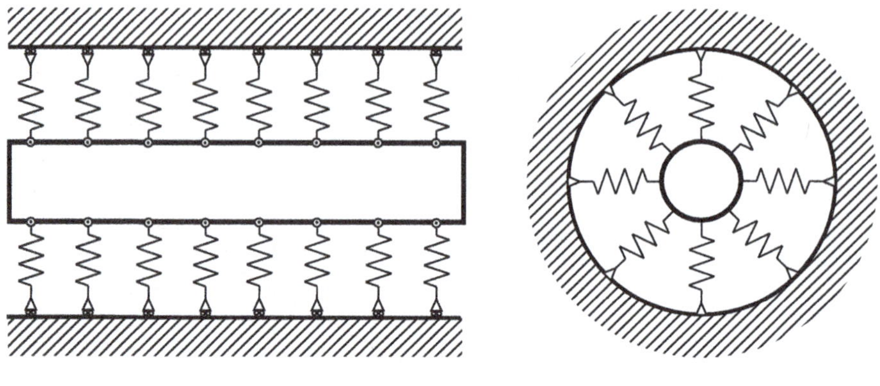

Fig. 6.15 Idealized schematic 2D views for a prismatic rod with lateral surface that is elastically constraint in the radial direction and prevented from circumferential displacement, but free to displace along the longitudinal direction

In the following analysis, we consider (i) intermediate case where the elastic constraint is at neither extremes, (ii) the influence of Poisson's ratio with special emphasis on the case where the Poisson's ratio is negative, and (iii) the influence from the longitudinal strain. Assumptions include: (a) the rod being elastically isotropic and (b) the absence of twisting. The latter assumption may fail in very thin rods made of chiral or anti-chiral layers oriented perpendicularly to the rod axis.

With reference to Fig. 6.16, the equation of motion for longitudinal wave in a prismatic rod, allowing for variable cross-sectional area A and variable density ρ, is

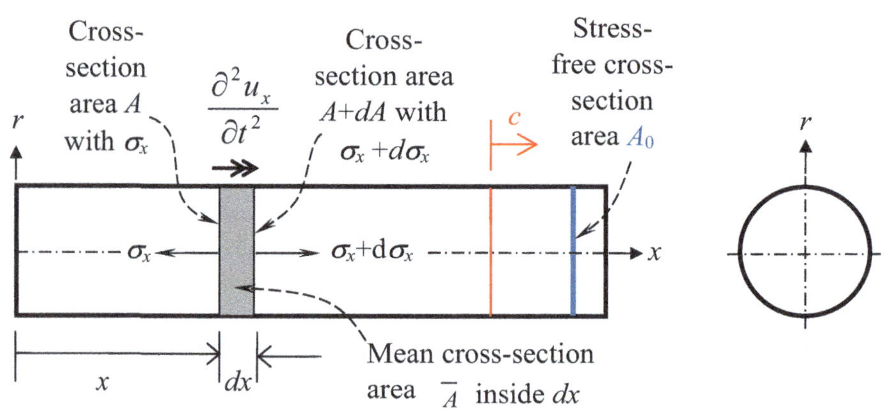

Fig. 6.16 Schematic view for analysis of longitudinal wave motion in a cylindrical rod considering changes to the cross-section and density with the lateral boundary condition specified in Fig. 6.15

generally written as Eq. (6.2.2), whereby $(A, A + dA)$ are the cross-sectional areas corresponding to stresses $(\sigma_x, \sigma_x + d\sigma_x)$ at longitudinal distances $(x, x + dx)$ from the origin. This removes the simplifying assumptions of constancy in the rod's cross-sectional area and density. The mean cross-sectional area of the rod confined within the elemental volume of length dx is assumed to be a simple average of the cross-sectional areas at x and $x + dx$ as described by Eq. (6.2.3). Substituting $\varepsilon_r = -v\varepsilon_x$ into Eq. (6.2.15) for a cylindrical isotropic rod, we have

$$\frac{A}{A_0} = 1 - 2v\varepsilon_x + v^2\varepsilon_x^2 \tag{6.6.1}$$

To implement the elastic constraint in the radial direction, as illustrated in Fig. 6.15, we impose a spring constant per unit area k_s such that the two extremes are $k_s = 0$ and $k_s \to \infty$ for achieving longitudinal wave speeds of unconstrained rods and plane waves of dilatation, respectively. It is more convenient, however, to adopt a modified quantity, herein termed as the boundary parameter acting on the curved surface along the radial direction

$$b_r = \frac{k_c}{k_c + k_s} = \frac{1}{1 + (k_s/k_c)} \tag{6.6.2}$$

where k_c is a constant that possesses a similar unit as k_s. As such, the extreme values of $k_s = 0$ and $k_s \to \infty$ can be described by $b_r = 1$ (rod with no lateral direction constraint) and $b_r = 0$ (plane wave of dilatation or rod with no lateral displacement allowed), respectively. This boundary parameter is multiplied to the lateral strains

$$\varepsilon_r = \varepsilon_\theta = -b_r v\varepsilon_x \tag{6.6.3}$$

so that a value of $b_r = 1$ permits full lateral strain, i.e., the elastic constraint vanishes, while a value of $b_r = 0$ diminishes the lateral strain. Implementing this boundary parameter on Eq. (6.6.1), we have

$$\frac{A}{A_0} = 1 - 2b_r v\varepsilon_x + b_r^2 v^2\varepsilon_x^2 \tag{6.6.4}$$

Writing the changes in the cross-sectional area and longitudinal strains as $A \to A + dA$ and $\varepsilon_x \to \varepsilon_x + d\varepsilon_x$, respectively, from x to $x + dx$, we extend from Eq. (6.6.4) to obtain

$$\frac{A + dA}{A_0} = 1 - 2b_r v(\varepsilon_x + d\varepsilon_x) + b_r^2 v^2(\varepsilon_x + d\varepsilon_x)^2 \tag{6.6.5}$$

or its expanded form

$$\frac{A + \mathrm{d}A}{A_0} = 1 - 2b_r v\left(\varepsilon_x + \frac{\partial \varepsilon_x}{\partial x}\mathrm{d}x\right) + b_r^2 v^2\left[\varepsilon_x^2 + 2\varepsilon_x \frac{\partial \varepsilon_x}{\partial x}\mathrm{d}x + \left(\frac{\partial \varepsilon_x}{\partial x}\mathrm{d}x\right)^2\right] \tag{6.6.6}$$

The incremental change in the cross-sectional area, dA, along the infinitesimal thickness dx can then be obtained by subtracting A/A_0 from $(A + \mathrm{d}A)/A_0$ to give

$$\frac{\mathrm{d}A}{A_0} = -2b_r v \frac{\partial \varepsilon_x}{\partial x}\mathrm{d}x + 2b_r^2 v^2 \varepsilon_x \frac{\partial \varepsilon_x}{\partial x}\mathrm{d}x + b_r^2 v^2\left(\frac{\partial \varepsilon_x}{\partial x}\mathrm{d}x\right)^2 \tag{6.6.7}$$

Substituting A/A_0 and $\mathrm{d}A/A_0$ into Eq. (6.2.3) allows the mean cross-sectional area within dx to be written in terms of longitudinal strain as

$$\frac{\overline{A}}{A_0} = 1 - 2b_r v\varepsilon_x + b_r^2 v^2 \varepsilon_x^2 - b_r v \frac{\partial \varepsilon_x}{\partial x}\mathrm{d}x + b_r^2 v^2 \varepsilon_x \frac{\partial \varepsilon_x}{\partial x}\mathrm{d}x + \frac{1}{2} b_r^2 v^2\left(\frac{\partial \varepsilon_x}{\partial x}\mathrm{d}x\right)^2 \tag{6.6.8}$$

Substituting Eq. (6.6.3) into the constitutive relation of Eq. (6.2.20), with subscripts $(y, z) \to (r, \theta)$, gives

$$\sigma_x = \frac{E\left(1 - v - 2b_r v^2\right)}{(1 + v)(1 - 2v)}\varepsilon_x \tag{6.6.9}$$

and, therefore,

$$\mathrm{d}\sigma_x = \frac{E\left(1 - v - 2b_r v^2\right)}{(1 + v)(1 - 2v)}\mathrm{d}\varepsilon_x = \frac{E\left(1 - v - 2b_r v^2\right)}{(1 + v)(1 - 2v)}\frac{\partial \varepsilon_x}{\partial x}\mathrm{d}x \tag{6.6.10}$$

Substituting Eq. (6.6.3) for isotropic rod into Eq. (6.2.18)

$$\rho = \frac{\rho_0}{(1 + \varepsilon_x)(1 - b_r v\varepsilon_x)^2} \tag{6.6.11}$$

Substituting A/A_0, $\mathrm{d}A/A_0$, $\overline{A}/A_0$, ρ, σ_x, and $\mathrm{d}\sigma_x$ into Eq. (6.2.2) would give an equation with second order of ε_x. Neglecting its highest order, we have

$$\left[1 - 2b_r v\varepsilon_x + b_r^2 v^2 \varepsilon_x^2\right]\frac{E\left(1 - v - 2b_r v^2\right)}{(1 + v)(1 - 2v)}\frac{\partial \varepsilon_x}{\partial x}\mathrm{d}x + \frac{E\left(1 - v - 2b_r v^2\right)}{(1 + v)(1 - 2v)}\varepsilon_x\left[-2b_r v \frac{\partial \varepsilon_x}{\partial x}\mathrm{d}x + 2b_r^2 v^2 \varepsilon_x \frac{\partial \varepsilon_x}{\partial x}\mathrm{d}x\right.$$

$$
+b_{\mathrm{r}}^2 v^2 \left(\frac{\partial \varepsilon_x}{\partial x} \mathrm{d}x \right)^2 \Bigg]
$$

$$
= \frac{\rho_0}{(1+\varepsilon_x)(1-b_{\mathrm{r}} v \varepsilon_x)^2} \Bigg[1 - 2b_{\mathrm{r}} v \varepsilon_x + b_{\mathrm{r}}^2 v^2 \varepsilon_x^2 - b_{\mathrm{r}} v \frac{\partial \varepsilon_x}{\partial x} \mathrm{d}x
$$

$$
+b_{\mathrm{r}}^2 v^2 \varepsilon_x \frac{\partial \varepsilon_x}{\partial x} \mathrm{d}x + \frac{1}{2} b_{\mathrm{r}}^2 v^2 \left(\frac{\partial \varepsilon_x}{\partial x} \mathrm{d}x \right)^2 \Bigg] \mathrm{d}x \frac{\partial^2 u_x}{\partial t^2} \tag{6.6.12}
$$

Diving Eq. (6.6.12) by dx on both sides and let

$$
\frac{\partial \varepsilon_x}{\partial x} = \frac{\partial^2 u_x}{\partial x^2} \tag{6.6.13}
$$

while

$$
\frac{\partial \varepsilon_x}{\partial x} \mathrm{d}x \to 0 \tag{6.6.14}
$$

we have

$$
\frac{E\left(1 - v - 2b_{\mathrm{r}} v^2\right)}{(1+v)(1-2v)} \left[1 - 4b_{\mathrm{r}} v \varepsilon_x + 3b_{\mathrm{r}}^2 v^2 \varepsilon_x^2\right] \frac{\partial^2 u_x}{\partial x^2}
$$

$$
= \frac{\rho_0}{(1+\varepsilon_x)(1-b_{\mathrm{r}} v \varepsilon_x)^2} \left[1 - 2b_{\mathrm{r}} v \varepsilon_x + b_{\mathrm{r}}^2 v^2 \varepsilon_x^2\right] \frac{\partial^2 u_x}{\partial t^2} \tag{6.6.15}
$$

This gives the longitudinal wave speed (Lim 2019a)

$$
c = \sqrt{\left(\frac{\partial^2 u_x}{\partial t^2}\right) / \left(\frac{\partial^2 u_x}{\partial x^2}\right)} = \sqrt{\frac{E(1+\varepsilon_x)(1-b_{\mathrm{r}} v \varepsilon_x)(1-3b_{\mathrm{r}} v \varepsilon_x)\left(1 - v - 2b_{\mathrm{r}} v^2\right)}{\rho_0 (1+v)(1-2v)}} \tag{6.6.16}
$$

The longitudinal wave speed described in Eq. (6.6.16) can be simplified under some special cases. Substituting $b_{\mathrm{r}} = 1$ for the case of unconstrained rod gives the longitudinal wave speed in prismatic rods with corrections as indicated in Eq. (6.3.14). Substituting $b_{\mathrm{r}} = 0$ or the case of rod that is fully constrained in the radial direction but free to move in the longitudinal direction, we obtain the speed of plane wave of dilatation with density correction described in Eq. (6.5.3).

Within the category of negligible strain $\varepsilon_x \to 0$, we have

$$
c = \sqrt{\frac{E\left(1 - v - 2b_{\mathrm{r}} v^2\right)}{\rho_0 (1+v)(1-2v)}} \tag{6.6.17}
$$

which can be further abridged to $c = \sqrt{E/\rho_0}$, i.e., the elementary longitudinal wave speed in prismatic rods for $b_r = 1$, while the speed of the elementary plane wave of dilatation given in Eq. (6.1.9) is recovered for $b_r = 0$. In addition, $c = \sqrt{E/\rho_0}$ for any value of b_r if $v = 0$; this is unsurprising since the elementary model neglects the lateral strain, which is exact when $v = 0$.

Special cases with $\varepsilon_x \to 0$, i.e., Eq. (6.6.17), have been briefly discussed for $b_r = 0, 1$. For other values of $0 < b_r < 1$, the wave speed can be visualized by plotting against the Poisson's ratio with normalization against the elementary longitudinal wave speed in rods

$$c\sqrt{\frac{\rho_0}{E}} = \sqrt{\frac{1 - v - 2b_r v^2}{(1 + v)(1 - 2v)}} \tag{6.6.18}$$

as shown in Fig. 6.17 (left). Due to the denominator at the RHS of Eq. (6.6.18), the wave speed goes to infinity as the Poisson's ratio approaches both limits. As such, Eq. (6.6.17) can be alternatively normalized against the elementary plane wave of dilatation speed

$$c\sqrt{\frac{\rho_0(1 + v)(1 - 2v)}{E(1 - v)}} = \sqrt{\frac{1 - v - 2b_r v^2}{1 - v}} \tag{6.6.19}$$

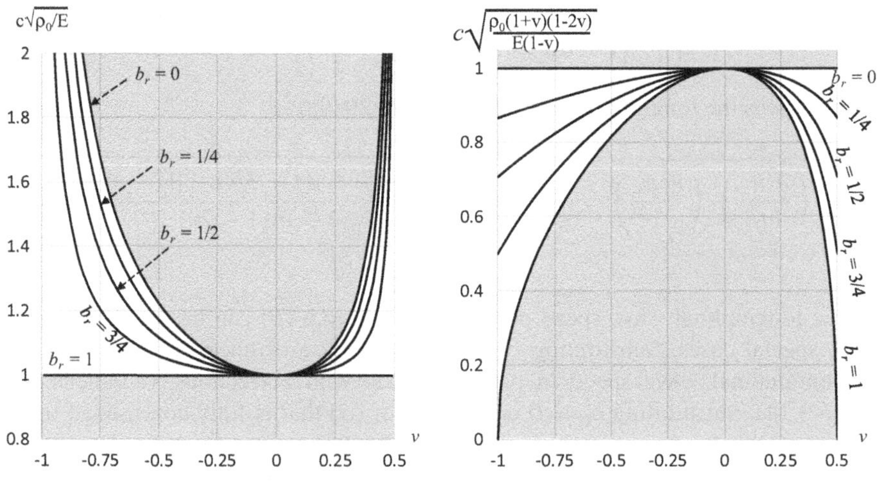

Fig. 6.17 Dimensionless longitudinal wave speed in prismatic rods for infinitesimal strain with normalization against the elementary longitudinal wave speed in prismatic rods (left) and the elementary wave speed in plane waves of dilatation (right). The shaded parts indicate theoretically impossible region within the discussed framework

as plotted in Fig. 6.17 (right). Both plots suggest that the wave speed is more easily controlled when the Poisson's ratio is negative rather than positive.

In consideration of the strain—alongside the Poisson' ratio and the boundary parameter—the wave speed as indicated in Eq. (6.6.16) can be normalized against $c = \sqrt{E/\rho_0}$ so as to provide a dimensionless wave speed

$$c\sqrt{\frac{\rho_0}{E}} = \sqrt{\frac{(1+\varepsilon_x)(1-b_r v\varepsilon_x)(1-3b_r v\varepsilon_x)\left(1-v-2b_r v^2\right)}{(1+v)(1-2v)}} \tag{6.6.20}$$

This is plotted in Fig. 6.18, which shows that at $b_r = 1$ the Poisson's ratio of conventional materials plays an insignificant role on the wave speed. However, the Poisson's ratio plays an increasing role in auxetic materials. In auxetic rods that are not laterally constrained, the tensile wave speed is greater than compressive wave speed, and the difference is accentuated when the Poisson's ratio of the rod becomes more negative. As the lateral boundary elastic constraint increases (i.e., b_r reduces), the chosen dimensionless wave speed increases at both limits of the Poisson's ratio, thereby forming minimum points. In the case of $0 < b_r < 1$, the minimum point for $\varepsilon_x = 0$ is at $v = 0$, while the minimum points for tensile and compressive waves

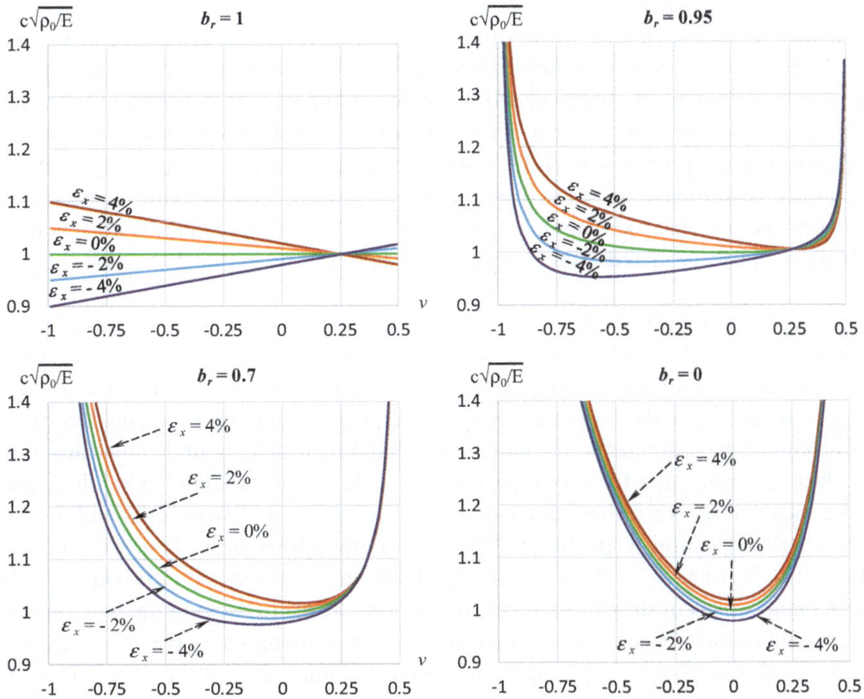

Fig. 6.18 Plots of dimensionless wave speed $c\sqrt{\rho_0/E}$ against Poisson's ratio for various longitudinal strains at different values of b_r

are in the conventional and auxetic regions, respectively. As the boundary parameter b_r reduces toward zero, their minimum points shift toward $v = 0$. The choice of dimensionless wave speed shown in Eq. (6.6.20) is useful for larger values of b_r; for smaller values of b_r, this choice of dimensionless wave speed is still useful for observing wave speed for materials in which the magnitude of Poisson's ratio is very small.

Due to the sharp rise of c or $c\sqrt{\rho_0/E}$ toward infinity at both limits of the Poisson's ratio for $0 \leq b_r < 1$, an alternative normalization of the wave speed can be made against the classical speed of plane waves of dilatation so as to give another dimensionless wave speed

$$c\sqrt{\frac{\rho_0(1+v)(1-2v)}{E(1-v)}} = \sqrt{\frac{(1+\varepsilon_x)(1-b_r v\varepsilon_x)(1-3b_r v\varepsilon_x)\left(1-v-2b_r v^2\right)}{1-v}} \tag{6.6.21}$$

This is plotted in Fig. 6.19, in which for the range $0 < b_r \leq 1$, the dimensionless wave speed drops at both limits of Poisson's ratio. For this range of b_r, there exists a maximum point such that it occurs at $v = 0$ when $\varepsilon_x = 0$; for $\varepsilon_x \neq 0$, the maximum points take place in the auxetic and conventional regions under tensile and compressive waves, respectively. As the boundary parameter b_r reduces, the curvatures reduce such that these dimensionless wave speed become constant at $b_r = 0$. Either way of normalization, both sets of dimensionless wave speeds show that, to a large extent, the wave speed is enhanced and slowed down by tensile and compressive impact loads, respectively. The choice of dimensionless wave speed as shown in Eq. (6.6.21) is useful for smaller values of b_r; for larger values of b_r, this choice of dimensionless wave speed is still useful for observing wave speed for materials in which the magnitude of Poisson's ratio is very small. Both schemes of making the wave speed dimensionless are shown to be applicable when the rod material is auxetic.

An overview is furnished in Fig. 6.20 for relating the longitudinal wave speed cylindrical rods subjected to elastic restraint in radial direction with other known wave speeds. In summary, the interlacing effect from the lateral boundary parameter, material auxeticity, and strain on the longitudinal wave speed has been considered. A lateral boundary parameter acting along the radial direction b_r, bounded by 0 and 1, is introduced so as to cater for varying extent of elastic constraint in the radial direction on the rods surface such that values of $b_r = 1$ and $b_r = 0$ fully permits and totally prohibits radial strain, respectively. When the longitudinal strain is negligible, the values of $b_r = 1$ and $b_r = 0$ correspond to elementary wave speed in prismatic rods and in plane waves of dilatation, respectively. Two forms of making the wave speed dimensionless have been implemented, one by normalizing the wave speed against the classical wave speed in prismatic rods and the other against the classical wave speed for plane waves of dilatation; the former is more advantageous for visualizing wave speed when the lateral constraint is non-existent or very small while the latter is more suitable for when the elastic constraint in the radial direction is very large or

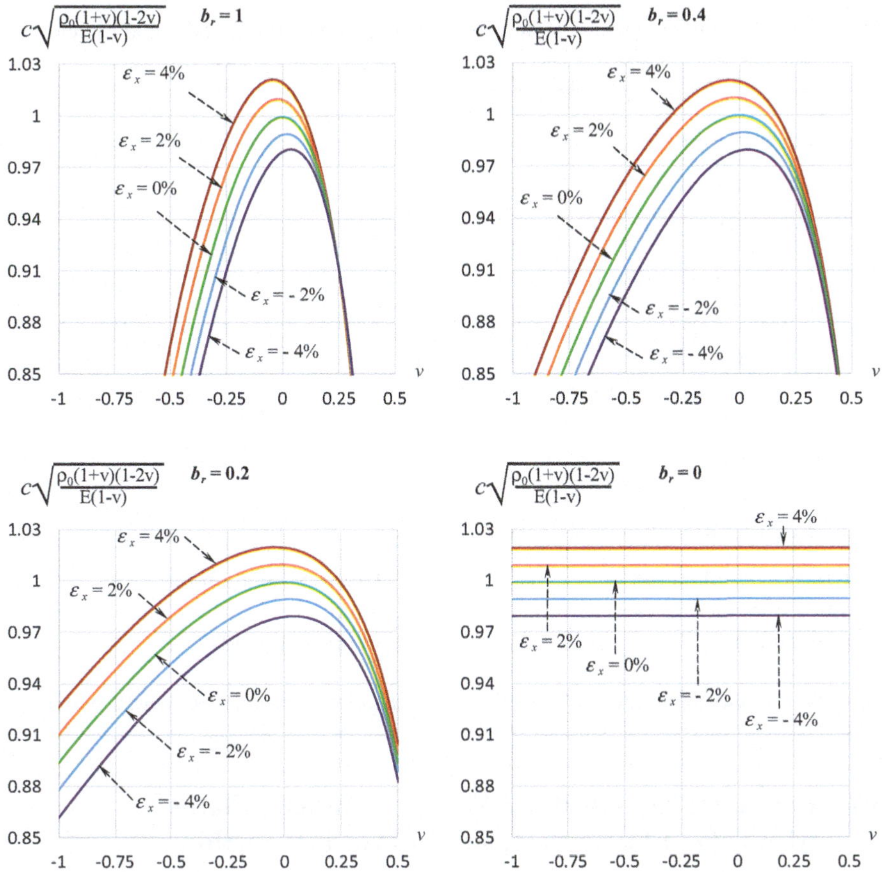

Fig. 6.19 Plots of dimensionless wave speed—normalized against Eq. (6.1.9)—versus Poisson's ratio for various longitudinal strains at different values of b_r

when the radial strain is zero. Both types of dimensionless wave speeds are suitable for auxetic materials, as they are able to clearly discriminate the plotted curves for various longitudinal strains when the Poisson's ratio is negative. Results from the developed wave speed herein that takes the longitudinal strain into consideration reveal that in most cases, especially in auxetic range, tensile waves travel faster than compressive waves. An understanding on these factors enables the wave speed to be controlled by adjusting the strain and selecting appropriate material properties for a given boundary condition.

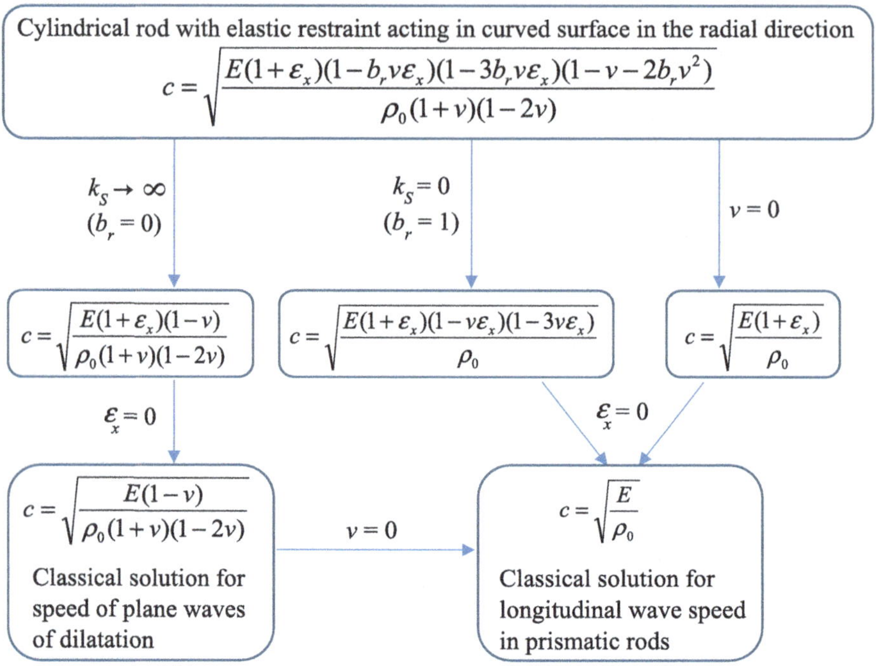

Fig. 6.20 Overview of the longitudinal wave speed cylindrical rods subjected to elastic restraint in radial direction with classical solutions and those from Sects. 6.3 and 6.5

6.7 Example: Longitudinal Waves in Auxetic Plates with Edge Elastic Restraint in Width Direction

Consider a flat plate of Young's modulus E and Poisson's ratio v with width y and thickness z, measured along the y- and z-axes as illustrated in Fig. 6.21, in which a longitudinal wave propagates along the x-axis. Let the cross-sectional area of the unstressed portion of the plate be A_0 while in the stressed portion the cross-sectional areas are $A_x = A$ and $A_{x+\mathrm{d}x} = A + \mathrm{d}A$ at locations x and $x + \mathrm{d}x$, respectively, being measured along the x-axis, as shown in Fig. 6.21. Taking the equation of motion for the elemental volume between x and $x + \mathrm{d}x$, we have Eq. (6.2.2) where u_x is the displacement parallel to the x-axis while the mean cross-sectional area between x and $x + \mathrm{d}x$ is averaged as described by Eq. (6.2.3).

Based on the constitutive relation of Eq. (6.2.20), we impose the conditions of plane stress $\sigma_z = 0$ and elastically restraint width with spring stiffness k_s at the plate side acting along the plate width direction or parallel to the y-axis, i.e.,

$$\varepsilon_y = -b_w v \varepsilon_x \tag{6.7.1}$$

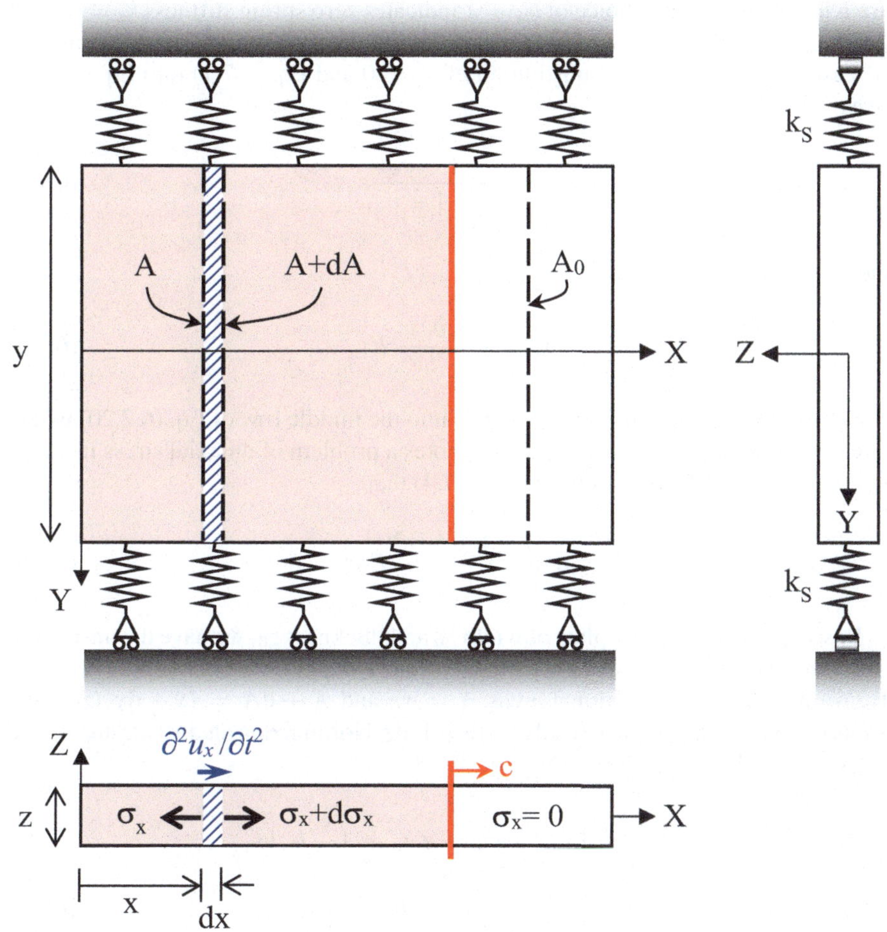

Fig. 6.21 Schematic views of a plate with longitudinal wave motion parallel to the x-axis and elastic restraint at two opposing sides in the y-axis

where the imposed elastic constraint is more conveniently executed by introducing a boundary parameter

$$b_{\mathrm{w}} = \frac{k_{\mathrm{c}}}{k_{\mathrm{c}} + k_{\mathrm{s}}} = \frac{1}{1 + (k_{\mathrm{s}}/k_{\mathrm{c}})} \tag{6.7.2}$$

An arbitrary spring constant k_{c} is introduced to non-dimensionalize the boundary parameter so as to confine it within $0 \leq b_{\mathrm{w}} \leq 1$. The choice of $b_{\mathrm{w}} = 0$ denotes infinite spring stiffness $k_{\mathrm{s}} \to \infty$, thereby preventing expansion or contraction of the width. It is conveniently shown that the substitution of $b_{\mathrm{w}} = 0$ into Eq. (6.7.1) gives $\varepsilon_y = 0$, thereby indicating a problem of plane stress in x–y plane as well as plane strain in x–z plane. The choice of $b_w = 1$ indicates zero spring stiffness $k_s = 0$, such that the plate width is allowed to freely expand or contract without restraint, akin to the case of prismatic rods. Substitution of $\sigma_z = 0$ and Eq. (6.7.1) into Eq. (6.2.20) gives

$$\varepsilon_z = -\upsilon \frac{1 - b_w \upsilon}{1 - \upsilon} \varepsilon_x \tag{6.7.3}$$

and

$$\sigma_x = \frac{1 - b_w \upsilon^2}{1 - \upsilon^2} E \varepsilon_x \tag{6.7.4}$$

Substitution of Eqs. (6.7.1) and (6.7.3) into the middle row of Eq. (6.2.20) would give $\sigma_y = 0$ if $b_w = 1$. Hence, $b_w = 1$ denotes a problem of uniaxial stress in x-axis and triaxial strain. It follows from Eq. (6.7.4) that

$$\mathrm{d}\sigma_x = \frac{1 - b_w \upsilon^2}{1 - \upsilon^2} E \mathrm{d}\varepsilon_x \tag{6.7.5}$$

Based on the unstressed plate of width y_0 and thickness z_0, we have the unstressed cross-sectional area $A_0 = y_0 z_0$ parallel to the y–z plane; at locations x and $x + \mathrm{d}x$, we have the cross-sectional areas $A = yz$ and $A + \mathrm{d}A = (y + \mathrm{d}y)(z + \mathrm{d}z)$, respectively, which gives $\mathrm{d}A = z\mathrm{d}y + y\mathrm{d}z + \mathrm{d}y\mathrm{d}z$. Normalizing against the unstressed cross-sectional area, we have from Eq. (6.2.8)

$$\begin{aligned}\frac{A}{A_0} &= 1 + \varepsilon_y + \varepsilon_z + \varepsilon_y \varepsilon_z = 1 - b_w \upsilon \varepsilon_x \\ &\quad - \upsilon \frac{1 - b_w \upsilon}{1 - \upsilon} \varepsilon_x + b_w \upsilon^2 \frac{1 - b_w \upsilon}{1 - \upsilon} \varepsilon_x^2\end{aligned} \tag{6.7.6}$$

which implies

$$\begin{aligned}\frac{A + \mathrm{d}A}{A_0} &= 1 - b_w \upsilon (\varepsilon_x + \mathrm{d}\varepsilon_x) - \upsilon \frac{1 - b_w \upsilon}{1 - \upsilon} (\varepsilon_x + \mathrm{d}\varepsilon_x) \\ &\quad + b_w \upsilon^2 \frac{1 - b_w \upsilon}{1 - \upsilon} \left(\varepsilon_x^2 + 2\varepsilon_x \mathrm{d}\varepsilon_x + (\mathrm{d}\varepsilon_x)^2\right)\end{aligned} \tag{6.7.7}$$

The above area expressions give

$$\frac{\mathrm{d}A}{A_0} = -b_w \upsilon \mathrm{d}\varepsilon_x - \upsilon \frac{1 - b_w \upsilon}{1 - \upsilon} \mathrm{d}\varepsilon_x + b_w \upsilon^2 \frac{1 - b_w \upsilon}{1 - \upsilon} \left(2\varepsilon_x \mathrm{d}\varepsilon_x + (\mathrm{d}\varepsilon_x)^2\right) \tag{6.7.8}$$

In addition to the variation in the cross-sectional area, the change in density can be established from Eq. (6.2.11) on the basis of mass conservation to give

$$\rho = \frac{\rho_0}{(1+\varepsilon_x)(1-b_w v\varepsilon_x)\left(1 - v\frac{1-b_w v}{1-v}\varepsilon_x\right)} \tag{6.7.9}$$

Substituting σ_x, $\mathrm{d}\sigma_x$, A/A_0, $\mathrm{d}A/A_0$ and ρ into Eq. (6.2.2) leads to

$$\begin{aligned}
&\frac{1-b_w v^2}{1-v^2}E\varepsilon_x\left[-b_w v\mathrm{d}\varepsilon_x - v\frac{1-b_w v}{1-v}\mathrm{d}\varepsilon_x + b_w v^2\frac{1-b_w v}{1-v}\left(2\varepsilon_x\mathrm{d}\varepsilon_x + (\mathrm{d}\varepsilon_x)^2\right)\right]\\
&\quad+\left[1 - b_w v\varepsilon_x - v\frac{1-b_w v}{1-v}\varepsilon_x + b_w v^2\frac{1-b_w v}{1-v}\varepsilon_x^2\right]\frac{1-b_w v^2}{1-v^2}E\mathrm{d}\varepsilon_x\\
&\quad+\frac{1-b_w v^2}{1-v^2}E\mathrm{d}\varepsilon_x\left[-b_w v\mathrm{d}\varepsilon_x - v\frac{1-b_w v}{1-v}\mathrm{d}\varepsilon_x\right.\\
&\quad\left.+b_w v^2\frac{1-b_w v}{1-v}\left(2\varepsilon_x\mathrm{d}\varepsilon_x + (\mathrm{d}\varepsilon_x)^2\right)\right]\\
&=\rho\left\{1 - b_w v\left(\varepsilon_x + \frac{\mathrm{d}\varepsilon_x}{2}\right) - v\frac{1-b_w v}{1-v}\left(\varepsilon_x + \frac{\mathrm{d}\varepsilon_x}{2}\right)\right.\\
&\quad\left.+ b_w v^2\frac{1-b_w v}{1-v}\left(\varepsilon_x^2 + \varepsilon_x\mathrm{d}\varepsilon_x + \frac{(\mathrm{d}\varepsilon_x)^2}{2}\right)\right\}\mathrm{d}x\frac{\partial^2 u_x}{\partial t^2}
\end{aligned} \tag{6.7.10}$$

where ρ is described in Eq. (6.7.9). Rearranging Eq. (6.7.10), such that only one $\mathrm{d}\varepsilon_x$ is a common multiplier on the LHS, we have

$$\begin{aligned}
&\frac{1-b_w v^2}{1-v^2}E\mathrm{d}\varepsilon_x\left[1 - b_w v(2\varepsilon_x + \mathrm{d}\varepsilon_x) - v\frac{1-b_w v}{1-v}(2\varepsilon_x + \mathrm{d}\varepsilon_x)\right.\\
&\quad\left.+b_w v^2\frac{1-b_w v}{1-v}\left(3\varepsilon_x^2 + 3\varepsilon_x\mathrm{d}\varepsilon_x + (\mathrm{d}\varepsilon_x)^2\right)\right]\\
&=\rho\left\{1 - b_w v\left(\varepsilon_x + \frac{\mathrm{d}\varepsilon_x}{2}\right) - v\frac{1-b_w v}{1-v}\left(\varepsilon_x + \frac{\mathrm{d}\varepsilon_x}{2}\right)\right.\\
&\quad\left.+b_w v^2\frac{1-b_w v}{1-v}\left(\varepsilon_x^2 + \varepsilon_x\mathrm{d}\varepsilon_x + \frac{(\mathrm{d}\varepsilon_x)^2}{2}\right)\right\}\mathrm{d}x\frac{\partial^2 u_x}{\partial t^2}
\end{aligned} \tag{6.7.11}$$

Substituting Eq. (6.2.21) for the common $\mathrm{d}\varepsilon_x$ on the LHS while setting the remaining $\mathrm{d}\varepsilon_x \to 0$, as well as substituting Eq. (6.7.9) into the above, we arrive at

$$\begin{aligned}
&\frac{1-b_w v^2}{1-v^2}E\left[1 - 2b_w v\varepsilon_x - 2v\frac{1-b_w v}{1-v}\varepsilon_x + 3b_w v^2\frac{1-b_w v}{1-v}\varepsilon_x^2\right]\frac{\partial^2 u_x}{\partial x^2}\\
&=\frac{\rho_0}{(1+\varepsilon_x)(1-b_w v\varepsilon_x)\left(1 - v\frac{1-b_w v}{1-v}\varepsilon_x\right)}\left\{1 - b_w v\varepsilon_x - v\frac{1-b_w v}{1-v}\varepsilon_x\right.\\
&\quad\left.+b_w v^2\frac{1-b_w v}{1-v}\varepsilon_x^2\right\}\frac{\partial^2 u_x}{\partial t^2}
\end{aligned} \tag{6.7.12}$$

after canceling dx on both sides. Based on the longitudinal wave speed is defined in Eq. (6.1.6)

we have from Eq. (6.7.12) (Lim 2019b)

$$c = \sqrt{\frac{E}{\rho_0}\left(\frac{1-b_{\mathrm{w}}v^2}{1-v^2}\right)(1+\varepsilon_x)(1-b_{\mathrm{w}}v\varepsilon_x)\left(1-v\frac{1-b_{\mathrm{w}}v}{1-v}\varepsilon_x\right)\frac{1-2b_{\mathrm{w}}v\varepsilon_x-2v\frac{1-b_{\mathrm{w}}v}{1-v}\varepsilon_x+3b_{\mathrm{w}}v^2\frac{1-b_{\mathrm{w}}v}{1-v}\varepsilon_x^2}{1-b_{\mathrm{w}}v\varepsilon_x-v\frac{1-b_{\mathrm{w}}v}{1-v}\varepsilon_x+b_{\mathrm{w}}v^2\frac{1-b_{\mathrm{w}}v}{1-v}\varepsilon_x^2}} \tag{6.7.13}$$

The general equation described above can be simplified under a few special cases. For a plate with fully constrained width ($b_{\mathrm{w}} = 0$), Eq. (6.7.13) reduces to

$$c = \sqrt{\frac{E}{\rho_0}\left(\frac{1+\varepsilon_x}{1-v^2}\right)\left(1-\frac{2v}{1-v}\varepsilon_x\right)} \tag{6.7.14}$$

Recognizing that

$$\left(1-\frac{v}{1-v}\varepsilon_x\right)^2 \approx 1-\frac{2v}{1-v}\varepsilon_x \tag{6.7.15}$$

since $|\varepsilon_x| \ll 1$, Eq. (6.7.14) can also be written as Eq. (6.4.13), which is the longitudinal wave speed in plates with constrained width but unconstrained thickness. If the plate's width is totally unconstrained $b_{\mathrm{w}} = 1$, Eq. (6.7.13) abridges to Eq. (6.3.14), which is the longitudinal wave speed in prismatic rods with corrections. In the case where the strain is infinitesimal $\varepsilon_x \to 0$, Eq. (6.7.13) becomes

$$c = \sqrt{\frac{E}{\rho_0}\left(\frac{1-b_{\mathrm{w}}v^2}{1-v^2}\right)} \tag{6.7.16}$$

This neglect of ε_x is typical of elementary solutions. Selecting $b_{\mathrm{w}} = 0$ for Eq. (6.7.16) reduces it to the elementary longitudinal wave speed in plates of infinite width or plates of constrained width described by Eq. (6.1.8) while the choice of $b_{\mathrm{w}} = 1$ greatly simplifies Eq. (6.7.16) to the case of elementary longitudinal wave speed in prismatic rods given in Eq. (6.1.7). Suppose the plate's Poisson's ratio is $v = 0$, Eq. (6.7.13) simplifies to the first of Eq. (6.3.15), which is independent from b_{w}.

In what follows, we shall firstly consider the case of infinitesimal strain as laid out in Eq. (6.7.16). While the extreme cases $b_{\mathrm{w}} = 0, 1$ have been pointed out, it remains to observe how the wave speed varies with this boundary parameter. This is best achieved by portraying the wave speed as the parameter b_{w} varies between these extremes. To furnish a family of dimensionless plots, the wave speed is normalized against the elementary longitudinal wave speed in prismatic rods described by Eq. (6.1.7). This is shown in Fig. 6.22 (left). An alternate non-dimensionalization is attainable by normalizing the wave speed against the elementary longitudinal wave speed in plates of infinite width or of constrained width described in Eq. (6.1.8), as

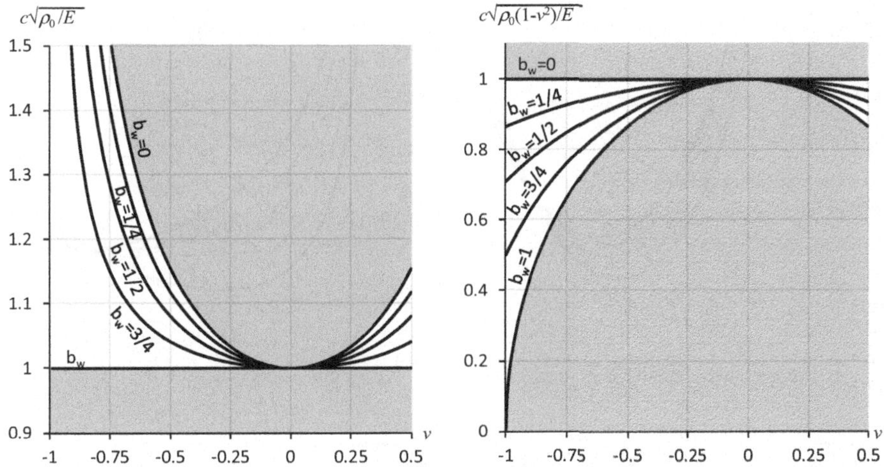

Fig. 6.22 Plots of dimensionless wave speed in plates with elastically constrained width based on normalization against the elementary longitudinal wave speed in prismatic rods (left) and the elementary longitudinal wave speed in plates of infinite width (right). The shaded loci indicate physically inadmissible region

displayed in Fig. 6.22 (right). The former and latter non-dimensionalizations allow the dimensionless wave speeds to be constant at $b_w = 1$ and $b_w = 0$, respectively. The latter non-dimensionalization, however, exhibits more uniform variation of the wave speed with b_w.

While Fig. 6.22 displays plots of dimensionless wave speeds under a special case of infinitesimal strain, the following discussion takes into consideration the strain effect. More specifically, Fig. 6.23 exhibits the culminating interlacing effect arising from the boundary parameter, Poisson's ratio, and longitudinal strain on the wave speed normalized against the elementary longitudinal wave speed in prismatic rods. The plots generally indicate that tensile waves travel faster through conventional plates with lower Poisson's ratio, and especially so in auxetic plates, but compressive waves travel faster in conventional plates of higher Poisson's ratio. This choice of normalization gives a minimum value approximately at $\upsilon = 0$ for $\varepsilon_x \approx 0$ (the minimum falls exactly at $\upsilon = 0$ for $\varepsilon_x = 0$) when $b_w = 0$, and linear variation with the Poisson's ratio for $b_w = 1$. The gradual transition of the wave speed between these two values of boundary parameters is furnished in Fig. 6.23, where by the scale on the vertical axis remain consistent to facilitate better comparison.

An alternative non-dimensionalization of the wave speed, again with the strain being accounted for, is plotted in Fig. 6.24 by normalizing it against the longitudinal wave speed in width-constrained plates. As before, tensile waves propagate faster through conventional plates of lower Poisson's ratio, and especially so through auxetic plates, while compressive waves move faster in conventional plates of higher Poisson's ratio. Unlike the other non-dimensionalization, this normalization does not lead to linear plots. Instead, the plots for $b_w = 0$ indicate negative and positive

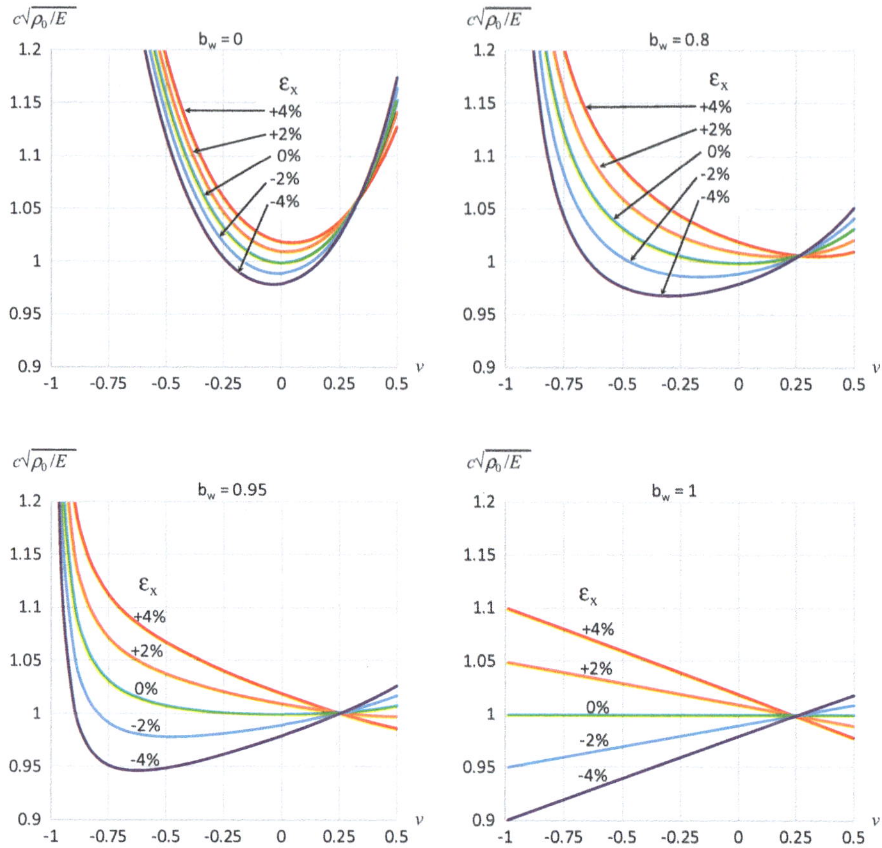

Fig. 6.23 Combined influence from the boundary parameter, Poisson's ratio, and longitudinal strain on the wave speed normalized against the elementary longitudinal wave speed in prismatic rods

curvatures for tensile and compressive waves, respectively. At $b_w = 1$, the normalized wave speed exhibits a maximum value approximately at $v = 0$ for $\varepsilon_x \approx 0$ (the maximum occurs exactly at $v = 0$ for $\varepsilon_x = 0$). The manner in which this dimensionless wave speed progresses from $b_w = 0$ to $b_w = 1$ is also displayed in Fig. 6.24, whereby the scale on the vertical axis remains consistent to facilitate better comparison. In both Figs. 6.23 and 6.24, it can be seen that the dimensionless wave speed can be more effectively controlled in the auxetic region than in the conventional region. The availability for the two types of non-dimensionalization allows better comparison to be made. Normalization of the wave speed against Eqs. (6.1.7) and (6.1.8) are more suitable for high b_w (or low elastic restraint) and low b_w (or high elastic restraint), respectively.

An overview on how the longitudinal wave speed in plates with elastic restraint along the width direction is related to other longitudinal wave speed models is furnished in Fig. 6.25. The classical or elementary models are placed inside a dashed

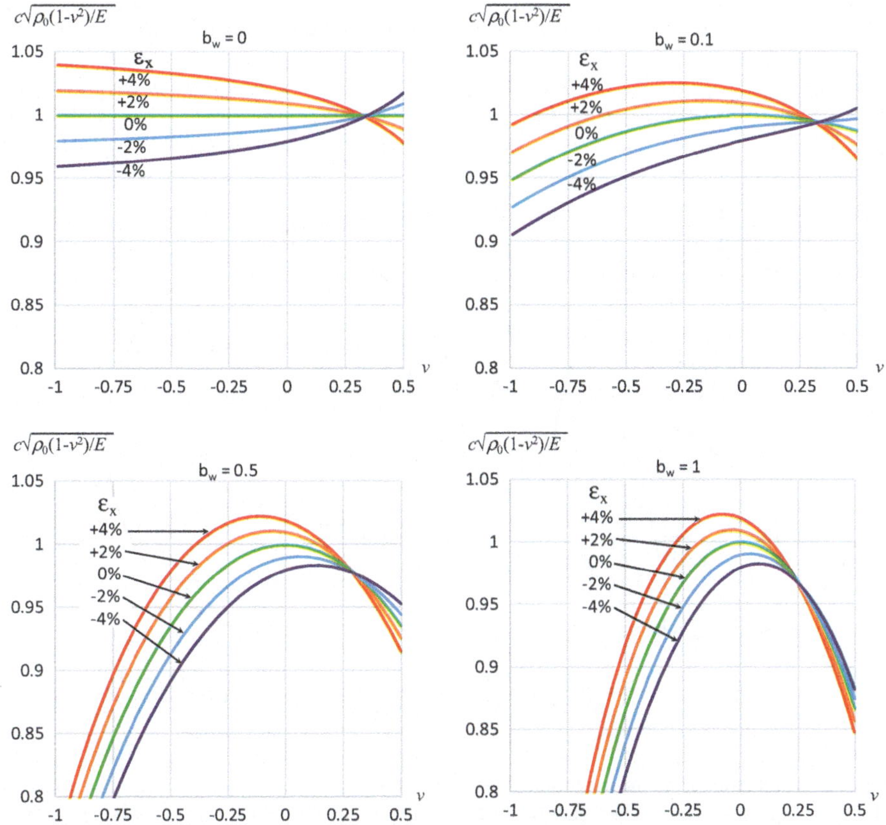

Fig. 6.24 Interlacing effect from the boundary parameter, Poisson's ratio, and longitudinal strain on the wave speed normalized against the elementary longitudinal wave speed in plates of infinite width or of constrained width

oval. In summary, the generalized speed of longitudinal waves in plates with elastically restrained sides along the width direction, taking into account the change in lateral dimensions and density, has been established with special consideration on the effect from material auxeticity. In addition to the well-known approach of selecting the Young's modulus and density for controlling the wave speed, results indicate that the wave propagation can be slowed down by decreasing the width elastic restraint. In the case of high elastic restraint, the speed of both tensile and compressive waves can be minimized by selecting plate materials with Poisson's ratio of low magnitude. In the case of low elastic restraint, the speed of tensile and compressive waves can be greatly reduced by selecting plate materials with large positive and large negative Poisson's ratio, respectively.

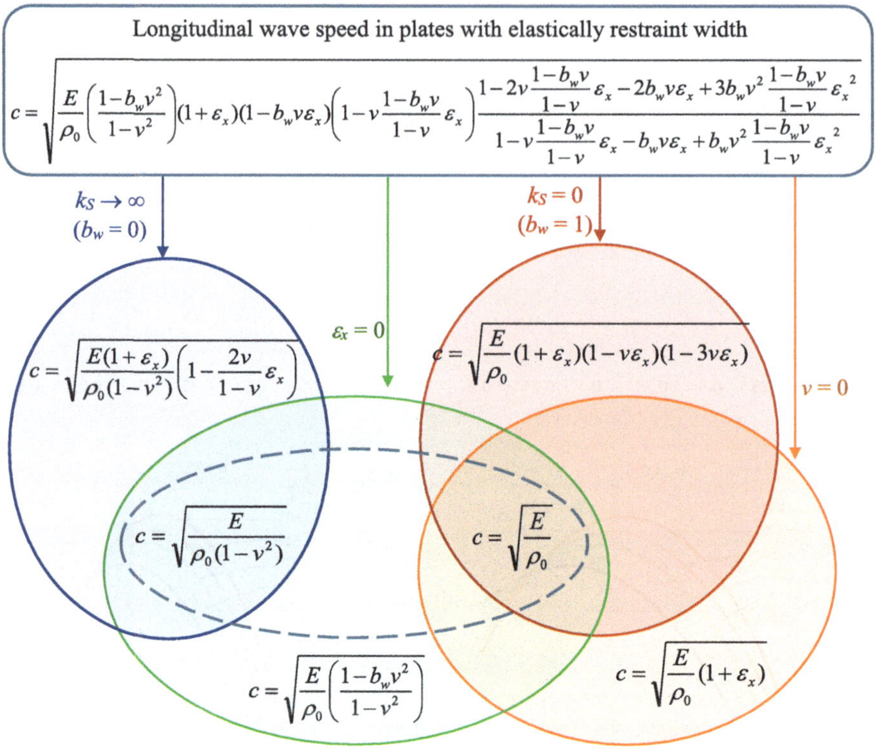

Fig. 6.25 Reduction from the longitudinal wave speed in plates with elastically restrained width to other wave speed models, including the classical solutions for longitudinal wave speeds in prismatic rods and in infinite plates inside the dashed oval

6.8 Example: Longitudinal Waves in Rectangular Slabs with Mixed Restraints in Lateral Directions

Consider the propagation of longitudinal wave in the x-direction, through a rectangular slab of original dimensions y_0 and z_0 measure along the y- and z-axes, respectively, as illustrated in Fig. 6.26. Taking the equation of motion in the x-direction for an elemental volume $\mathrm{d}V$ confined within

- $x \in [x, x + \mathrm{d}x]$
- $y \in [0, y_0]$
- $z \in [0, z_0]$

with the following boundary conditions

- $u_y = 0; \quad y = 0, y_0$
- $u_z = -F_s/k_s; \quad z = 0, z_0$

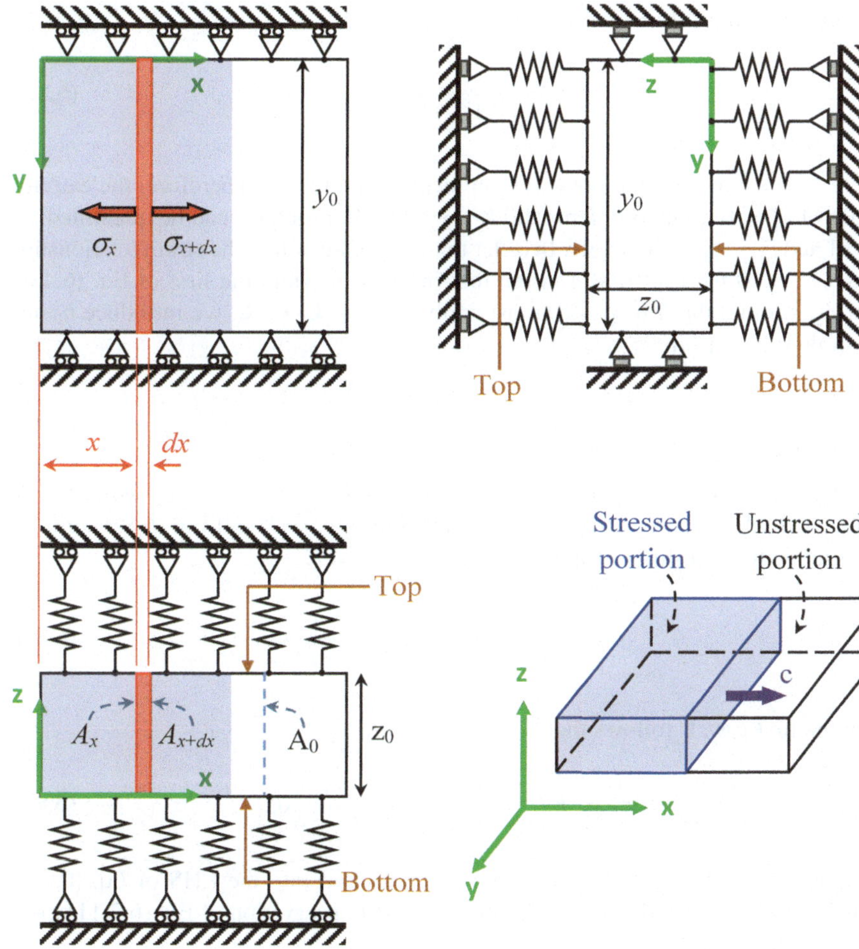

Fig. 6.26 Schematics for analysis of wave propagation in an auxetic rectangular slab with combined fixed and elastic constraints, viewed from various angles

where F_s and k_s are the overall restraint force and equivalent spring stiffness, respectively, acting along the z-direction at the surfaces $z = 0, z_0$, we arrive at Eq. (6.2.2) after performing non-dimensionalization on the cross-sectional area.

Based on the first boundary condition $u_y = 0$ at $y = 0, y_0$, substitution of $\varepsilon_y = 0$ into the constitutive relation of Eq. (6.2.19) gives

$$\begin{Bmatrix} \varepsilon_x \\ \varepsilon_z \end{Bmatrix} = \frac{1+v}{E} \begin{bmatrix} 1-v & -v \\ -v & 1-v \end{bmatrix} \begin{Bmatrix} \sigma_x \\ \sigma_z \end{Bmatrix} \tag{6.8.1}$$

A relation between strains in the x- and z-directions can be easily established by introducing a dimensionless boundary parameter b_t acting along the thickness direction, i.e., the z-direction, as

$$b_t = \frac{1}{1 + (k_s/k_c)} \tag{6.8.2}$$

where k_c is a constant that possesses an identical unit as k_s. Therefore, the extreme values of $k_s = 0$ (plate problem) and $k_s \to \infty$ (bulk problem) can be quantified via $b_t = 1$ and $b_t = 0$, respectively. In order to incorporate b_t into the strain relationship, we set $\sigma_z = 0$ for Eq. (6.8.1) in the first instance to yield the first of Eq. (6.4.6). Bearing in mind that Eq. (6.4.6) must apply for $0 \le k_s \le \infty$, we introduce b_t into the strain relationship as

$$\varepsilon_z = -\frac{b_t v}{1 - v}\varepsilon_x \tag{6.8.3}$$

such that the first of Eq. (6.4.6) is recovered when $k_s = 0$ (or $b_t = 1$), while $\varepsilon_z = 0$ is obtained when $k_s \to \infty$ (or $b_t = 0$). Substituting $\varepsilon_y = 0$ and Eq. (6.8.3) into the constitutive relationship of Eq. (6.2.20) gives rise to

$$\sigma_x = E\frac{1 - 2v + v^2(1 - b_t)}{(1 + v)(1 - 2v)(1 - v)}\varepsilon_x \tag{6.8.4}$$

from the first row. It follows that

$$\mathrm{d}\sigma_x = E\frac{1 - 2v + v^2(1 - b_t)}{(1 + v)(1 - 2v)(1 - v)}\mathrm{d}\varepsilon_x \tag{6.8.5}$$

Hence, Eqs. (6.8.4) and (6.8.5) can be substituted into the LHS of Eq. (6.2.2). Substituting $\varepsilon_y = 0$ and Eq. (6.8.3) into the mass conservation of Eq. (6.2.11) gives

$$\rho = \frac{\rho_0}{(1 + \varepsilon_x)\left(1 - \frac{b_t v}{1 - v}\varepsilon_x\right)} \tag{6.8.6}$$

for subsequent substitution into the RHS of Eq. (6.2.2). Finally, the expressions of A/A_0 and $\mathrm{d}A/A_0$, which appear on both sides of Eq. (6.2.2), can be incorporated by recognizing that the change in the cross-sectional area is solely due to the change in thickness, since $\varepsilon_y = 0$. Hence, $A/A_0 = 1 + \varepsilon_z$ or

$$\frac{A}{A_0} = 1 - \frac{b_t v}{1 - v}\varepsilon_x \tag{6.8.7}$$

by virtue of Eq. (6.8.3). It therefore follows that

$$\frac{A + \mathrm{d}A}{A_0} = 1 - \frac{b_t v}{1 - v}(\varepsilon_x + \mathrm{d}\varepsilon_x) \tag{6.8.8}$$

thereby leading to

$$\frac{\mathrm{d}A}{A_0} = -\frac{b_t v}{1 - v}\mathrm{d}\varepsilon_x \tag{6.8.9}$$

Substituting σ_x, $\mathrm{d}\sigma_x$, ρ, A/A_0, and $\mathrm{d}A/A_0$ into Eq. (6.2.2) leads to

$$\begin{aligned} &E\frac{1 - 2v + v^2(1 - b_t)}{(1 + v)(1 - 2v)(1 - v)}\varepsilon_x\left(-\frac{b_t v}{1 - v}\mathrm{d}\varepsilon_x\right) \\ &+ \left(1 - \frac{b_t v}{1 - v}\varepsilon_x\right)E\frac{1 - 2v + v^2(1 - b_t)}{(1 + v)(1 - 2v)(1 - v)}\mathrm{d}\varepsilon_x \\ &+ E\frac{1 - 2v + v^2(1 - b_t)}{(1 + v)(1 - 2v)(1 - v)}\mathrm{d}\varepsilon_x\left(-\frac{b_t v}{1 - v}\mathrm{d}\varepsilon_x\right) \\ &= \frac{\rho_0}{(1 + \varepsilon_x)\left(1 - \frac{b_t v}{1 - v}\varepsilon_x\right)}\left[1 - \frac{b_t v}{1 - v}\left(\varepsilon_x + \frac{\mathrm{d}\varepsilon_x}{2}\right)\right]\mathrm{d}x\frac{\partial^2 u_x}{\partial t^2} \end{aligned} \tag{6.8.10}$$

Neglecting the highest order terms on both sides of Eq. (6.8.10) and substituting Eq. (6.2.21), we have

$$E\frac{1 - 2v + v^2(1 - b_t)}{(1 + v)(1 - 2v)(1 - v)}\left(1 - \frac{2b_t v}{1 - v}\varepsilon_x\right)\frac{\partial^2 u_x}{\partial x^2} = \frac{\rho_0}{1 + \varepsilon_x}\frac{\partial^2 u_x}{\partial t^2} \tag{6.8.11}$$

which leads to the longitudinal wave speed (Lim 2019c)

$$c = \sqrt{\left(\frac{\partial^2 u_x}{\partial t^2}\right)/\left(\frac{\partial^2 u_x}{\partial x^2}\right)} = \sqrt{\frac{E}{\rho_0}\left(\frac{1 + \varepsilon_x}{1 - v^2}\right)\left(1 + \frac{v^2(1 - b_t)}{1 - 2v}\right)\left(1 - \frac{2b_t v}{1 - v}\varepsilon_x\right)} \tag{6.8.12}$$

The wave speed furnished in Eq. (6.8.12) can now be reduced to a few special cases. For Special Case I, where the spring stiffness diminishes ($k_s = 0$), substitution of $b_t = 1$ simplifies it to the case of plates with width constraint $(\varepsilon_y = \sigma_z = 0)$ but with unconstrained thickness described in Eq. (6.7.14). For the opposing Special Case II, where the spring stiffness becomes infinitely large ($k_s \to \infty$), substitution of $b_t = 0$ into Eq. (6.8.12) reduces it to the case of plane waves of dilatation $(\varepsilon_y = \varepsilon_z = 0)$ with due consideration of the change in density described in Eq. (6.5.3). Finally, in the Special Case III, where the Poisson's ratio of the rectangular slab is $v = 0$, Eq. (6.8.12) reduces to the first of Eq. (6.3.15), which is independent from the boundary parameter b_t.

Although the longitudinal wave speeds at the extreme cases of boundary parameter $b_t = 1$ and $b_t = 0$ have been shown to be simplified to the wave speed in plates with constrained width and in plane waves of dilatation, respectively, it is of interest to observe how the wave speed varies with b_t. To observe the effect of the dimensionless boundary parameter b_t and the Poisson's ratio v of the rectangular slab at infinitesimal deformation, we set $\varepsilon_x = 0$ for Eq. (6.8.12) to give

$$c = \sqrt{\frac{E}{\rho_0(1 - v^2)}\left(1 + \frac{v^2(1 - b_t)}{1 - 2v}\right)} \tag{6.8.13}$$

This permits the wave speed to be observed as a material property, which is independent from whether the stress wave is tensile or compressive. It is expedient to normalize this wave speed so as to give it a dimensionless form. One way to do so is to normalize it against the classical longitudinal wave speed in prismatic rods described in Eq. (6.1.7), to give rise to

$$c\sqrt{\frac{\rho_0}{E}} = \sqrt{\frac{1}{1 - v^2}\left(1 + \frac{v^2(1 - b_t)}{1 - 2v}\right)} \tag{6.8.14}$$

Under this dimensionless wave speed, it is clear that the Poisson's ratio v and dimensionless boundary parameter b_t constitute the primary and secondary influencing factors, respectively, as evident from Fig. 6.27. Specifically, this dimensionless wave speed can be greatly enhanced by increasing the Poisson's ratio magnitude

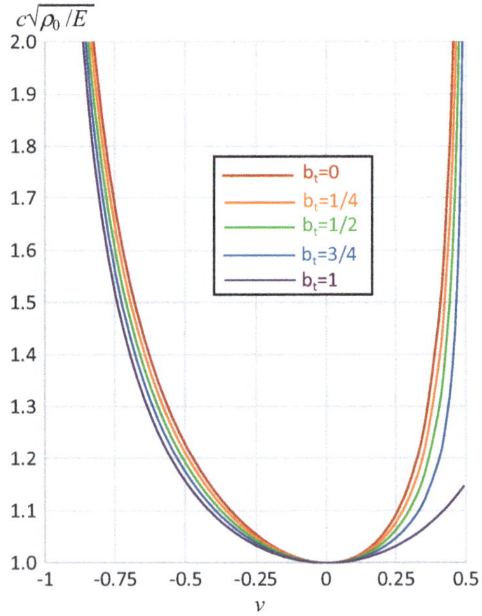

Fig. 6.27 Family of wave speed curves—normalized against the wave speed in prismatic rods—versus the Poisson's ratio for various boundary parameters when the strain in insignificant

$|v|$, and moderately incremented by lowering b_t, i.e., increasing the spring stiffness acting in the thickness direction.

As the wave speed falls in between that of plates with constrained width ($b_t = 1$) and that of plane waves of dilatation ($b_t = 0$), it follows that there are two physically meaningful ways to normalize the wave speed in such a manner that facilitates a clearer observation on the effect of b_t. Hence, the second way to non-dimensionalize the wave speed is to normalize it with respect to the classical wave speed in width-constrained plates to yield

$$c\sqrt{\frac{\rho_0\left(1-v^2\right)}{E}} = \sqrt{1+\frac{v^2(1-b_t)}{1-2v}} \tag{6.8.15}$$

The third way to non-dimensionalize Eq. (6.8.13) is to normalize it with reference to the classical wave speed in plane waves of dilatation to give

$$c\sqrt{\frac{\rho_0(1+v)(1-2v)}{E(1-v)}} = \sqrt{1-b_t\left(\frac{v}{1-v}\right)^2} \tag{6.8.16}$$

Based on the second and third types of non-dimensionalization, it can be seen that the following inequality holds

$$c\sqrt{\frac{\rho_0(1+v)(1-2v)}{E(1-v)}} \le 1 \le c\sqrt{\frac{\rho_0\left(1-v^2\right)}{E}} \tag{6.8.17}$$

More importantly,

$$\lim_{b_t\to 1} c\sqrt{\frac{\rho_0\left(1-v^2\right)}{E}} = \lim_{b_t\to 0} c\sqrt{\frac{\rho_0(1+v)(1-2v)}{E(1-v)}} = 1 \tag{6.8.18}$$

It is therefore not surprising that these two sets of dimensionless wave speeds can be combined in the manner furnished in Fig. 6.28, whereby each of the dimensionless wave speed appears as an almost symmetrical image of the other.

Since the effect from the boundary parameter is clearer when the wave speed is non-dimensionalized against Eqs. (6.1.8) and (6.1.9), these two schemes of normalizations are extended to the more general case, i.e., Eq. (6.8.12), when the strain in not negligible. The interlacing influence from the (i) tensile and compressive longitudinal strains, (ii) Poisson's ratio of the rectangular slab, and (iii) the boundary parameter is furnished in Fig. 6.29 (left) for the wave speed being normalized against Eq. (6.1.8), and in Fig. 6.29 (right) for the wave speed being normalized against Eq. (6.1.9). Generally, tensile and compressive strains increase and decrease the wave speed, respectively. In addition, the wave speed is more easily controlled when the Poisson's ratio is negative. In comparison with the longitudinal wave speed in

Fig. 6.28 Family of wave speed curves—normalized against the wave speeds in infinite plates (upper half) and against the plane waves of dilatation (bottom half)—versus the Poisson's ratio for various boundary parameters when the strain in insignificant

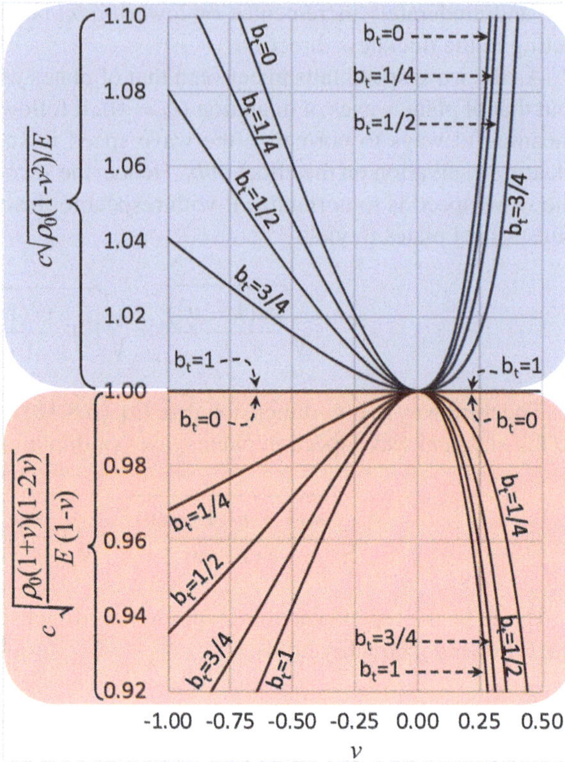

width-constrained plates, Fig. 6.29 (left) suggests that the increasing spring stiffness acting in the thickness direction (or decreasing b_t) enhances the wave speed in the rectangular slabs when $v \neq 0$. With reference to the plane wave of dilatation, Fig. 6.29 (right) indicates that the decreasing spring stiffness acting in the thickness direction (or increasing b_t) slows down the wave speed when $v \neq 0$.

In conclusion, the longitudinal wave speed through a rectangular slab—with the top and bottom surfaces being elastically restrained in the thickness direction, while the other two opposing side surfaces are constrained from motion in the width direction—has been established herein with due consideration to the change in cross-sectional area and density of the rectangular slab. A dimensionless boundary parameter has been introduced for quantifying a normalized spring compliance along the slab thickness direction. Three types of dimensionless wave speeds have been explored by normalizing the wave speed against some classical wave speeds. When normalized against the longitudinal wave speed in prismatic rods, results show that the Poisson's ratio magnitude and spring stiffness exert major and minor influences, respectively, on the wave speed. A clearer visualization on the effect of the boundary parameter is exhibited when the wave speed is normalized either against the classical wave speed in plates of constrained width and against the classical plane waves

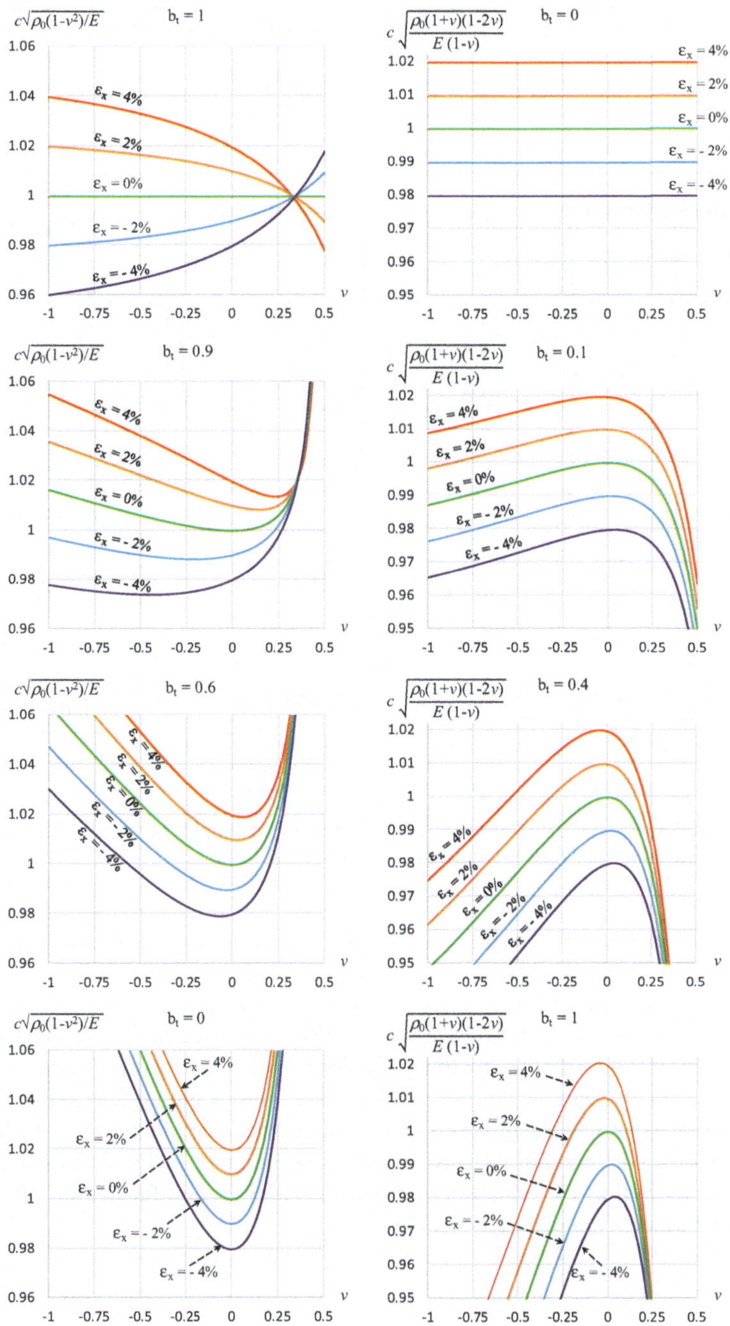

Fig. 6.29 Plots of dimensionless longitudinal wave speeds—being normalized against longitudinal waves in width-constrained plates (left) and against plane waves of dilatation (right)—versus Poisson's ratio of the rectangular slab for various longitudinal strains and boundary parameters

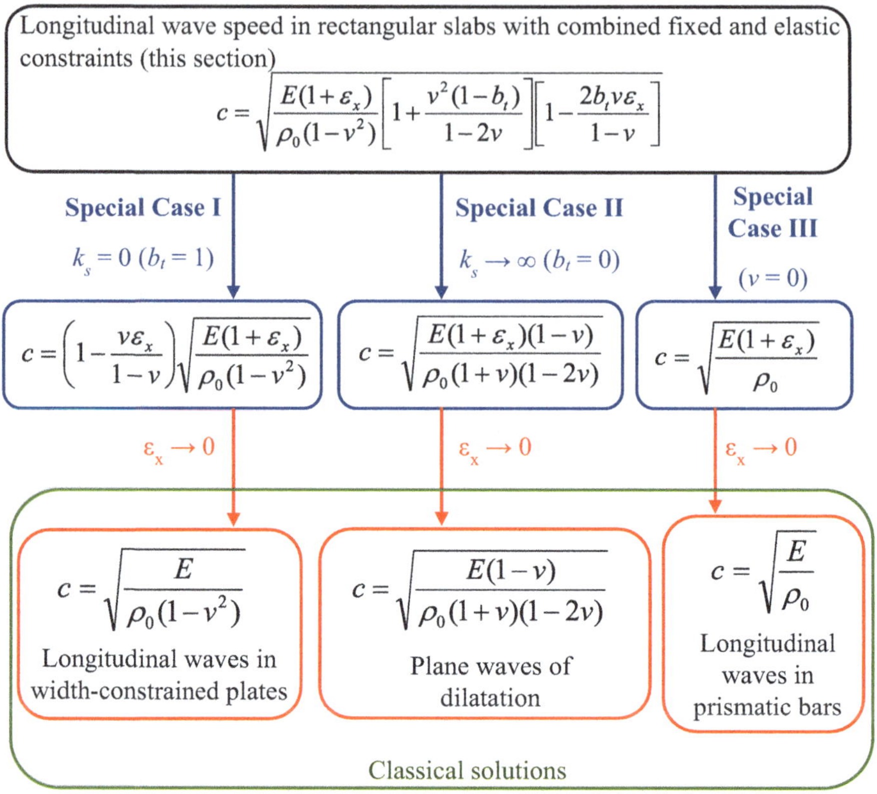

Fig. 6.30 Various special cases of wave speeds arising from simplifying assumptions as well as further simplifications that lead to elementary classical solutions

of dilatation speed. Results show that tensile and compressive strains quicken and depress, respectively, in comparison with the classical solutions. In addition, the wave speed under the current boundary condition shows that the wave speed is higher and lower than those in longitudinal wave speed in width-constrained plates and in plane waves of dilatation, respectively, unless $v = 0$. The incorporation of dimensional and density changes not only facilitates a more accurate wave speed model, but also permits greater design options for the engineering practitioner. A summary of the established longitudinal wave speed in rectangular slabs with mixed lateral constraints as well as its reduction to three special cases and further simplifications to classical models are displayed in Fig. 6.30.

References

Bianchi M, Scarpa F (2013) Vibration transmissibility and damping behaviour for auxetic and conventional foams under linear and nonlinear regimes. Smart Mater Struct 22(8):084010

Boldrin L, Hummel S, Scarpa F, Di Maio D, Lira C, Ruzzene M, Remillat CDL, Lim TC, Rajasekaran R, Patsias S (2016) Dynamic behaviour of auxetic gradient composite hexagonal honeycombs. Compos Struct 149:114–124

Chen CP, Lakes RS (1996) Micromechanical analysis of dynamic behavior of conventional and negative Poisson's ratio foams. J Eng Mater Technol 118(3):285–288

Graff KF (1975) Wave motion in elastic solids. Clarendon Press, Oxford

Goldstein RV, Gorodtsov VA, Lisovenko DS (2014) Rayleigh and love surface waves in isotropic media with negative Poisson's ratio. Mech Solids 49(4):422–434

He JH, Huang HH (2018) Tunable acoustic wave propagation through planar auxetic metamaterial. J Mech 34(2):113–122

Hou X, Deng Z, Zhou J (2011) Symplectic analysis for the wave propagation properties of conventional and auxetic cellular structures. Int J Num Anal Model Ser B 2(4):298–314

Koenders MA (2009) Wave propagation through elastic granular and granular auxetic materials. Phys Status Solidi B 246(9):2083–2088

Kolat P, Maruszewski BM, Wojciechowski KW (2010) Solitary waves in auxetic plates. J Non-Cryst Solids 356(37–40):2001–2009

Kolat P, Maruszewski BT, Tretiakov KV, Wojciechowski KW (2011) Solitary waves in auxetic rods. Phys Status Solidi B 248(1):148–157

Kolsky H (1963) Stress waves in solids. Dover Publications, New York

Lim TC (2013) Stress wave transmission and reflection through auxetic solids. Smart Mater Struct 22(8):084002

Lim TC, Alderson A, Alderson KL (2014a) Experimental studies on the impact properties of auxetic materials. Phys Status Solidi B 251(2):307–313

Lim TC, Cheang P, Scarpa F (2014b) Wave motion in auxetic solids. Phys Status Solidi B 251(2):388–396

Lim TC (2015a) Auxetic materials and structures. Springer, Singapore

Lim TC (2015b) Longitudinal wave velocity in auxetic rods. J Eng Mater Technol 137(2):024502

Lim TC (2016a) Longitudinal wave motion in width-constrained auxetic plates. Smart Mater Struct 25(5):054008

Lim TC (2016b) Plane waves of dilatation in auxetic bulk solids. Mater Sci Forum 866:206–210

Lim TC (2019a) Longitudinal wave speed in cylindrical auxetic rods with elastic constraint in radial direction. Euro J Mech A Solids 75:443–449

Lim TC (2019b) Longitudinal wave speed in auxetic plates with elastic constraint in width direction. Arch Appl Mech 89(4):659–668

Lim TC (2019c) Longitudinal wave speed in rectangular slabs with mixed restraints in lateral directions. Eng Comput Mech 172(4):153–161

Malischewsky PG, Lorato A, Scarpa F, Ruzzene M (2012) Unusual behaviour of wave propagation in auxetic structures: P-waves on free surface and S-waves in chiral lattices with piezoelectrics. Phys Status Solidi B 249(7):1339–1346

Reda H, Rahali Y, Ganghoffer JF, Lakiss H (2016) Wave propagation in 3D viscoelastic auxetic and textile materials by homogenized continuum micropolar models. Compos Struct 141:328–345

Ruzzene M, Scarpa F (2003) Control of wave propagation in sandwich beams with auxetic core. J Intell Mater Syst Struct 14(7):443–453

Scarpa F, Malischewsky PG (2008) Some new considerations concerning the Rayleigh-wave velocity in auxetic materials. Phys Status Solidi B 245(3):578–583

Sobieszczyk P, Majka M, Kuźma D, Lim TC, Zieliński P (2015) Effect of longitudinal stress on wave propagation in width-constrained elastic plates with arbitrary Poisson's ratio. Phys Status Solidi B 252(7):1615–1619

Tee KF, Spadoni A, Scarpa F, Ruzzene M (2010) Wave propagation in auxetic tetrachiral honeycombs. J Vib Acoust 132(3):031007
Timoshenko S, Goodier JN (1951) Theory of elasticity, 2nd edn. McGraw-Hill, New York
Trzupek D, Zielinski P (2009) Isolated true surface wave in a radiative band on a surface of a stressed auxetic. Phys Rev Lett 103(7):075504

Chapter 7
Elasticity of Auxetic Beams

Abstract This chapter employs elasticity models to understand the effect of Poisson's ratio negativity. Some discussions include how the use of classical elasticity models differs from the refined models as applied to auxetic beams.

Keywords Auxetic beams · Correction terms · Elasticity models

7.1 Fundamentals

Unlike the mechanics of materials (previously known as the strength of materials) approach for beam theory—which oversimplifies the beam cross section upon deformation—and its shear deformation theories—the first-order shear deformation theory (FSDT) and the third-order shear deformation theory (TSDT)—which makes less simplifying assumptions, the elasticity theory removes these simplifying assumptions. As a result, the elasticity solutions are good candidates as verifier of the mechanics of materials and the shear deformation theories. This chapter considers only the elasticity solutions of some beams and how they are related to their corresponding mechanics of materials models. Special emphasis is given to the effect of auxeticity on their solutions. For this reason, four examples are furnished in this chapter based on 2D problems in rectangular coordinates. Unlike the elementary beam theory according to the Euler–Bernoulli hypotheses that plane sections remain plane and normal to the axis of the beam, the theory of elasticity removes such simplifying assumptions in order to permit a more realistic stress and displacement fields.

For a 2D problem in rectangular coordinates, the Airy stress function $\phi = \phi(x, y)$ takes the form of the biharmonic equation

$$\frac{\partial^4 \phi}{\partial x^4} + 2\frac{\partial^4 \phi}{\partial x^2 \partial y^2} + \frac{\partial^4 \phi}{\partial y^4} = 0 \tag{7.1.1}$$

in the absence of body force, whereby

$$\sigma_x = \frac{\partial^2 \phi}{\partial y^2}; \quad \sigma_y = \frac{\partial^2 \phi}{\partial x^2}; \quad \tau_{xy} = -\frac{\partial^2 \phi}{\partial x \partial y} \tag{7.1.2}$$

T. Lim, *Mechanics of Metamaterials with Negative Parameters*, Engineering Materials,
https://doi.org/10.1007/978-981-15-6446-8_7

The effect of Poisson's ratio, and hence its negativity, is not apparent. In the presence of body force acting in the x- and y-directions

$$F_x = -\frac{\partial V}{\partial x}; \quad F_y = -\frac{\partial V}{\partial y} \tag{7.1.3}$$

written in the form of a potential function V, the biharmonic equation becomes

$$\frac{\partial^4 \phi}{\partial x^4} + 2\frac{\partial^4 \phi}{\partial x^2 \partial y^2} + \frac{\partial^4 \phi}{\partial y^4} = -\frac{1-2v}{1-v}\left(\frac{\partial^2 V}{\partial x^2} + \frac{\partial^2 V}{\partial y^2}\right) \tag{7.1.4}$$

under plane strain condition, and

$$\frac{\partial^4 \phi}{\partial x^4} + 2\frac{\partial^4 \phi}{\partial x^2 \partial y^2} + \frac{\partial^4 \phi}{\partial y^4} = -(1-v)\left(\frac{\partial^2 V}{\partial x^2} + \frac{\partial^2 V}{\partial y^2}\right) \tag{7.1.5}$$

under plane stress condition. Both equations become identical when $v = 0$

$$\frac{\partial^4 \phi}{\partial x^4} + 2\frac{\partial^4 \phi}{\partial x^2 \partial y^2} + \frac{\partial^4 \phi}{\partial y^4} = -\left(\frac{\partial^2 V}{\partial x^2} + \frac{\partial^2 V}{\partial y^2}\right) \tag{7.1.6}$$

which demarcates the auxetic region from the conventional region. Very little information on the effect of auxeticity can be extracted unless one considers specific elasticity problems, in which the Airy stress function must be determined.

The assumed solution for the Airy stress function can be taken in the form of

$$\phi = \sum_{m=0}^{\infty}\sum_{n=0}^{\infty} A_{mn} x^m y^n \tag{7.1.3}$$

where A_{mn} are constant coefficients to be evaluated, or in the form of $\phi = e^{\alpha x} e^{\beta y}$, i.e.,

$$\begin{aligned}
\phi = {} & \sin \beta x[(A + C\beta y) \sinh \beta y + (B + D\beta y) \cosh \beta y] \\
& + \cos \beta x\left[\left(A' + C'\beta y\right) \sinh \beta y + \left(B' + D'\beta y\right) \cosh \beta y\right] \\
& + \sin \alpha y[(E + G\alpha x) \sinh \alpha x + (F + H\alpha x) \cosh \alpha x] \\
& + \cos \alpha y\left[\left(E' + G'\alpha x\right) \sinh \alpha x + \left(F' + H'\alpha x\right) \cosh \alpha x\right] \\
& + C_0 + C_1 x + C_2 x^2 + C_3 x^3 + C_4 y + C_5 y^2 + C_6 y^3 \\
& + C_7 xy + C_8 x^2 y + C_9 xy^2
\end{aligned} \tag{7.1.4}$$

where the constant coefficients are obtained from the boundary conditions. Examples on the effect of auxeticity on transversely loaded beams with supports on both ends are furnished in Sects. 7.2 and 7.3 for uniform and sinusoidal loads, respectively. For the uniform load in Sect. 7.2, the polynomial solution to the trial Airy stress function is taken from Eq. (7.1.3) to give

$$\phi = A_{20}x^2 + A_{21}x^2y + A_{03}y^3 + A_{23}x^2y^3 - \frac{1}{5}A_{23}y^5 \tag{7.1.5}$$

where the last term is required to satisfy the biharmonic equation, while in the case of sinusoidal load in Sect. 7.3, the trial solution is taken from Eq. (7.1.4) to give

$$\phi = \sin\beta x[(A + C\beta y)\sinh\beta y + (B + D\beta y)\cosh\beta y] \tag{7.1.6}$$

For the examples of end-loaded cantilever in Sects. 7.4 and 7.5, the trial solution is again taken from Eq. (7.1.3) to give

$$\phi = A_{11}xy + A_{13}xy^3 \tag{7.1.7}$$

7.2 Example: Auxetic Beams Under Uniform Load

As a first example, we consider a loaded beam of length $2l$ and thickness $2c$ as measured along the x-axis and y-axis, respectively, with unit width. As shown in Fig. 7.1, a uniform load of

$$q = q_0 \tag{7.2.1}$$

is prescribed such that the overall load on each ends of the beam is q_0l.

Recall that the displacement field for this problem is (Timoshenko and Goodier 1970; Saad 2014)

$$u_x = \frac{q_0x}{2EI}\left[\left(l^2 - \frac{x^2}{3}\right)y + \left(\frac{2y^3}{3} - \frac{2c^2y}{5}\right) + v\left(\frac{y^3}{3} - c^2y + \frac{2c^3}{3}\right)\right] \tag{7.2.2}$$

and

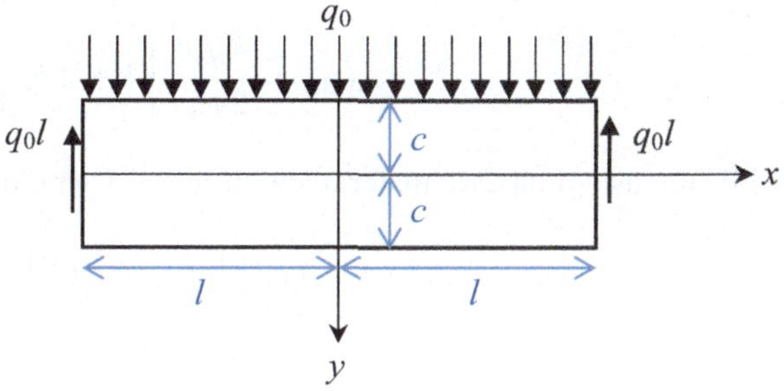

Fig. 7.1 Uniformly loaded beam

$$u_y = -\frac{q_0}{2EI}\left\{\frac{y^4 - x^4}{12} - \frac{c^2y^2}{2} + \frac{2c^3y}{3} + v\left[\frac{(l^2 - x^2)y^2}{2} + \frac{y^4}{6} - \frac{c^2y^2}{5}\right] + \left[\frac{l^2}{2} + \left(\frac{4}{5} + \frac{v}{2}\right)c^2\right]x^2\right\} + \frac{5q_0l^4}{24EI}\left[1 + \frac{12}{5}\left(\frac{4}{5} + \frac{v}{2}\right)\frac{c^2}{l^2}\right] \tag{7.2.3}$$

The displacement field simplifies to

$$u_x = \frac{q_0xy}{2EI}\left[\left(l^2 - \frac{x^2}{3}\right) + \left(\frac{2y^2}{3} - \frac{2c^2}{5}\right)\right] \tag{7.2.4}$$

and

$$u_y = -\frac{q_0}{2EI}\left\{\frac{y^4 - x^4}{12} - \frac{c^2y^2}{2} + \frac{2c^3y}{3} + \left[\frac{l^2}{2} + \frac{4c^2}{5}\right]x^2\right\} + \frac{5q_0l^4}{24EI}\left[1 + \frac{48}{25}\frac{c^2}{l^2}\right] \tag{7.2.5}$$

when Poisson's ratio is zero, which demarcates the auxetic region from the conventional region. Substituting $(x, y) = (0, 0)$ into Eq. (7.2.3) for the maximum deflection gives

$$u_{y,\max} = \frac{5q_0l^4}{24EI}\left[1 + \frac{12}{5}\left(\frac{4}{5} + \frac{v}{2}\right)\frac{c^2}{l^2}\right] \tag{7.2.6}$$

which is comparable to the mechanics of materials solution

$$u_{y,\max} = \frac{5q_0l^4}{24EI} \tag{7.2.7}$$

with the difference being the correction term due to the presence of shear force in the beam. This shows that the beam's dimensionless thickness c/l plays a major role in contrasting the elasticity solution from the mechanics of materials solution, i.e., for thick beams, the use of the mechanics of materials solution underestimates the deflection. From the materials viewpoint, the use of materials with high Poisson's ratio creates greater difference between the elasticity and the mechanics of materials solution. For example, the use of incompressible materials ($v = 0.5$) would give

$$u_{y,\max} = \frac{5q_0l^4}{24EI}\left[1 + 2.52\frac{c^2}{l^2}\right] \tag{7.2.8}$$

while the use of auxetic materials with $v = -1$ would give

$$u_{y,\max} = \frac{5q_0l^4}{24EI}\left[1 + 0.72\frac{c^2}{l^2}\right] \sim \frac{5q_0l^4}{24EI} \tag{7.2.9}$$

In other words, a geometrically thick beam approximates a mechanically thin beam if the negativity of Poisson's ratio is sufficient to reduce the correction term.

7.3 Example: Auxetic Beams Under Sinusoidal Load

Figure 7.2 shows a schematic for a simply supported beam of Young's modulus E, Poisson's ratio v, length l, and thickness $2c$ subjected to a sinusoidal load

$$q = q_0 \sin \frac{\pi x}{l} \tag{7.3.1}$$

in which the displacement fields $u_x(x, y)$ and $u_y(x, y)$ are functions of the coordinate position (x, y) of the beam. In the classical elasticity solution to this problem (Saad 2014; Soutas-Little 1999), the mid-plane transverse deflection of the beam is obtained as

$$u_y(x, 0) = \frac{D\beta}{E} \sin \beta x [2 + (1 + v)\beta c \tanh \beta c] \tag{7.3.2}$$

where $\beta = \pi / l$ and

$$D = \frac{q_0 \cosh \frac{\pi c}{l}}{2 \frac{\pi^2}{l^2} \left(\frac{\pi c}{l} - \sinh \frac{\pi c}{l} \cosh \frac{\pi c}{l} \right)} \tag{7.3.3}$$

In typical elasticity textbooks, it is assumed that $l \gg c$, and so substituting

$$\cosh \frac{\pi c}{l} \approx 1 \tag{7.3.4}$$

and

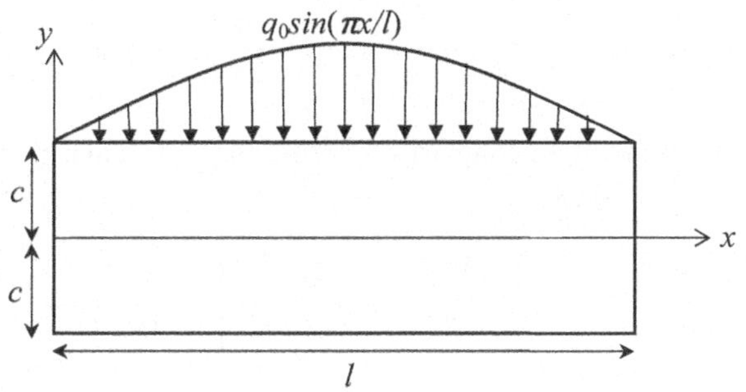

Fig. 7.2 Schematics of a beam under transverse sinusoidal load with simple supports at both ends

$$\sinh\frac{\pi c}{l}\cosh\frac{\pi c}{l}\approx\left(\frac{\pi c}{l}+\frac{1}{3!}\frac{\pi^3c^3}{l^3}\right)\left(1+\frac{1}{2!}\frac{\pi^2c^2}{l^2}\right)\approx\frac{\pi c}{l}+\frac{2}{3}\frac{\pi^3c^3}{l^3}\qquad(7.3.5)$$

into the numerator and denominator of Eq. (7.3.3), respectively, gives

$$D\approx-\frac{3q_0l^5}{4c^3\pi^5}\qquad(7.3.6)$$

Using Eq. (7.3.6) and $\beta=\pi/l$, Eq. (7.3.2) becomes

$$u_y(x,0)=-\frac{3q_0l^4}{2c^3\pi^4E}\sin\frac{\pi x}{l}\left[1+\left(\frac{1+\upsilon}{2}\right)\frac{\pi c}{l}\tanh\frac{\pi c}{l}\right]\qquad(7.3.7)$$

Hence, Eq. (7.3.7) is typically taken to be the typical "textbook" elasticity solution for the transverse deflection of a sinusoidally loaded beam, which can then be reduced to the mechanics of materials solution

$$u_y(x,0)=-\frac{3q_0l^4}{2c^3\pi^4E}\sin\frac{\pi x}{l}\qquad(7.3.8)$$

by considering $l\gg c$ again such that the second term in Eq. (7.3.7) diminishes. The premature assumption of $l\gg c$ that leads to Eq. (7.3.7) therefore disqualifies this elasticity solution as a verifier for the mechanics of materials deflection model of this beam problem, except for cases where the beam thickness is very small compared to its length, as the corresponding textbooks correctly state (Saad 2014; Soutas-Little 1999). In referring to some monographs (Fung 1965; Landau and Lifshitz 1959; Lim 2015) pertaining to the range of Poisson's ratio being $-1\le\upsilon\le 1/2$ for isotropic solids, Eq. (7.3.7) gives a false impression that the use of auxetic materials of $\upsilon=-1$ reduces the elasticity solution to the mechanics of materials solution. This interpretation is untrue, and it will be proven that the use of materials with $\upsilon=-1$ will not reduce the deflection model by elasticity theory to that by the elementary theory. From Eq. (7.3.3) and $\beta=\pi/l$, we have

$$\frac{D\beta}{E}=\frac{q_0l}{2\pi E}\frac{\cosh\frac{\pi c}{l}}{\frac{\pi c}{l}-\frac{1}{2}\sinh\frac{2\pi c}{l}}\qquad(7.3.9)$$

To facilitate comparison with Eqs. (7.3.7) and (7.3.8), it is expedient to rewrite Eq. (7.3.9) as

$$\frac{D\beta}{E}=-\frac{3q_0l^4}{2c^3\pi^4E}\left\{-\frac{\pi^3c^3}{3l^3}\frac{\cosh\frac{\pi c}{l}}{\frac{\pi c}{l}-\frac{1}{2}\sinh\frac{2\pi c}{l}}\right\}\qquad(7.3.10)$$

which, upon substitution into Eq. (7.3.2), leads to (Lim 2018)

$u_y(x,0)$

$$= -\frac{3q_0 l^4}{2c^3\pi^4 E} \sin\frac{\pi x}{l}\left[1 + \left(\frac{1+v}{2}\right)\frac{\pi c}{l}\tanh\frac{\pi c}{l}\right]\left\{-\frac{2}{3}\left(\frac{\pi c}{l}\right)^3 \frac{\cosh\frac{\pi c}{l}}{\frac{\pi c}{l} - \frac{1}{2}\sinh\frac{2\pi c}{l}}\right\} \tag{7.3.11}$$

It can be seen that the {...} term in Eq. (7.3.11) can be alternatively expressed as

$$\left\{-\frac{2}{3}\left(\frac{\pi c}{l}\right)^3 \frac{\cosh\frac{\pi c}{l}}{\frac{\pi c}{l} - \frac{1}{2}\sinh\frac{2\pi c}{l}}\right\} = \frac{1 + \frac{1}{2}\left(\frac{\pi c}{l}\right)^2 + \frac{1}{24}\left(\frac{\pi c}{l}\right)^4 + \cdots}{1 + \frac{1}{5}\left(\frac{\pi c}{l}\right)^3 + \frac{2}{105}\left(\frac{\pi c}{l}\right)^5 + \cdots} \tag{7.3.12}$$

so as to show that $\{...\} = 1$ as $c/l \to 0$. Nevertheless, the form of {...} furnished in Eq. (7.3.11) is adopted herein instead of that shown in Eq. (7.3.12) as the former is exact. The substitution of $v = -1$ into Eq. (7.3.11) gives

$$u_y(x, 0) = -\frac{3q_0 l^4}{2c^3\pi^4 E} \sin\frac{\pi x}{l}\left\{-\frac{2}{3}\left(\frac{\pi c}{l}\right)^3 \frac{\cosh\frac{\pi c}{l}}{\frac{\pi c}{l} - \frac{1}{2}\sinh\frac{2\pi c}{l}}\right\} \tag{7.3.13}$$

which is different from the mechanics of materials model. To put in a systematic perspective, one may write the mid-plane deflection profile of the given beam as

$$u_y(x, 0) = -\frac{3q_0 l^4}{2c^3\pi^4 E} \sin\frac{\pi x}{l} f\left(\frac{c}{l}, v\right) g\left(\frac{c}{l}\right) \tag{7.3.14}$$

where the correction functions

$$f\left(\frac{c}{l}, v\right) = 1 + \left(\frac{1+v}{2}\right)\frac{\pi c}{l}\tanh\frac{\pi c}{l} \tag{7.3.15}$$

and

$$g\left(\frac{c}{l}\right) = -\frac{2}{3}\left(\frac{\pi c}{l}\right)^3 \frac{\cosh\frac{\pi c}{l}}{\frac{\pi c}{l} - \frac{1}{2}\sinh\frac{2\pi c}{l}} \tag{7.3.16}$$

can be reduced to the mechanics of materials and the simplified elasticity models, as furnished in Table 7.1.

Table 7.1 Comparison of mechanics of materials, simplified elasticity, and exact elasticity models for the given problem

Models for $u_y(x, 0)$	$f\left(\frac{c}{l}, v\right)$	$g\left(\frac{c}{l}\right)$
Mechanics of materials, $u_y^{\text{Mech Mater}}$	1	1
Simplified elasticity, $u_y^{\text{simpl Elast}}$	$1 + \left(\frac{1+v}{2}\right)\frac{\pi c}{l}\tanh\frac{\pi c}{l}$	1
Exact elasticity, $u_y^{\text{exact Elast}}$	$1 + \left(\frac{1+v}{2}\right)\frac{\pi c}{l}\tanh\frac{\pi c}{l}$	$-\frac{2}{3}\left(\frac{\pi c}{l}\right)^3 \frac{\cosh\frac{\pi c}{l}}{\frac{\pi c}{l} - \frac{1}{2}\sinh\frac{2\pi c}{l}}$

To give an appreciation on the exact model, the deflection profiles based on the mechanics of materials, simplified elasticity and exact elasticity models can be plotted, preferably in the form of dimensionless deflection $uE\pi/q_0 l$, as shown in Fig. 7.3 for the typical Poisson's ratio of $v = 0.3$, as well as at the upper and lower limits of Poisson's ratio for isotropic solids ($v = 0.5, -1$) with $c/l = 0.2$.

Plots of the dimensionless deflection profile are useful to show that the gaps between the three models widen and narrow when Poisson's ratio of the beam increases and decreases, respectively. More importantly, plots of the deflection profile for the mechanics of materials model and the simplified elasticity model collapse into a single curve when $v = -1$, which underestimate the extent of deflection predicted by the exact elasticity model.

Reference to Eqs. (7.3.7), (7.3.8), and (7.3.11) shows that the ratio of elasticity's (both simplified and exact) mid-plane deflection to the mechanics of materials deflection model is independent from the location x. Figure 7.4 displays these ratios using the case of a "thin" beam of $c/l = 0.1$ and a "thick" beam of $c/l = 0.2$, clearly showing that the deflection by the simplified elasticity model underestimates that by the exact elasticity model, especially for thick beams. Elasticity textbooks that adopt this beam problem correctly declare upfront in their derivation that Eq. (7.3.7) is obtained by assuming $l \gg c$. On what situation, then, is Eq. (7.3.7) applicable? This question is best answered in a quantitative manner. Suppose we let the acceptable level of error be confined within 5%, then Eq. (7.3.7) is valid for $c/l \leq 0.135$, as indicated by the curve of the simplified-to-exact elasticity ratio plotted against the c/l ratio in Fig. 7.5. The error exceeds 10% for $c/l = 0.2$.

In summary, an exact elasticity model for the deflection of a simply supported beam under a sinusoidal load has been furnished in this section. By ending the exact formulation for the beam deflection at Eq. (7.3.2) in current textbooks, one may get the impression that the exact deflection profile is not in the form that can be readily reduced to the mechanics of materials deflection profile unless a first $l \gg c$ assumption is made to simplify the expression of D from Eq. (7.3.3) to Eq. (7.3.6), and thereafter, a second $l \gg c$ assumption is made on the simplified elasticity solution, Eq. (7.3.7), to recover the mechanics of materials solution, Eq. (7.3.8). However, this section shows that it is not only possible, but also very convenient, to express the exact elasticity solution of Eq. (7.3.2) in the form of Eq. (7.3.11), which is easily reduced to both Eqs. (7.3.7) and (7.3.8). In a practical sense, the effect from the first $l \gg c$ assumption is quantified in terms of the simplified elasticity's deflection percentage error vis-à-vis the exact elasticity model, which stands at 5% for $c/l = 0.135$. Unlike the current simplified elasticity model, the exact model does not reduce to the mechanics of materials model when $v = -1$.

7.4 Example: Deflections of Auxetic Cantilever Beams

The example that is considered in this section is that of end-loaded cantilever beam, as shown in Fig. 7.6 (top left), which has been discussed in a number of elasticity

Fig. 7.3 Dimensionless deflection profile, $uE\pi/q_0l$ along the dimensionless beam length, x/l for a beam of $c/l = 0.2$ with $\nu = 0.5$ (top), $\nu = 0.3$ (middle), and $\nu = -1$ (bottom)

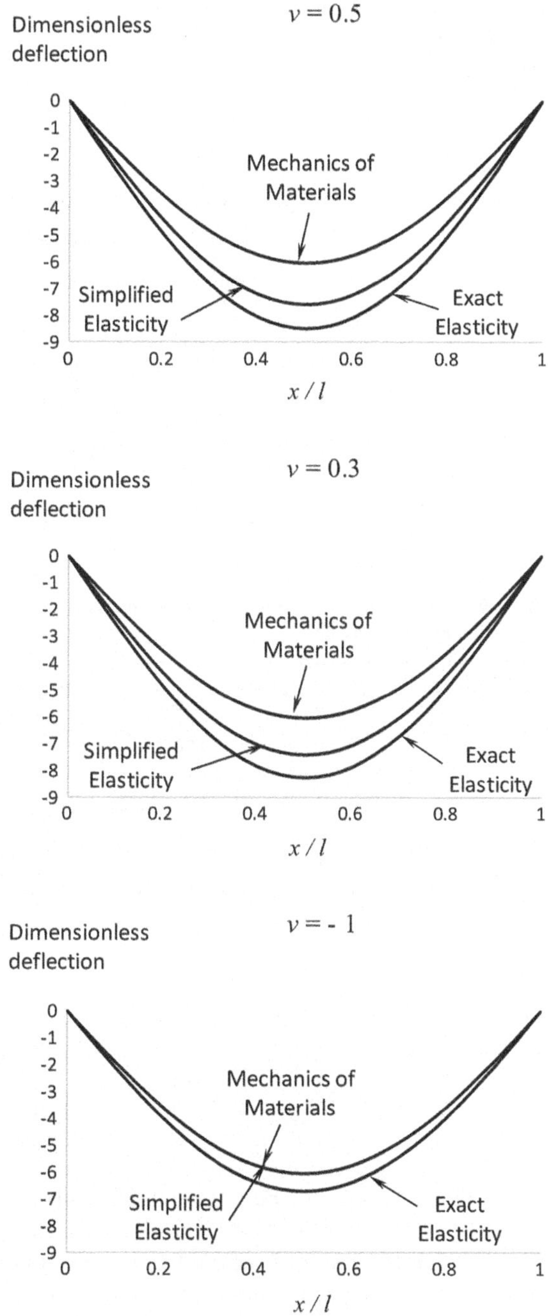

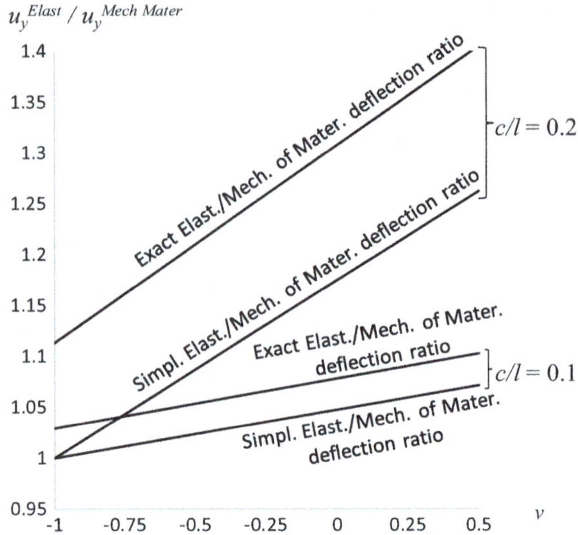

Fig. 7.4 Ratio of exact and simplified elasticity to mechanics of materials deflection versus the entire range of Poisson's ratio for a typical "thick" beam ($c/l = 0.2$) and a relatively "thin" beam ($c/l = 0.1$)

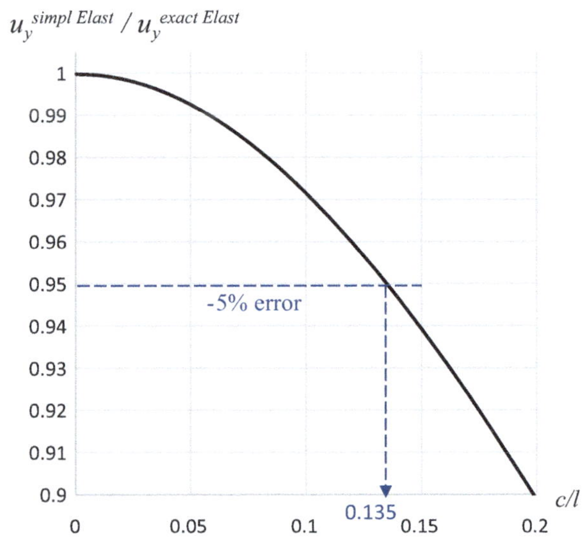

Fig. 7.5 Deficiency of the simplified elasticity model vis-à-vis the exact elasticity model

textbooks (Sokolnikoff 1956; Atanankovic and Guran 2000; Saada 2009; Boresi et al. 2011; Gould 2013; Saad 2014). The discourse by Love (1906) on this subject matter suggests that the condition of the fixed end is that illustrated in Fig. 7.6 (top right), which is the mechanics of materials solution. In addition to this profile, Timoshenko and Goodier (1970) and Housner and Vreeland (1965) explained the effect of shearing force on the beam deflection, which results in the fixed end profile furnished in Fig. 7.6 (bottom left). Having demonstrated these two profiles, Barber (2010) suggested the

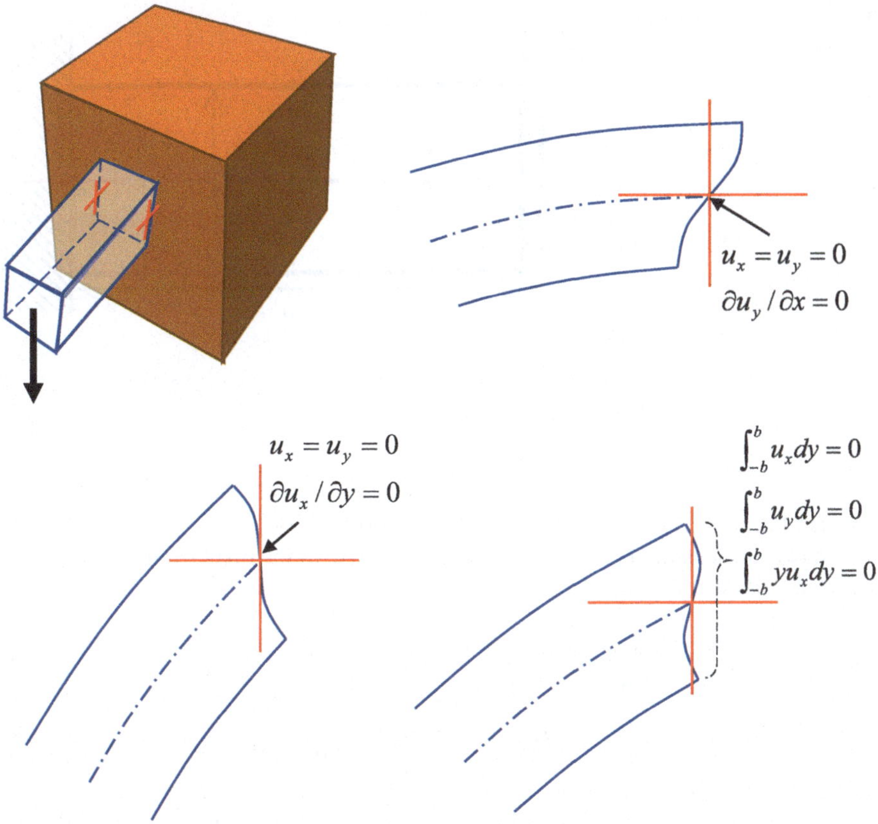

Fig. 7.6 Isometric view of an end-loaded cantilever (top left) with strong boundary conditions akin to the elementary mechanics of materials approach (top right), strong boundary conditions that account for the effect of shearing force on the deflection (bottom left), and weak boundary conditions (bottom right)

weak boundary conditions that lead to the end condition indicated in Fig. 7.6 (bottom right) as the best solution.

Figure 7.7 shows a cantilever of Young's modulus E and Poisson's ratio v with rectangular cross section of length a, thickness $2b$, and unit width that is loaded by a force P at the free end, i.e., the boundary conditions are

$$\begin{aligned} \sigma_x &= 0; \quad x = 0 \\ \sigma_y &= 0; \quad y = \pm b \\ \tau_{xy} &= 0; \quad y = \pm b \end{aligned} \tag{7.4.1}$$

and

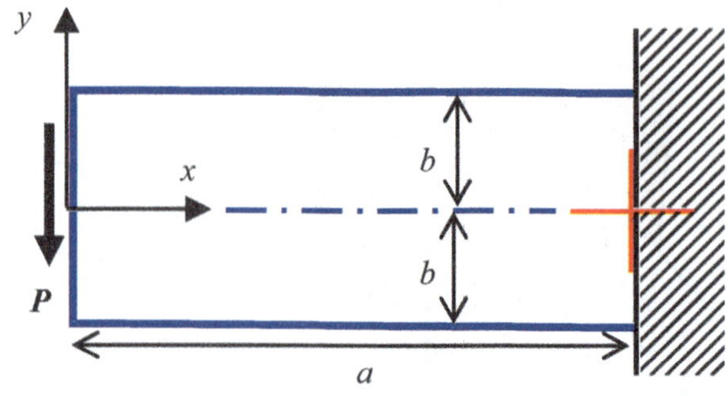

Fig. 7.7 Schematic view of an end-loaded cantilever

$$\int_{-b}^{b} \tau_{xy} \mathrm{d}y = P; \quad x = 0 \tag{7.4.2}$$

in which Eq. (7.4.2) is prescribed in the weak form.

The two-dimensional elasticity solution gives rise to the displacement field (e.g., Barber 2010) described as

$$u_x = \frac{3Px^2y}{4Eb^3} + \frac{3P(1+v)y}{2Eb} - \frac{P(2+v)y^3}{4Eb^3} + A - Cy \tag{7.4.3}$$

$$u_y = -\frac{3vPxy^2}{4Eb^3} - \frac{Px^3}{4Eb^3} + B + Cx \tag{7.4.4}$$

To obtain the constants *A*, *B*, and *C*, the required three boundary conditions at the built-in end, as illustrated in Fig. 7.6 (top left), are ideally $u_x = u_y = 0$ at $x = a$ and $-b \leq y \leq b$. However, the displacement field does not permit such boundary condition to be implemented. In the mechanics of materials approach, the following strong boundary condition imposed

$$u_x = u_y = \frac{\partial u_y}{\partial x} = 0; \quad (x, y) = (a, 0) \tag{7.4.5}$$

is illustrated in Fig. 7.6 (top right). The substitution of Eq. (7.4.5) into Eqs. (7.4.3) and (7.4.4) gives

$$A = 0; \quad B = -\frac{Pa^3}{2Eb^3}; \quad C = \frac{3Pa^2}{4Eb^3} \tag{7.4.6}$$

If one were to implement an alternate strong boundary condition whereby

$$u_x = u_y = \frac{\partial u_x}{\partial y} = 0; \quad (x, y) = (a, 0) \tag{7.4.7}$$

then substitution of Eq. (7.4.7) into Eqs. (7.4.3) and (7.4.4) leads to

$$A = 0;\quad B = -\frac{Pa^3}{2Eb^3}\left(1 + 3(1+v)\frac{b^2}{a^2}\right);\quad C = \frac{3Pa^2}{4Eb^3}\left(1 + 2(1+v)\frac{b^2}{a^2}\right) \tag{7.4.8}$$

As opposed to the two strong boundary conditions, a set of weak boundary conditions (Barber 2010)

$$\int_{-b}^{b} u_x \mathrm{d}y = \int_{-b}^{b} u_y \mathrm{d}y = \int_{-b}^{b} y u_x \mathrm{d}y = 0;\quad x = a \tag{7.4.9}$$

yields the following coefficients

$$A = 0;\quad B = -\frac{Pa^3}{2Eb^3}\left(1 + \frac{(12+11v)}{5}\frac{b^2}{a^2}\right);\quad C = \frac{3Pa^2}{4Eb^3}\left(1 + \frac{(8+9v)}{5}\frac{b^2}{a^2}\right) \tag{7.4.10}$$

The displacement of the loaded end at the mid-plane $(x, y) = (0, 0)$ can therefore be obtained by substituting these coefficients into Eqs. (7.4.3) and (7.4.4) to give the horizontal displacement

$$u_x = A = 0 \tag{7.4.11}$$

regardless of the type of boundary conditions selected, while the vertical deflection at the same location $(x, y) = (0, 0)$, $u_{y,\max} = B$ is dependent on the chosen boundary conditions, i.e.,

$$u_{y,\max} = -\frac{Pa^3}{2Eb^3} \tag{7.4.12}$$

based on the mechanics of materials' strong boundary conditions set out in Eq. (7.4.5),

$$u_{y,\max} = -\frac{Pa^3}{2Eb^3}\left(1 + 3(1+v)\frac{b^2}{a^2}\right) \tag{7.4.13}$$

based on the alternate strong boundary condition described in Eq. (7.4.7), and

$$u_{y,\max} = -\frac{Pa^3}{2Eb^3}\left(1 + \frac{(12+11v)}{5}\frac{b^2}{a^2}\right) \tag{7.4.14}$$

based on the weak boundary condition indicated in Eq. (7.4.9).

The deflection model described in Eq. (7.4.13) takes into consideration the shear deflection. Note that if we let $v = -1$ in Eq. (7.4.13), then the effect of shear deflection diminishes such that it reduces to Eq. (7.4.12). In other words, the shear deflection vanishes for the choice of $v = -1$ material.

Since the term containing $3(1+v)$ has been identified as the correction due to shear deflection, one may express Eq. (7.4.14) in the form

$$u_{y,\max} = -\frac{Pa^3}{2Eb^3}\left[1 + \left(3(1+v) - \frac{3+4v}{5}\right)\frac{b^2}{a^2}\right] \quad (7.4.15)$$

so that by isolating the shear correction term, the other term containing $(3+4v)/5$ is identified as the remaining end correction. This remaining end correction diminishes if $v = -3/4$. Again, the choice of auxetic materials diminishes one type of correction.

7.5 Example: Deflection of Beams with Partially Built-in End

While the method that adopts weak boundary conditions appears to be the best solution for a cantilever that is transversely loaded at its free end, an exact solution using strong boundary conditions remains elusive. For this reason, a modified example is introduced herein so that an exact solution using strong boundary conditions can be achieved. Recall that to obtain the constants A, B, and C, the required three boundary conditions at the built-in end, as illustrated in Fig. 7.6 (top left), are ideally $u_x = u_y = 0$ at $x = a$ and $-b \leq y \leq b$. However, the displacement field as described by Eqs. (7.4.3) and (7.4.4) does not permit such boundary condition to be implemented.

An exact result using strong boundary conditions can be obtained from the given displacement field if one were to select a different boundary condition from that of totally built end, i.e., by defining a different problem in which the end conditions permit displacement so that the imposed strong boundary condition is exact. This modified boundary condition is illustrated in Fig. 7.8 (left), which shows that at $x \geq a$ the cantilever is perfectly bonded to the wall at top and bottom but the left and right sides are free surfaces. In other words, the boundary condition at $x = a$ prevents displacement at the upper and lower surfaces ($y = \pm b$) while permitting displacement for $-b < y < b$.

Implementing the boundary conditions

$$u_x = u_y = 0; \quad (x, y) = (a, \pm b) \quad (7.5.1)$$

as indicated in Fig. 7.8 (right), we have

$$B + Ca = \frac{3vPab^2}{4Eb^3} + \frac{Pa^3}{4Eb^3} \quad (7.5.2)$$

and

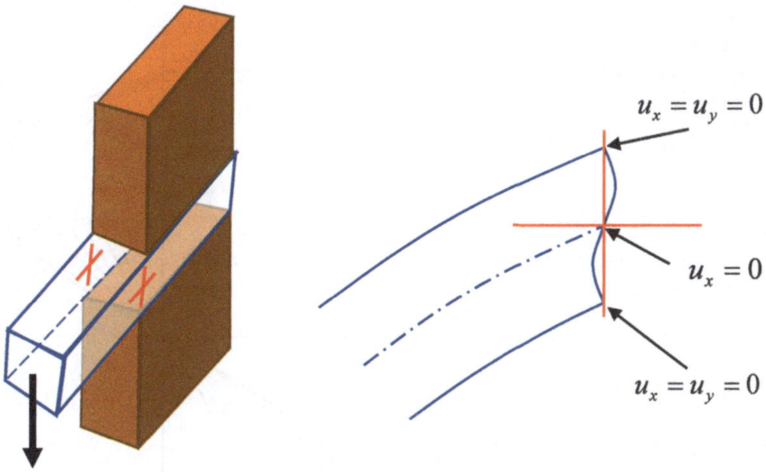

Fig. 7.8 Variant of cantilever, showing partially built-in end, which is clamped at the top and bottom surfaces but free at the other two side surfaces (left), and the corresponding strong boundary condition (right)

$$
\begin{aligned}
+Cb - A &= +\frac{3Pa^2}{4Eb^2} + \frac{3P(1+v)}{2E} - \frac{P(2+v)}{4E} \\
-Cb - A &= -\frac{3Pa^2}{4Eb^2} - \frac{3P(1+v)}{2E} + \frac{P(2+v)}{4E}
\end{aligned}
\tag{7.5.3}
$$

which can be solved to give (Lim 2021)

$$
A = 0; \quad B = -\frac{Pa^3}{2Eb^3}\left(1 + (2+v)\frac{b^2}{a^2}\right); \quad C = \frac{3Pa^2}{4Eb^3}\left(1 + \frac{(4+5v)}{3}\frac{b^2}{a^2}\right)
\tag{7.5.4}
$$

It can be deduced, by symmetry, that the following boundary condition holds

$$
u_x = 0; \quad (x, y) = (a, 0)
\tag{7.5.5}
$$

However, this boundary condition need not be imposed since it is intrinsically attained by the imposition of Eq. (7.5.1) on the displacement field. At the partially built-in end, the horizontal displacement profile is

$$
(u_x)_{x=a} = \frac{P}{E}\left(\frac{1}{2} + \frac{v}{4}\right)\left(1 - \frac{y^2}{b^2}\right)\frac{y}{b}
\tag{7.5.6}
$$

wherein its maximum magnitude of horizontal displacement occurs at

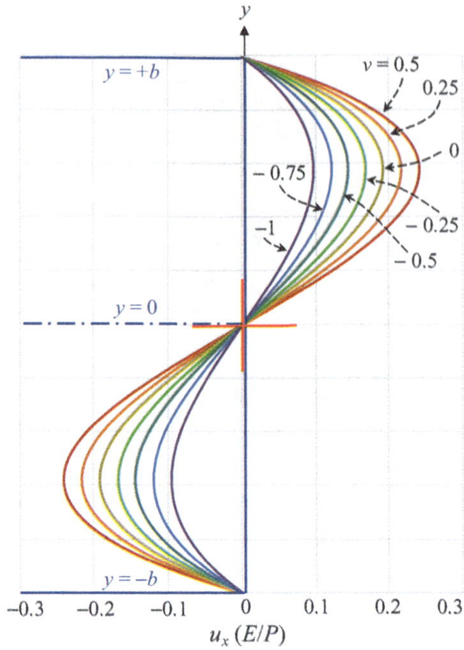

Fig. 7.9 Dimensionless horizontal displacement profile at $x = a$ for various Poisson's ratio of the cantilever material

$$y = \pm\frac{b}{\sqrt{3}} \tag{7.5.7}$$

Figure 7.9 shows the dimensionless horizontal displacement, of the cantilever, defined as $(E/P)u_x$, at $x = a$ for $-1 \le v \le 1/2$. It can be seen that the use of cantilever with $v = -1$ gives the least horizontal displacement at the partially built-in end.

The vertical displacement profile at the partially built-in end

$$\left(u_y\right)_{x=a} = \frac{3vPa}{4Eb}\left(1 - \frac{y^2}{b^2}\right) \tag{7.5.8}$$

is reduced to zero if Poisson's ratio of the cantilever is $v = 0$. Since the horizontal and vertical displacement profiles are minimized at $v = -1$ and $v = 0$, respectively, it follows that the use of auxetic materials, i.e., $-1 \le v < 0$, is more effective than conventional materials in reducing the amount of displacement at the partially built-in end. The reduced displacement at the fixed end for auxetic cantilevers implies that while the ideal boundary condition of $u_x = u_y = 0$ at $x = a$ and $-b \le y \le b$ is still elusive, it is better approximated in comparison with conventional materials. The deflection at the mid-point of the free end can be obtained by substituting $x = y = 0$ into Eq. (7.4.4) to give

$$u_{y,\max} = B = -\frac{Pa^3}{2Eb^3}\left(1 + (2+v)\frac{b^2}{a^2}\right) \tag{7.5.9}$$

Perusal to Eq. (7.5.9) suggests that the use of auxetic materials can reduce the maximum deflection, but this reduction is negligible since $a > 2b$. To compare the new solution, based on the partially clamed boundary condition, against the standard approach for fully clamped end, we recall that the displacement field indicated by Eqs. (7.4.3) and (7.4.4) holds. Comparing Eqs. (7.4.14) and (7.5.9), the difference between both maximum deflections is

$$\Delta u_{y,\max} = u_{y,\max}^{\text{fully - clamped}} - u_{y,\max}^{\text{partially - clamped}} = -\frac{P}{E}\left(\frac{1+3v}{5}\right)\frac{a}{b} \tag{7.5.10}$$

Since the displacement field expressed in Eqs. (7.4.3) and (7.4.4) is 2D in the x-y plane, the width effect is undefined. The restriction of displacements at $(x, y) = (a, \pm b)$ gives a lesser displacement than the standard model's weak boundary condition, which permits displacements at the built-in corners. However, both maximum displacements coincide when $v = -1/3$, and that the partially clamped beam gives a greater maximum deflection than the fully clamped beam if $-1 \leq v < -1/3$.

In summary, an exact solution to the end-loaded cantilever problem with the use of strong boundary conditions has been proposed not by a different route of formulation, but by changing the problem into one in which the existing displacement field can satisfactorily fit the boundary conditions. Specifically, the boundary condition is one in which clamping exists, but provision is made so that displacement is prohibited only at the top and bottom surfaces. The results confirm that the horizontal displacement is zero at $(x, y) = (a, 0)$, as one would expect, and that the use of auxetic cantilevers helps to reduce horizontal displacement at $x = a$ for $-b \leq y \leq b$.

References

Atanackovic TM, Guran A (2000) Theory of elasticity for scientists and engineers. Birkhauser, Boston

Barber JR (2010) Elasticity, 3rd edn. Springer, Dordrecht

Boresi AP, Chong KP Lee JD (2011) Elasticity in engineering mechanics, 3rd edn. Wiley, New Jersey

Fung YC (1965) Foundations of solid mechanics. Prentice-Hall, Englewood Cliffs

Gould PL (2013) Introduction to linear elasticity. Springer, New York

Housner GW, Vreeland TJ (1965) The analysis of stress and deformation. Housner and Vreeland, California

Landau LD, Lifshitz EM (1959) Theory of elasticity. Pergamon Press, Oxford

Lim TC (2015) Auxetic materials and structures. Springer, Singapore

Lim TC (2018) Revisiting the elasticity solution for a simply supported beam under sinusoidal load. Int J Mech Eng Edu 46(1):41–49

Lim TC (2021) An exact deflection solution to a type of cantilever with partially built-in end using strong boundary conditions. J Mech Eng Edu, Int. https://doi.org/10.1177/0306419019851363
Love AEH (1906) A treatise on the mathematical theory of elasticity, 2nd edn. Cambridge University Press, Cambridge
Saada AS (2009) Elasticity: theory and applications, 2nd edn. J Ross Publishing, Plantation
Saad M (2014) Elasticity: theory, applications, and numerics, 3rd edn. Elsevier, Oxford
Sokolnikoff IS (1956) Mathematical theory of elasticity, 2nd edn. Tata-McGraw-Hill, Bombay
Soutas-Little RW (1999) Elasticity. Dover, New York
Timoshenko SP, Goodier JN (1970) Theory of elasticity, 3rd edn. McGraw-Hill, Auckland

Chapter 8
Auxetic Composites with Mixed Auxeticity

Abstract This chapter considers composites containing phases with opposing Poisson's ratio signs. Special emphasis is placed on the overall sign of the composite, which differs depending on the mode of loading.

Keywords Auxetic composites · Loading mode dependency · Mixed auxeticity

8.1 Fundamentals

This chapter deals with the phenomenon of composites that possess both overall auxetic and conventional properties depending on the mode of loading. Consider a concentric cylindrical core and shell shown in Fig. 8.1 (top left), whereby the cross-sectional area of the cylindrical core is greater than the cross-sectional area of the cylindrical shell, as shown in Fig. 8.1 (top right). Therefore, under the prescription of uniform axial strain, as illustrated in Fig. 8.1 (bottom left), the cylindrical core material properties have greater influence than the cylindrical shell material properties. However, under the prescription of an angular twist, as shown in Fig. 8.1 (bottom right), the cylindrical shell encounters a greater extent of shear strain in comparison with the cylindrical core. As a result, under the action of twisting, the cylindrical shell material properties exert a greater influence over the cylindrical core material properties. Now if Poisson's ratio of the cylindrical core and cylindrical shell possesses opposite signs, there is a possibility that the overall Poisson's ratio of the composite cylindrical assembly possesses a sign similar to the core under the action of axial load, but exhibits a sign similar to the shell under the action of twisting load. The example illustrated in Sect. 8.2 includes the effects arising from Young's modulus and shear modulus of the cylindrical core and shell. Thereafter, the theory is extended into an assembly of trilayered concentric core and shell assembly in Sect. 8.3 to account for the influence of interface adhesive material.

Section 8.4 considers a sandwich structure consisting of a core and two facesheets as depicted in Fig. 8.2 (top left), whereby the cross-sectional area of the core is greater than the cross-sectional area of both facesheets added together. Since the width of the core and facesheets is common, their cross-sectional areas can also be represented by their thicknesses, as shown in Fig. 8.2 (top right). By similar

T. Lim, *Mechanics of Metamaterials with Negative Parameters*, Engineering Materials,
https://doi.org/10.1007/978-981-15-6446-8_8

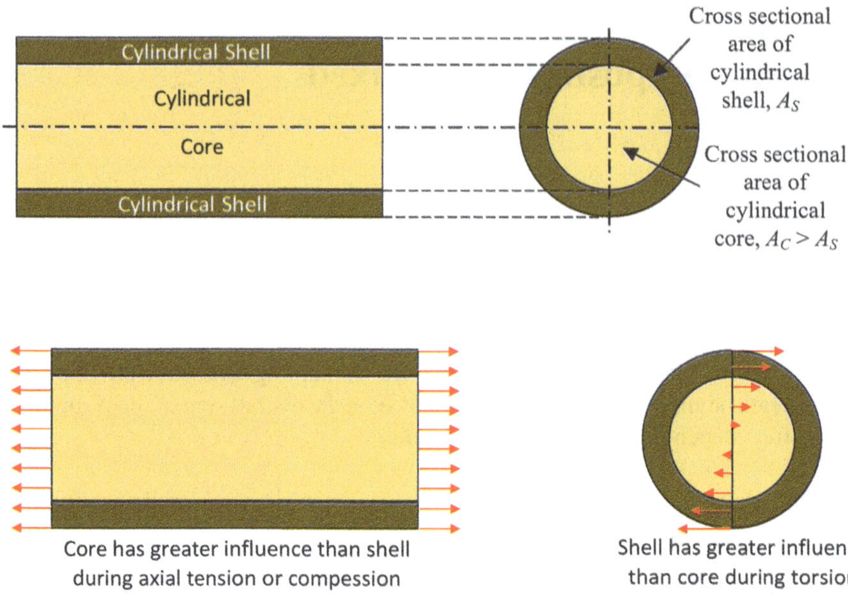

Fig. 8.1 Opposing overall Poisson's ratio signs for a concentric cylindrical core and shell assembly under axial and torsional loading modes

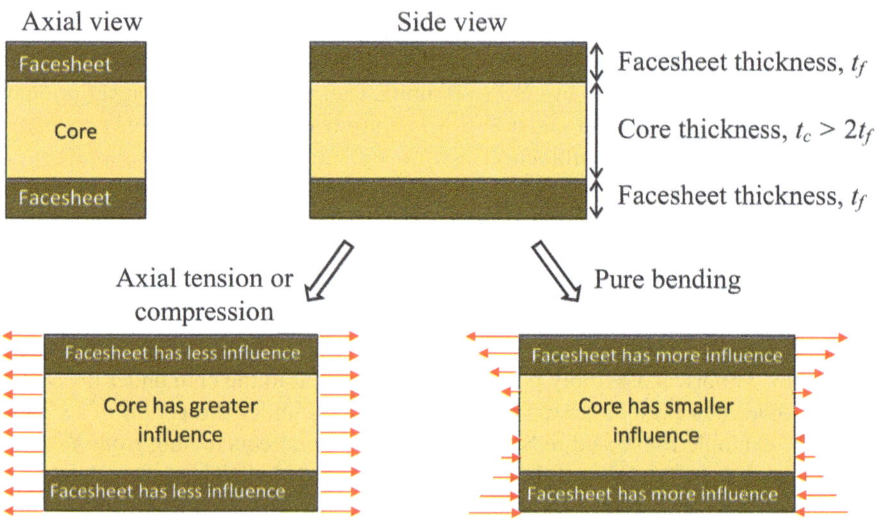

Fig. 8.2 Opposing effective Poisson's ratio signs for a sandwich structure under axial and bending loading modes

reasoning, under the prescription of uniform axial strain, as indicated in Fig. 8.2 (bottom left), the core material properties have stronger influence than the facesheet material properties. On the other hand, under the action of bending in the manner described by the linearly varying axial strain in Fig. 8.2 (bottom right), the facesheets encounter a greater amount of axial strain in comparison with the core. As a result, under the action of bending, the facesheets material properties exert greater influence over the core material properties. Suppose Poisson's ratio of the core and facesheets possesses opposite signs, there is a similar possibility that the overall Poisson's ratio of the sandwich structure possesses a sign similar to the core under axial load, but displays a sign similar to the facesheets under the action of pure bending. The example furnished in Sect. 8.4 includes the influence arising from Young's moduli of the core and facesheet materials. Finally, an example is given in Sect. 8.5 for the case of sandwich structure under the action of axial load, bending and twisting on their effective auxeticity arising from the three loading modes.

8.2 Example: Semi-auxetic Rods

This example demonstrates a way to evaluate the effective Poisson's ratio of a rod consisting of a core and shell of similar shape but possessing opposite Poisson's ratio signs. Consider a rod of length L made from N number of disks in a series arrangement as shown in Fig. 8.3a and another rod also of length L made from N number of hollow rods of similar shape in concentric arrangement as shown in Fig. 8.3b.

With reference to Fig. 8.3a, a rod consisting of different material properties arranged in series will experience common torsion while the twist angle distribution is piecewise. Therefore, a single or constant torque is imposed

$$T = T_n \tag{8.2.1}$$

for $n = 1, 2, \ldots, N$ while the overall angular twist over the entire length of the rod is summed up from individual twist angles from every disk, i.e.,

$$\phi = \sum_{n=1}^{N} \phi_n \tag{8.2.2}$$

Conversely, the torsion of a rod made from concentrically similar-shaped rods, as shown in Fig. 8.3b, produces common angular twist from one end to the other, while the torsional load is piecewise distributed radially. Therefore, a single or constant angular twist is imposed

$$\phi = \phi_n \tag{8.2.3}$$

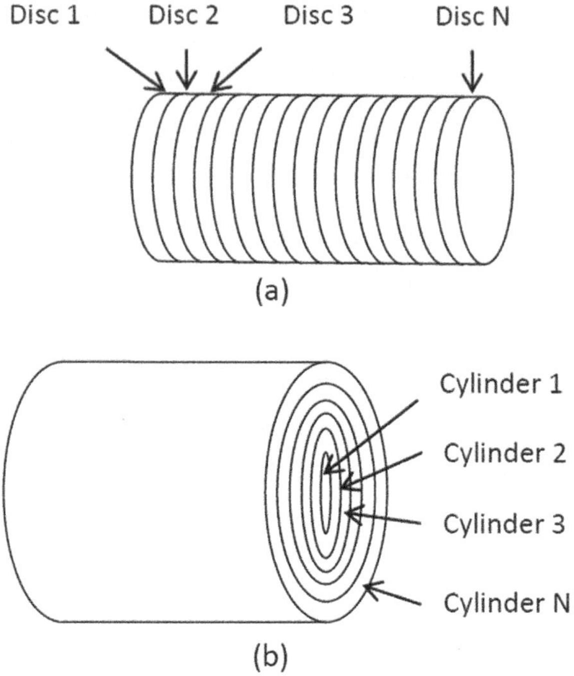

Fig. 8.3 Rod made from N different materials in **a** series arrangement and **b** concentric arrangement

for $n = 1, 2, \ldots, N$ while the total torsional load is summed from individual torsional load from each concentric rod.

$$T = \sum_{n=1}^{N} T_n \tag{8.2.4}$$

The framework specified by Eqs. (8.2.3) and (8.2.4) is adopted in the proceeding analysis. It follows that this model is valid for perfect bonding at the interfaces of adjacent cylinders, and that the elastic properties from the bonding adhesive are negligible if its modulus is in the same order as those of the concentric cylinders with the adhesive thickness being very small in comparison with the radial dimension of each cylinder. The relation between the applied torsion, rod length, shear modulus, polar moment area perpendicular to the torsional axis, and the angular twist for a solid prismatic rod is given by

$$\frac{TL}{\phi} = GJ = GCD^4 \tag{8.2.5}$$

whereby the polar moment area of the solid rod's cross section is $J = CD^4$, in which D is a characteristic dimension of the rod cross section and C is cross-sectional shape-dependent coefficient. Using $E/G = 2(1 + v)$, we have the following relation

$$1 + v = \frac{CD^4 E\phi}{2LT} \tag{8.2.6}$$

whereby the RHS term is a dimensionless group. Since the polar moment of area is

$$J = C\left(D_o^4 - D_i^4\right) \tag{8.2.7}$$

we have

$$1 + v = \frac{CD_o^4 E\phi}{2LT}\left[1 - \left(\frac{D_i}{D_o}\right)^4\right] \tag{8.2.8}$$

whereby a similar dimensionless group appears on the RHS. It can be seen for both cases that the effective Poisson's ratio of a rod consisting of an inner core of arbitrary cross section and a similarly shaped outer shell undergoing torsion is a function of two dimensionless groups

$$v_{\text{TOR}} = f\left(\frac{CD_o^4 E_n \phi}{2LT}, \frac{D_i}{D_o}\right) \tag{8.2.9}$$

for $n = 1, 2$. The overall torsion is combined from both components while the angular twist is common

$$T = T_i + T_o = \frac{\phi}{L}(G_i J_i + G_o J_o) \tag{8.2.10}$$

such that

$$\frac{TL}{\phi} = \frac{CD_i^4 E_i}{2(1 + v_i)} + \frac{C\left(D_o^4 - D_i^4\right)E_o}{2(1 + v_o)} \tag{8.2.11}$$

To pave a way for comparison and illustration, normalization is imposed for both the inner core and outer shell materials. Normalizing Young's modulus $E_i = E_0 = E$, we obtain an expression for the inverse of the dimensionless group,

$$\frac{2TL}{CD_o^4 E\phi} = \frac{\left(\frac{D_i}{D_o}\right)^4}{1 + v_i} + \frac{1 - \left(\frac{D_i}{D_o}\right)^4}{1 + v_o} \tag{8.2.12}$$

On the basis of the effective torsional Poisson's ratio being functions of the identified dimensionless groups as described in Eq. (8.2.6), the effective Poisson's ratio under torsional load can be inferred as

$$1 + v_{\text{TOR}} = \frac{CD_o^4 E\phi}{2LT} \tag{8.2.13}$$

such that substituting Eq. (8.2.12) into Eq. (8.2.13) gives (Lim 2011)

$$v_{\mathrm{TOR}} = \frac{(1+v_o)(1+v_i)}{(1+v_o)\left(\frac{D_i}{D_o}\right)^4 + (1+v_i)\left[1-\left(\frac{D_i}{D_o}\right)^4\right]} - 1 \tag{8.2.14}$$

For comparison with Poisson's ratio under axial loading, we have its corresponding effective Poisson's ratio

$$v_{\mathrm{AX}} = \left(\frac{D_i}{D_o}\right)^2 v_i + \left[1-\left(\frac{D_i}{D_o}\right)^2\right] v_o \tag{8.2.15}$$

To illustrate the change in the overall auxeticity of a semi-auxetic rod, we consider a special category whereby Poisson's ratio of the core and shell possesses equal magnitude but opposite signs. Since Poisson's ratio of an isotropic material is within the range $-1 \le v \le 0.5$, we select two cases: (i) auxetic core, in which the inner core Poisson's ratio is $v_i = -0.5$ and the outer shell Poisson's ratio is $v_o = 0.5$, and (ii) auxetic shell in which $v_i = 0.5$ and $v_o = -0.5$. See Fig. 8.4.

For the special case of auxetic core rod, the effective Poisson's ratio under axial loading mode and torsional loading mode is

$$v_{\mathrm{AX}} = \frac{1}{2} - \left(\frac{D_i}{D_o}\right)^2 \tag{8.2.16}$$

and

$$v_{\mathrm{TOR}} = \frac{3}{6\left(\frac{D_i}{D_o}\right)^4 + 2\left[1-\left(\frac{D_i}{D_o}\right)^4\right]} - 1 \tag{8.2.17}$$

respectively. Plots of the effective Poisson's ratios with reference to the inner-to-outer characteristic dimension ratio, D_i/D_0, are shown in Fig. 8.5.

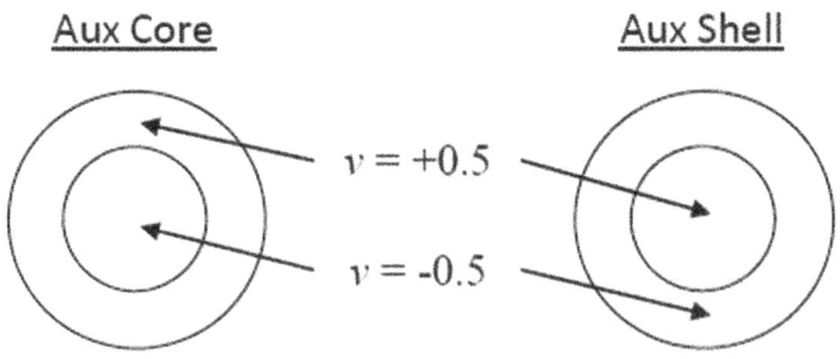

Fig. 8.4 Semi-auxetic rod based on auxetic core (left) and auxetic shell (right)

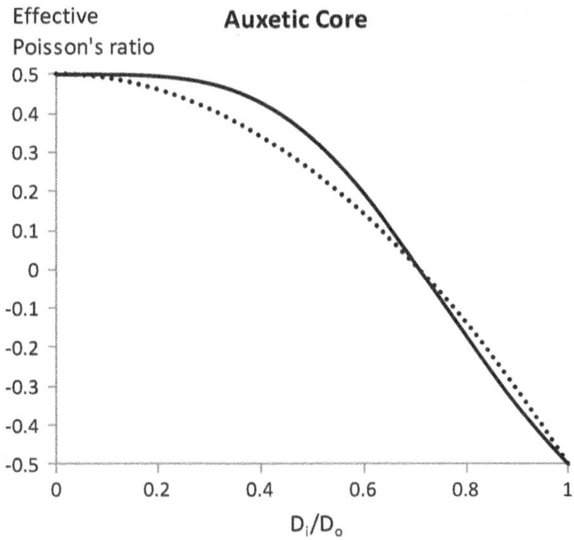

Fig. 8.5 Effective Poisson's ratio of a semi-auxetic rod with an auxetic core under axial load (dashed curve) and torsional load (solid curve)

Perusal to this figure, as well as to Eqs. (8.2.16) and (8.2.17), shows that such rods possess zero Poisson's ratio at $D_i/D_0 = 0.5^{0.5}$ such that they exhibit an overall conventional behavior for $D_i/D_0 < 0.5^{0.5}$ and overall auxetic behavior for $D_i/D_0 > 0.5^{0.5}$. More strikingly, the effective Poisson's ratio is of higher magnitude when the rod undergoes torsion as compared to axial loading.

In the special case of auxetic shell, the effective Poisson's ratio for axial and torsional loading modes is

$$\upsilon_{\mathrm{AX}} = -\frac{1}{2} + \left(\frac{D_i}{D_o}\right)^2 \tag{8.2.18}$$

and

$$\upsilon_{\mathrm{TOR}} = \frac{3}{2\left(\frac{D_i}{D_o}\right)^4 + 6\left[1 - \left(\frac{D_i}{D_o}\right)^4\right]} - 1 \tag{8.2.19}$$

respectively. Variations of the effective Poisson's ratios with respect to the inner-to-outer characteristic dimension ratio, D_i/D_0, are plotted in Fig. 8.6.

As in the previous section, the effective Poisson's ratio for axial loading is zero at $D_i/D_0 = 0.5^{0.5}$. Unlike the previous case, the effective Poisson's ratio, under the influence of torsional load, is zero at $D_i/D_0 = 0.75^{0.25}$. As a result, the rod exhibits an overall auxetic behavior for $D_i/D_0 < 0.5^{0.5}$ and an overall conventional behavior for $D_i/D_0 > 0.75^{0.25}$ regardless of the loading mode. For the range $0.5^{0.5} < D_i/D_0 < 0.75^{0.25}$, the rod exhibits a mixed behavior which is loading mode dependant, i.e., it

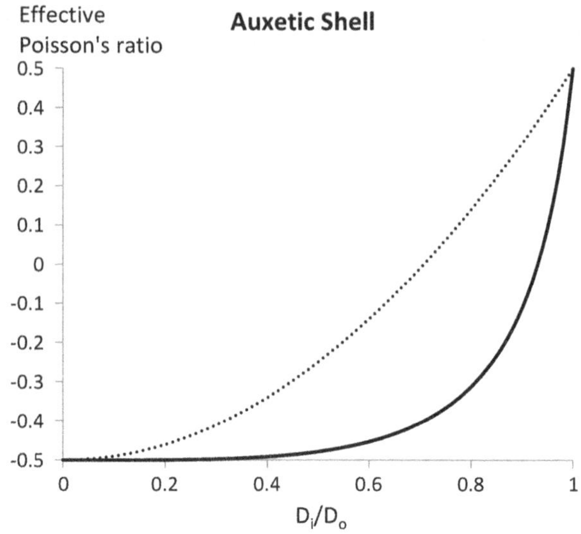

Fig. 8.6 Effective Poisson's ratio of a semi-auxetic rod with an auxetic shell under axial load (dashed curve) and torsional load (solid curve)

behaves as a conventional rod when an axial load is imposed but as an auxetic rod during twisting (Lim 2011).

For the purpose of symmetry, the plots of effective Poisson's ratio of the combined rods in Figs. 8.5 and 8.6 are replotted against the inner-to-outer characteristic dimensions raised to the second and fourth powers, respectively, as depicted in Figs. 8.7 and 8.8.

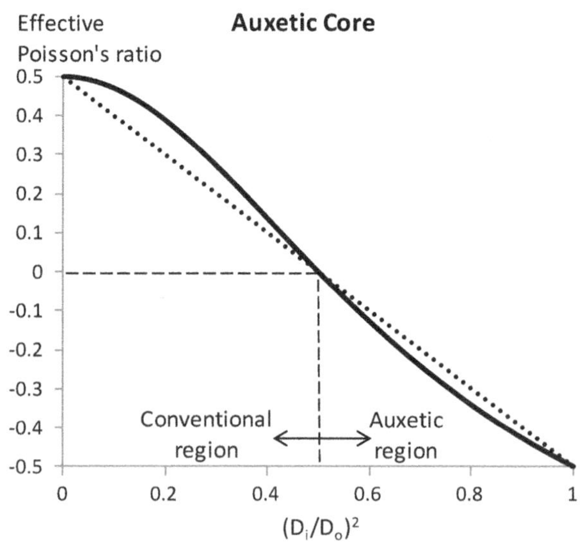

Fig. 8.7 Symmetric plot of the effective Poisson's ratio of a semi-auxetic rod with an auxetic core under axial load (dashed curve) and torsional load (solid curve)

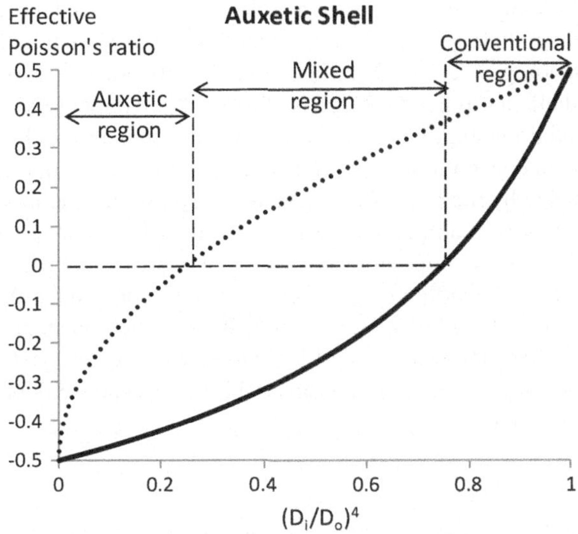

Fig. 8.8 Symmetric plot of the effective Poisson's ratio of a semi-auxetic rod with an auxetic shell under axial load (dashed curve) and torsional load (solid curve)

The range of the inner-to-outer rod dimension that gives rise to different auxeticity levels, i.e.,

$$\left.\begin{array}{l} 0 < \upsilon_{\mathrm{AX}} < \upsilon_{\mathrm{TOR}};\, 0 < \left(\frac{D_i}{D_o}\right)^2 < \frac{1}{2} \\ \upsilon_{\mathrm{TOR}} < \upsilon_{\mathrm{AX}} < 0;\, \frac{1}{2} < \left(\frac{D_i}{D_o}\right)^2 < 1 \end{array}\right\} \Leftrightarrow \left\{\begin{array}{c} \upsilon_i \\ \upsilon_o \end{array}\right\} = \frac{1}{2}\left\{\begin{array}{c} -1 \\ +1 \end{array}\right\} \tag{8.2.20}$$

for the auxetic core and

$$\left.\begin{array}{l} \upsilon_{\mathrm{TOR}} < \upsilon_{\mathrm{AX}} < 0;\, 0 < \left(\frac{D_i}{D_o}\right)^4 < \frac{1}{4} \\ \upsilon_{\mathrm{TOR}} < 0 < \upsilon_{\mathrm{AX}};\, \frac{1}{4} < \left(\frac{D_i}{D_o}\right)^4 < \frac{3}{4} \\ 0 < \upsilon_{\mathrm{TOR}} < \upsilon_{\mathrm{AX}};\, \frac{3}{4} < \left(\frac{D_i}{D_o}\right)^4 < 1 \end{array}\right\} \Leftrightarrow \left\{\begin{array}{c} \upsilon_i \\ \upsilon_o \end{array}\right\} = \frac{1}{2}\left\{\begin{array}{c} +1 \\ -1 \end{array}\right\} \tag{8.2.21}$$

for the auxetic shell may well be elucidated by the area property of the rod's cross section. In the case of auxetic shell, the higher auxeticity that is manifested during torsional loading mode as compared to axial loading mode is attributed to the greater influence of the outermost rod material. This results in a more negative effective Poisson's ratio during torsion than during axial loading for any value of D_i/D_0. In the case of auxetic core, the lower auxeticity for torsional loading as compared to that of axial loading for a relatively small core is due to the strong influence of the conventional shell. As D_i/D_0 increases, the overall auxeticity of the rod in the context of torsional loading catches up and exceeds the overall auxeticity in the context of axial loading.

Within the category whereby the core and shell possess equal Poisson's ratio magnitude, plotted results reveal different torsional loading auxeticity but equal axial loading auxeticity. For a concentric compound rod with auxetic core and conventional shell, the sign of the effective Poisson ratio under torsional loading follows that under axial loading, but with slightly greater magnitude. For a concentric compound rod with conventional core and auxetic shell, there is a range of relative volume fraction whereby the rod exhibits overall conventional and auxetic behavior under axial and torsional loading modes, respectively. This phenomenon has been elucidated herein from the standpoint of cross-sectional area properties of rods undergoing axial and torsional loading modes. Hence, a compound rod with auxetic shell exhibits greater extent of loading mode-dependent auxeticity than that with auxetic core. The present results suggest the use of a compound rod with conventional core and auxetic shell as a smart structure that gives different response depending on the type of loading imposed on it.

8.3 Example: Concentric Auxetic Cylinders with Interface Adhesives

This example models a single trilayered cylindrical rod consisting of a solid inner foam rod adhered to a hollow outer foam rod, with significant adhesive modulus as the interface of the two foam rods. The overall auxeticity of a concentric rod whereby the core and shell possess Poisson's ratio of opposite signs is influenced by the mode of loading. It is obvious that during axial loading both the core and shell experience equal longitudinal strain, but during torsional loading the shell undergoes greater shear strain than the core. In this concentrically bilayered rod system, the auxeticity during axial and torsional loading modes changes at different rate with respect to the ratio of the inner-to-outer diameters, thereby implying a range of this ratio whereby the concentric rod exhibits overall Poisson's ratio of different signs depending on the loading mode. However, the assumption of a bilayered concentric foam system is no longer valid when a layer of high modulus adhesive exists between the two concentric rods. The adhesive layer can be thought of as an intermediate layer that fills up the gap between the two concentric cylinders. Even in the case where the surfaces of the inner and outer cylinders are in contact with each other, the slight seepage of adhesive fluid into the foam before its solidification gives rise to a high modulus intermediate layer. The effect of the adhesive elastic properties on the concentric foam rods of opposing Poisson's ratio signs under torsional load is investigated in this section as a trilayered concentric rod system.

Unlike uniaxial loading whereby the diameter changes for nonzero Poisson's ratio, there is no salient change in the rod diameter under torsional load. With reference to Fig. 8.9, a component of radial increase due to one principal strain is canceled by a component of radial decrease due to another principal strain in the same plane. Hence, there is no change in the rod diameter under torsion. Nevertheless, it is obvious that

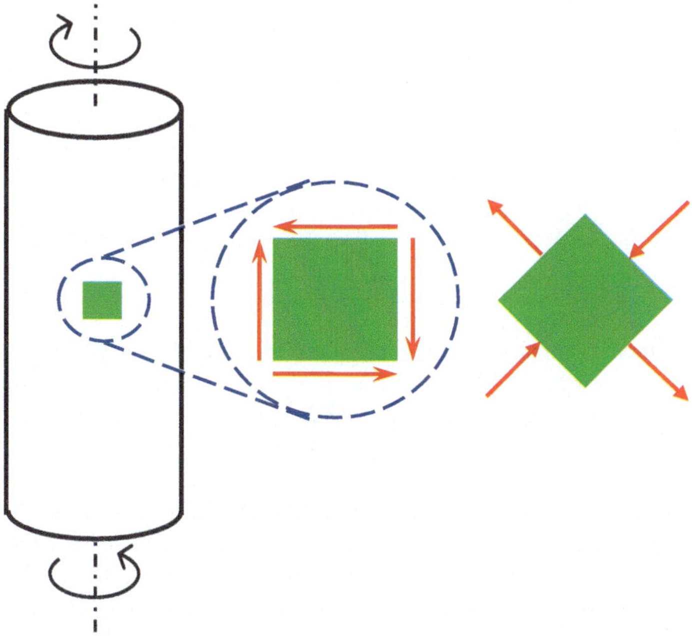

Fig. 8.9 Schematic for a rod under torsion (left), and elemental volume under pure shearing (middle), and an equivalent stress state at 45° showing in-plane principal strains (right)

the auxeticity of the rod must be related to that of the material that constitutes the rod. The analysis consists of three parts, namely the single solid rods, the bilayered concentric rods, and the trilayered concentric rods. In the trilayered concentric rods, we let the middle material be the adhesive material.

Consider the case of a single solid rod. Since an observation on rod diameter yields no change regardless of the rod auxeticity, we herein consider the rod auxeticity—under torsional loading—in terms of the moduli ratio G/E (or E/G) using the elastic relation

$$G = \frac{E}{2(1+v)} \tag{8.3.1}$$

for isotropic solids. By virtue of Eq. (8.3.1), the moduli ratio indicates Poisson's ratio and hence its auxeticity. Therefore, the auxeticity of a material can be inferred from the moduli ratio, as an alternative to the usual way of measuring change in dimension during axial loading. For a single solid rod of diameter D and length L undergoing torsion T, the angular twist ϕ is given as

$$\phi = \frac{TL}{GJ} \tag{8.3.2}$$

whereby the polar moment area of the circular cross section J is

$$J = \frac{\pi D^4}{32} \tag{8.3.3}$$

Substituting Eqs. (8.3.1) and (8.3.3) into Eq. (8.3.2), we have Poisson's ratio of the rod as

$$v = \frac{D^4 E \pi \phi}{64TL} - 1 \tag{8.3.4a}$$

or, for convenient comparison with subsequent formulation, we write

$$v_{\text{Single}} = \left(\frac{64TL}{D^4 E \pi \phi} \right)^{-1} - 1 \tag{8.3.4b}$$

We now turn our attention to the case of bilayered concentric rods. As opposed to summative angular twist and common transmitted torsional load for two rods arranged in series, the case of two concentrically arranged rods is governed by common angular twist with summative torsional loads as depicted in Fig. 8.10 (top) and (bottom), respectively. Hence, the common angular twist for the inner and outer rods

$$\phi_i = \phi_o \tag{8.3.5}$$

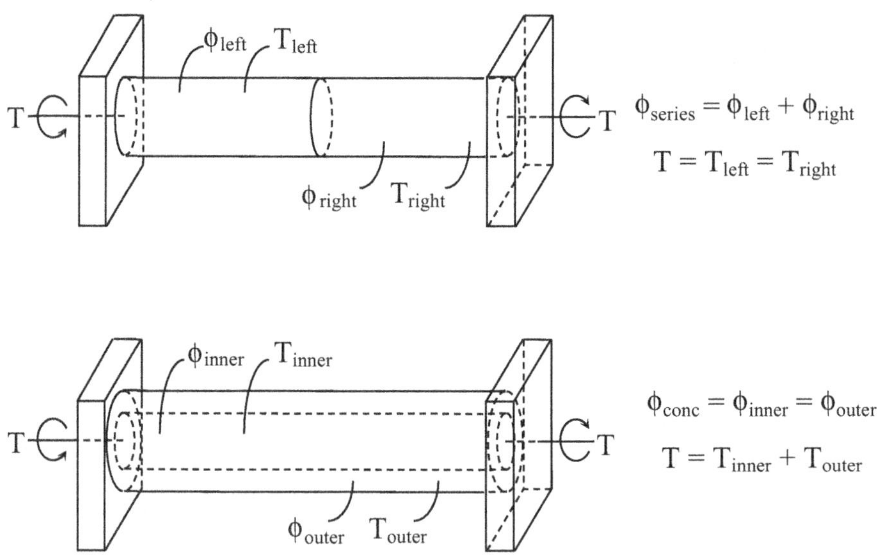

Fig. 8.10 Comparison between two rods in series arrangement (top) and concentric arrangement (bottom)

and the summative torsional load

$$T = T_i + T_o = \frac{\phi_i G_i J_i}{L} + \frac{\phi_o G_o J_o}{L} \tag{8.3.6}$$

give

$$\frac{TL}{\phi} = G_i J_i + G_o J_o \tag{8.3.7}$$

Substituting Eqs. (8.3.1) and (8.3.3) into Eq. (8.3.7) leads to

$$\frac{64TL}{D_o^4 E_o \pi \phi} = \frac{\frac{E_i}{E_o}\left(\frac{D_i}{D_o}\right)^4}{1 + \upsilon_i} + \frac{1 - \left(\frac{D_i}{D_o}\right)^4}{1 + \upsilon_o} \tag{8.3.8}$$

It can be easily seen that substituting $D_i = 0$ (or $D_i = D_0$ with $E_i = E_0$) into Eq. (8.3.8) reduces it to Eq. (8.3.4). By virtue of Eq. (8.3.4) and neglecting adhesive layer, the effective Poisson's ratio for two perfectly bonded concentric cylinders under torsional loading mode is

$$\upsilon_{\text{Bi-layered}} = \left(\frac{64TL}{D_o^4 E \pi \phi}\right)^{-1} - 1 \tag{8.3.9}$$

where the term in the parenthesis is given by Eq. (8.3.8) in the case of a bilayered concentric rod.

Finally, for this example, we consider the case of trilayered concentric rod. Perusal to Fig. 8.11 shows the adhesive layer being the interlayer, thereby extending the bilayered concentric rod into a more realistic trilayered concentric rod, with the thickness of the adhesive layer, δ being

$$\delta = \frac{D_A - D_i}{2} \tag{8.3.10}$$

As in the case of bilayered concentric rod, the angular twist for a trilayered rod is common

$$\phi_i = \phi_A = \phi_o \tag{8.3.11}$$

while the torsional load is carried by all three layers

$$T = T_i + T_A + T_o \tag{8.3.12}$$

Proceeding similarly as in the case of bilayered concentric rod, the effective Poisson's ratio for a trilayered concentric rod under torsional loading mode is (Lim 2018)

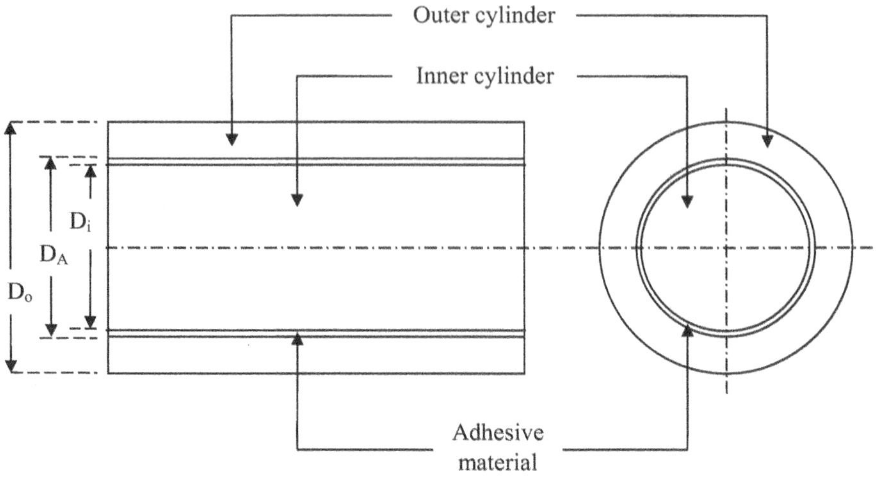

Fig. 8.11 Side (left) and axial (right) views of a rod made from two concentric foams with opposite Poisson's ratio sign, in which the adhesive material is assumed to take on the shape of a thin cylindrical shell

$$v_{\text{Tri-layered}} = \left(\frac{64TL}{D_o^4 E \pi \phi} \right)^{-1} - 1 \tag{8.3.13}$$

with

$$\frac{64TL}{D_o^4 E_o \pi \phi} = \frac{\frac{E_i}{E_o}\left(\frac{D_i}{D_o}\right)^4}{1+v_i} + \frac{\frac{E_A}{E_o}\left[\left(\frac{D_A}{D_o}\right)^4 - \left(\frac{D_i}{D_o}\right)^4\right]}{1+v_A} + \frac{1-\left(\frac{D_A}{D_o}\right)^4}{1+v_o} \tag{8.3.14}$$

where

$$\left(\frac{D_A}{D_o}\right)^4 - \left(\frac{D_i}{D_o}\right)^4 = 8\frac{\delta}{D_o}\left(\frac{D_i}{D_o}\right)^3 + 24\left(\frac{\delta}{D_o}\right)^2\left(\frac{D_i}{D_o}\right)^2 + 32\left(\frac{\delta}{D_o}\right)^3\frac{D_i}{D_o} + 16\left(\frac{\delta}{D_o}\right)^4 \tag{8.3.15}$$

and

$$1 - \left(\frac{D_A}{D_o}\right)^4 = 1 - \left(\frac{D_i}{D_o}\right)^4 - 8\frac{\delta}{D_o}\left(\frac{D_i}{D_o}\right)^3 - 24\left(\frac{\delta}{D_o}\right)^2\left(\frac{D_i}{D_o}\right)^2 - 32\left(\frac{\delta}{D_o}\right)^3\frac{D_i}{D_o} - 16\left(\frac{\delta}{D_o}\right)^4 \tag{8.3.16}$$

It can be seen from Eqs. (8.3.15) and (8.3.16) that as $\delta \to 0$, Eq. (8.3.14) reduces to Eq. (8.3.8). If the adhesive layer is very small in comparison with other radial dimensions and the adhesive modulus is in the same order as that of the foam material, then the following simplifications

$$\left(\frac{D_A}{D_o}\right)^4 - \left(\frac{D_i}{D_o}\right)^4 \approx 8\frac{\delta}{D_o}\left(\frac{D_i}{D_o}\right)^3 \tag{8.3.17}$$

and

$$1 - \left(\frac{D_A}{D_o}\right)^4 \approx 1 - \left(\frac{D_i}{D_o}\right)^4 \tag{8.3.18}$$

for Eqs. (8.3.15) and (8.3.16), respectively, are valid. A direct consequence of this simplification is that Eq. (8.3.14) resembles Eq. (8.3.8), i.e.,

$$\frac{64TL}{D_o^4 E_o \pi \phi} = \frac{\frac{E_i}{E_o}\left(\frac{D_i}{D_o}\right)^4}{1+v_i} + \frac{8\frac{E_A}{E_o}\frac{\delta}{D_o}\left(\frac{D_i}{D_o}\right)^3}{1+v_A} + \frac{1-\left(\frac{D_i}{D_o}\right)^4}{1+v_o} \tag{8.3.19}$$

such that the influence from the adhesive material is confined to only one term, i.e., on the second term on the RHS of Eq. (8.3.19). Here, the relative modulus of the adhesive-to-outer foam material E_A/E_0 and the relative adhesive thickness δ/D_0 play equal importance. The simplification suggested in Eq. (8.3.17), however, is no longer valid when the relative modulus of the adhesive material E_A/E_0 is several orders higher. Hence, the retention of Eq. (8.3.15) with the use of Eq. (8.3.18) provides a balanced simplification and accuracy.

The influence of the intermediate layer on the overall Poisson's ratio of a rod made from foams of opposing Poisson's ratio signs can be observed by considering the adhesive layer's elastic and geometrical properties listed below:

1. Elastic properties
   a. adhesive modulus relative to that of the foam, E_A/E_{Foam}
   b. adhesive Poisson's ratio, ν_A
2. Geometrical properties
   a. adhesive thickness relative to the rod diameter, δ/D_o
   b. adhesive diameter relative to that of the rod, D_i/D_o

In the plotted results of the effective Poisson's ratio that follow, we adopt the concentrically trilayered cylinder using Eqs. (8.3.13)–(8.3.15) and (8.3.17), for the case of equal inner and outer foam modulus

$$E_i = E_o = E_{Foam} \tag{8.3.20}$$

and equal the Poisson's ratio magnitudes for the inner and outer foams

$$\pm v_i = \mp v_0 = \frac{1}{2} \tag{8.3.21}$$

The effect of relative moduli ratio E_A/E_{Foam} on the variation of the combined rod's effective Poison's ratio, v_{eff} with the inner-to-outer cylinder diameters, D_i/D_0 is plotted in Fig. 8.12 with relative adhesive thickness at $\delta/D_0 = 0.001$, Poisson's ratio of solidified adhesive at $v_A = 0$ and relative adhesive modulus at $E_A/E_{\text{Foam}} = 10^n$ for $n = 1, 2, 3, 4, 5$. Figure 8.12 (top and bottom) corresponds to the use of auxetic core ($-v_i = v_0 = 0.5$) and auxetic shell ($v_i = -v_0 = 0.5$), respectively.

As expected, the rod auxeticity increases (or the overall Poisson's ratio decreases) with the relative size of the auxetic core, as shown in Fig. 8.12(top). However, an unexpected trend is observed whereby the overall Poisson's ratio approaches -1 as the inner-to-outer diameter approaches 1, although none of the material component possesses any Poisson's ratio lower than -0.5. Plotted results also reveal that the rod auxeticity increases with the use of higher adhesive modulus. When the position of the auxetic and conventional parts is swopped, only the case of moderate relative adhesive modulus gives an intuitive trend, i.e., increasing Poisson's ratio with increasing

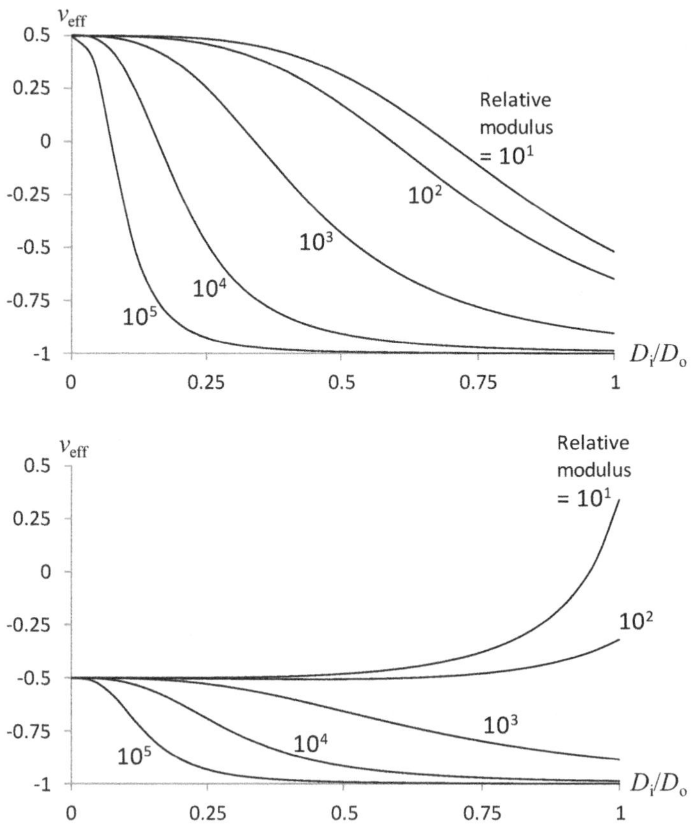

Fig. 8.12 Effect of relative adhesive modulus on the compound rod auxeticity with auxetic core (top) and auxetic shell (bottom)

conventional inner cylinder size. However, the trend reverses for extremely large adhesive modulus. This may well be due to the high torsional stiffness that translates into high G/E ratio, which is associated with auxeticity.

The effect of relative thickness ratio δ/D_0 on the variation of the combined rod's effective Poison's ratio, v_{eff} with the inner-to-outer cylinder diameters, D_i/D_0 is plotted in Fig. 8.13 with relative adhesive modulus at $E_A/E_{\mathrm{Foam}} = 1000$, Poisson's ratio of solidified adhesive at $v_A = 0$ and relative adhesive thickness at $\delta/D_0 = 10^n$ for $n = -1, -2, -3, -4, -5$. Figure 8.13 (top and bottom) corresponds to the use of auxetic core ($-v_i = v_0 = 0.5$) and auxetic shell ($v_i = -v_0 = 0.5$), respectively.

The trends obtained in Fig. 8.13 are somewhat similar to those of Fig. 8.12, signifying an almost similar effect of adhesive thickness with adhesive modulus. The similarity is attributed to the increasing stiffness contributed by the intermediate layer's increasing modulus and increasing thickness.

The effect of adhesive Poisson's ratio, v_A on the variation of the combined rod's effective Poison's ratio, v_{eff} with the inner-to-outer cylinder diameters, D_i/D_0 is

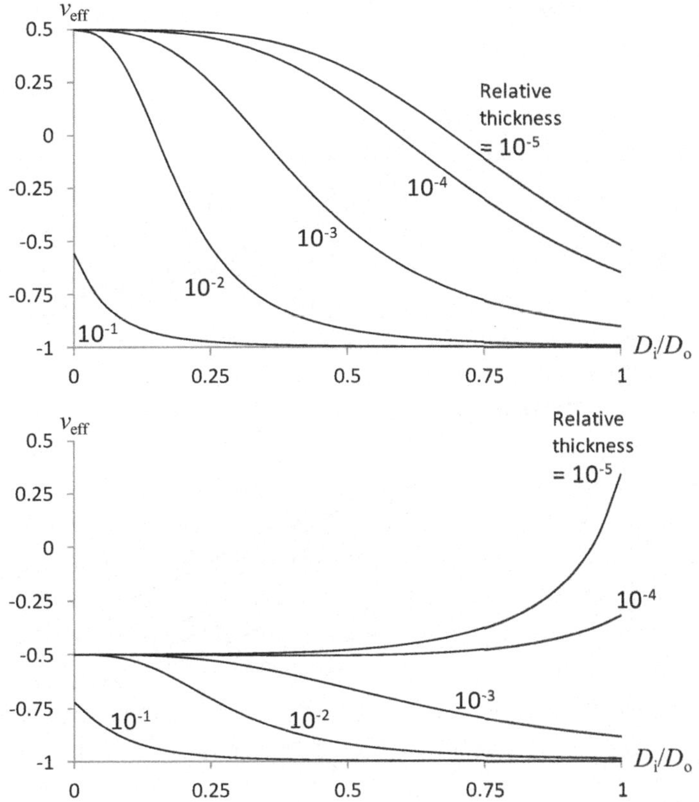

Fig. 8.13 Effect of relative adhesive thickness on the compound rod auxeticity with auxetic core (top) and auxetic shell (bottom)

plotted in Fig. 8.14 with relative adhesive modulus at $E_A/E_{\text{Foam}} = 1000$, relative adhesive thickness at $\delta/D_O = 0.001$, and adhesive Poisson's ratio at $v_A = -0.5, -0.3, -0.1, 0.1, 0.3, 0.5$. Figure 8.14 (top and bottom) corresponds to the use of auxetic core ($-v_i = v_O = 0.5$) and auxetic shell ($v_i = -v_O = 0.5$), respectively. As with Figs. 8.12 and 8.13, a drop in the overall Poisson's ratio toward -1 is obtained with increasing adhesive ring diameter. As expected, the overall Poisson's ratio is influenced by Poisson's ratio of the adhesive material, thereby causing an upward or downward shift to the overall Poisson's ratio arising from the positive and negative signs of the adhesive material's Poisson's ratio.

In summary, an indirect way for inferring Poisson's ratio of a concentrically multilayered rod was obtained. Results show that the following factors increase the auxeticity of the rod under consideration: (a) adhesive modulus, (b) adhesive Poisson's ratio, (c) adhesive thickness, and (d) adhesive radius. Plotted results also suggest that, even with each components possessing Poisson's ratio not lower than −

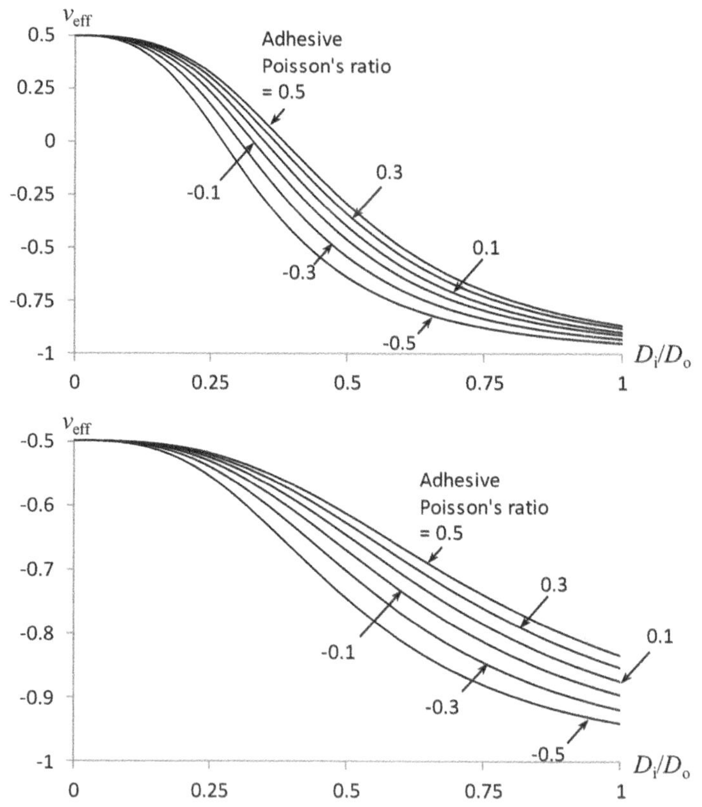

Fig. 8.14 Effect of adhesive Poisson's ratio on the compound rod auxeticity with auxetic core (top) and auxetic shell (bottom)

0.5, the overall Poisson's ratio within the framework of torsional loading approaches -1 as the outer shell thins (Lim 2018).

8.4 Example: Semi-auxetic Sandwich Plates

Consider a three-layered laminate as shown in Fig. 8.15. Based on this layup, a symmetric laminate can be obtained by assigning the facesheet and core portions to isotropic materials possessing Poisson's ratio of opposite signs. It follows that we can have two broad categories: (i) the positive–negative–positive (PNP) layup whereby the facesheets and core have positive and negative Poisson's ratio, respectively, and (ii) the negative–positive–negative (NPN) layup whereby the facesheets and the core possess negative and positive Poisson's ratio, respectively.

For k number of plies within a laminate of thickness t whose plane is perpendicular to the z-axis, the force N_{ij} and moment M_{ij} resultants of the laminate can be related to the reference plane strains ε_{ij}^0 and curvatures κ_{ij} via the ABD matrices

$$\begin{Bmatrix} N_{11} \\ N_{22} \\ N_{12} \\ M_{11} \\ M_{22} \\ M_{12} \end{Bmatrix} = \begin{bmatrix} A_{11} & A_{12} & A_{16} & B_{11} & B_{12} & B_{16} \\ & A_{22} & A_{26} & B_{12} & B_{22} & B_{26} \\ & & A_{66} & B_{16} & B_{26} & B_{66} \\ & & & D_{11} & D_{12} & D_{16} \\ & \mathrm{sym} & & & D_{22} & D_{26} \\ & & & & & D_{66} \end{bmatrix} \begin{Bmatrix} \varepsilon_{11}^0 \\ \varepsilon_{22}^0 \\ \gamma_{12}^0 \\ \kappa_{11} \\ \kappa_{22} \\ \kappa_{12} \end{Bmatrix} \tag{8.4.1}$$

where the stiffness matrices of the laminate can be obtained from the transformed reduced stiffness matrix as

$$\begin{Bmatrix} [A_{ij}] \\ [B_{ij}] \\ [D_{ij}] \end{Bmatrix} = \sum_{k=1}^{n} \left[\overline{Q}_{ij}\right]_k \begin{Bmatrix} \left(z_k^1 - z_{k-1}^1\right)/1 \\ \left(z_k^2 - z_{k-1}^2\right)/2 \\ \left(z_k^3 - z_{k-1}^3\right)/3 \end{Bmatrix} \tag{8.4.2}$$

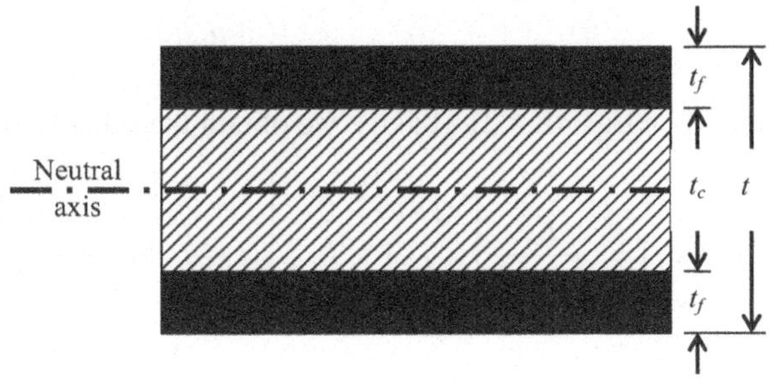

Fig. 8.15 Generalized 3-ply symmetrical laminate consisting of facesheets (subscript f) and core (subscript c) for analysis

based on Classical Lamination Theory (CLT). In the case of isotropic plies considered herein,

$$\left[\overline{Q}_{ij}\right] = \left[Q_{ij}\right] \tag{8.4.3}$$

where by the reduced stiffness matrix $\left[Q_{ij}\right]$ can be expressed in terms of Young's modulus E and Poisson's ratio v as

$$\left[Q_{ij}\right] = \frac{E}{1-v^2}\begin{bmatrix} 1 & v & 0 \\ v & 1 & 0 \\ 0 & 0 & \frac{1-v}{2} \end{bmatrix} \tag{8.4.4}$$

The transformed reduced stiffness matrix $\left[\overline{Q}_{ij}\right]$ for an orthotropic ply rotated at an angle θ clockwise about the z-axis can be obtained from the reduced stiffness matrix $\left[Q_{ij}\right]$ as

$$\overline{Q}_{11} = c^4 Q_{11} + s^4 Q_{22} + 2c^2 s^2 (Q_{12} + 2Q_{66}) \tag{8.4.5}$$

$$\overline{Q}_{22} = s^4 Q_{11} + c^4 Q_{22} + 2c^2 s^2 (Q_{12} + 2Q_{66}) \tag{8.4.6}$$

$$\overline{Q}_{12} = c^2 s^2 (Q_{11} + Q_{22} - 4Q_{66}) + \left(c^4 + s^4\right) Q_{12} \tag{8.4.7}$$

$$\overline{Q}_{16} = c^3 s (Q_{11} - Q_{12} - 2Q_{66}) - cs^3 (Q_{22} - Q_{12} - 2Q_{66}) \tag{8.4.8}$$

$$\overline{Q}_{26} = cs^3 (Q_{11} - Q_{12} - 2Q_{66}) - c^3 s (Q_{22} - Q_{12} - 2Q_{66}) \tag{8.4.9}$$

$$\overline{Q}_{66} = c^2 s^2 (Q_{11} + Q_{22} - 2Q_{12}) + \left(c^2 - s^2\right)^2 Q_{66} \tag{8.4.10}$$

where $c = \cos\theta$ and $s = \sin\theta$. Due to the symmetric layup, the force–strain and moment–curvature relations are decoupled since

$$\left[B_{ij}\right] = 0 \tag{8.4.11}$$

based on Eqs. (8.4.2)–(8.4.4), we obtain the other two stiffness matrices

$$\left[A_{ij}\right] = \frac{E_f}{1-v_f^2}\begin{bmatrix} 1 & v_f & 0 \\ v_f & 1 & 0 \\ 0 & 0 & \frac{1-v_f}{2} \end{bmatrix}(t - t_c) + \frac{E_c}{1-v_c^2}\begin{bmatrix} 1 & v_c & 0 \\ v_c & 1 & 0 \\ 0 & 0 & \frac{1-v_c}{2} \end{bmatrix}(t_c) \tag{8.4.12}$$

$$[D_{ij}] = \frac{E_f}{1-v_f^2}\begin{bmatrix} 1 & v_f & 0 \\ v_f & 1 & 0 \\ 0 & 0 & \frac{1-v_f}{2} \end{bmatrix}\left(\frac{t^3 - t_c^3}{12}\right) + \frac{E_c}{1-v_c^2}\begin{bmatrix} 1 & v_c & 0 \\ v_c & 1 & 0 \\ 0 & 0 & \frac{1-v_c}{2} \end{bmatrix}\left(\frac{t_c^3}{12}\right) \tag{8.4.13}$$

where subscripts f and c refer to the facesheet and core properties, respectively. We herein consider a special case in which Young's modulus for both the facesheet and the core materials are equal

$$E_f = E_c = E_0 \tag{8.4.14}$$

and that Poisson's ratio for both materials is of equal magnitude but opposite signs such that

$$\left|v_f\right| = |v_c| = v_0 \tag{8.4.15}$$

whereby both E_0 and v_0 are positive. As such, we arrive at the following laminate stiffness

$$\frac{1-v_0^2}{E_0}\left(\frac{1}{t}\right)[A_{ij}] = \begin{bmatrix} 1 & v_f + \left(\frac{t_c}{t}\right)(v_c - v_f) & 0 \\ v_f + \left(\frac{t_c}{t}\right)(v_c - v_f) & 1 & 0 \\ 0 & 0 & \frac{1-\left[v_f + \left(\frac{t_c}{t}\right)(v_c - v_f)\right]}{2} \end{bmatrix} \tag{8.4.16}$$

and

$$\frac{1-v_0^2}{E_0}\left(\frac{12}{t^3}\right)[D_{ij}] = \begin{bmatrix} 1 & v_f + \left(\frac{t_c}{t}\right)^3(v_c - v_f) & 0 \\ v_f + \left(\frac{t_c}{t}\right)^3(v_c - v_f) & 1 & 0 \\ 0 & 0 & \frac{1-\left[v_f + \left(\frac{t_c}{t}\right)^3(v_c - v_f)\right]}{2} \end{bmatrix} \tag{8.4.17}$$

which can be readily compared with the normalized laminate stiffness matrices

$$\left[\overline{A}_{ij}\right] = \begin{bmatrix} 1 & v_{\text{eff}}^A & 0 \\ v_{\text{eff}}^A & 1 & 0 \\ 0 & 0 & \frac{1-v_{\text{eff}}^A}{2} \end{bmatrix} \tag{8.4.18}$$

and

$$\left[\overline{D}_{ij}\right] = \begin{bmatrix} 1 & v_{\text{eff}}^D & 0 \\ v_{\text{eff}}^D & 1 & 0 \\ 0 & 0 & \frac{1-v_{\text{eff}}^D}{2} \end{bmatrix} \tag{8.4.19}$$

The effective laminate Poisson's ratios are v_{eff}^{A} and v_{eff}^{D} as a result of the application of axial loading and bending moment, respectively. Comparison of terms among Eqs. (8.4.16)–(8.4.19) implies (Lim 2007)

$$\begin{Bmatrix} v_{\text{eff}}^{A} \\ v_{\text{eff}}^{D} \end{Bmatrix} = \begin{bmatrix} 1 & (t_c/t)^1 \\ 1 & (t_c/t)^3 \end{bmatrix} \begin{Bmatrix} v_f \\ v_c - v_f \end{Bmatrix} \quad (8.4.20)$$

The obtained [A] and [D] matrices can then be extended for application to plates via the consideration of plate governing equations—(a) equilibrium equations and (b) boundary conditions. By substituting $v_f = -v_c = v_0$ for the PNP layup, we have $v_{\text{eff}}^{A} = 0$ and $v_{\text{eff}}^{D} = 0.75v_0 > 0$ for $t_c/t = 1/2$, while $v_{\text{eff}}^{A} = -0.5874v_0 < 0$ and $v_{\text{eff}}^{D} = 0$ for $t_c/t = (1/2)^{1/3}$. In the same manner, by substituting $-v_f = v_c = v_0$ for the NPN layup, we have $v_{\text{eff}}^{A} = 0$ and $v_{\text{eff}}^{D} = -0.75v_0 < 0$ for $t_c/t = 1/2$, with $v_{\text{eff}}^{A} = 0.5874v_0 > 0$ and $v_{\text{eff}}^{D} = 0$ for $t_c/t = (1/2)^{1/3}$. The implications of these effective laminate Poisson's ratios are shown in Table 8.1. If the relative core thickness falls between $1/2$ and $(1/2)^{1/3}$, then one may expect the properties to also fall in between the properties corresponding to the abovementioned relative core thickness. Since the intermediary between zero and a particular value takes the same sign as the latter, then it can be inferred that for the type of 3-ply symmetric laminate considered herein, the sign of the effective Poisson's ratio is determined by the type of loading—whether axial loading or bending moment (see Table 8.1).

To provide a graphical view for the change in the effective Poisson's ratios with different core thickness, curves of the relative effective Poisson's ratio v_{eff}/v_0 versus the relative core thickness t_c/t are plotted in Fig. 8.16 (top) and (bottom) for PNP and NPN layups, respectively.

Since v_{eff}^{A} and v_{eff}^{D} correspond to $[A_{ij}]$ and $[D_{ij}]$, respectively, any opposing signs in the mixed region (See Fig. 8.16) do not cancel off but coexist. This implies the manifestation of both auxetic and non-auxetic behavior as a result of combined loading modes.

The occurrence of interfacial stresses between laminas is inherent in almost all composite laminates, and there is no exception in the case of Poisson's ratio mismatch. The consequence of opposing Poisson's ratio signs is schematically shown in Fig. 8.17. The CLT is formulated on the assumption that the laminate and all

Table 8.1 Effective laminate Poisson's ratio

Relative core thickness	Loading modes	PNP layup	NPN layup
$\frac{t_c}{t} = \frac{1}{2}$	Axial	0	0
	Pure bending	Positive	Negative
$\frac{t_c}{t} = \left(\frac{1}{2}\right)^{\frac{1}{3}}$	Axial	Negative	Positive
	Pure bending	0	0
$\frac{1}{2} < \frac{t_c}{t} < \left(\frac{1}{2}\right)^{\frac{1}{3}}$	Axial	Negative	Positive
	Pure bending	Positive	Negative

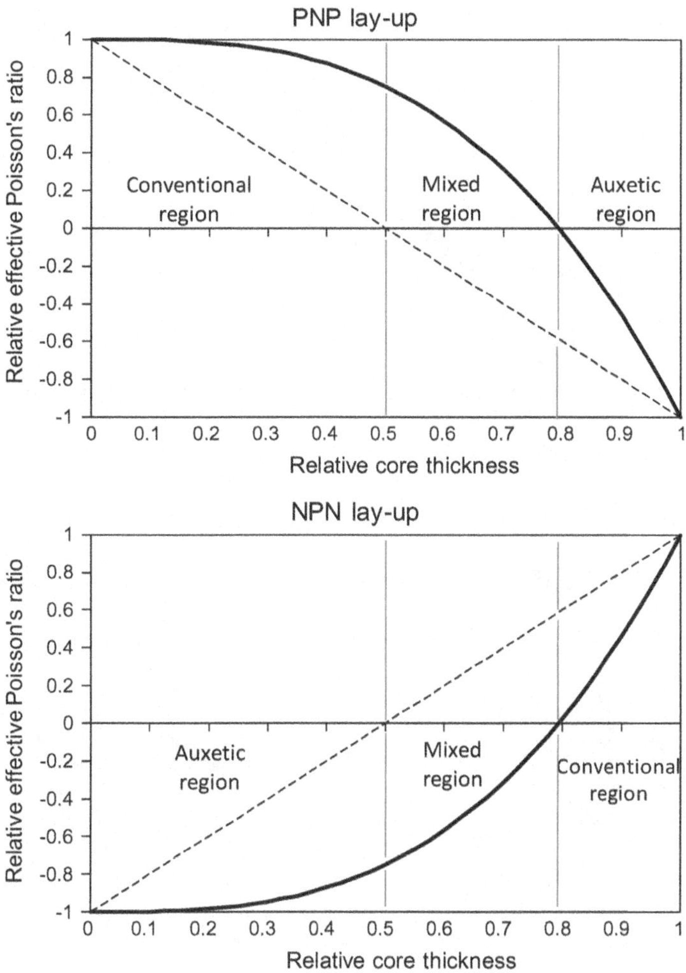

Fig. 8.16 Changes in the effective Poisson's ratio for axial loading (dashed lines) and bending moment (thick curves) based on the PNP lay-up (top) and NPN lay-up (bottom)

its layers are in-plane stress, thereby neglecting all the out-of-plane stresses. This assumption is valid in regions away from the free edges. Near the edges, however, shear stresses may arise between the neighboring layers, as shown in Fig. 8.17f. Therefore, the use of good bonding material is important for preventing interlaminar delamination.

It can be concluded that for the special case where Young's moduli are equal for all plies and the core ply possesses a Poisson's ratio of equal magnitude but opposite sign compared to the facesheet plies, the CLT shows that a region of relative core thickness exists which exhibit loading mode dependency on the laminate's effective Poisson's ratio. For combined loading, such as stretch bending, the PNP laminate

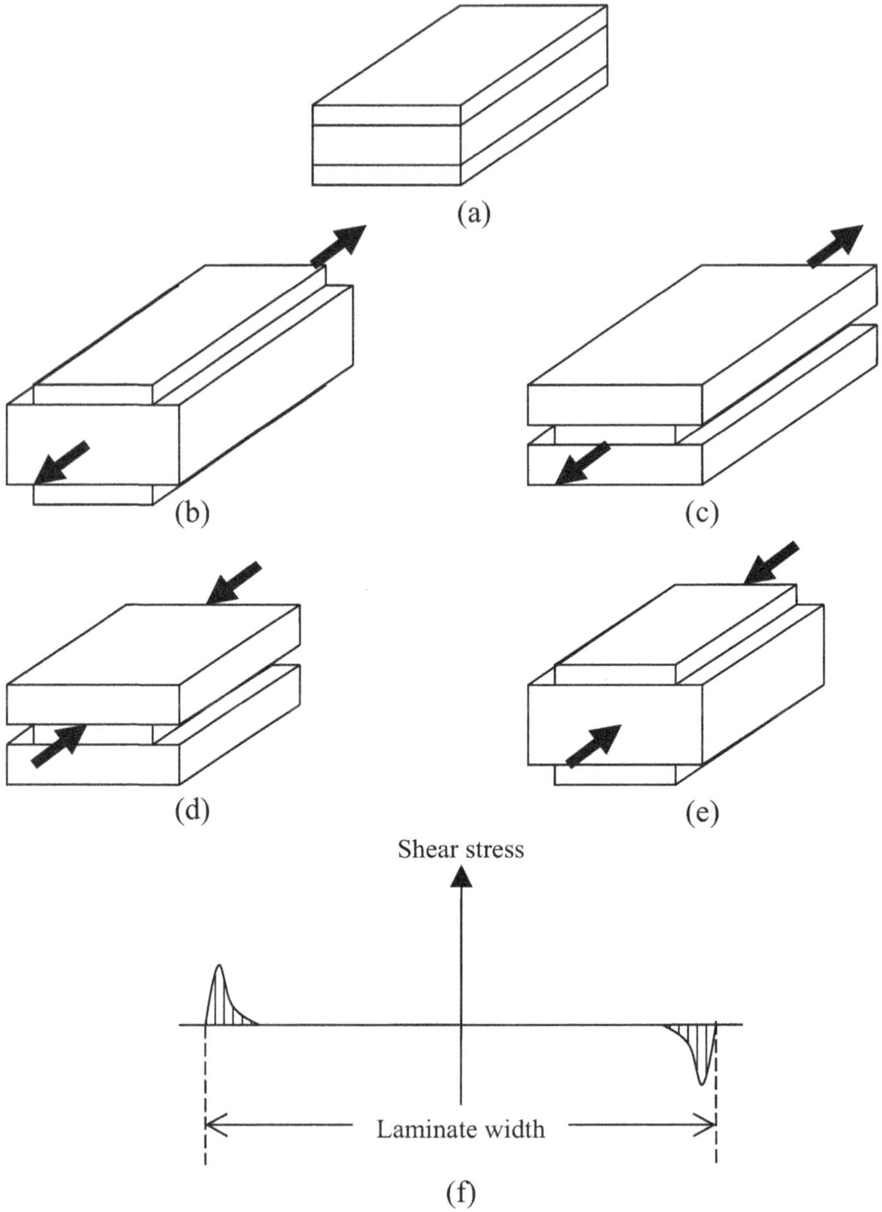

Fig. 8.17 Effect of opposing Poisson's ratio on in-plane transverse deformation without interlaminar bonding for **a** a 3-ply laminate with: **b** PNP in tension, **c** NPN in tension, **d** PNP in compression, **e** NPN in compression, and **f** a typical interlaminar shear stress profile with perfect interlaminar bonding

exhibits lateral expansion (due to negative Poisson's ratio) and anti-clastic shape (due to positive Poisson's ratio) while the NPN laminate gives lateral shrinkage (due to positive Poisson's ratio) and synclastic shape (due to negative Poisson's ratio). Therefore, a semi-auxetic structure can be designed to exhibit a positive, negative, or both signs for the effective Poisson's ratio as a result of loading modes (Lim 2007).

8.5 Example: Mixed Auxeticity of Semi-auxetic Sandwich Structures

Unlike axial loading and bending, which indicates the sign of the overall Poisson's ratio from the transverse deformation and the shape of the bent plate, respectively, as shown in Fig. 8.18, no geometrical observation can be obtained for identifying the effective Poisson's ratio for torsional loading. This can again be inferred from Fig. 8.9, where the resulting state of pure shear on an element can be viewed as principal stresses in a rotated element with equal in-plane tensile and compressive stress. Here, the out-of-plane deformation of the rotated element (i.e., radial deformation) is canceled, thereby preventing any indication arising from the sign of the overall Poisson's ratio to be ascertained on the basis of the deformed geometry. This example demonstrates a more physical approach for obtaining the overall Poisson's ratio of a sandwich structure than an earlier approach for that of a compound rod. Specifically, the case of sandwich structure where both the core and the facesheets are isotropic but possess Poisson's ratio of opposite signs is considered, for comparison with the overall Poisson's ratio of the same structure during axial loading and bending.

	Conventional behavior during axial loading (upper picture)	Auxetic behavior during axial loading (upper picture)
Conventional behavior during bending (lower picture)	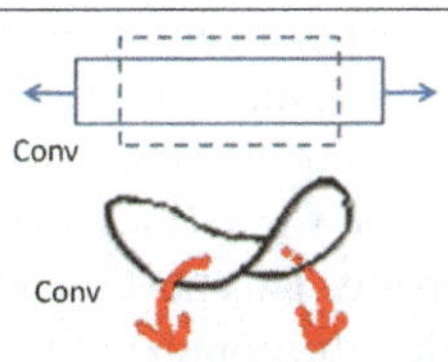	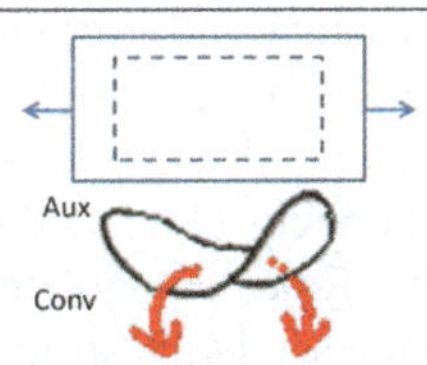
Auxetic behavior during bending (lower picture)	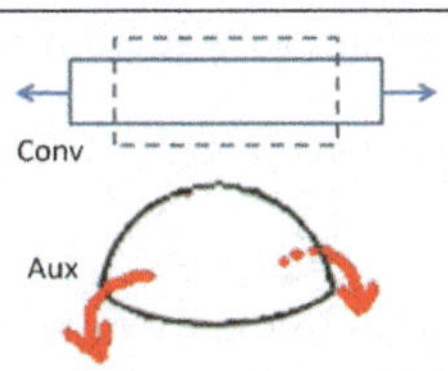	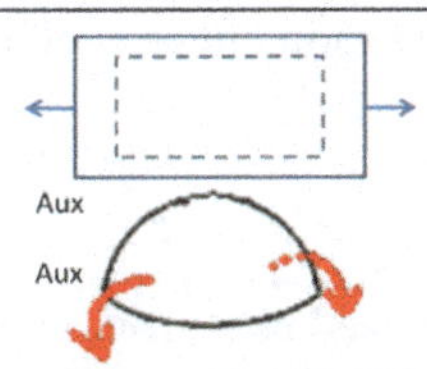

Fig. 8.18 Signs of overall Poisson's ratio during axial loading and bending are determined from transverse deformation and the resulting shell shape from the bent plate, respectively

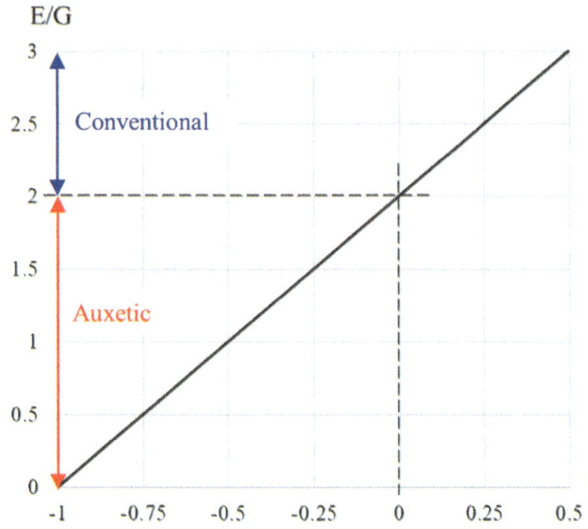

Fig. 8.19 Although torsion of a conventional or auxetic rod does not show any difference in terms of the deformed size nor shape, as shown in Fig. 8.9, auxeticity can still be inferred from the E/G ratio

Based on the relation $E/G = 2(1 + \upsilon)$ for isotropic solids, substitution of $\nu = -1$, 0, and 0.5 gives $E/G = 0$, 2, and 3, respectively. It follows that the range

$$0 < \frac{E}{G} < 2 \Rightarrow \upsilon < 0 \tag{8.5.1}$$

or $G/E > 1/2$ indicate auxetic behavior, while the range

$$2 < \frac{E}{G} < 3 \Rightarrow \upsilon > 0 \tag{8.5.2}$$

or $1/3 < G/E < 1/2$ imply conventional behavior. See Fig. 8.19. This is the first basis on which the overall Poisson's ratio can be inferred for a sandwich structure undergoing only torsion loading, which may differ under other modes of loading.

The second basis is to obtain two sets of torsional load per twisted angle, one for a homogeneous block and another one for a sandwich structure with similar overall geometry, for a comparison to be made. This paves a way for an equivalent Poisson's ratio of the sandwich structure within the context of torsional load. For convenience, the polar moment area of a rectangular cross section of width w and thickness t

$$J = \frac{wt}{12}\left(w^2 + t^2\right) \tag{8.5.3}$$

may be used as a gross approximation to the torsional polar moment area described in Eq. (8.5.30). The torsional load, T, on a homogenous rectangular block of length L has been given in Eq. (8.2.5) where ϕ is the angular twist and G is the shear modulus of the rectangular block material. Substituting $E/G = 2(1 + \upsilon)$ and Eq. (8.5.3) into Eq. (8.2.5) gives

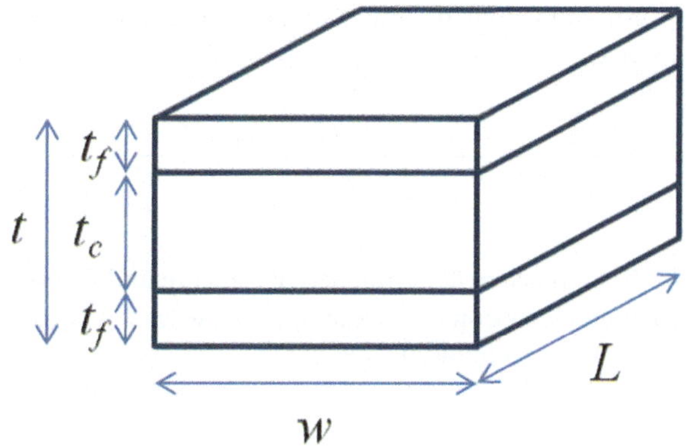

Fig. 8.20 Sandwich structure of core and facesheet thickness t_c and t_f, respectively

$$\frac{TL}{\phi} = \frac{E_{\text{equiv}} w^3 t}{24(1 + v_{\text{equiv}})}\left(1 + \frac{t^2}{w^2}\right) \tag{8.5.4}$$

where the subscript equiv refers to the homogeneous equivalence for comparison with the sandwich structure.

For a sandwich structure shown in Fig. 8.20, the total torsional load is carried by the core and facesheets

$$T = T_c + T_f \tag{8.5.5}$$

or

$$\frac{TL}{\phi} = \frac{T_c L}{\phi} + \frac{T_f L}{\phi} = G_c J_c + G_f J_f \tag{8.5.6}$$

whereby subscripts c and f refer to the core and both facesheets, respectively. Substituting the component shear modulus

$$\begin{Bmatrix} G_c \\ G_f \end{Bmatrix} = \frac{1}{2}\begin{Bmatrix} E_c/(1 + v_c) \\ E_f/(1 + v_f) \end{Bmatrix} \tag{8.5.7}$$

and their corresponding simplified polar moment area

$$\begin{Bmatrix} J_c \\ J_f \end{Bmatrix} = \frac{1}{12}\begin{Bmatrix} wt_c(w^2 + t_c^2) \\ wt(w^2 + t^2) - wt_c(w^2 + t_c^2) \end{Bmatrix} \tag{8.5.8}$$

into Eq. (8.5.6) yields

$$\frac{TL}{\phi} = \frac{E_c w^3 t_c}{24(1 + v_c)}\left(1 + \frac{t_c^2}{w^2}\right) + \frac{E_f w^3 t}{24(1 + v_f)}\left[\left(1 + \frac{t^2}{w^2}\right) - \frac{t_c}{t}\left(1 + \frac{t_c^2}{w^2}\right)\right] \tag{8.5.9}$$

An equivalent or effective Poisson's ratio for this sandwich structure can be approximated by equating Eq. (8.5.4) with Eq. (8.5.9) to give (Lim 2012)

$$\frac{1}{1+v_{\text{equiv}}}=\frac{1}{1+v_f}\left(\frac{E_f}{E_{\text{equiv}}}\right)+\frac{t_c}{t}\left(\frac{w^2+t_c^2}{w^2+t^2}\right)\left[\frac{1}{1+v_c}\left(\frac{E_c}{E_{\text{equiv}}}\right)-\frac{1}{1+v_f}\left(\frac{E_f}{E_{\text{equiv}}}\right)\right] \tag{8.5.10}$$

To emphasize the influence of the core and facesheet Poisson's ratios, the subsequent analysis considers a broad case whereby both the core and facesheets possess equal Young's modulus, i.e.,

$$E_c=E_f=E_{\text{equiv}} \tag{8.5.11}$$

Under this condition, Eq. (8.5.10) reduces to

$$\frac{1}{1+v_{\text{equiv}}}=\frac{1}{1+v_f}+\frac{t_c}{t}\left(\frac{w^2+t_c^2}{w^2+t^2}\right)\left[\frac{1}{1+v_c}-\frac{1}{1+v_f}\right] \tag{8.5.12}$$

or

$$v_{\text{equiv}}=-1+\frac{(1+v_f)(1+v_c)}{(1+v_c)+(v_f-v_c)\frac{t_c}{t}\left(\frac{w^2+t_c^2}{w^2+t^2}\right)} \tag{8.5.13}$$

In calculating the special case where the core and facesheets possess Poisson's ratio of equal magnitude but opposite signs, the torsional Poisson's ratio is proposed as

$$v_T=\frac{1}{2}\left\{\begin{matrix} v_{\text{equiv}}(v_f=+v_0,v_c=-v_0)-v_{\text{equiv}}(v_f=-v_0,v_c=+v_0);\ v_c<0<v_f \\ v_{\text{equiv}}(v_f=-v_0,v_c=+v_0)-v_{\text{equiv}}(v_f=+v_0,v_c=-v_0);\ v_c>0>v_f \end{matrix}\right\} \tag{8.5.14}$$

where $v_0>0$, i.e.,

$$v_T=\frac{1-v_0^2}{2}\left[\frac{1}{1-v_0+2v_0\frac{t_c}{t}\left(\frac{w^2+t_c^2}{w^2+t^2}\right)}-\frac{1}{1+v_0-2v_0\frac{t_c}{t}\left(\frac{w^2+t_c^2}{w^2+t^2}\right)}\right] \tag{8.5.15}$$

for auxetic core, and

$$v_T=\frac{1-v_0^2}{2}\left[\frac{1}{1+v_0-2v_0\frac{t_c}{t}\left(\frac{w^2+t_c^2}{w^2+t^2}\right)}-\frac{1}{1-v_0+2v_0\frac{t_c}{t}\left(\frac{w^2+t_c^2}{w^2+t^2}\right)}\right] \tag{8.5.16}$$

for auxetic facesheets.

For a square cross section, $w = t$, or

$$\lim_{\frac{w}{t}\to 1}\left(\frac{w^2+t_c^2}{w^2+t^2}\right) = \frac{1}{2}\left(1+\frac{t_c^2}{t^2}\right) \tag{8.5.17}$$

This leads to

$$\upsilon_T = \frac{1-\upsilon_0^2}{2}\left[\frac{1}{1-\upsilon_0\left[1-\frac{t_c}{t}-\left(\frac{t_c}{t}\right)^3\right]} - \frac{1}{1+\upsilon_0\left[1-\frac{t_c}{t}-\left(\frac{t_c}{t}\right)^3\right]}\right] \tag{8.5.18}$$

and

$$\upsilon_T = \frac{1-\upsilon_0^2}{2}\left[\frac{1}{1+\upsilon_0\left[1-\frac{t_c}{t}-\left(\frac{t_c}{t}\right)^3\right]} - \frac{1}{1-\upsilon_0\left[1-\frac{t_c}{t}-\left(\frac{t_c}{t}\right)^3\right]}\right] \tag{8.5.19}$$

for auxetic core and auxetic facesheets, respectively. In comparison with similar condition whereby $E_f = E_c$ and $\upsilon_f/\upsilon_c = -1$, the effective Poisson's ratios under axial loading and bending were established in the previous example as

$$\upsilon_A = \upsilon_f + \frac{t_c}{t}(\upsilon_c - \upsilon_f) \tag{8.5.20}$$

and

$$\upsilon_B = \upsilon_f + \left(\frac{t_c}{t}\right)^3(\upsilon_c - \upsilon_f) \tag{8.5.21}$$

respectively. Hence,

$$\begin{Bmatrix} \upsilon_A \\ \upsilon_B \end{Bmatrix} = \upsilon_0 \begin{Bmatrix} +1 - 2(t_c/t)^1 \\ +1 - 2(t_c/t)^3 \end{Bmatrix} \tag{8.5.22}$$

and

$$\begin{Bmatrix} \upsilon_A \\ \upsilon_B \end{Bmatrix} = \upsilon_0 \begin{Bmatrix} -1 + 2(t_c/t)^1 \\ -1 + 2(t_c/t)^3 \end{Bmatrix} \tag{8.5.23}$$

for auxetic core and auxetic facesheets, respectively, with equal Poisson's ratio magnitude. Figure 8.21 shows the variation of loading modes on the auxeticity with reference to the relative core thickness for the case of auxetic core, i.e., $\upsilon_c = -\upsilon_f = -0.5$, using Eqs. (8.5.18) and (8.5.22), and auxetic facesheets, i.e., $\upsilon_c = -\upsilon_f = +0.5$, using Eqs. (8.5.19) and (8.5.23). As expected, the curves fall within $-1/2 \le \upsilon \le 1/2$ for $0 \le t_c/t \le 1$.

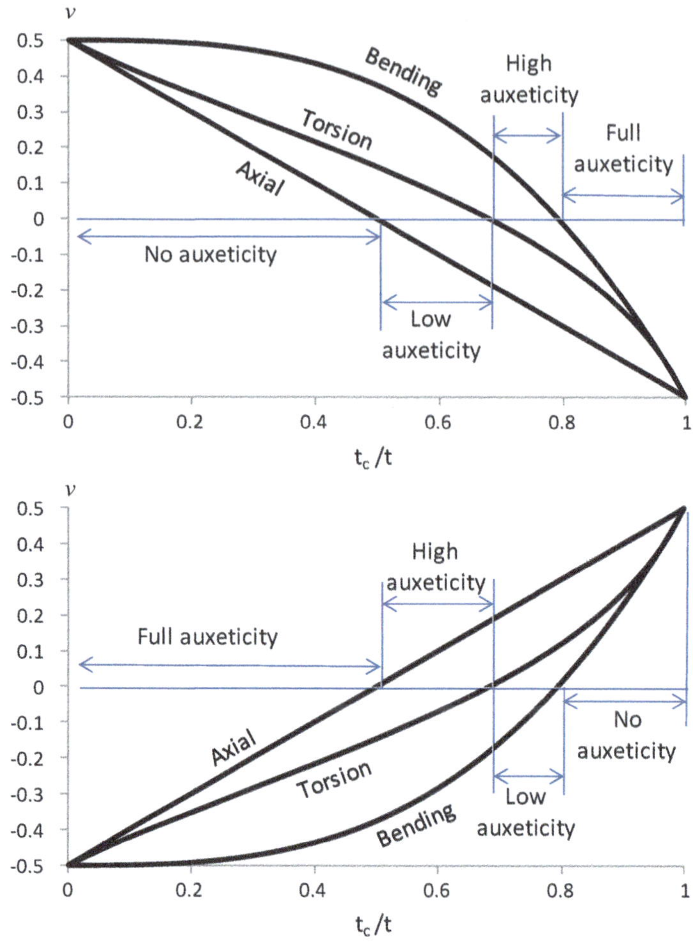

Fig. 8.21 Effect of loading modes on the auxeticity of semi-auxetic sandwich structure with square cross section for the case of auxetic core (top) and auxetic facesheets (bottom), where t_c/t refers to the relative core thickness

Based on these plots, four levels of auxeticity are identified (Lim 2012). They are as follows:

1. full auxeticity (FA) if the structure behaves as an auxetic structure under all three modes of loading
2. high auxeticity (HA) if the structure behaves as an auxetic structure in two of the loading modes
3. low auxeticity (LA) if the structure behaves as an auxetic structure in only one of the loading modes, and
4. no auxeticity (NA) if the structure behaves as a conventional structure under all of the three loading modes.

The relative core thickness at which the switch from one level of auxeticity to the next, in the context of Fig. 8.21, is $t_c/t = 0.5, 0.682328$, and 0.793701. The relative core thickness at which change in the level of auxeticity occurs will be different if the Poisson's ratio of the core and facesheets is altered because of the accompanying shift in the curves.

Unlike the axial loading and bending modes, the auxeticity of torsion modes is influenced by the width-to-thickness aspect ratio, *w*/*t*. To observe the effect of aspect ratio, we consider the limiting states

$$\lim_{\frac{w}{t}\to\infty}\left(\frac{w^2+t_c^2}{w^2+t^2}\right)=1 \tag{8.5.24}$$

and

$$\lim_{\frac{w}{t}\to 0}\left(\frac{w^2+t_c^2}{w^2+t^2}\right)=\frac{t_c^2}{t^2} \tag{8.5.25}$$

The limiting state described by Eq. (8.5.24) refers to a very wide sandwich plate which leads to

$$v_T=\frac{1-v_0^2}{2}\left[\frac{1}{1-v_0\left[1-2\left(\frac{t_c}{t}\right)\right]}-\frac{1}{1+v_0\left[1-2\left(\frac{t_c}{t}\right)\right]}\right] \tag{8.5.26}$$

and

$$v_T=\frac{1-v_0^2}{2}\left[\frac{1}{1+v_0\left[1-2\left(\frac{t_c}{t}\right)\right]}-\frac{1}{1-v_0\left[1-2\left(\frac{t_c}{t}\right)\right]}\right] \tag{8.5.27}$$

for auxetic core and auxetic facesheets, respectively. The limiting state described by Eq. (8.5.25) refers to a very narrow sandwich beam, which leads to

$$v_T=\frac{1-v_0^2}{2}\left[\frac{1}{1-v_0\left[1-2\left(\frac{t_c}{t}\right)^3\right]}-\frac{1}{1+v_0\left[1-2\left(\frac{t_c}{t}\right)^3\right]}\right] \tag{8.5.28}$$

and

$$v_T=\frac{1-v_0^2}{2}\left[\frac{1}{1+v_0\left[1-2\left(\frac{t_c}{t}\right)^3\right]}-\frac{1}{1-v_0\left[1-2\left(\frac{t_c}{t}\right)^3\right]}\right] \tag{8.5.29}$$

for auxetic core and auxetic facesheets, respectively. Figure 8.22 shows the torsional Poisson's ratio plotted against the relative core thickness for various aspect ratios in the case of auxetic core, i.e., $v_c = -v_f = -0.5$, using Eqs. (8.5.18), (8.5.26) and

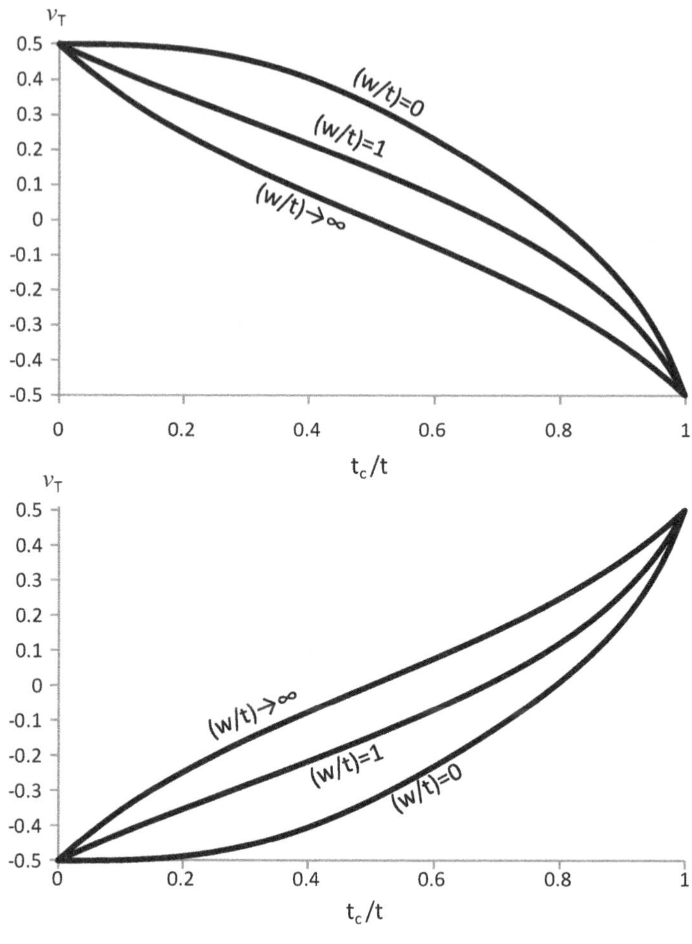

Fig. 8.22 Effect of cross-sectional aspect ratio on the auxeticity of semi-auxetic sandwich structure in the case of auxetic core (top) and auxetic facesheets (bottom), where t_c/t refers to the relative core thickness

(8.5.28), and auxetic facesheets, i.e., $v_c = -v_f = +0.5$, using Eqs. (8.5.19), (8.5.27) and (8.5.29). As expected, the curves fall within $-1/2 \leq v \leq 1/2$ for $0 \leq t_c/t \leq 1$.

The plots of torsional Poisson's ratio, v_T versus relative core thickness, t_c/t for $w/t = 0$ and $w/t\infty$, in Fig. 8.22 further imply that v_T approaches its limits when the cross-sectional aspect ratio tends to extreme values. However, the curves will shift when core and facesheets of other Poisson's ratio are used. To illustrate two special cases that give antisymmetric curves, the relative core thickness that gives zero torsional Poisson's ratio was solved from either Eqs. (8.5.15) or (8.5.16) to give $t_c/t = 0.682328$ for $w/t = 1$. Based on this value of relative core thickness, variation of the torsional Poisson's ratio with respect to the cross-sectional aspect ratio can be plotted as shown in Fig. 8.23. The plot also shows the rate at which υ_T varies with respect to w/t at $t_c/t = 0.682328$. Unlike Fig. 8.22, the plots in Fig. 8.23 show clearly that υ_T varies rapidly with respect to w/t for the range $10^{-1} < w/t < 10^{+1}$. Outside this range, υ_T is almost constant for $t_c/t = 0.682328$. The two curves are antisymmetric when the horizontal axis is in logarithmic scale.

As a further comment in regard to non-antisymmetric curves, the equivalent torsional Poisson's ratio curves for auxetic core and auxetic facesheets will become more positive and more negative, respectively, for lower relative core thickness. Similarly, the equivalent torsional Poisson's ratio curves for auxetic core and auxetic facesheets will become more negative and more positive, respectively, for higher relative core thickness. As a numerical example, the equivalent torsional Poisson's ratio of $\upsilon_T = +1/4$ for auxetic core and $\upsilon_T = -1/4$ for auxetic facesheets is obtained at $w/t = 1$ if $t_c/t = 0.351146$. This gives the range of equivalent torsional Poisson's ratio $0.114 < \upsilon < 0.433$ and $-0.433 < \upsilon < -0.114$ for auxetic core and auxetic facesheets, respectively, as shown in Fig. 8.24 (top). Likewise, the equivalent torsional Poisson's ratio of $\upsilon_T = -1/4$ for auxetic core and $\upsilon_T = +1/4$ for auxetic facesheets is obtained at $w/t = 1$ if $t_c/t = 0.892831$. This gives the range of equivalent torsional Poisson's ratio $-0.348 < \upsilon < -0.166$ and $0.166 < \upsilon < 0.348$ for auxetic core and auxetic facesheets, respectively, as shown in Fig. 8.24 (bottom).

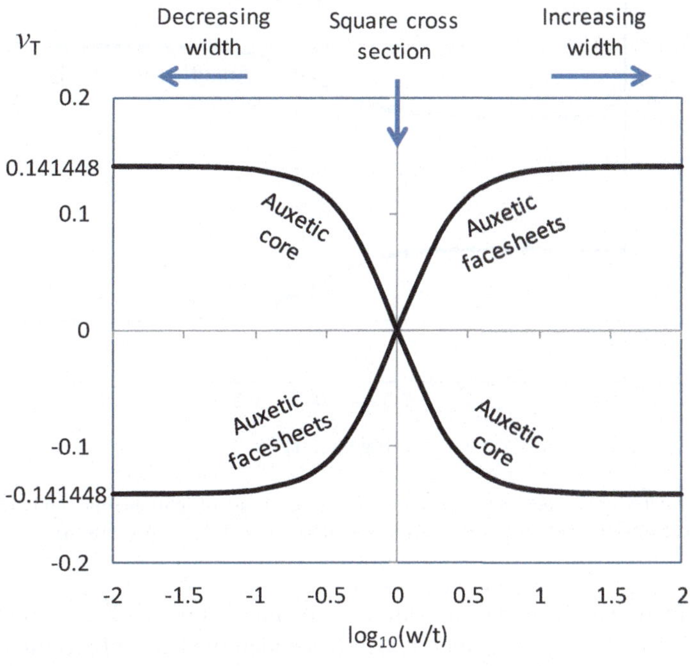

Fig. 8.23 Effect of cross-sectional aspect ratio on the auxeticity of semi-auxetic sandwich structure at relative core thickness of $t_c/t = 0.682328$

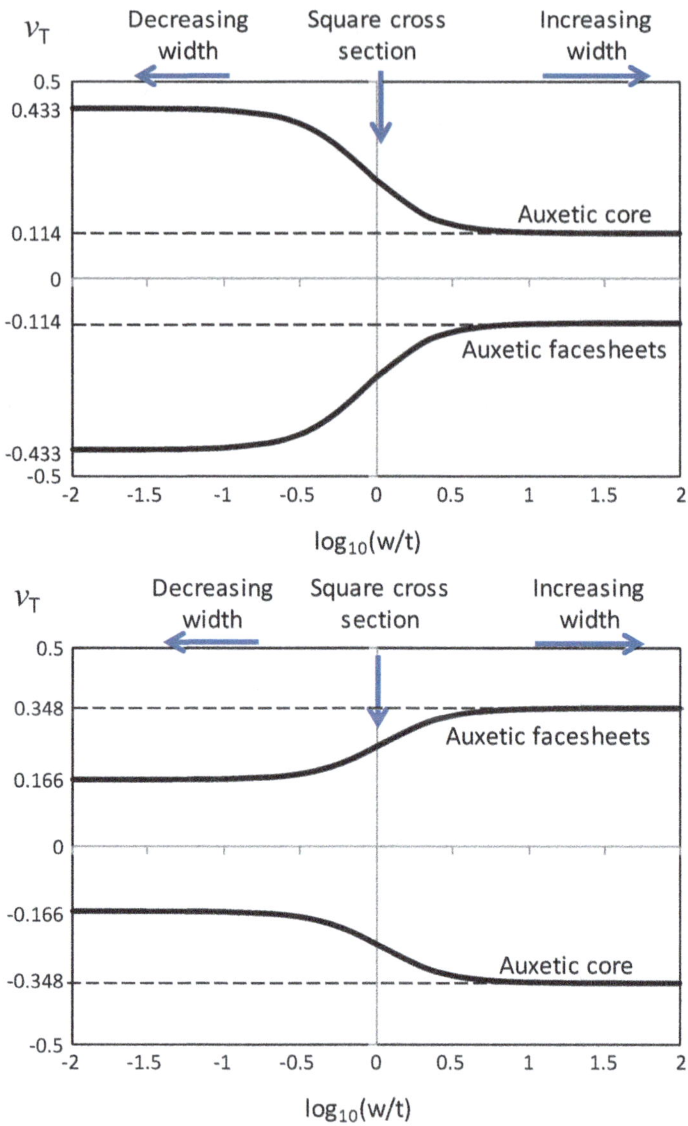

Fig. 8.24 Effect of cross-sectional aspect ratio on the auxeticity of semi-auxetic sandwich structure at relative core thickness of $t_c/t = 0.351146$ (top) and $t_c/t = 0.892831$ (bottom)

In considering three modes of loading on a semi-auxetic sandwich structure, four categories of loading mode auxeticity have been identified, namely (i) full auxeticity (FA), (ii) high auxeticity (HA), (iii) low auxeticity (LA), and (iv) no auxeticity (NA). It was found that, unlike axial loading and bending auxeticities, the cross-sectional aspect ratio affects torsional auxeticity and that the change in the torsional

Poisson's ratio is largest in the range $10^{-1} < w/t < 10^{+1}$. In addition, the torsional auxeticity approaches its limits for extreme cross-sectional aspect ratios. Therefore, the overall Poisson's ratio at structural level can be designed to respond differently under different loading modes as a means to achieve structural optimization. The polar moment area given in Eq. (8.5.3) is highly simplistic for convenient calculation. Further refinement can be achieved by using (Young et al. 2011)

$$J = wt^3\left[\frac{1}{3} - 0.21\frac{t}{w}\left(1 - \frac{t^4}{12w^4}\right)\right] \tag{8.5.30}$$

for $w > t$.

8.6 Advanced Topics

Strek and Jopek (2012) performed computational modeling and simulation to provide efficient alternatives to predict the parameters of the composites where one of the phases is auxetic. They used a finite element method o find the engineering constants (Young's modulus and Poisson's ratio) of auxetic composites consisting of concentric cylindrical inclusions made of combinations of auxetic and classic (non-auxetic) materials and observed that not only the mechanical properties of the different composite phases influence the effective mechanical properties of the whole composite, but also the location of the same material phases do matter (Strek and Jopek 2012). Thereafter, Strek et al. (2014) analyzed sandwich-structured composites consisting of two different materials: auxetic and structural steel. The optimization criterion adopted was minimum compliance for the load case where the frame's top boundary was downward loaded. Outer layers are made of steel while the middle layer is two-phase solid material composite. Only the middle layer is optimized by means of minimization of the objective function defined as the internal strain energy. In the first part of this paper, Strek et al. (2014) studied the application of the solid isotropic material with penalization (SIMP) model to find the optimal distribution of a given amount of materials in sandwich-structured composite. In the second part, they proposed a multilayered composite structure in which internal layers surfaces are wavy. In both cases, the total energy strain was analyzed. Subsequently, Strek et al. (2015) analyzed the effective properties and dynamic response of a sandwich panel made of two facesheets and auxetic core by computer simulations, whereby the inner composite layer is made of a cellular auxetic structure immersed in a filler material of a given Poisson's ratio (filler material fills the voids in structure), and each cell is composed of an auxetic structure (reentrant honeycomb or rotating square), i.e., exhibiting negative Poisson's ratio without any filler. The effect of filler material on the effective properties of the sandwich panel was investigated, and their proposed structure showed interesting structural characteristics and dynamic properties. Strek et al. (2015) clearly demonstrated that it is possible to create auxetic sandwich panels made of two solid materials with positive Poisson's ratio. They showed that it is even possible if the filler material is nearly incompressible, but can move in out-of-plane direction, and that the effective Young's modulus of such sandwich panels becomes very large if Poisson's ratio of the filler material tends to -1.

References

Lim TC (2007) On simultaneous positive and negative Poisson's ratio laminates. Phys Status Solidi B 244(3):910–918

Lim TC (2011) Torsion of semi-auxetic rods. J Mater Sci 46(21):6904–6909

Lim TC (2012) Mixed auxeticity of auxetic sandwich structures. Phys Status Solidi B 249(7):1366–1372

Lim TC (2018) Auxeticity of concentric auxetic-conventional foam rods with high modulus interface adhesive. Materials 11(2):223

Strek T, Jopek H (2012) Effective mechanical properties of concentric cylindrical composites with auxetic phase. Phys Status Solidi B 249(7):1359–1365

Strek T, Jopek H, Maruszewski BT, Nienartowicz M (2014) Computational analysis of sandwich-structured composites with an auxetic phase. Phys Status Solidi B 251(2):354–366

Strek T, Jopek H, Nienartowicz M (2015) Dynamic response of sandwich panels with auxetic cores. Phys Status Solidi B 252(7):1540–1550

Young WC, Budymas RG, Sadegh AM (2011) Roark's formulas for stress and strains, 8th edn. McGraw-Hill, New York

Chapter 9
Auxetic Composites with Enhanced Moduli

Abstract This chapter evaluates the extent of moduli increase, especially beyond the simple rule-of-mixture mode, for composites consisting of positive and negative Poisson's ratio phases. Specific topics include fiber composites, laminates, and particle composites. The refined moduli models include correction terms or functions to cater for the increased stiffness.

Keywords Auxetic composites · Correction terms · Enhanced moduli

9.1 Fundamentals

Arising from opposing Poisson's ratio signs in constituents of composite materials, the requirement for retaining geometrical compatibility translates into higher modulus. In this chapter, we consider the extent of moduli increase beyond the Voigt rule-of-mixture.

9.2 Example: Semi-Auxetic Unidirectional Fiber Composites

The 1D micromechanical model for the longitudinal Young's modulus of a continuous unidirectional fiber composite follows the Voigt formula

$$E_L = E_{Lf} V_f + E_{Lm}\left(1 - V_f\right) \tag{9.2.1}$$

The 3D model incorporates Poisson's ratios of the individual phases, such as that given by the self-consistent method

$$\begin{aligned} E_L = {} & E_{Lf} V_f + E_{Lm}\left(1 - V_f\right) \\ & + \frac{4\left(\upsilon_{LTm} - \upsilon_{LTf}\right)^2 K_{Tf} K_{Tm} G_{TTm}\left(1 - V_f\right) V_f}{\left(K_{Tf} + G_{TTm}\right) K_{Tm} + \left(K_{Tf} - K_{Tm}\right) G_{TTm} V_f} \end{aligned} \tag{9.2.2}$$

T. Lim, *Mechanics of Metamaterials with Negative Parameters*, Engineering Materials,
https://doi.org/10.1007/978-981-15-6446-8_9

In this example, we assume isotropic phases, i.e., $E_{Lf} = E_f$, $E_{Lm} = E_m$, $\upsilon_{LTm} = \upsilon_m$, $\upsilon_{LTf} = \upsilon_f$, $K_{Tf} = K_f$, $K_{Tm} = K_m$ and $G_{TTm} = G_m$ so that

$$\frac{E_L}{E_m} = 1 + \left(\frac{E_f}{E_m} - 1\right)V_f + \frac{4(\upsilon_m - \upsilon_f)^2(1 - V_f)V_f}{\left[\left(\frac{1}{G_m} + \frac{1}{K_f}\right) + \left(\frac{1}{K_m} - \frac{1}{K_f}\right)V_f\right]E_m} \tag{9.2.3}$$

where the following moduli relations for isotropic phases

$$G_m = \frac{E_m}{2(1 + \upsilon_m)}, \quad K_m = \frac{E_m}{3(1 - 2\upsilon_m)}, \quad K_f = \frac{E_f}{3(1 - 2\upsilon_f)} \tag{9.2.4}$$

are valid. This leads to

$$\frac{E_L}{E_m} = 1 + \left(\frac{E_f}{E_m} - 1\right)V_f + \frac{4(\upsilon_m - \upsilon_f)^2(1 - V_f)V_f}{2(1 + \upsilon_m) + 3(1 - 2\upsilon_f)\frac{E_m}{E_f} + \left(3(1 - 2\upsilon_m) - 3(1 - 2\upsilon_f)\frac{E_m}{E_f}\right)V_f} \tag{9.2.5}$$

such that one may consider two special cases: (i) Special Case I in which the constituents possess equal Poisson's ratio magnitudes but opposite signs and (ii) Special Case II whereby the phases have equal Young's modulus and volume fractions.

For Special Case I where each phase's Poisson's ratio is equal

$$|\upsilon_f| = |\upsilon_m| = \upsilon_0 \tag{9.2.6}$$

but with opposite signs

$$\upsilon_m < 0 < \upsilon_f \tag{9.2.7}$$

then

$$\frac{E_L}{E_m} = 1 + \left(\frac{E_f}{E_m} - 1\right)V_f + \frac{(4\upsilon_0)^2(1 - V_f)V_f}{2(1 - \upsilon_0) + 3(1 + 2\upsilon_0)V_f + 3(1 - 2\upsilon_0)(1 - V_f)(E_m/E_f)} \tag{9.2.8}$$

For illustration, we select a typical value of Poisson's ratio magnitude of $\upsilon_0 = 0.3$ to give

$$\frac{E_L}{E_m} = 1 + \left(\frac{E_f}{E_m} - 1\right)V_f + \frac{1.44(1 - V_f)V_f}{1.4 + 4.8V_f + 1.2(1 - V_f)(E_m/E_f)} \tag{9.2.9}$$

This illustration was performed for various E_f/E_m ratio, as furnished in Fig. 9.1.

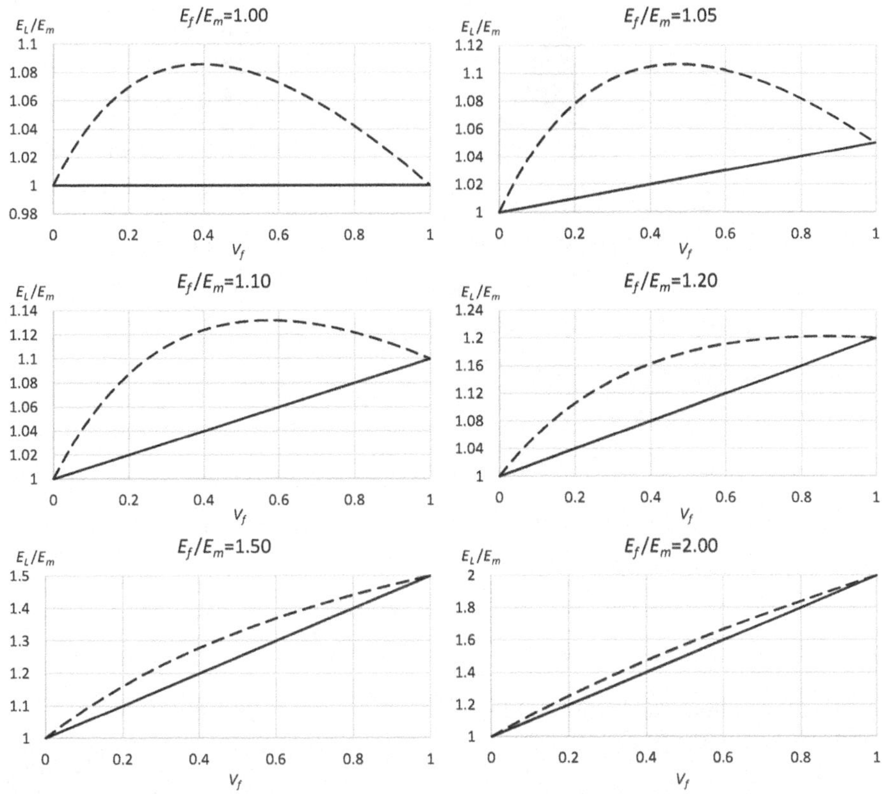

Fig. 9.1 Dimensionless longitudinal Young's modulus, E_L/E_m, versus fiber volume fraction of continuous unidirectional fiber composites considering constituent Poisson's ratios (dashed curves) and based on rule-of-mixture (straight lines) at $v_f = 0.3$, $v_m = -0.3$ for various E_f/E_m ratio

Reference to Fig. 9.1 shows that the mismatch in Poisson's ratio cannot be neglected when the Young's moduli ratio of the constituent materials fall between 0.5 and 2. Specifically when the difference in Young's moduli is less than 10%, the effect of Poisson's ratio mismatches, which result in Young's modulus that exceeds the rule-of-mixture description, is significant for a semi-auxetic composite.

For Special Case II, where the phases possess of equal Young's modulus $E_f = E_m$, we have (Lim and Acharya 2010)

$$\frac{E_L}{E_m} = 1 + \frac{4(v_m - v_f)^2(1 - V_f)V_f}{5 + 2(v_m - 3v_f) + 6(v_f - v_m)V_f} \tag{9.2.10}$$

For illustration, we let the volume fraction of each phase to be equal, i.e., $V_f = 0.5$, such that

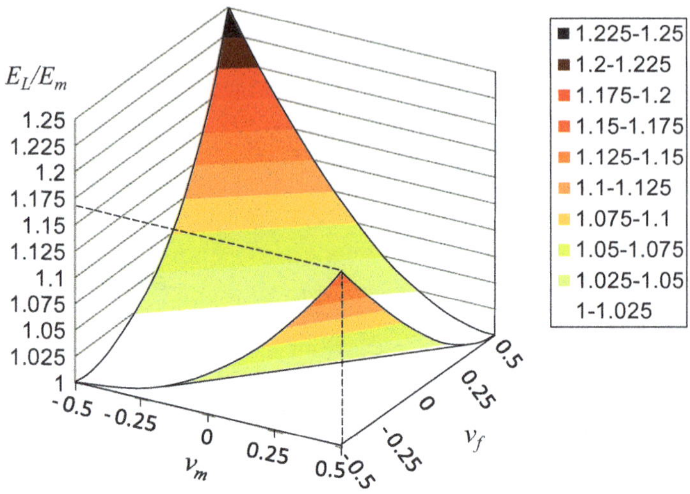

Fig. 9.2 Plot of E_L/E_m with $E_f = E_m$ and $V_f = V_m$ for various combinations of v_f and v_m

$$\frac{E_L}{E_m} = 1 + \frac{\left(v_m - v_f\right)^2}{5 - 3v_f - v_m} \tag{9.2.11}$$

This illustration was performed for $-1/2 \le v_f \le 1/2$ and $-1/2 \le v_m \le 1/2$ and furnished in Fig. 9.2.

It can be seen that the surpassing of the rule-of-mixture is insignificant for a fully positive and a fully negative Poisson's ratio phases, i.e., $v_f v_m > 0$. However, the overshooting of Young's modulus above the rule-of-mixture becomes significant for a semi-auxetic composite, whereby $v_f v_m < 0$.

It can be concluded that the surpassing of the rule-of-mixture relation is significant when the phases possess Poisson's ratios of opposing signs and where the difference of Young's moduli between the phases is insignificant. As a design practice, it is suggested that the rule-of-mixtures equation be replaced by other descriptions, which take into account Poisson's ratios of the constituents, when dealing with composites that possess positive and negative Poisson's ratio phases.

9.3 Example: Out-of-Plane Modulus of Semi-auxetic Laminates

With reference to Fig. 9.3, the effective in-plane Young's modulus, i.e., in the X-direction or Y-direction, obeys the simple rule-of-mixture by Voigt (1889, 1910)

$$E_{\text{Voigt}} = V_A E_A + V_B E_B \tag{9.3.1}$$

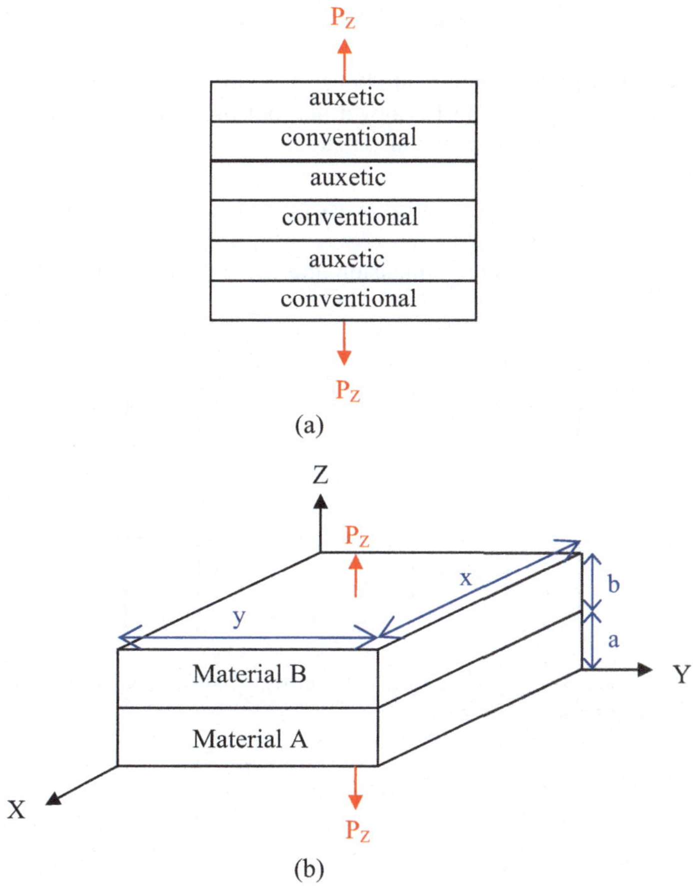

Fig. 9.3 **a** Schematics for a semi-auxetic laminate, and **b** idealized representative volume element for analysis

while the effective out-of-plane Young's modulus, i.e., in the *Z*-direction, obeys the inverse rule-of-mixture by Reuss (1929)

$$\frac{1}{E_{\text{Reuss}}} = \frac{V_A}{E_A} + \frac{V_B}{E_B} \tag{9.3.2}$$

where the volume fractions for materials A and B obey

$$V_A + V_B = 1 \tag{9.3.3}$$

Although it is widely believed that the effective Young's modulus of any composite E_C is bounded by the simple and inverse rules-of-mixtures

$$E_{\text{Reuss}} < E_C < E_{\text{Voigt}} \tag{9.3.4}$$

and that the out-of-plane Young's modulus of a laminate consisting of isotropic laminas is $E_C = E_{\text{Reuss}}$, it will be shown that not only Poisson's ratio mismatch gives $E_C > E_{\text{Reuss}}$ for the latter, but also that

$$E_C > E_{\text{Voigt}} \tag{9.3.5}$$

Taking Hooke's Law in three dimension, the stress–strain relations for materials A and B are

$$\begin{Bmatrix} \varepsilon_{AX} \\ \varepsilon_{AY} \\ \varepsilon_{AZ} \end{Bmatrix} = \frac{1}{E_A} \begin{bmatrix} 1 & -v_A & -v_A \\ -v_A & 1 & -v_A \\ -v_A & -v_A & 1 \end{bmatrix} \begin{Bmatrix} \sigma_{AX} \\ \sigma_{AY} \\ \sigma_{AZ} \end{Bmatrix} \tag{9.3.6}$$

and

$$\begin{Bmatrix} \varepsilon_{BX} \\ \varepsilon_{BY} \\ \varepsilon_{BZ} \end{Bmatrix} = \frac{1}{E_B} \begin{bmatrix} 1 & -v_B & -v_B \\ -v_B & 1 & -v_B \\ -v_B & -v_B & 1 \end{bmatrix} \begin{Bmatrix} \sigma_{BX} \\ \sigma_{BY} \\ \sigma_{BZ} \end{Bmatrix} \tag{9.3.7}$$

respectively. Prescribing a normal load P_Z on the Z-surface as shown in Fig. 9.3, equilibrium consideration leads to

$$\sigma_{AZ} = \sigma_{BZ} = \sigma_Z = \frac{P_Z}{xy} \tag{9.3.8}$$

and

$$\sigma_{BX} = -\frac{a}{b}\sigma_{AX}, \quad \sigma_{BY} = -\frac{a}{b}\sigma_{AY} \tag{9.3.9}$$

Substituting Eqs. (9.3.8) and (9.3.9) into Eqs. (9.3.6) and (9.3.7), and imposing common in-plane deformation, i.e.,

$$\varepsilon_{AX} = \varepsilon_{BX}, \quad \varepsilon_{AY} = \varepsilon_{BY} \tag{9.3.10}$$

give

$$\left(1 + \frac{aE_A}{bE_B}\right)\sigma_{AX} - \left(v_A + v_B\frac{aE_A}{bE_B}\right)\sigma_{AY} = \frac{P_Z}{xy}\left(v_A - v_B\frac{E_A}{E_B}\right) \tag{9.3.11}$$

and

$$-\left(v_A + v_B\frac{aE_A}{bE_B}\right)\sigma_{AX} + \left(1 + \frac{aE_A}{bE_B}\right)\sigma_{AY} = \frac{P_Z}{xy}\left(v_A - v_B\frac{E_A}{E_B}\right) \tag{9.3.12}$$

Comparison of terms in Eqs. (9.3.11) and (9.3.12) suggests $\sigma_{AX} = \sigma_{AY}$, which leads to

$$\sigma_{AX} = \sigma_{AY} = \frac{\frac{P_Z}{xy}\left(v_A - v_B \frac{E_A}{E_B}\right)\left[(1 + v_A) + (1 + v_B)\frac{aE_A}{bE_B}\right]}{\left(1 - v_A^2\right) + 2(1 - v_A v_B)\frac{aE_A}{bE_B} + \left(1 - v_B^2\right)\left(\frac{aE_A}{bE_B}\right)^2} \tag{9.3.13}$$

Substituting

$$\varepsilon_{AZ} = \frac{1}{E_A}[\sigma_{AZ} - v_A(\sigma_{AX} + \sigma_{AY})] = \frac{1}{E_A}\left[\frac{P_Z}{xy} - 2v_A\sigma_{AX}\right] \tag{9.3.14}$$

and

$$\varepsilon_{BZ} = \frac{1}{E_B}[\sigma_{BZ} - v_B(\sigma_{BX} + \sigma_{BY})] = \frac{1}{E_B}\left[\frac{P_Z}{xy} + 2v_B\frac{a}{b}\sigma_{AX}\right] \tag{9.3.15}$$

into

$$\varepsilon_Z = \frac{a}{a+b}\varepsilon_{AZ} + \frac{b}{a+b}\varepsilon_{BZ} \tag{9.3.16}$$

gives

$$\varepsilon_Z = \frac{P_Z}{xy}\left[\frac{a}{a+b}\left(\frac{1}{E_A}\right) + \frac{b}{a+b}\left(\frac{1}{E_B}\right) - C\right] \tag{9.3.17}$$

where

$$C = \frac{\frac{2}{a+b}\left(\frac{v_A}{E_A} - \frac{v_B}{E_B}\right)^2\left(\frac{1+v_A}{aE_A} + \frac{1+v_B}{bE_B}\right)}{\frac{1-v_A^2}{(aE_A)^2} + \frac{2(1-v_A v_B)}{(aE_A)(bE_B)} + \frac{1-v_B^2}{(bE_B)^2}} \tag{9.3.18}$$

Since $E_C = \sigma_Z/\varepsilon_Z$ and

$$V_A = \frac{a}{a+b}, \quad V_B = \frac{b}{a+b} \tag{9.3.19}$$

for the laminate under consideration, therefore

$$\frac{1}{E_C} = \frac{\varepsilon_Z}{P_Z/(xy)} = \frac{V_A}{E_A} + \frac{V_B}{E_B} - C \tag{9.3.20}$$

where

$$C = \frac{2\left(\frac{v_A}{E_A} - \frac{v_B}{E_B}\right)^2 \left(\frac{1+v_A}{V_A E_A} + \frac{1+v_B}{V_B E_B}\right)}{\frac{1-v_A^2}{(V_A E_A)^2} + \frac{2(1-v_A v_B)}{(V_A E_A)(V_B E_B)} + \frac{1-v_B^2}{(V_B E_B)^2}} \tag{9.3.21}$$

When $E_A = E_B$ and $v_A = v_B$, we have $C = 0$, therefore with reference to Eq. (9.3.20), we have $E_C = E_{\text{Reuss}}$ under this specific condition. When $E_A = E_B$ but $v_A \neq v_B$, it is possible to have $C > 0$, therefore implying $E_C > E_{\text{Reuss}}$. Hence, this example illustrates the out-of-plane Young's modulus exceeding the inverse rule-of-mixture when there is a mismatch in Poisson's ratios of the individual laminas.

To visually observe the influence of Poisson's ratios of the individual phases, we consider a special case whereby Young's moduli of both phases are equal, i.e., $E_A = E_B = E_0$. With this condition, both the simple and inverse rules-of-mixture collapse into a constant independent of the volume fraction of constituents. When $V_A = V_B = 0.5$, the dimensionless out-of-plane Young's modulus can be expressed as (Lim 2009)

$$\frac{E_C}{E_0} = \left(1 - \frac{(2 + v_A + v_B)(v_A - v_B)^2}{4 - (v_A + v_B)^2}\right)^{-1} \tag{9.3.22}$$

Figure 9.4a shows the plot of Eq. (9.3.22) for each phase with individual Poisson's ratios ranging between $-1/2$ and $1/2$ (Lim 2013). It is observed that Poisson's ratios of the individual phases do not significantly influence the out-of-plane Young's modulus of a conventional laminate (i.e., both v_A and v_B are positive) and also for the case of fully auxetic laminate (i.e., both v_A and v_B are negative), as evident from the symmetric moduli profile about $v_A = v_B$ but not so about $v_A = -v_B$, as shown in Fig. 9.4b. However, Poisson's ratios of individual phases significantly influence the out-of-plane Young's moduli for a semi-auxetic laminate (i.e., $v_A v_B < 0$). This is an example whereby both the fully conventional and fully auxetic structures share common characteristics, while a semi-auxetic structure possesses behavior that is unique from fully conventional and fully auxetic structures.

To investigate the effect of varying Young's moduli of the constituent materials, we plot E_C/E_A based on Eqs. (9.3.20) and (9.3.21) at $v_A v_B = -0.1$ whereby v_A and v_B are of equal magnitude but opposite signs. The magnitude $|v_A| = |v_B| = \sqrt{0.1} = 0.316$ would be realistic for illustration since most materials possess Poisson's ratio between 0.3 and 1/3. Substituting $\pm v_A = \mp v_B = \sqrt{0.1}$, and hence $v_A v_B = -0.1$ and $v_A^2 = v_B^2 = 0.1$, into Eqs. (9.3.20) and (9.3.21) leads to (Lim 2009)

$$\frac{E_C}{E_A} = \left(V_A(1-d) + V_B\frac{E_A}{E_B}\right)^{-1} \tag{9.3.23}$$

where

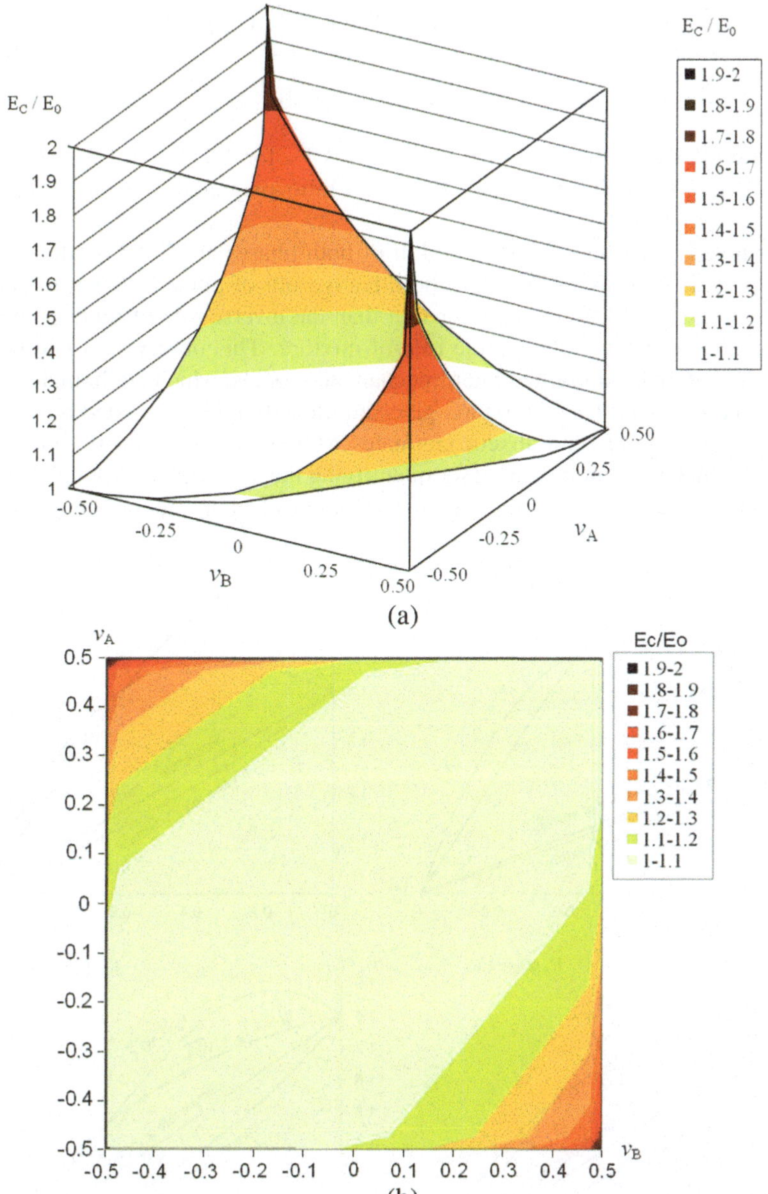

Fig. 9.4 Out-of-plane Young's modulus profile of a laminate with laminas of equal volume fractions and equal Young's moduli for $-1/2 \leq v_i \leq 1/2$ where $i = A, B$, showing: **a** large out-of-plane modulus when Poisson's ratio of alternating laminas possesses opposite Poisson's ratio signs, and **b** a contour plot showing symmetry about $v_A = v_B$ but no symmetry about $v_A = -v_B$

$$d = \frac{0.2\left(1 + \frac{E_A}{E_B}\right)^2\left(\left(1 \pm \sqrt{0.1}\right) + \left(1 \mp \sqrt{0.1}\right)\frac{V_A E_A}{V_B E_B}\right)}{0.9 + 2.2\frac{V_A E_A}{V_B E_B} + 0.9\left(\frac{V_A E_A}{V_B E_B}\right)^2} \quad (9.3.24)$$

Figure 9.5 shows the plots of E_C/E_A versus V_A for $E_B/E_A = 1.2, 1.4, 1.6, 1.8$ with alternating laminas of equal Poisson's ratio magnitude but opposing signs $\pm\upsilon_A = \mp\upsilon_B = \sqrt{0.1}$ (Lim 2013).

It is noted that when Young's moduli of both phases are equal or almost equal $E_B/E_A \approx 1$, such as in Fig. 9.5, the effective out-of-plane Young's moduli for $\pm\upsilon_A = \mp\upsilon_B = \sqrt{0.1}$ are not only greater than the inverse rule-of-mixture, but also significantly overshoots the simple rule-of-mixture. This observation is crucial as it reveals that many simplified micromechanical models, which are bounded by the simple and inverse rules-of-mixture, give significantly underestimated out-of-plane Young's modulus of a semi-auxetic laminates. More importantly, the inverse rule-of-mixture, which is commonly used for quantifying the out-of-plane Young's modulus of laminates, gives the worst estimation. The out-of-plane Young's modulus of a

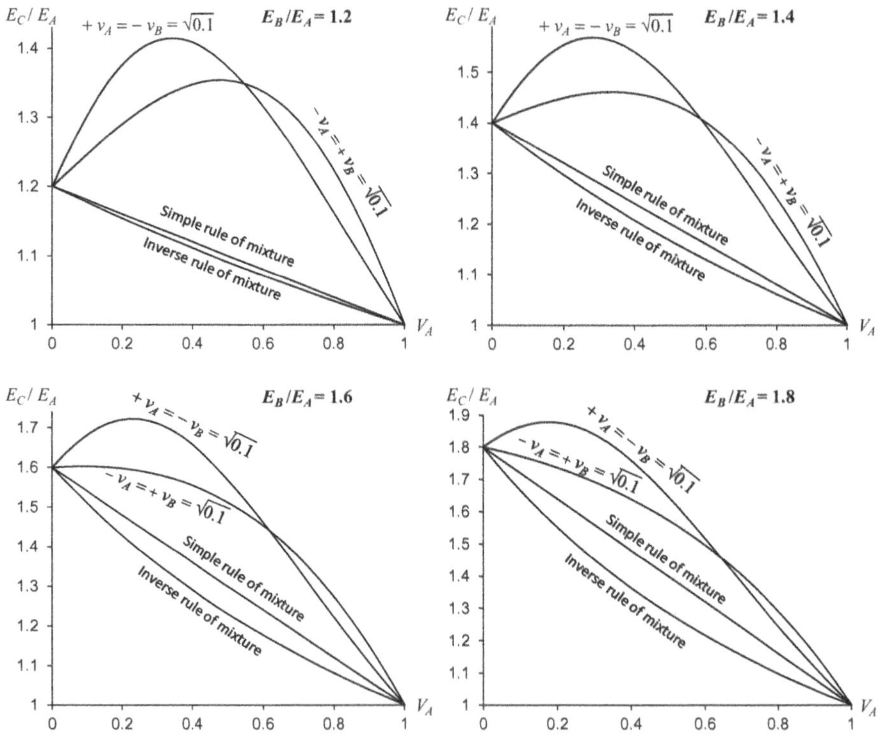

Fig. 9.5 Plots of dimensionless out-of-plane Young's moduli with respect to the modulus of material A versus the volume fraction of material A with $\pm\upsilon_A = \mp\upsilon_B = \sqrt{0.1}$ in comparison with the simple rule-of-mixture and inverse rule-of-mixture for various E_B/E_A ratio

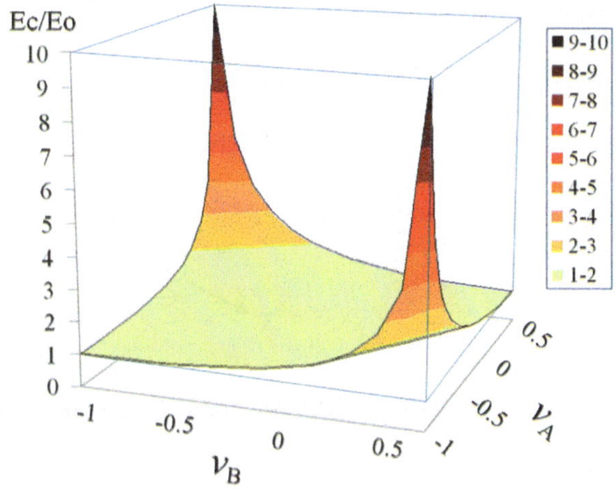

Fig. 9.6 Out-of-plane Young's modulus profile for a laminate with laminas of equal volume fractions and equal Young's moduli for $-1 \leq \upsilon_i \leq 1/2$ where $i = A, B$

semi-auxetic laminate reduces for higher E_B/E_A ratio; however, both out-of-plane Young's moduli for $\pm\upsilon_A = \mp\upsilon_B = \sqrt{0.1}$ are still significantly above the inverse rule-of-mixture. Therefore, the use of laminas of alternating Poisson's ratio signs gives rise to very high out-of-plane modulus that not only exceeds the inverse rule-of-mixture, but also the simple rule-of-mixture especially when the Young's moduli of the laminas are very close.

Reference to Eq. (9.3.22) for laminates of isotropic laminas with equal Young's moduli and equal thicknesses, computed results for the effective out-of-plane modulus at $\pm\upsilon_A = \mp\upsilon_B = 1/2$ is $E_C = 2E_0$, i.e., twice the modulus of the lamina. When extended to either $(\upsilon_A, \upsilon_B) = (-1, 1/2)$ or $(\upsilon_A, \upsilon_B) = (1/2, -1)$, the computed results give $E_C = 10E_0$, or ten times the modulus of the lamina. A plot of dimensionless out-of-plane modulus E_C/E_0 for the entire range of Poisson's ratio is shown in Fig. 9.6.

9.4 Example: In-Plane Modulus of Semi-auxetic Laminates

Extending the Young's modulus according to Voigt (1889, 1910) for n number of phases, we have

$$E_{\text{Voigt}} = V_1 E_1 + V_2 E_2 + \cdots + V_n E_n \tag{9.4.1}$$

where $V_1 + V_2 + \cdots + V_n = 1$, which is applicable for describing the in-plane modulus of a laminate of isotropic laminas. Alternatively, Eq. (9.4.1) can be rewritten in a dimensionless form as

$$\frac{E_C}{E_0} = \frac{V_1 E_1 + V_2 E_2 + \cdots + V_n E_n}{\sqrt[n]{E_1 E_2 \ldots E_n}} \tag{9.4.2}$$

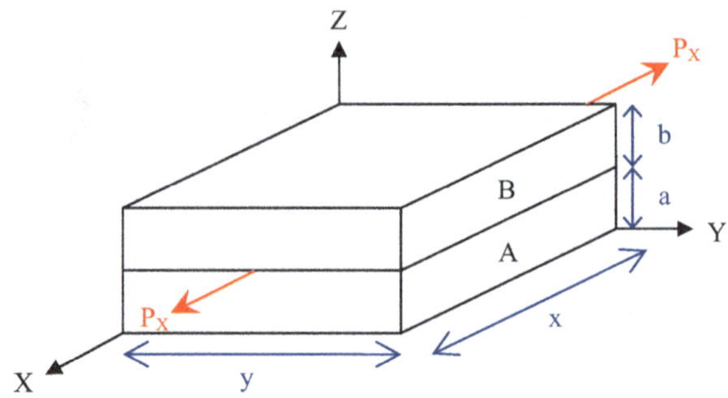

Fig. 9.7 Representative volume element for a laminate of alternating laminas

where $E_0 = \sqrt[n]{E_1 E_2 \ldots E_n}$. For a two-phase composite, the dimensionless modulus form reduces to

$$\frac{E_C}{E_0} = V_A \sqrt{\frac{E_A}{E_B}} + V_B \sqrt{\frac{E_B}{E_A}} \tag{9.4.3}$$

where

$$E_0 = \sqrt{E_A E_B} \tag{9.4.4}$$

in paving a way for comparison with a model that considers Poisson's ratios

$$\frac{E_C}{E_0} = V_A \sqrt{\frac{E_A}{E_B}} f_A(V_A, V_B, E_A, E_B, \upsilon_A, \upsilon_B) + V_B \sqrt{\frac{E_B}{E_A}} f_B(V_A, V_B, E_A, E_B, \upsilon_A, \upsilon_B) \tag{9.4.5}$$

With reference to Fig. 9.7, the application of Hooke's law gives the constitutive relations

$$\begin{Bmatrix} \varepsilon_{AX} \\ \varepsilon_{AY} \\ \varepsilon_{AZ} \end{Bmatrix} = \frac{1}{E_A} \begin{bmatrix} 1 & -\upsilon_A & -\upsilon_A \\ & 1 & -\upsilon_A \\ sym & & 1 \end{bmatrix} \begin{Bmatrix} \sigma_{AX} \\ \sigma_{AY} \\ \sigma_{AZ} \end{Bmatrix} \tag{9.4.6}$$

and

$$\begin{Bmatrix} \varepsilon_{BX} \\ \varepsilon_{BY} \\ \varepsilon_{BZ} \end{Bmatrix} = \frac{1}{E_B} \begin{bmatrix} 1 & -\upsilon_B & -\upsilon_B \\ & 1 & -\upsilon_B \\ sym & & 1 \end{bmatrix} \begin{Bmatrix} \sigma_{BX} \\ \sigma_{BY} \\ \sigma_{BZ} \end{Bmatrix} \tag{9.4.7}$$

for laminas *A* and *B*, respectively.

Prescribing an applied load P_X on the x-surface, we have

$$\sigma_{AX}(ay) + \sigma_{BX}(by) = P_X \tag{9.4.8}$$

and

$$\sigma_{AY}(ax) + \sigma_{BY}(bx) = 0 \tag{9.4.9}$$

based upon equilibrium in the x- and y-directions, respectively. Imposing equal in-plane normal strains as indicated by Eq. (9.3.10), plane stress condition

$$\sigma_{AZ} = \sigma_{BZ} = 0 \tag{9.4.10}$$

zero shear strains

$$\tau_{Aij} = \tau_{Bij} = 0 \quad (i, j = X, Y, Z) \tag{9.4.11}$$

and the equilibrium relations on the constitutive relations give

$$\begin{aligned} \sigma_{AX}\left(\frac{E_B}{E_A} + \frac{a}{b}\right) - \sigma_{AY}\left(\upsilon_A\frac{E_B}{E_A} + \upsilon_B\frac{a}{b}\right) &= \frac{P_X}{by} \\ \sigma_{AY}\left(\frac{E_B}{E_A} + \frac{a}{b}\right) - \sigma_{AX}\left(\upsilon_A\frac{E_B}{E_A} + \upsilon_B\frac{a}{b}\right) &= -\upsilon_B\frac{P_X}{by} \end{aligned} \tag{9.4.12}$$

The stresses σ_{AX} and σ_{AY} can be simultaneously solved from Eq. (9.4.12) to give

$$\begin{aligned} \sigma_{AX} &= \frac{\frac{P_X}{by}}{\left(\frac{E_B}{E_A}+\frac{a}{b}\right)^2 - \left(\upsilon_A\frac{E_B}{E_A}+\upsilon_B\frac{a}{b}\right)^2}\left[+\left(\frac{E_B}{E_A}+\frac{a}{b}\right) - \upsilon_B\left(\upsilon_A\frac{E_B}{E_A}+\upsilon_B\frac{a}{b}\right)\right] \\ \sigma_{AY} &= \frac{\frac{P_X}{by}}{\left(\frac{E_B}{E_A}+\frac{a}{b}\right)^2 - \left(\upsilon_A\frac{E_B}{E_A}+\upsilon_B\frac{a}{b}\right)^2}\left[-\upsilon_B\left(\frac{E_B}{E_A}+\frac{a}{b}\right) + \left(\upsilon_A\frac{E_B}{E_A}+\upsilon_B\frac{a}{b}\right)\right] \end{aligned} \tag{9.4.13}$$

Substituting the expression of σ_{AX} from Eq. (9.4.13) into Eq. (9.4.8),we have

$$\begin{aligned} \sigma_{BX} = & \frac{\frac{P_X}{by}}{\left(\frac{E_B}{E_A}+\frac{a}{b}\right)^2 - \left(\upsilon_A\frac{E_B}{E_A}+\upsilon_B\frac{a}{b}\right)^2} \\ & \left[\frac{E_B}{E_A}\left(\frac{E_B}{E_A}+\frac{a}{b}\right) - \upsilon_A\frac{E_B}{E_A}\left(\upsilon_A\frac{E_B}{E_A}+\upsilon_B\frac{a}{b}\right)\right] \end{aligned} \tag{9.4.14}$$

These three stresses would be sufficient for calculating the in-plane Young's modulus, defined as $E_X = \sigma_X/\varepsilon_X$ or

$$E_C = \frac{\frac{P_X}{(a+b)y}}{\varepsilon_{AX}} = \frac{\sigma_{AX}\left(\frac{a}{a+b}\right) + \sigma_{BX}\left(\frac{b}{a+b}\right)}{\frac{1}{E_A}[\sigma_{AX} - \upsilon_A \sigma_{AY}]} \tag{9.4.15}$$

Substituting Eqs. (9.4.13) and (9.4.14) into Eq. (9.4.15) leads to (Lim 2010)

$$\begin{aligned} \frac{E_C}{E_0} &= V_A \sqrt{\frac{E_A}{E_B}} \frac{aE_A\left(1-\upsilon_B^2\right) + bE_B(1-\upsilon_A\upsilon_B)}{aE_A\left(1-\upsilon_B^2\right) + bE_B\left(1-\upsilon_A^2\right)} \\ &+ V_B \sqrt{\frac{E_B}{E_A}} \frac{aE_A(1-\upsilon_A\upsilon_B) + bE_B\left(1-\upsilon_A^2\right)}{aE_A\left(1-\upsilon_B^2\right) + bE_B\left(1-\upsilon_A^2\right)} \end{aligned} \tag{9.4.16}$$

Comparing Eq. (9.4.16) with Eq. (9.4.5) implies

$$\begin{aligned} f_A &= \frac{V_A E_A\left(1-\upsilon_B^2\right) + V_B E_B(1-\upsilon_A\upsilon_B)}{V_A E_A\left(1-\upsilon_B^2\right) + V_B E_B\left(1-\upsilon_A^2\right)}, \\ f_B &= \frac{V_A E_A(1-\upsilon_A\upsilon_B) + V_B E_B\left(1-\upsilon_A^2\right)}{V_A E_A\left(1-\upsilon_B^2\right) + V_B E_B\left(1-\upsilon_A^2\right)} \end{aligned} \tag{9.4.17}$$

The following may then be inferred from Eq. (9.4.17)

(i) $f_A = f_B = 1$ when $\upsilon_A = \upsilon_B$, therefore $E_C = E_{\text{Voigt}}$
(ii) $f_A > 1$, $f_B > 1$ when $\upsilon_A\upsilon_B < 0$, therefore $E_C > E_{\text{Voigt}}$
(iii) $f_A > f_B = 1$ when $\upsilon_A \neq \upsilon_B = 0$, therefore $E_C > E_{\text{Voigt}}$
(iv) $f_B > f_A = 1$ when $\upsilon_B \neq \upsilon_A = 0$, therefore $E_C > E_{\text{Voigt}}$

Of the four special cases considered, the first case reduces the in-plane Young's modulus of the semi-auxetic laminate into the rule-of-mixture. The other three conditions ($\upsilon_A\upsilon_B < 0$, $\upsilon_A \neq \upsilon_B = 0$, and $\upsilon_B \neq \upsilon_A = 0$) give descriptions of Young's modulus surpassing that of the rule-of-mixture. The variation of the relative constituents' volume fraction, moduli, and Poisson's ratios are separately discussed.

To investigate the effect of Poisson's ratio product, we let $E_A = E_B = E_0$ and $\upsilon_A = -\upsilon_B$ in order to observe the change in the in-plane Young's modulus when both the constituent material possess equal Young's moduli. The dimensionless Young's moduli are hence reduced to

$$\frac{E_C}{E_0} = V_A f_A + V_B f_B \tag{9.4.18}$$

where

$$f_A = \frac{V_A\left(1-\upsilon_0^2\right) + V_B\left(1+\upsilon_0^2\right)}{1-\upsilon_0^2}, \quad f_B = \frac{V_A\left(1+\upsilon_0^2\right) + V_B\left(1-\upsilon_0^2\right)}{1-\upsilon_0^2} \tag{9.4.19}$$

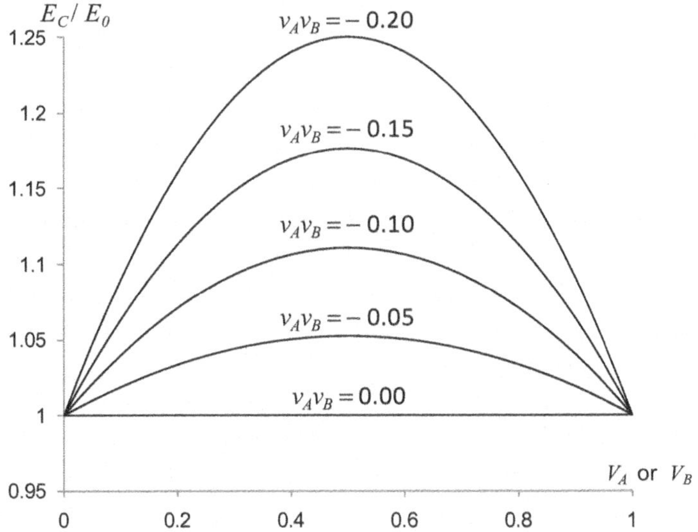

Fig. 9.8 Variation in the dimensionless in-plane Young's modulus with the volume fraction for increasing magnitude of Poisson's ratio product $\upsilon_A\upsilon_B$ whereby $E_A = E_B$ and $\upsilon_A = -\upsilon_B$

in which υ_0 is a non-negative number defined as

$$\upsilon_0 = |\upsilon_A| = |\upsilon_B| \tag{9.4.20}$$

Figure 9.8 shows the variation of E_C/E_0 with one of the constituent's volume fraction for incremental product of the constituents' Poisson's ratios $\upsilon_A\upsilon_B = -n/20$ for $n = 0, 1, 2, 3, 4$. It can be seen that the E_C/E_0 profile is symmetric due to equal magnitudes of the Poisson's ratios, and that the increase in E_C/E_0 is up to 25% at $V_A = V_B = 0.5$ for $\upsilon_A\upsilon_B = -0.2$. This significance is overlooked when the rule-of-mixture is used.

The effect of relative constituent moduli, i.e., E_A/E_B, on E_C/E_0 can be evaluated for constituents of opposing Poisson's ratio signs. The composite's in-plane Young's modulus remains as the form given in Eq. (9.4.5) with the functions in Eq. (9.4.17) reduced to

$$f_A = \frac{V_A E_A\left(1 - \upsilon_0^2\right) + V_B E_B\left(1 + \upsilon_0^2\right)}{E_{\text{Voigt}}\left(1 - \upsilon_0^2\right)}, \quad f_B = \frac{V_A E_A\left(1 + \upsilon_0^2\right) + V_B E_B\left(1 - \upsilon_0^2\right)}{E_{\text{Voigt}}\left(1 - \upsilon_0^2\right)} \tag{9.4.21}$$

Figure 9.9 shows the plots of E_C/E_0 curves at relative constituent moduli of $E_A/E_B = 2^{2n-1}$ with $n = -1, 0, 1, 2$, i.e., geometric increment of 2^2, with $\upsilon_A = -\upsilon_B$ and $\upsilon_A\upsilon_B = -0.1$. The straight lines that connect the two ends of each curve are E_{Voigt}/E_0.

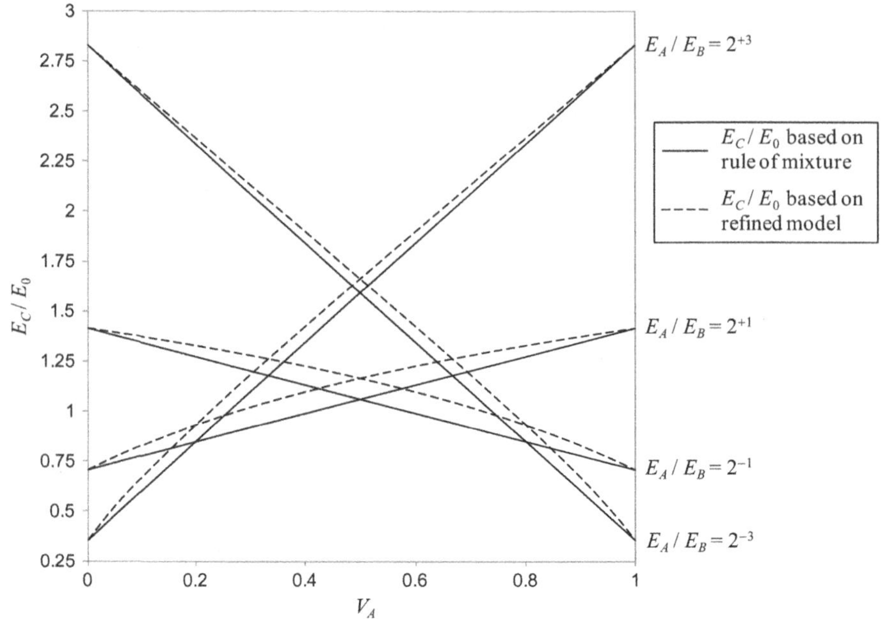

Fig. 9.9 Variation in the semi-auxetic's in-plane Young's modulus with volume fraction of material A for various E_A/E_B ratio at $v_A v_B = -0.1$ with $v_A = -v_B$

The E_C/E_0 profiles with varying v_A and v_B at $V_A = V_B = 0.5$ are considered next to observe the effect of constituents' relative Poisson's ratio. The constituents' Young's moduli are set equal for this illustration. For this category, the dimensionless in-plane Young's modulus simplifies to

$$\frac{E_C}{E_0} = \frac{1}{2}(f_A + f_B) \tag{9.4.22}$$

with

$$f_A = \frac{2 - v_A v_B - v_B^2}{2 - v_A^2 - v_B^2}, \quad f_B = \frac{2 - v_A v_B - v_A^2}{2 - v_A^2 - v_B^2} \tag{9.4.23}$$

Figure 9.10 shows the plot of E_C/E_0 for $-1/2 \leq v_i \leq 1/2$ with $i = A, B$ fixed at $E_A/E_B = V_A/V_B = 1$. The results range from $E_C/E_0 = 1$ at $v_A = v_B$ to $E_C/E_0 = 4/3$ at $v_A v_B = -1/4$, exhibiting symmetry about $v_A = v_B$ and $v_A = -v_B$.

Based on the plot of Fig. 9.10, it is evident that the magnitude of difference between the constituents' Poisson's ratios, $|v_A - v_B|$, contributes significantly toward the laminate's in-plane Young's modulus. The summary in Fig. 9.11 reiterates the effect of combined positive and negative Poisson's ratio laminas. The attainment of in-plane Young's modulus beyond the upper limit set by the rule-of-mixture is salient only

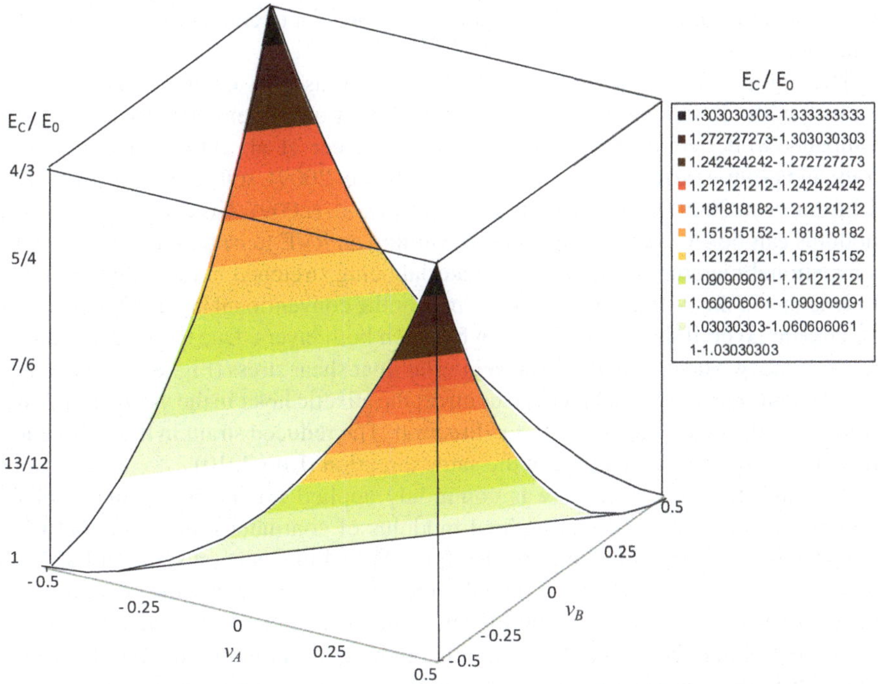

Fig. 9.10 Variation of the semi-auxetic's in-plane Young's modulus with the constituents' Poisson's ratios at equal volume fractions and $E_A = E_B$

Fig. 9.11 Map summarizing the extent of in-plane Young's modulus of a laminate that surpasses the rule-of-mixture theory

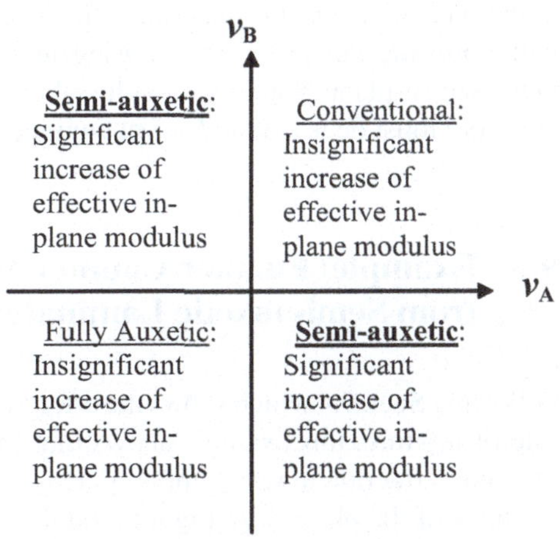

when both conventional and auxetic laminas are present to constitute a semi-auxetic laminate.

The higher than rule-of-mixture in-plane modulus of a laminate consisting of laminas with alternating Poisson's ratio sign based on present analytical approach compares well with finite element approach (Kocer et al. 2009), Cosserat elasticity and homogenization approach (Donescu et al. 2009), self-consistent approach (Chirima et al. 2009), and the formulation by Liu et al. (2009). The enhanced effective modulus can be elucidated as follows. Assuming an RVE to be defined by a layer of conventional material and an auxetic material being stretched in the x-direction, the dimensions in the y- and z-directions shrink for the conventional layer but expand for the auxetic layer, if there is no bonding between both layers. Due to perfect bonding between the two layers, there exists an interlaminar shear stress (Fig. 8.4.3) that tends to expand the conventional layer and contract the auxetic layer in the y-direction. This results in reduced elongation in the x-direction. The reduced strain in the x-direction translates to increased modulus in the same direction (Lim 2010).

Although the rule-of-mixture is commonly applied for predicting the in-plane modulus of laminates and longitudinal modulus of continuous unidirectional fiber composites, and hence known to be the upper bound in a general sense, it has been shown that the rule-of-mixture is actually the lower bound within the context of in-plane properties of semi-auxetic composite laminates. For a fixed volume fraction of constituent materials, the in-plane modulus reduces to a minimum when Poisson's ratios for both constituents are zero. Laminates with isotropic laminas possess in-plane properties that overshoot the modulus by rule-of-mixture when there is a difference in the laminas' Poisson's ratios. The characteristic of surpassing the rule-of-mixture formula is therefore generally existent, albeit not significant, in conventional and fully auxetic materials. However, for laminates that consist of laminas with opposing Poisson's ratio signs arranged in alternative sequence, the effects of increased in-plane Young's modulus above the rule-of-mixture are significant and therefore must be accounted for in engineering design.

9.5 Example: Further Counter-Intuitive Modulus from Semi-auxetic Laminates

Following Sect. 9.4 which shows the effective Young's modulus exceeding the simple rule-of-mixture, this example shows that for a composite whereby there is a range of volume fraction in which the in-plane composite modulus, E_C, is greater than the modulus of the phase with higher modulus, i.e.,

$$E_A < E_B < E_C \tag{9.5.1}$$

whereby phase B has greater modulus than phase A, under certain circumstances. This is one counter-intuitive property as it is normally taken for granted that the composite modulus falls between the moduli of the constituent phases, i.e., $E_A < E_C < E_B$. The second counter-intuitive property is that, under limited cases, the maximum point of the in-plane composite modulus occurs with the volume fraction of the stiffer phase being less than that of the other phase, i.e.,

$$\frac{\partial E_C}{\partial V_B} = 0; \quad (V_A > V_B, E_A < E_B) \tag{9.5.2}$$

This characteristic is counter-intuitive because one would intuitively expect that if a maximum point was to occur, it would take at $V_A < V_B$ if $E_A < E_B$.

Recall that for a multiphase composite, its effective modulus E_C on the basis of one-dimensional analysis falls within the range

$$\left(\sum_{i=1}^{n} \frac{V_i}{E_i}\right)^{-1} \le E_C \le \sum_{i=1}^{n} V_i E_i, \quad \sum_{i=1}^{n} V_i = 1 \tag{9.5.3}$$

where n is the total number of phases, while V_i and E_i are the volume fraction and Young's modulus, respectively, of the ith phase. This can be reduced for the case of two-phase composite

$$\left(\frac{V_A}{E_A} + \frac{V_B}{E_B}\right)^{-1} \le E_C \le V_A E_A + V_B E_B, V_A + V_B = 1 \tag{9.5.4}$$

whereupon the upper and lower limits are easily recognizable as the direct rule-of-mixture, or Voigt model (1889, 1910), and the inverse rule-of-mixture, or the Reuss model (1929), respectively. However, the analysis in the previous section gave the in-plane modulus of laminates as

$$E_C = V_A E_A f_{AB} + V_B E_B f_{BA} \tag{9.5.5}$$

where

$$f_{ij} = \frac{V_i E_i\left(1 - v_j^2\right) + V_j E_j\left(1 - v_i v_j\right)}{V_i E_i\left(1 - v_j^2\right) + V_j E_j\left(1 - v_i^2\right)} \tag{9.5.6}$$

for $i, j = A, B$. Since $f_{ij} = 1$ if and only if $v_i = v_j$, it follows that $E_C \ge V_A E_A + V_B E_B$. In the case of laminates with layers of alternating Poisson's ratio, the in-plane modulus always exceeds the rule-of-mixture. Due to the 3D nature of Eq. (9.5.5) while retaining its resemblance to the rule-of-mixture, this equation is adapted for obtaining a criterion by which the laminate's in-plane modulus exceeds the stiffness of the stiffer phase. To pave a way for this analysis, Eqs. (9.5.5) and (9.5.6) are rewritten as

$$E_C = \frac{E_A + V_B\left[E_B\left(\frac{1-\upsilon_A\upsilon_B}{1-\upsilon_B^2}\right) - E_A\right]}{1 + \frac{V_B}{1-V_B}\frac{E_B}{E_A}\left(\frac{1-\upsilon_A^2}{1-\upsilon_B^2}\right)} + \frac{E_A\left(\frac{1-\upsilon_A\upsilon_B}{1-\upsilon_A^2}\right) + V_B\left[E_B - E_A\left(\frac{1-\upsilon_A\upsilon_B}{1-\upsilon_A^2}\right)\right]}{1 + \frac{1-V_B}{V_B}\frac{E_A}{E_B}\left(\frac{1-\upsilon_B^2}{1-\upsilon_A^2}\right)} \tag{9.5.7}$$

so as to obtain the first derivative with respect to the volume fraction of the stiffer phase

$$\begin{aligned}\frac{\mathrm{d}E_C}{\mathrm{d}V_B} = {} & \frac{(1-V_B)\left[E_B\left(\frac{1-\upsilon_A\upsilon_B}{1-\upsilon_B^2}\right) - E_A\right]}{1 - V_B + V_B\frac{E_B}{E_A}\left(\frac{1-\upsilon_A^2}{1-\upsilon_B^2}\right)} \\ & - \frac{\left\{E_A + V_B\left[E_B\left(\frac{1-\upsilon_A\upsilon_B}{1-\upsilon_B^2}\right) - E_A\right]\right\}\frac{E_B}{E_A}\left(\frac{1-\upsilon_A^2}{1-\upsilon_B^2}\right)}{\left[1 - V_B + V_B\frac{E_B}{E_A}\left(\frac{1-\upsilon_A^2}{1-\upsilon_B^2}\right)\right]^2} \\ & + \frac{V_B\left[E_B - E_A\left(\frac{1-\upsilon_A\upsilon_B}{1-\upsilon_A^2}\right)\right]}{V_B + (1-V_B)\frac{E_A}{E_B}\left(\frac{1-\upsilon_B^2}{1-\upsilon_A^2}\right)} \\ & + \frac{\left\{E_A\left(\frac{1-\upsilon_A\upsilon_B}{1-\upsilon_A^2}\right) + V_B\left[E_B - E_A\left(\frac{1-\upsilon_A\upsilon_B}{1-\upsilon_A^2}\right)\right]\right\}\frac{E_A}{E_B}\left(\frac{1-\upsilon_B^2}{1-\upsilon_A^2}\right)}{\left[V_B + (1-V_B)\frac{E_A}{E_B}\left(\frac{1-\upsilon_B^2}{1-\upsilon_A^2}\right)\right]^2}\end{aligned} \tag{9.5.8}$$

Imposing the boundary condition

$$\frac{\mathrm{d}E_C}{\mathrm{d}V_B} = 0; \quad (V_B = 1) \tag{9.5.9}$$

on Eq. (9.5.8), we have the threshold moduli ratio

$$\left(\frac{E_B}{E_A}\right)^* = \frac{2(1-\upsilon_A\upsilon_B) - \left(1-\upsilon_B^2\right)}{1-\upsilon_A^2} \tag{9.5.10}$$

As such, there will be certain range of volume fraction in which $E_C > E_B$, whereby lamina B is the stiffer phase if $E_B/E_A < (E_B/E_A)^*$. Since we consider $E_B > E_A$, then there exists a maximum point when the following range condition

$$1 < \frac{E_B}{E_A} < \frac{2(1-\upsilon_A\upsilon_B) - \left(1-\upsilon_B^2\right)}{1-\upsilon_A^2} \tag{9.5.11}$$

is complied with (Lim and Acharya 2011).

To obtain the range of volume fraction whereby the in-plane laminate modulus is higher than that of the stiffer lamina, we substitute $E_C = E_B$ into Eq. (9.5.7) to give

$$V_B^2 - [1 + g(A, B)]V_B + g(A, B) = 0 \tag{9.5.12}$$

where

$$g(A, B) = \frac{E_A(E_A - E_B)(1 - v_B^2)}{(E_A - E_B)^2 - (E_A v_B - E_B v_A)^2} \tag{9.5.13}$$

and that the solutions to Eq. (9.5.12) are

$$V_B = \frac{1}{2}\left\{[1 + g(A, B)] \pm \sqrt{[1 - g(A, B)]^2}\right\} \tag{9.5.14}$$

Since the upper solution

$$V_B^{\text{upp}} = 1 \tag{9.5.15}$$

is trivial, the range for $E_C > E_B$ is defined by the lower solution

$$V_B^{\text{low}} = g(A, B) \tag{9.5.16}$$

i.e., (Lim and Acharya 2011)

$$\frac{\left(1 - \frac{E_B}{E_A}\right)(1 - v_B^2)}{\left(1 - \frac{E_B}{E_A}\right)^2 - \left(v_B - \frac{E_B}{E_A} v_A\right)^2} < V_B < 1 \tag{9.5.17}$$

for $E_C > E_B > E_A$.

To demonstrate the criterion for the in-plane laminate modulus that exceeds the modulus of the stiffer phase, we use a typical case whereby $|v_A| = |v_B| = 1/3$ such that $v_A = -v_B$. This gives the threshold moduli ratio $(E_B/E_A)^*$ described in Eq. (9.5.10) as 1.5. Figure 9.12 depicts a plot of dimensionless in-plane modulus, E_C/E_A, against the volume fraction of the stiffer phase, V_B, for moduli ratio E_B/E_A of 1.2, 1.5, and 1.8, based on Eq. (9.5.7). It can be seen that for $E_B/E_A < 1.5$, there exists a range of V_B where $E_C > E_B$ but not when $E_B/E_A > 1.5$, while a threshold point is observed when $E_B/E_A = 1.5$. Figure 9.12 (bottom right) combines the other three plots by replacing E_C/E_A with the normalized in-plane modulus $(E_C - E_A)/(E_B - E_A)$. Subsequent illustrations adopt this normalized modulus instead of the dimensionless modulus in order to collapse the end points (at $V_B = 1$) into a single point for better comparison.

Figure 9.13 shows a 3D surface plot of the threshold moduli ratio described in Eq. (9.5.10). This diagram shows a U-shaped surface that determines the existence

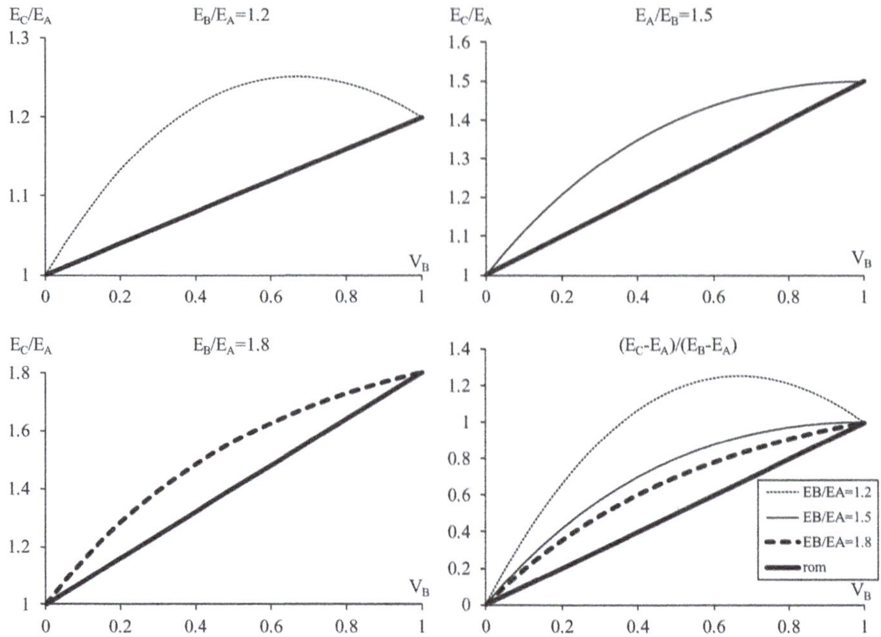

Fig. 9.12 Plots of dimensionless laminate in-plane modulus versus volume fraction of the stiffer phase with $|\upsilon_A| = |\upsilon_B| = 1/3$ and $\upsilon_A = -\upsilon_B$ for various E_B/E_A ratio

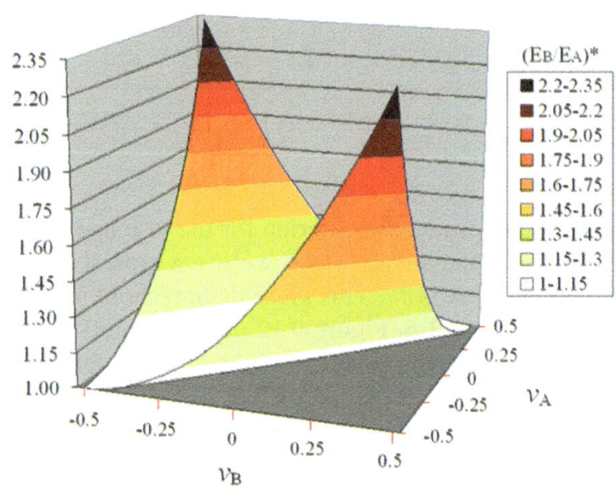

Fig. 9.13 Distribution of threshold moduli ratio $(E_B/E_A)^*$ for $-1/2 \le \upsilon_i \le 1/2$ where $i = A, B$

of a maximum point for the in-plane modulus within $0 < V_B < 1$. One can expect the in-plane modulus to exceed that of the stiffer phase for a certain combination of Poisson's ratio of both phases if the moduli ratio E_B/E_A is trapped between the plane E_B/E_A and the curved surface. Since the surface touches the plane at $\upsilon_A = \upsilon_B$, it follows that there are two separate regions (one on each side of the line $v_A = v_B$) by which the in-plane modulus can exceed the modulus of the stiffer phase. The nature of this surface shows that the in-plane modulus is more likely to exceed the modulus of the stiffer phase when the two phases possess Poisson's ratio of opposite signs.

Although the present model allows plotting of the in-plane modulus for maximum Poisson's ratio of 0.5, it does not cater to Poisson's ratio of exactly −1. As such, illustration is made for cases where Poisson's ratio of the laminas is being alternated at 0.5 and −0.99. Figure 9.14 (top) shows the plots of normalized in-plane modulus against the volume fraction of the stiffer phase at $v_A = -0.99$ and $v_B = 1/2$ for $E_B/E_A = 2$, 10, 50, and 250. The rule-of-mixture is incorporated for comparison. Since the threshold moduli ratio is $(E_B/E_A)^* = 112.6$, there exists a range of V_B where $E_C > E_B$ for $E_B/E_A = 2$, 10, and 50 but not for $E_B/E_A = 250$. When

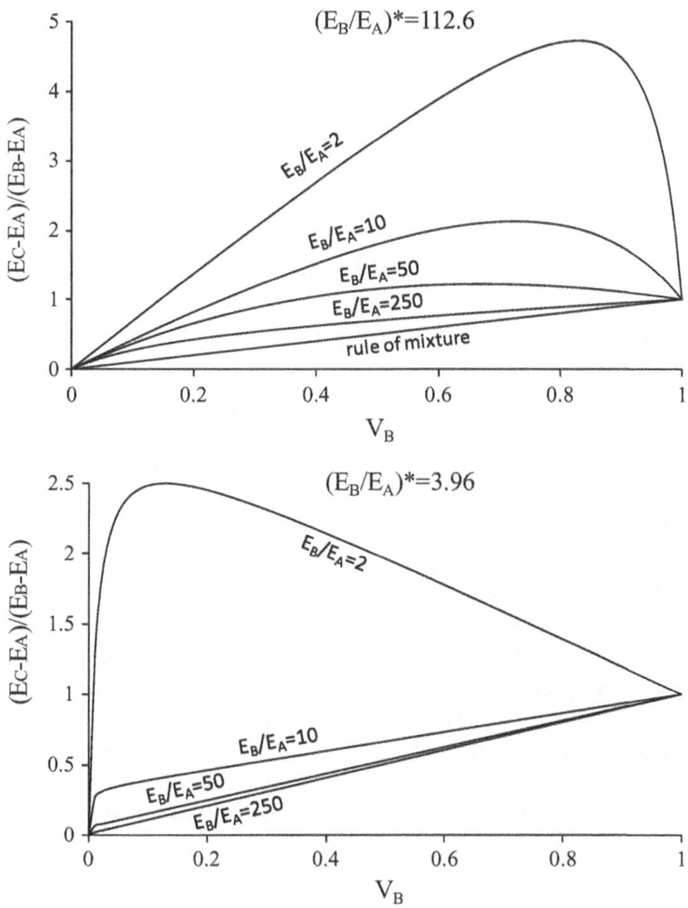

Fig. 9.14 Plots of normalized in-plane laminate modulus against the volume fraction of the stiffer phase for $v_A = -0.99$ and $v_B = 1/2$ (top), and $v_A = 1/2$ and $v_B = -0.99$ (bottom)

Poisson's ratios are switched to $\upsilon_A = 1/2$ and $\upsilon_B = -0.99$ as in Fig. 9.14 (bottom), only the $E_B/E_A = 2$ curve exhibits a range of V_B where $E_C > E_B$. This is not surprising since the corresponding threshold ratio is $(E_B/E_A)^* = 3.96$. What is counter-intuitive, however, is that the maximum point occurs at $V_B = 0.13$. In other words, the maximum modulus can occur even when the volume fraction of the more compliant phase is greater than that of the stiffer phase.

Perusal to Fig. 9.14 (top) again shows that as the moduli ratio increases from $E_B/E_A = 2$ to 10 and then to 50, the V_B of the maximum point drops from 0.83 to 0.73 and then to 0.67, respectively. Correspondingly, the drop in the moduli ratio widens the range in which $E_C > E_B$. Reference to Fig. 9.14 (top) reveals the range as $0.37 < V_B < 1$, $0.25 < V_B < 1$ and $0.15 < V_B < 1$ for $E_B/E_A = 50$, 10, and 2, respectively. In Fig. 9.14 (bottom), the range is extremely wide, i.e., $0.0067 < V_B < 1$ for $E_B/E_A = 2$.

The range of V_B at which the in-plane modulus exceeds the modulus of the stiffer phase can be visualized from Eq. (9.5.16), which is the lower solution for V_B for $E_C = E_B$, since it is known that the upper solution is $V_B = 1$. The curves of moduli ratio which gives rise to this property, $(E_B/E_A)^{**}$, against the corresponding volume fraction of the stiffer phase

$$V_B^{\text{low}} = \frac{\left(1 - \left(\frac{E_B}{E_A}\right)^{**}\right)\left(1 - \upsilon_B^2\right)}{\left(1 - \left(\frac{E_B}{E_A}\right)^{**}\right)^2 - \left(\upsilon_B - \left(\frac{E_B}{E_A}\right)^{**}\upsilon_A\right)^2} \tag{9.5.18}$$

are plotted in Fig. 9.15 for $\upsilon_A = -0.99$ with $\upsilon_B = 1/2$ (top) and $\upsilon_A = 1/2$ with $\upsilon_B = -0.99$ (bottom).

For the extreme case where $\upsilon_A = -0.99$ and $\upsilon_B = 1/2$, a range of V_B where $E_C > E_B$ begins to exist from the threshold point of $V_B = 1$ at $E_B/E_A = 112.6$, and this range widens with decreasing moduli ratio with diminishing rate until about $E_B/E_A \approx 20$ and thereafter with increasing rate until $E_B/E_A = 1$. As for the other extreme condition of $\upsilon_A = 1/2$ and $\upsilon_B = -0.99$, the range of V_B where $E_C > E_B$ begins to exist from the threshold point of $V_B = 1$ at $E_B/E_A = 3.96$, and this range widens rapidly with decreasing moduli ratio until $E_B/E_A = 3.5$ and thereafter very little change in the V_B range is seen until $E_B/E_A = 1$.

We may now conclude that there exist certain conditions that lead to further counter-intuitive properties whereby the in-plane laminate modulus exceeds the modulus of the stiffer phase. The conditions that enable this phenomenon to take place have been established together with the corresponding range of volume fraction. In addition, it has been shown for one of the extreme cases that the maximum point of the laminate modulus can occur even when the volume fraction of the stiffer phase is lower than the volume fraction of the more compliant phase (Lim and Acharya, 2011).

Fig. 9.15 Plots of moduli ratio versus the lower limits of the V_B range whereby $E_C = E_B$ for $\upsilon_A = -0.99$ and $\upsilon_B = 1/2$ (top), and $\upsilon_A = 1/2$ and $\upsilon_B = -0.99$ (bottom)

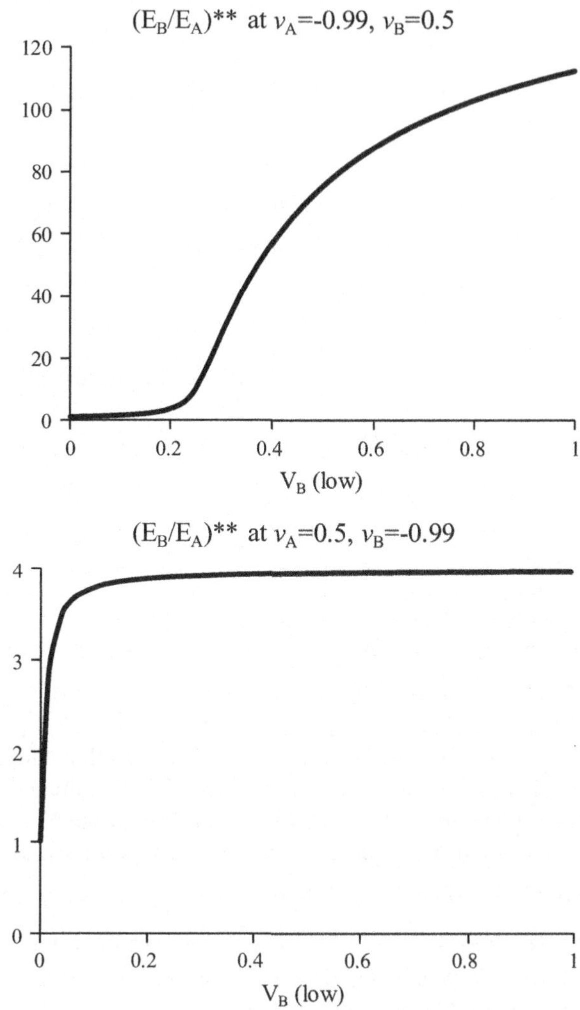

9.6 Example: Comparison Between In-Plane and Out-of-Plane Modulus of Semi-auxetic Laminates

Since most materials possess Poisson's ratio of between 0.3 and 1/3, the magnitude $|\upsilon_A| = |\upsilon_B| = \sqrt{0.1} \approx 0.316$ was selected in Sect. 9.3 such that the adoption of $\upsilon_A \upsilon_B = -0.1$ and $\upsilon_A^2 = \upsilon_B^2 = 0.1$ gave Eqs. (9.3.23) and (9.3.24). Using lamina modulus ratio of $E_B/E_A = 1.2$, 1.4, 1.6, and 1.8, it was shown in Fig. 9.5 that the maximum point of E_C/E_A at $(\upsilon_A, \upsilon_B) = \left(+\sqrt{0.1}, -\sqrt{0.1}\right)$ is higher than the maximum point of E_C/E_A at $(\upsilon_A, \upsilon_B) = \left(-\sqrt{0.1}, +\sqrt{0.1}\right)$, and that the maximum

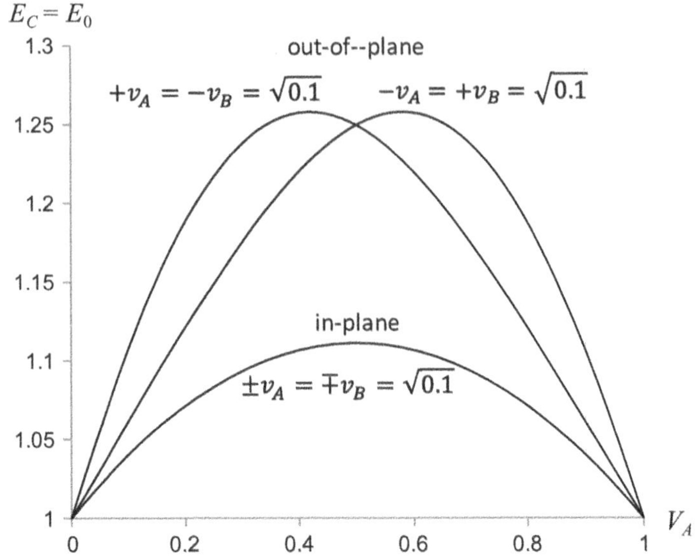

Fig. 9.16 Plots of dimensionless effective Young's modulus with respect to the moduli of materials A and B, E_C/E_0, versus the volume fraction of materials A, V_A, whereby $E_A = E_B = E_0$ with $+v_A = -v_B = \sqrt{0.1}$ and $-v_A = +v_B = \sqrt{0.1}$ for out-of-plane direction and with $\pm v_A = \mp v_B = \sqrt{0.1}$ for in-plane direction

point of E_C/E_A becomes significantly higher than the modulus of the stiffer material when the difference in modulus between both lamina decreases. What happens when both the laminas possess equal Young's modulus?

For such a unique case, the same maximum out-of-plane modulus of $E_C = 1.258E_0$ occurs at $V_A = 0.42$ for $(v_A, v_B) = \left(+\sqrt{0.1}, -\sqrt{0.1}\right)$ and at $V_A = 0.58$ for $(v_A, v_B) = \left(-\sqrt{0.1}, +\sqrt{0.1}\right)$, as shown in Fig. 9.16. A uniqueness exhibited in Fig. 9.16 is that the maximum out-of-plane modulus is higher than the maximum in-plane modulus of semi-auxetic laminate under the condition of equal constituent Young's modulus $E_A = E_B$ and equal Poisson's ratio magnitude $v_A = -v_B$. Normally, the out-of-plane modulus is lower than that in the in-plane direction. It can therefore be concluded that the out-of-plane modulus of a laminate is higher than that in the in-plane direction under the conditions of equal Young's modulus if the laminate is semi-auxetic with opposite Poisson's ratio signs in alternating laminas.

9.7 Example: Particulate Auxetic Composites

Zhu et al. (2015) established the design of single-level two-phase composite materials, whereby the designed single-level composite materials are assumed to be

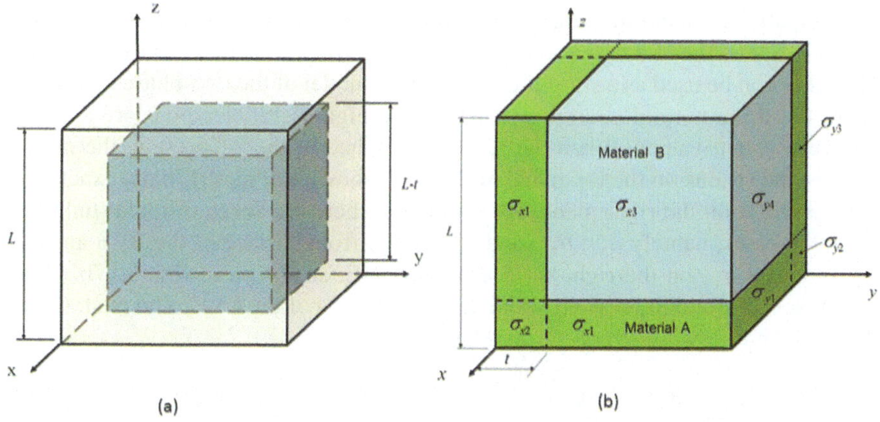

Fig. 9.17 Cubic periodic representative volume element (RVE) of the two-phase composite material (Zhu et al. 2015): **a** a cubic periodic unit RVE, and **b** cubic periodic mechanics model

composed of a large number of identical cubic periodic cells, as shown in Fig. 9.17a which is one representative volume element (RVE) of the composite. In the RVE, material A is a hollow cubic box which has square walls of uniform thickness $t/2$ and an external edge length *L*; material *B* is a solid cube which is inside the hollow cubic box of material *A* and has an edge length $L - t$. The interfaces of materials *A* and *B* are assumed to be perfectly bonded.

In the two-phase composite, the volume fraction of material *A* is

$$f_A = 1 - \frac{(L-t)^3}{L^3} \tag{9.7.1}$$

and the volume fraction of material *B* is thus

$$f_B = \frac{(L-t)^3}{L^3} \tag{9.7.2}$$

The designed composite material has a cubic symmetry and thus has only up to three independent elastic constants, namely E_{xx}, G_{xy}, and υ_{xy}. To obtain the effective Young's modulus E_{xx} and the Poisson's ratio υ_{xy} for the composite material, the cubic periodic RVE shown in Fig. 9.17b, Zhu et al. (2015) prescribed strain ε_x in the *x*-direction by an effective uniaxial tension. The periodic boundary conditions and the symmetry of the applied load require that all the six outside planes of the cubic periodic unit RVE in Fig. 9.17a remain plane after deformation.

To simplify the analysis, Zhu et al. (2015) divided the RVE into eight parallelepipeds, as can be seen in Fig. 9.17b. In order to carry out analytical solution, they considered only the normal stresses within each of the eight parallelepipeds in the RVE and the periodic conditions (i.e., compatibility conditions) on the outside surfaces of the RVE and ignore the shear stresses inside the parallelepipeds and the compatibility conditions on the interfaces between the parallelepipeds inside the RVE. Thus, the cubic periodic representative volume element (RVE) shown in Fig. 9.17b can be used as a simplified mechanics model of the two-phase composite, where the three normal stresses in each of the eight parallelepipeds are assumed to have constant values. When the RVE is stretched in the x-direction, the normal stresses and strains on the top plane of the RVE shown in Fig. 9.17b are exactly the same as those on the right plane. By symmetry, there are seven different unknown normal stresses, namely σ_{x1}, σ_{x2}, and σ_{x3} on the front surface of the RVE and σ_{y1}, σ_{y2}, σ_{y3}, and σ_{y4} on the right surface of the RVE, as shown in Fig. 9.17b. Based on Hooke's law and the periodic boundary conditions of the RVE, Zhu et al. (2015) obtained the following stress–strain relations

$$\frac{L-t}{L}\left(\frac{\sigma_{x1}-v_A\sigma_{y1}-v_A\sigma_{y4}}{E_A}\right)+\frac{t}{L}\left(\frac{\sigma_{x1}-v_A\sigma_{y2}-v_A\sigma_{y3}}{E_A}\right)=\varepsilon_x \tag{9.7.3}$$

$$\frac{L-t}{L}\left(\frac{\sigma_{x2}-2v_A\sigma_{y1}}{E_A}\right)+\frac{t}{L}\left(\frac{\sigma_{x2}-2v_A\sigma_{y2}}{E_A}\right)=\varepsilon_x \tag{9.7.4}$$

$$\frac{L-t}{L}\left(\frac{\sigma_{x3}-2v_B\sigma_{y4}}{E_B}\right)+\frac{t}{L}\left(\frac{\sigma_{x3}-2v_A\sigma_{y3}}{E_A}\right)=\varepsilon_x \tag{9.7.5}$$

$$\frac{L-t}{L}\left(\frac{\sigma_{y1}-v_A\sigma_{x1}-v_A\sigma_{y4}}{E_A}\right)+\frac{t}{L}\left(\frac{\sigma_{y1}-v_A\sigma_{x2}-v_A\sigma_{y1}}{E_A}\right)=\varepsilon_y \tag{9.7.6}$$

$$\frac{L-t}{L}\left(\frac{\sigma_{y2}-v_A\sigma_{x1}-v_A\sigma_{y3}}{E_A}\right)+\frac{t}{L}\left(\frac{\sigma_{y2}-v_A\sigma_{y2}-v_A\sigma_{x2}}{E_A}\right)=\varepsilon_y \tag{9.7.7}$$

$$\frac{L-t}{L}\left(\frac{\sigma_{y3}-v_A\sigma_{x3}-v_A\sigma_{y3}}{E_A}\right)+\frac{t}{L}\left(\frac{\sigma_{y3}-v_A\sigma_{x1}-v_A\sigma_{y2}}{E_A}\right)=\varepsilon_y \tag{9.7.8}$$

$$\frac{L-t}{L}\left(\frac{\sigma_{y4}-v_B\sigma_{y4}-v_B\sigma_{x3}}{E_B}\right)+\frac{t}{L}\left(\frac{\sigma_{y4}-v_A\sigma_{x1}-v_A\sigma_{y1}}{E_A}\right)=\varepsilon_y \tag{9.7.9}$$

In addition, the zero total force in the normal direction of the top or right plane of the RVE in Fig. 9.17b requires

$$(L-t)^2\sigma_{y4}+(L-t)t\sigma_{y1}+(L-t)t\sigma_{y3}+t^2\sigma_{y2}=0 \tag{9.7.10}$$

For a given value of the tensile strain ε_x, we have in total only eight unknowns to be determined: σ_{x1}, σ_{x2}, σ_{x3}, σ_{y1}, σ_{y2}, σ_{y3}, σ_{y4}, and ε_y. They can be solved from the eight simultaneous linear Eqs. (9.7.3)–(9.7.10). Thus, the effective Young's modulus and Poisson's ratio of the composite material can be obtained as

$$E_{xx}=\frac{t^2\sigma_{x2}+2(L-t)t\sigma_{x1}+(L-t)^2\sigma_{x3}}{L^2\varepsilon_x} \tag{9.7.11}$$

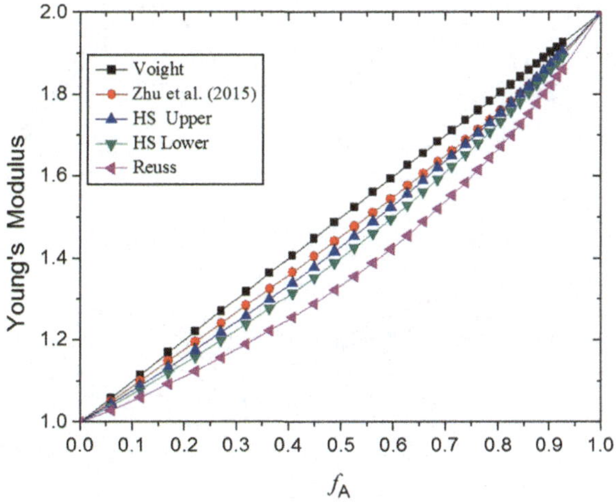

Fig. 9.18 Young's modulus of the two-phase composite with the cubic periodic RVE structure shown in Fig. 9.17 and with $E_A = 2E_B$ and $\upsilon_A = \upsilon_B$ versus the volume fraction of material *A* (Zhu et al. 2015), compared with the Voigt limit, the Reuss limit, and the Hashin–Shtrikman upper and lower limits. Young's moduli are normalized by E_B

and $\upsilon_{xy} = -\varepsilon_y/\varepsilon_x$. For single-level two-phase composite materials with the cubic periodic RVE structure shown in Fig. 9.17 and with $E_A = 2E_B$ and $\upsilon_A = \upsilon_B = 0.3$, the relationship between the effective Young's modulus E_{xx} and the volume fraction f_A is plotted in Fig. 9.18. The Voigt bound, the Reuss bound, and the Hashin–Shtrikman (1963) upper and lower bounds are also presented for comparison, with Young's moduli being normalized by E_B. As can be seen from Fig. 9.18, the effective Young's modulus of the composite material predicted from our mechanics model shown in Fig. 9.17b is larger than the Hashin–Shtrikman upper limit when the volume fraction f_A is smaller than 82%. As the possible effect of the Poisson's ratios of materials *A* and *B* is completely absent in Fig. 9.18, the enhancement of the effective Young's modulus (i.e., larger than the Hashin–Shtrikman upper limit) can be attributed to the geometrical structure. Zhu et al. (2015) have also tested cases of $\upsilon_A = \upsilon_B = 0$ and other values and found that as long as $\upsilon_A = \upsilon_B$, the results of the effective Young's modulus of the composite materials remain unchanged.

An exploration was then attempted to make Young's modulus of a single-level composite material larger than the Voigt limit. For a single-level two-phase composite material with the cubic periodic RVE structure shown in Fig. 9.17, the effects of different combinations of Young's moduli and Poisson's ratios of materials *A* and *B* on the relationship between the effective Young's modulus of the composite and the volume fraction f_A are illustrated in Fig. 9.19, where Young's modulus of the composite is normalized by the Voigt limit $(E_C)_{\text{upp}} = E_A f_A + E_B f_B$. As the Voigt limit normalized by itself is constantly 1.0, a value above 1.0 in Fig. 9.19 indicates that Young's modulus of the composite material is larger than the Voigt limit.

It can be observed from Fig. 9.19 that when $E_A = E_B$ (i.e., when the possible effects of the difference between E_A and E_B are absent), the difference between υ_A and υ_B can make the normalized Young's modulus of the composite material larger than 1.0 (i.e., exceeding the Voigt limit). Moreover, the larger the difference between υ_A and υ_B, the larger the Young's modulus of the two-phase composite

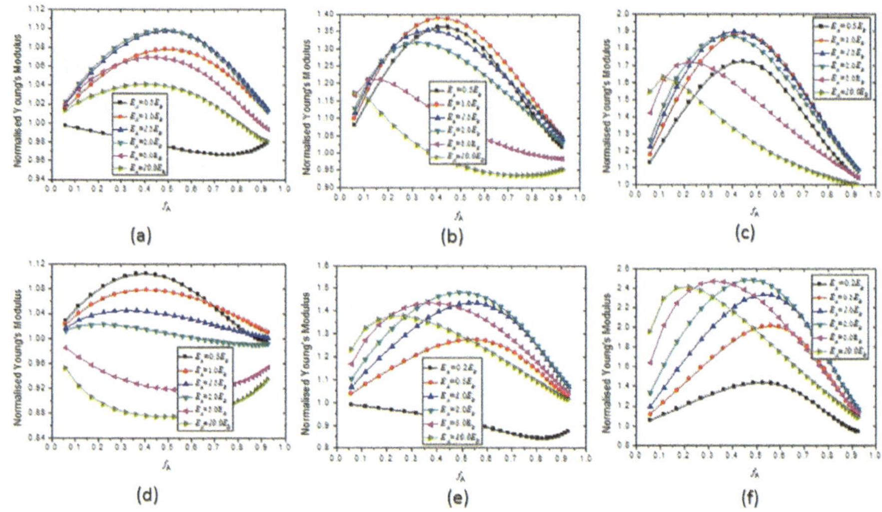

Fig. 9.19 Effects of the value of E_A/E_B on the relationship between the normalized Young's modulus of the composites and the volume fraction of material A (Zhu et al. 2015): **a** $\upsilon_A = 0.05$ and $\upsilon_B = 0.495$; **b** $\upsilon_A = 0.45$ and $\upsilon_B = -0.5$; **c** $\upsilon_A = 0.45$ and $\upsilon_B = -0.8$; **d** $\upsilon_A = 0.495$ and $\upsilon_B = 0.05$; **e** $\upsilon_A = -0.5$ and $\upsilon_B = 0.45$; **f** $\upsilon_A = -0.8$ and $\upsilon_B = 0.45$

material. Comparing Fig. 9.19b, c to e, f, it can be found that if υ_A is negative and υ_B is positive, the composite material has a larger Young's modulus than the case when υ_A is positive and υ_B is negative. In the case when $\upsilon_A = -0.8$ and $\upsilon_B = 0.45$, Young's modulus of the composite material is about 150% larger than the Voigt limit. Figure 9.20 shows that by properly choosing Young's moduli and Poisson's ratios of materials A and B, Poisson's ratio of a two-phase composite material can be designed to have a desired value, e.g., positive, or negative, or zero.

To validate the analytical results for the effective Young's moduli and Poisson's ratios of the two-phase composite materials obtained from Eqs. (9.7.3) to (9.7.10), Zhu et al. (2015) used the commercial finite element software ABAQUS to perform a number of simulations (i.e., to do numerical experiments) for the cubic periodic RVE structural model shown in Fig. 9.17a. The RVE was partitioned into 8000 C3D8 elements. Periodic boundary conditions were used in all the finite element simulations, and the obtained simulation results can be assumed to be the exact results. Zhu et al. (2015) presented in Table 9.1 the analytical results and the finite element simulation results for the two-phase composite materials with different combinations among the values of E_A, E_B, υ_A, υ_B, and f_A, where the effective Young's moduli of the composites are normalized by the Voigt limit $(E_C)_{\text{upp}}$.

Table 9.1 shows that the analytical results for Young's modulus of the single-level composite materials obtained from Eqs. (9.7.3) to (9.7.10) are always smaller than the simulation results, suggesting that the analytical results always tend to underestimate Young's modulus of the composite materials. Zhu et al. (2015) pointed out that this is consistent with the mechanics principle because any additional restraint

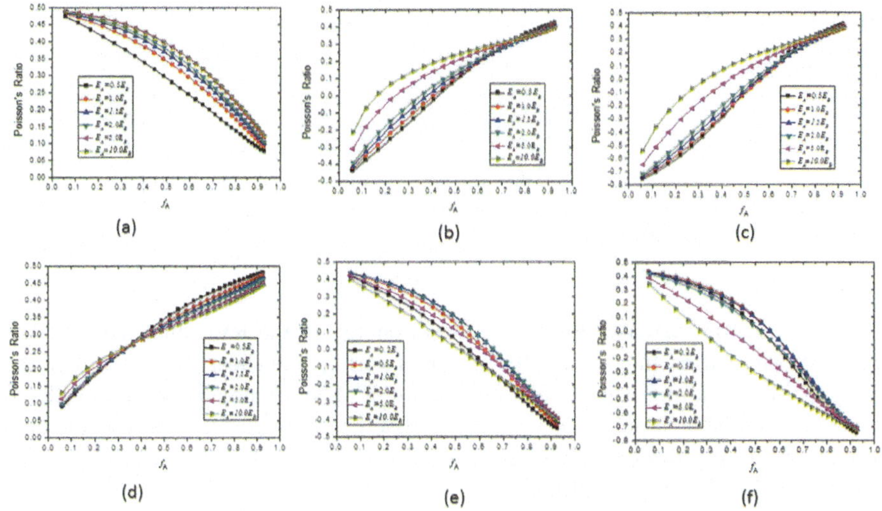

Fig. 9.20 Effects of the value of E_A/E_B on the relationship between Poisson's ratio of the composite and the volume fraction of material A (Zhu et al. 2015): **a** $\upsilon_A = 0.05$ and $\upsilon_B = 0.495$; **b** $\upsilon_A = 0.45$ and $\upsilon_B = -0.5$; **c** $\upsilon_A = 0.45$ and $\upsilon_B = -0.8$; (**d**) $\upsilon_A = 0.495$ and $\upsilon_B = 0.05$; (**e**) $\upsilon_A = -0.5$ and $\upsilon_B = 0.45$; (**f**) $\upsilon_A = -0.8$ and $\upsilon_B = 0.45$

Table 9.1 Comparison between the analytical results and finite element simulation results where $E_A = 2E_B$ by Zhu et al. (2015)

Volume fraction and constituents' Poisson's ratio			Analytical results		Simulation results	
f_A	υ_A	υ_B	$E_{xx}/(E_C)_{\text{upp}}$	υ_{xy}	$E_{xx}/(E_C)_{\text{upp}}$	υ_{xy}
0.271	0.05	0.495	1.0794	0.4453	1.0858	0.4550
0.271	0.45	−0.8	1.8128	−0.3981	1.8930	−0.3792
0.271	0.495	−0.99	2.9665	−0.8896	3.5342	−0.9175
0.488	0.05	0.495	1.0986	0.3828	1.1156	0.3920
0.488	0.45	−0.8	1.8216	−0.0680	1.9637	−0.0237
0.488	0.495	−0.9	2.5586	−0.2554	2.9617	−0.1282
0.488	0.495	−0.95	3.0174	−0.4095	3.9841	−0.2998
0.488	0.495	−0.99	3.6526	−0.4976	5.5141	−0.5817

always makes a material or structure stiffer. In the analysis of Eqs. (9.7.3)–(9.7.10), only normal stresses in the RVE and periodic conditions on the outside boundaries of the RVE are considered, while all the possible shear stresses and all the compatibility conditions inside the RVE are ignored. This could result in possible gaps or overlaps between the eight deformed parallelepipeds inside the RVE. To remove the gaps and overlaps (i.e., to make the interfaces between the eight deformed parallelepipeds inside the RVE perfectly bonded), additional work has to be done, and

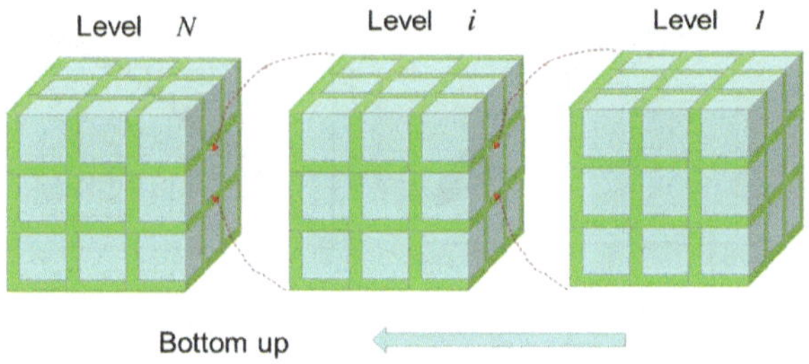

Fig. 9.21 Bottom-up structure of hierarchical composites (Zhu et al. 2015)

this consequently increases the stored strain energy in the RVE and hence makes the composite stiffer. In contrast, all the actual normal and shear stresses and all the compatibility conditions inside and outside the RVE have already been considered in the finite element simulations using the ABAQUS software. As the finite element simulations have considered much more restraints between the interfaces of the eight parallelepipeds than the simplified mechanics model shown in Fig. 9.17b, the exact results for the effective Young's modulus obtained from the finite element simulations are consequently always larger than the analytical results obtained from Eqs. (9.7.3)–(9.7.10). Table 9.1 shows that when $v_B \geq -0.8$, the difference between the effective Young's modulus of the composite materials obtained from Eqs. (9.7.3)–(9.7.10) and that obtained from the ABAQUS finite element simulation is constantly less than 8%, indicating that the analytical results shown in Figs. 9.19 and 9.20 are quite accurate and hence reliable. When $v_B \to -1$, although the error of the analytical results becomes larger, the predicted trend of the effects remains correct.

Zhu et al. (2015) also demonstrated how structure hierarchy could further enhance the elastic properties of a two-phase composite material. The two-phase hierarchical composite material is assumed to be made of isotropic materials A and B with Young's moduli E_A and E_B, Poisson's ratios v_A and v_B, and volume fraction f_B. At each hierarchical level n, the composite material is assumed to be composed of a large number of identical RVEs, as shown in Fig. 9.21, and each of the cubic fillers/inclusions (i.e., equivalent to material 'B' in Fig. 9.17) in the RVEs is also made of a large number of identical lower level (i.e., level $n-1$) cubic periodic RVEs. For simplicity, Zhu et al. (2015) assumed that the hierarchical composite material is self-similar in structure and the volume fraction of the cubic fillers/inclusions (i.e., material 'B') in the RVEs is assumed to remain fixed at all hierarchical levels,

$$f_B(n) = (f_B)^{1/N}, n = 1, 2, 3, \ldots, N \tag{9.7.12}$$

where n is the specific hierarchical level and N is the total number of the hierarchical levels.

For a given material volume fraction f_B and a given number of the total hierarchical levels N, the volume fraction of the cubic fillers/inclusions in the RVEs at each hierarchical level, $f_{B(n)}$, can be obtained from Eq. (9.7.12) and Young's modulus $E_{(n)}$ and Poisson's ratio $v_{(n)}$ at each hierarchical level can be obtained from Eqs. (9.7.3) to (9.7.11) and $v_{xy} = -\varepsilon_y/\varepsilon_x$. Figures 9.22 and 9.23 show the analytical results of

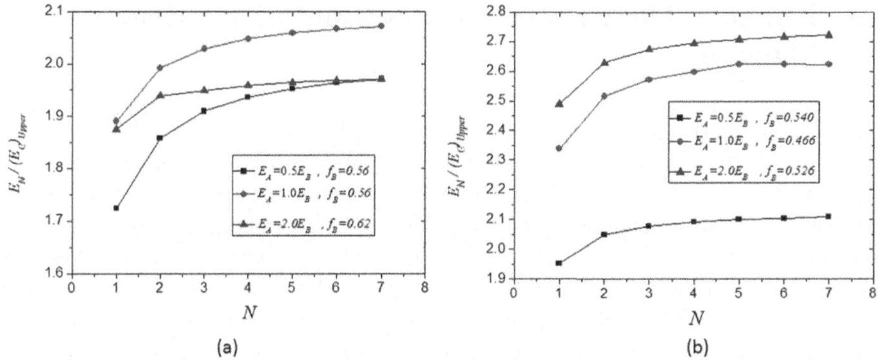

Fig. 9.22 Dimensionless Young's modulus of hierarchical composites as a function of the total number of the hierarchical levels (Zhu et al. 2015): **a** $\upsilon_A = 0.45$ and $\upsilon_B = -0.8$; **b** $\upsilon_A = -0.8$ and $\upsilon_B = 0.45$

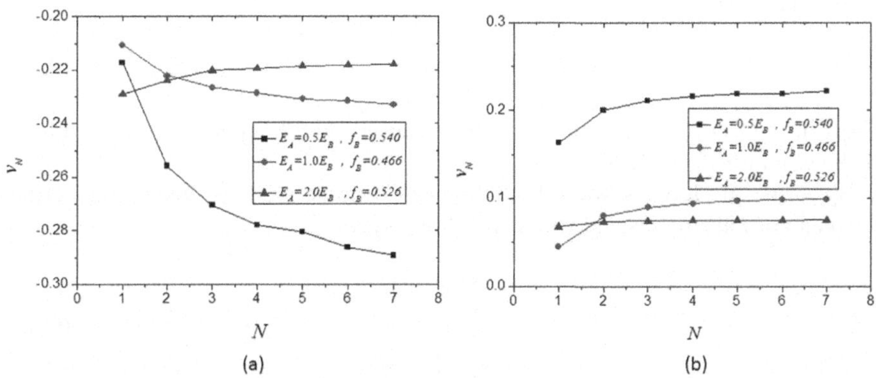

Fig. 9.23 Poisson's ratio of hierarchical composites as a function of the total number of the hierarchical levels (Zhu et al. 2015): **a** $\upsilon_A = 0.45$ and $\upsilon_B = -0.8$; **b** $\upsilon_A = -0.8$ and $\upsilon_B = 0.45$

Young's modulus E_N and Poisson's ratio υ_N for a few hierarchical and self-similar composite materials as functions of the number of the total hierarchical levels N, where Young's modulus is normalized by the Voigt limit $(E_C)_{upp} = E_A f_A + E_B f_B$. In Figs. 9.22 and 9.23, the results of the case $N = 1$ are those of the single-level composites, which can also be seen from Figs. 9.19c, f and 9.20c, f. The results in Fig. 9.22 indicate that increasing the number of hierarchical levels tends to enhance the stiffness of composite materials.

9.8 Example: Three-Layered Plates with Cubic Auxetic Materials

This section reports a study on three-layered cubic plates in which the cubic layers are sequenced as non-auxetic–auxetic–non-auxetic and auxetic–non-auxetic–auxetic. Cubic materials belong to a broad category of anisotropic solids such as body-centered cubic (BCC) and face-centered cubic (FCC) crystals. The constitutive relationship of cubic materials can be described by three material constants: Young's modulus E, shear modulus G, and Poisson's ratio ν. However, the usual relationship that relates these three material constants in the case of isotropic materials does not apply for cubic materials. The analyses consist of two parts: the stretching analysis and the bending analysis.

The stretching of a thin three-layered plate, lying on a plane defined by x- and y-axes, with layers of cubic crystals in the approximation of a plane-stressed state in the absence of forces on external horizontal surfaces is characterized by the following simplifying relations

$$\sigma_z^1 = \sigma_z^2 = \sigma_z^3 = 0 \tag{9.8.1}$$

where the subscript z is the out-of-plane direction and the superscripts 1,2,3 refer to the layer numbers, as defined in Fig. 9.24.

By virtue of Hooke's law for small elastic strains, we have linear connections between normal stresses and strains for three layers

$$\begin{aligned} \varepsilon_x^k &= s_{11}^k \sigma_x^k + s_{12}^k \sigma_y^k \\ \varepsilon_y^k &= s_{11}^k \sigma_y^k + s_{12}^k \sigma_x^k \\ \varepsilon_z^k &= s_{12}^k \sigma_x^k + s_{12}^k \sigma_y^k \end{aligned} \tag{9.8.2}$$

where $k = 1, 2, 3$ are the layer numbers. The conditions for the common in-plane strains $\varepsilon_x^1 = \varepsilon_x^2 = \varepsilon_x^3$ and $\varepsilon_y^1 = \varepsilon_y^2 = \varepsilon_y^3$ of three perfectly bonded layers for strains are rewritten in the following conditions for stresses

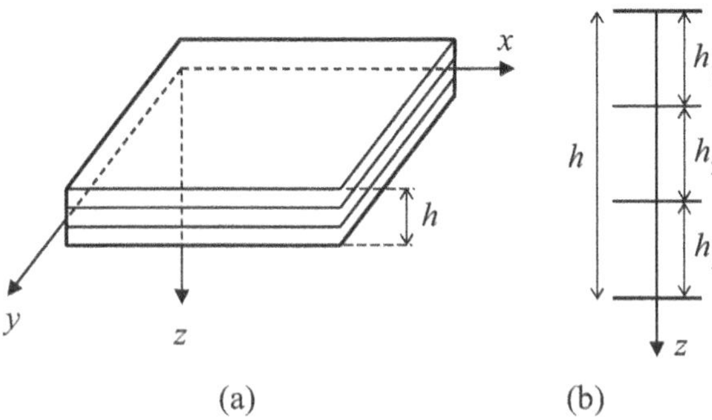

Fig. 9.24 Geometrical nomenclature of the three-layered plate

$$\begin{aligned} s_{11}^1\sigma_x^1 + s_{12}^1\sigma_y^1 &= s_{11}^2\sigma_x^2 + s_{12}^2\sigma_y^2 = s_{11}^3\sigma_x^3 + s_{12}^3\sigma_y^3 \\ s_{11}^1\sigma_y^1 + s_{12}^1\sigma_x^1 &= s_{11}^2\sigma_y^2 + s_{12}^2\sigma_x^2 = s_{11}^3\sigma_y^3 + s_{12}^3\sigma_x^3 \end{aligned} \tag{9.8.3}$$

The equilibrium conditions when the plate is stretched in the direction of the x-axis by force P when the transverse width of the plate is assumed to be equal to unity have the form

$$\begin{aligned} h_1\sigma_x^1 + h_2\sigma_x^2 + h_3\sigma_x^3 &= P \\ h_1\sigma_y^1 + h_2\sigma_y^2 + h_3\sigma_y^3 &= 0 \end{aligned} \tag{9.8.4}$$

where h_1, h_2, and h_3 are the thicknesses of layers $1, 2$, and 3, respectively, as indicated by Fig. 9.24b.

Equations (9.8.3) and (9.8.4) allow the following expressions for the stress components to be obtained

$$\begin{aligned} \sigma_x^1 &= \frac{P}{2}\left(\frac{1}{(s_{11}^1+s_{12}^1)S_+} + \frac{1}{(s_{11}^1-s_{12}^1)S_-}\right) \\ \sigma_x^2 &= \frac{P}{2}\left(\frac{1}{(s_{11}^2+s_{12}^2)S_+} + \frac{1}{(s_{11}^2-s_{12}^2)S_-}\right) \\ \sigma_x^3 &= \frac{P}{2}\left(\frac{1}{(s_{11}^3+s_{12}^3)S_+} + \frac{1}{(s_{11}^3-s_{12}^3)S_-}\right) \end{aligned} \tag{9.8.5a}$$

$$\begin{aligned} \sigma_y^1 &= \frac{P}{2}\left(\frac{1}{(s_{11}^1+s_{12}^1)S_+} - \frac{1}{(s_{11}^1-s_{12}^1)S_-}\right) \\ \sigma_y^2 &= \frac{P}{2}\left(\frac{1}{(s_{11}^2+s_{12}^2)S_+} - \frac{1}{(s_{11}^2-s_{12}^2)S_-}\right) \\ \sigma_y^3 &= \frac{P}{2}\left(\frac{1}{(s_{11}^3+s_{12}^3)S_+} - \frac{1}{(s_{11}^3-s_{12}^3)S_-}\right) \end{aligned} \tag{9.8.5b}$$

where

$$S_\pm = \sum_{k=1}^{3} \frac{h_k}{s_{11}^k \pm s_{12}^k} \tag{9.8.6}$$

We have then, in accordance with Hooke's law, the following expressions for the strains

$$\begin{aligned} \varepsilon_x^1 = \varepsilon_x^2 = \varepsilon_x^3 &= \frac{P}{2}\left(\frac{1}{S_+} + \frac{1}{S_-}\right) = \varepsilon_x \\ \varepsilon_y^1 = \varepsilon_y^2 = \varepsilon_y^3 &= \frac{P}{2}\left(\frac{1}{S_+} - \frac{1}{S_-}\right) = \varepsilon_y \end{aligned} \tag{9.8.7a}$$

$$\begin{aligned} \varepsilon_z^1 &= P\frac{s_{12}^1}{(s_{11}^1+s_{12}^1)S_+} \\ \varepsilon_z^2 &= P\frac{s_{12}^2}{(s_{11}^2+s_{12}^2)S_+} \\ \varepsilon_z^3 &= P\frac{s_{12}^3}{(s_{11}^3+s_{12}^3)S_+} \end{aligned} \tag{9.8.7b}$$

Using these expressions for strains, one can now obtain the results for effective Poisson's ratios and Young's modulus for a thin three-layered plate by means of constraints

$$\upsilon_{xy} = -\frac{\varepsilon_y}{\varepsilon_x} = \upsilon_{xy}^1 = \upsilon_{xy}^2 = \upsilon_{xy}^3 \tag{9.8.8a}$$

$$\begin{aligned} \upsilon_{xz}^1 &= -\tfrac{\varepsilon_z^1}{\varepsilon_x}, \quad \upsilon_{xz}^2 = -\tfrac{\varepsilon_z^2}{\varepsilon_x}, \quad \upsilon_{xz}^3 = -\tfrac{\varepsilon_z^3}{\varepsilon_x} \\ \upsilon_{xz} &= \tfrac{1}{h_1+h_2+h_3} \sum_{k=1}^{3} h_k \upsilon_{xz}^k \end{aligned} \tag{9.8.8b}$$

$$E_x = \frac{P}{(h_1 + h_2 + h_3)\varepsilon_x} \tag{9.8.8c}$$

to get

$$\upsilon_{xy} = \frac{S_+ - S_-}{S_+ + S_-} \tag{9.8.9a}$$

$$\begin{aligned} \upsilon_{xz}^1 &= -\tfrac{2s_{12}^1}{s_{11}^1+s_{12}^1}\left(\tfrac{S_-}{S_++S_-}\right) \\ \upsilon_{xz}^2 &= -\tfrac{2s_{12}^2}{s_{11}^2+s_{12}^2}\left(\tfrac{S_-}{S_++S_-}\right) \\ \upsilon_{xz}^3 &= -\tfrac{2s_{12}^3}{s_{11}^3+s_{12}^3}\left(\tfrac{S_-}{S_++S_-}\right) \\ \upsilon_{xz} &= -\tfrac{2}{h_1+h_2+h_3}\left(\tfrac{S_-}{S_++S_-}\right)\sum_{k=1}^{3}\tfrac{h_k s_{12}^k}{s_{11}^k+s_{12}^k} \end{aligned} \tag{9.8.9b}$$

$$E_x = \frac{2}{h_1 + h_2 + h_3}\left(\frac{S_+ S_-}{S_+ + S_-}\right) \tag{9.8.9c}$$

These results are invariant with respect to permutations of the layers. By this symmetry, the case of a three-layered plate with any pair of identical layers is reduced to the case of a two-layered plate. If all the layers have the same thickness, then the results for effective Poisson's ratios and Young's modulus are independent of the thickness of the layers (Gorodtsov et al. 2018)

$$\upsilon_{xy} = -\frac{\sum_k \frac{s_{12}^k}{\left(s_{11}^k\right)^2 - \left(s_{12}^k\right)^2}}{\sum_k \frac{s_{11}^k}{\left(s_{11}^k\right)^2 - \left(s_{12}^k\right)^2}} \tag{9.8.10a}$$

$$\begin{aligned}
\upsilon_{xz}^{1} &= -\frac{\frac{s_{12}^{1}}{s_{11}^{1}+s_{12}^{1}}\sum_k \frac{1}{s_{11}^{k}-s_{12}^{k}}}{\sum_k \frac{s_{11}^{k}}{\left(s_{11}^{k}\right)^2-\left(s_{12}^{k}\right)^2}} \\
\upsilon_{xz}^{2} &= -\frac{\frac{s_{12}^{2}}{s_{11}^{2}+s_{12}^{2}}\sum_k \frac{1}{s_{11}^{k}-s_{12}^{k}}}{\sum_k \frac{s_{11}^{k}}{\left(s_{11}^{k}\right)^2-\left(s_{12}^{k}\right)^2}} \\
\upsilon_{xz}^{3} &= -\frac{\frac{s_{12}^{3}}{s_{11}^{3}+s_{12}^{3}}\sum_k \frac{1}{s_{11}^{k}-s_{12}^{k}}}{\sum_k \frac{s_{11}^{k}}{\left(s_{11}^{k}\right)^2-\left(s_{12}^{k}\right)^2}} \\
\upsilon_{xz} &= -\frac{1}{3}\frac{\sum_k \frac{1}{s_{11}^{k}-s_{12}^{k}}\sum_k \frac{s_{12}^{k}}{s_{11}^{k}+s_{12}^{k}}}{\sum_k \frac{s_{11}^{k}}{\left(s_{11}^{k}\right)^2-\left(s_{12}^{k}\right)^2}}
\end{aligned} \tag{9.8.10b}$$

$$E_x = \frac{1}{3}\frac{\sum_k \frac{1}{s_{11}^{k}+s_{12}^{k}}\sum_k \frac{1}{s_{11}^{k}-s_{12}^{k}}}{\sum_k \frac{s_{11}^{k}}{\left(s_{11}^{k}\right)^2-\left(s_{12}^{k}\right)^2}} \tag{9.8.10c}$$

All these relations for the three-layered plates remain valid also in the case of plates with more layers if we assume that the index k ranges from 1 to n. The uniform longitudinal stretching of the layers when the crystallographic axes coincide with the edges of the rectangular plate makes it possible to use the following substitutions of the compliance coefficients s_{11}^{k}, s_{12}^{k} through original Young's modulus E_k and Poisson's ratios υ_k for the individual layers

$$s_{11}^{k} = \frac{1}{E_k}, s_{12}^{k} = -\frac{\upsilon_k}{E_k} \tag{9.8.11}$$

Cubic crystals with negative Poisson's ratio when oriented in a crystallographic direction coinciding with the direction of the rectangular plate edge will be termed below as in-plane auxetics. The term in-plane non-auxetic will be used in the opposite situation.

So far, the definition of Poisson's ratio has been defined from the stretching of a rod. One can also obtain Poisson's ratio of a plate by pure bending, specifically by applying pure bending moments on two opposing sides of a rectangular plate. Recall that for an isotropic plate made from a single layer, the bending moments are as follows:

$$\begin{aligned} M_x &= -D(\kappa_x + \upsilon\kappa_y) \\ M_y &= -D(\kappa_y + \upsilon\kappa_x) \end{aligned} \tag{9.8.12}$$

where curvatures along the x- and y-axes are defined as $\kappa_x = \partial^2 w/\partial x^2$ and $\kappa_y = \partial^2 w/\partial y^2$ in which w is the plate deflection, while D is the plate's flexural rigidity. For application of M_x only (i.e., $M_y = 0$), effective in-plane Poisson's ratio can be obtained from the second of Eq. (9.8.12) as

$$\upsilon_{xy} = -\frac{\kappa_y}{\kappa_x} \tag{9.8.13}$$

It can be seen that Eq. (9.8.13) is a bending mode analogy to the usual Poisson's ratio definition $\upsilon_{xy} = -\varepsilon_y/\varepsilon_x$ under stretching mode. For a plate made of a single material, Poisson's ratios are the same under both the stretching and bending modes, but it will later be shown that for the three-layered plate, the effective in-plane Poisson's ratios are different. The effective in-plane Poisson's ratio for the three-layered plate under the bending mode can be obtained by using the Classical Lamination Theory (CLT), in which the in-plane normal and shearing loads (N_x, N_y, N_{xy}) and bending and twisting loads (M_x, M_y, M_{xy}) are related to the mid-plane strains $(\varepsilon_x^0, \varepsilon_y^0, \gamma_{xy}^0)$ and curvatures $(\kappa_x, \kappa_y, \kappa_{xy})$ as

$$\begin{Bmatrix} N_x \\ N_y \\ N_{xy} \\ M_x \\ M_y \\ M_{xy} \end{Bmatrix} = \begin{bmatrix} A_{11} & A_{12} & A_{16} & B_{11} & B_{12} & B_{16} \\ & A_{22} & A_{26} & B_{12} & B_{22} & B_{26} \\ & & A_{66} & B_{16} & B_{26} & B_{66} \\ & & & D_{11} & D_{12} & D_{16} \\ & \text{sym} & & & D_{22} & D_{26} \\ & & & & & D_{66} \end{bmatrix} \begin{Bmatrix} \varepsilon_x^0 \\ \varepsilon_y^0 \\ \gamma_{xy}^0 \\ \kappa_x \\ \kappa_y \\ \kappa_{xy} \end{Bmatrix} \tag{9.8.14}$$

where the extensional $[A]$, coupling $[B]$, and bending $[D]$ stiffness matrices are defined as

$$\begin{aligned} [A_{ij}] &= \sum_{k=1}^{n} [\bar{Q}_{ij}]_k (z_k - z_{k-1}) \\ [B_{ij}] &= \tfrac{1}{2} \sum_{k=1}^{n} [\bar{Q}_{ij}]_k (z_k^2 - z_{k-1}^2) \\ [D_{ij}] &= \tfrac{1}{3} \sum_{k=1}^{n} [\bar{Q}_{ij}]_k (z_k^3 - z_{k-1}^3) \end{aligned} \tag{9.8.15}$$

with $i, j = 1, 2, 6$ and, in the case of three-layered plate $n = 3$, the elements of the transformed reduced stiffness matrix $[\bar{Q}]$ for each layer are related to the reduced stiffness matrix coefficients $[Q]$ as

$$\begin{aligned} \bar{Q}_{11} &= Q_{11}\cos^4\theta + Q_{22}\sin^4\theta + 2(Q_{12} + 2Q_{66})\cos^2\theta\sin^2\theta \\ \bar{Q}_{22} &= Q_{11}\sin^4\theta + Q_{22}\cos^4\theta + 2(Q_{12} + 2Q_{66})\cos^2\theta\sin^2\theta \\ \bar{Q}_{12} &= (Q_{11} + Q_{22} - 4Q_{66})\cos^2\theta\sin^2\theta + Q_{12}(\cos^4\theta + \sin^4\theta) \\ \bar{Q}_{16} &= (Q_{11} - Q_{12} - 2Q_{66})\cos^3\theta\sin\theta - (Q_{22} - Q_{12} - 2Q_{66})\cos\theta\sin^3\theta \\ \bar{Q}_{26} &= (Q_{11} - Q_{12} - 2Q_{66})\cos\theta\sin^3\theta - (Q_{22} - Q_{12} - 2Q_{66})\cos^3\theta\sin\theta \\ \bar{Q}_{66} &= (Q_{11} + Q_{22} - 2Q_{12})\cos^2\theta\sin^2\theta + Q_{66}(\cos^2\theta - \sin^2\theta)^2 \end{aligned} \tag{9.8.16}$$

where

$$Q_{11} = Q_{22} = \frac{E}{1 - v^2}, \; Q_{12} = \frac{vE}{1 - v^2}, \quad Q_{66} = G \tag{9.8.17}$$

Simplifications can be obtained if (a) the principal axes of the cubic material are rotated such that they are aligned to the sides of the plates, i.e., $\theta = 0$, so that $[\bar{Q}] = [Q]$, and (b) the plates are layered symmetrically with the first and third layers being $Q_{ij}^1 = Q_{ij}^3$ and $h_1 = h_3$ so that $[B] = 0$. Let $h_2 = xh_1 = xh_3$, then

$$\begin{aligned} D_{11} = D_{22} &= \frac{h_1^3}{3}\left[\left(1 + \frac{3}{2}x + \frac{3}{4}x^2\right)\frac{E_1}{1 - v_1^2} + \frac{x^3}{4}\frac{E_2}{1 - v_2^2} \right. \\ &\left. + \left(1 + \frac{3}{2}x + \frac{3}{4}x^2\right)\frac{E_3}{1 - v_3^2}\right] \\ D_{12} &= \frac{h_1^3}{3}\left[\left(1 + \frac{3}{2}x + \frac{3}{4}x^2\right)\frac{v_1 E_1}{1 - v_1^2} + \frac{x^3}{4}\frac{v_2 E_2}{1 - v_2^2} + \left(1 + \frac{3}{2}x + \frac{3}{4}x^2\right)\frac{v_3 E_3}{1 - v_3^2}\right] \end{aligned} \tag{9.8.18}$$

In the absence of prescribed strain and twisting, the bending moments are related to the curvatures as

$$\begin{aligned} M_x &= D_{11}\kappa_x + D_{12}\kappa_y \\ M_y &= D_{12}\kappa_x + D_{22}\kappa_y \end{aligned} \tag{9.8.19}$$

For our present problem ($M_y = 0$, $D_{22} = D_{11}$), we obtain information regarding the curvatures

$$\begin{aligned} \kappa_x &= \frac{D_{11}}{D_{11}^2 - D_{12}^2} M_x \\ \kappa_y &= \frac{D_{12}}{D_{12}^2 - D_{11}^2} M_x \end{aligned} \tag{9.8.20}$$

by solving Eqs. (9.8.19) simultaneously. Substituting Eqs. (9.8.20) into Eq. (9.8.13), we therefore obtain effective in-plane Poisson's ratio by bending mode, through prescription of M_x, as

$$v_{\text{eff}}^B = \frac{D_{12}}{D_{11}} = \frac{\left(1 + \frac{3}{2}x + \frac{3}{4}x^2\right)\frac{v_1 E_1}{1 - v_1^2} + \frac{x^3}{4}\frac{v_2 E_2}{1 - v_2^2} + \left(1 + \frac{3}{2}x + \frac{3}{4}x^2\right)\frac{v_3 E_3}{1 - v_3^2}}{\left(1 + \frac{3}{2}x + \frac{3}{4}x^2\right)\frac{E_1}{1 - v_1^2} + \frac{x^3}{4}\frac{E_2}{1 - v_2^2} + \left(1 + \frac{3}{2}x + \frac{3}{4}x^2\right)\frac{E_3}{1 - v_3^2}} \tag{9.8.21}$$

Equation (9.8.21) can also be obtained directly, without obtaining the curvatures κ_x and κ_y in Eq. (9.8.20); substituting $M_y = 0$ and $D_{22} = D_{11}$ into the second of Eq. (9.8.19) gives $D_{12}\kappa_x = -D_{11}\kappa_y$, by which the Poisson's ratio by bending mode can be inferred from Eq. (9.8.13). In the case of effective Poisson's ratio of the same three-layered plate via stretching mode, Eqs. (9.8.9a) can be expressed as

$$\upsilon_{\text{eff}}^{S} = \frac{\frac{\upsilon_1 E_1}{1-\upsilon_1^2} + x\frac{\upsilon_2 E_2}{1-\upsilon_2^2} + \frac{\upsilon_3 E_3}{1-\upsilon_3^2}}{\frac{E_1}{1-\upsilon_1^2} + x\frac{E_2}{1-\upsilon_2^2} + \frac{E_3}{1-\upsilon_3^2}} \tag{9.8.22}$$

for comparison with Eq. (9.8.21). If the layers are of equal thickness $x = 1$, then

$$\begin{aligned} D_{11} = D_{22} &= \frac{h_1^3}{3}\left[3.25\frac{E_1}{1-\upsilon_1^2} + 0.25\frac{E_2}{1-\upsilon_2^2} + 3.25\frac{E_3}{1-\upsilon_3^2}\right] \\ D_{12} &= \frac{h_1^3}{3}\left[3.25\frac{\upsilon_1 E_1}{1-\upsilon_1^2} + 0.25\frac{\upsilon_2 E_2}{1-\upsilon_2^2} + 3.25\frac{\upsilon_3 E_3}{1-\upsilon_3^2}\right] \end{aligned} \tag{9.8.23}$$

which leads to effective in-plane Poisson's ratio under bending mode (Gorodtsov et al. 2018)

$$\upsilon_{\text{eff}}^{B} = \frac{13\frac{\upsilon_1 E_1}{1-\upsilon_1^2} + \frac{\upsilon_2 E_2}{1-\upsilon_2^2} + 13\frac{\upsilon_3 E_3}{1-\upsilon_3^2}}{13\frac{E_1}{1-\upsilon_1^2} + \frac{E_2}{1-\upsilon_2^2} + 13\frac{E_3}{1-\upsilon_3^2}} \tag{9.8.24}$$

while from Eq. (9.8.10a), or more conveniently from Eq. (9.8.22), we obtain effective in-plane Poisson's ratio under stretching mode

$$\upsilon_{\text{eff}}^{S} = \frac{\frac{\upsilon_1 E_1}{1-\upsilon_1^2} + \frac{\upsilon_2 E_2}{1-\upsilon_2^2} + \frac{\upsilon_3 E_3}{1-\upsilon_3^2}}{\frac{E_1}{1-\upsilon_1^2} + \frac{E_2}{1-\upsilon_2^2} + \frac{E_3}{1-\upsilon_3^2}} \tag{9.8.25}$$

for comparison with Eq. (9.8.24). Regardless of whether the layers are of equal or unequal thicknesses, bending-based Poisson's ratio for the symmetrically arranged three-layered plates gives more influence to layers 1 and 3, and less influence to layer 2.

As mentioned earlier, the usual relationship that connects Young's modulus E, the shear modulus G, and Poisson's ratio v for isotropic materials is not valid for cubic materials. Specifically, its elastic compliances are related as $s_{ii} = 1/E$, $s_{ij} = -\upsilon/E$, and $s_{kk} = 1/G$ where $i, j = 1, 2, 3$ and $k = 4, 5, 6$ with the remaining elastic compliances being zero. As such, a measure of the cubic material's anisotropy is

$$A = 2(1+\upsilon)\frac{G}{E} = 2\frac{s_{ii} - s_{ij}}{s_{kk}} \tag{9.8.26}$$

where a value of $A = 1$ indicates isotropy. Coefficient A is called Zener's anisotropy factor.

Some results on the effective elastic properties of the three-layered plates made from different cubic materials are listed in Table 9.2. The non-auxetic–auxetic–non-auxetic plates are sub-categorized into the (i) NiTi–$Sm_{0.75}Y_{0.25}S$–non-auxetic plates, the (ii) Al–$Sm_{0.75}Y_{0.25}S$–non-auxetic plates, and the (iii) LiBr–$Sm_{0.75}Y_{0.25}S$–non-auxetic plates where the third layers of cubic non-auxetic materials are KI, LiBr, Al, Ge, and Mo, W, Ir in the first case and KI, NaBr, Mg_2Pb, Ge and Mo, W, Ir in the second and third cases. All the layers are equal thicknesses $h_1 = h_2 = h_3$, and the

Table 9.2 Effective elastic properties results for three-layered non-auxetic–auxetic–non-auxetic plates category consisting cubic materials of equal thicknesses

Non-auxetics	E_1, GPa	E_2, GPa	E_3, GPa	E_{eff}, GPa	υ_1	υ_2	υ_3	υ_{xy}	υ_{xz}
NiTi-$Sm_{0.75}Y_{0.25}S$-non-auxetics									
KI	47.8	55.6	26.2	60.0	0.44	−0.67	0.14	−0.20	0.22
LiBr			27.2	62.1			0.33	−0.17	0.34
Al			62.5	77.3			0.36	−0.07	0.34
Ge			103	90.5			0.27	−0.04	0.27
Mo			380	187			0.26	0.11	0.22
W			408	198			0.28	0.14	0.23
Ir			446	211			0.30	0.16	0.23
Al–$Sm_{0.75}Y_{0.25}S$–non-auxetics									
KI	62.5	55.6	26.2	64.2	0.36	−0.67	0.14	- 0.19	0.13
NaBr			35.6	68.3			0.21	- 0.16	0.17
Mg_2Pb			61.3	78.4			0.23	- 0.11	0.17
Ge			103	94.6			0.27	- 0.04	0.19
Mo			380	191			0.26	0.11	0.15
W			408	202			0.28	0.13	0.16
Ir			446	216			0.30	0.16	0.17
LiBr–$Sm_{0.75}Y_{0.25}S$–non-auxetics									
KI	27.2	55.6	26.2	46.5	0.33	−0.67	0.14	−0.34	0.11
NaBr			61.3	51.3			0.23	−0.30	0.15
NiTi			47.8	62.1			0.44	−0.17	0.34
Mg_2Pb			61.3	62.4			0.23	−0.22	0.16
Ge			103	79.8			0.27	−0.12	0.17
Mo			380	178			0.26	0.09	0.13
W			408	189			0.28	0.12	0.14
Ir			446	203			0.30	0.14	0.15

properties of the layers are denoted by subscripts 1, 2, and 3 for the first, middle, and last layers, respectively. In-plane Poisson's ratio υ_{xy}, out-of-plane Poisson's ratio υ_{xz}, and Young's modulus E_{eff} were calculated using Eq. (9.8.10). There are no cases when a three-layered plate exhibited auxeticity in the out-of-plane direction, i.e., $\upsilon_{xz} > 0$. Calculated in-plane Poisson's ratio υ_{xy} reveals auxeticity in the case of relatively soft non-auxetic layers. As Young's modulus of the third layer increases, in-plane Poisson's ratio υ_{xy} increases. This coefficient becomes positive for such high stiffness non-auxetics of the third layer as Mo, W, and Ir. Some plates (with relatively soft original layers) show that effective Young's modulus is higher than Young's moduli of all the layers. An example of such a situation, together with the dependence of effective Young's modulus of the three-layered plate NiTi–$Sm_{0.75}Y_{0.25}S$–Al on the thickness parameter $x = h_2/h_1 = h_2/h_3$, is given in Fig. 9.25 (left). Effec-

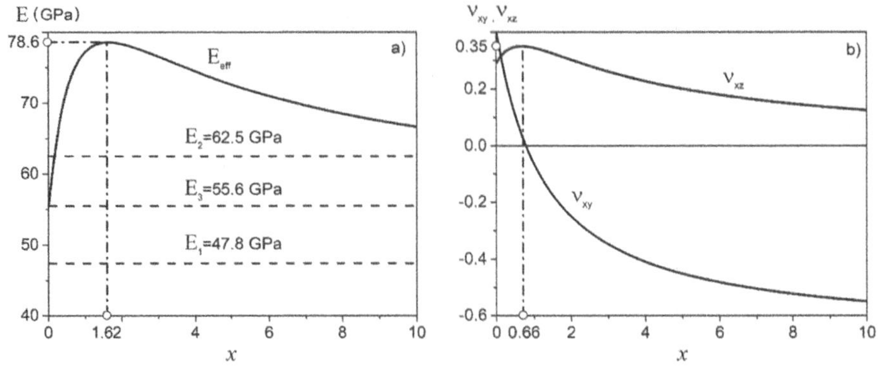

Fig. 9.25 Dependence of effective Young's modulus E_{eff} (left) and Poisson's ratios υ_{xy} and υ_{xz} (right) on the ratio of the thicknesses of the layers $x = h_2/h_1 = h_2/h_3$ for three-layered NiTi–$Sm_{075}Y_{0.25}S$–Al plate

tive Young's modulus is larger than Young's modulus of the layer with the greatest modulus at $x > 0.2$ and the maximum is attained at $x = 1.62$. At the same time, out-of-plane Poisson's ratio υ_{xz} remains always positive, reaching a maximum at $x = 0.66$, and in-plane Poisson's ratio υ_{xy} decreases rapidly with increasing x and becomes negative at $x > 0.77$ (Fig. 9.25 right).

The effective Young's moduli and Poisson's ratios of three-layered NiTi–$Sm_{0.75}Y_{0.25}S$–non-auxetic plates, Al–$Sm_{0.75}Y_{0.25}S$–non-auxetic plates, LiBr–$Sm_{0.75}Y_{0.25}S$–non-auxetic plates, and Mo–$Sm_{0.75}Y_{0.25}S$–non-auxetic plates at the thickness parameter $x = 1$ for more than seven hundred different non-auxetics had been calculated. It was found that effective Young's moduli for NiTi–$Sm_{0.75}Y_{0.25}S$–non-auxetic plates and Al–$Sm_{0.75}Y_{0.25}S$–non-auxetic plates exceeded Young's moduli of the three original layers in the case of more than two hundred non-auxetics. Whereas out-of-plane Poisson's ratio is always positive and in-plane Poisson's ratio for such three-layered plates is negative for half of the original non-auxetics of the third layer. One-third among such three-layered plates is weak auxetics ($-1 < \upsilon_{xy} < 0$). It was found that effective Young's modulus of the three-layered LiBr–$Sm_{0.75}Y_{0.25}S$–non-auxetic plates exceeds all three original Young's moduli in a smaller number of cases (about 50), while in-plane Poisson's ratio is more often negative than for the two previous three-layered plates. However, out-of-plane Poisson's ratio is still always positive. Finally, the three-layered Mo–$Sm_{0.75}Y_{0.25}S$–non-auxetic plates are always non-auxetics ($\upsilon_{xy} > 0$, $\upsilon_{xz} > 0$) because of the overwhelming influence of the high stiffness non-auxetic Mo for the first layer.

The category auxetic–non-auxetic–auxetic plates was considered on the examples of $Sm_{0.75}Y_{0.25}S$–non-auxetic–$Sm_{0.7}Y_{0.3}S$ plates at the thickness parameter $x = 1$ with the same numerous set (about 700) of non-auxetics. It was found that effective Young's moduli for three-layered plates exceeded Young's moduli of the three original layers for one-tenth non-auxetics. Most of these three-layered plates were auxetics both in the ratio of in-plane Poisson's ratio and out-of-plane Poisson's ratio.

A small selection of such three-layered plates is given in Table 9.3. This table also shows the results for $Sm_{0.65}La_{0.35}S$–non-auxetic–$Sm_{0.75}Tm_{0.25}S$ plates with two other auxetics and non-auxetics from the same set. It demonstrates the role of stiffness for non-auxetics in the negativity of in-plane Poisson's ratio υ_{xy} for the auxetic–non-auxetic–auxetic plate category. Poisson's ratio υ_{xy} increases with increasing Young's modulus of the original non-auxetics, similar to the case of the auxetic–non-auxetic–auxetics category discussed above. The change in Poisson's ratios υ_{xy} and υ_{xz} with an increase in the relative thickness of the inner layer x for the three-layered $Sm_{0.75}Y_{0.25}S$–KI–$Sm_{0.7}Y_{0.3}S$ plate is shown in Fig. 9.26.

For plates of equal layer thicknesses, comparison between Eqs. (9.8.24) and (9.8.25) shows that in-plane Poisson's ratio is contributed equally by all layers under stretching mode but in the case of bending mode each of the outer layers has a factor

Table 9.3 Effective elastic properties results for three-layered of auxetic–non-auxetic–auxetic plate category consisting cubic materials of equal thicknesses

Non-auxetics	E_1, GPa	E_2, GPa	E_3, GPa	E_{eff}, GPa	υ_1	υ_2	υ_3	υ_{xy}	υ_{xz}
$Sm_{0.75}Y_{0.25}S$-non-auxetics-$Sm_{0.7}Y_{0.3}S$									
$BaCa_2(C_2H_5CO_2)_6$	55.6	4.31	76.3	47.2	−0.67	0.37	−0.59	−0.60	−0.10
KI		26.2		57.4		0.14		−0.54	−0.32
LiBr		27.2		61.2		0.33		−0.51	−0.15
NiTi		47.8		78.0		0.44		−0.40	0.01
Al		62.5		82.6		0.36		−0.38	−0.09
Ge		103		98.1		0.27		−0.32	−0.18
Si		129		111		0.29		−0.27	−0.16
V		148		123		0.34		−0.21	−0.10
Mo		380		208		0.26		−0.05	−0.15
W		408		220		0.28		−0.02	−0.13
Ir		446		236		0.30		0.01	−0.11
$Sm_{0.65}La_{0.35}S$–non-auxetics–$Sm_{0.75}Tm_{0.25}S$									
$BaCa_2(C_2H_5CO_2)_6$	50.8	4.31	65.4	41.3	−0.35	0.37	−0.46	−0.39	0.01
KI		26.2		49.8		0.14		−0.33	−0.18
LiBr		27.2		52.5		0.33		−0.28	−0.04
NiTi		47.8		65.1		0.44		−0.16	0.09
Al		62.5		69.3		0.36		−0.15	0.00
Ge		103		82.8		0.27		−0.11	−0.07
Si		129		93.2		0.29		−0.07	−0.07
V		148		103		0.34		0.00	−0.02
Mo		380		181		0.26		0.09	−0.07
W		408		192		0.28		0.11	−0.05
Ir		446		206		0.30		0.14	−0.04

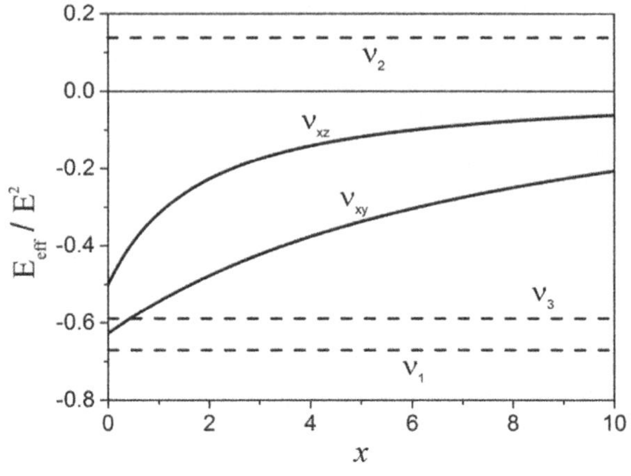

Fig. 9.26 Dependence of effective Poisson's ratios υ_{xy} and υ_{xz} on the ratio of the thicknesses of the layers $x = h_2/h_1 = h_2/h_3$ for three-layered $Sm_{0.75}Y_{0.25}S$–KI–$Sm_{0.7}Y_{0.3}S$ plate

of 13. Table 9.4 shows sample results for three groups of symmetric three-layered plates in which in-plane auxeticity is dependent on the loading mode. Non-auxetics and complete auxetics used in it have positive or negative values of Poisson's ratios for any orientations of such crystals. However, the original single-layer plates are characterized by Poisson's ratios υ_k for crystallographic orientations corresponding to the edges of the plates. Since effective Poisson's ratios υ_{xy} are calculated for these same particular directions in the three-layered plates, we will use the terms in-plane auxetics and in-plane non-auxetics for them in the cases $\upsilon_{xy} < 0$ and $\upsilon_{xy} > 0$, respectively. The first group of symmetric three-layered plates consists of non-auxetic–$Sm_{0.75}Y_{0.25}S$–non-auxetic symmetric plates that are in-plane auxetic under stretching mode but in-plane non-auxetic under bending mode; the second group consists of $Sm_{0.65}La_{0.35}S$–non-auxetic–$Sm_{0.65}La_{0.35}S$ symmetric plates that are in-plane non-auxetic under stretching mode but in-plane auxetic under bending mode; the third group consists of $Sm_{0.75}Y_{0.25}S$–non-auxetic–$Sm_{0.75}Y_{0.25}S$ symmetric plates that are in-plane auxetics for both modes. Examples for the cases of unequal layer thickness with $x = h_2/h_1 = h_2/h_3$ are plotted in Figs. 9.27, 9.28, and 9.29.

In-plane Poisson's ratio of symmetrical Al–$Sm_{0.75}Y_{0.25}S$–Al plates is plotted in Fig. 9.27. Specifically, Fig. 9.27 (left) reveals that this plate is fully in-plane non-auxetic when $x < 0.766$ and fully in-plane auxetic when $x > 3.743$; within the range $0.766 \le x \le 3.743$ this plate behaves as an in-plane auxetic plate when the loading mode is stretching but behaves as an in-plane non-auxetic plate when the loading mode is bending. Figure 9.27 (right) shows a zoomed out view to reveal that the in-plane Poisson's ratios under both modes of loading converge for extreme values of x. This is not surprising because the plate becomes 100% Al and 100% $Sm_{0.75}Y_{0.25}S$ when $x = 0$ and $x \to \infty$, respectively. As mentioned earlier in the bending analysis, Poisson's ratios of a plate are the same under both the stretching and bending modes if the plate is made of a single material.

Table 9.4 Effective in-plane Poisson's ratio of three-layered symmetric with equal layer thickness under stretching and bending modes

Symmetric three-layered plate of equal layer thickness	E_1, E_3 GPa	E_2 GPa	E_{eff} GPa	v_1, v_3	v_2	v_{xy} (stretch)	v_{xy} (bend)	Remarks
KI–$Sm_{0.75}Y_{0.25}S$–KI	26.2	55.6	43.6	0.14	−0.67	−0.39	0.03	In-plane auxetic behavior under axial load, but in-plane non-auxetic behavior under bending load
LiBr–$Sm_{0.75}Y_{0.25}S$–LiBr	27.2		49.3	0.33		−0.30	0.21	
Mg_2Pb–$Sm_{0.75}Y_{0.25}S$–Mg_2Pb	61.3		75.0	0.23		−0.16	0.18	
Al–$Sm_{0.75}Y_{0.25}S$–Al	62.5		81.4	0.36		−0.07	0.31	
Ge–$Sm_{0.75}Y_{0.25}S$–Ge	103		108	0.27		−0.02	0.24	
$Sm_{0.65}La_{0.35}S$–Si-$Sm_{0.65}La_{0.35}S$	50.8	129	85.2	−0.35	0.29	0.00	−0.29	In-plane non-auxetic by axial load, in-plane auxetic by bending load
$Sm_{0.65}La_{0.35}S$–Cr–$Sm_{0.65}La_{0.35}S$		328	151		0.16	0.03	−0.25	
$Sm_{0.65}La_{0.35}S$–Mo–$Sm_{0.65}La_{0.35}S$		380	171		0.26	0.12	−0.22	
$Sm_{0.65}La_{0.35}S$–W–$Sm_{0.65}La_{0.35}S$		408	182		0.28	0.15	−0.20	
$Sm_{0.65}La_{0.35}S$–Ir–$Sm_{0.65}La_{0.35}S$		446	196		0.30	0.18	−0.19	
$Sm_{0.75}Y_{0.25}S$–Al–$Sm_{0.75}Y_{0.25}S$	55.6	62.5	76.8	−0.67	0.36	−0.40	−0.64	In-plane auxetic by axial load, in-plane auxetic by bending load
$Sm_{0.75}Y_{0.25}S$–Ge–$Sm_{0.75}Y_{0.25}S$		103	92.6		0.27	−0.34	−0.63	
$Sm_{0.75}Y_{0.25}S$–Si–$Sm_{0.75}Y_{0.25}S$		129	105		0.29	−0.28	−0.62	
$Sm_{0.75}Y_{0.25}S$–Cr–$Sm_{0.75}Y_{0.25}S$		328	176		0.16	−0.15	−0.58	
$Sm_{0.75}Y_{0.25}S$–Mo–$Sm_{0.75}Y_{0.25}S$		380	203		0.26	−0.05	−0.55	
$Sm_{0.75}Y_{0.25}S$–W–$Sm_{0.75}Y_{0.25}S$		408	215		0.28	−0.02	−0.53	

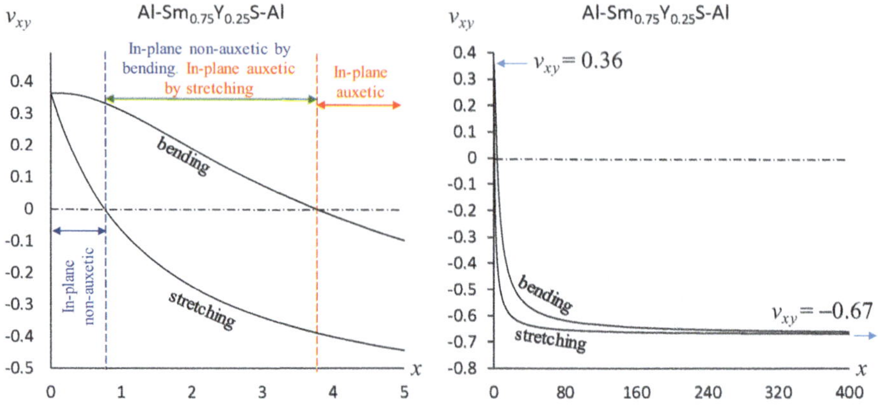

Fig. 9.27 Mode dependency of in-plane Poisson's ratio υ_{xy} in symmetric Al–$Sm_{0.75}Y_{0.25}S$–Al plates, showing: fully in-plane auxetic region, fully in-plane non-auxetic region, and mode-dependent mixed region in zoomed-in view (left), and convergence of bending and stretching-based υ_{xy} at extreme values of x (right)

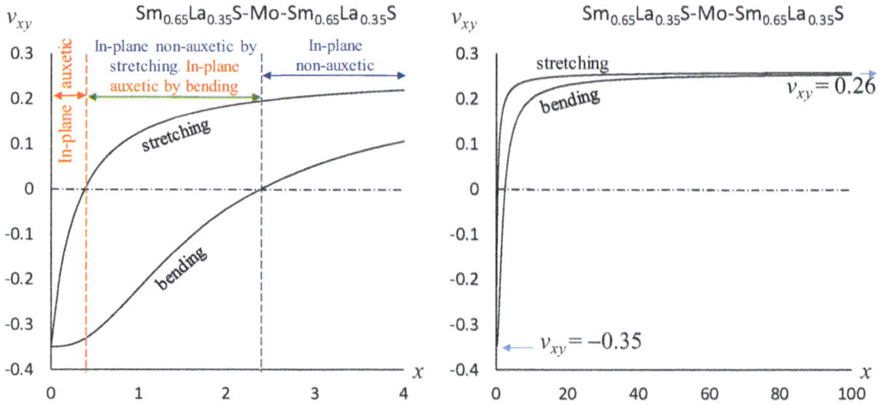

Fig. 9.28 Mode dependency of in-plane Poisson's ratio υ_{xy} in symmetric $Sm_{0.65}La_{0.35}S$–Mo–$Sm_{0.65}La_{0.35}S$ plates, showing: fully in-plane auxetic region, fully in-plane non-auxetic region, and mode-dependent mixed region in zoomed-in view (left), and convergence of bending and stretching-based υ_{xy} at extreme values of x (right)

Curves of Poisson's ratio υ_{xy} for symmetrically layered $Sm_{0.65}La_{0.35}S$–Mo–$Sm_{0.65}La_{0.35}S$ plates are plotted in Fig. 9.25 (left), in which the plate is fully in-plane auxetic when $x \leq 0.382$ but fully in-plane non-auxetic when $x \geq 2.381$. More interestingly, the plate exhibits in-plane auxeticity dependency based on loading mode when the middle layer's relative thickness is within $0.382 < x < 2.381$. As with the other plate, Poisson's ratios for both loading modes converge at extreme values of x for the same reason. See Fig. 9.28 (right). Comparison between Figs. 9.27 and 9.28

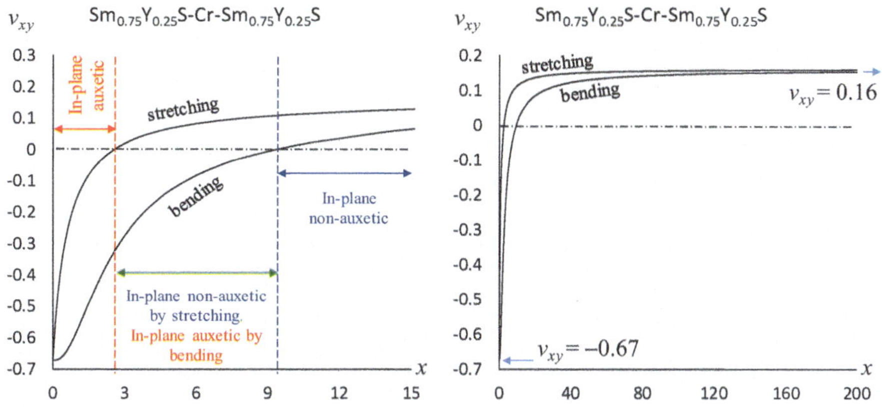

Fig. 9.29 Mode dependency of in-plane Poisson's ratio υ_{xy} in symmetric $Sm_{0.75}Y_{0.25}S$–Cr–$Sm_{0.75}Y_{0.25}S$ plates, showing: fully in-plane auxetic region, fully in-plane non-auxetic region, and mode-dependent mixed region in zoomed-in view (left), and convergence of bending and stretching-based υ_{xy} at extreme values of x (right)

suggests that in the case of non-auxetic–auxetic–non-auxetic plates, the effective in-plane Poisson's ratio is more positive under bending mode than in stretching mode, while in the case of auxetic–non-auxetic–auxetic plate the υ_{xy} is more negative under bending mode than in stretching mode.

Symmetric three-layered plates of the $Sm_{0.75}Y_{0.25}S$–non-auxetic–$Sm_{0.75}Y_{0.25}S$ type with equal thickness of the layers from Table 9.4 are characterized by negative in-plane Poisson's ratios for both stretching mode and bending mode. The magnitude of in-plane Poisson's ratios for both types increases with increasing stiffness of non-auxetics. The modal dependence of in-plane Poisson's ratios on the relative thickness of the inner layer x on the example of symmetrical $Sm_{0.75}Y_{0.25}S$–Cr–$Sm_{0.75}Y_{0.25}S$ plate is shown in Fig. 9.29. In this case, the region of the fully in-plane auxetic is realized at $x < 2.530$, and the region of the fully in-plane non-auxetics is reached when $x > 9.332$. Mixed behavior takes place within the region $2.530 \leq x \leq 9.332$.

In-plane Poisson's ratio of symmetrical Ca–$Sm_{0.65}La_{0.35}S$–Ca plates is plotted in Fig. 9.30. Specifically, Fig. 9.30 (left) reveals that this plate is fully in-plane non-auxetic when $x \leq 0.452$ and fully in-plane auxetic when $x > 2.642$; within the range $0.452 < x \leq 2.642$, this plate behaves as an in-plane auxetic plate when the loading mode is stretching but behaves as an in-plane non-auxetic plate when the loading mode is bending. Figure 9.30 (right) shows a zoomed out view to reveal that in-plane Poisson's ratios under both modes of loading converge for extreme values of x.

Many cubic crystals are partial auxetics (about 300), i.e., they have negative or positive Poisson's ratio, depending on their orientation. Nevertheless, often in-plane Poisson's coefficients υ_k for the original single-layer plates are positive (or negative), so that one can speak of in-plane non-auxetics (or in-plane auxetics). Examples of symmetrical three-layered plates with such crystals and complete auxetics,

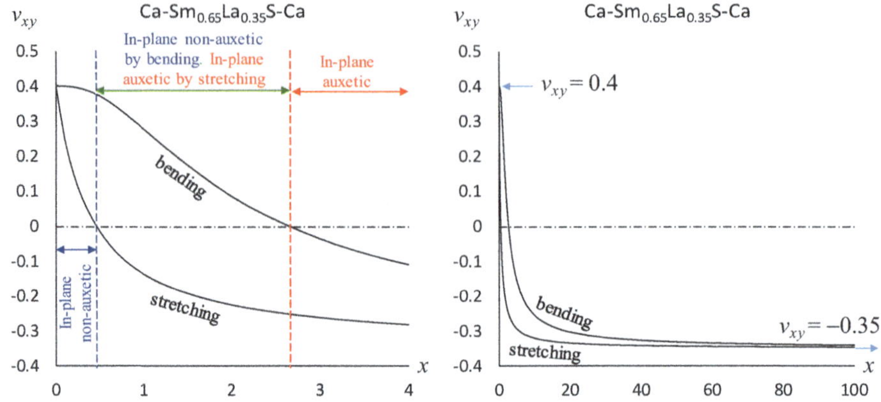

Fig. 9.30 Mode dependency of in-plane Poisson's ratio υ_{xy} in symmetric Ca–$Sm_{0.65}La_{0.35}S$–Ca plates, showing: fully in-plane non-auxetic region, fully in-plane auxetic region, and mode-dependent mixed region in zoomed-in view (left), and convergence of bending and stretching-based υ_{xy} at extreme values of x (right)

$Sm_{0.65}La_{0.35}S$ in particular, are given in Table 9.5. Since the following cubic crystals that were discussed—$Sm_{0.65}La_{0.35}S$, $Sm_{0.75}Y_{0.25}S$, $Sm_{0.7}Y_{0.3}S$, and $Sm_{0.75}Tm_{0.25}S$—are complete auxetics, i.e., their Poisson's ratio sign does not change with orientation, and they are herein simply referred to as auxetics.

In summary, the effective Young's moduli of the laminated plates were larger than Young's moduli of all layers in many cases except for composites with very stiffness crystals. Effective in-plane Poisson's ratios for stretching of three-layered plates often increase with increasing rigidity of non-auxetics stiffness, changing the negative sign to positive. In many cases, the three-layered non-auxetic–auxetic–non-auxetic plates are characterized by positive effective out-of-plane Poisson's ratios, whereas both effective Poisson's ratios may be negative in the three-layered auxetic–non-auxetic–auxetic plates. In the case for bending of symmetric plates, the outer layers impose stronger influence under the bending mode than in stretching mode, thereby resulting three regions of characteristics: (a) fully in-plane auxetic, (b) fully in-plane non-auxetic, and (c) combination of in-plane auxetic and non-auxetic, depending on the loading mode applied. Since it is well known that both auxetic and non-auxetic materials have their own advantages and disadvantages, the possibility to allow coexistence of dual properties—both auxetic and non-auxetic—means that materials can be tailor-made so that the structures can behave as an auxetic material when it is more advantageous to be auxetic under one type of loading mode, and to also behave as a non-auxetic material when it is disadvantageous to be auxetic under another type of loading mode. An example application is in the case where a sheet needs to be wrapped into a dome of double curvature, which undergoes stretching. For such an application, negative Poisson's ratio under bending mode will generate less bending stresses when wrapped onto the dome. Poisson's ratio of low magnitude will be beneficial in preserving the sheet thickness in the presence

Table 9.5 Effective in-plane Poisson's ratio of three-layered symmetric with equal layer thickness under stretching and bending modes

Symmetric three-layered plate of equal layer thickness	E_1, E_3 GPa	E_2 GPa	E_{eff} GPa	υ_1, υ_3	υ_2	υ_{xy} (stretch)	υ_{xy} (bend)	Remarks
Ca–TmSe–Ca	9.62	115	50.6	0.4	−0.54	−0.43	0.07	In-plane auxetic behavior under axial load, but in-plane non-auxetic behavior under bending load
Ce–TmSe–Ce	15.9		56.8	0.36		−0.38	0.13	
La–TmSe–La	19.3		60.9	0.37		−0.34	0.17	
Ag–TmSe–Ag	43.5		87.3	0.43		−0.16	0.32	
Cu–TmSe–Cu	66.7		108	0.42		−0.06	0.35	
$Sm_{0.65}La_{0.35}S$–Co–$Sm_{0.65}La_{0.35}S$	50.8	114	83.2	−0.35	0.40	0.05	−0.29	In-plane non-auxetic by axial load, in-plane auxetic by bending load
$Sm_{0.65}La_{0.35}S$–Fe–$Sm_{0.65}La_{0.35}S$		130	88.5		0.37	0.06	−0.28	
$Sm_{0.65}La_{0.35}S$–Ni–$Sm_{0.65}La_{0.35}S$		130	89.0		0.38	0.07	−0.28	
$Sm_{0.65}La_{0.35}S$–SiC–$Sm_{0.65}La_{0.35}S$		272	136		0.29	0.11	−0.24	
$Sm_{0.65}La_{0.35}S$–TiC–$Sm_{0.65}La_{0.35}S$		476	201		0.17	0.07	−0.22	

of in-plane stretching. Hence, in this application, a low magnitude of Poisson's ratio for in-plane stretching and a large negative Poisson's ratio for pure stretching will reduce both the stretching and bending stresses.

References

Chirima GT, Zied KM, Ravirala N, Alderson KL, Alderson A (2009) Numerical and analytical modelling of multi-layer adhesive–film interface systems. Phys Status Solidi B 246(9):2072–2082

Donescu S, Chiroiu V, Munteanu L (2009) On the Young's modulus of a auxetic composite structure. Mech Res Commun 36(3):294–301

Gorodtsov VA, Lisovenko DS, Lim TC (2018) Three-layered plate exhibiting auxeticity based on stretching and bending modes. Compos Struct 194:643–651

Hashin Z, Shtrikman S (1963) A variational approach to the elastic behavior of multiphase minerals. J Mech Phys Solids 11(2):127–140

Kocer C, McKenzie DR, Bilek MM (2009) Elastic properties of a material composed of alternating layers of negative and positive Poisson's ratio. Mater Sci Eng A 505(1–2):111–115

Lim TC (2009) Out-of-plane modulus of semi-auxetic laminates. Eur J Mech A Solids 28(4):752–756

Lim TC (2010) In-plane stiffness of semiauxetic laminates. ASCE J Eng Mech 136(9):1176–1180

Lim TC (2013) Corrigendum to "Out-of-plane modulus of semi-auxetic laminates". Eur J Mech A Solids 37(1):379–380

Lim TC, Acharya UR (2010) Longitudinal modulus of semi-auxetic unidirectional fiber composites. J Reinf Plast Compos 29(10):1441–1445

Lim TC, Acharya UR (2011) Counterintuitive modulus from semi-auxetic laminates. Phys Status Solidi B 248(1):60–65

Liu B, Feng X, Zhang SM (2009) The effective Young's modulus of composites beyond the Voigt estimation due to the Poisson effect. Compos Sci Technol 69(13):2198–2204

Reuss A (1929) Berechnung der Fließgrenze von Mischkristallen auf Grund der Plastizitätsbedingung für Einkristalle. Zeitschrift für Angewandte Mathematik und Mechanik 9(1):49–58

Voigt W (1889) Über die Beziehung zwischen den beiden Elastizitätskonstanten isotroper Körper. Wied Ann 38:573–589

Voigt W (1910) Lehrbuch der Kristallphysik. Teubner, Berlin

Zhang W, Evans KE (1992) A Fortran program for the design of laminates with required mechanical properties. Comput Struct 45(5–6):919–939

Zhang R, Yeh HL, Yeh HY (1999) A discussion of negative Poisson's ratio design for composites. J Reinf Plast Compos 18(17):1546–1556

Zhu HX, Fan TX, Zhang D (2015) Composite materials with enhanced dimensionless Young's modulus and desired Poisson's ratio. Sci Rep 5:14103

Chapter 10
Auxetic Membranes

Abstract This chapter evaluates the deflection and stretching stresses occurring in large deflection of auxetic membranes using strain energy minimization approach. Results show that auxeticity tends to increase the deflection. Optimization study shows that the membrane stresses can be minimized by controlling the Poisson's ratio and aspect ratio of the rectangular membranes.

Keywords Auxetic membranes · Circular membranes · Rectangular membranes · Large deflection

10.1 Preliminaries

In assessing the effect of auxeticity, we have so far considered infinitesimal deformation is previous chapters, as well as in a previous monograph (Lim 2015). In soft materials as well as in very thin rods and plates, one encounters large deformation. This chapter deals with large deflection of membranes that possess negative Poisson's ratio. Unlike plates, membranes do not carry bending moments. Hence, bending stresses are absent in membranes and stresses are evaluated based on stretching at the mid-surface of the membrane. Unlike plates where a fully clamped edge requires zero slope therein in the direction perpendicular to the side, the absence of bending moment permits free rotation at the clamped side akin to simply supported plates. In this chapter, we consider circular and rectangular membranes subjected to pressure on one side with fully clamped sides.

Consider an originally linear element in the radial direction as illustrated in Fig. 10.1, and the radial and tangential strains under large deflection are

$$\varepsilon_r = \frac{\mathrm{d}u_r}{\mathrm{d}r} + \frac{1}{2}\left(\frac{\mathrm{d}w}{\mathrm{d}r}\right)^2 \tag{10.1.1}$$

and

$$\varepsilon_\theta = \frac{u_r}{r} \tag{10.1.2}$$

T. Lim, *Mechanics of Metamaterials with Negative Parameters*, Engineering Materials,
https://doi.org/10.1007/978-981-15-6446-8_10

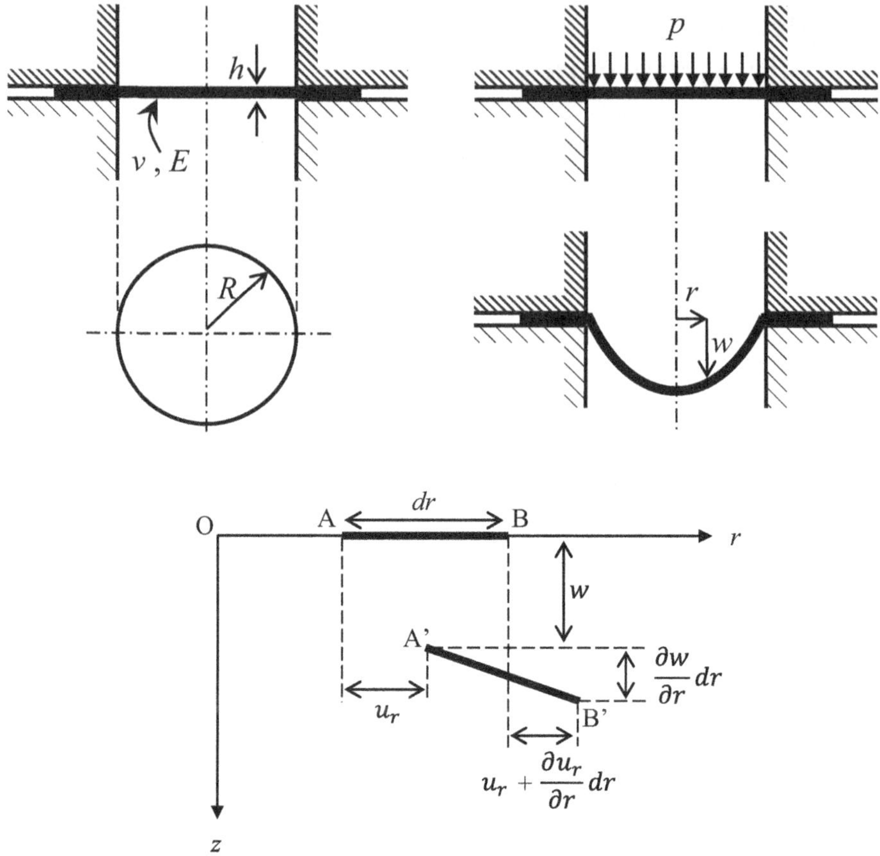

Fig. 10.1 A circular membrane of Young's modulus E, Poisson's ratio ν, thickness h, and effective radius R (top left) being subjected to a net pressure p on one surface to produce deflection w at a radial distance r from the membrane center (top right), and schematics for description of displacements and strains (bottom)

respectively. For *very* large deflection, Reissner (1949) showed that

$$\varepsilon_r = \frac{\mathrm{d}u_r}{\mathrm{d}r} + \frac{1}{2}\left[\left(\frac{\mathrm{d}u_r}{\mathrm{d}r}\right)^2 + \left(\frac{\mathrm{d}w}{\mathrm{d}r}\right)^2\right] \tag{10.1.3}$$

In this chapter, we shall adopt the large strain, rather than the very large strain, model.

We note that $\sigma_z = -p$ on the surface subjected to pressure p while $\sigma_z = 0$ on the opposing side for being a free surface. Since σ_z is relatively insignificant in comparison to in-plane stresses σ_x and σ_y, it is reasonable to assume that the out-of-plane stress is negligible $\sigma_z = 0$. Based on constitutive relations in the absence of shearing as given in Eq. (6.2.20) and substituting $\sigma_x = \sigma_r$, $\sigma_y = \sigma_\theta$ and $\sigma_z = 0$, we have

$$\begin{Bmatrix} \sigma_r \\ \sigma_\theta \\ 0 \end{Bmatrix} = \frac{E}{(1+v)(1-2v)} \begin{bmatrix} 1-v & v & v \\ v & 1-v & v \\ v & v & 1-v \end{bmatrix} \begin{Bmatrix} \varepsilon_r \\ \varepsilon_\theta \\ \varepsilon_z \end{Bmatrix} \tag{10.1.4}$$

from which

$$\varepsilon_z = -\frac{v}{1-v}(\varepsilon_r + \varepsilon_\theta) \tag{10.1.5}$$

This gives

$$\begin{aligned} \sigma_r &= \frac{E}{1-v^2}(\varepsilon_r + v\varepsilon_\theta) \\ \sigma_\theta &= \frac{E}{1-v^2}(v\varepsilon_r + \varepsilon_\theta) \end{aligned} \tag{10.1.6}$$

At the center of the circular membrane ($r = 0$), $\sigma_\theta = \sigma_r$ and $\varepsilon_\theta = \varepsilon_r$ such that

$$\sigma_r = \frac{E\varepsilon_r}{1-v} \tag{10.1.7}$$

wherein

$$\frac{E}{2} \leq \frac{\sigma_r}{\varepsilon_r} \leq 2E \; ; \; -1 \leq v \leq \frac{1}{2} \tag{10.1.8}$$

In other words, at the center of isotropic circular membranes, we have the range $E/2 \leq \sigma_r/\varepsilon_r \leq E$ for auxetic membranes while $E \leq \sigma_r/\varepsilon_r \leq 2E$ for conventional membranes. Writing in terms of tensile force per unit length $(N_r, N_\theta) = (\sigma_r h, \sigma_\theta h)$ and using Eqs. (10.1.1) and (10.1.2), we have from Eq. (10.1.6)

$$\begin{aligned} N_r &= \frac{Eh}{1-v^2}\left(\frac{du_r}{dr} + \frac{1}{2}\left(\frac{dw}{dr}\right)^2 + v\frac{u_r}{r}\right) \\ N_\theta &= \frac{Eh}{1-v^2}\left(v\frac{du_r}{dr} + \frac{v}{2}\left(\frac{dw}{dr}\right)^2 + \frac{u_r}{r}\right) \end{aligned} \tag{10.1.9}$$

which will be used in Sect. 10.2.

While the polar coordinate system is expedient for circular membranes, the Cartesian coordinate system is obvious choice for rectangular membranes. As such the large strain descriptions for rectangular membranes are

$$\begin{aligned}\varepsilon_x &= \frac{\partial u_x}{\partial x} + \frac{1}{2}\left(\frac{\partial w}{\partial x}\right)^2\\ \varepsilon_y &= \frac{\partial u_y}{\partial y} + \frac{1}{2}\left(\frac{\partial w}{\partial y}\right)^2\\ \gamma_{xy} &= \frac{\partial u_x}{\partial y} + \frac{\partial u_y}{\partial x} + \frac{\partial w}{\partial x}\frac{\partial w}{\partial y}\end{aligned} \tag{10.1.10}$$

With reference to the constitutive relations given in Eq. (6.2.19), we again apply $\sigma_z = 0$. In addition, the out-of-plane shear stress is also zero $\gamma_{yz} = \gamma_{zx} = 0$. Therefore,

$$\begin{aligned}\varepsilon_x &= \frac{1}{E}\left(\sigma_x - \upsilon\sigma_y\right)\\ \varepsilon_y &= \frac{1}{E}\left(\sigma_y - \upsilon\sigma_x\right)\\ \gamma_{xy} &= \frac{\tau_{xy}}{G} = \frac{2(1+\upsilon)\tau_{xy}}{E}\end{aligned} \tag{10.1.11}$$

It is worth noticing that

$$\begin{aligned}\lim_{\upsilon\to-1}\varepsilon_x = \lim_{\upsilon\to-1}\varepsilon_y = \lim_{\upsilon\to0}\left(\varepsilon_x + \varepsilon_y\right) = \frac{\sigma_x+\sigma_y}{E}\\ \lim_{\upsilon\to-1}\frac{\gamma_{xy}E}{\tau_{xy}} = 0\end{aligned} \tag{10.1.12}$$

The first of Eq. (10.1.12) can also be observed for the strains and stresses in polar coordinate. Writing in terms of tensile force per unit length $\left(N_x, N_y, N_{xy}\right) = \left(\sigma_x h, \sigma_y h, \tau_{xy} h\right)$, we have

$$\begin{aligned}N_x &= \frac{Eh}{1-\upsilon^2}\left(\varepsilon_x + \upsilon\varepsilon_y\right)\\ N_y &= \frac{Eh}{1-\upsilon^2}\left(\varepsilon_y + \upsilon\varepsilon_x\right)\\ N_{xy} &= \frac{Eh}{2(1+\upsilon)}\gamma_{xy}\end{aligned} \tag{10.1.13}$$

such that substituting Eq. (10.1.10) into the above gives

$$\begin{aligned}N_x &= \frac{Eh}{1-\upsilon^2}\left(\frac{\partial u_x}{\partial x} + \frac{1}{2}\left(\frac{\partial w}{\partial x}\right)^2 + \upsilon\frac{\partial u_y}{\partial y} + \frac{\upsilon}{2}\left(\frac{\partial w}{\partial y}\right)^2\right)\\ N_y &= \frac{Eh}{1-\upsilon^2}\left(\frac{\partial u_y}{\partial y} + \frac{1}{2}\left(\frac{\partial w}{\partial y}\right)^2 + \upsilon\frac{\partial u_x}{\partial x} + \frac{\upsilon}{2}\left(\frac{\partial w}{\partial x}\right)^2\right)\\ N_{xy} &= \frac{Eh}{1-\upsilon^2}\left(\frac{1-\upsilon}{2}\right)\left(\frac{\partial u_x}{\partial y} + \frac{\partial u_y}{\partial x} + \frac{\partial w}{\partial x}\frac{\partial w}{\partial y}\right)\end{aligned} \tag{10.1.14}$$

which will be used in Sect. 10.3.

10.2 Example: Approximate Approach for Circular Membranes by Strain Energy of Stretching

Suppose a circular membrane of radius R is imposed with uniform load on one surface, one may, for the sake of simplicity, assume that the deflection takes the profile

$$w = w_0\left(1 - \frac{r^2}{R^2}\right)^2 \tag{10.2.1}$$

where w_0, the maximum deflection, occurs at $r = 0$. As for the radial displacement, one may adopt an expression of the form

$$u_r = r(R-r)\sum_{n=1}^{\infty} C_n r^{n-1} = r(R-r)\left(C_1 + C_2 r + C_3 r^2 + \cdots\right) \tag{10.2.2}$$

wherein the condition $u_r = 0$ must be satisfied at $r = 0$ and $r = R$. The stretching energy based on the mid-plane of the membrane can be obtained via

$$U = 2\pi\int_0^R\left(\frac{1}{2}N_r\varepsilon_r + \frac{1}{2}N_\theta\varepsilon_\theta\right)r\mathrm{d}r = \pi\int_0^R\left(N_r\varepsilon_r + N_\theta\varepsilon_\theta\right)r\mathrm{d}r \tag{10.2.3}$$

where the strains and tensile forces per unit length are obtained from Eqs. (10.1.1), (10.1.2), and (10.1.9) to give

$$\begin{aligned} N_r\varepsilon_r &= \tfrac{Eh}{1-\upsilon^2}\left(\left(\tfrac{du_r}{dr}\right)^2 + \tfrac{du_r}{dr}\left(\tfrac{dw}{dr}\right)^2 + \tfrac{1}{4}\left(\tfrac{dw}{dr}\right)^4 + \upsilon\tfrac{u_r}{r}\tfrac{du_r}{dr} + \tfrac{\upsilon}{2}\tfrac{u_r}{r}\left(\tfrac{dw}{dr}\right)^2\right) \\ N_\theta\varepsilon_\theta &= \tfrac{Eh}{1-\upsilon^2}\left(\upsilon\tfrac{u_r}{r}\tfrac{du_r}{dr} + \tfrac{\upsilon}{2}\tfrac{u_r}{r}\left(\tfrac{dw}{dr}\right)^2 + \left(\tfrac{u_r}{r}\right)^2\right) \end{aligned} \tag{10.2.4}$$

Considering only the first two terms of the series in Eq. (10.2.2), we have from Eq. (10.2.3)

$$\begin{aligned} U = \frac{\pi EhR^2}{1-\upsilon^2}\Big(&0.25C_1^2R^2 + 0.3C_1C_2R^3 + 0.1167C_2^2R^4 \\ &-0.06768\frac{C_1w_0^2}{R} + 0.05456C_2w_0^2 + 0.30528\frac{w_0^4}{R^4}\Big) \end{aligned} \tag{10.2.5}$$

The constants C_1 and C_2 are evaluated from the condition that the total stretching energy of the membrane is a minimum at equilibrium. Prescribing

$$\frac{\partial U}{\partial C_1} = \frac{\partial U}{\partial C_2} = 0 \tag{10.2.6}$$

leads to

$$C_1 = 1.185\frac{w_0^2}{R^3}\ ;\ C_2 = -1.75\frac{w_0^2}{R^4} \tag{10.2.7}$$

such that

$$U = 0.2159\frac{\pi Ehw_0^4}{(1 - \upsilon^2)R^2} \tag{10.2.8}$$

In other words, the range of stretching energy for moderately auxetic $\upsilon = -0.5$ to incompressibility $\upsilon = 0.5$ is

$$0.216\frac{\pi Ehw_0^4}{R^2} \leq U \leq 0.288\frac{\pi Ehw_0^4}{R^2} \tag{10.2.9}$$

while the stretching energy for highly auxetic range $-1 \leq \upsilon < -0.5$ is

$$U > 0.288\frac{\pi Ehw_0^4}{R^2} \tag{10.2.10}$$

The membrane deflection can be evaluated by applying the principle of virtual displacements, i.e.,

$$\frac{\mathrm{d}U}{\mathrm{d}w_0}\delta w_0 = 2\pi\int_0^R p\delta wr\mathrm{d}r \tag{10.2.11}$$

Substituting Eq. (10.2.8) into the LHS of Eq. (10.2.11) gives

$$\frac{\mathrm{d}U}{\mathrm{d}w_0}\delta w_0 = 0.8636\frac{\pi Ehw_0^3}{(1 - \upsilon^2)R^2}\delta w_0 \tag{10.2.12}$$

while the substitution of Eq. (10.2.1) into the RHS of Eq. (10.2.11) yields

$$2\pi p\delta w_0\int_0^R \left(1 - 2\frac{r^2}{R^2} + \frac{r^4}{R^4}\right)r\mathrm{d}r = \frac{\pi}{3}p\delta w_0 R^2 \tag{10.2.13}$$

Therefore, substituting Eqs. (10.2.12) and (10.2.13) into Eq. (10.2.11) leads to

$$w_0 = \left(\frac{pR^4(1 - \upsilon^2)}{2.591Eh}\right)^{\frac{1}{3}} \tag{10.2.14}$$

Reference to Eq. (10.2.14) implies that if the Young's modulus remains constant with the change in Poisson's ratio, the deflection is highest when the Poisson's ratio of the membrane material is zero, and the deflection decreases with the magnitude of the Poisson's ratio. It must be borne in mind that, in addition to being an approximated approach, the above analysis applies only for membranes, which carry no bending moment. If such analysis is to be extended for large deflection of circular plates, then the strain energy of bending must be included.

10.3 Example: Approximate Approach for Rectangular Membranes by Strain Energy of Stretching

The approximate strain energy of stretching in membranes of other shapes follow the same approach as with the circular membrane, and a brief outline is furnished in this section for the case of rectangular membranes to highlight the differences by virtue of the different geometry. Figure 10.2 shows a rectangular membrane of sides $2a$ by $2b$ so as to place the origin of the coordinate system at the center of the square membrane. Again, the membrane possesses a Young's modulus of E, Poisson's ratio of ν, and thickness h. This membrane is being subjected to the following boundary conditions:

- $x = -a, a, y \in \langle -b, b\rangle$ **and** $z \in \langle -h/2, h/2\rangle$: $u_x = u_y = w = 0$
- $y = -b, b, x \in \langle -a, a\rangle$ **and** $z \in \langle -h/2, h/2\rangle$: $u_x = u_y = w = 0$
- $x \in \langle -a, a\rangle, y \in \langle -b, b\rangle, z = h/2 : \sigma_z = -p$
- $x \in \langle -a, a\rangle, y \in \langle -b, b\rangle, z = -h/2 : \sigma_z = 0$

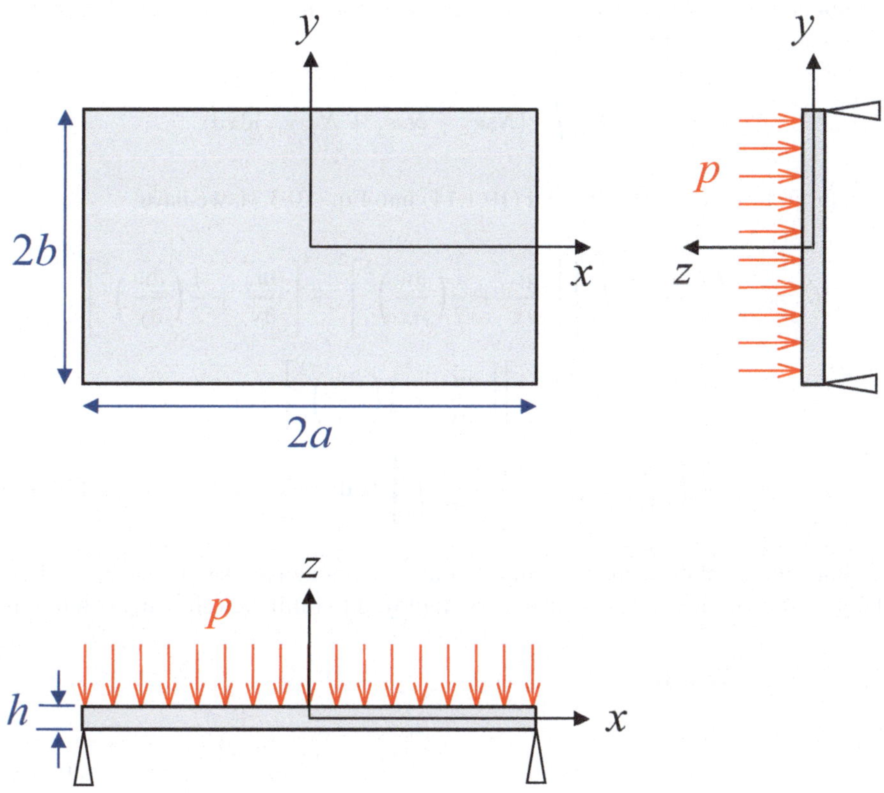

Fig. 10.2 Schematic view of rectangular membrane for analysis

As with the circular membrane, there is a need to assume a deflection profile that gives zero slope at the center and zero deflection at all four sides. This can be achieved by assuming the deflection profile of the form

$$w = w_0 \cos\frac{\pi x}{2a}\cos\frac{\pi y}{2b} \tag{10.3.1}$$

while the in-plane displacements can be well approximated by

$$\begin{aligned} u_x &= c_x \sin\tfrac{\pi x}{a}\cos\tfrac{\pi y}{2b} \\ u_y &= c_y \sin\tfrac{\pi y}{b}\cos\tfrac{\pi x}{2a} \end{aligned} \tag{10.3.2}$$

Unlike the case of circular membranes being subjected to pressure where the choice of polar coordinates removes the need for shear strain components, in the case of rectangular membranes one must add the stretching strain due to in-plane shearing. Hence, the strain energy for rectangular membranes consists of three terms, namely $N_x\varepsilon_x$, $N_y\varepsilon_y$ and $N_{xy}\gamma_{xy}$. Again, unlike circular membranes whereby only a single integral is performed due to the axisymmetrical profile of the deflected membrane, double integral needs to be performed for obtaining the stretching energy of rectangular membranes. Hence

$$U = \frac{1}{2}\int_{-b}^{b}\int_{-a}^{a}\left(N_x\varepsilon_x + N_y\varepsilon_y + N_{xy}\gamma_{xy}\right)\mathrm{d}x\mathrm{d}y \tag{10.3.3}$$

Substituting Eqs. (10.1.10) and (10.1.14) into Eq. (10.3.3), we have

$$\begin{aligned} U = \frac{Eh}{2(1-v^2)}\int_{-b}^{b}\int_{-a}^{a}\Bigg\{&\left[\frac{\partial u_x}{\partial x} + \frac{1}{2}\left(\frac{\partial w}{\partial x}\right)^2\right]^2 + \left[\frac{\partial u_y}{\partial y} + \frac{1}{2}\left(\frac{\partial w}{\partial y}\right)^2\right]^2 \\ &+ 2v\left[\frac{\partial u_x}{\partial x} + \frac{1}{2}\left(\frac{\partial w}{\partial x}\right)^2\right]\left[\frac{\partial u_y}{\partial y} + \frac{1}{2}\left(\frac{\partial w}{\partial y}\right)^2\right] \\ &+ \frac{1-v}{2}\left(\frac{\partial u_x}{\partial y} + \frac{\partial u_y}{\partial x} + \frac{\partial w}{\partial x}\frac{\partial w}{\partial y}\right)^2\Bigg\}\mathrm{d}x\,\mathrm{d}y \end{aligned} \tag{10.3.4}$$

Since the displacements contain arbitrary magnitude parameters w_0, c_x, and c_y, the principle of virtual displacements is employed to fulfil the following conditions

$$\frac{\partial U}{\partial c_x} = \frac{\partial U}{\partial c_y} = 0$$

$$\frac{\partial U}{\partial w_0}\delta w_0 = \int_{-b}^{b}\int_{-a}^{a} p\delta w\,\mathrm{d}x\,\mathrm{d}y = p\delta w_0\int_{-b}^{b}\int_{-a}^{a}\cos\frac{\pi x}{2a}\cos\frac{\pi y}{2b}\mathrm{d}x\,\mathrm{d}y \tag{10.3.5}$$

from which the arbitrary parameters can be solved to yield the strains and stresses. A special case of this problem has been furnished by Timoshenko and Woinowsky-Krieger (1959) for a square membrane $a = b$ (and hence $c_x = c_y = c$) with $v = 0.25$ to give

$$U = \frac{Eh}{7.5}\left[\frac{5\pi^4}{64}\frac{w_0^4}{a^2} - \frac{17\pi^2}{6}\frac{cw_0^2}{a} + c^2\left(\frac{35\pi^2}{4} + \frac{80}{9}\right)\right] \tag{10.3.6}$$

wherein the application of the principle of virtual displacements leads to

$$\begin{aligned} c &= 0.147\tfrac{w_0^2}{a} \\ w_0 &= 0.802\left(\tfrac{pa^4}{Eh}\right)^{\frac{1}{3}} \end{aligned} \tag{10.3.7}$$

The in-plane tensile strain $\varepsilon_x = \varepsilon_y$ at the center of the membrane is obtained from either the second or third of Eq. (10.3.2)

$$\varepsilon = \frac{\partial u_x}{\partial x} = \frac{\partial u_y}{\partial y} = \frac{\pi c}{a} = 0.462\frac{w_0^2}{a^2} \tag{10.3.8}$$

From Hooke's law in 3D with $\sigma_z \sim 0$, we have the in-plane stress $\sigma_x = \sigma_y = E\varepsilon/(1 - v^2)$. Since this special case applies only for $v = 0.25$, we have

$$\sigma = 0.616E\frac{w_0^2}{a^2} \tag{10.4.9}$$

The more generic cases where $a \neq b$ with other Poisson's ratio (including negative values) are discussed in Sect. 10.5.

10.4 Example: Circular Auxetic Membranes

Consider again a circular membrane of radius R and thickness h with Young's modulus E and Poisson's ratio v as shown in Fig. 10.1 (top), the deflection w due to a uniform load p has been given as a function of the radial distance r from the membrane center according to Hencky (1915) as

$$\frac{w}{R} = \left(\frac{pR}{Eh}\right)^{\frac{1}{3}}\sum_{n=0}^{\infty} a_{2n}\left[1 - \left(\frac{r}{R}\right)^{2n+2}\right] \tag{10.4.1}$$

where the coefficients are

$$a_0 = b_0^{-1}$$

$$
\begin{aligned}
a_2 &= \frac{1}{2} b_0^{-4} \\
a_4 &= \frac{5}{9} b_0^{-7} \\
a_6 &= \frac{55}{72} b_0^{-10} \\
a_8 &= \frac{7}{6} b_0^{-13} \\
a_{10} &= \frac{205}{108} b_0^{-16} \\
a_{12} &= \frac{17,051}{5292} b_0^{-19} \\
a_{14} &= \frac{2,864,485}{508,032} b_0^{-22} \\
a_{16} &= \frac{103,863,265}{10,287,648} b_0^{-25} \\
a_{18} &= \frac{27,047,983}{1,469,664} b_0^{-28} \\
a_{20} &= \frac{42,367,613,873}{1,244,805,408} b_0^{-31}
\end{aligned}
\tag{10.4.2}
$$

in which b_0 is numerically solved from

$$
\begin{aligned}
&(1-v)b_0 - \frac{3-v}{b_0^2} - \frac{2(5-v)}{3b_0^5} - \frac{13(7-v)}{18b_0^8} - \frac{17(9-v)}{18b_0^{11}} - \frac{37(11-v)}{27b_0^{14}} \\
&- \frac{1205(13-v)}{567b_0^{17}} - \frac{219,241(15-v)}{63,504b_0^{20}} - \frac{6,634,069(17-v)}{1,143,072b_0^{23}} \\
&- \frac{51,523,763(19-v)}{5,143,824b_0^{26}} - \frac{998,796,305(21-v)}{56,582,064b_0^{29}} - \cdots = 0
\end{aligned}
\tag{10.4.3}
$$

A major setback of this analytical model to compute the membrane deflection is the need to perform three stages of calculations. In the first stage, a Poisson's ratio is selected so as to numerically solve for b_0 from Eq. (10.4.3). Thereafter, the coefficients a_{2n} in Eq. (10.4.2) are directly calculated from the obtained b_0. Finally, the membrane deflection is calculated from Eq. (10.4.1) using the values of a_{2n}, bearing in mind that this solution is based on the particular Poisson's ratio that was selected earlier for calculating b_0. In order to facilitate convenient calculation of circular membrane deflection for all Poisson's ratio within the bounds $-1 \le v \le 0.5$ for isotropic membrane materials, a closed-form semi-empirical solution is proposed as (Lim 2016)

$$
\frac{w}{R} = \left(\frac{pR}{Eh}\right)^{\frac{1}{3}} \left\{ c_0 - c_2\left(\frac{r}{R}\right)^2 - c_4\left(\frac{r}{R}\right)^4 \right\} \tag{10.4.4}
$$

where the coefficients c_i $(i = 0, 2, 4)$ are to be obtained by curve-fitting.

Hencky (1915) used only the first seven terms of Eq. (10.4.3) to solve b_0 for $v = 0.3$. The underestimated value is, however, due to a computation error, in which his last term's coefficient was $407/189$ instead of $1205/567$. This error was corrected by Campbell (1956), whose analysis includes initial tension. Both Hencky (1915) and Campbell (1956) calculated b_0 for $v = 0.3$ only, hence the validity of their subsequent results are confined to circular membranes of this Poisson's ratio only. By extending Eq. (10.4.3) up to the 11th term, Fichter (1997) obtained a more accurate value of b_0 for $v = 0.3$; a comparison between the b_0 results of Campbell (1956) and Fichter (1997) confirms the accuracy of the former up to three decimal places. In addition, Fichter (1997) obtained results of b_0 for $v = 0.2$ and $v = 0.4$, thereby allowing interpolated values of deflection for $0.2 \leq v \leq 0.4$ as well as extrapolated values of deflection for Poisson's ratio slightly outside this range but still within the confines of positive Poisson's ratio. The deflection results based on $v = 0.2$, $v = 0.3$ and $v = 0.4$, however, are not applicable for extrapolating the deflection of circular membranes into the auxetic range. As such the values of b_0 are computed for the entire range of Poisson's ratio of isotropic solids $-1 \leq v \leq 0.5$ as shown in Table 10.1. It can be seen that as the Poisson's ratio of the membrane material becomes more negative, the value of b_0 decreases at diminishing rate.

Suppose the deflection is expressed in the following dimensionless form

Table 10.1 Comparison of computed b_0 results

Poisson's ratio	Hencky (1915)	Campbell (1956)	Fichter (1997)	Lim (2016)	Remarks
0.5				1.845309	Conventional
0.4			1.7769	1.776866	Conventional
0.3	1.713	1.724	1.7244	1.724393	Conventional
0.2			1.6827	1.682711	Conventional
0.1				1.648699	Conventional
0				1.620353	Conventional
−0.1				1.596323	Auxetic
−0.2				1.575662	Auxetic
−0.3				1.557686	Auxetic
−0.4				1.541886	Auxetic
−0.5				1.527875	Auxetic
−0.6				1.515357	Auxetic
−0.7				1.504095	Auxetic
−0.8				1.493902	Auxetic
−0.9				1.484627	Auxetic
−1				1.476147	Auxetic

$$w^* = \frac{w}{R}\left(\frac{Eh}{pR}\right)^{\frac{1}{3}} \tag{10.4.5}$$

then the exact analytical model can be expressed as

$$\begin{aligned} w^* = &\frac{1}{b_0}\left(1-\frac{r^2}{R^2}\right)+\frac{1}{2b_0^4}\left(1-\frac{r^4}{R^4}\right)+\frac{5}{9b_0^7}\left(1-\frac{r^6}{R^6}\right)\\ &+\frac{55}{72b_0^{10}}\left(1-\frac{r^8}{R^8}\right)+\frac{7}{6b_0^{13}}\left(1-\frac{r^{10}}{R^{10}}\right)+\frac{205}{108b_0^{16}}\left(1-\frac{r^{12}}{R^{12}}\right)\\ &+\frac{17,051}{5292b_0^{19}}\left(1-\frac{r^{14}}{R^{14}}\right)+\frac{2,864,485}{508,032b_0^{22}}\left(1-\frac{r^{16}}{R^{16}}\right)\\ &+\frac{103,863,265}{10,287,648b_0^{25}}\left(1-\frac{r^{18}}{R^{18}}\right)+\frac{27,047,983}{1,469,664b_0^{28}}\left(1-\frac{r^{20}}{R^{20}}\right)\\ &+\frac{42,367,613,873}{1,244,805,408b_0^{31}}\left(1-\frac{r^{22}}{R^{22}}\right) \end{aligned} \tag{10.4.6}$$

where b_0 is solved from Eq. (10.4.3) for various Poisson's ratio. Using results from this exact solution, a first stage curve-fitting was performed to give the coefficients in Eq. (10.4.4), for each value of Poisson's ratio from $v = -1$ to $v = 0.5$ at increment of 0.1, as listed in Table 10.2.

Table 10.2 List of semi-empirical coefficients for various Poisson's ratio

v	c_0	c_2	c_4
−1.0	0.8473	0.6298	0.2136
−0.9	0.8375	0.6288	0.2051
−0.8	0.8273	0.6275	0.1964
−0.7	0.8165	0.6259	0.1875
−0.6	0.8050	0.6239	0.1783
−0.5	0.7929	0.6214	0.1689
−0.4	0.7799	0.6184	0.1592
−0.3	0.7660	0.6148	0.1492
−0.2	0.7511	0.6104	0.1389
−0.1	0.7349	0.6050	0.1283
0.0	0.7173	0.5986	0.1174
0.1	0.6981	0.5907	0.1062
0.2	0.6767	0.5812	0.0946
0.3	0.6529	0.5694	0.0828
0.4	0.6258	0.5547	0.0706
0.5	0.5946	0.5360	0.0582

From this result, a second-stage curve-fitting to fit the variation of the coefficients across the Poisson's ratio was then performed, thereby completing the surface fit, to express the coefficients of Eq. (10.4.4) in terms of the Poisson's ratio as

$$\begin{Bmatrix} c_0 \\ c_2 \\ c_4 \end{Bmatrix} = \frac{1}{10,000}\begin{bmatrix} 7176 & 1881 & 932 & 356 \\ 5989 & 754 & 782 & 346 \\ 1173 & 1107 & 146 & 0 \end{bmatrix}\begin{Bmatrix} 1 \\ -v \\ -v^2 \\ -v^3 \end{Bmatrix} \tag{10.4.7}$$

It should also be pointed out at this juncture that the solution in the form given by Eq. (10.4.4) is also a truncation of Eq. (10.4.1) with summation up to the term containing r^4/R^4, i.e.

$$\frac{w}{R} = \left(\frac{pR}{Eh}\right)^{\frac{1}{3}}\left\{(a_0 + a_2 + a_4 + \cdots) - a_0\left(\frac{r}{R}\right)^2 - a_2\left(\frac{r}{R}\right)^4\right\} \tag{10.4.8}$$

or, with reference to Eqs. (10.4.2) and (10.4.5),

$$w^* = \left\{\frac{1}{b_0} + \frac{1}{2b_0^4} + \frac{5}{9b_0^7} + \cdots\right\} - \frac{1}{b_0}\left(\frac{r}{R}\right)^2 - \frac{1}{2b_0^4}\left(\frac{r}{R}\right)^4 \tag{10.4.9}$$

In considering only the first three terms of a_{2n} ($n = 0, 1, 2$), the corresponding first three terms of Eq. (10.4.3)

$$(1 - v)b_0^6 - (3 - v)b_0^3 - \frac{2}{3}(5 - v) = 0 \tag{10.4.10}$$

can be solved to give several mathematical solutions but in which only one root is physically possible, which is

$$b_0 = \left[\frac{1}{2}\left(\frac{3 - v}{1 - v}\right)\left(1 + \sqrt{1 + \frac{8}{3}\frac{(1 - v)(5 - v)}{(3 - v)^2}}\right)\right]^{\frac{1}{3}} \tag{10.4.11}$$

To display the comparative validity of the semi-empirical solution described by Eqs. (10.4.4) and (10.4.7) with the truncated analytical solution described by Eqs. (10.4.9) and (10.4.11), the dimensionless deflection distribution by the exact method, Eq. (10.4.6), was indirectly solved and compared with the dimensionless deflection distribution by the semi-empirical method

$$\begin{aligned} w^* = & \frac{7176 - 1881v - 932v^2 - 356v^3}{10,000} - \frac{5989 - 754v - 782v^2 - 346v^3}{10,000}\left(\frac{r}{R}\right)^2 \\ & - \frac{1173 - 1107v - 146v^2}{10,000}\left(\frac{r}{R}\right)^4 \end{aligned} \tag{10.4.12}$$

and the dimensionless deflection distribution by the truncated analytical method

$$\begin{aligned} w^* = &\left\{ \left[\frac{1}{2}\left(\frac{3-v}{1-v}\right)\left(1+\sqrt{1+\frac{8}{3}\frac{(1-v)(5-v)}{(3-v)^2}}\right)\right]^{-\frac{1}{3}} \right. \\ &+\frac{1}{2}\left[\frac{1}{2}\left(\frac{3-v}{1-v}\right)\left(1+\sqrt{1+\frac{8}{3}\frac{(1-v)(5-v)}{(3-v)^2}}\right)\right]^{-\frac{4}{3}} \\ &\left. +\frac{5}{9}\left[\frac{1}{2}\left(\frac{3-v}{1-v}\right)\left(1+\sqrt{1+\frac{8}{3}\frac{(1-v)(5-v)}{(3-v)^2}}\right)\right]^{-\frac{7}{3}} \right\} \\ &-\left[\frac{1}{2}\left(\frac{3-v}{1-v}\right)\left(1+\sqrt{1+\frac{8}{3}\frac{(1-v)(5-v)}{(3-v)^2}}\right)\right]^{-\frac{1}{3}}\left(\frac{r}{R}\right)^2 \\ &-\frac{1}{2}\left[\frac{1}{2}\left(\frac{3-v}{1-v}\right)\left(1+\sqrt{1+\frac{8}{3}\frac{(1-v)(5-v)}{(3-v)^2}}\right)\right]^{-\frac{4}{3}}\left(\frac{r}{R}\right)^4 \end{aligned} \quad (10.4.13)$$

both of which take the form described in Eq. (10.4.4) and can be solved directly.

To assess the accuracy of the semi-empirical model vis-à-vis the truncated analytical model, a comparison is made between the dimensionless deflection distributions of these two models with that of the exact analytical model, in terms of w^* versus r/R, at a typical Poisson's ratio value of $v = 0.3$. Figure 10.3 shows that the truncated analytical model overestimates the exact analytical model while the semi-empirical model agrees well with the exact analytical model. Figure 10.3 further suggests that the semi-empirical model, in addition to being more convenient than the truncated analytical model, gives superior accuracy at $v = 0.3$. The better accuracy of the semi-empirical model in comparison to the truncated analytical model is attributed to the former taking more terms from the exact analytical model into consideration as compared to the truncated analytical model.

Having established the accuracy of the semi-empirical model, there is now a need to investigate the consistency of its accuracy for the entire range of Poisson's ratio of the isotropic membrane material. Figure 10.4 compares the distribution of dimensionless deflection, w^*, versus dimensionless radial distance, r/R; for clarity, the families of $w*$ curves with even and odd values of $10v$ are separately plotted. A very good agreement can be observed between the exact and the semi-empirical solutions, thereby verifying the validity of the proposed semi-empirical model across the entire Poisson's ratio range for isotropic solids. It follows that there exists a separation of auxetic and conventional regions bounded by $v = -1, 0, 0.5$, as shown in Fig. 10.5. Although the Poisson's ratio range in the auxetic region ($-1 \le v < 0$) is twice that in the conventional region ($0 \le v \le 0.5$), the area bounded by both regions in Fig. 10.5 is almost equal. This is due to the diminishing increment of the membrane deflection as the Poisson's ratio becomes more negative.

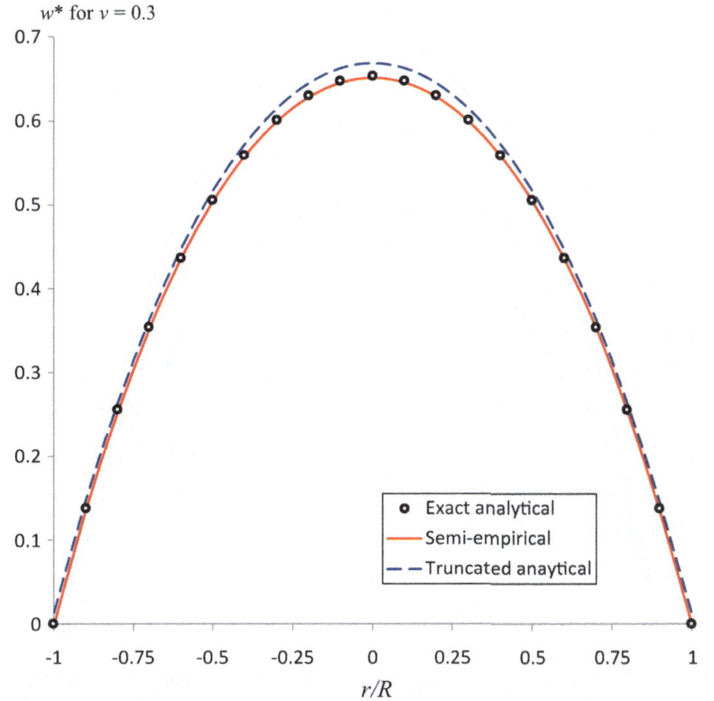

Fig. 10.3 Comparison of the truncated analytical deflection (dashed curve) and the semi-empirical deflection (continuous curve) against the exact analytical deflection (circles)

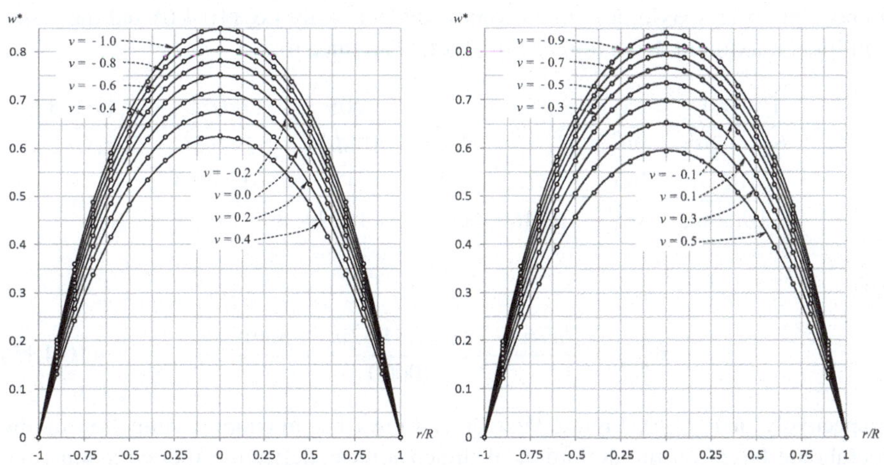

Fig. 10.4 Comparison between the dimensionless deflection distributions between the exact model (discrete data points) and the semi-empirical model (continuous curves) for $\upsilon = -1, -0.8, -0.6, -0.4, -0.2, 0, 0.2\&0.4$ (left), and $\upsilon = -0.9, -0.7, -0.5, -0.3, -0.1, 0.1, 0.3$ and 0.5 (right)

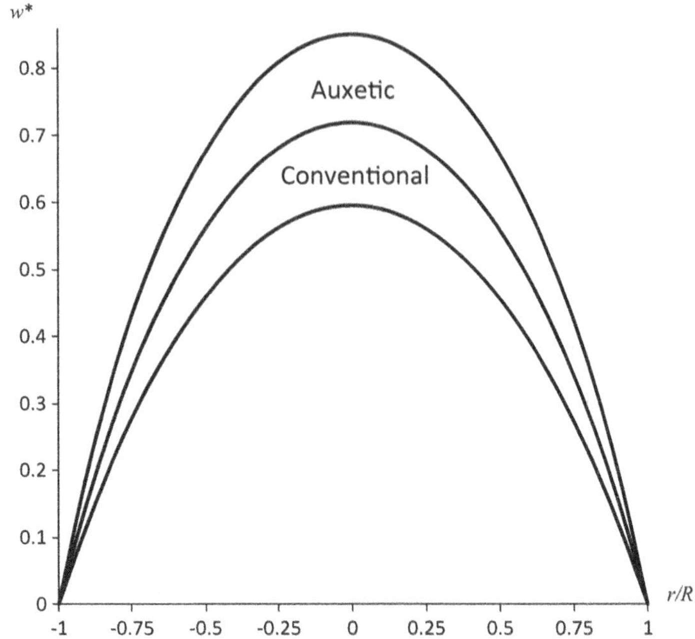

Fig. 10.5 Loci of dimensionless deflection versus dimensionless radial distance for the auxetic and conventional regions

The deflection is obviously maximum at the center of the membrane, $r = 0$. Hence, the exact maximum dimensionless deflection by Eq. (10.4.6) and the semi-empirical maximum deflection of Eq. (10.4.12) reduce to

$$w^*_{\max} = \frac{1}{b_0} + \frac{1}{2b_0^4} + \frac{5}{9b_0^7} + \frac{55}{72b_0^{10}} + \frac{7}{6b_0^{13}} + \frac{205}{108b_0^{16}} + \frac{17,051}{5292b_0^{19}} + \frac{2,864,485}{508,032b_0^{22}} + \frac{103,863,265}{10,287,648b_0^{25}} + \frac{27,047,983}{1,469,664b_0^{28}} + \frac{42,367,613,873}{1,244,805,408b_0^{31}} \quad (10.4.14)$$

and

$$w^*_{\max} = \frac{7176 - 1881v - 932v^2 - 356v^3}{10000} \quad (10.4.15)$$

respectively at $r = 0$. Figure 10.6 (left) shows the maximum deflection of the membrane based on the definition of dimensionless deflection plotted against the dimensionless radial distance. The exact and semi-empirical plots are denoted by discrete points and discontinuous curve, respectively. There appears to be a slight underestimation of the maximum deflection by the semi-empirical model when $r = 0$ is simply substituted into Eq. (10.4.12) to yield Eq. (10.4.15), especially in the auxetic

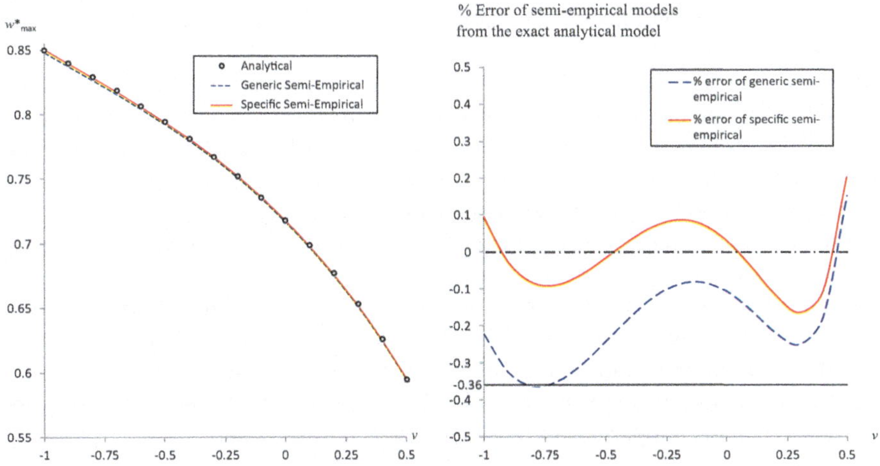

Fig. 10.6 Plots of dimensionless maximum membrane deflection versus the dimensionless radial distance for generic and specific semi-empirical results compared against exact solution (left), and percentage error of generic and specific semi-empirical with reference to the exact solution (right)

region. When measured as a percentage error with reference to the exact solution, the maximum percentage error is approximately -0.36% at $\upsilon = -0.8$, as shown by the discontinuous curve in Fig. 10.6 (right). Although this error magnitude is very small and can be neglected, it is of interest to understand the reason for its overall trend.

In the curve-fitting process for obtaining the semi-empirical model, the dimensionless deflection was fitted as a function of both the Poisson's ratio and the dimensionless radial distance, with the error being generally underestimated and overestimated at the central and outer portions of the membrane. As a result, the mere substitution of $r = 0$ into Eq. (10.4.12) inherently leads to deflection underestimation therein. Reference to Fig. 10.6 (right) shows that the percentage error is very small especially in the conventional region. To obtain a more accurate maximum deflection model for auxetic membranes, it is useful to curve-fit the maximum deflection directly from the exact solution. The resulting dimensionless semi-empirical maximum deflection is thus

$$w^*_{\max} = \frac{7186 - 1896\upsilon - 930\upsilon^2 - 356\upsilon^3}{10,000} \tag{10.4.16}$$

To differentiate both semi-empirical models for the maximum deflection, we define Eq. (10.4.15) as the generic semi-empirical model, for it comes from Eq. (10.4.12) which includes the radial distance, and the specific semi-empirical model, for it is specifically curve-fitted at the maximum deflection. Reference to Fig. 10.6 (left) shows the extremely good agreement between the specific semi-empirical model for the maximum deflection and the exact model. When assessed in terms of the percentage error, the maximum error of 0.2% takes place in the conventional region, or $v = 0.5$ to be precise. In the auxetic range, the error magnitude is less than 0.1%. Perusal to Fig. 10.6 (left) again provides evidence that as the Poisson's ratio of the membrane becomes more negative the maximum deflection increases, but this change takes place with diminishing effect. This observation is not surprising, as biaxial stretching is facilitated by negative Poisson's ratio. Consequently, the deflection of the entire membrane increases with auxeticity.

Finally, it is of interest to evaluate the combined effect of auxeticity on large deformation vis-à-vis infinitesimal deformation theory of thin plates. To do so, a normalized deflection distribution is introduced as $w/w_{r=v=0}$; this ratio normalizes the deflection distribution against the deflection at the pole ($r = 0$) and, to observe the effect of Poisson's ratio sign, at zero Poisson's ratio. For the infinitesimal theory, the flexural rigidity described in Eq. (4.1.3) is substituted into the plate deflection profile of a simply supported circular plate under uniform load

$$w = \frac{p}{64D}\left(R^2 - r^2\right)\left(\frac{5+v}{1+v}R^2 - r^2\right) \tag{10.4.17}$$

to give

$$w = \frac{3p\left(1 - v^2\right)}{16Eh^3}\left(R^2 - r^2\right)\left(\frac{5+v}{1+v}R^2 - r^2\right) \tag{10.4.18}$$

and hence

$$\frac{w}{w_{r=v=0}} = \frac{1}{5}\left(1 - \frac{r^2}{R^2}\right)\left[(5+v)(1-v) - \left(1 - v^2\right)\frac{r^2}{R^2}\right] \tag{10.4.19}$$

while the corresponding ratio for large deflection of membranes is based on Eq. (10.4.12)

$$\begin{aligned}\frac{w}{w_{r=v=0}} = \frac{1}{0.7176}\Bigg[&\frac{7176 - 1881v - 932v^2 - 356v^3}{10,000} \\ &- \frac{5989 - 754v - 782v^2 - 346v^3}{10,000}\left(\frac{r}{R}\right)^2 \\ &- \frac{1173 - 1107v - 146v^2}{10,000}\left(\frac{r}{R}\right)^4\Bigg]\end{aligned} \tag{10.4.20}$$

due to its impeccable accuracy. The comparison of membrane deflection is made herein against that of the simply supported circular plates due to the similarity of boundary condition for both problems. Figure 10.7 shows that while auxeticity increases deflection, the increment of deflection is lower for large deformation of membranes than for infinitesimal deformation of thin plates when normalization is performed against their respective maximum deflection ($r = 0$) at $v = 0$. In other

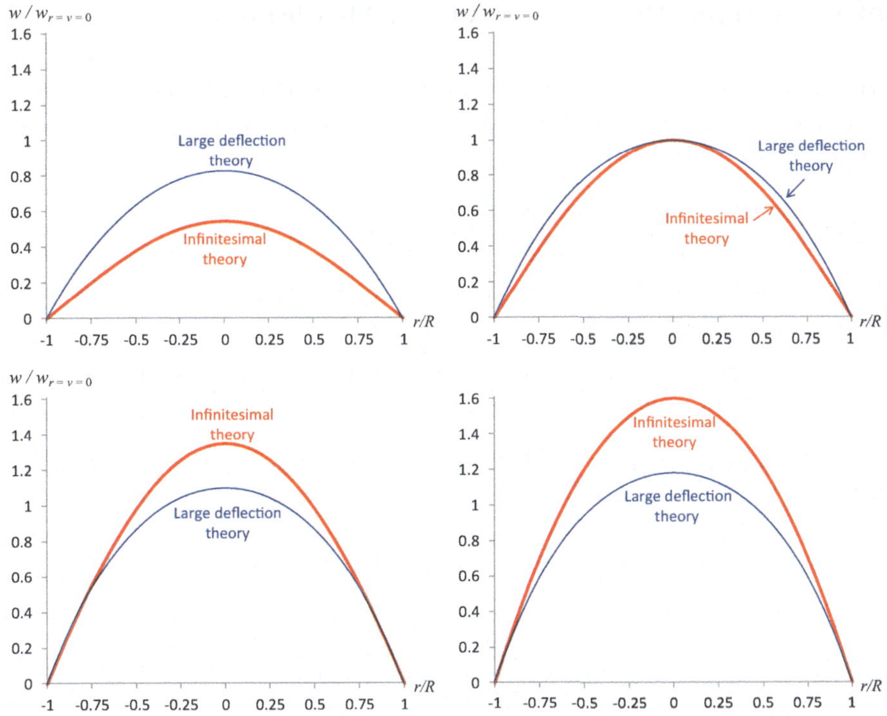

Fig. 10.7 Comparison of normalized deflection distributions between infinitesimal deformation theory of plates with the large deformation theory of membranes with $v = 0.5$ (top left), $v = 0$ (top right), $v = -0.5$ (bottom left), and (d) $v = -1$ (bottom right)

words, an auxetic membrane performs better than an auxetic plate under the selected normalization criterion.

In summary, the plotted deflection results reveal that the negativity of Poisson's ratio facilitates deflection but with diminishing effect. Using these exact results, a semi-empirical model has been developed to facilitate convenient and direct calculation of the membrane deflection. The generic semi-empirical model exhibits excellent agreement with the exact result, with the highest error magnitude at 0.36% at the point of maximum deflection, and this reduces to 0.2% for the specific semi-empirical model. It is known that membranes are compliant and are therefore not used for structural support functions; they are used as diaphragms or other extremely thin-walled structures that function by large deflection. Hence, auxetic membranes exhibit greater flexibility and stretching (bio)functions in comparison conventional membranes, at equal Young's modulus.

10.5 Example: Rectangular Auxetic Membranes

With reference to the rectangular membrane furnished in Fig. 10.2 and its boundary conditions, the strain energy has been obtained from the assumed displacements given in Eqs. (10.3.1) and (10.3.2) as (Conway, 1946)

$$U = \frac{Eh\pi^2}{2(1-v^2)} \left\{ \frac{\pi^2 w_0^4}{1024ab} \left(9\frac{a^2}{b^2} + 9\frac{b^2}{a^2} + 2 \right) - \frac{w_0^2 c_x}{6} \left(\frac{2b}{a^2} + \frac{1-3v}{2b} \right) - \frac{w_0^2 c_y}{6} \left(\frac{2a}{b^2} + \frac{1-3v}{2a} \right) + c_x^2 \left(\frac{b}{a} + \frac{1-v}{8}\frac{a}{b} \right) + c_y^2 \left(\frac{a}{b} + \frac{1-v}{8}\frac{b}{a} \right) + \frac{16(1+v)}{9\pi^2} c_x c_y \right\} \quad (10.5.1)$$

The arbitrary parameters c_x and c_y are evaluated by the minimization of U, as per the first of Eq. (10.3.5), to give

$$c_x = 3\pi^2 \frac{w_0^2}{ab} \frac{\begin{vmatrix} \frac{a(16a^2+b^2)}{9\pi^2(32a^2+3b^2)} & \frac{b(16b^2+a^2)}{320ab} \end{vmatrix}}{\begin{vmatrix} \frac{320ab}{9\pi^2(32b^2+3a^2)} & \frac{9\pi^2(32a^2+3b^2)}{320ab} \end{vmatrix}} \quad (10.5.2)$$

and

$$c_y = 3\pi^2 \frac{w_0^2}{ab} \frac{\begin{vmatrix} \frac{b(16b^2+a^2)}{9\pi^2(32b^2+3a^2)} & \frac{a(16a^2+b^2)}{320ab} \end{vmatrix}}{\begin{vmatrix} \frac{320ab}{9\pi^2(32b^2+3a^2)} & \frac{9\pi^2(32a^2+3b^2)}{320ab} \end{vmatrix}} \quad (10.5.3)$$

Applying the principle of virtual displacement as described in the second of Eq. (10.3.5), from which

$$\frac{\partial U}{\partial w_0} = p\frac{16ab}{\pi^2} \quad (10.5.4)$$

leads to the deflection at the membrane center

$$w_0 = \gamma \left(\frac{pa^4}{Eh} \right)^{\frac{1}{3}} \quad (10.5.5)$$

where γ is dependent on the membrane aspect ratio a/b. For convenience, Conway (1946) introduced coefficients α and β for selected values of c_x and c_y in the form

Table 10.3 Coefficients α, β and γ for various membrane aspect ratios by Conway (1946)

a/b	α	β	γ
1.0	0.1468	0.1468	0.818
1.5	0.1238	0.2217	0.569
2.0	0.0982	0.3254	0.411
3.0	0.0503	0.4964	0.248
4.0	0.0194	0.6650	0.171

$$\begin{Bmatrix} c_x \\ c_y \end{Bmatrix} = \frac{w_0^2}{a} \begin{Bmatrix} \alpha \\ \beta \end{Bmatrix} \tag{10.5.6}$$

while the values of γ were calculated from the abovementioned principle of virtual displacement; these coefficients were calculated for $a/b = 1, 1.5, 2, 3$ and 4 (see Table 10.3).

To facilitate convenient analysis of such rectangular membranes at other membrane aspect ratios, empirical modeling from Table 10.3 gives the following simplified relations

$$\alpha = 0.1873 - 0.0433\frac{a}{b} \tag{10.5.7}$$

$$\beta = 0.1745\frac{a}{b} - 0.03031 \tag{10.5.8}$$

$$\gamma = 0.8684\frac{b}{a} - 0.034 \tag{10.5.9}$$

that are applicable for $1 \leq a/b \leq 4$. The validity of the empirical models can be established by comparing them to the data by Conway (1946), as shown in Fig. 10.8. The use of Eqs. (10.5.7)–(10.5.9) allows calculations of c_x, c_y and w_0 that are simpler and more direct than Eqs. (10.5.2)–(10.5.4), respectively.

Having established the correlation between the empirical models and the exact values of the coefficient α, β, and γ, we now apply them for evaluating the extent of stresses

$$\begin{Bmatrix} \sigma_x \\ \sigma_y \end{Bmatrix}_{x=y=0} = \frac{\pi E}{1 - v^2} \begin{bmatrix} 1 & v \\ v & 1 \end{bmatrix} \begin{Bmatrix} c_x/a \\ c_y/b \end{Bmatrix} \tag{10.5.10}$$

at the center of the rectangular membranes. Where stresses are concerned, Conway (1946) evaluated the maximum membrane stress only for the case where $v = 0.25$. Since $(\sigma_y)_{x=y=0} > (\sigma_x)_{x=y=0}$ for $a > b$, we evaluate the maximum membrane stresses in terms of σ_y at the membrane center. As it is well known that the membrane stresses are aggravated by their Young's modulus and the applied load, a dimensionless maximum stress is introduced as

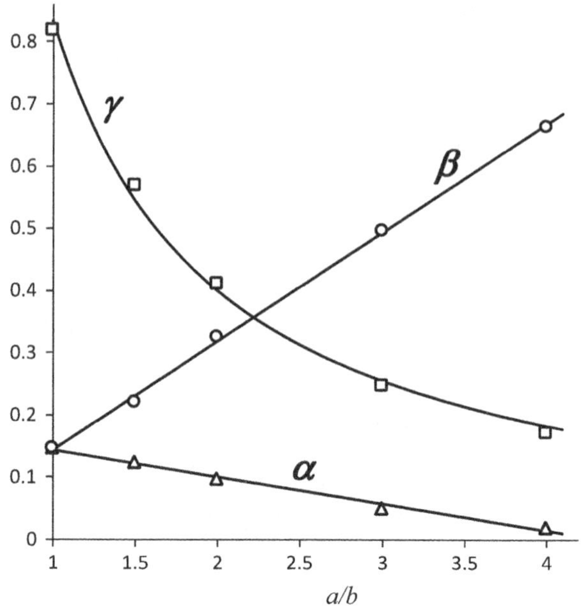

Fig. 10.8 Comparison between exact (points) and empirical (lines) values of coefficients α, β and γ

$$\sigma^*_{\max} = \frac{(\sigma_y)_{x=y=0}}{E}\frac{ab}{\pi}\left(\frac{Eh}{pa^4}\right)^{\frac{2}{3}} \tag{10.5.11}$$

to give

$$\sigma^*_{\max} = \frac{\gamma^2}{1-v^2}\left(\beta + v\alpha\frac{b}{a}\right) \tag{10.5.12}$$

so as to allow direct observation on the effects of Poisson's ratio and the aspect ratio of the membrane. Substitution of Eqs. (10.5.7)–(10.5.9) into Eq. (10.5.12) gives

$$\sigma^*_{\max} = \frac{1}{1-v^2}\left(0.8684\frac{b}{a} - 0.034\right)^2 \left\{0.1745\frac{a}{b} - 0.03031 + v\left(0.1873\frac{b}{a} - 0.0433\right)\right\} \tag{10.5.13}$$

To find the optimum Poisson's ratio, that minimizes the dimensionless maximum stress, we impose

$$\frac{\partial \sigma^*_{\max}}{\partial v} = 0 \tag{10.5.14}$$

on Eq. (10.5.12) to give two possible mathematical solutions on the optimal Poisson's ratio (Lim 2020)

$$v_{\mathrm{opt}} = -\frac{a}{b}\frac{\beta}{\alpha} \pm \sqrt{\left(\frac{a}{b}\frac{\beta}{\alpha}\right)^2 - 1} \tag{10.5.15}$$

Since only one solution exists, the choice of physically admissible solution from Eq. (10.5.15) has to be made based on the bounds of Poisson's ratio for isotropic materials, as will be shown later in Eq. (10.5.16).

A comparison on the effect of Poisson's ratio negativity can be made using Eq. (10.5.13) for $-0.5 \leq v \leq 0.5$. This is the range of Poisson's ratio at which one can compare mechanical responses for Poisson's ratio of opposite signs while retaining the magnitude, hence comparison is therefore made on purely the effect of Poisson's ratio negativity but not its magnitude. Figure 10.9 shows that for $v = 0$, the dimensionless maximum stress exhibits an exponential-like decay as the membrane's aspect ratio increases. Perusal to the maximum stress plots for $v = \pm 0.1, \pm 0.3 and \pm 0.5$ shows that the trend remains the same for conventional membranes, in which the dimensionless maximum stress increases with the Poisson's ratio. However, in the case of auxetic membranes, it is observed that there is no clear increase in the dimensionless maximum stress within $-0.5 \leq v < 0$ and that a buildup of maximum point becomes more evident as the Poisson's ratio becomes more negative. More importantly, the dimensionless maximum stress is lower when the membrane is auxetic in comparison to conventional membranes for the same Poisson's ratio magnitude. This is attributed to the ease of forming synclastic deformation when the Poisson's ratio is negative. The remaining Poisson's ratio range $-0.999 \leq v < -0.6$, for which there is no counterpart in the conventional range, is also furnished for the sake of completeness. Due to the presence of $1 - v^2$ at the denominator of Eq. (10.5.13), the plot of $v = -0.999$ was used instead of $v = -1$. As the aspect ratio increases, the portion of the membrane away from the short side approximates a cylindrical shape. When a flat surface transforms into a surface of single (not double) curvature, the Poisson's ratio sign becomes insignificant. An alternate view of Fig. 10.9 is given in Fig. 10.10, for emphasizing the variation of the dimensionless maximum stress for gradual change in Poisson's ratio.

To observe the effect from a wider range of auxeticity, $-1 < v < 0$, a family of dimensionless maximum stress curves is plotted against the membrane aspect ratio from $v = -0.1$ to $v = -0.9$. Plotted results in Fig. 10.11 (top) show that the peak dimensionless maximum stress increases and occurs at higher membrane aspect ratio (indicated by a dashed curve) as the membrane's Poisson's ratio becomes more negative. These dimensionless maximum stress curves also suggest that stresses can be minimized by judicious choice of Poisson's ratio indicated by the lower boundary formed in Fig. 10.11 (top). A zoomed in view of this boundary is shown in Fig. 10.11 (bottom), indicating that the dimensionless maximum stress can be minimized by controlling the Poisson's ratio. Specifically, the optimal Poisson's ratio is $v = 0$ as the membrane becomes very long and narrow; the optimal Poisson's ratio tends to

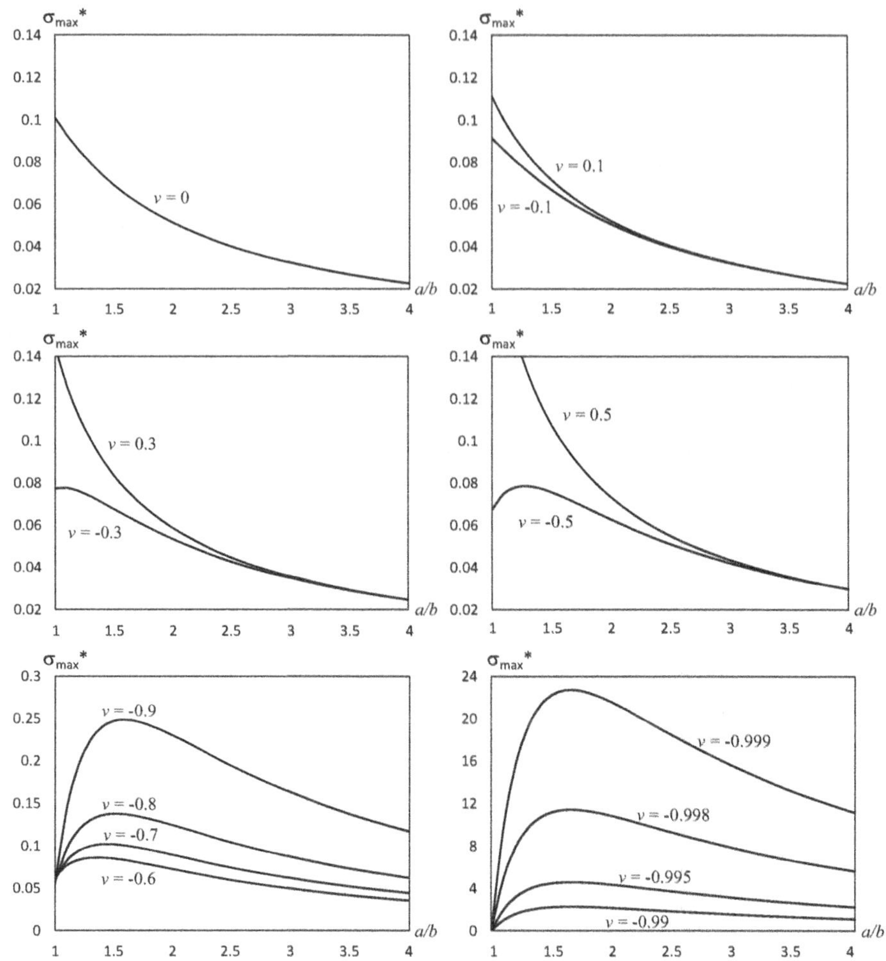

Fig. 10.9 Effect from the negativity of the Poisson's ratio on the dimensionless maximum membrane stresses

$\upsilon \to -1$ for square membranes. The former observation is attributed to the tendency for very long and narrow membranes to deform into a cylindrical-like shell, hence a Poisson's ratio of $\upsilon = 0$ would give the least stress. The latter observation is attributed to the tendency for the square membrane to approximate a spherical shell at its central portion, such that the value of $\upsilon = -1$ would give the lowest stress. These two extremes suggest that a negative Poisson's ratio is beneficial in lowering the membrane stresses of rectangular membranes.

While the optimal Poisson's ratio in Fig. 10.11 (bottom) is based on piecewise increments of Poisson's ratio of 0.1, a continuous and smooth relationship between this optimal Poisson's ratio and its corresponding minimized dimensionless

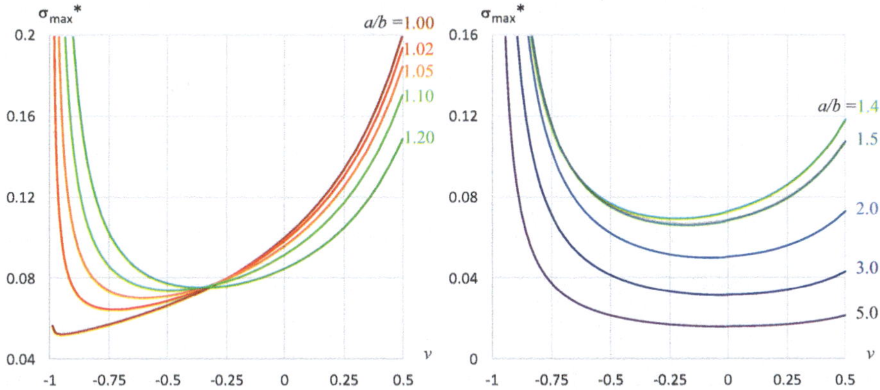

Fig. 10.10 Variation of the dimensionless maximum stress with the Poisson's ratio for square and almost square membranes (left), and rectangle membranes of higher aspect ratio (right)

maximum stress can be obtained as follows. Substituting Eqs. (10.5.7) and (10.5.8) into Eq. (10.5.15) gives

$$-\infty < v_{\text{opt}}^{\text{low}} < -1 < v_{\text{opt}}^{\text{upp}} < 0 \tag{10.5.16}$$

for $1 \leq a/b \leq 4$, hence the lower solution is discarded in favor of the upper solution. Based on the latter, the optimal Poisson's ratio

$$v_{\text{opt}} = -f(a,b) + \sqrt{(f(a,b))^2 - 1} \tag{10.5.17}$$

where

$$f(a,b) = \frac{0.1745\frac{a}{b} - 0.03031}{0.1873\frac{b}{a} - 0.0433} \tag{10.5.18}$$

can be substituted into Eq. (10.5.13) to give the minimized dimensionless maximum stress (Lim 2020)

$$\min_{\frac{a}{b}\in[1,4]} \sigma_{\max}^* = \left(\sigma_{\max}^*\right)_{v=v_{\text{opt}}} \tag{10.5.19}$$

both of which are plotted in Fig. 10.12 with reference to the membrane aspect ratio. While minimization of the dimensionless maximum stress is effectively carried out for square membranes as well as long and narrow membranes via $v = -1$ and $v = 0$, respectively, the minimization is less effective for slightly rectangular membranes of aspect ratio $1.1 < a/b \leq 1.2$ whereby the required Poisson's ratio would be about $v \sim -0.4$.

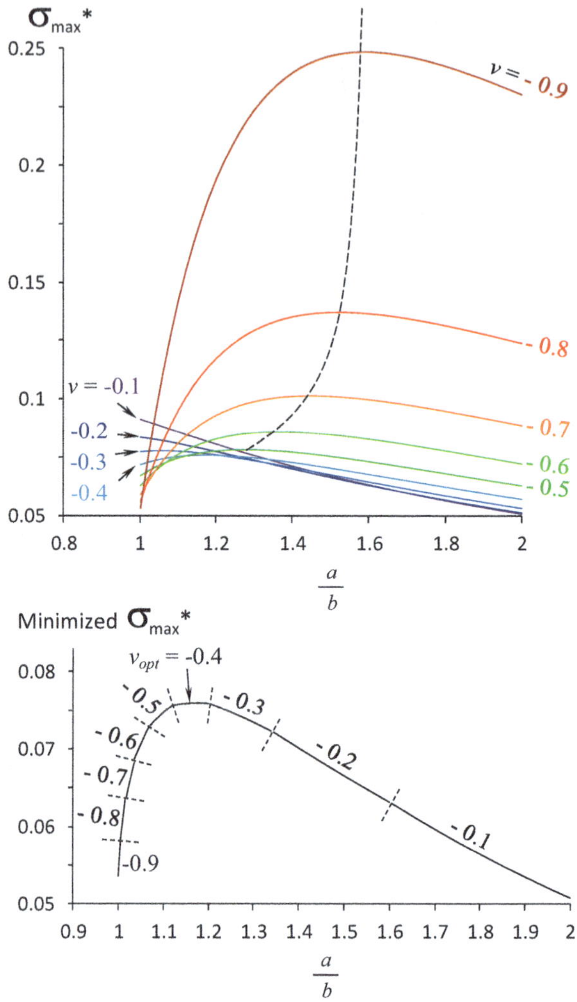

Fig. 10.11 Dimensionless maximum stresses for rectangular auxetic membranes, showing: locus of maximum stress indicated by dashed curve (top) and optimal Poisson's ratio, that minimizes the maximum stresses for various membrane aspect ratio (bottom)

Since the influence of the membrane's Young's modulus, relative thickness and load are obvious, a dimensionless maximum stress description has been introduced to observe the less obvious effects stemming from the Poisson's ratio and aspect ratio of the membrane. Results from the empirical model reveal that:

(a) as the membrane aspect ratio increases, the dimensionless maximum stress decreases monotonically and non-monotonically for conventional and auxetic membranes, respectively

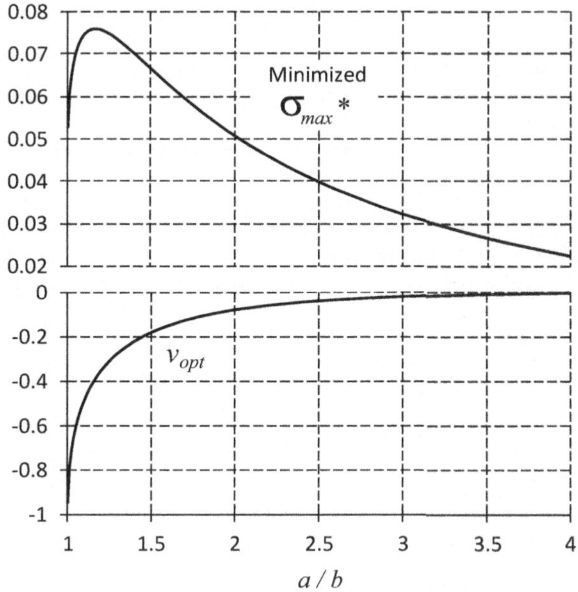

Fig. 10.12 Curves of dimensionless maximum stress, minimized with respect to the Poisson's ratio (top), and the corresponding optimal Poisson's ratio (bottom) plotted against the membrane aspect ratio

(b) for the same Poisson's ratio magnitude, the dimensionless maximum stress is lower in auxetic membranes than in conventional ones
(c) minimization of the dimensionless maximum stress is very effective for square membranes and very long and narrow membranes with the use of $\upsilon = -1$ and $\upsilon = 0$, but less effective for slightly rectangular membranes ($1.1 < a/b \leq 1.2$) and
(d) it is easier to control the dimensionless maximum stress for $a/b \geq 1.2$ due to the more gradual change in the optimal Poisson's ratio.

In other words, the use of auxetic materials with low Young's modulus is recommended for lowering the extent of stresses in rectangular membranes under uniform load.

References

Campbell JD (1956) On the theory of initially tensioned circular membranes subjected to uniform pressure. Q J Mech Appl Mech 9(1):84–93
Conway HD (1946) The large deflections of rectangular membranes and plates. Lond Edinb Dublin Philos Mag J Sci 37(274):767–778
Fichter WB (1997) Some solutions for the large deflections of uniformly loaded circular membranes. NASA Technical paper 3658

Hencky H (1915) On the stress state in circular plates with vanishing bending stiffnesses. Zeitschrift für Mathematik und Physik 63:311–317

Lim TC (2015) Auxetic materials and structures. Springer, Singapore

Lim TC (2016) Large deflection of circular auxetic membranes under uniform load. J Eng Mater Technol 138(4):041011

Lim TC (2020) Maximum stresses in rectangular auxetic membranes. Physica Status Solidi B 257:2000300

Reissner E (1949) On finite deflections of circular plate. In: Reissner E, Prager W, Stoker JJ (eds) Proceedings of symposia in applied mathematics, vol 1. American Mathematical Society, pp 213–219. https://doi.org/10.1090/psapm/001

Timoshenko SP, Woinowsky-Krieger S (1959) Theory of plates and shells, 2nd edn. McGraw-Hill, New York

Chapter 11
Negative Thermal Expansion

Abstract This chapter reviews the various 2D NTE systems, including those constructed from bimaterial strips, laminates (of various stiffness disparity), trusses (of triangular cells, Y-shaped elements, and Hoberman circle), meshes, rigid unit modes, and ring-rod assemblies (both 2D and 3D). Finally, a few examples of 3D NTE structures are briefly mentioned.

Keyword Negative thermal expansion

11.1 Fundamentals and Historical Development

For an original length L_0 of a solid at a reference temperature T_0, an increase in temperature $\mathrm{d}T = T - T_0$ brings about an increase to the length by an amount $\mathrm{d}L = L - L_0$. The thermal strain in terms of dimensional change is no different from the strain due to uniaxial loading

$$\varepsilon_{\mathrm{L}}^{(T)} = \frac{\mathrm{d}L}{L_0} \tag{11.1.1}$$

while the thermal strain in terms of temperature change is

$$\varepsilon_{\mathrm{L}}^{(T)} = \alpha_{\mathrm{L}}^{(T)} \mathrm{d}T \tag{11.1.2}$$

where $\alpha_{\mathrm{L}}^{(T)}$, the linear coefficient of thermal expansion (CTE) is a proportionality constant that describes the extent of strain per unit change in temperature. From Eqs. (11.1.1) and (11.1.2), the linear CTE can be written as

$$\alpha_{\mathrm{L}}^{(T)} = \frac{1}{L_0} \frac{\mathrm{d}L}{\mathrm{d}T} \tag{11.1.3}$$

Typically, an increase in dimension $\mathrm{d}L > 0$ results from an increase in the temperature of the solid $\mathrm{d}T > 0$, or a decrease in dimension $\mathrm{d}L < 0$ results from a decrease in the temperature of the solid $\mathrm{d}T < 0$. For this reason, the linear CTE is normally

T. Lim, *Mechanics of Metamaterials with Negative Parameters*, Engineering Materials,
https://doi.org/10.1007/978-981-15-6446-8_11

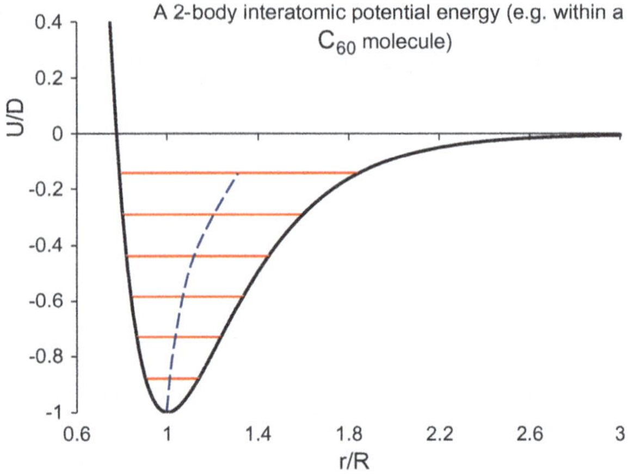

Fig. 11.1 An example of anharmonicity (see blue-dashed curve) passing through the center of the mean bond length for various energy levels (red lines) for a 2-body bond energy between bonded carbon atoms within each C_{60} molecules. Adapted from Lim (2010)

positive because $dL/dT > 0$, and this can be attributed to the anharmonicity of the lattice vibration. See Fig. 11.1 for example. Suppose the changes in the lattice vibration frequency arising from anharmonicity is linearly correlated to the infinitesimal change in the volume V, then the linear CTE is described by

$$\alpha_{\mathrm{L}}^{(T)} = \frac{\gamma_{\mathrm{G}} C_{\mathrm{Deb}}}{3\mathrm{KV}} \tag{11.1.4}$$

where γ_{G} is the Grüneisen parameter, C_{Deb} is the Debye specific heat, and K is the bulk modulus.

Two illustrations of local vibrational modes responsible for NTE are illustrated in Fig. 11.2 for single-atom linkage (left) and diatomic linkage (right). The larger circles (in black) refer to heavier atoms such as metals, while the smaller circles (in white) correspond with lighter atoms such as C, N, or O. The vibration of the lighter atoms in transverse direction to the bond, due to heating, causes the heavier atoms

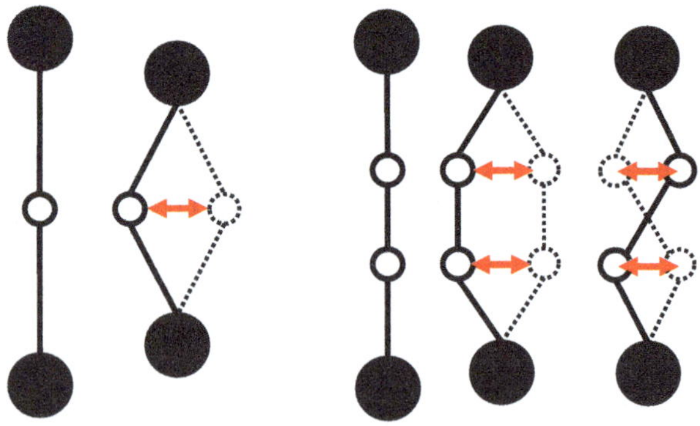

Fig. 11.2 Local vibrational modes responsible for with single-atom (left) and diatomic (right) linkages

to move closer to one another. This results in NTE along the line formed by the two heavier atoms.

A solid is said to exhibit negative thermal expansion (NTE) when a decrease in dimension $dL < 0$ results from an increase in the temperature of the solid $dT > 0$, or an increase in dimension $dL > 0$ results from a decrease in the temperature of the solid $dT < 0$. Under such a condition the solid is said to possess negative thermal expansion (NTE), and consequently its CTE is negative, $\alpha_L^{(T)} < 0$. Some solids may exhibit NTE in one direction but conventional, or positive thermal expansion (PTE), in another direction. The overall negativity of thermal expansion for these solids is therefore assessed by considering its volumetric CTE

$$\alpha_V^{(T)} = \frac{1}{V_0}\frac{dV}{dT} \tag{11.1.5}$$

for a change in volume per unit temperature change dV/dT in a solid of original volume V_0 at reference temperature T_0. For comparison with gases, recall that its CTE is

$$\alpha_V^{(T)} = \frac{1}{V_0}\left(\frac{dV}{dT}\right)_{P=\text{const.}} \tag{11.1.6}$$

at constant pressure. In an anisotropic solid that exhibits NTE and PTE in different directions, overall NTE is defined by $\alpha_V^{(T)} < 0$. For completeness' sake, one may also define an areal CTE in a similar fashion for an area change per temperature change dA/dT

$$\alpha_A^{(T)} = \frac{1}{A_0}\frac{dA}{dT} \tag{11.1.7}$$

where A_0 can the original area for a 2D material or an original area on a plane of consideration. If both NTE and PTE exist at right angles from one another within the plane of consideration, then overall planar NTE is said to be demonstrated if $\alpha_A^{(T)} < 0$ for that plane. Historically, the shrinkage of a solid was reported by Scheel (1907a, b) for quartz and vitreous silica at low temperatures at the turn of the previous century, and further documentations of materials that contract over various temperature ranges appeared in the literature throughout the years, including work on lithium aluminum silicates by Hummel (1948, 1951) half a century later.

A number of landmark reports on NTE have been made before the turn of the millennium; these include, but not limited to, the works by Chu et al. (1987), Sleight (1995, 1998a, b), Mary et al. (1996), Evans et al. (1999). A recent review on NTE by Takenaka (2018) classifies NTE materials into two broad categories: "conventional NTE" and phase transition-type of NTE, as outlined in Fig. 11.3. The working mechanisms of these NTE systems are illustrated by Takenaka (2018) as follows. The schematic of anisotropic thermal expansion in the silicates is shown in Fig. 11.4 (top).

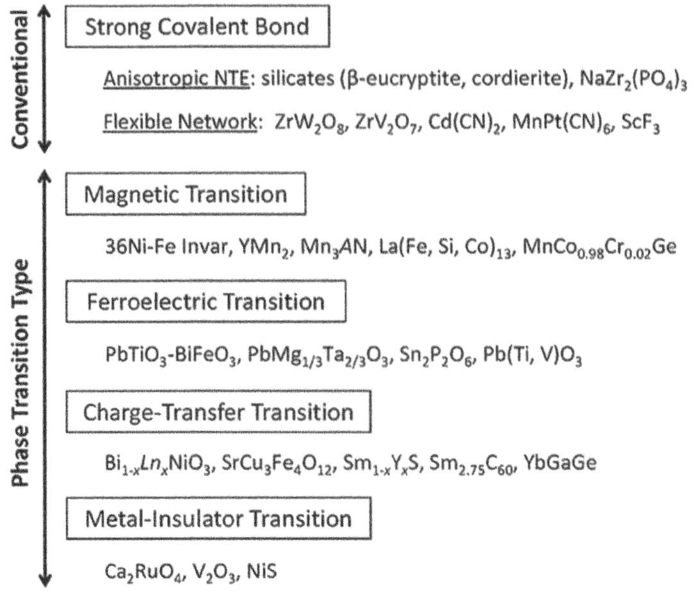

Fig. 11.3 Classification of negative thermal expansion materials by Takenaka (2018)

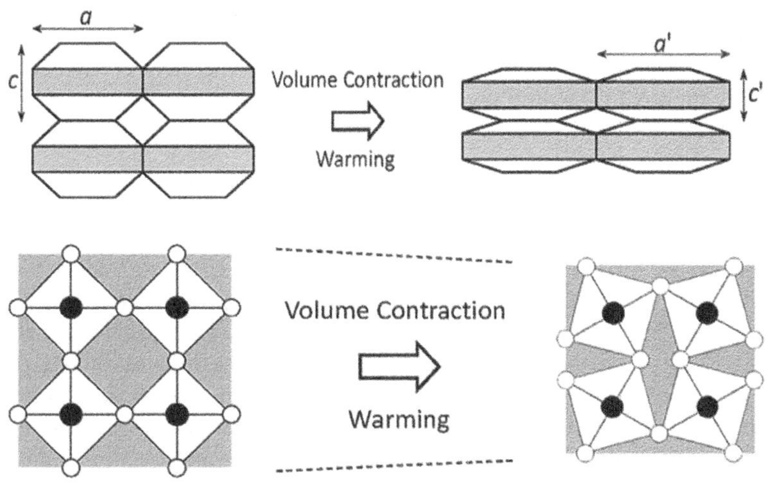

Fig. 11.4 Schematics of NTE in silicates (top) and in flexible network (bottom), as reviewed by Takenaka (2018)

As the shaded layers undergo thermal expansion, they are pulled closer together in the direction perpendicular to the layer. This causes significant thermal contraction in the perpendicular direction and yields slight net volumetric thermal contraction. The schematic of NTE in a flexible network is furnished in Fig. 11.4 (bottom). A vibrational mode consuming an open space in a crystal lattice is thermally excited, which yields net volumetric thermal contraction.

The concept of total volume change $\Delta V/V$ related to NTE is employed for elucidating the case of phase transition-type NTE materials. With reference to Fig. 11.5 (top), the "slope" $\alpha_V^{(T)}$ (coefficient of volumetric thermal expansion) shares a tradeoff relation with the width of the operating-temperature window ΔT, roughly described as $\Delta V/V \sim \left|\alpha_V^{(T)}\right| \Delta T$. As an isotropic material, a linear CTE $\alpha_L^{(T)}$ is related to the volumetric CTE $\alpha_V^{(T)}$ as $\alpha_V^{(T)} = 3\alpha_L^{(T)}$. Therefore, coefficients $\alpha_V^{(T)}$ and $\alpha_L^{(T)}$ are not intrinsic for such phase transition-type materials. Instead, the total volume change $\Delta V/V$ is the intrinsic index indicating the potential of NTE. Figure 11.5 (bottom) shows a schematic explanation of microstructural effect for bulk negative thermal expansion. The ceramic body consists of crystal grains with anisotropic thermal expansion and pores. Upon an increase in temperature, the crystal grain expands in one direction but contracts in a perpendicular direction. If there exists open space in

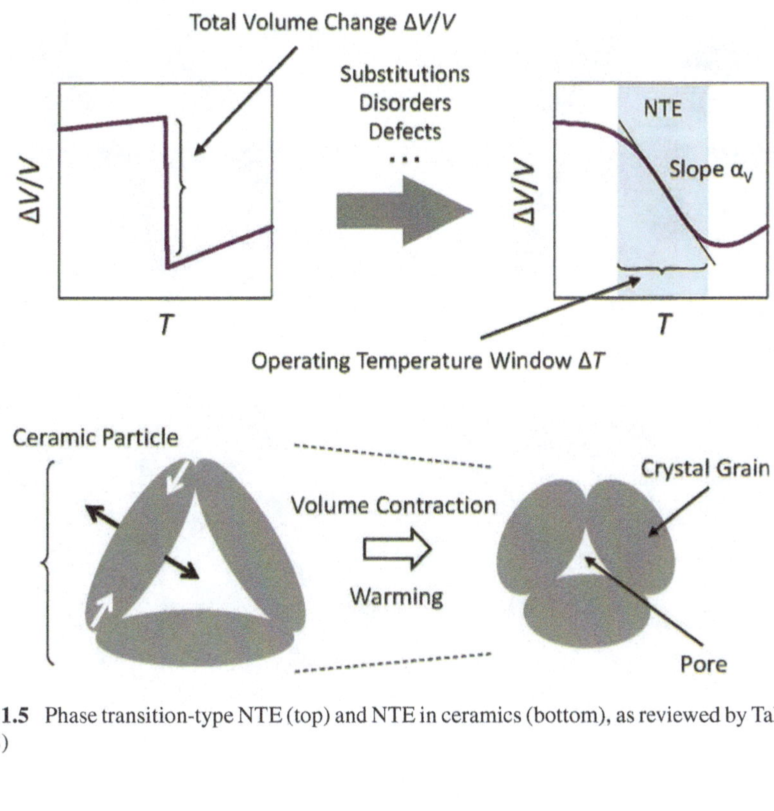

Fig. 11.5 Phase transition-type NTE (top) and NTE in ceramics (bottom), as reviewed by Takenaka (2018)

the direction in which the crystal grain expands, then this ceramic body encounters overall contraction (Takenaka 2018), thereby exhibiting NTE.

With reference to Fig. 11.4 (bottom), which is an idealized 2D representation, we shall now consider its 3D equivalences as examples from Lind (2012); Fig. 11.6 shows the crystal structures of α-ZrW_2O_8 (top left), ZrV_2O_7, an ideal high-temperature structure, (top right), chabazite (bottom left), and AlPO-17 (bottom right). ZrW_2O_8 was first discovered by Graham et al. (1959), and its crystal structure was determined by Auray et al. (1995). This led Mary et al. (1996) to elucidate the NTE behavior of ZrW_2O_8 from 0.3 to 1050 K based on its structure and the rigid units mode (RUM) of vibration. Negative thermal expansion in the zirconium vanadate family was first reported by Korthuis et al. (1995). Unlike ZrW_2O_8, vibrational modes always involve distortions of the polyhedral, and hence, the vibration is described as quasi-rigid unit modes, or qRUM (Tao and Sleight 2003). ZrV_2O_7 exhibits NTE with CTE values between $-7 \times 10^{-6}K^{-1}$ and $-10 \times 10^{-6}K^{-1}$ for $T > 375$ K, but shows PTE at room temperature (Khosrovani et al. 1997; Evans et al. 1998; Withers et al. 1998). With reference to Fig. 11.6 (bottom), the CTE of chabazite and AlPO-17 have been

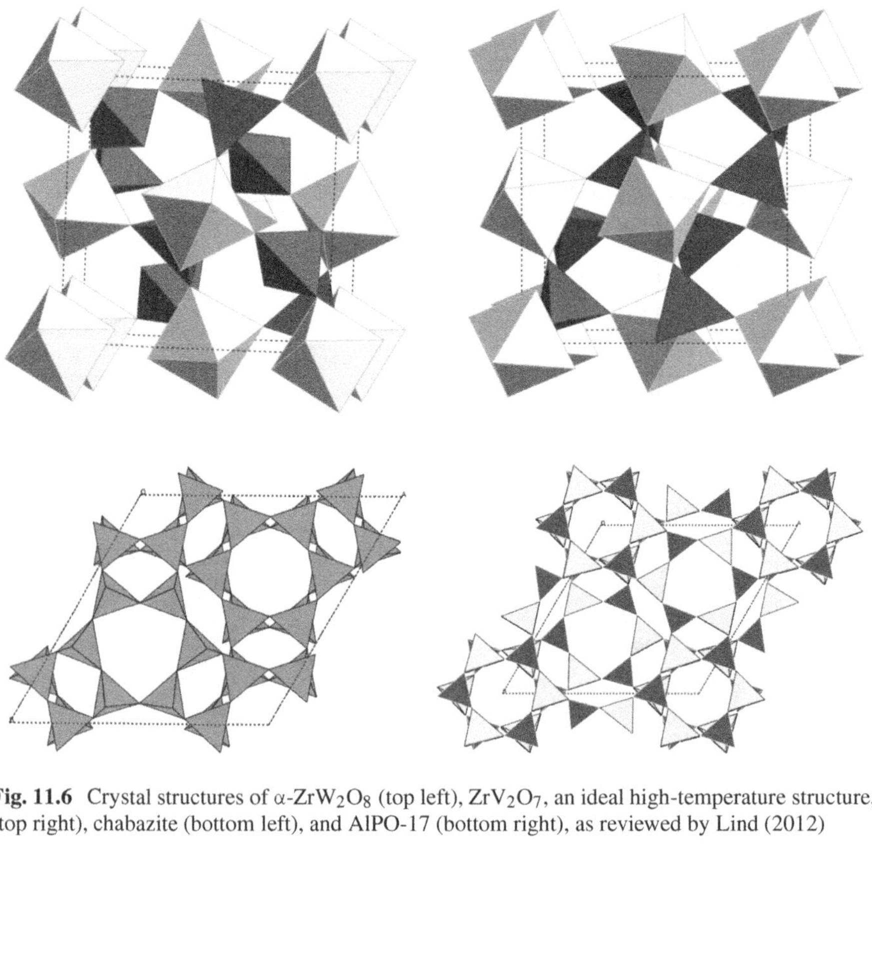

Fig. 11.6 Crystal structures of α-ZrW_2O_8 (top left), ZrV_2O_7, an ideal high-temperature structure, (top right), chabazite (bottom left), and AlPO-17 (bottom right), as reviewed by Lind (2012)

measured to be about zero and $-11.7 \times 10^{-6}\mathrm{K}^{-1}$, respectively (Attfield and Sleight 1998; Woodcock et al. 1999; Lightfoot et al. 2001)

A comprehensive survey of NTE materials has been given by Fisher (2018). Subsequent sections consider examples of composites consisting of phases with non-negative CTEs, but whose microarchitectures are designed such that they exhibit overall NTE properties.

11.2 Example: NTE Systems via Bimaterial Strips

An early account on an innovative approach for producing NTE and PTE microstructure was proposed in the form of cellular structures made from bimaterial strips (Lakes 1996). See Fig. 11.7 (left). It is known that bimaterial strips are made from materials (normally metals) with vastly different CTEs, i.e., $\alpha_1^{(T)}$ and $\alpha_2^{(T)}$ for layers 1 and 2, respectively. Arising from their differential strain and the perfect bond at the interface, the straight bimaterial strip transforms into a curve during a change in temperature $\mathrm{d}T$. For a bimaterial layer thicknesses h_1 and h_2, this curvature κ has been obtained by Timoshenko (1925) as

$$\kappa = \frac{6\left(\alpha_2^{(T)} - \alpha_1^{(T)}\right)\mathrm{d}T\left(1 + \frac{h_1}{h_2}\right)^2}{(h_1 + h_2)\left[3\left(1 + \frac{h_1}{h_2}\right)^2 + \left(1 + \frac{E_1 h_1}{E_2 h_2}\right)\left(\frac{h_1^2}{h_2^2} + \frac{E_2 h_2}{E_1 h_1}\right)\right]} \tag{11.2.1}$$

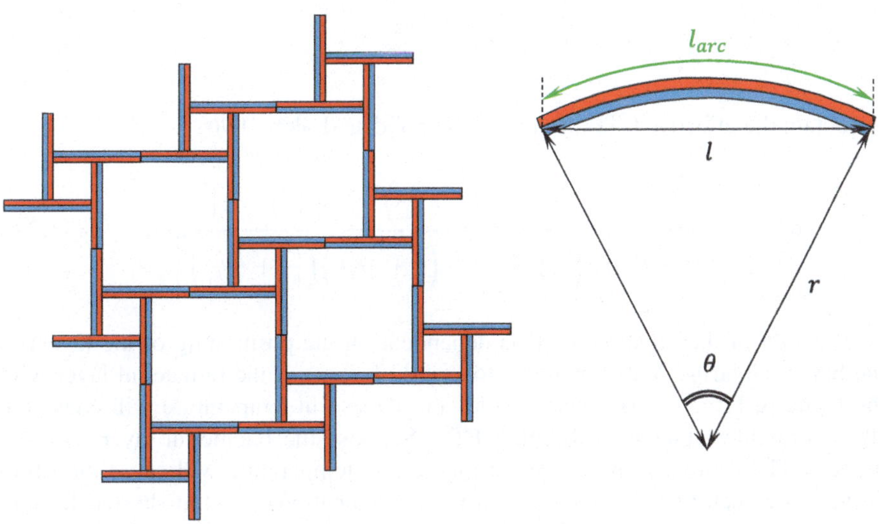

Fig. 11.7 An example of cellular structure consisting of bimaterial strips (left) and a bimaterial strip with initial curvature (right) studied by Lakes (1996)

where E_1 and E_2 are Young's moduli of layers 1 and 2, respectively. While Eq. (11.2.1) applies for the curvature formed from an initially straight bimaterial, it is also valid for describing the change in curvature for an initially curved bimaterial strip. Suppose the temperature increases, the curving of the bimaterial strips brings the junctions closer together. This essentially causes contraction. In other words, the microstructure exhibits an effective CTE with a negative value for increasing temperature.

Consider a curved rib of arc length l_{arc} with radius of curvature r and an included angle of θ. With reference to Fig. 11.7 (right), the rib length l, defined as the straight line connecting both ends, is

$$l = l_{\mathrm{arc}} \frac{2}{\theta} \sin\left(\frac{\theta}{2}\right) \tag{11.2.2}$$

The thermal strain as measured by the distance between the ends of the rib is

$$\varepsilon_{\mathrm{T}} = \frac{\mathrm{d}l}{l} \tag{11.2.3}$$

while the effect of curvature change due to thermal bending is

$$\varepsilon_{\mathrm{T}} = \left[\frac{1}{2}\cot\left(\frac{\theta}{2}\right) - \frac{1}{\theta}\right] l_{\mathrm{arc}} \mathrm{d}\kappa \tag{11.2.4}$$

Since the CTE is defined as

$$\alpha^{(T)} = \frac{\varepsilon_{\mathrm{T}}}{\mathrm{d}T} \tag{11.2.5}$$

we obtain the effective CTE of the cellular solid as (Lakes 1996)

$$\alpha_{\mathrm{eff}}^{(T)} = \frac{6 l_{\mathrm{arc}} \left(\alpha_2^{(T)} - \alpha_1^{(T)}\right) \left(1 + \frac{h_1}{h_2}\right)^2 \left[\frac{1}{2}\cot\left(\frac{\theta}{2}\right) - \frac{1}{\theta}\right]}{(h_1 + h_2)\left[3\left(1 + \frac{h_1}{h_2}\right)^2 + \left(1 + \frac{E_1 h_1}{E_2 h_2}\right)\left(\frac{h_1^2}{h_2^2} + \frac{E_2 h_2}{E_1 h_1}\right)\right]} \tag{11.2.6}$$

The sign of the effective CTE is dependent on the positioning of the layers in the binaterial strips. With reference to Fig. 11.8 (top), if the bimaterial layer with the higher CTE is on its concave side, an increase in temperature will cause the rib to straighten, thereby exhibiting PTE. Suppose the bimaterial layer with the higher CTE is on its convex side, an increase in temperature will cause the rib to curve more such that the distance between the junctions become shorter, leading to NTE. With reference to Fig. 11.8 (bottom) whereby the bimaterial strips are initially straight, an increase or decrease in temperature will cause the bimaterial strips to curve in opposing manner. Now, if the difference in the curved length of

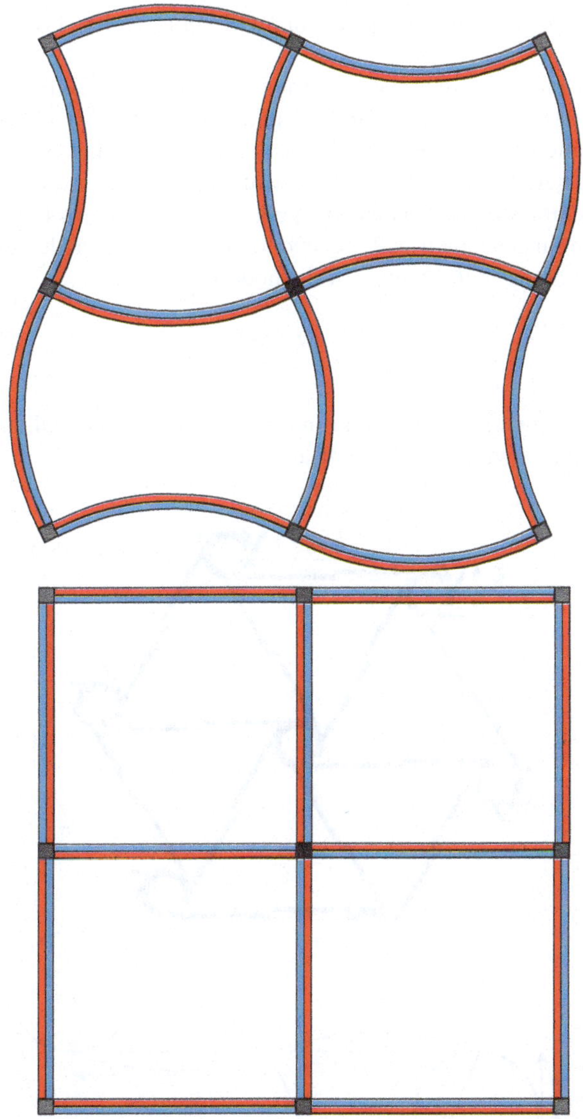

Fig. 11.8 Square array of cells where the walls are initially curved (top) and initially straight (bottom)

the bimaterial strip under the effect of increasing and decreasing temperature is negligible, then the cellular solid encounters contraction regardless of increasing or decreasing temperature. In other words, this cellular solid behaves as an NTE material with increasing temperature, but a PTE material with decreasing temperature. Such sign-toggling of expansion coefficients has been investigated recently by Lim (2019a, b, c, d, e; 2020a, b, c) as a follow up on work with sign-toggling Poisson's ratio (Lim 2019f, g).

Further work on the use of bimaterial strips for developing cellular microstructures with NTE and PTE by this group can be found in Lakes (2007) and Ha et al. (2015, 2017). In the alternate arrangement by Ha et al. (2015), the bimaterial strips of alternating orientation are adopted for constructing in-plane isotropic NTE system that resembles chiral structure, as shown in Fig. 11.9 (top). With reference to Fig. 11.9 (bottom), the strain is geometrically related to the rotation ϕ, node outer radius r_{node} and the spacing between the centers of the nodes R as

$$\varepsilon = \frac{r_{\text{node}}\phi}{R} \tag{11.2.7}$$

A change in temperature $\mathrm{d}T$ in an initially straight bimaterial strip produces a radius of curvature r with included half-angle ϕ of

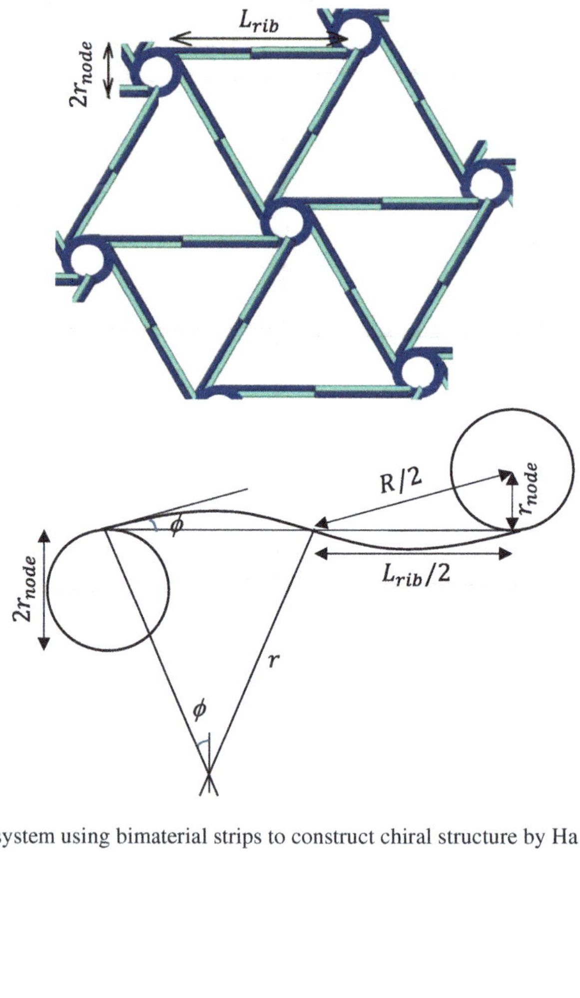

Fig. 11.9 NTE system using bimaterial strips to construct chiral structure by Ha et al. (2015)

$$\phi = \frac{L_{\text{rib}}}{4r} \tag{11.2.8}$$

where L_{rib} is the rib length. From Pythagoras' theorem, we have

$$R = \sqrt{L_{\text{rib}}^2 + (2r_{\text{node}})^2} \tag{11.2.9}$$

Substituting Eqs. (11.2.8) and (11.2.9) into Eq. (11.2.7) gives

$$\varepsilon = \frac{r_{\text{node}}}{4r} \frac{1}{\sqrt{1 + \left(\frac{2r_{\text{node}}}{L_{\text{rib}}}\right)^2}} = \alpha_{\text{eff}}^{(T)} \mathrm{d}T \tag{11.2.10}$$

where $r^{-1} = \kappa$ as given in Eq. (11.2.1).

11.3 Example: NTE System via Multilayered Systems with Large Stiffness Disparity

An early work on the use of multilayered systems to produce NTE has been established by Grima et al. (2010) through the use of thin layers with high modulus and high CTE and thick layers of low modulus and low CTE; the latter possessing significantly higher Poisson's ratio than the former. Both materials are isotropic. With reference to Fig. 11.10 (top left), let the CTE of materials A and B be $\alpha_A^{(T)}$ and $\alpha_B^{(T)}$, respectively, whereby $\alpha_A^{(T)} > \alpha_B^{(T)}$ so that upon heating, material A has a propensity to expand to a much greater extent than material B if they are not bonded to each other. Let Young's moduli of materials A and B be E_A and E_B, respectively, whereby $E_A \gg E_B$ so that the middle layer encounters an in-plane stretching much greater than the in-plane contraction of the upper and lower layers in fulfillment of geometrical compatibility at the boundaries on the condition of the perfectly bonded interface. Finally, let v_A and v_B be Poisson's ratio of materials A and B, respectively, whereby $v_B > v_A$ so that the magnitude of the negative thickness strain of the middle layer is greater than the thickness strain of the upper and lower layers. The NTE effect is also enhanced by increasing the thickness of material B vis-à-vis material A.

The subsequent analysis by Grima et al. (2010) further assumes no necking. With reference to Fig. 11.11 (bottom), where total thickness is

$$z = 2z_A + z_B \tag{11.3.1}$$

the change in the total thickness is accordingly

$$\mathrm{d}z = 2\mathrm{d}z_A + \mathrm{d}z_B \tag{11.3.2}$$

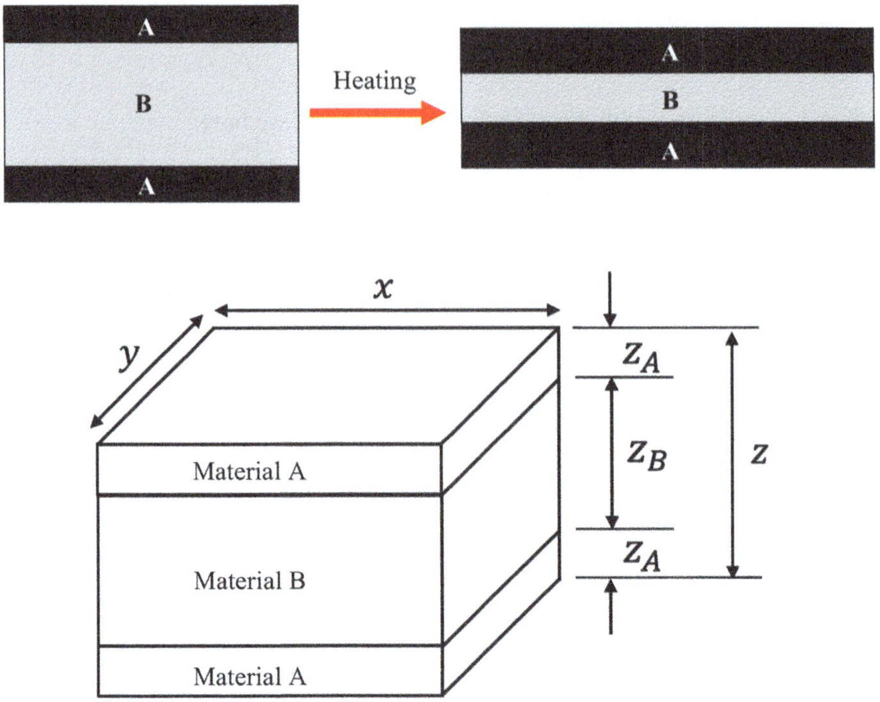

Fig. 11.10 NTE in the direction orthogonal to the layers (top) and schematics for analysis (bottom) by Grima et al. (2010)

Fig. 11.11 Comparison between Eq. (11.3.5) (line) with finite element results (circles) for $dT = 100\,K$ by Grima et al. (2010)

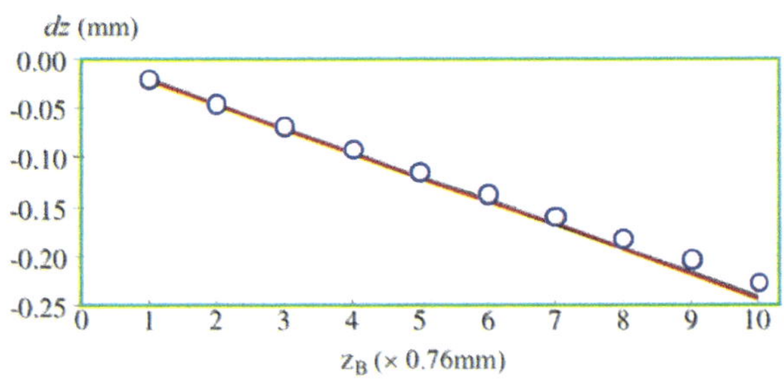

where the change of thickness in material A is

$$dz_A = z_A \alpha_A^{(T)} dT \tag{11.3.3}$$

while the thickness change in material B is given by Rees (1990) as

$$dz_B = z_B \left(\alpha_B^{(T)} dT - \frac{2\upsilon_B \left(\alpha_A^{(T)} - \alpha_B^{(T)} \right) dT}{1 - \upsilon_B} \right) \tag{11.3.4}$$

These descriptions of dz_A and dz_B in Eqs. (11.3.3) and (11.3.4), respectively, are valid only if $E_A \gg E_B$ in that the in-plane dimension of the material A in not influenced by material B, whereas the in-plane dimension of material B when unbonded must be stretched to match the in-plane dimensions of material A. This results in a contraction of material B thickness due to its Poisson's ratio by the amount stated in the second term on the RHS of Eq. (11.3.4). Substituting Eqs. (11.3.3) and (11.3.4) into Eq. (11.3.2) gives

$$dz = \left(2z_A \alpha_A^{(T)} + z_B \alpha_B^{(T)} - 2z_B \upsilon_B \frac{\alpha_A^{(T)} - \alpha_B^{(T)}}{1 - \upsilon_B} \right) dT \tag{11.3.5}$$

With reference to Eq. (11.3.5), the NTE condition is satisfied if

$$2z_B \upsilon_B \frac{\alpha_A^{(T)} - \alpha_B^{(T)}}{1 - \upsilon_B} > 2z_A \alpha_A^{(T)} + z_B \alpha_B^{(T)} \tag{11.3.6}$$

A comparison between the Eq. (11.3.5) and finite element results are furnished in Fig. 11.11 for validating the former. In the FE simulation, Grima et al. (2010) adopted $x = y = 200\,\text{mm}$, $z_A = 0.076\,\text{mm}$, z_B in multiples of z_A (see horizontal axis of Fig. 11.11) and the material properties are listed in Table 11.1. The finite element results in Fig. 11.11 reveals that the CTE is less negative when compared against Eq. (11.3.5). This observation can be attributed to the assumption of $E_B/E_A \to 0$, which enhances the negativity of the out-of-plane CTE. Since the out-of-plane CTE is defined by the ratio of the thermal strain in that direction to the change in temperature $\alpha_Z^{(T)} = \varepsilon_Z/dT$, we obtain from Eqs. (11.3.1) and (11.3.5)

$$\alpha_Z^{(T)} = \frac{2z_A \alpha_A^{(T)} + z_B \alpha_B^{(T)} - 2z_B \upsilon_B \frac{\alpha_A^{(T)} - \alpha_B^{(T)}}{1 - \upsilon_B}}{2z_A + z_B} \tag{11.3.7}$$

Note that the relative thicknesses of the layers can be expressed in terms of their volume fractions V_A and V_B for materials A and B, respectively. Therefore,

Table 11.1 Material properties of material A and material B layers

	Material i	Material B
Young's modulus	$E_A = 0.552\,\text{GPa}$	$E_B = 0.000517\,\text{GPa}$
CTE	$\alpha_A^{(T)} = 190 \times 10^{-6} K^{-1}$	$\alpha_B^{(T)} = 19 \times 10^{-6} K^{-1}$
Poisson's ratio	$\upsilon_A = 0.333$	$\upsilon_B = 0.499$

substituting

$$\frac{2z_A}{2z_A + z_B} = V_A$$
$$\frac{z_B}{2z_A + z_B} = V_B \tag{11.3.8}$$

into Eq. (11.3.7) yields

$$\alpha_Z^{(T)} = V_A\alpha_A^{(T)} + V_B\alpha_B^{(T)} - 2V_B\left(\alpha_A^{(T)} - \alpha_B^{(T)}\right)\frac{\upsilon_B}{1-\upsilon_B} \tag{11.3.9}$$

In other words, the out-of-plane CTE description in Eq. (11.3.9) simplifies to the rule-of-mixture model, also known as the Voigt model, if the effect of Poisson's ratio for material B is neglected. The use of Eq. (11.3.9) is, therefore, more accurate than the Voigt model because it takes into account the elastic properties of material B. Since it has been assumed that $E_A \gg E_B$, the in-plane CTE is dictated by the CTE of material A, i.e.,

$$\alpha_X^{(T)} = \alpha_Y^{(T)} = \alpha_A^{(T)} \tag{11.3.10}$$

If Young's moduli for both materials A and B are of the same order, then the elastic properties of both materials must be taken into account. This is discussed in the next example which, in addition to establishing the out-of-plane CTE, also includes the in-plane CTE.

11.4 Example: NTE System via Multilayered Systems with Comparable Stiffness

In this example, the CTE of laminates, with isotropic laminas, are developed for predicting the effective in-plane and out-of-plane laminate CTE. The laminate consists of 2 types of laminas arranged in alternate layers, in which both materials possess Young's moduli of equal order. Thereafter, 3 special cases are discussed for the laminate. They are: (a) alternating conventional (positive Poisson's ratio and positive thermal expansion) with non-conventional (auxetic and NTE) laminas, (b) alternating auxetic (with positive thermal expansion) and NTE (with positive Poisson's ratio) laminas, and (c) simplification to the models developed in Sect. 11.3. The CTE of composites is well-established, especially in the engineering design field. The effective in-plane CTE of a laminate with two types of isotropic laminas is given as

$$\alpha_{\text{in}} = \frac{\alpha_f^{(T)} V_f E_f + \alpha_m^{(T)} V_m E_m}{V_f E_f + V_m E_m} \tag{11.4.1}$$

This equation was developed based on 1D approach, i.e., with the imposition of equal strain whereby the laminas are considered as elements in parallel, without the consideration of Poisson's ratio. Since most engineering materials possess the range of Poisson's ratio of between 1/4 and 1/3, then Eq. (11.4.1) is valid. A similar equation can be arrived when one considers the effective CTE of a unidirectional (UD) fiber composite in the longitudinal direction based on 1D analysis by Schapery (1968). The effective out-of-plane CTE of the same laminate is easily obtained as

$$\alpha_{\text{out}} = \alpha_f^{(T)} V_f + \alpha_m^{(T)} V_m \tag{11.4.2}$$

based on 1D approach. Equation (11.4.2) is based on the summation of individual lamina thickness expansion, whereby the laminas are considered as elements in series. The validity of this equation is justified when the difference in the CTE of both materials is negligible such that the difference in the unrestrained thermal in-plane deformation is insignificant.

In this section, correction factors f_i and g_i (for $i = A, B$), with A and B representing the two lamina materials, are incorporated into Eq. (11.4.1) to give the effective in-plane CTE of a laminate

$$\alpha_{\text{in}} = \frac{\alpha_A^{(T)} V_A E_A f_B g_A + \alpha_B^{(T)} V_B E_B f_A g_B}{V_A E_A f_B g_A + V_B E_B f_A g_B} \tag{11.4.3}$$

such that Eq. (11.4.3) reduces to Eq. (11.4.1) when Poisson's ratio and CTE of both lamina materials are negligible. These correction factors—together with the volume fraction, modulus and CTE of individual phases—are incorporated into Eq. (11.4.2) to give the effective out-of-plane CTE of the same laminate

$$\alpha_{\text{out}} = \alpha_A^{(T)} V_A + \alpha_B^{(T)} V_B - 2\left(\alpha_A^{(T)} - \alpha_B^{(T)}\right) \frac{v_B E_A g_A - v_A E_B g_B}{E_A V_B^{-1} f_B g_A + E_B V_A^{-1} f_A g_B} \tag{11.4.4}$$

so that Eq. (11.4.4) simplifies to Eq. (11.4.2) when Poisson's ratio and CTE of both phases are negligible, or when both phases possess equal CTE. In addition, if $E_A \gg E_B$ such that the terms containing E_B vanish, then Eq. (11.4.4) reduces to Eq. (11.3.9) if $f_B = 1 - v_B$. This will be proven later in Eq. (11.4.30). Likewise if $E_A \gg E_B$ such that the terms containing E_B in Eq. (11.4.3) vanish, then Eq. (11.4.3) simplifies to Eq. (11.3.10).

Formulation of the effective laminate CTE in the in-plane and out-of-plane direction is made in two stages. In the first stage, it is assumed that there is no bonding between adjacent laminas such that the contact surfaces of adjacent laminas are allowed to freely slide relative to one another in the in-plane direction. Geometrical compatibility is then imposed in the second stage by application of load on the sides of every laminas such that (a) the in-plane dimensions are made equal, and that (b) the net load is zero. This sideload approximates the interlaminar shear stresses that arise following the unrestrained thermal expansion in the first stage. Figure 11.12(top) shows an RVE that consists of two laminas—materials *A* and *B* with original thickness a_0 and b_0, respectively, both with original side dimensions l_0. For clarity, we let material A possess positive CTE while material B be an NTE material such that an increment of temperature alters the dimensions of both laminas as shown in Fig. 11.12(middle), considering no restraint.

From the definition of thermal strain

$$\varepsilon_{\mathrm{T}} = \alpha^{(T)} \mathrm{d}T \equiv \frac{\mathrm{d}l}{l_0} = \frac{l_1 - l_0}{l_0} \tag{11.4.5}$$

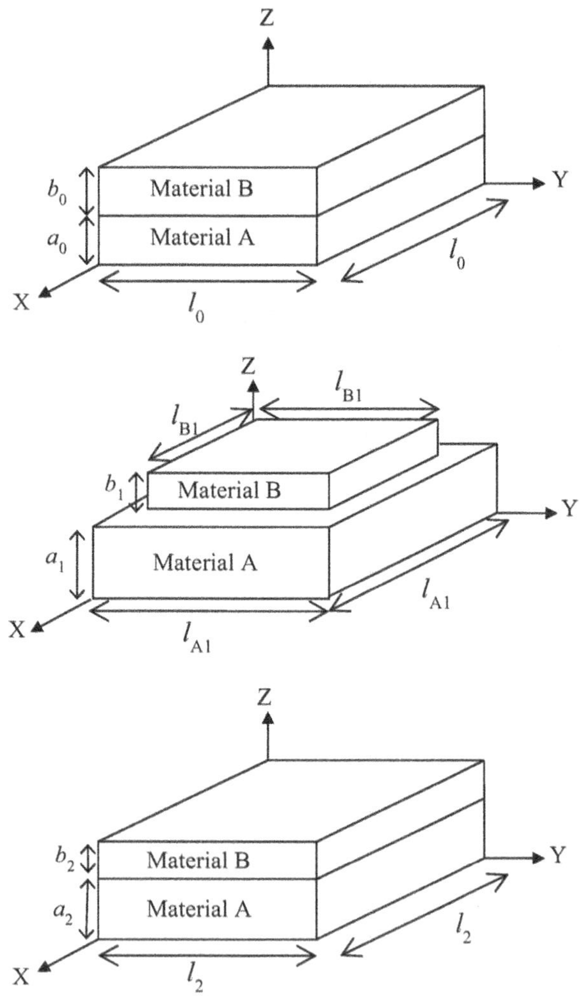

Fig. 11.12 Representative volume element of laminate before thermal strain (top), unrestrained thermal strain (middle), and restrained thermal strain (bottom)

we have the unrestrained resultant dimension, l_1, as a function of original dimension, l_0, as

$$l_1 = l_0\left(1 + \alpha^{(T)} \mathrm{d}T\right) \tag{11.4.6}$$

This translates to the unrestrained dimensions for material A

$$\begin{aligned} a_1 &= a_0\left(1 + \alpha_A^{(T)} \mathrm{d}T\right) \\ l_{A1} &= l_0\left(1 + \alpha_A^{(T)} \mathrm{d}T\right) \end{aligned} \tag{11.4.7}$$

and those of material B

$$\begin{aligned} b_1 &= b_0\left(1 + \alpha_B^{(T)} \mathrm{d}T\right) \\ l_{B1} &= l_0\left(1 + \alpha_B^{(T)} \mathrm{d}T\right) \end{aligned} \tag{11.4.8}$$

For perfect bonding between laminas A and B, there exists interlaminar shear stresses acting on the contact surfaces of the laminas, whereby the shear stress on the z-surfaces on materials A and B acts inward and outward, respectively, in the x–y plane. In this analysis, normal forces are prescribed such that the load on the sides of lamina A and B acts inward and outward, respectively, to bridge the gap $(l_{A1} - l_{B1})$. Since the interlaminar shear stresses are internal stresses, prescribed normal stresses are governed by zero net force in the following equilibrium equation

$$\begin{aligned} \sigma_{AX}(a_1 l_{A1}) + \sigma_{BX}(b_1 l_{B1}) = 0 \\ \sigma_{AY}(a_1 l_{A1}) + \sigma_{BY}(b_1 l_{B1}) = 0 \end{aligned} \tag{11.4.9}$$

such that the final in-plane dimensions are common, i.e.,

$$l_{A2} = l_{B2} \equiv l_2 \tag{11.4.10}$$

while incurring further changes to the out-of-plane dimensions, as shown in Fig. 11.12 (bottom).

Considering isotropic laminas, the constitutive relations for materials A and B are described by Eqs. (9.4.6) and (9.4.7), respectively. Since the in-plane dimensions along the x- and y-axes have been set equal, and that the individual laminas are isotropic, it follows that the in-plane stresses and strains along both axes for each lamina are identical by virtue of symmetry, i.e.,

$$\begin{bmatrix} \sigma_{AX} & \sigma_{BX} \\ \varepsilon_{AX} & \varepsilon_{BX} \end{bmatrix} = \begin{bmatrix} \sigma_{AY} & \sigma_{BY} \\ \varepsilon_{AY} & \varepsilon_{BY} \end{bmatrix} \tag{11.4.11}$$

The above relation reduces the number of variables for analysis; i.e., its first row combines both the first and second of Eq. (11.4.9) into one common equation. Further reduction of variables is made possible by assuming plane stress condition described in Eq. (9.4.10) by virtue of free z-surface of the laminate. This free surface further implies zero out-of-plane internal stresses, and that any internal out-of-plane stresses would cancel out one another. Variable reduction as specified by Eq. (11.4.11) and the plane stress condition simplify the constitutive relations to

$$\begin{Bmatrix} \varepsilon_{AX2} \\ \varepsilon_{AY2} \\ \varepsilon_{AZ2} \end{Bmatrix} = \frac{\sigma_{AX}}{E_A} \begin{Bmatrix} 1 - \upsilon_A \\ 1 - \upsilon_A \\ -2\upsilon_A \end{Bmatrix} \tag{11.4.12a}$$

and

$$\begin{Bmatrix} \varepsilon_{BX2} \\ \varepsilon_{BY2} \\ \varepsilon_{BZ2} \end{Bmatrix} = \frac{\sigma_{BX}}{E_B} \begin{Bmatrix} 1 - \upsilon_B \\ 1 - \upsilon_B \\ -2\upsilon_B \end{Bmatrix} \tag{11.4.12b}$$

for laminas A and B, respectively. Substituting from the first rows of Eqs. (11.4.12a) and (11.4.12b)

$$\begin{aligned} \sigma_{AX} &= \frac{E_A \varepsilon_{AX2}}{1 - \upsilon_A} \\ \sigma_{BX} &= \frac{E_B \varepsilon_{BX2}}{1 - \upsilon_B} \end{aligned} \tag{11.4.13}$$

as well as Eqs. (11.4.7) and (11.4.8) into the first of Eq. (11.4.9) leads to

$$\frac{E_A \varepsilon_{AX2} a_0 \left(1 + \alpha_A^{(T)} \mathrm{d}T\right)^2}{1 - \upsilon_A} + \frac{E_B \varepsilon_{BX2} b_0 \left(1 + \alpha_B^{(T)} \mathrm{d}T\right)^2}{1 - \upsilon_B} = 0 \tag{11.4.14}$$

The second stage of in-plane strain, being defined as

$$\begin{aligned} \varepsilon_{AX2} &= \frac{l_{A2} - l_{A1}}{l_{A1}} \\ \varepsilon_{BX2} &= \frac{l_{B2} - l_{B1}}{l_{B1}} \end{aligned} \tag{11.4.15}$$

gives the final in-plane dimension

$$l_2 \equiv l_{A1}(1 + \varepsilon_{AX2}) = l_{B1}(1 + \varepsilon_{BX2}) \tag{11.4.16a}$$

Using the second rows of Eqs. (11.4.7) and (11.4.8), the above equation can be written as

$$\left(1+\alpha_A^{(T)}\mathrm{d}T\right)(1+\varepsilon_{AX2}) = \left(1+\alpha_B^{(T)}\mathrm{d}T\right)(1+\varepsilon_{BX2}) \tag{11.4.16b}$$

The second stage of out-of-plane strain, being defined as

$$\begin{aligned} \varepsilon_{AZ2} &= \frac{a_2 - a_1}{a_1} \\ \varepsilon_{BZ2} &= \frac{b_2 - b_1}{b_1} \end{aligned} \tag{11.4.17}$$

together with the first rows of Eqs. (11.4.7) and (11.4.8), give the final thickness of laminas A and B as

$$\begin{aligned} a_2 &= a_0\left(1+\alpha_A^{(T)}\mathrm{d}T\right)(1+\varepsilon_{AZ2}) \\ b_2 &= b_0\left(1+\alpha_B^{(T)}\mathrm{d}T\right)(1+\varepsilon_{BZ2}) \end{aligned} \tag{11.4.18}$$

Based on the criterion of zero net force on the lamina sides as furnished in Eq. (11.4.14), we have the second stage in-plane strain of lamina B

$$\varepsilon_{BX2} = -\varepsilon_{AX2}\frac{E_A a_0}{E_B b_0}\left(\frac{1-v_B}{1-v_A}\right)\left(\frac{1+\alpha_A^{(T)}\mathrm{d}T}{1+\alpha_B^{(T)}\mathrm{d}T}\right)^2 \tag{11.4.19}$$

Based on the criterion of equal in-plane dimensions as shown in Eq. (11.4.16b), we have the second stage in-plane strain of lamina B in a different form,

$$\varepsilon_{BX2} = (1+\varepsilon_{AX2})\frac{1+\alpha_A^{(T)}\mathrm{d}T}{1+\alpha_B^{(T)}\mathrm{d}T} - 1 \tag{11.4.20}$$

The second stage in-plane strain of lamina A can then be obtained by equating Eqs. (11.4.19) and (11.4.20) to give

$$\varepsilon_{AX2} = \frac{1 - \frac{1+\alpha_A^{(T)}\mathrm{d}T}{1+\alpha_B^{(T)}\mathrm{d}T}}{\frac{1+\alpha_A^{(T)}\mathrm{d}T}{1+\alpha_B^{(T)}\mathrm{d}T}\left[1+\frac{a_0 E_A}{b_0 E_B}\left(\frac{1-v_B}{1-v_A}\right)\left(\frac{1+\alpha_A^{(T)}\mathrm{d}T}{1+\alpha_B^{(T)}\mathrm{d}T}\right)\right]} \tag{11.4.21}$$

which is a function of Young's modulus, CTE, Poisson's ratio, and thickness of individual laminas. With reference to Eqs. (11.4.6) and (11.4.16a), as well as the definition of overall in-plane thermal strain

$$\varepsilon_{\mathrm{in}} = \alpha_{\mathrm{in}}^{(T)}\mathrm{d}T \equiv \frac{l_2 - l_0}{l_0} = \frac{l_0\left(1+\alpha_A^{(T)}\mathrm{d}T\right)(1+\varepsilon_{AX2}) - l_0}{l_0}, \quad (i = A, B) \tag{11.4.22}$$

we arrive at the effective in-plane CTE of the laminate (Lim 2011)

$$\alpha_{\text{in}}^{(T)} = \frac{\alpha_A^{(T)} V_A E_A (1 - \upsilon_B)\left(1 + \alpha_A^{(T)} \mathrm{d}T\right) + \alpha_B^{(T)} V_B E_B (1 - \upsilon_A)\left(1 + \alpha_B^{(T)} \mathrm{d}T\right)}{V_A E_A (1 - \upsilon_B)\left(1 + \alpha_A^{(T)} \mathrm{d}T\right) + V_B E_B (1 - \upsilon_A)\left(1 + \alpha_B^{(T)} \mathrm{d}T\right)} \tag{11.4.23}$$

where the volume fractions are defined as

$$\begin{aligned} V_A &= \frac{a_0}{a_0 + b_0} \\ V_B &= \frac{b_0}{a_0 + b_0} \end{aligned} \tag{11.4.24}$$

Defining the overall out-of-plane strain as

$$\varepsilon_{\text{out}} = \frac{(a_2 + b_2) - (a_0 + b_0)}{(a_0 + b_0)} \tag{11.4.25a}$$

and considering the relations given in Eq. (11.4.18), then Eq. (11.4.25a) can be expressed as

$$\varepsilon_{\text{out}} = \frac{a_0\left(1 + \alpha_A^{(T)} \mathrm{d}T\right)(1 + \varepsilon_{AZ2}) + b_0\left(1 + \alpha_B^{(T)} \mathrm{d}T\right)(1 + \varepsilon_{BZ2})}{a_0 + b_0} - 1 \tag{11.4.25b}$$

Substituting the expressions for σ_{AX} and σ_{BX} from the first rows of Eqs. (11.4.12a) and (11.4.12b) into the expressions for ε_{AZ2} and ε_{BZ2} in the last rows of the same equation, we have the relation

$$\begin{aligned} \varepsilon_{AZ2} &= -2\varepsilon_{AX2} \frac{\upsilon_A}{1 - \upsilon_A} \\ \varepsilon_{BZ2} &= -2\varepsilon_{BX2} \frac{\upsilon_B}{1 - \upsilon_B} \end{aligned} \tag{11.4.26}$$

This out-of-plane strain, being functions of ε_{AZ2} and ε_{BZ2}, is hence in terms of ε_{AX2} and ε_{BX2}. With reference to the expression of ε_{AX2} in Eq. (11.4.21), the expression for ε_{BX2} can be inferred as

$$\varepsilon_{BX2} = \frac{1 - \frac{1 + \alpha_B^{(T)} \mathrm{d}T}{1 + \alpha_A^{(T)} \mathrm{d}T}}{\frac{1 + \alpha_B^{(T)} \mathrm{d}T}{1 + \alpha_A^{(T)} \mathrm{d}T}\left[1 + \frac{b_0 E_B}{a_0 E_A}\left(\frac{1 - \upsilon_A}{1 - \upsilon_B}\right)\left(\frac{1 + \alpha_B^{(T)} \mathrm{d}T}{1 + \alpha_A^{(T)} \mathrm{d}T}\right)\right]} \tag{11.4.27}$$

Hence, the out-of-plane strain in Eq. (11.4.25b) can be expressed purely in terms of temperature increment and individual lamina properties such as thickness, CTE,

Poisson's ratio, and Young's modulus by substituting Eqs. (11.4.21) and (11.4.27) into Eq. (11.4.26), which is then substituted into Eq. (11.4.25b). The overall out-of-plane thermal strain, being defined as

$$\varepsilon_{\text{out}} = \alpha_{\text{out}}^{(T)} \mathrm{d}T \tag{11.4.28}$$

can then be equated with Eq. (11.4.25b) to give the effective out-of-plane CTE of the laminate as (Lim 2011)

$$\alpha_{\text{out}}^{(T)} = V_A \alpha_A^{(T)} + V_B \alpha_B^{(T)} - 2\left(\alpha_A^{(T)} - \alpha_B^{(T)}\right) \frac{\upsilon_B E_A\left(1 + \alpha_A^{(T)} \mathrm{d}T\right) - \upsilon_A E_B\left(1 + \alpha_B^{(T)} \mathrm{d}T\right)}{E_A V_B^{-1}(1 - \upsilon_B)\left(1 + \alpha_A^{(T)} \mathrm{d}T\right) + E_B V_A^{-1}(1 - \upsilon_A)\left(1 + \alpha_B^{(T)} \mathrm{d}T\right)} \tag{11.4.29}$$

Comparing the effective in-plane CTE, Eq. (11.4.23), and the effective out-of-plane CTE, Eq. (11.4.29), with Eqs. (11.4.3) and (11.4.4), respectively, leads to the expressions of the correction factors

$$\begin{aligned} f_i &= 1 - \upsilon_i \\ g_i &= 1 + \alpha_i^{(T)} \mathrm{d}T \end{aligned} \tag{11.4.30}$$

for $i = A, B$. Here, the correction factor f_i takes into account Poisson's ratio, while the correction factor g_i considers the CTE of each lamina. The coupling of Poisson's ratio and CTE in the laminate's effective thermal expansion, that embodies large temperature change, is generally described in Eqs. (11.4.23) and (11.4.29).

There exist two forms of CTE nonlinearity, namely (a) material nonlinearity and (b) geometrical nonlinearity. Material nonlinearity refers to the individual phase's CTE not being a constant but varies either with the change in temperature, i.e., $\alpha_i^{(T)} = \alpha_i^{(T)}(\mathrm{d}T)$ for significant thermal fluctuation, or, within the context of infinitesimal change in temperature, the CTE is a function of the instantaneous temperature, i.e., $\alpha_i^{(T)} = \alpha_i^{(T)}(T)$. Geometrical nonlinearity refers to the existence of CTE nonlinearity of the entire laminate for significant change in temperature even if every lamina possesses constant CTE with respect to the temperature. Thus, Eqs. (11.4.23) and (11.4.29) per se accommodate geometrical nonlinearity, while material nonlinearity can be incorporated by substituting the individual lamina's CTE with curve-fitted CTEs as functions of temperature.

The general expressions of the considered laminate CTE can be further simplified under specific conditions. In the case where the temperature change is insignificant, we take the limits of Eqs. (11.4.23) and (11.4.29) as the temperature change tends to zero, to give

$$\alpha_{\text{in}}^{(T)} = \frac{\alpha_A^{(T)} V_A E_A (1 - \upsilon_B) + \alpha_B^{(T)} V_B E_B (1 - \upsilon_A)}{V_A E_A (1 - \upsilon_B) + V_B E_B (1 - \upsilon_A)} \tag{11.4.31}$$

and

$$\alpha_{\text{out}}^{(T)} = V_A \alpha_A^{(T)} + V_B \alpha_B^{(T)} - 2\left(\alpha_A^{(T)} - \alpha_B^{(T)}\right) \frac{\upsilon_B E_A - \upsilon_A E_B}{E_A V_B^{-1}(1 - \upsilon_B) + E_B V_A^{-1}(1 - \upsilon_A)} \tag{11.4.32}$$

respectively. For the case where Poisson's ratios of both phases are equal or insignificant, then Eq. (11.4.23) reduces to

$$\alpha_{\text{in}}^{(T)} = \frac{\alpha_A^{(T)} V_A E_A \left(1 + \alpha_A^{(T)} \mathrm{d}T\right) + \alpha_B^{(T)} V_B E_B \left(1 + \alpha_B^{(T)} \mathrm{d}T\right)}{V_A E_A \left(1 + \alpha_A^{(T)} \mathrm{d}T\right) + V_B E_B \left(1 + \alpha_B^{(T)} \mathrm{d}T\right)} \tag{11.4.33}$$

while Eq. (11.4.29) simplifies to

$$\alpha_{\text{out}}^{(T)} = \alpha_{\text{ROM}}^{(T)} - 2V_A V_B \left(\alpha_A^{(T)} - \alpha_B^{(T)}\right)\left(\frac{\upsilon_i}{1 - \upsilon_i}\right) \frac{(E_A - E_B) + \left(E_A \alpha_A^{(T)} - E_B \alpha_B^{(T)}\right)\mathrm{d}T}{E_{\text{ROM}} + \left(E_A V_A \alpha_A^{(T)} + E_B V_B \alpha_B^{(T)}\right)\mathrm{d}T} \tag{11.4.34}$$

if $\upsilon_i \equiv \upsilon_A = \upsilon_B \neq 0$, whereby $\alpha_{\text{ROM}}^{(T)}$ and E_{ROM} denote the simple rule-of-mixtures for the CTE and Young's modulus, respectively. This out-of-plane CTE reduces to simple rule-of-mixtures when $\upsilon_A = \upsilon_B = 0$, thereby giving rise to the relation

$$\alpha_{\text{in}}^{(T)} \alpha_{\text{out}}^{(T)} = \frac{\alpha_{\text{ROM}}^{(T)}}{E_{\text{ROM}}} \sum_{i=A,B,\ldots} \alpha_i^{(T)} V_i E_i \tag{11.4.35}$$

if $\mathrm{d}T = 0$, which can be extended to more than 2 types of isotropic laminas. At this juncture, we consider a group of special cases for illustration under the following conditions

$$\frac{E_A}{E_B} = \frac{\left|\alpha_A^{(T)}\right|}{\left|\alpha_B^{(T)}\right|} = \frac{|\upsilon_A|}{|\upsilon_B|} = 1 \tag{11.4.36}$$

whereby $|\upsilon_A| = |\upsilon_B| = 1/3$ and $\mathrm{d}T \to 0$. Further, we consider two sub-cases illustrated below, in which the property $\alpha_0^{(T)}$ is positive.

In the first sub-case, we let $\alpha_A^{(T)} = -\alpha_B^{(T)} = \alpha_0^{(T)}$ and $\upsilon_A = -\upsilon_B = 1/3$. Here, we have the following normalized in-plane CTE

$$\frac{\alpha_{\text{in}}^{(T)}}{\alpha_0^{(T)}} = \frac{2 - 3V_B}{2 - V_B} \tag{11.4.37}$$

and normalized out-of-plane CTE

$$\frac{\alpha_{\text{out}}^{(T)}}{\alpha_0^{(T)}} = 1 - 2V_B + \frac{4V_B(1 - V_B)}{2 - V_B} \tag{11.4.38}$$

In the second sub-case, we let $\alpha_A^{(T)} = -\alpha_B^{(T)} = \alpha_0^{(T)}$ and $v_A = -v_B = -1/3$. For this sub-case, we have

$$\frac{\alpha_{\text{in}}^{(T)}}{\alpha_0^{(T)}} = \frac{1 - 3V_B}{1 + V_B} \tag{11.4.39}$$

and

$$\frac{\alpha_{\text{out}}^{(T)}}{\alpha_0^{(T)}} = 1 - 2V_B - \frac{4V_B(1 - V_B)}{1 + V_B} \tag{11.4.40}$$

Figure 11.13 shows the plots of normalized in-plane and out-of-plane CTEs as described in Eqs. (11.4.37)–(11.4.40).

When temperature change is substantial, there are two types of nonlinearities as previously mentioned. In the following illustration, we consider a simplified case of geometrical nonlinearity only with

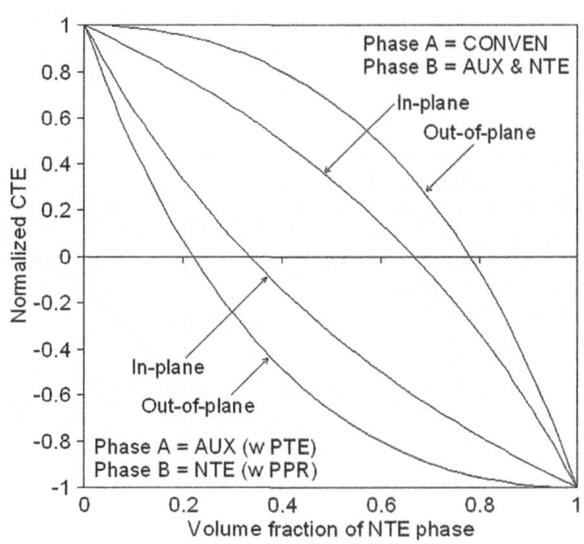

Fig. 11.13 Plots of normalized CTE with respect to NTE volume fraction

$$\frac{E_A}{E_B} = \frac{V_A}{V_B} = \frac{\left|\alpha_A^{(T)}\right|}{\left|\alpha_B^{(T)}\right|} = \frac{|\upsilon_A|}{|\upsilon_B|} = 1 \tag{11.4.41}$$

in order to illustrate the interlacing effect of Poisson's ratio and CTE. Again we consider the two sub-cases mentioned earlier. For the first sub-case whereby $\alpha_A^{(T)} = -\alpha_B^{(T)} = \alpha_0^{(T)}$ and $\upsilon_A = -\upsilon_B = 1/3$, the normalized in-plane and normalized out-of-plane laminate CTEs are

$$\frac{\alpha_{\text{in}}^{(T)}}{\alpha_0^{(T)}} = \frac{1 + 3\alpha_0^{(T)}\text{d}T}{3 + \alpha_0^{(T)}\text{d}T} \tag{11.4.42}$$

and

$$\frac{\alpha_{\text{out}}^{(T)}}{\alpha_0^{(T)}} = \frac{2}{3 + \alpha_0^{(T)}\text{d}T} \tag{11.4.43}$$

respectively. Based on Eqs. (11.4.42) and (11.4.43), the normalized effective in-plane and out-of-plane CTE versus the dimensionless change in temperature, $\alpha_0^{(T)}\text{d}T$, are plotted in Fig. 11.14. The in-plane and out-of-plane CTEs increase and decrease, respectively, with respect to the dimensionless temperature change—with decreasing slope. Essentially, the change in CTE with the temperature change is gradual.

The corresponding normalized CTEs for the second sub-case whereby $\alpha_A^{(T)} = -\alpha_B^{(T)} = \alpha_0^{(T)}$ and $\upsilon_A = -\upsilon_B = -1/3$ are

$$\frac{\alpha_{\text{in}}^{(T)}}{\alpha_0^{(T)}} = -\frac{1 - 3\alpha_0^{(T)}\text{d}T}{3 - \alpha_0^{(T)}\text{d}T} \tag{11.4.44}$$

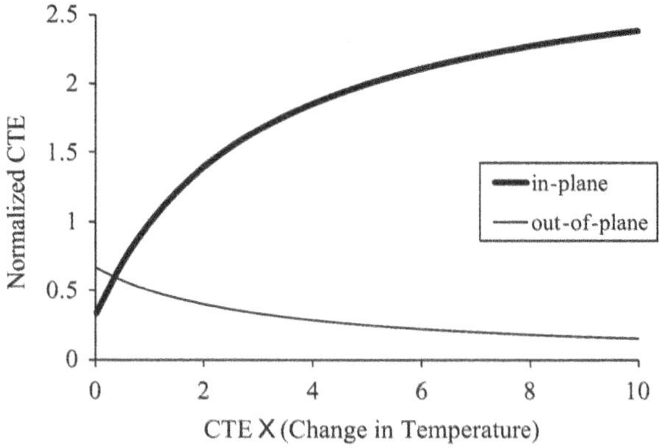

Fig. 11.14 Plots of normalized effective CTE against dimensionless temperature change for alternating conventional and non-conventional (auxetic and NTE) laminas

and

$$\frac{\alpha_{\text{out}}^{(T)}}{\alpha_0^{(T)}} = -\frac{2}{3 - \alpha_0^{(T)}\mathrm{d}T} \tag{11.4.45}$$

On the basis of Eqs. (11.4.44) and (11.4.45), plots of normalized effective CTEs for this case are shown in Fig. 11.15. For this sub-case whereby $\alpha_A^{(T)} = -\alpha_B^{(T)} = \alpha_0^{(T)}$ and $v_A = -v_B = -1/3$, the effective in-plane and out-of-plane CTEs possess extreme values at $\alpha_0^{(T)}\mathrm{d}T = |v_A|^{-1} = |v_B|^{-1}$. This is an interesting phenomenon in that the alternating layup of auxetic lamina with NTE lamina gives extremal thermal expansion, i.e., very extreme positive and negative values of laminate CTE when the increase in temperature reaches certain values.

It should be borne in mind that the plots of normalized CTEs in Figs. 11.14 and 11.5 are based on positive $\mathrm{d}T$. To reflect the variation in CTE with decreasing temperature, i.e., negative $\mathrm{d}T$, the range of dimensionless temperature change is extended to -10, as shown in Fig. 11.16. The results exhibit rotational symmetry of the plotted CTE curves about the origin.

Finally, the CTE models developed in this section can be related to those furnished in Sect. 11.3, whereby $E_A \gg E_B$. Substituting $E_B = 0$ into Eq. (11.4.23) reduces it to Eq. (11.3.10). Similarly, if we let $E_B = 0$ then Eq. (11.4.29) abridges to Eq. (11.3.9).

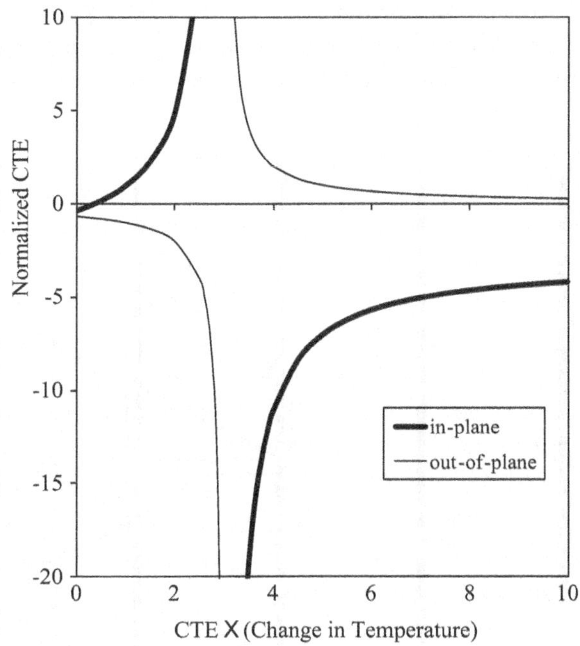

Fig. 11.15 Plots of normalized effective CTE against dimensionless temperature change for alternating auxetic and NTE laminas

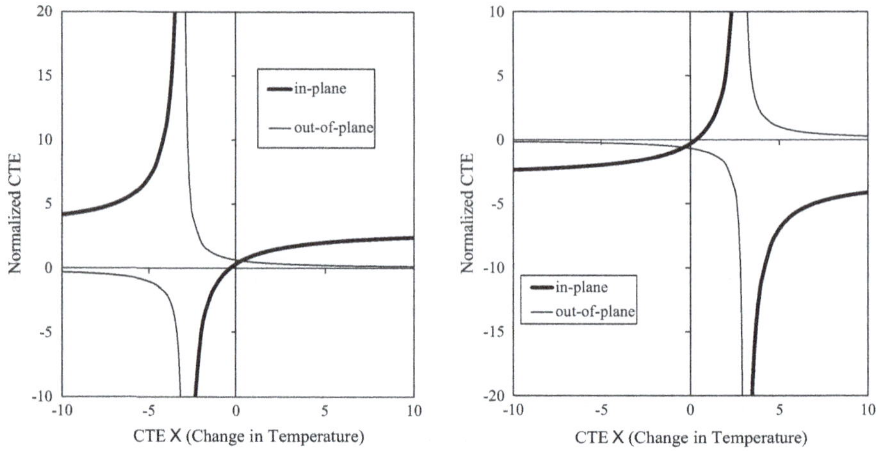

Fig. 11.16 Plots of normalized in-plane and out-of-plane CTE with respect to dimensionless temperature change for alternating conventional (positive v and PTE) laminas with non-conventional (negative v and NTE) laminas (left), as well as alternating positive v and NTE laminas with negative v and PTE laminas (right)

11.5 Example: 2D Truss System with Triangular Cells

An early 2D truss system with pin joints was introduced by Grima et al. (2007). With reference to Fig. 11.17, the lattice is made from three types of rod materials; rods made from the same materials are aligned in the same direction.

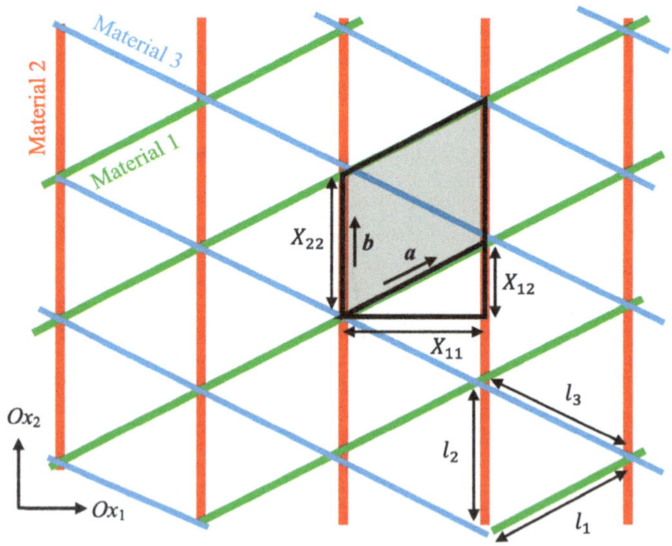

Fig. 11.17 Two-dimensional truss system with pin joints by Grima et al. (2007)

The geometry of the construct by Grima et al. (2007) can be described using a parallelogram unit cell consisting of 2 adjacent triangles. With reference to Fig. 11.17, one may identify two unit cell vectors

$$\begin{aligned} \boldsymbol{a} &= (X_{11}, X_{12}) \\ \boldsymbol{b} &= (0, X_{22}) \end{aligned} \tag{11.5.1}$$

where

$$\begin{aligned} X_{11} &= \frac{1}{2l_2}\sqrt{(l_1+l_2+l_3)(-l_1+l_2+l_3)(l_1-l_2+l_3)(l_1+l_2-l_3)} \\ X_{22} &= l_2 \\ X_{12} &= \sqrt{l_1^2 - X_{11}^2} = \frac{1}{2}\left(\frac{l_1^2-l_3^2}{l_2} + l_2\right) \end{aligned} \tag{11.5.2}$$

Since the three types of rods, in general, possess different CTEs of $\alpha_{\mathrm{Sm}}^{(T)}$ for rods $m = 1, 2, 3$, then a change in temperature by $\mathrm{d}T$ will change the lengths by

$$\mathrm{d}l_m = l_m \alpha_{\mathrm{Sm}}^{(T)} \mathrm{d}T \tag{11.5.3}$$

and the anisotropic CTE $\alpha_{ij}^{(T)}$ is described by the thermal strain

$$\varepsilon_{ij} = \alpha_{ij}^{(T)} \mathrm{d}T \tag{11.5.4}$$

where ε_{11} and ε_{22} are the strain components parallel to the Ox_1 and Ox_2 directions, respectively, while ε_{12} and ε_{21} are each half of the shear strain γ. These strain components are expressed as

$$\begin{aligned} \varepsilon_{11} &= \frac{\mathrm{d}X_{11}}{X_{11}} \\ \varepsilon_{22} &= \frac{\mathrm{d}X_{22}}{X_{22}} \\ \gamma = 2\varepsilon_{12} = 2\varepsilon_{21} &= \frac{1}{X_{11}}\left[\mathrm{d}X_{12} - \left(\frac{X_{12}}{X_{22}}\right)\mathrm{d}X_{22}\right] \end{aligned} \tag{11.5.5}$$

Since

$$\mathrm{d}X_{ij} = \sum_{m=1}^{3} \frac{\partial X_{ij}}{\partial l_m}\mathrm{d}l_m = \sum_{m=1}^{3} \frac{\partial X_{ij}}{\partial l_m} l_m \alpha_{\mathrm{Sm}}^{(T)} \mathrm{d}T \quad (i, j = 1, 2) \tag{11.5.6}$$

the elements of thermal expansion tensor $\alpha_{ij}^{(T)}$ can be simplified to

$$\alpha_{11}^{(T)} = \frac{\varepsilon_{11}}{\mathrm{d}T} = \left[l_1^2(l_2^2+l_3^2)\alpha_{S1}^{(T)} + l_2^2(l_1^2+l_3^2)\alpha_{S2}^{(T)} + l_3^2(l_1^2+l_2^2)\alpha_{S3}^{(T)}\right]\frac{1}{2l_2^2 X_{11}^2}$$

$$-\left[2l_2^2X_{11}^2\alpha_{S2}^{(T)}+l_1^4\alpha_{S1}^{(T)}+l_2^4\alpha_{S2}^{(T)}+l_3^4\alpha_{S3}^{(T)}\right]\frac{1}{2l_2^2X_{11}^2} \tag{11.5.7}$$

$$\alpha_{22}^{(T)}=\frac{\varepsilon_{22}}{\mathrm{d}T}=\alpha_{S2}^{(T)} \tag{11.5.8}$$

$$\alpha_{12}^{(T)}=\alpha_{21}^{(T)}=\frac{\varepsilon_{12}}{\mathrm{d}T}=\frac{1}{2}\frac{\gamma}{\mathrm{d}T}=\frac{l_1^2\left(\alpha_{S1}^{(T)}-\alpha_{S2}^{(T)}\right)-l_3^2\left(\alpha_{S3}^{(T)}-\alpha_{S2}^{(T)}\right)}{2X_{11}l_2} \tag{11.5.9}$$

By adopting the standard axis transformation techniques by Nye (1957), the CTE at a direction subtending by an angle ζ to the Ox_1 axis was obtained as

$$\alpha^{(T)}(\zeta)=\alpha_{11}^{(T)}\cos^2(\zeta)+2\alpha_{12}^{(T)}\sin(\zeta)\cos(\zeta)+\alpha_{22}^{(T)}\sin^2(\zeta) \tag{11.5.10}$$

while using the standard theory of principal strains (Gere 2001) the maximum and minimum CTEs obtained

$$\alpha^{(T)}(\zeta)_{\max/\min}=\frac{\alpha_{11}^{(T)}+\alpha_{22}^{(T)}}{2}\pm\sqrt{\left(\frac{\alpha_{11}^{(T)}-\alpha_{22}^{(T)}}{2}\right)^2+\left(\alpha_{12}^{(T)}\right)^2} \tag{11.5.11}$$

are at mutually orthogonal directions with

$$\zeta_{\max/\min}=\frac{1}{2}\tan^{-1}\left(\frac{2\alpha_{12}^{(T)}}{\alpha_{11}^{(T)}-\alpha_{22}^{(T)}}\right) \tag{11.5.12}$$

Thereafter, Grima et al. (2007) gave solutions and plotted results for the following special cases: equilateral triangles with $\alpha_{S1}^{(T)}=\alpha_{S3}^{(T)}\neq\alpha_{S2}^{(T)}$, equilateral triangles with $\alpha_{S1}^{(T)}=\alpha_{S2}^{(T)}\neq\alpha_{S3}^{(T)}$, equilateral triangles with $\alpha_{S1}^{(T)}\neq\alpha_{S3}^{(T)}\neq\alpha_{S2}^{(T)}$, isosceles triangles with $l_1=l_3\neq l_2$ and $\alpha_{S1}^{(T)}=\alpha_{S3}^{(T)}\neq\alpha_{S2}^{(T)}$, and more general cases. Results are plotted in Fig. 11.18 (equilateral triangles) and Fig. 11.19 (triangles of other shapes).

11.6 Example: 2D Truss System with Y-Shaped Elements

Another 2D microstructure that exhibits NTE was recently proposed by Cabras et al. (2019) by the joining of Y-shapes elements of lower CTEs $\alpha_{\text{low}}^{(T)}$ with rod elements of higher CTEs $\alpha_{\text{high}}^{(T)}$ as shown in Fig. 11.20 (top left) such that a temperature increase deforms the microstructure into that illustrated in Fig. 11.20 (top right). All connections are pin joints so as to permit free rotations. When considering the analysis

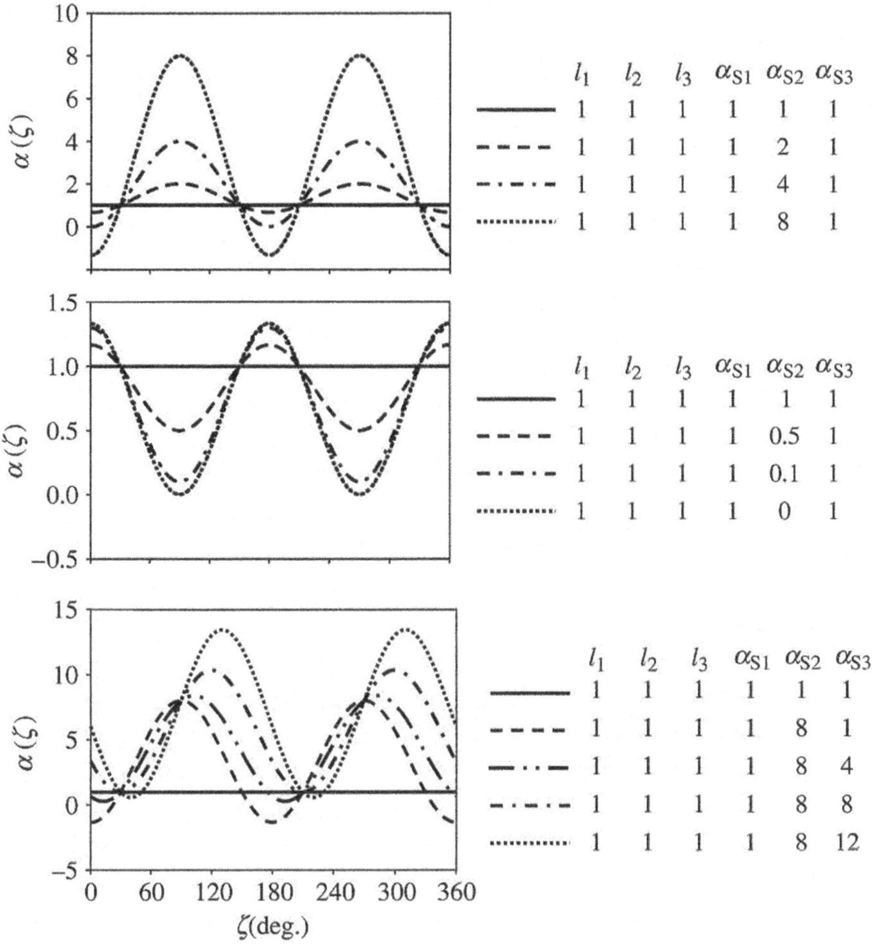

Fig. 11.18 Effective CTE for equilateral triangles by Grima et al. (2007). Reprinted with permission from the Royal Society

shown in Fig. 11.20 (bottom), one observes an essentially triangular cell, although arranged in a different manner from Sect. 11.5.

A description of the triangle formed by two Y-shaped elements and a connecting rod can be established by introducing two geometrical parameters; l and θ, as shown in Fig. 11.20 (bottom left). This gives the base length $2l\ \cos\theta$ and the height $L_0 = l\ \sin\theta$. The CTE in the direction of the connecting rod is none other than the CTE of the connecting rod. The movement of the apex can be obtained by vector addition of the thermally induced displacement. Assuming $\alpha_{\text{high}}^{(T)} = 0$, an increase of temperature by $\mathrm{d}T$ would elongate the inclined rod by an amount $l\alpha_{\text{low}}^{(T)}\mathrm{d}T$ such that, when resolved along the vertical axis gives an upward displacement of

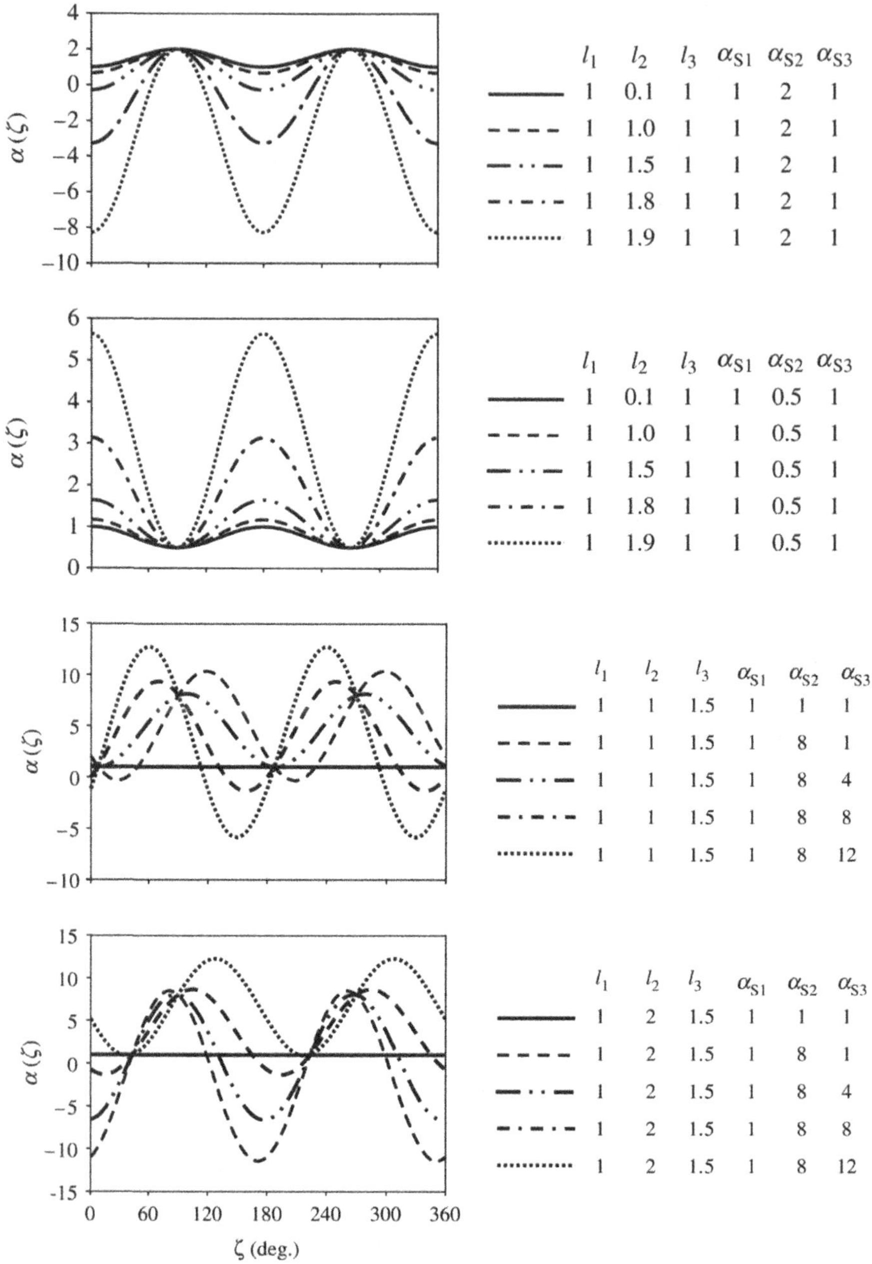

Fig. 11.19 Effective CTE for triangles of other shapes by Grima et al. (2007). Reprinted with permission from the Royal Society

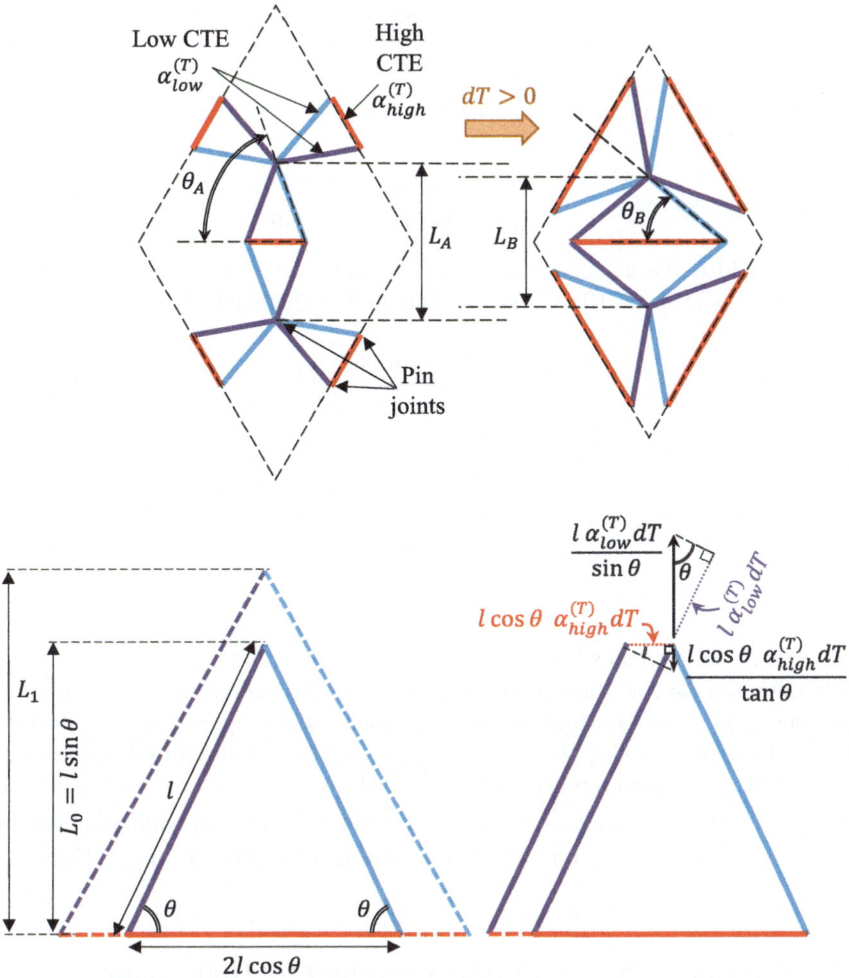

Fig. 11.20 Illustrations on the deformation of an NTE microstructure (top) and schematics for analysis (bottom) by Cabras et al. (2019)

$$dL_1' = \frac{l\alpha_{\text{low}}^{(T)}dT}{\sin\theta} \tag{11.6.1}$$

as shown in Fig. 11.20 (bottom right). Assuming $\alpha_{\text{low}}^{(T)} = 0$, an increase of temperature by dT would elongate the horizontal rod by an amount $2l\ \cos\theta\alpha_{\text{high}}^{(T)}dT$, which would cause a horizontal displacement of each inclined rod by $l\ \cos\theta\alpha_{\text{high}}^{(T)}dT$, as shown in Fig. 11.20 (bottom right). When resolved along the vertical axis, the apex displacement is downward by the amount

$$\mathrm{d}L_1'' = \frac{l \cos\theta \alpha_{\text{high}}^{(T)} \mathrm{d}T}{\tan\theta} \tag{11.6.2}$$

The overall displacement of the apex is thus

$$L_1 - L_0 = \mathrm{d}L_1' - \mathrm{d}L_1'' = \frac{l\alpha_{\text{low}}^{(T)} \mathrm{d}T}{\sin\theta} - \frac{l \cos\theta \alpha_{\text{high}}^{(T)} \mathrm{d}T}{\tan\theta} \tag{11.6.3}$$

Equation (11.6.3) is a linearization in $\mathrm{d}T$, and the exact, nonlinear in $\mathrm{d}T$, has been given by Cabras et al. (2019), thereby leading to the effective CTE

$$\alpha_{\text{eff}}^{(T)} = \frac{\left(1 - \eta \cos^2\theta\right)}{\sin^2\theta} \alpha_{\text{low}}^{(T)} \tag{11.6.4}$$

where

$$\eta = \frac{\alpha_{\text{high}}^{(T)}}{\alpha_{\text{low}}^{(T)}} \tag{11.6.5}$$

in which NTE is obtained when $\eta > 1/\cos^2\theta$.

In addition to the analytical model, Cabras et al. (2019) performed simulation as shown in Fig. 11.21 (left), which was substantiated by their experimental verification shown in Fig. 11.21 (right). Taking the difference between the initial temperature $T_0 = 26.4\ ^\circ\text{C}$ and a final temperature $T_f = 55.74\ ^\circ\text{C}$, Cabras et al. (2019) showed that the effective CTEs are $\alpha_{\text{eff}}^{(T)} = -2048\,\mu\text{m}/(\text{m}\ ^\circ\text{C})$ based on simulation results and $\alpha_{\text{eff}}^{(T)} = -2162\,\mu\text{m}/(\text{m}\ ^\circ\text{C})$ from their experimental results. See Fig. 11.22.

11.7 Example: 2D Truss System Inspired by Hoberman Sphere

An NTE system inspired by the Hoberman sphere has been introduced by Li et al. (2018). Its 2D version—the Hoberman circle—is shown in Fig. 11.23. Unlike the usual Hoberman sphere or circle, an additional connecting rod element is incorporated. By joining the outer and inner end points, the rod elements are therefore aligned radially. In order to exhibit NTE, the CTE of the joining rods must be higher than the CTE of the rod elements in the Hoberman circle. As temperature increases, the connecting rods expand to a greater extent than the rods in the Hoberman circle, which causes the circumference to contract. The opposite effect takes place with decreasing temperature. See Figs. 11.24 and 11.25.

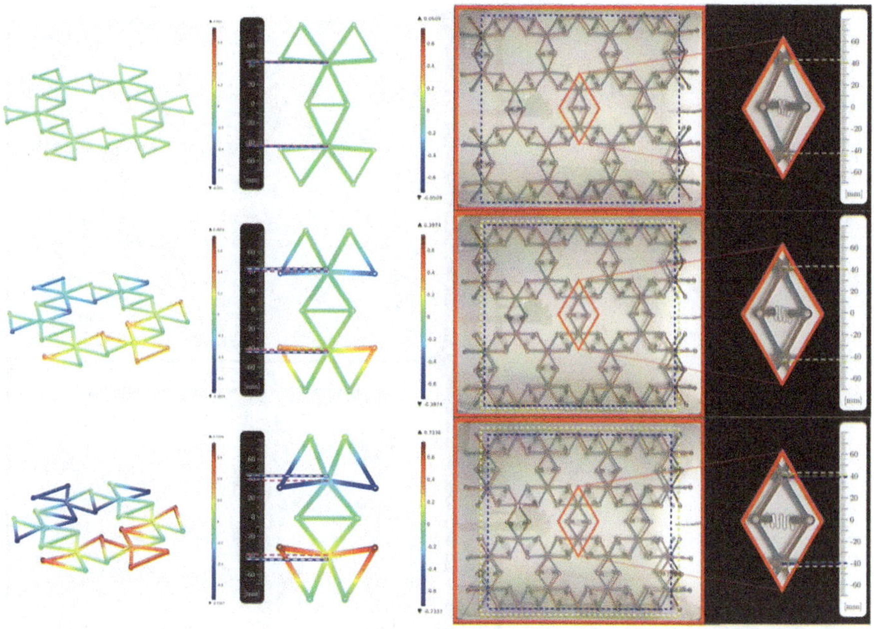

Fig. 11.21 Simulated (left) and experimental (right) deformation of an NTE microstructure by Cabras et al. (2019) at 22 °C (top row), 39°C (middle row) and 56 °C (bottom row)

Due to symmetry, only a triangle is required for analysis, as shown in Fig. 11.26 and, by symmetry again, the apex of the triangle is arrested from moving circumferentially; i.e., it can only displace radially. Let each Hoberman circle's rod element be of length m_0 and the length of each connecting rods is l_0. The distance of the apex from the origin is R_0, and the half-angle formed by the apex within the triangle is θ_0, while the distance of the outer joint from the origin is s_0. With a change in temperature by $\mathrm{d}T$, the Hoberman circle's rod length changes to m_1 while the length of the connecting rod changes to l_1. During the course of the changing rod lengths, both the connecting rod and the apex displace radially. In the selected triangle shown in Fig. 11.26, the connecting rod displaces along the x-axis while the apex moves along a straight line that subtends at an angle φ to the x-axis. As a consequence, the apex moves a location at a distance R_1 from the origin and the half-angle formed by the apex changes to θ_1 while the outer joint is displaced to a location s_1 from the origin.

By geometrical consideration involving the rod lengths

$$\left.\begin{aligned} h_i &= m_i \cos\theta_i = R_i \sin\varphi \\ l_i &= 2R_i \sin\varphi \, \tan\theta_i \end{aligned}\right\} \quad i = 0, 1 \tag{11.7.1}$$

we have the distances of the outer joints

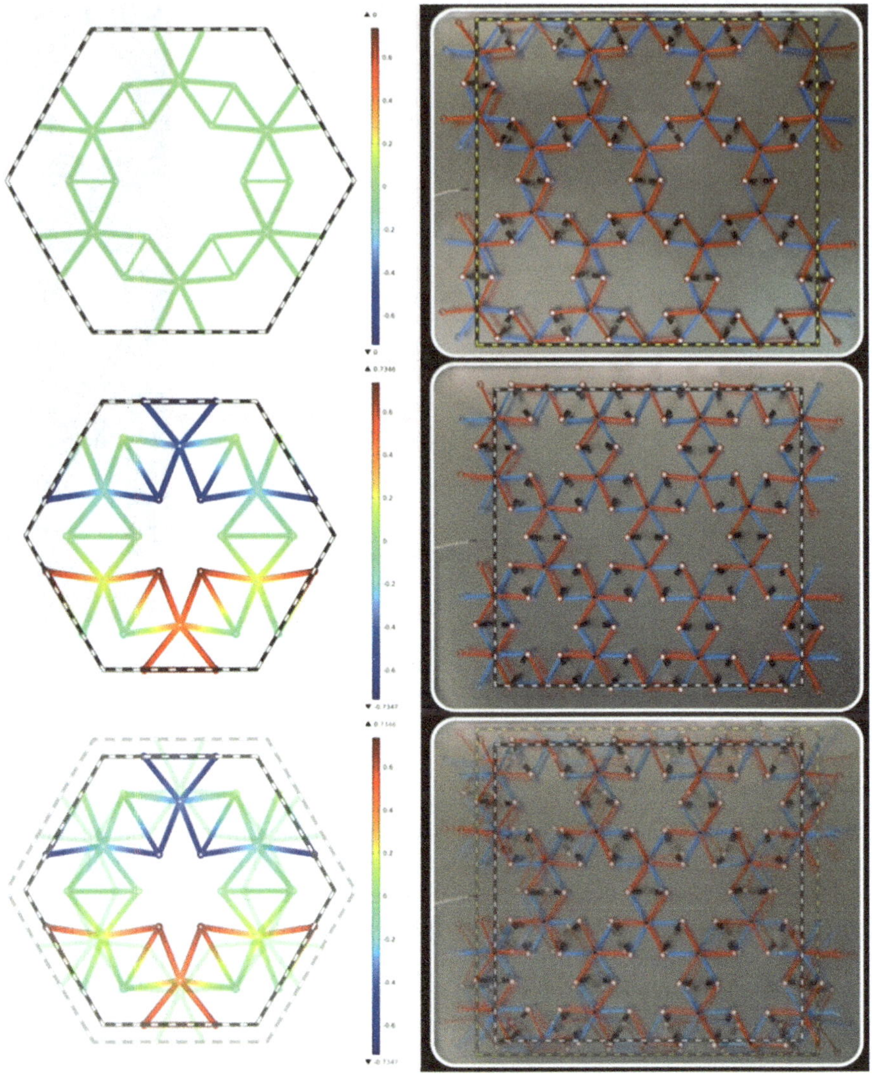

Fig. 11.22 Simulated (left) and experimental (right) deformation of an NTE microstructure by Cabras et al. (2019) at 26.4 °C (top row) and 55.74 °C (middle row), giving the effective CTE as $-2048\,\mu m/(m\ °C)$ by simulation and $-2162\,\mu m/(m\ °C)$ by experiment (bottom row)

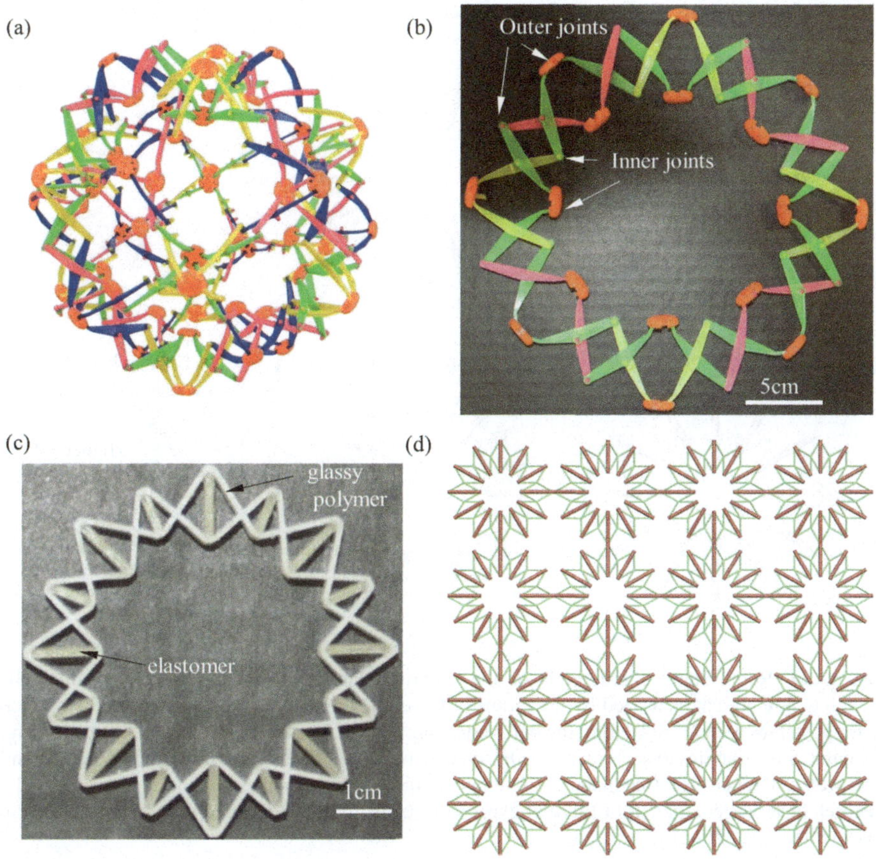

Fig. 11.23 Design of the lattice metamaterials with tunable thermal expansion: **a** Hoberman expandable sphere, **b** the planar element, **c** the unit cell fabricated by 3D printing technique, and **d** the lattice metamaterial with 4 × 4 unit cells by Li et al. (2018). Reprinted with permission from Elsevier

$$s_i = R_i \cos\varphi + m_i \sin\theta_i = R_i(\cos\varphi + \sin\varphi \ \tan\theta_i) \tag{11.7.2}$$

before ($i = 0$) and after ($i = 1$) temperature change. Thus, thermal strain $\varepsilon = (s_1 - s_0)/s_0$ leads to the effective CTE (Li et al. 2018)

$$\alpha_{\text{eff}}^{(T)} = \frac{1}{\mathrm{d}T} \frac{s_1 - s_0}{R_0(\cos\varphi + \sin\varphi \ \tan\theta_0)} \tag{11.7.3}$$

By solving for $s_1 - s_0$ computationally, Li et al. (2018) gave solutions of the effective CTE based on different combinations of rod materials. In addition to analytical model, Li et al. (2018) performed computational modeling for the same problem, on the basis of Fig. 11.27.

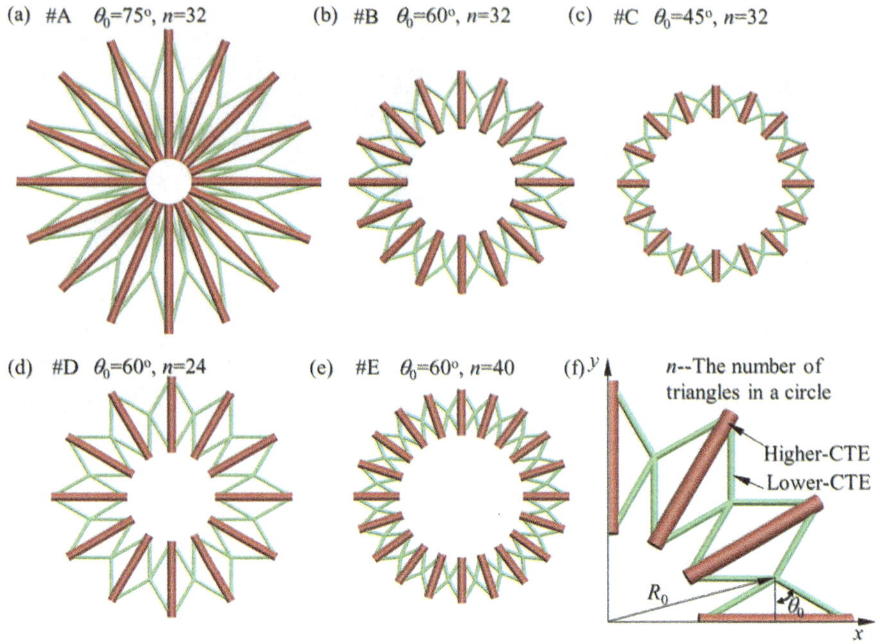

Fig. 11.24 Representative unit cells of the proposed lattice metamaterials. Those unit cells including **a–e** are characterized by the radius of the circle (distance between the center and green cross-point) R_0 with $R_0 = 10$ mm, the half vertex angle between blue beams in the triangle $\theta 0$, and the number of triangles in a circle, n. *(f)* One quarter unit cell with definitions including R_0, θ_0, and n, as well as Higher-CTE (the material of higher coefficient of thermal expansion), Lower-CTE (the material of lower coefficient of thermal expansion) constituents. Reprinted with permission from Elsevier (Li et al. 2018)

11.8 Example: A 2D Anepectic Mesh System

Raminhos et al. (2019) introduced the term "anepectic" from the Greek root "Επέκταση" (Epéktasi'), meaning expansion, for materials capable of simultaneously demonstrating both auxetic and NTE behavior. In this section, emphasis is on its NTE behavior. An illustration on this microstructure and its deformation with temperature change is shown in Fig. 11.28. The extension of the rods with greater CTE causes the inclined rods, of lower CTE, to reorientate such that the star-shape voids appear sharper, thereby shortening the distances between the four sharp corners of each star. A set of experiments on this microstructure using various combinations of materials have been reported by Raminhos et al. (2019).

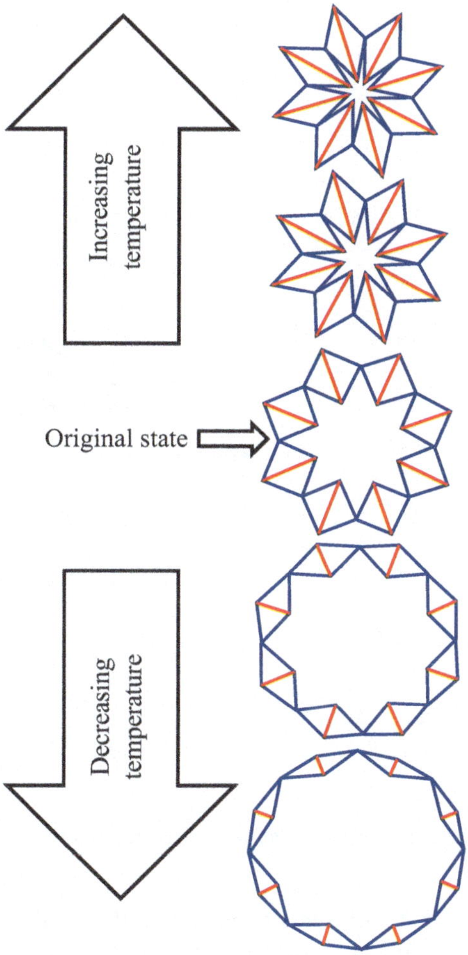

Fig. 11.25 An NTE system based on Hoberman circle by Li et al. (2018)

11.9 Example: 2D Thermal Shearing with Opposing CTE Signs in Orthogonal Directions

This example is a definite anisotropic system with no possibility of reducing into an isotropic system based on the stipulated geometry which ensures that the CTE is positive (maximum) and negative (minimum) in orthogonal directions, thereby leading to thermal shearing. Consider a cellular microstructure shown in Fig. 11.29 and its idealized representative volume element (RVE), as shown in Fig. 11.30, which consists of thermally inexpandible rod elements (thick lines) and thermally expandible rod elements (thin lines). The connections are in the form of pin joints to permit free rotation except at the cross-junction of the thermally inexpandible rod

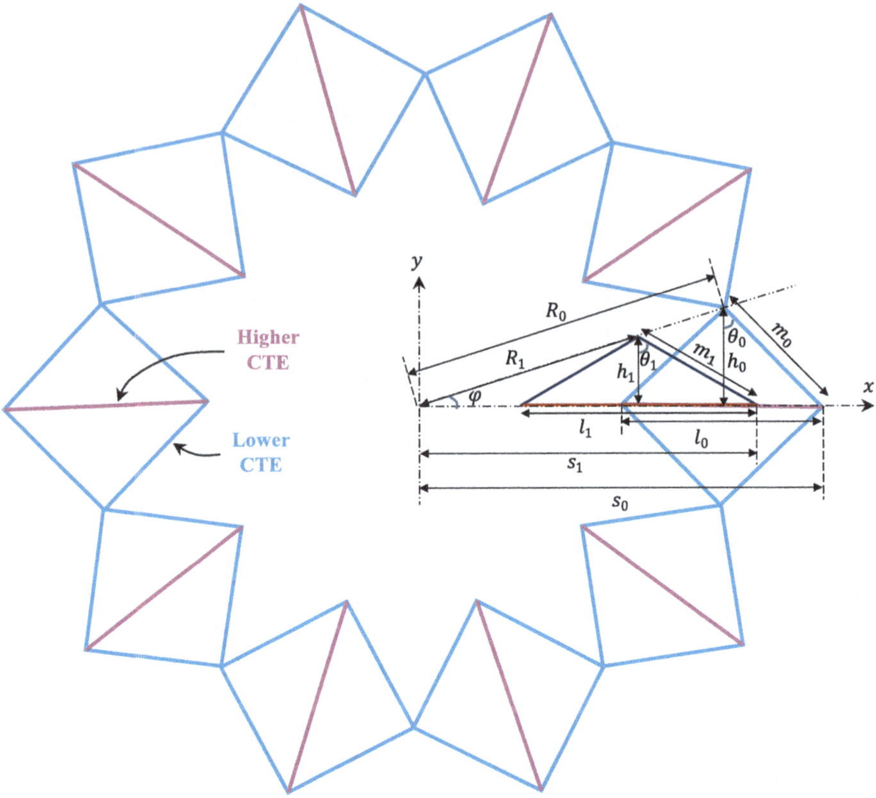

Fig. 11.26 A schematics for Hoberman sphere-inspired NTE system by Li et al. (2018) for analysis

elements. The sides of the RVE remain mutually orthogonal under thermal loading such that the RVE dies not to undergo a distortion leading to repeat-cell angles departing from 90° in the plane. However, unequal principal normal strains lead to maximum shearing that can be observed for an element rotated by 45° from the principal axes. We let the inexpandible rod be rigidly fixed at the central junction (i.e., θ_3 and θ_4 being constant). The assumption of rigid rods at the central junction is highly essential to ensure that point C remains fixed to enable RVE contraction with increase in temperature—hence a negative thermal expansion—when both θ_1 and θ_2 are less than $\pi/4$. The magnitude of negative thermal expansion is greatly reduced when the central junction angles and rods are rotatable and expandible, respectively.

The thermal strains for every rod elements in this system are thus

$$\begin{aligned} \varepsilon_{\text{expand}} &= \alpha^{(T)} \mathrm{d}T \\ \varepsilon_{\text{inexpand}} &= 0 \end{aligned} \tag{11.9.1}$$

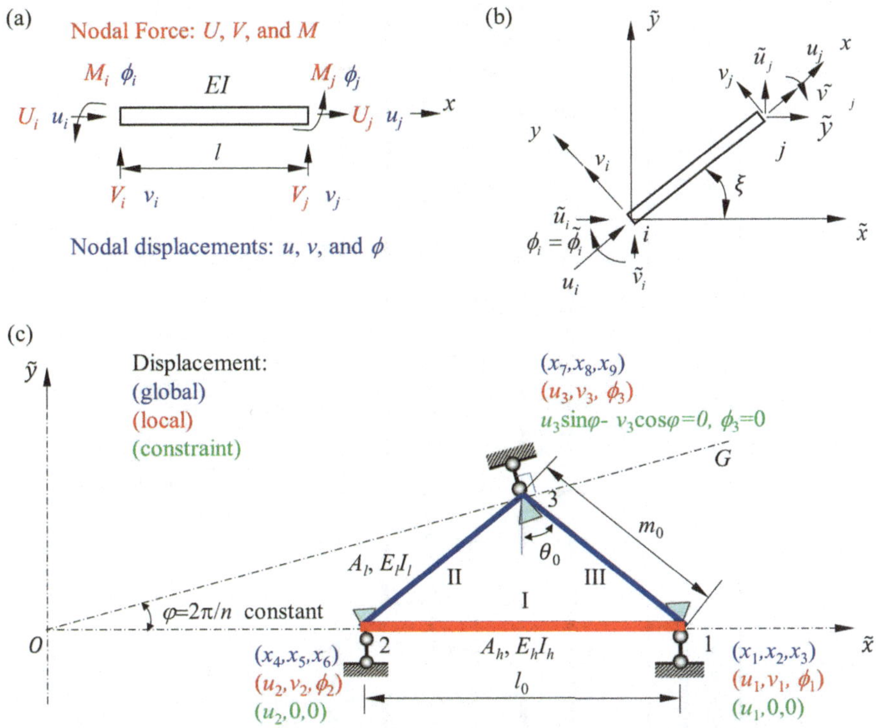

Fig. 11.27 Beam element and representative isosceles triangle frame with boundary conditions. **a** Beam element with nodal displacements, rotations, forces, and moments. **b** Conversion from the local system to the global system. **c** Elements, nodes, and constraint conditions. The global elements are I, II, and III. The global nodes are designated as 1, 2, and 3. The global nodal displacements are $x_1, x_2, x_3, x_4, x_5, x_6, x_7, x_8,$ and x_9. A_l, E_l, and I_l are the sectional area, Young's Modulus, and the inertia moment of the lower-CTE beam, respectively. A_h, E_h, and I_h are the sectional area, Young's Modulus, and the inertia moment of the higher-CTE beam, respectively. Reprinted with permission from Elsevier (Li et al. 2018)

where ΔT refers to the change in temperature, we define effective relationship for the entire RVE as

$$\varepsilon_{\text{eff}} = \alpha_{\text{eff}}^{(T)} \text{d}T \tag{11.9.2}$$

such that the effective property is obtained by considering the geometry of the RVE and every elements contained therein. The effective shear strain for the RVE as a result of temperature change can be written in a form analogous to Eq. (11.9.2), that is

$$\gamma = \beta \, \text{d}T \tag{11.9.3}$$

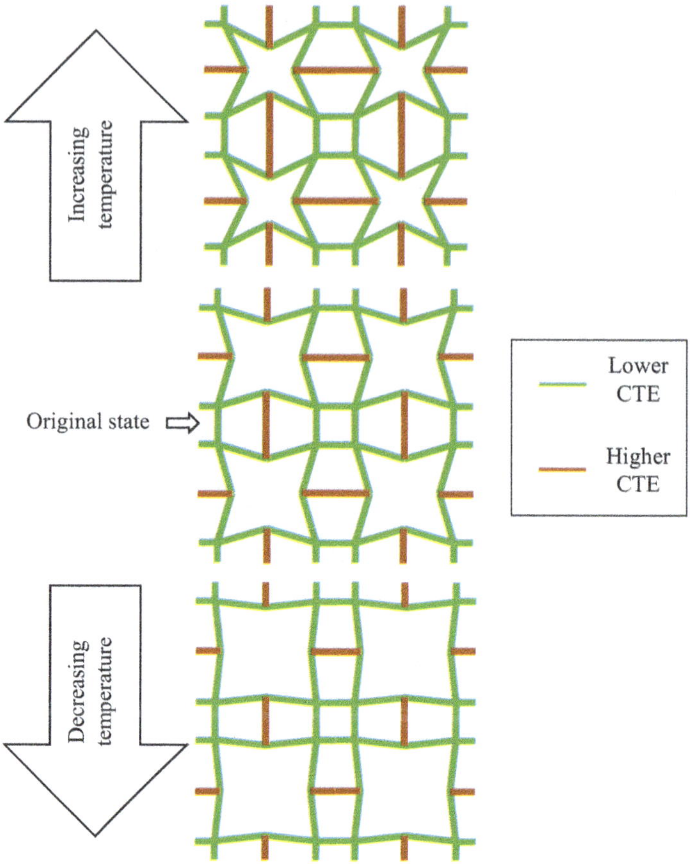

Fig. 11.28 Thermal deformation of the anepectic microstructure by Raminhos et al. (2019)

where β is herein defined as the coefficient of thermal shearing. The following analysis expresses the shear strain in terms of temperature induced geometrical alteration of the proposed RVE. With reference to Eq. (11.9.1), the new lengths of l_1 and l_2 are

$$l_i' = l_i\left(1 + \alpha^{(T)}\mathrm{d}T\right) \;\; ; \;\; (i = 1, 2) \tag{11.9.4}$$

with a change in environmental temperature by dT, as shown in Fig. 11.31.

Taking equal projected length, we have the new angles of θ_1 and θ_2 as

$$\theta_i' = \sin^{-1}\left(\frac{\sin\theta_i}{1 + \alpha^{(T)}\mathrm{d}T}\right) \tag{11.9.5}$$

This gives the displacements of points A and B as

Fig. 11.29 A 2D thermal shearing microstructure exhibiting PTE and NTE in the vertical and horizontal directions (Lim 2005)

$$dx_i = l_i \cos\theta_i - l_i\left(1 + \alpha^{(T)} dT\right) \cos\left[\sin^{-1}\left(\frac{\sin\theta_i}{1 + \alpha^{(T)} dT}\right)\right] \tag{11.9.6}$$

along axes x_1 and x_2 respectively. Perusal to Fig. 11.31 shows that the half widths of the RVE are

$$X_i = l \cos\theta_{i+2} - l_i \cos\theta_i + l_{i+2} \tag{11.9.7}$$

such that the principal strains are

$$\varepsilon_i = \frac{dx_i}{X_i} \tag{11.9.8}$$

Since the relationship between the inexpandible and expandible rods of lengths l and $l_i (i = 1, 2)$, respectively, is

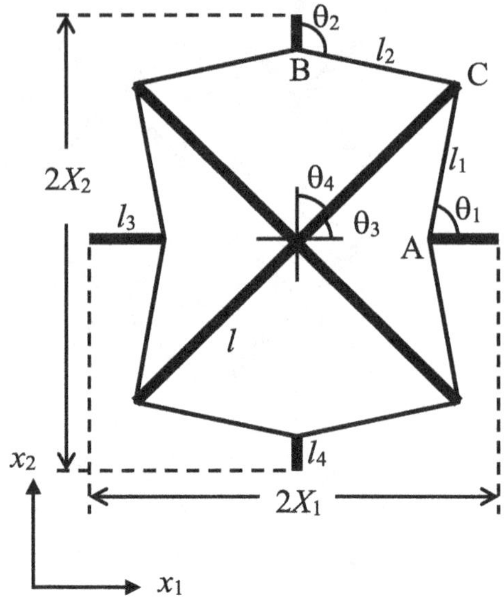

Fig. 11.30 Geometrical properties for one RVE of the thermal shearing system

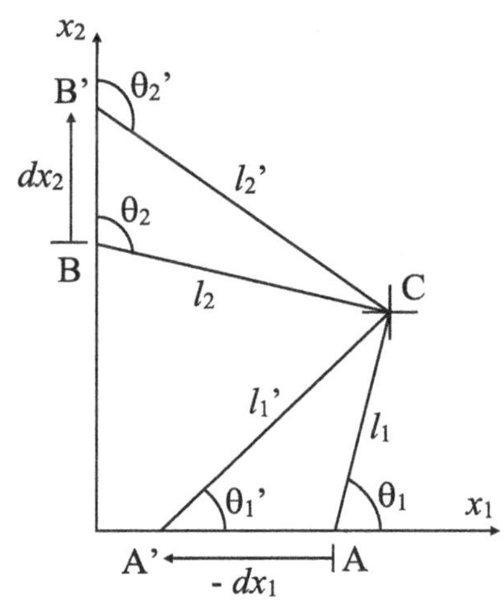

Fig. 11.31 Changes in lengths and angles in one-quarter of RVE

$$\frac{l}{l_i} = \frac{\sin\theta_i}{\sin\theta_{i+2}} \tag{11.9.9}$$

the coefficient of thermal shearing is therefore

$$\beta = \frac{\varepsilon_1 - \varepsilon_2}{\mathrm{d}T} \tag{11.9.10}$$

where the principal strains for the RVE are

$$\varepsilon_i = \frac{\cos\theta_i - \left(1 + \alpha^{(T)}\mathrm{d}T\right)\cos\left[\sin^{-1}\left(\frac{\sin\theta_i}{1+\alpha^{(T)}\mathrm{d}T}\right)\right]}{\frac{\sin\theta_i}{\tan\theta_{i+2}} - \cos\theta_i + \frac{l_{i+2}}{l_i}} \tag{11.9.11}$$

For illustration purposes, a special case is considered whereby the criss-cross inexpandible rod element is symmetrical about diagonal axes $\theta_3 = \theta_4 = \pi/4$; the extendible rod elements are of equal length $l_1 = l_2$; the RVE is a square $X_1 = X_2$, the angles $\theta_{1,2} = 90^\circ \mp \phi$ such that $\theta_1 + \theta_2 = \pi$, and $l_4 = 0$. For such a geometry, the principal thermal strains and the coefficient of thermal shearing reduce to (Lim 2005)

$$\varepsilon_1 = -\varepsilon_2 = -\frac{\cos\theta_2 - \left(1 + \alpha^{(T)}\mathrm{d}T\right)\cos\left[\sin^{-1}\left(\frac{\sin\theta_2}{1+\alpha^{(T)}\mathrm{d}T}\right)\right]}{\sin\theta_2 - \cos\theta_2} \tag{11.9.12}$$

and

$$\beta = \frac{2}{\mathrm{d}T}\left(\frac{(-1)^{i+1}\cos\theta_i}{\sin\theta_i + (-1)^{i+1}\cos\theta_i}\right) \tag{11.9.13}$$

respectively. Plots of principal strains and shear strain versus θ_1 are furnished in Fig. 11.32 to demonstrate the influence of microstructural geometry on the thermally induced shear strain.

11.10 A 2D NTE System by Rigid Unit Modes

Grima et al. (2015) introduced a group of 2D NTE mechanism based on rotating unit modes (RUM) shown in Fig. 11.33. With reference to the Type I connected rectangles shown in Fig. 11.33 (b), the unit cell projections along the Ox_1 and Ox_2 axes are

$$\begin{aligned} X_{11} &= 2\left[a \cos\tfrac{\theta}{2} + b \sin\tfrac{\theta}{2}\right] \\ X_{22} &= 2\left[a \sin\tfrac{\theta}{2} + b \cos\tfrac{\theta}{2}\right] \end{aligned} \tag{11.10.1}$$

from which one obtains the unit cell area at any angle θ

$$A(\theta) = X_{11}X_{22} = 2\left(a^2 + b^2\right)\sin\theta + 4ab \tag{11.10.2}$$

Writing the instantaneous angle θ in terms of its original angle θ_0 and its change $\Delta\theta$, then the area at an angle $\theta = \theta_0 \pm \Delta\theta$ is

$$A(\theta_0 \pm \Delta\theta) = 2\left(a^2 + b^2\right)\sin(\theta_0 \pm \Delta\theta) + 4ab \tag{11.10.3}$$

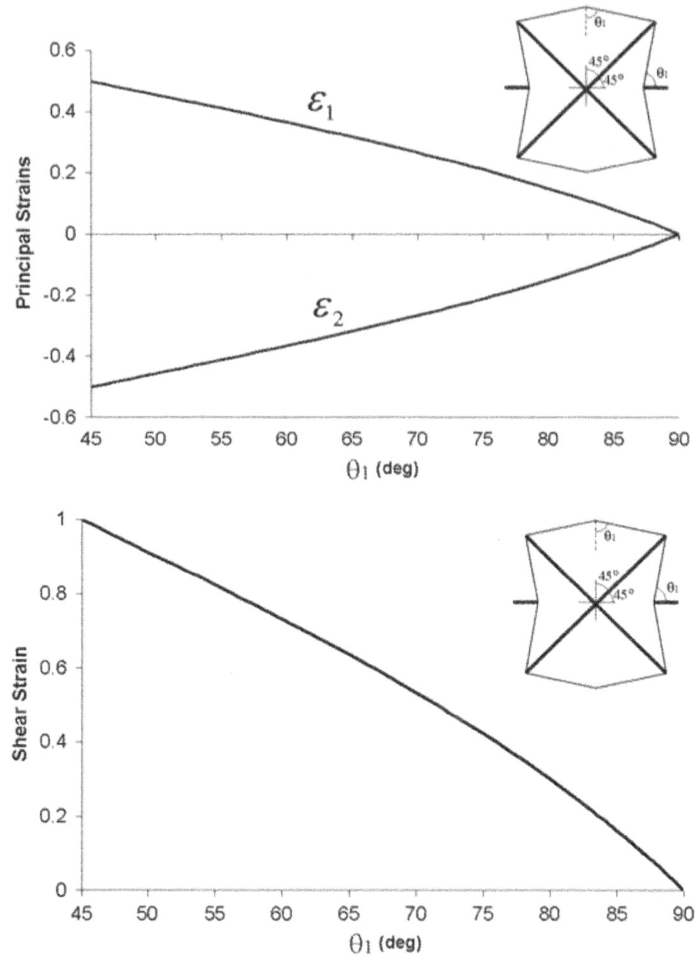

Fig. 11.32 Effect of special case RVE geometry on the thermally induced principal strains (top) and shear strain (bottom)

If the connected rectangles are in its open state in its original condition, i.e., $\theta_0 = \pi/2$, then for a given $\Delta\theta$ we have $\sin(\pi/2 \pm \Delta\theta) = \cos\Delta\theta$. One may express the average unit cell area at temperature T, $\langle A \rangle_{\mathrm{T}}$, as

$$\langle A \rangle_{\mathrm{T}} = 2\left(a^2 + b^2\right)\cos\Delta\theta + 4ab \approx 2\left(a^2 + b^2\right)\left(1 - \frac{\left\langle \Delta\theta^2 \right\rangle_{\mathrm{T}}}{2}\right) + 4ab \quad (11.10.4)$$

where $\left\langle \Delta\theta^2 \right\rangle_{\mathrm{T}}$ is the thermal average of $\Delta\theta^2$. The potential energy due to rotations arising from thermal fluctuations of $\Delta\theta$ is (Dove et al. 1998).

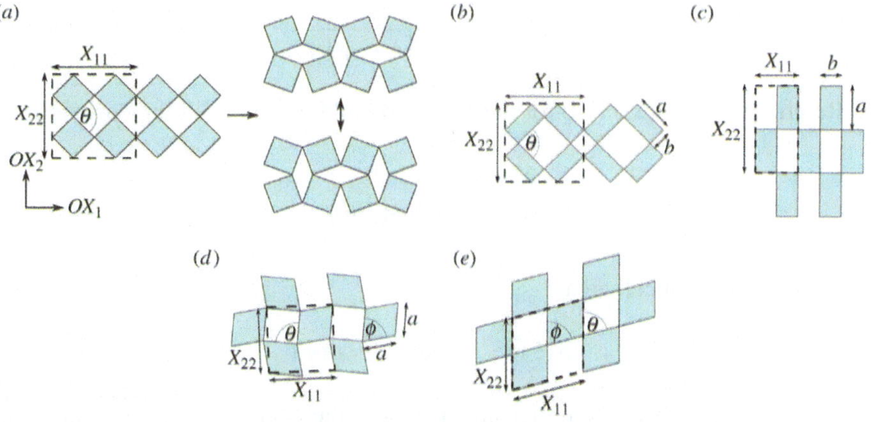

Fig. 11.33 Connected networks composed of **a** squares, **b** Type I rectangles, **c** Type II rectangles, **d** Type α rhombi, and **e** Type β rhombi by Grima et al. (2015). The respective RUM for the square-based network is also shown in **a**; it is assumed that the other geometries follow a similar mode of motion. Reprinted with permission from the Royal Society

$$\frac{1}{2}I\omega^2\langle\Delta\theta^2\rangle_{\mathrm{T}} = \frac{1}{2}k_B T \tag{11.10.5}$$

where I is the moment of inertia for each rigid unit, ω is the angular frequency of the rotational motion, and k_B is the Boltzmann constant. Substituting $\langle\Delta\theta^2\rangle_{\mathrm{T}}$ from Eq. (11.10.5) into Eq. (11.10.4) gives

$$\langle A\rangle_{\mathrm{T}} = 2\left(a^2+b^2\right)\left(1-\frac{k_B T}{2I\omega^2}\right)+4ab \tag{11.10.6}$$

from which the area CTE is obtained as

$$\alpha_A^{(T)} = \frac{1}{\langle A\rangle_{\mathrm{T}}}\frac{\partial\langle A\rangle_{\mathrm{T}}}{\partial T} = -\frac{k_B\left(a^2+b^2\right)}{I\omega^2\left[-\left(a^2+b^2\right)\frac{k_B T}{I\omega^2}+2\left(a^2+2ab+b^2\right)\right]} \tag{11.10.7}$$

With reference to the Type II connected rectangles shown in Fig. 11.33 (c), the unit cell projections along the Ox_1 and Ox_2 axes are

$$\begin{aligned} X_{11} &= 2b\,\sin\left(\frac{\theta}{2}+\frac{\pi}{4}\right) \\ X_{22} &= 2a\,\sin\left(\frac{\theta}{2}+\frac{\pi}{4}\right) \end{aligned} \tag{11.10.8}$$

from which, through a similar approach to that of Type I, gives the area CTE wherein $\theta_0 = \pi/2$

$$\alpha_A^{(T)} = -\frac{k_B}{I\omega^2\left[4 - \frac{k_B T}{I\omega^2}\right]} \tag{11.10.9}$$

Referring to Fig. 11.33 (d) for the Type α rhombi, the unit cell dimensions are

$$\begin{aligned} X_{11} &= 2a\ \sin\left(\tfrac{\phi+\theta}{2}\right) \\ X_{22} &= 2a\ \cos\left(\tfrac{\phi-\theta}{2}\right) \end{aligned} \tag{11.10.10}$$

which gives the unit cell area

$$A(\theta) = 2a^2(\sin\theta + \sin\phi) \tag{11.10.11}$$

Assuming again that the initial state is that of the open state $\theta_0 = \pi/2$ and that the angular displacement $\pm\Delta\theta$ gives a symmetric function such that $A(\theta_0 + \Delta\theta) = A(\theta_0 - \Delta\theta)$, then using similar approach

$$\begin{aligned} \langle A\rangle_\mathrm{T} &= 2a^2(\sin(\theta_0 + \Delta\theta) + \sin\phi) = 2a^2(\cos\Delta\theta + \sin\phi) \\ &\approx 2a^2\left(1 - \frac{\langle\Delta\theta^2\rangle_\mathrm{T}}{2}\right) + 2a^2\sin\phi = 2a^2\left(1 - \frac{k_B T}{2I\omega^2}\right) + 2a^2\sin\phi \end{aligned} \tag{11.10.12}$$

which leads to the area CTE

$$\alpha_A^{(T)} = -\frac{k_B}{I\omega^2\left[2 - \frac{k_B T}{I\omega^2} + 2\sin\phi\right]} \tag{11.10.13}$$

In the case of the Type β rhombi shown in Fig. 11.33 (e), the most open conformation in the "cold state" is $\theta_0 = \pi - \phi$, from which the area CTE is given by Eq. (11.10.9). See Fig. 11.34 for plotted CTE results, as well as illustrations for extreme shapes.

11.11 Example: 2D and 3D NTE from Ring-Rod Assemblies

Following up from Sect. 2.5 on the single-ring and double-ring structures that give auxetic behavior, this section explores their NTE properties (Lim 2017). See Fig. 11.35a, wherein the ring has positive CTE and that the rods, being rigid, possess zero CTE. Suppose there is an increase to the environmental temperature, the ring expands such that the rods are drawn toward into the ring as shown in Fig. 11.35b, thereby decreasing the dimension along the rod alignment direction while the dimension perpendicular to the rod direction expands. As shown in Fig. 11.35c, the ring

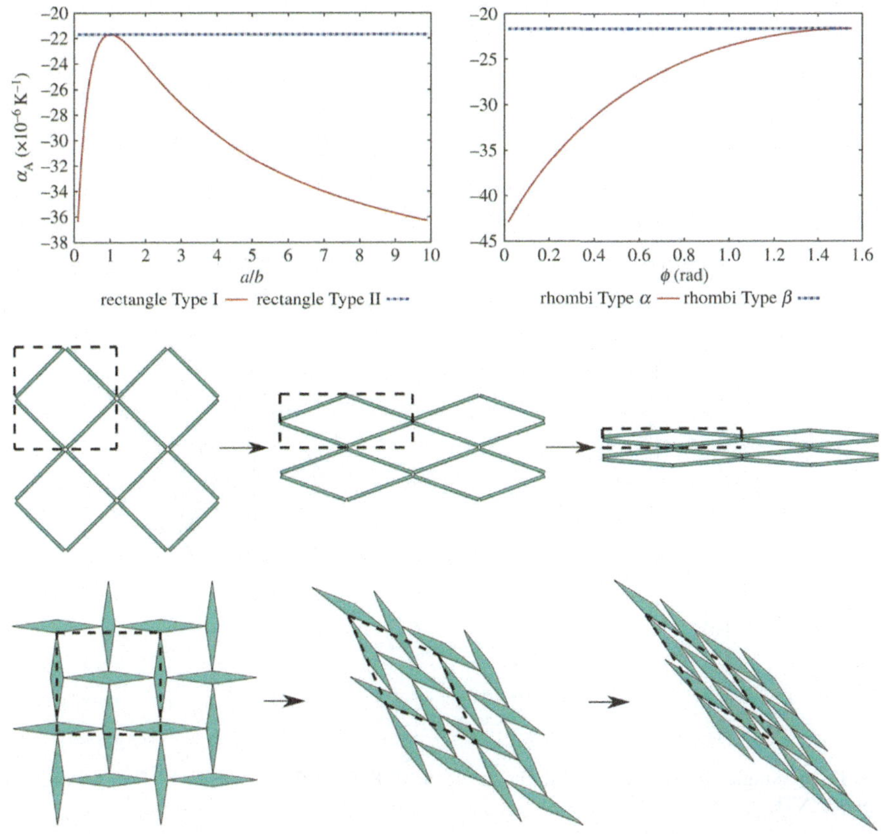

Fig. 11.34 Results of area CTE for the four types of RUMs for varying shapes (top), and illustrations on deformation mechanism of Type I rectangles (middle) and Type α rhombi (bottom) with extreme aspect ratio Grima et al. (2015). Reprinted with permission from the Royal Society

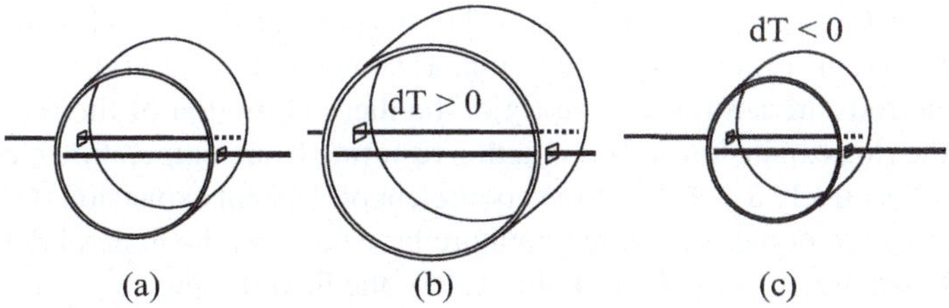

Fig. 11.35 **a** A ring structure with a pair of sliding rods before temperature change, **b** increase in temperature causes increase in ring diameter and draw-in of sliding rods, and **c** decrease of temperature contracts the ring and pushes out the sliding rods

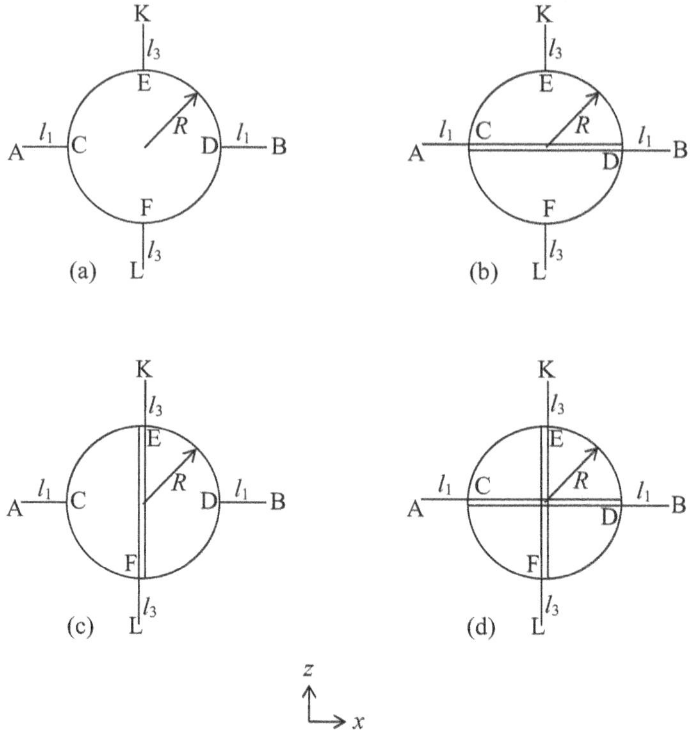

Fig. 11.36 Single-ring structure exhibiting **a** planar PTE, **b** NTE in x-axis, **c** NTE in y-axis, and **d** planar NTE

deformation and rod movement reverse with a drop in temperature. Hence, this ring-rod structure exhibits both NTE and positive thermal expansion (PTE) parallel and perpendicular, respectively, to the rod alignment. See Fig. 11.36a–c. In follows that an equi-biaxial NTE structure can be obtained if two pairs of sliding rods perpendicular to each other are incorporated into the ring, as shown in Fig. 11.36d.

Since the rods are assumed to be rigid, the thermal motion of the rods is determined by the ring expansion and contraction resulting from temperature increase and decrease, respectively. Let $\alpha_r^{(T)}$ be the coefficient of thermal expansion (CTE) of the ring and consider an increase in temperature by $\mathrm{d}T$. From the general definition of the thermal strain $\varepsilon = \alpha^{(T)}\mathrm{d}T$ or, in the case of the flexible ring,

$$\frac{2\pi(R + \mathrm{d}R) - 2\pi R}{2\pi R} = \alpha_r^{(T)}\mathrm{d}T \tag{11.11.1}$$

we have the change in radius $R\alpha_r^{(T)}\mathrm{d}T$ or the change in diameter $2R\alpha_r^{(T)}\mathrm{d}T$. Considering increasing temperature, there is a thermal expansion by $2R\alpha_r^{(T)}\mathrm{d}T$ in the direction where the short rods are attached to the outer surface to the ring and a thermal contraction of $2R\alpha_r^{(T)}\mathrm{d}T$ in the direction where the long sliding rods are attached to the inner surface of the ring. Writing the effective CTE of the ring and rod structure as $\varepsilon/\mathrm{d}T$, we have

$$\begin{aligned} \alpha_x^{(T)} &= \frac{1}{\mathrm{d}T}\frac{\Delta\overline{AB}}{\overline{AB}} \\ \alpha_z^{(T)} &= \frac{1}{\mathrm{d}T}\frac{\Delta\overline{KL}}{\overline{KL}} \end{aligned} \tag{11.11.2}$$

or

$$\begin{aligned} \alpha_x^{(T)} &= \frac{1}{\mathrm{d}T}\frac{\Delta\overline{AB}}{2R+2l_1} \\ \alpha_z^{(T)} &= \frac{1}{\mathrm{d}T}\frac{\Delta\overline{KL}}{2R+2l_3} \end{aligned} \tag{11.11.3}$$

where

$$\left|\Delta\overline{AB}\right| = \left|\Delta\overline{KL}\right| = 2R\alpha_r^{(T)}\mathrm{d}T \tag{11.11.4}$$

The calculations of $\alpha_x^{(T)}$ and $\alpha_z^{(T)}$ are summarized in Table 11.2.

Figure 11.37a shows further details on the use of two pairs of sliding rods to give a fully planar NTE ring structure. Expansion of the ring due to temperature increase draws in the sliding rods while contraction of the ring due to temperature decrease pushes out the sliding rods, as indicated in Fig. 11.37b, c, respectively. Hence, the use of two pairs of sliding rods gives fully planar NTE characteristics but this structure results in non-auxetic property, as discussed in Sect. 2.5. The turning propensity of the rings about the axes can be arrested by arranging the sliding rods as suggested in Fig. 11.37d, in which the possibility of turning in one ring is canceled by its neighboring ring with a propensity of turning in the opposite direction.

An appreciation on the effective structural CTE is given in Fig. 11.38. Essentially, the CTE is positive and negative along the axes where the sliding rods are absent and present, respectively. With that, the case of purely double-ring, i.e., no sliding rods as in Fig. 11.38a, exhibits positive thermal expansion. Where one pair of sliding rods is incorporated, the structure displays NTE in the direction of the sliding rods while positive thermal expansion remains on the plane perpendicular to the sliding rods. With reference to the structure shown in Fig. 11.38b, NTE and PTE are evident in the x-axis and y-z plane, respectively. Since the NTE behavior is confined to one axis while the PTE behavior applies for a plane (i.e., two axes), we say that this structure has "low NTE." Suppose two pairs of sliding rods are inserted, the structure exhibits NTE in the plane where the sliding rods lie, while the other direction remains PTE. Perusal to Fig. 11.38c therefore suggests that this structure demonstrates NTE and PTE in the x–y plane and the z-axis, respectively. Since the structure gives NTE

Table 11.2 Summary of effective CTE for the ring structure

Schematics. See Fig. 11.35 for details	Effective CTE	Remarks
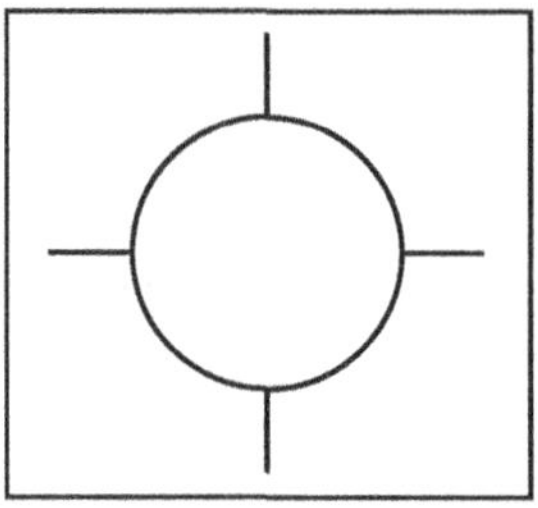	$\Delta\overline{AB} = +2R\alpha_r^{(T)}\mathrm{d}T$ $\Delta\overline{KL} = +2R\alpha_r^{(T)}dT$ $\therefore \begin{matrix} \alpha_x^{(T)} = +\alpha_r^{(T)}\frac{R}{R+l_1} \\ \alpha_x^{(T)} = +\alpha_r^{(T)}\frac{R}{R+l_3} \end{matrix}$	PTE
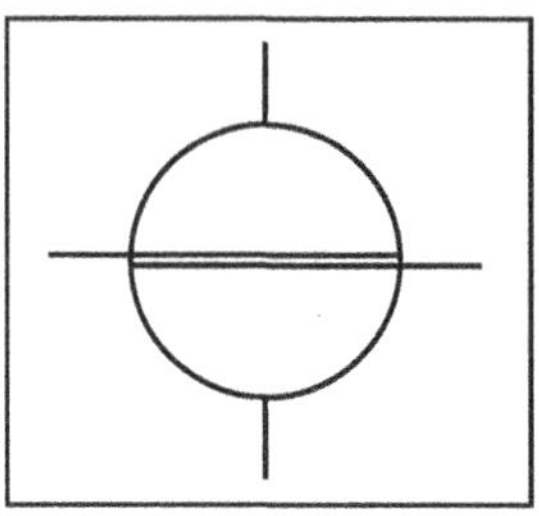	$\Delta\overline{AB} = -2R\alpha_r^{(T)}\mathrm{d}T$ $\Delta\overline{KL} = +2R\alpha_r^{(T)}\mathrm{d}T$ $\therefore \begin{matrix} \alpha_x^{(T)} = -\alpha_r^{(T)}\frac{R}{R+l_1} \\ \alpha_x^{(T)} = +\alpha_r^{(T)}\frac{R}{R+l_3} \end{matrix}$	1D NTE
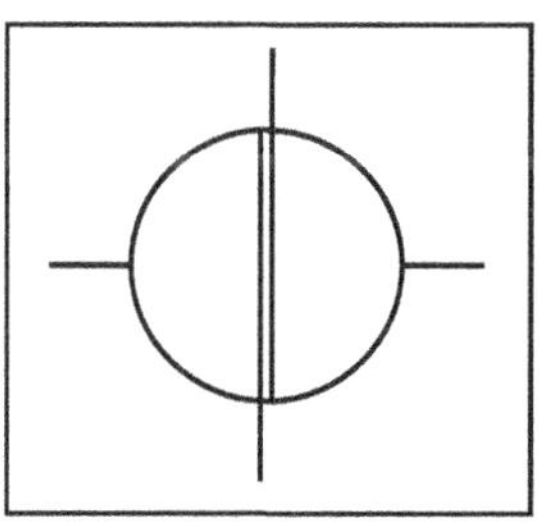	$\Delta\overline{AB} = +2R\alpha_r^{(T)}\mathrm{d}T$ $\Delta\overline{KL} = -2R\alpha_r^{(T)}\mathrm{d}T$ $\therefore \begin{matrix} \alpha_x^{(T)} = +\alpha_r^{(T)}\frac{R}{R+l_1} \\ \alpha_x^{(T)} = -\alpha_r^{(T)}\frac{R}{R+l_3} \end{matrix}$	1D NTE
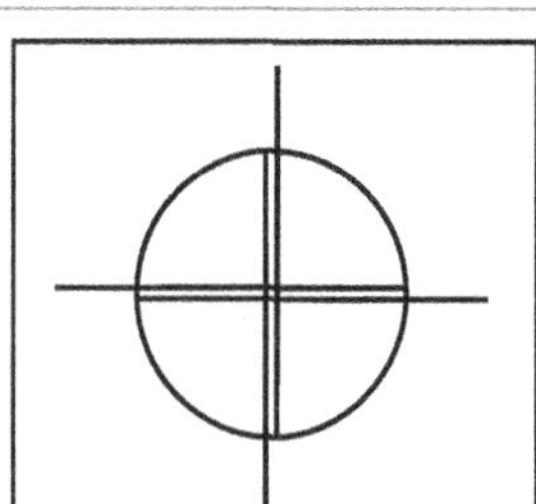	$\Delta\overline{AB} = -2R\alpha_r^{(T)}\mathrm{d}T$ $\Delta\overline{KL} = -2R\alpha_r^{(T)}\mathrm{d}T$ $\therefore \begin{matrix} \alpha_x^{(T)} = -\alpha_r^{(T)}\frac{R}{R+l_1} \\ \alpha_x^{(T)} = -\alpha_r^{(T)}\frac{R}{R+l_3} \end{matrix}$	2D NTE

characteristic in two axes, we shall term it "high NTE." In the final example shown in Fig. 11.38d, pairs of sliding rods are incorporated in all three axes, and hence, we have a fully NTE structure.

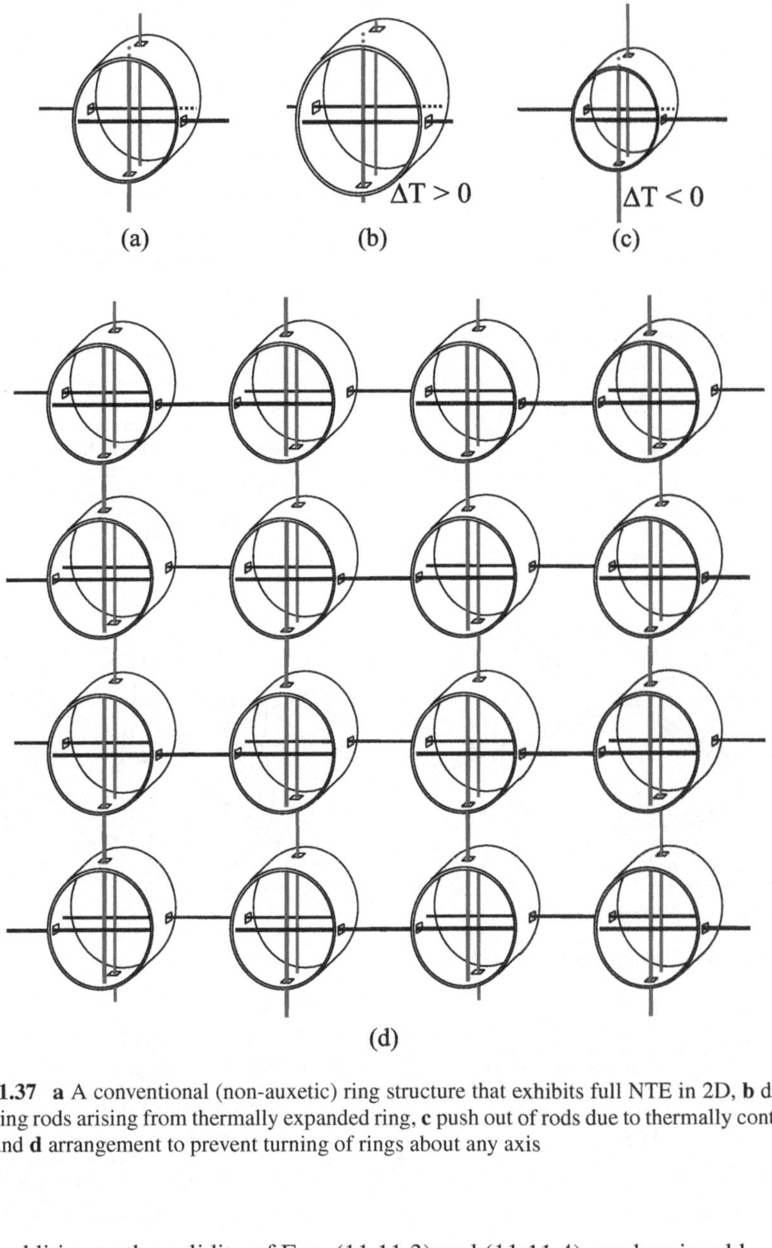

Fig. 11.37 **a** A conventional (non-auxetic) ring structure that exhibits full NTE in 2D, **b** draw-in of sliding rods arising from thermally expanded ring, **c** push out of rods due to thermally contracted ring, and **d** arrangement to prevent turning of rings about any axis

In addition to the validity of Eqs. (11.11.3) and (11.11.4), we herein add

$$\alpha_y^{(T)} = \frac{1}{\mathrm{d}T}\frac{\Delta \overline{IJ}}{2R + 2l_2} \tag{11.11.5}$$

where

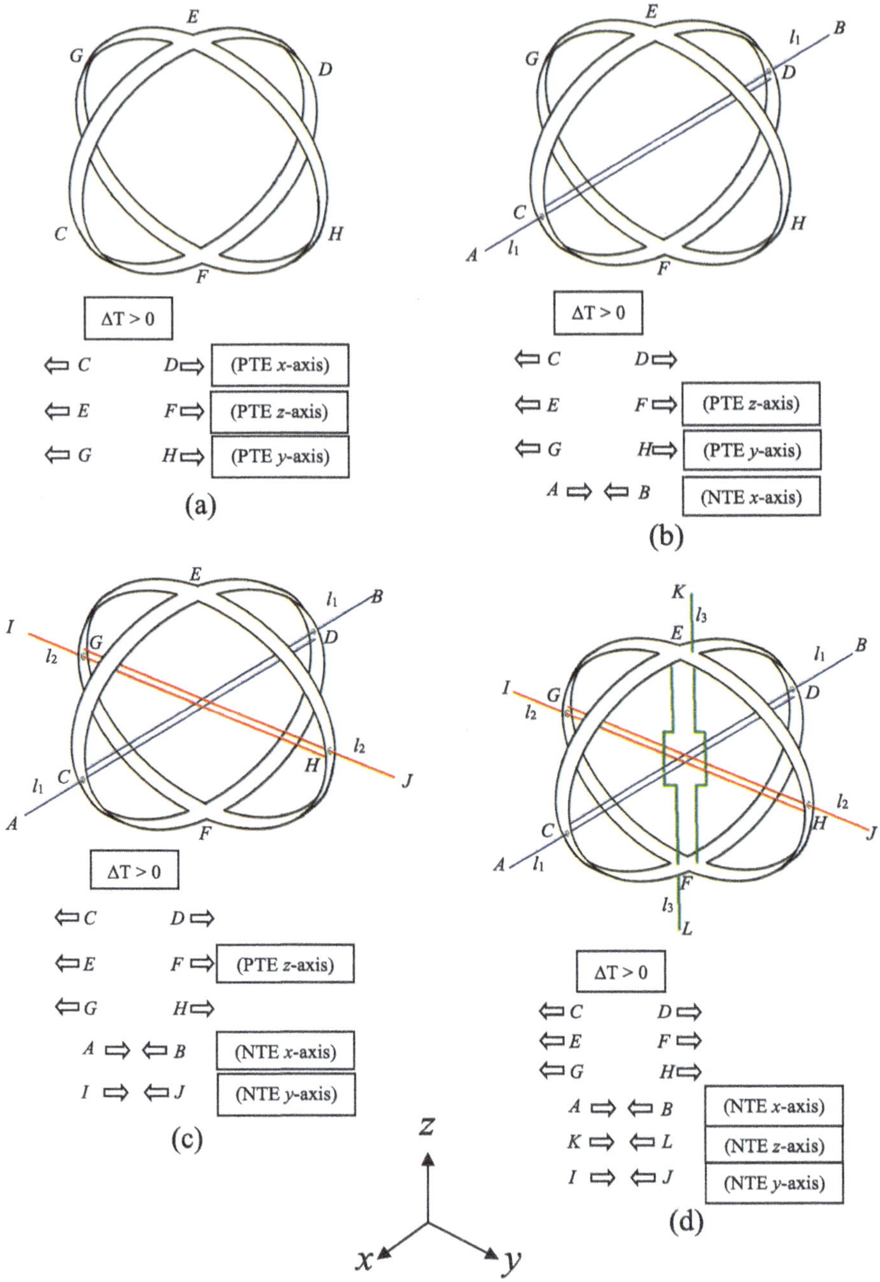

Fig. 11.38 NTE evaluation on double-rings: **a** fully PTE, **b** NTE in *x*-axis, **c** NTE in *x*- and *y*-axes, and **d** fully NTE

$$\left|\Delta\overline{IJ}\right| = 2R\alpha_r^{(T)}\mathrm{d}T \tag{11.11.6}$$

Specific CTEs for all the 8 combinations of double-rings with fixed and/or sliding rods are listed in Table 11.3. The following terms are introduced for the double-ring structure and are listed in the remarks column of Table 11.3: (i) "PTE" for the double-ring structure that has no sliding rods due to its positive thermal expansion in all three axes, (ii) "1D NTE" refers to double-ring structures that possess one pair of sliding rods because they exhibit NTE in one axis, (iii) "2D NTE" term is used for double-ring structures with two pairs of sliding rods due to their manifestation of NTE is two axes, and (iv) "3D NTE" for the double-ring structure that exhibits NTE in all three axes as a consequence of three pairs of sliding rods. See Table 11.2.

For the first special case of $l_1 = l_2 = l_3$, i.e., the short fixed rods are of equal length l and the long sliding rods are of equal length $2R + l$ such that the single-ring and the double-ring structures are arranged in square and cube arrays, respectively, the effective CTEs reduce to

$$\left|\alpha_i^{(T)}\right| = \alpha_r^{(T)}\frac{R}{R+l} \quad ; \quad i = x, y, z \tag{11.11.7}$$

So far the rings and pairs of rods have been arranged in such a manner that the thermal expansion and contraction of the rings do not alter the distance between the ring center. Recall that the expansion and contraction of the rings occur simultaneously with the rod draw-in and push-out of, respectively, from the rings. Consider now the case where the rings are arranged in such a way that the holes from 4 neighboring rings face one another, as shown in Fig. 11.39. A consequence of this arrangement is the shift of ring centers toward one another as well as away from each other. This is shown in Fig. 11.40a, in which contraction of the rings causes the rods to be pushed out. When the rings expand due to temperature rise, as shown in Fig. 11.40b, the rings move in the opposite direction as compared to Fig. 11.40a.

Reference to Fig. 11.40 also reveals that the change in ring diameter occurs in tandem with the lateral shifting of the rods. As such there are two factors in influencing the void size (excluding the voids confined within each ring), which is herein defined by the area confined by the rods and the outer surface of the rings. The first factor is the change in ring size, in which contracting ring diameter (due to $\mathrm{d}T < 0$) and expanding ring diameter (due to $\mathrm{d}T > 0$) lead to expanding and contracting void size, respectively. In other words, the increasing and decreasing void size in response to the decreasing and increasing temperature, respectively, suggests that the voids behave in an NTE manner. The second factor is the in-plane transverse motion of the rods, which transforms the original mid-sized squares voids into small squares, large squares, and rectangles.

Bearing in mind that the rings behave as PTE while the voids behave as NTE, we shall now see that the entire structure, being defined by its boundary, exhibits zero thermal expansion (ZTE) characteristic. Figure 11.41 provides an appreciation of this combined PTE, NTE, and ZTE characteristics with the usual PTE and NTE behavior, in which the top, middle, and bottom rows correspond to the usual PTE,

Table 11.3 Effect of microstructural arrangement in the double ring structure on the negativity of thermal expansion

Schematics	Effective CTE in x-, y-, and z-axes based on $dT > 0$	Remarks
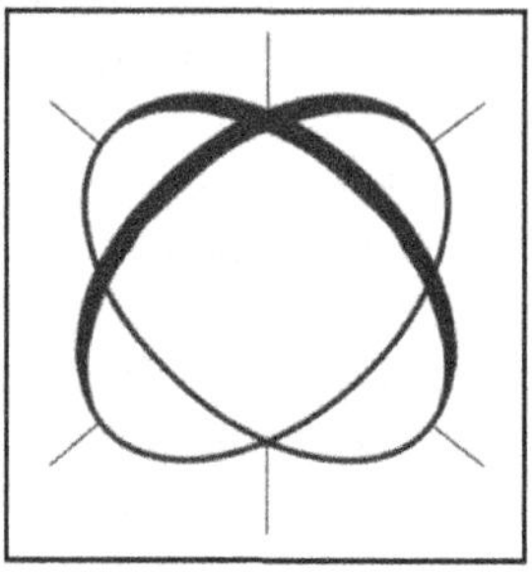	$\alpha_x^{(T)} = +\alpha_r^{(T)} \frac{R}{R+l_1}$ $\alpha_y^{(T)} = +\alpha_r^{(T)} \frac{R}{R+l_2}$ $\alpha_z^{(T)} = +\alpha_r^{(T)} \frac{R}{R+l_3}$	PTE
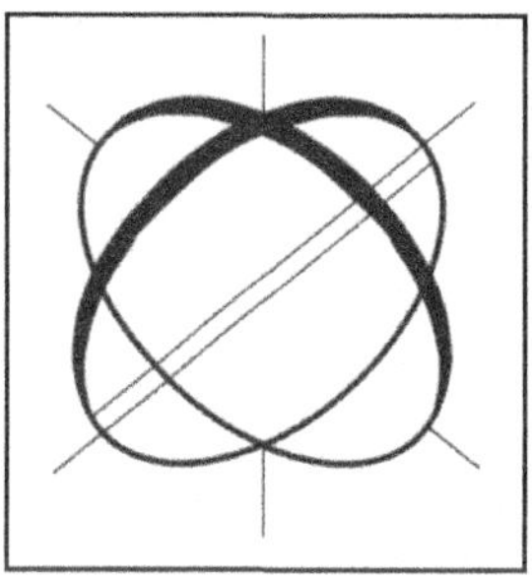	$\alpha_x^{(T)} = -\alpha_r^{(T)} \frac{R}{R+l_1}$ $\alpha_y^{(T)} = +\alpha_r^{(T)} \frac{R}{R+l_2}$ $\alpha_z^{(T)} = +\alpha_r^{(T)} \frac{R}{R+l_3}$	1D NTE
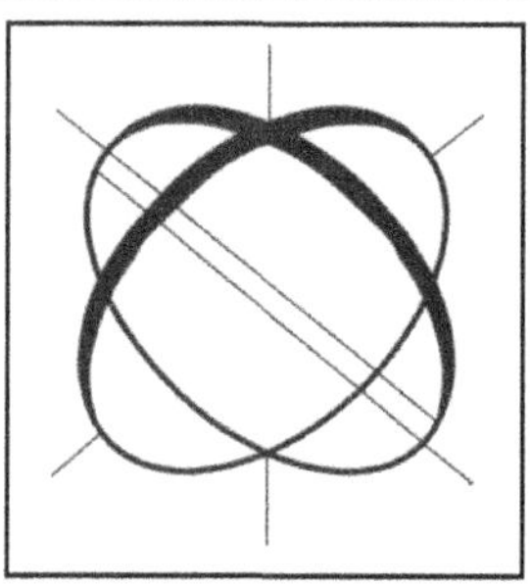	$\alpha_x^{(T)} = +\alpha_r^{(T)} \frac{R}{R+l_1}$ $\alpha_y^{(T)} = -\alpha_r^{(T)} \frac{R}{R+l_2}$ $\alpha_z^{(T)} = +\alpha_r^{(T)} \frac{R}{R+l_3}$	1D NTE
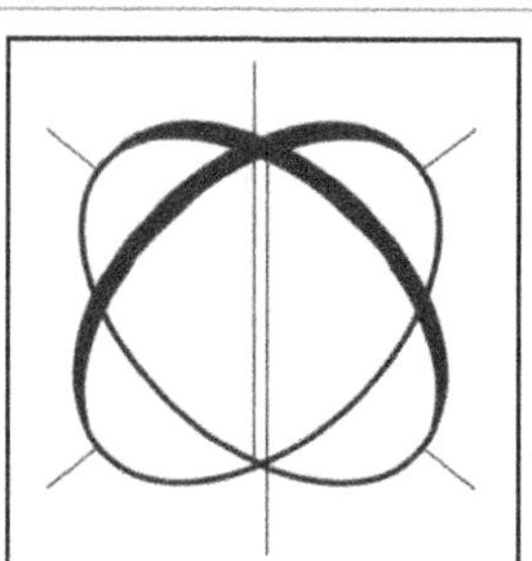	$\alpha_x^{(T)} = +\alpha_r^{(T)} \frac{R}{R+l_1}$ $\alpha_y^{(T)} = +\alpha_r^{(T)} \frac{R}{R+l_2}$ $\alpha_z^{(T)} = -\alpha_r^{(T)} \frac{R}{R+l_3}$	1D NTE

(continued)

Table 11.3 (continued)

Schematics	Effective CTE in x-, y-, and z-axes based on $\mathrm{d}T > 0$	Remarks
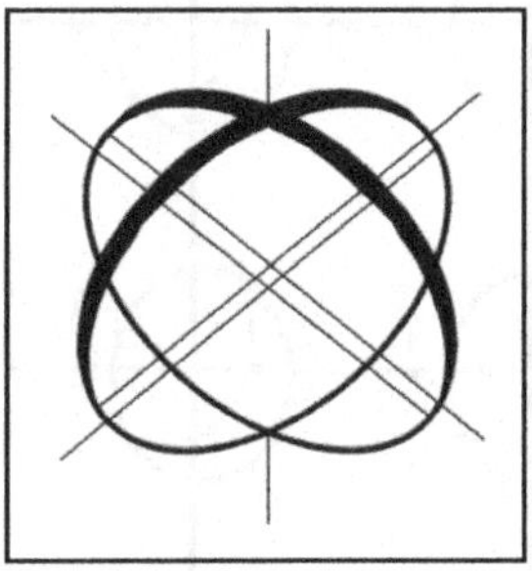	$\alpha_x^{(T)} = -\alpha_r^{(T)} \frac{R}{R+l_1}$ $\alpha_y^{(T)} = -\alpha_r^{(T)} \frac{R}{R+l_2}$ $\alpha_z^{(T)} = +\alpha_r^{(T)} \frac{R}{R+l_3}$	2D NTE
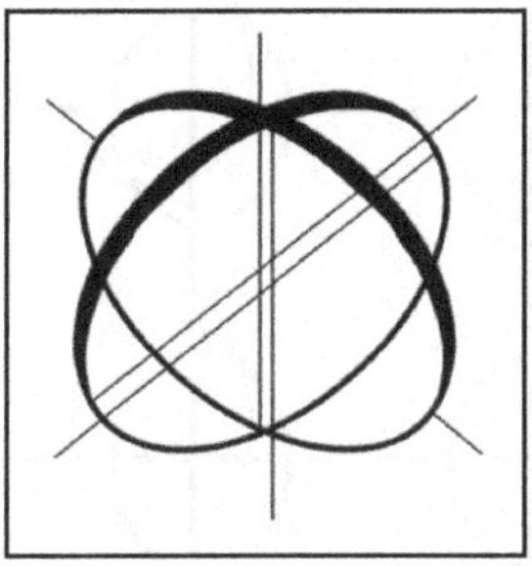	$\alpha_x^{(T)} = -\alpha_r^{(T)} \frac{R}{R+l_1}$ $\alpha_y^{(T)} = +\alpha_r^{(T)} \frac{R}{R+l_2}$ $\alpha_z^{(T)} = -\alpha_r^{(T)} \frac{R}{R+l_3}$	2D NTE
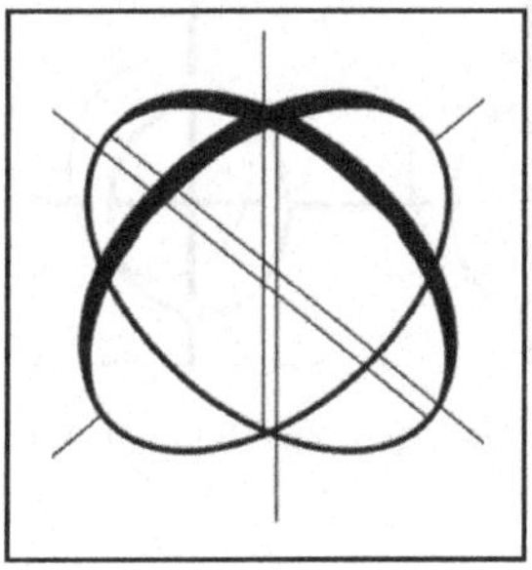	$\alpha_x^{(T)} = +\alpha_r^{(T)} \frac{R}{R+l_1}$ $\alpha_y^{(T)} = -\alpha_r^{(T)} \frac{R}{R+l_2}$ $\alpha_z^{(T)} = -\alpha_r^{(T)} \frac{R}{R+l_3}$	2D NTE
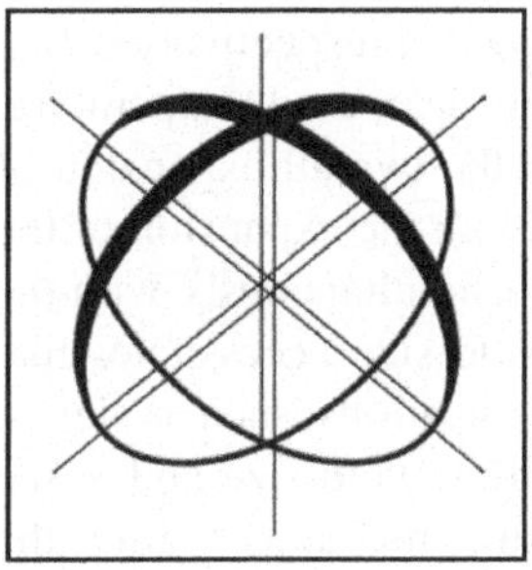	$\alpha_x^{(T)} = -\alpha_r^{(T)} \frac{R}{R+l_1}$ $\alpha_y^{(T)} = -\alpha_r^{(T)} \frac{R}{R+l_2}$ $\alpha_z^{(T)} = -\alpha_r^{(T)} \frac{R}{R+l_3}$	3D NTE

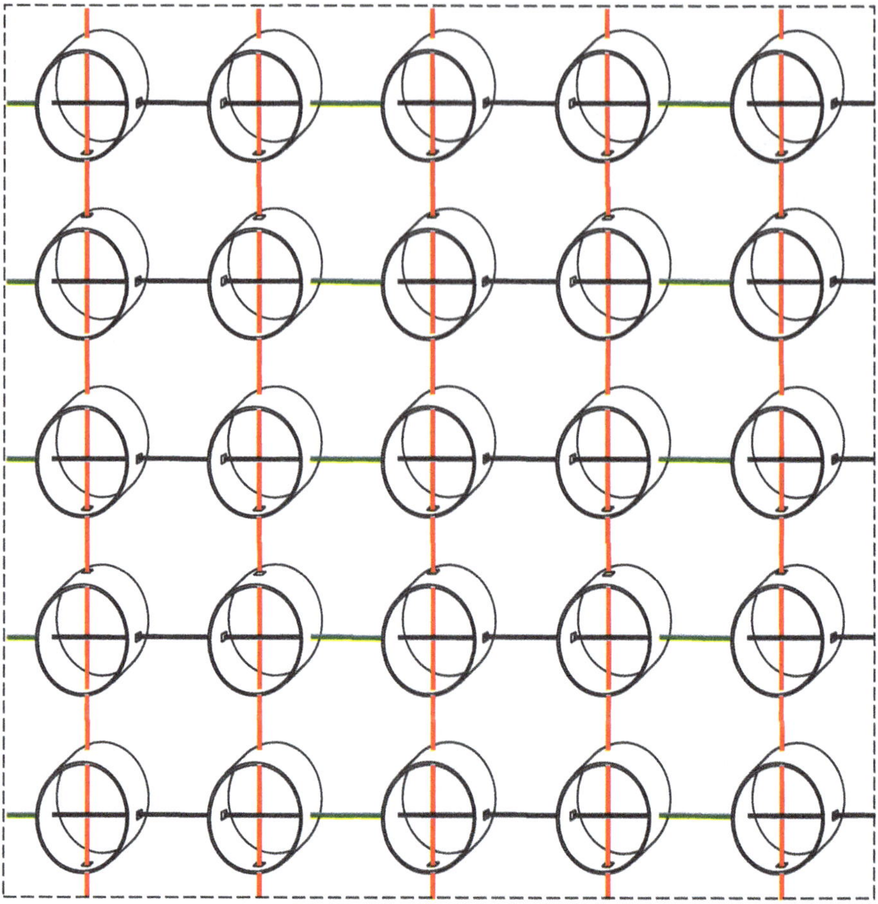

Fig. 11.39 An overall zero thermal expansion (ZTE) structure that exhibits internal NTE characteristics

usual NTE and the currently combined PTE/NTE/ZTE models, respectively, while the left, middle, and right columns refer to the original state, decreased temperature, and increased temperature, respectively. Needless to say, the overall expansion of the usual PTE and NTE structures occurs in conjunction with the expansion of the voids, and conversely, their overall contraction takes place simultaneously with the contraction of the voids. In other words, the change in void size necessitates like changes to the boundary of the entire structure. In some applications, such as sieves, it is desirable to have a design that can thermally control the void size and at the same time prevent any change to the size of the overall structure so as to limit the built-up of thermal stresses at the boundary and the consequential stresses within the boundary. It is instructive to incorporate the markings of "*X*" and "*Y*" at the bottom row of Fig. 11.41, to indicate the locations that experience significant and

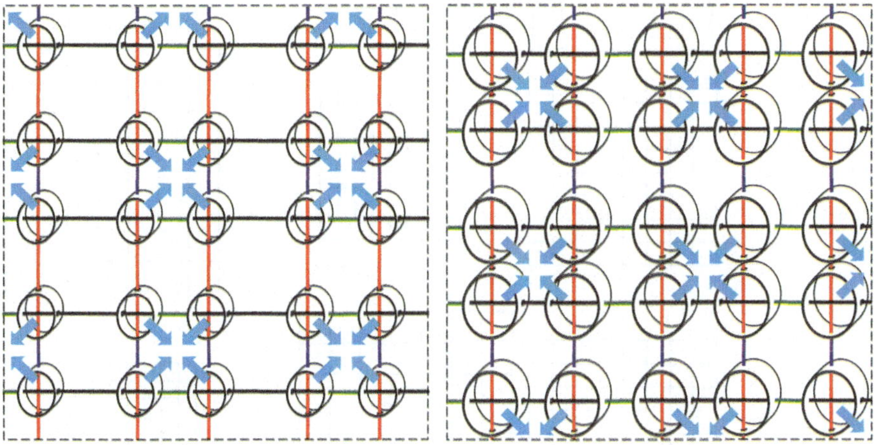

Fig. 11.40 Change in area bounded by the rods and ring outer surface as well as the in-plane transverse shift of the rods during **a** ring contraction, and **b** ring expansion

insignificant changes in void size; the former due to compounding effect from both factors and the latter due to the cancelation effect from both factors. More importantly, the last row of Fig. 11.41 visually shows that the alternating direction in which the rods move facilitates the preservation of the overall size of the structure.

Section 2.5 and this section report the extent of auxeticity and NTE in different combinations. To put into perspective, the extent and interrelation between auxeticity and NTE characteristics, it is useful to place each of these single-ring and double-ring structures on Euler diagrams displayed in Figs. 11.42 and 11.43, respectively (Lim 2017); the axes and loading directions for auxeticity evaluation are defined in Sect. 2.5.

11.12 Example: Three-Dimensional Truss Systems with Pin Joints

With reference to the 2D periodic network that consists of isosceles triangles shown in Fig. 11.44, it is easily seen that when the horizontal rod begins expanding (at fixed diagonal rod length) there is an increase in the triangular area until a maximum point is reached, whereupon further expansion of the horizontal rod is accompanied by area reduction until a state of zero area is attained when the triangle collapses into a horizontal line. However, an increase in the length of the diagonal rods (at fixed horizontal rod length) does not lead to area reduction.

On the basis of this understanding, the 3D version is identified as a tetrahedron consisting of 2 sets of rods: (a) the first set of 3 rods of length a and CTE $\alpha_a^{(T)}$ are

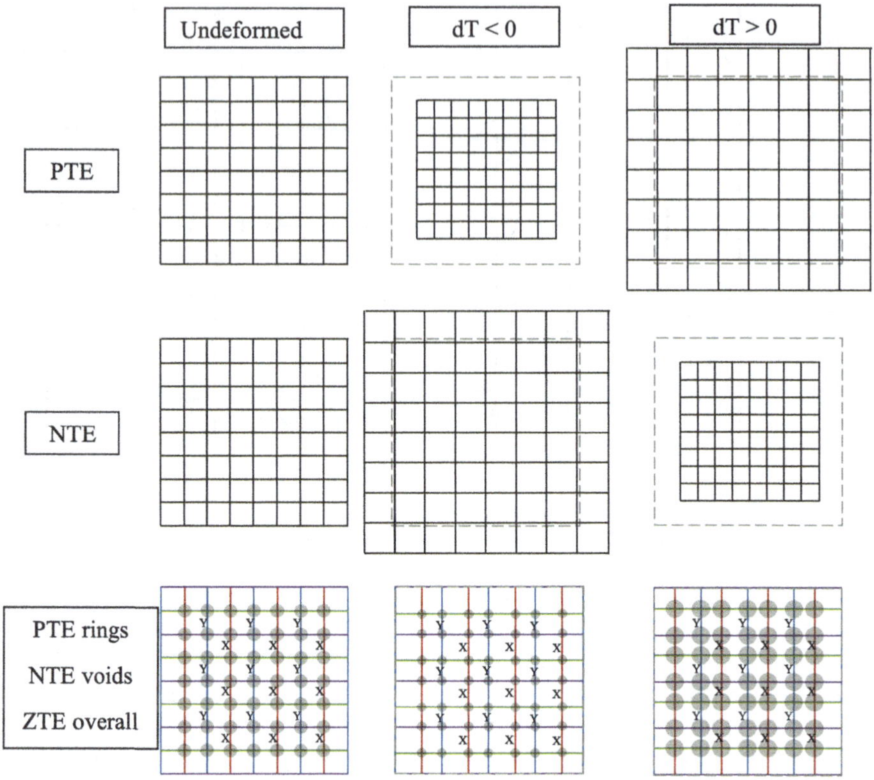

Fig. 11.41 A comparison between the overall ZTE characteristics of the ring-rod array with the usual PTE and NTE characteristics. The dashed squares indicate the original overall size for convenient comparison

branched from the tetrahedron apex, while (b) the second set of 3 rods of length b and CTE $\alpha_b^{(T)}$ forms a triangular loop that defines the boundary of the tetrahedron base (Lim 2012). This combination is selected due to its analogy with the 2D network mentioned; i.e., an increase in the "apex" rod length (at fixed "base" rod length) does not give any volumetric reduction. Increase in the "base" rod length (at fixed "apex" rod length) from the state of sharp tetrahedron is followed by volumetric increase. However, the same increment of base rod length from the state of tetrahedron is followed by volumetric decrease. A state of zero volume is finally attained when all 6 rods lie on the same plane. (Figure 11.45)

To establish the ratio of the apex-to-base rod length at the optimum point, we write the volume of the tetrahedron as

$$V = \frac{b^3}{12}\sqrt{3\frac{a^2}{b^2} - 1} \tag{11.12.1}$$

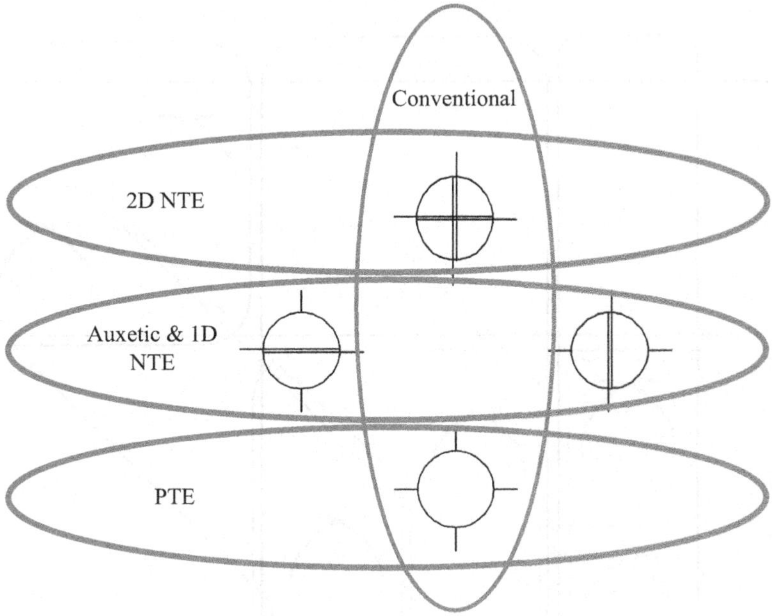

Fig. 11.42 Euler diagram for auxeticity and NTE mapping of single-ring structure with short fixed rods and/or long sliding rods. Ring, axes, and loading details are defined in Sect. 2.5

such that the condition

$$\frac{\mathrm{d}V}{\mathrm{d}a} = 0 \tag{11.12.2}$$

has no solution, thereby confirming positive thermal expansion (PTE) of the tetrahedron when $\alpha_a^{(T)} > \alpha_b^{(T)} = 0$. On the other hand, there is a solution for

$$\frac{\mathrm{d}V}{\mathrm{d}b} = 0 \Leftrightarrow \frac{a}{b} = \frac{1}{\sqrt{2}} \tag{11.12.3}$$

thereby quantifying the earlier qualitative elucidation for $\alpha_b^{(T)} > \alpha_a^{(T)} = 0$. Table 11.4 summarizes the geometrical condition that leads to NTE for the scope of tetrahedral structure considered herein.

As a solid undergoes thermal expansion from its original volume V_0 to its final volume V_f by an amount $\mathrm{d}V$ resulting from a temperature increment of $\mathrm{d}T$, i.e.,

$$V_f = V_0 + \mathrm{d}V \tag{11.12.4}$$

the volumetric strain by definition is

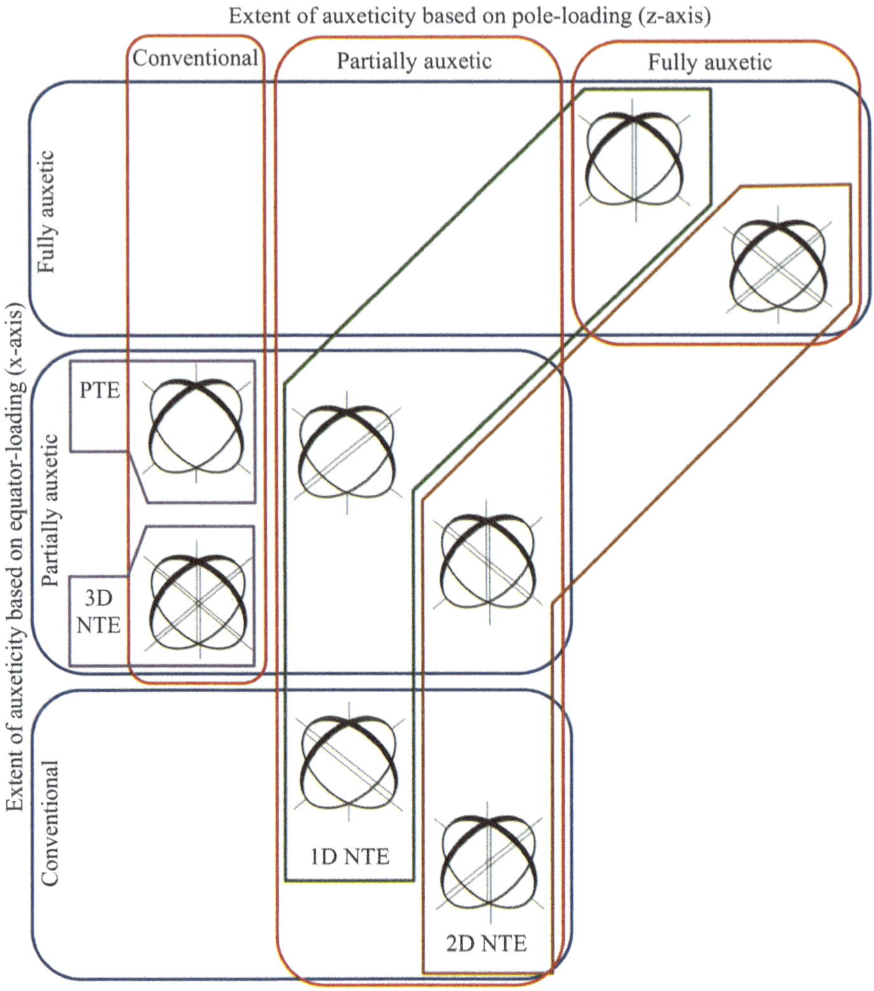

Fig. 11.43 Euler diagram for auxeticity and NTE mapping of double-ring structure with short fixed rods and/or long sliding rods. Double-ring, axes, and loading details are defined in Sect. 2.5

$$\varepsilon_{\mathrm{V}} = \frac{\mathrm{d}V}{V_0} \tag{11.12.5}$$

while its relationship with the coefficient of volumetric thermal expansion (CVTE), $\alpha_{\mathrm{V}}^{(T)}$ is

$$\varepsilon_{\mathrm{VT}} = \alpha_{\mathrm{V}}^{(T)}\mathrm{d}T \tag{11.12.6}$$

Equating both volumetric strains

Fig. 11.44 A triangular periodic rod/truss structure that gives variable CTE: 2D periodic structure (left), and a representative area element (right)

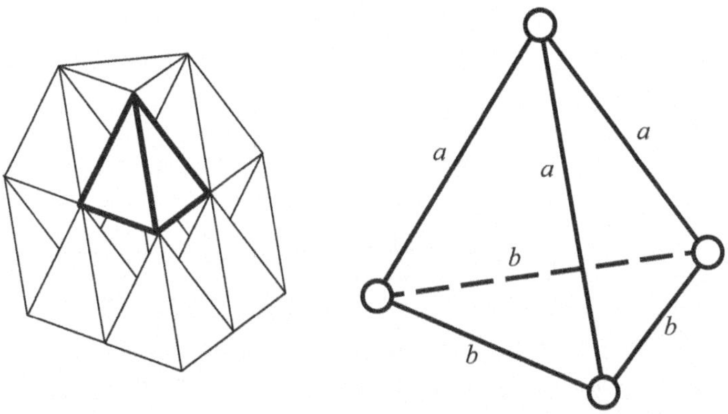

Fig. 11.45 A tetrahedral periodic rod/truss structure that gives variable CTE: 3D periodic structure (left), and a representative volume element (right) by Lim (2012)

Table 11.4 Geometrical condition for NTE

	$\alpha_b^{(T)} > \alpha_a^{(T)} = 0$	$\alpha_a^{(T)} > \alpha_b^{(T)} = 0$
$\frac{a}{b} < \frac{1}{\sqrt{2}}$	NTE	PTE
$\frac{a}{b} > \frac{1}{\sqrt{2}}$	PTE	PTE

$$\mathrm{d}V = \alpha_{\mathrm{V}}^{(T)} V_0 \mathrm{d}T \tag{11.12.7}$$

we have

$$V_f = V_0\left(1 + \alpha_{\mathrm{V}}^{(T)} \mathrm{d}T\right) \tag{11.12.8}$$

Hence, the volumetric strain resulting from thermal expansion

$$\varepsilon_{\mathrm{VT}} = \frac{V_f}{V_0} - 1 \tag{11.12.9}$$

can be expressed in terms of the rod dimensions before and after thermal expansions, i.e.,

$$\begin{aligned} V_f &= \tfrac{b_f^3}{12}\sqrt{3\tfrac{a_f^2}{b_f^2} - 1} \\ V_0 &= \tfrac{b_0^3}{12}\sqrt{3\tfrac{a_0^2}{b_0^2} - 1} \end{aligned} \tag{11.12.10}$$

Substituting

$$\begin{aligned} a_f &= a_0\left(1 + \alpha_a^{(T)}\mathrm{d}T\right) \\ b_f &= b_0\left(1 + \alpha_b^{(T)}\mathrm{d}T\right) \end{aligned} \tag{11.12.11}$$

we arrive at

$$\varepsilon_{\mathrm{VT}} = -1 + \left(1 + \alpha_b^{(T)}\mathrm{d}T\right)^2 \sqrt{\frac{3\frac{a_0^2}{b_0^2}\left(1 + \alpha_a^{(T)}\mathrm{d}T\right)^2 - \left(1 + \alpha_b^{(T)}\mathrm{d}T\right)^2}{3\frac{a_0^2}{b_0^2} - 1}} \tag{11.12.12}$$

It follows that, for a solid to exhibit NTE, the volumetric thermal strain must be less than zero. Hence, the condition

$$\left(1 + \alpha_b^{(T)}\mathrm{d}T\right)^2 \sqrt{\frac{3\frac{a_0^2}{b_0^2}\left(1 + \alpha_a^{(T)}\mathrm{d}T\right)^2 - \left(1 + \alpha_b^{(T)}\mathrm{d}T\right)^2}{3\frac{a_0^2}{b_0^2} - 1}} < 1 \tag{11.12.13}$$

must be fulfilled for the considered space frame structure to be NTE. The CVTE can be obtained by taking the first derivative of the volumetric thermal strain with respect to the thermal increment

$$\alpha_{\mathrm{V}}^{(T)} = \frac{\partial \varepsilon_{\mathrm{VT}}}{\partial(\mathrm{d}T)} \tag{11.12.14}$$

to give

$$\alpha_{\rm V}^{(T)} = 2\alpha_b^{(T)}\left(1+\alpha_b^{(T)}\mathrm{d}T\right)\sqrt{\frac{3\frac{a_0^2}{b_0^2}\left(1+\alpha_a^{(T)}\mathrm{d}T\right)^2-\left(1+\alpha_b^{(T)}\mathrm{d}T\right)^2}{3\frac{a_0^2}{b_0^2}-1}}$$
$$+\alpha_b^{(T)}\left(1+\alpha_b^{(T)}\mathrm{d}T\right)^2\frac{3\frac{\alpha_a^{(T)}}{\alpha_b^{(T)}}\frac{a_0^2}{b_0^2}\left(1+\alpha_a^{(T)}\mathrm{d}T\right)-\left(1+\alpha_b^{(T)}\mathrm{d}T\right)}{\sqrt{3\frac{a_0^2}{b_0^2}-1}\times\sqrt{3\frac{a_0^2}{b_0^2}\left(1+\alpha_a^{(T)}\mathrm{d}T\right)^2-\left(1+\alpha_b^{(T)}\mathrm{d}T\right)^2}} \quad (11.12.15)$$

Equation (11.12.15) implies that the CVTE of a structure composed of rods in tetrahedral arrangement is dependent on the magnitude of temperature change. This result is not surprising since it is understood that the structural geometry changes with changes to the rod length.

The CVTE as a material property that is independent from the change in temperature is obtained by taking the limits (Lim 2012)

$$\lim_{\mathrm{d}T\to 0}\alpha_{\rm V}^{(T)} = \alpha_b^{(T)}\left(2+\frac{3\frac{\alpha_a^{(T)}}{\alpha_b^{(T)}}\frac{a_0^2}{b_0^2}-1}{3\frac{a_0^2}{b_0^2}-1}\right) \quad (11.12.16)$$

Equation (11.12.16) refers to the CVTE at infinitesimal change in temperature and is hence valid for small change in temperature. It is also valid for moderate change in temperature under the condition that the changes in the tetrahedron volume and shape are insignificant. It follows that, for the structure to possess a negative CVTE, the condition

$$\left(2+\frac{\alpha_a^{(T)}}{\alpha_b^{(T)}}\right)\frac{a_0^2}{b_0^2} < 1 \quad (11.12.17)$$

must be met. The following discusses the results of variation in volumetric thermal strain and volumetric CTE with reference to a dimensionless rod CTE and the rod ratio, respectively. Figure 11.46 shows variation of the volumetric thermal strain with reference to a dimensionless rod CTE, $\alpha_b^{(T)}\mathrm{d}T$, under various rod length ratios. This parameter $\alpha_b^{(T)}\mathrm{d}T$ was selected over $\alpha_a^{(T)}\mathrm{d}T$ due to the role that the "base" rods play in giving rise to NTE. The effect of the "apex" rods' CTE is taken into consideration on Fig. 11.46 for $\alpha_a^{(T)}/\alpha_b^{(T)} = 0.0, 0.1, 0.2$.

Perusal to Fig. 11.46 shows that the extent of negative expansivity becomes more significant with

(a) decreasing rod length ratio a_0/b_0,
(b) increasing dimensionless rod CTE, $\alpha_b^{(T)}\mathrm{d}T$, and

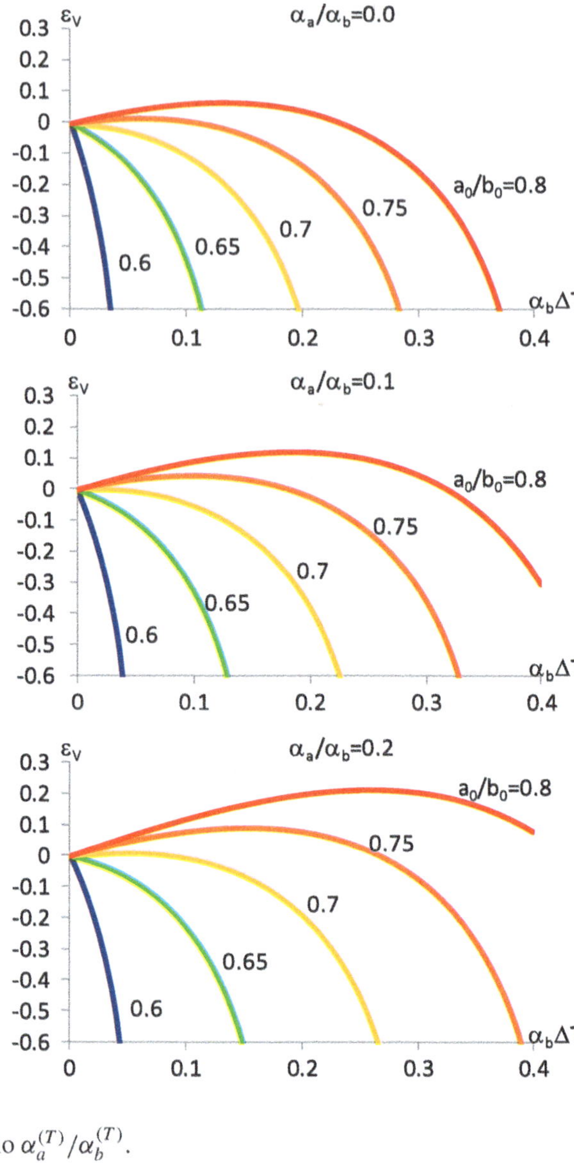

Fig. 11.46 Plots of volumetric thermal expansion versus dimensionless rod CTE for various rod ratios at $\alpha_a^{(T)}/\alpha_b^{(T)}$ of 0 (top), 0.1 (middle), and 0.2 (bottom)

(c) decreasing rod CTE ratio $\alpha_a^{(T)}/\alpha_b^{(T)}$.

In addition to the above, it can be seen that under the condition where there is a slight PTE, the overall volumetric thermal expansion goes into a negative region beyond a certain increase in temperature, hence temperature change-dependent NTE.

Figure 11.47 (top) shows general variation of the infinitesimal volumetric CTE with reference to the rod length ratio, under various ratios of base rod CTE to apex rod CTE. Due to the close variation in the curves, an appreciation on the influence of rod length ratio and rod CTE ratio is made in Fig. 11.47 (bottom) which zooms into a narrow range of dimensionless volumetric CTE.

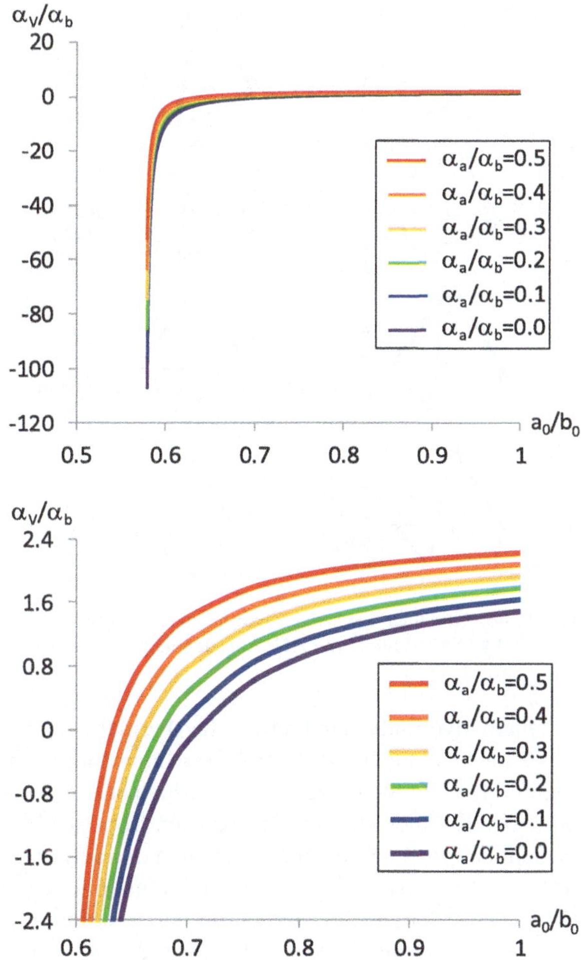

Fig. 11.47 Plots of dimensionless volumetric CTE versus rod ratio for various rod CTE ratios, showing the general trend (top) and zoomed in view at dimensionless volumetric CTE near zero (bottom)

As with Fig. 11.46, reference to Fig. 11.47 reveals that the degree of negative expansivity is more pronounced with decreasing rod length ratio a_0/b_0 and rod CTE ratio $\alpha_a^{(T)}/\alpha_b^{(T)}$. More striking, however, is that the extent of negative expansivity increases gradually with decreasing rod ratio for $a_0/b_0 > 0.6$, but the negative expansivity increases sharply with decreasing rod ratio for $a_0/b_0 < 0.6$. This may well be attributed to the limiting geometrical condition, in which the minimum limit for the rod ratio is $a_0/b_0 > 3^{-0.5} = 0.57735$.

To possess an overview of tetrahedrons with PTE and/or NTE characteristics within the context of nonnegative CTE rods, it is convenient to confine, in the first instance, our consideration into only two types of rods: (a) PTE rods (of equal CTE), and (b) zero thermal expansion (ZTE) rods. In so doing, it is possible to list up to 11 types of tetrahedron as shown in Fig. 11.48 (Lim 2013).

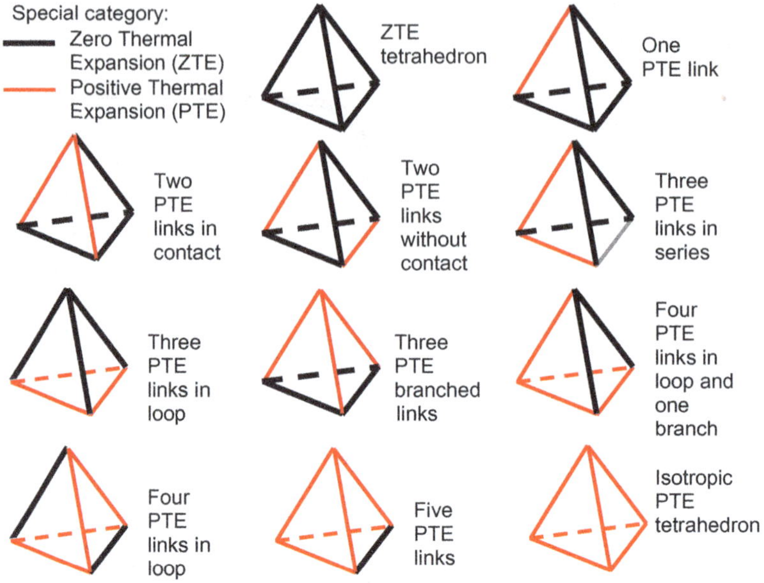

Fig. 11.48 Graphical list of tetrahedrons made from two types of rods: PTE rods (of equal CTE) denoted by fine red lines, and ZTE rods denoted by bold black lines

Qualitative understanding on the thermal expansion behavior of these tetrahedrons can be obtained by the following elucidations, with the obvious cases omitted. Figure 11.49 shows a tetrahedron in which one of the rods possess PTE while the rest possess ZTE, such that initial expansion from the former from zero length increases the tetrahedron's volume by the opening of an equilateral triangle into a tetrahedron, hence a volumetric PTE, until a maximum volume is attained. Further expansion of the PTE rod flattens the tetrahedron into a rhombus, thereby signifying an effective NTE for the tetrahedron. For this case, the ratio of PTE to ZTE rod lengths range from 0 to $\sqrt{3}$.

There are two possible ways by which two PTE and four ZTE rods can be arranged. Figure 11.50a, b shows the two variations, in which the PTE rods of the former are in contact in one of their vertices, while the PTE rods for the latter are not in contact. From a PTE to ZTE rod ratio in the form of the nested radical $\sqrt{2-\sqrt{3}}$ that gives

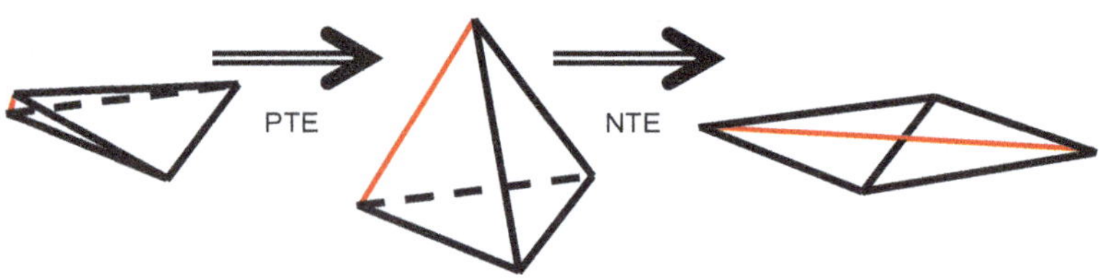

Fig. 11.49 PTE followed by NTE tetrahedral truss from one PTE and five ZTE rods

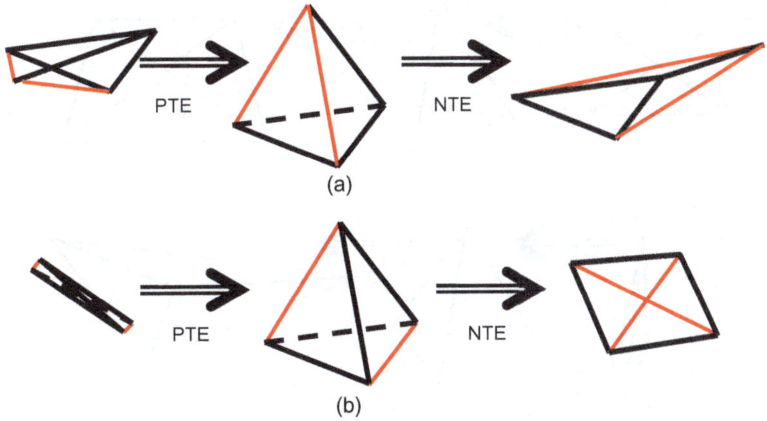

Fig. 11.50 PTE followed by NTE tetrahedral truss from two PTE and four ZTE rods: **a** PTE rods in contact, and **b** PTE rods not in contact (except in planar form)

the shape of a kite, thermal expansion of the PTE rod transforms the flat kite into a tetrahedron until a maximum volume is attained, as shown in Fig. 11.50a. Continued lengthening of the PTE rod reduces the tetrahedron volume so that the structure collapses into a bilateral triangle such that the ratio of PTE to ZTE rod lengths takes the value of the nested radical $\sqrt{2+\sqrt{3}}$. In the case depicted by Fig. 11.50b, the tetrahedron takes the shape of a rod whose length is equal to that of the ZTE when the PTE rod is at zero length. Increase in the PTE rod length transforms the rod into a tetrahedron, and further, expansion of the PTE rods causes the tetrahedron to collapse into a square in which the PTE to ZTE rod lengths ratio is $\sqrt{2}$.

Figure 11.51 shows three possible ways by which three PTE and three ZTE rods can be arranged. For the case where the PTE and ZTE rods are in series, as in Fig. 11.51a, an expansion of the PTE rods eventually leads to an isosceles trapezoid when the ratio of the PTE to ZTE rod length achieves the golden ratio $\phi = 1.61803\ldots$, which is the upper limit. A contraction of the PTE rods also leads to an isosceles trapezoid when the ratio of the PTE to ZTE rod length achieves the inverse of the golden ratio, i.e., $1/\phi = \phi - 1 = 0.61803\ldots$, which is the lower limit. In the arrangement shown in Fig. 11.51b, the tetrahedron takes the shape of a sharp needle when the PTE rods are at zero length such that lengthening of these rods increases the volume of the tetrahedron until a maximum volume is achieved. Further lengthening of the PTE rods collapses the tetrahedron into an equilateral triangle. As such, the ratio of the PTE to ZTE rod lengths ranges from 0 to $\sqrt{3}$. The opposite configuration, as depicted in Fig. 11.51c, shows that the minimal PTE to ZTE rod lengths ratio is $1/\sqrt{3}$. However, this arrangement does not lead to negative volumetric thermal expansion.

As a final qualitative discussion, we refer to the tetrahedron made from four PTE rods and two ZTE rods. The arrangement shown in Fig. 11.52a, b is conjugates of Fig. 11.50a, b, respectively. Starting from a bilateral triangle whose PTE to ZTE rod

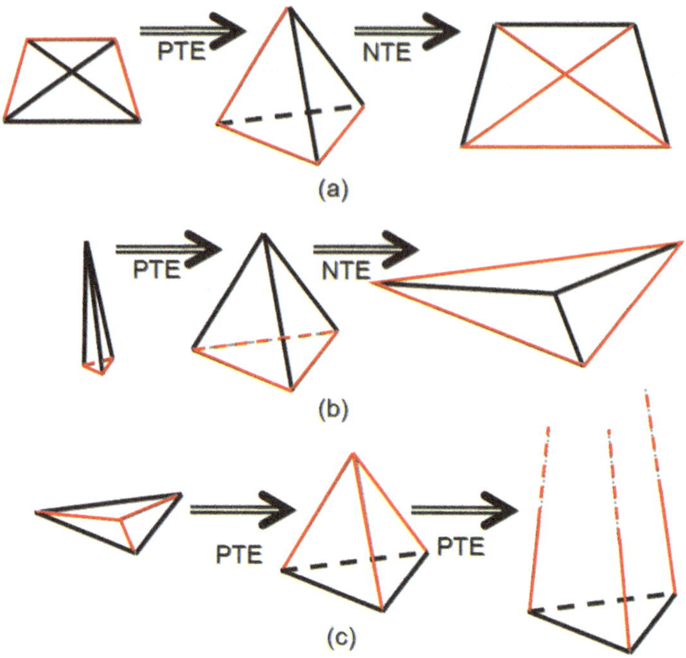

Fig. 11.51 Tetrahedral trusses from three PTE and three ZTE rods: **a** PTE followed by NTE for series PTE rods **b** PTE followed by NTE for looped PTE rods, and **c** PTE with branched PTE rods

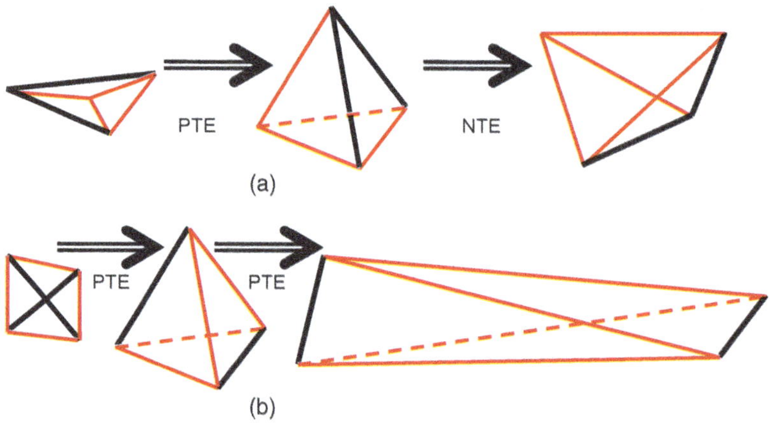

Fig. 11.52 Tetrahedral truss from four PTE and two ZTE rods: **a** PTE followed by NTE tetrahedron from PTE rods in end contact, and **b** PTE tetrahedron from skew PTE rods

lengths ratio is $\left(2+\sqrt{3}\right)^{-1/2}$ or $\sqrt{2-\sqrt{3}}$, the structure exhibits positive thermal expansion until its maximum volume is attained. Thereafter, further increase in the PTE rods reduces the tetrahedron volume until a zero volume is reached whereby the tetrahedron collapses into a kite-shaped planar structure whereby the PTE to ZTE rod lengths ratio is at $\left(2-\sqrt{3}\right)^{-1/2}$ or $\sqrt{2+\sqrt{3}}$. In the case depicted by Fig. 11.52b, the PTE to ZTE rod length ratio begins at $1/\sqrt{2}$ in the shape of a square such that lengthening the PTE rods increases the tetrahedron volume with no optimal condition; i.e., the structure does not exhibit negative volumetric thermal expansion.

Of the several 3D configurations outlined, only the case described by Fig. 11.51b is selected for analysis due to its analogy to the 2D version that gives NTE with two of its rods being equal in length and properties. These 2D and 3D arrangements share a similarity in its symmetric change in shape. Recall that the volumetric strain for this tetrahedral model is described by Eq. (11.12.12). Instead of plotting for temperature increase only, Fig. 11.53 shows the volumetric strains for both increasing and decreasing temperature.

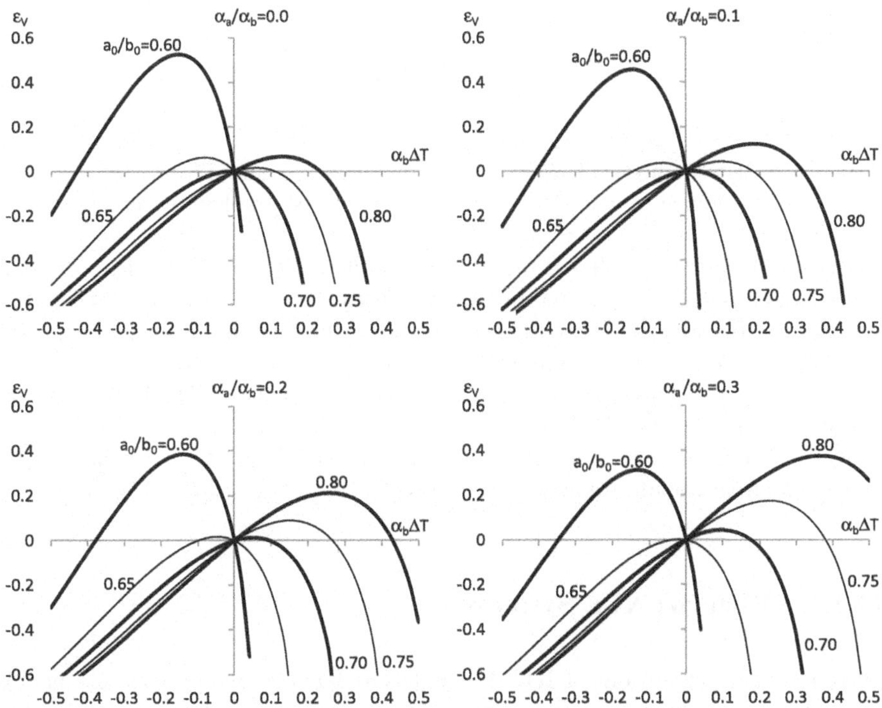

Fig. 11.53 Volumetric thermal strain versus dimensionless change in temperature in the base rods for various initial rod length ratios of 0.0 (top left), 0.1 (top right), 0.2 (bottom left), and 0.3 (bottom right)

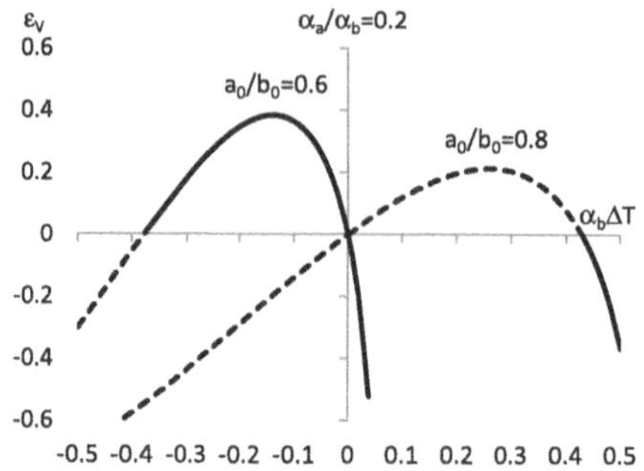

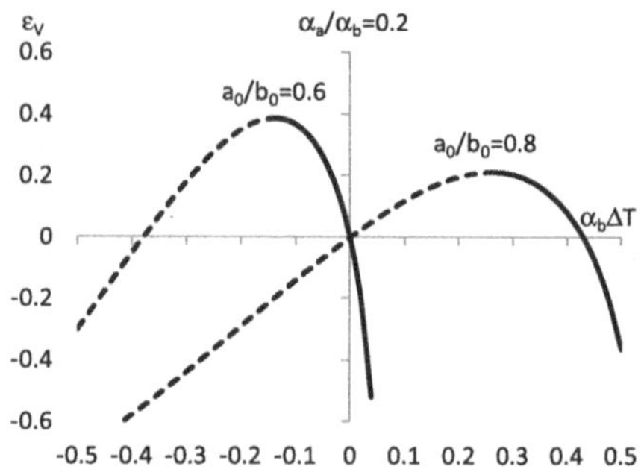

Fig. 11.54 Graphical representation of PTE (dashed curves) and NTE (continuous curves) of the tetrahedral space frame trusses based on overall definition of NTE (left) and incremental definition of NTE (right)

Recall that the condition of NTE described by Eq. (11.12.14) is on the basis of incremental change to the volumetric strain and hence defined by the negativity of the slopes in the curves plotted in Fig. 11.53. An alternative description of CTE can be conceptualized as

$$\alpha_{\mathrm{V}}^{(T)} = \frac{\varepsilon_{\mathrm{VT}}}{\mathrm{d}T} \tag{11.12.18}$$

from which the condition of NTE is satisfied when $\varepsilon_{\mathrm{VT}}$ and $\mathrm{d}T$ possess opposite signs. The difference in both definitions of NTE can be graphically represented in Fig. 11.54.

Arising from the overall and the incremental definitions of NTE, there exists a region which is NTE by one definition but PTE by the other definition. Where the space frame trusses are consistently NTE or PTE by both definitions, we say that the structure is definite NTE or definite PTE, respectively. When the trusses are NTE by only one of the definitions, then we say that it exhibits selective NTE. The conditions that lead to both PTE and NTE by virtue of different NTE definitions is hence called "selective NTE," because whether the truss is deemed NTE or PTE depends on the selected reference point. An example is shown in Fig. 11.55 (Lim 2013).

11.13 Other 3D NTE Systems

There have been extensions of 2D NTE models to 3D ones, and for each 2D model, there can be more than one version of its 3D counterpart. It is therefore impossible to cover all of them within this book. It suffices to consider a few examples and only briefly as the fundamentals is more easily understood from the 2D models. The first example of 3D model is that of tetrakaidecahedral structure, which resembles open cell foams shown in Fig. 11.56 envisaged by Lakes (2007) as a 3D extension of his

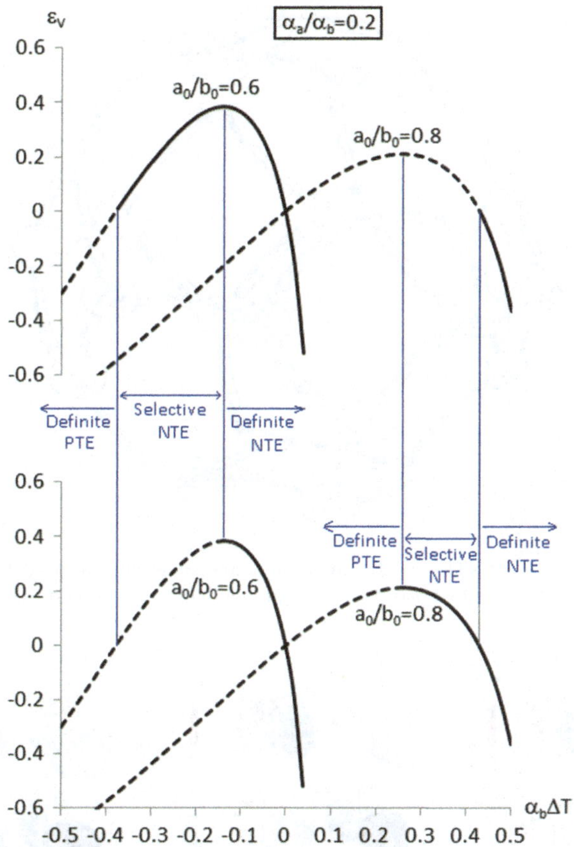

Fig. 11.55 Range of selective NTE as well as definite NTE and PTE of space frame trusses (NTE and PTE are denoted by continuous and dashed lines, respectively)

2D NTE models—discussed in Sect. 11.2 in Figs. 11.7 and 11.8—in which each rib is made by curved bimaterial strip. If the layer on the convex side possesses a greater CTE than that on the concave side, then an increase in temperature will cause the curved rib to curve even more; i.e., the junctions move closer together. The resulting shrinkage of the cell indicates NTE. Wu et al. (2016) investigated both 2D and 3D NTE models using anti-chiral structures. Specifically, Wu et al. (2016) performed experiment and simulation on 2D NTE based on anti-tetrachiral and anti-trichiral structures. The 2D anti-tetrachiral structure was then extended to its 3D version. See Fig. 11.57 for the simulation results of the total displacement, in mm, using node radii of 20 mm (top left), 25 mm (top center), 30 mm (top right), 35 mm (bottom center), and 40 mm (bottom right) with ligament length of 100 mm and thickness 2 mm, as furnished in the supplementary materials by Wu et al. (2016). A number of microlattice structural systems have been investigated by Xu and Pasini (2016), which the effective CTE, effective Young's modulus, and effective yield strength

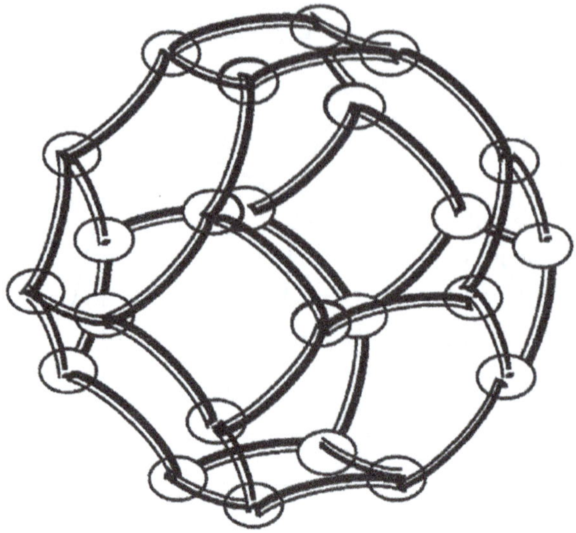

Fig. 11.56 Example of 3D NTE model in the form of tetrakaidecahedral structure with ribs made from curved bimaterial strips by Lakes (2007). Reprinted with permission from AIP

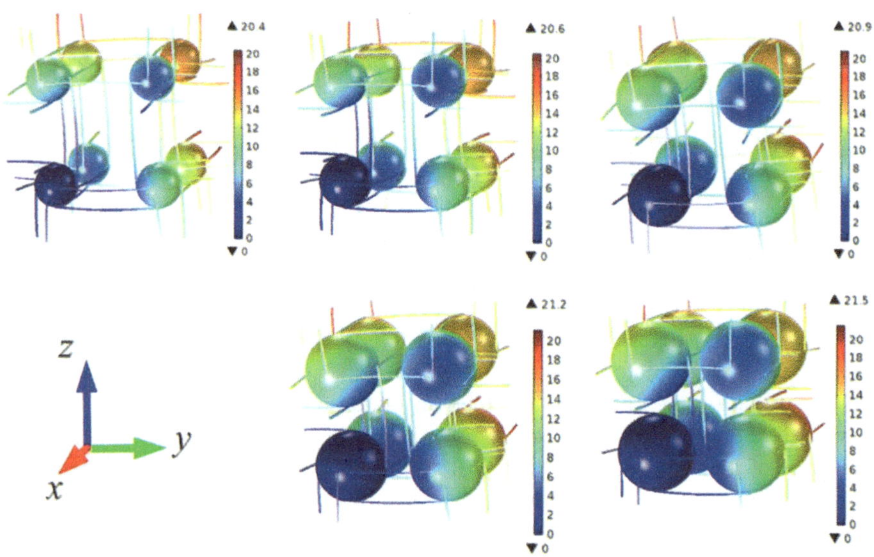

Fig. 11.57 Example of NTE structure using 3D anti-tetrachiral structure by Wu et al. (2016). Reprinted with permission from ACS

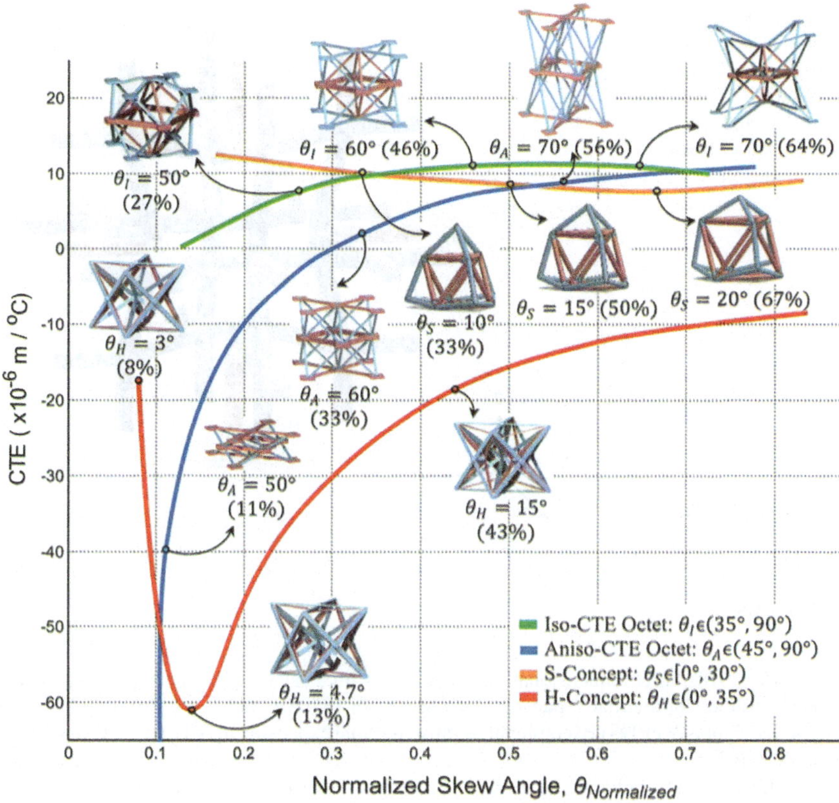

Fig. 11.58 Structurally efficient 3D metamaterials with controllable CTEs by Xu and Pasini (2016)

were evaluated. This work is an exhaustive study which approximately extends from earlier works, e.g., in Sects. 11.5–11.9 and Sect. 11.12. Plots of the effective CTEs by Xu and Pasini (2016) are furnished in Fig. 11.58. Finally, it can be seen that one of the 3D microstructures developed by Ai and Gao (2018), as shown in Fig. 11.59, is a 3D counterpart to the 2D version by Raminhos et al. (2019) discussed in Sect. 11.8.

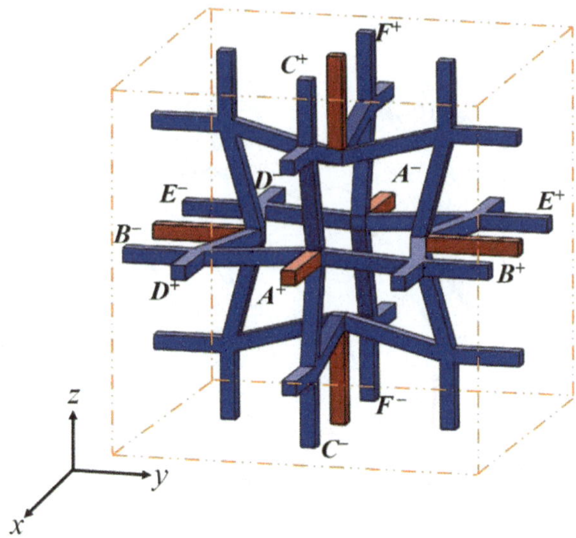

Fig. 11.59 One of the 3D metamaterials with auxetic and non-positive NTE investigated by Ai and Gao (2018). Reprinted with permission from Elsevier

References

Ai L, Gao XL (2018) Three-dimensional metamaterials with a negative Poisson's ratio and a non-positive coefficient of thermal expansion. Int J Mech Sci 135:101–113

Attfield MP, Sleight AW (1998) Exceptional negative thermal expansion in $AlPO_4$–17. Chem Mater 10(7):2013–2019

Auray M, Quarton M, Leblanc M (1995) Zirconium tungstate. Acta Crystallographica C 51(11):2210–2213

Cabras L, Brun M, Misseroni D (2019) Micro-structured medium with large isotropic negative thermal expansion. Proc Roy Soc A 475(2232):20190468

Chu CN, Saka N, Suh NP (1987) Negative thermal expansion ceramics: a review. Mater Sci Eng 95:303–308

Dove MT, Welche PRL, Heine V (1998) Negative thermal expansion in beta-quartz. Phys Chem Miner 26(1):63–77

Evans JSO, Hanson JC, Sleight AW (1998) Room-temperature superstructure of ZrV_2O_7. Acta Crystallographica B 54(6):705–713

Evans JSO, Jorgensen JD, Short S, David WIF, Ibberson RM, Sleight AW (1999) Thermal expansion in the orthorhombic γ phase of ZrW2O8. Phy Rev B 60(21):14643–14648

Fisher DJ (2018) Negative thermal expansion materials. Materials Research Forum LLC, Millersville

Gere JM (2001) Mechanics of materials, 5th edn. Nelson Thones Ltd, Cheltenham

Graham J, Wadsley AD, Weymouth JH, Williams LS (1959) A new ternary oxide, ZrW2O8. J Am Ceram Soc 42(11):570–570

Grima JN, Farrugia PS, Gatt R, Zammit V (2007) A system with adjustable positive or negative thermal expansion. Proc Roy Soc A 463(2082):1585–1596

Grima JN, Oliveri L, Ellul B, Gatt R, Attard D, Cicala G, Recca G (2010) Adjustable and negative thermal expansion from multilayered systems. Phys Status Solidi RRL 4(5–6):133–135

Grima JN, Bajada M, Attard D, Dudek KK, Gatt R (2015) Maximizing negative thermal expansion via rigid unit modes: a geometry-based approach. Proc Roy Soc A 471(2179):20150188

Ha CS, Hestekin E, Li J, Plesha ME, Lakes RS (2015) Controllable thermal expansion of large magnitude in chiral negative Poisson's ratio lattices. Phys Status Solidi B 252(7):1431–1434

Ha CS, Plesha ME, Lakes RS (2017) Simulations of thermoelastic triangular cell lattices with bonded joints by finite element analysis. Extreme Mech Lett 12:101–107

Hummel FA (1948) Thermal expansion properties of natural Lithia minerals. Foote Prints 20:3–11

Hummel FA (1951) Thermal expansion properties of some synthetic lithia minerals. J Am Ceram Soc 34:235–239

Khosrovani K, Sleight AW, Vogt T (1997) Structure of ZrV_2O_7 from −263 to 470 °C. J Solid State Chem 132(2):355–360

Korthuis V, Khosrovani N, Sleight AW, Roberts N, Dupree R, Warren WW (1995) Negative thermal expansion and phase transitions in the ZrV2 − xPxO7 series. Chem Mater 7(2):412–417

Lakes R (1996) Cellular solid structures with unbounded thermal expansion. J Mater Sci Lett 15(6):475–477

Lakes R (2007) Cellular solids with tunable positive or negative thermal expansion of unbounded magnitude. Appl Phys Lett 90(22):221905

Li Y, Chen Y, Li T, Cao S, Wang L (2018) Hoberman-sphere-inspired lattice metamaterials with tunable negative thermal expansion. Compos Struct 189:586–597

Lightfoot P, Woodcock DA, Maple MJ, Villaescusa LA, Wright PA (2001) The widespread occurrence of negative thermal expansion in zeolites. J Mater Chem 11(1):212–216

Lim TC (2005) Anisotropic and negative thermal expansion behavior in a cellular microstructure. J Mater Sci 40(12):3275–3277

Lim TC (2010) Preliminary assessment of a multifunctional potential energy function. Mol Phys 108(12):1589–1597

Lim TC (2011) Coefficient of thermal expansion of stacked auxetic and negative thermal expansion laminates. Phys Status Solidi B 248(1):140–147

Lim TC (2012) Negative thermal expansion structures constructed from positive thermal expansion trusses. J Mater Sci 47(1):368–373

Lim TC (2013) Negative thermal expansion in transversely isotropic space frame trusses. Phys Status Solidi B 250(10):2062–2069

Lim TC (2017) Auxetic and negative thermal expansion structure based on interconnected array of rings and sliding rods. Phys Status Solidi B 254(12):1600775

Lim TC (2019a) A 2D auxetikos system based on interconnected shurikens. SN Appl Sci 1(11):1383

Lim TC (2019b) 2D metamaterial with in-plane positive and negative thermal expansion and thermal shearing based on interconnected alternating biomaterials. Mater Res Express 6(11):115804

Lim TC (2019c) A class of shape-shifting composite metamaterial honeycomb structures with thermally-adaptive Poisson's ratio signs. Compos Struct 226:111256

Lim TC (2019d) Composite metamaterial with sign-switchable coefficients of hygroscopic, thermal and pressure expansions. Adv Compos Hybrid Mater 2(4):657–669

Lim TC (2019e) A composite metamaterial with sign switchable elastic and hygrothermal properties induced by stress direction and environmental change reversals. Compos Struct 220:185–193

Lim TC (2019f) Composite microstructures with Poisson's ratio sign switching upon stress reversal. Compos Struct 209:34–44

Lim TC (2019g) Metamaterials with Poisson's ratio sign toggling by means of microstructural duality. SN Appl Sci 1(2):176

Lim TC (2020a) Metacomposite with auxetic and in situ sign reversible thermal expansivity upon temperature fluctuation. Compos Commun 19:30–36

Lim TC (2020b) Composite metamaterial square grids with sign-flipping expansion coefficients leading to a type of Islamic design. SN Appl Sci 2(5):918

Lim TC (2020c) Metacomposite structure with sign-changing coefficients of hygrothermal expansions inspired by Islamic motif. Compos Struct 251:112660

Lind C (2012) Two decades of negative thermal expansion research: Where do we stand? Materials 5(6):1125–1154

Mary TA, Evans JSO, Vogt T, Sleight AW (1996) Negative thermal expansion from 0.3 to 1050 Kelvins in ZrW_2O_8. Science 272(5258):90–92

Nye JF (1957) Physical properties of crystals. Clarendon Press, Oxford, UK

Raminhos JS, Borges JP, Velhinho A (2019) Development of polymeric anepectic meshes: auxetic metamaterials with negative thermal expansion. Smart Mater Struct 28(4):045010

Rees DWA (1990) The mechanics of solids and structures. McGraw-Hill, Maidenhead

Scheel K (1907a) Versuche über die Ausdehnung fester Körper, insbesondere von Quarz in Richtung der Hauptachse, Platin, Palladium und Quarzglas bei der Temperatur der flüssigen Luft. Verh Dtsch Phys Ges 9:3–23

Scheel K (1907b) Über die ausdehnung des quarzglases. Verh Dtsch Phys Ges 9:719–721

Schapery RA (1968) Thermal expansion coefficients of composite materials based on energy principles. J Compos Mater 2(3):380–404

Sleight AW (1995) Thermal contraction. Endeavour 19(2):64–68

Sleight AW (1998a) Compounds that contract on heating. Inorg Chem 37(12):2854–2860

Sleight (1998b) Isotropic negative thermal expansion. Ann Rev Mater Sci 28:29–43

Takenaka K (2018) Progress of research in negative thermal expansion materials: paradigm shift in the control of thermal expansion. Front Chem 6:267

Tao JZ, Sleight AW (2003) The role of rigid unit modes in negative thermal expansion. J Solid State Chem 173(2):442–448

Timoshenko SP (1925) Analysis of bi-metal thermostats. J Opt Soc Am 11(3):233–235

Withers RL, Evans JSO, Hanson JC, Sleight AW (1998) An in-situ temperature dependent electron and X-ray diffraction study of the structural phase transitions in ZrV_2O_7. J Solid State Chem 137(1):161–167

Woodcock DA, Lightfoot P, Villaescusa LA, Diaz-Cabanas MJ, Camblor MA, Engberg D (1999) Negative thermal expansion in the siliceous zeolites chabazite and ITQ-4: a neutron powder diffraction study. Chem Mater 11(9):2508–2514

Wu L, Li B, Zhou J (2016) Isotropic negative thermal expansion metamaterials. ACS Appl Mater Interfaces 8(27):17721–17727

Xu H, Pasini D (2016) Structurally efficient three-dimensional metamaterials with controllable thermal expansion. Sci Rep 6:34924

Chapter 12
Negative Compressibility

Abstract This chapter surveys work done on negative compressibility (NC) systems, which can be broadly categorized into cellular system, bimaterial strip system, and interconnected membrane system. In the category of cellular NC system, topics include deformation solely by rib stretching and those solely by joint rotation, as well as those with combined modes of deformation.

Keywords Negative compressibility · Negative linear compressibility · Negative area compressibility · Negative volume compressibility

12.1 Introduction

For a typical solid, an increase in its surrounding pressure p, such as the environmental pressure or hydrostatic pressure, decreases its dimension at constant temperature. The compressibility—also known as the coefficient of compressibility, isothermal compressibility, or the coefficient of pressure expansion—can be defined for linear, areal, and volumetric compressibilities as (Baughman et al. 1998)

$$\begin{aligned} \alpha_L^{(P)} &= -\tfrac{1}{L}\left(\tfrac{\partial L}{\partial p}\right)_{T=\text{constant}} \\ \alpha_A^{(P)} &= -\tfrac{1}{A}\left(\tfrac{\partial A}{\partial p}\right)_{T=\text{constant}} \\ \alpha_V^{(P)} &= -\tfrac{1}{V}\left(\tfrac{\partial V}{\partial p}\right)_{T=\text{constant}} \end{aligned} \tag{12.1.1}$$

respectively. From there, one may recognize that

$$\begin{aligned} \tfrac{\partial L}{L} &= \varepsilon_L \\ \tfrac{\partial A}{A} &= \varepsilon_A \\ \tfrac{\partial V}{V} &= \varepsilon_V \end{aligned} \tag{12.1.2}$$

refer to the linear strain, areal strain, and volumetric strain, respectively, in response to the pressure change dp. Hence,

T. Lim, *Mechanics of Metamaterials with Negative Parameters*, Engineering Materials,
https://doi.org/10.1007/978-981-15-6446-8_12

$$\begin{aligned}\alpha_L^{(P)} &= -\left(\frac{\varepsilon_L}{\mathrm{d}p}\right)_{T=\text{constant}}\\ \alpha_A^{(P)} &= -\left(\frac{\varepsilon_A}{\mathrm{d}p}\right)_{T=\text{constant}}\\ \alpha_V^{(P)} &= -\left(\frac{\varepsilon_V}{\mathrm{d}p}\right)_{T=\text{constant}}\end{aligned} \tag{12.1.3}$$

For isotropic systems, the strain is uniform in all directions. Under this limited condition, $\varepsilon_A = 2\varepsilon_L$ and $\varepsilon_V = 3\varepsilon_L$, therefore

$$\alpha_L^{(P)} = \frac{1}{2}\alpha_A^{(P)} = \frac{1}{3}\alpha_V^{(P)} = -\left(\frac{\varepsilon_L}{\mathrm{d}p}\right)_{T=\text{constant}} \tag{12.1.4}$$

Since matter—gas, liquid, and solid—typically contract and expand with increasing and decreasing pressures, respectively, the strains and pressure change have opposing signs. Therefore, negative signs are required in Eqs. (12.1.1), (12.1.3), and (12.1.4) to produce compressibility with positive signs. Solids that behave in the opposing manner, i.e., expand and contract with increasing and decreasing pressures, respectively, are counter-intuitive. For such materials, the strains and pressure change possess equal signs. Using the above definition of compressibility, such solids give rise to compressibility values that are negative. These are negative compressibility (NC) materials. The occurrences of NC were first discovered in tellurium by Bridgman (1922), and has lately been found in methanol monohydrate (Fortes et al. 2011), silver (I) hexacyanocobaltate (III) (Goodwin et al. 2008), zinc dicyanoaurate (Cairns et al. 2013), wine-rack-like carbon allotropes and related poly(phenylacetylene) systems (Degabriele et al. 2019), and silver oxalate (Colmenero 2019), to name a few. The following sections provide examples of NC systems from the mechanics approach.

12.2 Generalized Compressibility Analysis in 3D Cellular Systems

This section furnishes a generalized compressibility model for 3D cellular systems based on the works of Grima et al. (2008, 2011, 2012, 2013), Attard et al. (2016), Dudek et al. (2016), Zhou et al. (2016, 2018), Ma et al. (2019), and Grima-Cornish et al. (2020). For any cellular structure, a representative volume element, a unit cell, or any repetitive unit, must be identified and geometrical quantities are then assigned to the rib lengths and their angular inclinations. The unit cell dimensions are indicated by (X_1, X_2, X_3) as measured along the axes (Ox_1, Ox_2, Ox_3), where O is the origin of the 3D Cartesian coordinate system. Given a number of ribs with various orientations, one may identify m number of ribs of lengths $l_1, l_2, \ldots, l_m$ as well as n number of rib inclinations $\theta_1, \theta_2, \ldots, \theta_n$ that are sufficient to describe the unit cell for analysis. The angles need not necessarily be measured from a particular direction; in some cases, rib orientations are measured with reference to other ribs.

Hence, the unit cell dimensions are functions of rib lengths and inclination angles

$$\begin{aligned} X_1 &= X_1(l_1, l_2, \ldots, l_m, \theta_1, \theta_2, \ldots, \theta_n) \\ X_2 &= X_2(l_1, l_2, \ldots, l_m, \theta_1, \theta_2, \ldots, \theta_n) \\ X_3 &= X_3(l_1, l_2, \ldots, l_m, \theta_1, \theta_2, \ldots, \theta_n) \end{aligned} \tag{12.2.1}$$

from which the application of load $\sigma^{[1]}$ in the Ox_1 direction gives strains in all three orthogonal directions parallel to the Ox_1, Ox_2 and Ox_3 axes

$$\begin{aligned} \varepsilon_1^{[1]} &= \frac{1}{X_1}\left(\frac{\partial X_1}{\partial l_1}\mathrm{d}l_1^{[1]} + \frac{\partial X_1}{\partial l_2}\mathrm{d}l_2^{[1]} + \cdots + \frac{\partial X_1}{\partial l_m}\mathrm{d}l_m^{[1]} + \frac{\partial X_1}{\partial \theta_1}\mathrm{d}\theta_1^{[1]} + \frac{\partial X_1}{\partial \theta_2}\mathrm{d}\theta_2^{[1]} + \cdots + \frac{\partial X_1}{\partial \theta_n}\mathrm{d}\theta_n^{[1]}\right) \\ \varepsilon_2^{[1]} &= \frac{1}{X_2}\left(\frac{\partial X_2}{\partial l_1}\mathrm{d}l_1^{[1]} + \frac{\partial X_2}{\partial l_2}\mathrm{d}l_2^{[1]} + \cdots + \frac{\partial X_2}{\partial l_m}\mathrm{d}l_m^{[1]} + \frac{\partial X_2}{\partial \theta_1}\mathrm{d}\theta_1^{[1]} + \frac{\partial X_2}{\partial \theta_2}\mathrm{d}\theta_2^{[1]} + \cdots + \frac{\partial X_2}{\partial \theta_n}\mathrm{d}\theta_n^{[1]}\right) \\ \varepsilon_3^{[1]} &= \frac{1}{X_3}\left(\frac{\partial X_3}{\partial l_1}\mathrm{d}l_1^{[1]} + \frac{\partial X_3}{\partial l_2}\mathrm{d}l_2^{[1]} + \cdots + \frac{\partial X_3}{\partial l_m}\mathrm{d}l_m^{[1]} + \frac{\partial X_3}{\partial \theta_1}\mathrm{d}\theta_1^{[1]} + \frac{\partial X_3}{\partial \theta_2}\mathrm{d}\theta_2^{[1]} + \cdots + \frac{\partial X_3}{\partial \theta_n}\mathrm{d}\theta_n^{[1]}\right) \end{aligned} \tag{12.2.2}$$

while the application of load $\sigma^{[2]}$ in the Ox_2 direction again gives strains in these three directions

$$\begin{aligned} \varepsilon_1^{[2]} &= \frac{1}{X_1}\left(\frac{\partial X_1}{\partial l_1}\mathrm{d}l_1^{[2]} + \frac{\partial X_1}{\partial l_2}\mathrm{d}l_2^{[2]} + \cdots + \frac{\partial X_1}{\partial l_m}\mathrm{d}l_m^{[2]} + \frac{\partial X_1}{\partial \theta_1}\mathrm{d}\theta_1^{[2]} + \frac{\partial X_1}{\partial \theta_2}\mathrm{d}\theta_2^{[2]} + \cdots + \frac{\partial X_1}{\partial \theta_n}\mathrm{d}\theta_n^{[2]}\right) \\ \varepsilon_2^{[2]} &= \frac{1}{X_2}\left(\frac{\partial X_2}{\partial l_1}\mathrm{d}l_1^{[2]} + \frac{\partial X_2}{\partial l_2}\mathrm{d}l_2^{[2]} + \cdots + \frac{\partial X_2}{\partial l_m}\mathrm{d}l_m^{[2]} + \frac{\partial X_2}{\partial \theta_1}\mathrm{d}\theta_1^{[2]} + \frac{\partial X_2}{\partial \theta_2}\mathrm{d}\theta_2^{[2]} + \cdots + \frac{\partial X_2}{\partial \theta_n}\mathrm{d}\theta_n^{[2]}\right) \\ \varepsilon_3^{[2]} &= \frac{1}{X_3}\left(\frac{\partial X_3}{\partial l_1}\mathrm{d}l_1^{[2]} + \frac{\partial X_3}{\partial l_2}\mathrm{d}l_2^{[2]} + \cdots + \frac{\partial X_3}{\partial l_m}\mathrm{d}l_m^{[2]} + \frac{\partial X_3}{\partial \theta_1}\mathrm{d}\theta_1^{[2]} + \frac{\partial X_3}{\partial \theta_2}\mathrm{d}\theta_2^{[2]} + \cdots + \frac{\partial X_3}{\partial \theta_n}\mathrm{d}\theta_n^{[2]}\right) \end{aligned} \tag{12.2.3}$$

and in similar fashion the application of $\sigma^{[3]}$ in the Ox_3 direction leads to the following strains

$$\begin{aligned} \varepsilon_1^{[3]} &= \frac{1}{X_1}\left(\frac{\partial X_1}{\partial l_1}\mathrm{d}l_1^{[3]} + \frac{\partial X_1}{\partial l_2}\mathrm{d}l_2^{[3]} + \cdots + \frac{\partial X_1}{\partial l_m}\mathrm{d}l_m^{[3]} + \frac{\partial X_1}{\partial \theta_1}\mathrm{d}\theta_1^{[3]} + \frac{\partial X_1}{\partial \theta_2}\mathrm{d}\theta_2^{[3]} + \cdots + \frac{\partial X_1}{\partial \theta_n}\mathrm{d}\theta_n^{[3]}\right) \\ \varepsilon_2^{[3]} &= \frac{1}{X_2}\left(\frac{\partial X_2}{\partial l_1}\mathrm{d}l_1^{[3]} + \frac{\partial X_2}{\partial l_2}\mathrm{d}l_2^{[3]} + \cdots + \frac{\partial X_2}{\partial l_m}\mathrm{d}l_m^{[3]} + \frac{\partial X_2}{\partial \theta_1}\mathrm{d}\theta_1^{[3]} + \frac{\partial X_2}{\partial \theta_2}\mathrm{d}\theta_2^{[3]} + \cdots + \frac{\partial X_2}{\partial \theta_n}\mathrm{d}\theta_n^{[3]}\right) \\ \varepsilon_3^{[3]} &= \frac{1}{X_3}\left(\frac{\partial X_3}{\partial l_1}\mathrm{d}l_1^{[3]} + \frac{\partial X_3}{\partial l_2}\mathrm{d}l_2^{[3]} + \cdots + \frac{\partial X_3}{\partial l_m}\mathrm{d}l_m^{[3]} + \frac{\partial X_3}{\partial \theta_1}\mathrm{d}\theta_1^{[3]} + \frac{\partial X_3}{\partial \theta_2}\mathrm{d}\theta_2^{[3]} + \cdots + \frac{\partial X_3}{\partial \theta_n}\mathrm{d}\theta_n^{[3]}\right) \end{aligned} \tag{12.2.4}$$

In Eqs. (12.2.2) to (12.2.4), the first m terms correspond to strain components resulting from rib stretching, while the last n terms refer to strain components arising from rib rotation. The stretching force F and the bending moment M that are built up from the changes in rib length δl and angle $\delta\theta$, respectively, are quantified as

$$\begin{aligned} F &= k_s \delta l \\ M &= k_h \delta\theta \end{aligned} \tag{12.2.5}$$

where k_s is stretching stiffness constant of the rib and k_h is the rotational stiffness constant of the hinge. From the prescribed stresses $\sigma^{[1]}$, $\sigma^{[2]}$, $\sigma^{[3]}$ and the generated strains described in Eqs. (12.2.2) to (12.2.4), the following effective elastic constants can be calculated for loading in the Ox_1 direction

$$\begin{aligned} E_1 &= \frac{\sigma^{[1]}}{\varepsilon_1^{[1]}} \\ \upsilon_{12} &= -\frac{\varepsilon_2^{[1]}}{\varepsilon_1^{[1]}} \\ \upsilon_{13} &= -\frac{\varepsilon_3^{[1]}}{\varepsilon_1^{[1]}} \end{aligned} \tag{12.2.6}$$

as well as for loading along the Ox_2

$$\begin{aligned} E_2 &= \frac{\sigma^{[2]}}{\varepsilon_2^{[2]}} \\ \upsilon_{21} &= -\frac{\varepsilon_1^{[2]}}{\varepsilon_2^{[2]}} \\ \upsilon_{23} &= -\frac{\varepsilon_3^{[2]}}{\varepsilon_2^{[2]}} \end{aligned} \tag{12.2.7}$$

and also for Ox_3 loading

$$\begin{aligned} E_3 &= \frac{\sigma^{[3]}}{\varepsilon_3^{[3]}} \\ \upsilon_{31} &= -\frac{\varepsilon_1^{[3]}}{\varepsilon_3^{[3]}} \\ \upsilon_{32} &= -\frac{\varepsilon_2^{[3]}}{\varepsilon_3^{[3]}} \end{aligned} \tag{12.2.8}$$

In some cases, it is more expedient to obtain the effective Young's modulus and the effective Poisson's ratio under separate modes of deformation, i.e., E_i^s and υ_{ij}^s under rib stretching mode, and E_i^h and υ_{ij}^h under hinge rotation mode. The overall Young's modulus E_i^{s+h} and Poisson's ratio υ_{ij}^{s+h}, incorporating both modes of deformation, can thus be obtained as

$$\begin{aligned} \frac{1}{E_i^{s+h}} &= \frac{1}{E_i^s} + \frac{1}{E_i^h} \\ \frac{\upsilon_{ij}^{s+h}}{E_i^{s+h}} &= \frac{\upsilon_{ij}^s}{E_i^s} + \frac{\upsilon_{ij}^h}{E_i^h} \end{aligned} ; \quad (i, j = 1, 2, 3) \tag{12.2.9}$$

From the definition of linear compressibility and recognizing that the strain in each orthogonal direction is a result of applied stress in all three orthogonal directions,

$$\begin{aligned} \alpha_1^{(P)} &= -\left(\frac{\varepsilon_1^{[1]}+\varepsilon_1^{[2]}+\varepsilon_1^{[3]}}{\mathrm{d}p}\right)_{T=\text{constant}} \\ \alpha_2^{(P)} &= -\left(\frac{\varepsilon_2^{[1]}+\varepsilon_2^{[2]}+\varepsilon_2^{[3]}}{\mathrm{d}p}\right)_{T=\text{constant}} \\ \alpha_3^{(P)} &= -\left(\frac{\varepsilon_3^{[1]}+\varepsilon_3^{[2]}+\varepsilon_3^{[3]}}{\mathrm{d}p}\right)_{T=\text{constant}} \end{aligned} \tag{12.2.10}$$

since the applied stress in all directions is equal to the pressure change

$$\sigma^{[1]} = \sigma^{[2]} = \sigma^{[3]} = \mathrm{d}p. \tag{12.2.11}$$

Having obtained the effective Young's modulus and effective Poisson's ratio, the linear compressibilities $\alpha_L^{(P)}$ along the Ox_1, Ox_2, and Ox_3 directions are thus

$$\begin{aligned}\alpha_1^{(P)} &= \frac{1}{E_1} - \left(\frac{\upsilon_{21}}{E_2} + \frac{\upsilon_{31}}{E_3}\right)\\ \alpha_2^{(P)} &= \frac{1}{E_2} - \left(\frac{\upsilon_{12}}{E_1} + \frac{\upsilon_{32}}{E_3}\right)\\ \alpha_3^{(P)} &= \frac{1}{E_3} - \left(\frac{\upsilon_{13}}{E_1} + \frac{\upsilon_{23}}{E_2}\right)\end{aligned} \quad (12.2.12)$$

Since the areal strain is the summation of the two principal strains in the plane of consideration, we similarly have the areal compressibilities $\alpha_A^{(P)}$ on the $Ox_1 - Ox_2$, the $Ox_2 - Ox_3$, and the $Ox_1 - Ox_3$ planes

$$\begin{aligned}\alpha_{12}^{(P)} &= \alpha_1^{(P)} + \alpha_2^{(P)}\\ \alpha_{23}^{(P)} &= \alpha_2^{(P)} + \alpha_3^{(P)}\\ \alpha_{31}^{(P)} &= \alpha_3^{(P)} + \alpha_1^{(P)}\end{aligned} \quad (12.2.13)$$

or, in terms of mechanical properties,

$$\begin{aligned}\alpha_{12}^{(P)} &= \frac{1}{E_1} + \frac{1}{E_2} - \left(\frac{\upsilon_{21}}{E_2} + \frac{\upsilon_{31}}{E_3} + \frac{\upsilon_{12}}{E_1} + \frac{\upsilon_{32}}{E_3}\right)\\ \alpha_{23}^{(P)} &= \frac{1}{E_2} + \frac{1}{E_3} - \left(\frac{\upsilon_{12}}{E_1} + \frac{\upsilon_{32}}{E_3} + \frac{\upsilon_{13}}{E_1} + \frac{\upsilon_{23}}{E_2}\right)\\ \alpha_{31}^{(P)} &= \frac{1}{E_3} + \frac{1}{E_1} - \left(\frac{\upsilon_{13}}{E_1} + \frac{\upsilon_{23}}{E_2} + \frac{\upsilon_{21}}{E_2} + \frac{\upsilon_{31}}{E_3}\right)\end{aligned} \quad (12.2.14)$$

Since the volumetric strain is the summation of all three principal strains, we have likewise the volumetric compressibility $\alpha_V^{(P)}$ as a summation of linear compressibilities in all three orthogonal directions

$$\alpha_V^{(P)} = \alpha_1^{(P)} + \alpha_2^{(P)} + \alpha_3^{(P)} \quad (12.2.15)$$

or in terms of mechanical properties

$$\alpha_V^{(P)} = \frac{1}{E_1} + \frac{1}{E_2} + \frac{1}{E_3} - 2\left(\frac{\upsilon_{12}}{E_1} + \frac{\upsilon_{23}}{E_2} + \frac{\upsilon_{31}}{E_3}\right) \quad (12.2.16)$$

The conditions for linear, areal, and volumetric negative compressibilities can then be extracted by solving for $\alpha_L^{(P)} < 0$, $\alpha_A^{(P)} < 0$, and $\alpha_V^{(P)} < 0$.

Simplifications to the formulation can be obtained when solving for microstructural systems that deform either by stretching only or by rotation only. An example of the former is encountered when the angle is constrained such that Eqs. (12.2.2) to (12.2.4) reduce to

$$\begin{aligned}
\varepsilon_1^{[1]} &= \frac{1}{X_1}\left(\frac{\partial X_1}{\partial l_1}\mathrm{d}l_1^{[1]} + \frac{\partial X_1}{\partial l_2}\mathrm{d}l_2^{[1]} + \cdots + \frac{\partial X_1}{\partial l_m}\mathrm{d}l_m^{[1]}\right) \\
\varepsilon_2^{[1]} &= \frac{1}{X_2}\left(\frac{\partial X_2}{\partial l_1}\mathrm{d}l_1^{[1]} + \frac{\partial X_2}{\partial l_2}\mathrm{d}l_2^{[1]} + \cdots + \frac{\partial X_2}{\partial l_m}\mathrm{d}l_m^{[1]}\right) \\
\varepsilon_3^{[1]} &= \frac{1}{X_3}\left(\frac{\partial X_3}{\partial l_1}\mathrm{d}l_1^{[1]} + \frac{\partial X_3}{\partial l_2}\mathrm{d}l_2^{[1]} + \cdots + \frac{\partial X_3}{\partial l_m}\mathrm{d}l_m^{[1]}\right) \\
\varepsilon_1^{[2]} &= \frac{1}{X_1}\left(\frac{\partial X_1}{\partial l_1}\mathrm{d}l_1^{[2]} + \frac{\partial X_1}{\partial l_2}\mathrm{d}l_2^{[2]} + \cdots + \frac{\partial X_1}{\partial l_m}\mathrm{d}l_m^{[2]}\right) \\
\varepsilon_2^{[2]} &= \frac{1}{X_2}\left(\frac{\partial X_2}{\partial l_1}\mathrm{d}l_1^{[2]} + \frac{\partial X_2}{\partial l_2}\mathrm{d}l_2^{[2]} + \cdots + \frac{\partial X_2}{\partial l_m}\mathrm{d}l_m^{[2]}\right) \\
\varepsilon_3^{[2]} &= \frac{1}{X_3}\left(\frac{\partial X_3}{\partial l_1}\mathrm{d}l_1^{[2]} + \frac{\partial X_3}{\partial l_2}\mathrm{d}l_2^{[2]} + \cdots + \frac{\partial X_3}{\partial l_m}\mathrm{d}l_m^{[2]}\right) \\
\varepsilon_1^{[3]} &= \frac{1}{X_1}\left(\frac{\partial X_1}{\partial l_1}\mathrm{d}l_1^{[3]} + \frac{\partial X_1}{\partial l_2}\mathrm{d}l_2^{[3]} + \cdots + \frac{\partial X_1}{\partial l_m}\mathrm{d}l_m^{[3]}\right) \\
\varepsilon_2^{[3]} &= \frac{1}{X_2}\left(\frac{\partial X_2}{\partial l_1}\mathrm{d}l_1^{[3]} + \frac{\partial X_2}{\partial l_2}\mathrm{d}l_2^{[3]} + \cdots + \frac{\partial X_2}{\partial l_m}\mathrm{d}l_m^{[3]}\right) \\
\varepsilon_3^{[3]} &= \frac{1}{X_3}\left(\frac{\partial X_3}{\partial l_1}\mathrm{d}l_1^{[3]} + \frac{\partial X_3}{\partial l_2}\mathrm{d}l_2^{[3]} + \cdots + \frac{\partial X_3}{\partial l_m}\mathrm{d}l_m^{[3]}\right)
\end{aligned} \tag{12.2.17}$$

while the latter is applied especially when dealing with rotating rigid units such that one has

$$\begin{aligned}
\varepsilon_1^{[1]} &= \frac{1}{X_1}\left(\frac{\partial X_1}{\partial \theta_1}\mathrm{d}\theta_1^{[1]} + \frac{\partial X_1}{\partial \theta_2}\mathrm{d}\theta_2^{[1]} + \cdots + \frac{\partial X_1}{\partial \theta_n}\mathrm{d}\theta_n^{[1]}\right) \\
\varepsilon_2^{[1]} &= \frac{1}{X_2}\left(\frac{\partial X_2}{\partial \theta_1}\mathrm{d}\theta_1^{[1]} + \frac{\partial X_2}{\partial \theta_2}\mathrm{d}\theta_2^{[1]} + \cdots + \frac{\partial X_2}{\partial \theta_n}\mathrm{d}\theta_n^{[1]}\right) \\
\varepsilon_3^{[1]} &= \frac{1}{X_3}\left(\frac{\partial X_3}{\partial \theta_1}\mathrm{d}\theta_1^{[1]} + \frac{\partial X_3}{\partial \theta_2}\mathrm{d}\theta_2^{[1]} + \cdots + \frac{\partial X_3}{\partial \theta_n}\mathrm{d}\theta_n^{[1]}\right) \\
\varepsilon_1^{[2]} &= \frac{1}{X_1}\left(\frac{\partial X_1}{\partial \theta_1}\mathrm{d}\theta_1^{[2]} + \frac{\partial X_1}{\partial \theta_2}\mathrm{d}\theta_2^{[2]} + \cdots + \frac{\partial X_1}{\partial \theta_n}\mathrm{d}\theta_n^{[2]}\right) \\
\varepsilon_2^{[2]} &= \frac{1}{X_2}\left(\frac{\partial X_2}{\partial \theta_1}\mathrm{d}\theta_1^{[2]} + \frac{\partial X_2}{\partial \theta_2}\mathrm{d}\theta_2^{[2]} + \cdots + \frac{\partial X_2}{\partial \theta_n}\mathrm{d}\theta_n^{[2]}\right) \\
\varepsilon_3^{[2]} &= \frac{1}{X_3}\left(\frac{\partial X_3}{\partial \theta_1}\mathrm{d}\theta_1^{[2]} + \frac{\partial X_3}{\partial \theta_2}\mathrm{d}\theta_2^{[2]} + \cdots + \frac{\partial X_3}{\partial \theta_n}\mathrm{d}\theta_n^{[2]}\right) \\
\varepsilon_1^{[3]} &= \frac{1}{X_1}\left(\frac{\partial X_1}{\partial \theta_1}\mathrm{d}\theta_1^{[3]} + \frac{\partial X_1}{\partial \theta_2}\mathrm{d}\theta_2^{[3]} + \cdots + \frac{\partial X_1}{\partial \theta_n}\mathrm{d}\theta_n^{[3]}\right) \\
\varepsilon_2^{[3]} &= \frac{1}{X_2}\left(\frac{\partial X_2}{\partial \theta_1}\mathrm{d}\theta_1^{[3]} + \frac{\partial X_2}{\partial \theta_2}\mathrm{d}\theta_2^{[3]} + \cdots + \frac{\partial X_2}{\partial \theta_n}\mathrm{d}\theta_n^{[3]}\right) \\
\varepsilon_3^{[3]} &= \frac{1}{X_3}\left(\frac{\partial X_3}{\partial \theta_1}\mathrm{d}\theta_1^{[3]} + \frac{\partial X_3}{\partial \theta_2}\mathrm{d}\theta_2^{[3]} + \cdots + \frac{\partial X_3}{\partial \theta_n}\mathrm{d}\theta_n^{[3]}\right)
\end{aligned} \tag{12.2.18}$$

The 3D models can be greatly simplified when reduced to 2D models. For the latter, we have the linear compressibilities for a cellular structure defined in the $Ox_1 - Ox_2$ plane

$$\begin{aligned}
\alpha_1^{(P)} &= \frac{1}{E_1} - \frac{\upsilon_{21}}{E_2} \\
\alpha_2^{(P)} &= \frac{1}{E_2} - \frac{\upsilon_{12}}{E_1}
\end{aligned} \tag{12.2.19}$$

and the areal compressibility

$$\alpha_{12}^{(P)} = \frac{1}{E_1} + \frac{1}{E_2} - 2\frac{\upsilon_{12}}{E_1} \tag{12.2.20}$$

The following sections consider examples of NC microstructures based on rib stretching mode of deformation (Sect. 12.3), hinge rotation mode of deformation (Sect. 12.4), and combined stretching and rotation modes of deformation (Sect. 12.5).

12.3 Examples of Negative Compressibility in Microstructures with Rib Stretching Mode of Deformation

There are two sub-categories of NC in microstructures with rib stretching mode of deformation: (a) network of rods or ribs joined at the junctions with pin joints to permit free rotation about the hinges without incurring bending moment and (b) network of rods or ribs joined at connectors that permit only sliding translational motion. The first three examples that were given by Grima et al. (2008) consider two types of 2D NC system (Fig. 12.1) and a 3D NC system (Fig. 12.2).

In the case of the rotationally constrained system by Grima et al. (2013), the connectors are designed to permit only translational motion of the rods in response to fluctuating pressure, as shown in Fig. 12.3, where rods of length l and stretching stiffness constant k_s^l are inclined at an angle θ to the Ox_1 axis, while rods of height h and stretching stiffness constant k_s^h are oriented parallel to the Ox_2 axis. Grima et al. (2013) obtained the linear compressibilities along the Ox_1 and Ox_2 axes

$$\begin{aligned} \alpha_1^{(P)} &= \tfrac{\cos\theta}{k_s^l}\left(\tfrac{h}{l} + 2\sin\theta\right) \\ \alpha_2^{(P)} &= \tfrac{\cos\theta}{k_s^l}\left(\sin\theta + \frac{2\frac{k_s^l}{k_s^h} + \sin^2\theta}{\frac{h}{l} + \sin\theta}\right) \end{aligned} \tag{12.3.1}$$

which gives the area compressibility in the $Ox_1 - Ox_2$ plane

$$\alpha_{12}^{(P)} = \frac{\cos\theta}{k_s^l}\left(\frac{h}{l} + 3\sin\theta + \frac{2\frac{k_s^l}{k_s^h} + \sin^2\theta}{\frac{h}{l} + \sin\theta}\right) \tag{12.3.2}$$

A plot of $\alpha_2^{(P)}$ versus θ for various h/l ratio with $k_s^l = 10$ and $\frac{k_s^l}{k_s^h}/\frac{h}{l} = 0.001$ is shown in Fig. 12.4.

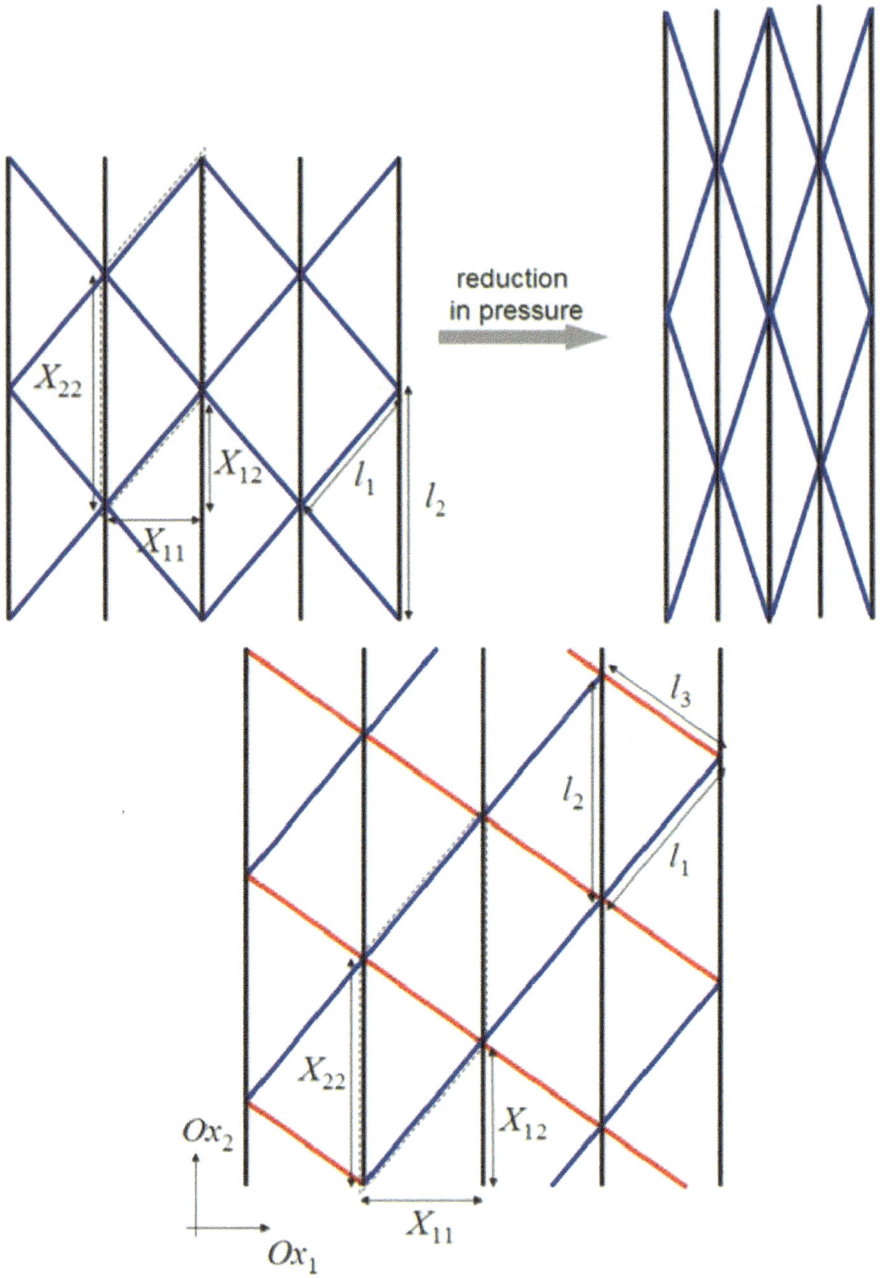

Fig. 12.1 2D truss-like system with negative compressibility made from two types of materials (top) and three types of materials (bottom) by Grima et al. (2008). Reprinted with permission from John Wiley and Sons

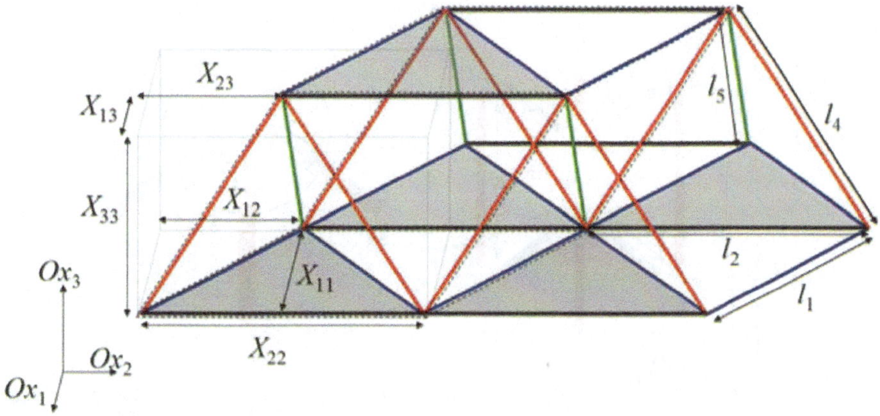

Fig. 12.2 3D truss-like system with negative compressibility made from four types of materials by Grima et al. (2008). Reprinted with permission from John Wiley and Sons

12.4 Examples of Negative Compressibility in Microstructures with Rotational Mode of Deformation

As with the previous section, two sub-categories are identified in this section. These sub-categories are: (a) deformation by rotation of rigid ribs and (b) deformation by rotation of rigid units. The first example is taken from the works of Grima et al. (2012), who proposed a unit cell in the form of hexagonal dodecahedron. Based on the geometrical details furnished in Fig. 12.5 with k_h being the rotational stiffness constant at the rib junctions, Grima et al. (2012) established the compressibilities in the three orthogonal directions as

$$\begin{aligned}
\alpha_1^{(P)} &= \frac{X_2 l_1^2}{4k_h}\left(\frac{X_3}{X_1}\sin^2\theta_1 - \cos\theta_1 \sin\theta_1\right) \\
\alpha_2^{(P)} &= \frac{X_1 l_2^2}{4k_h}\left(\frac{X_3}{X_2}\sin^2\theta_2 - \cos\theta_2 \sin\theta_2\right) \\
\alpha_3^{(P)} &= \frac{1}{4k_h}\left\{\frac{X_1 X_2}{X_3}\left[l_1^2\cos^2\theta_1 + l_2^2\cos^2\theta_2\right] - X_2 l_1^2 \sin\theta_1\cos\theta_1 - X_1 l_2^2 \sin\theta_2 \cos\theta_2\right\}
\end{aligned} \tag{12.4.1}$$

where

$$\begin{aligned}
X_1 &= 2l_1\cos\theta_1 \\
X_2 &= 2l_2\cos\theta_2 \\
X_3 &= 2(l_3 + l_1\sin\theta_1 + l_2\sin\theta_2)
\end{aligned} \tag{12.4.2}$$

For the special case where $l_1 = l_2 = l$ and $\theta_1 = \theta_2 = \theta$, and hence $X_1 = X_2 = X$ the compressibilities greatly simplify to

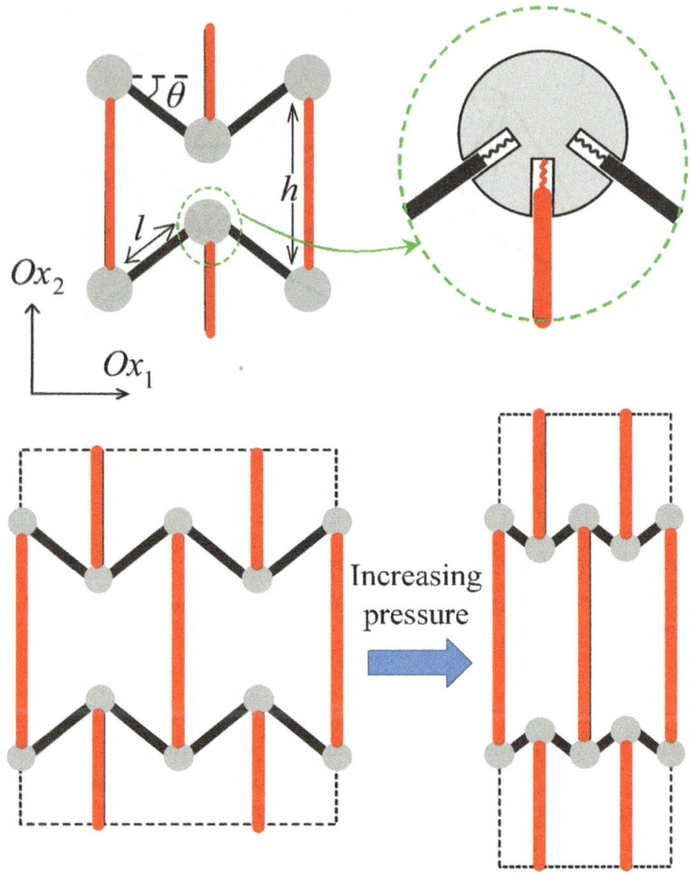

Fig. 12.3 An NC system with rotational constraint by Grima et al. (2013)

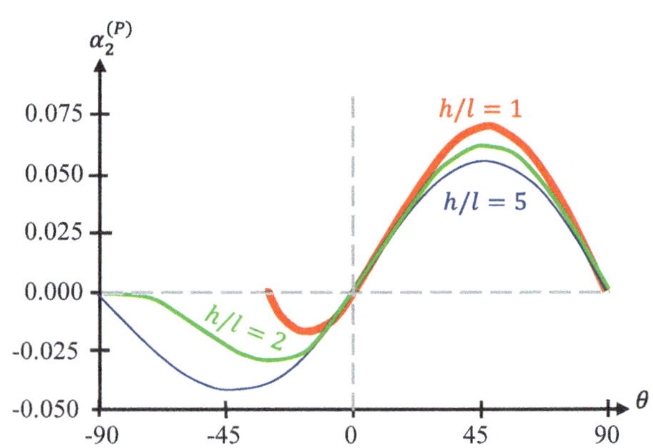

Fig. 12.4 A plot of $\alpha_2^{(P)}$ versus θ for various h/l ratio with $k_s^l = 10$ and $\frac{k_s^l}{k_s^h} / \frac{h}{l} = 0.001$ for the NC system shown in Fig. 12.3

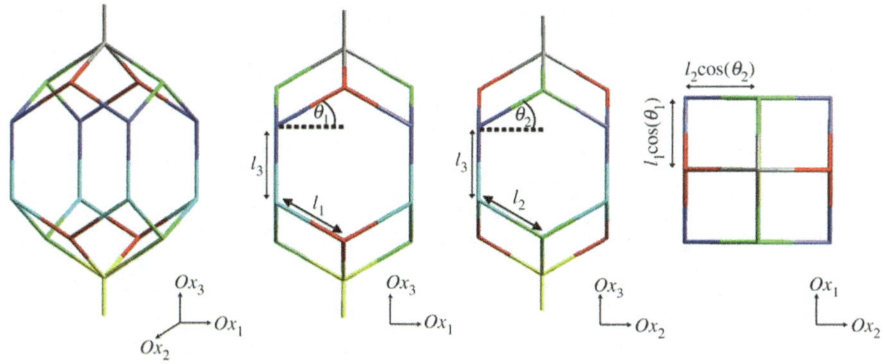

Fig. 12.5 Hexagonal dodecahedron model for NC analysis by Grima et al. (2012). Reprinted with permission from the Royal Society

$$\begin{aligned} \alpha_1^{(P)} = \alpha_2^{(P)} &= \tfrac{l^2}{4k_h}\left\{2(l_3 + 2l\sin\theta)\sin^2\theta - 2l\cos^2\theta\sin\theta\right\} \\ \alpha_3^{(P)} &= \tfrac{l^2\cos^2\theta}{2k_h}\left\{\tfrac{4l^2\cos^2\theta}{2(l_3+2l\sin\theta)} - 2l\sin\theta\right\} \end{aligned} \tag{12.4.3}$$

Figure 12.6 shows the contour maps of linear and area NC for this hexagonal dodecahedron model with $l_1 = l_2 = 0.2$ nm and $l_3 = 0.5$ nm. From the equation of on-axis linear compressibility

$$\alpha_j^{(P)} = \frac{1}{E_j} - \left(\sum_{i=1}^{3}\frac{v_{ij}}{E_i}(1-\delta_{ij})\right); \quad \delta_{ij} = \begin{cases} 0 \text{ if } i \neq i \\ 1 \text{ if } i = j \end{cases} \tag{12.4.4}$$

linear NC is attainable when $v_{ij} > E_i/E_j$. For the special case where $l_1 = l_2 = l$ and $\theta_1 = \theta_2 = \theta$, linear NC is observed along the Ox_1 and Ox_2 directions when

$$\alpha_1^{(P)} = \alpha_2^{(P)} < 0 \Rightarrow \frac{l\cos\theta}{l_3 + 2l\sin\theta} > \tan\theta \tag{12.4.5}$$

and along the Ox_3 direction when

$$\alpha_3^{(P)} < 0 \Rightarrow \frac{l\cos\theta}{l_3 + 2l\sin\theta} < \tan\theta \tag{12.4.6}$$

which indicates that zero compressibility is attained in all three orthogonal directions when

$$\alpha_1^{(P)} = \alpha_2^{(P)} = \alpha_3^{(P)} = 0 \Rightarrow \frac{l\cos\theta}{l_3 + 2l\sin\theta} = \tan\theta \tag{12.4.7}$$

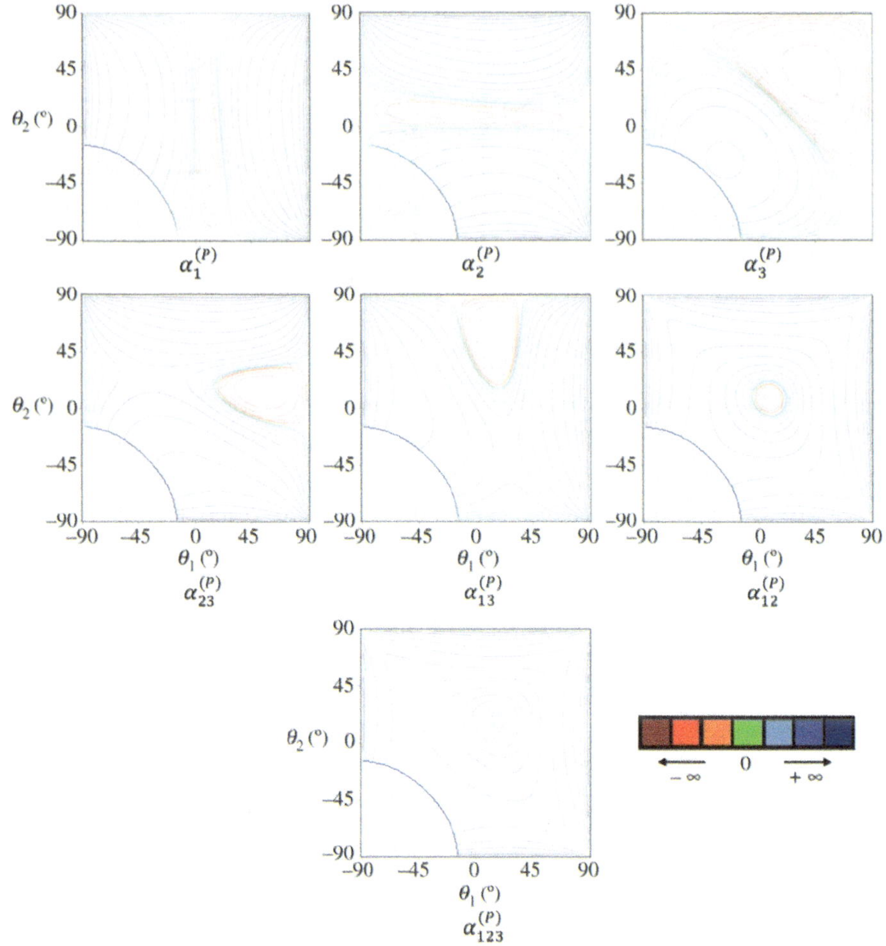

Fig. 12.6 Contour maps of linear and area NC for the hexagonal dodecahedron model shown in Fig. 12.5 with $l_1 = l_2 = 0.2$ nm and $l_3 = 0.5$ nm by Grima et al. (2012). Reprinted with permission from the Royal Society

Since the area compressibility in the $Ox_1 - Ox_2$ plane is obtained from $\alpha_{12}^{(P)} = \alpha_1^{(P)} + \alpha_2^{(P)}$, it follows that area NC on this plane occurs when

$$\alpha_{12}^{(P)} < 0 \Rightarrow \frac{l \cos\theta}{l_3 + 2l \sin\theta} > \tan\theta \tag{12.4.8}$$

For the area compressibilities on the $Ox_1 - Ox_3$ and $Ox_2 - Ox_3$ planes, the area NC can be identified by solving for the boundary between positive and negative area compressibilities

$$\alpha_i^{(P)} + \alpha_3^{(P)} = 0; \quad (i = 1, 2) \tag{12.4.9}$$

to give the roots

$$\begin{aligned} \frac{l\cos\theta}{l_3 + 2l\sin\theta} &= \tan\theta \\ \frac{2l\cos\theta}{l_3 + 2l\sin\theta} &= \tan\theta \end{aligned} \tag{12.4.10}$$

while the solution for $\alpha_V^{(P)} < 0$ does not exist (Grima et al. 2012).

Similar method on rotation mode of deformation has been taken for the NC study of cellular hexahedron, octahedron, and dodecahedron microstructures by Zhou et al. (2016) and elongated cellular octahedron microstructure by Ma et al. (2019). An investigation on the linear NC of hexagonal honeycomb under different layouts has been furnished by Zhou et al. (2018) by the same approach. The following examples involve NC studies on rigid mode rotation. The first of these examples is by Dudek et al. (2016) shown in Fig. 12.7, from which the unit cell dimensions are

$$\begin{aligned} X_1 &= 2\left(\max\left[0, a\cos\left(\tfrac{\theta}{2}\right), b\cos\left(\gamma + \tfrac{\theta}{2}\right)\right] - \min\left[0, a\cos\left(\tfrac{\theta}{2}\right), b\cos\left(\gamma + \tfrac{\theta}{2}\right)\right]\right) \\ X_2 &= 2\left(\max\left[0, a\sin\left(\tfrac{\theta}{2}\right), b\sin\left(\gamma + \tfrac{\theta}{2}\right)\right] - \min\left[0, a\sin\left(\tfrac{\theta}{2}\right), b\sin\left(\gamma + \tfrac{\theta}{2}\right)\right]\right) \end{aligned} \tag{12.4.11}$$

The in-plane Poisson's ratio and Young's moduli are thus

$$\begin{aligned} \upsilon_{12} &= \frac{1}{\upsilon_{21}} = -\frac{X_1}{X_2}\left(\frac{\mathrm{d}X_2}{\mathrm{d}\theta}\right)\cdot\left(\frac{\mathrm{d}X_1}{\mathrm{d}\theta}\right)^{-1} \\ E_1 &= \frac{k_h}{z}\frac{X_1}{X_2}\left(\frac{\mathrm{d}X_1}{\mathrm{d}\theta}\right)^{-2} \\ E_2 &= \frac{k_h}{z}\frac{X_2}{X_1}\left(\frac{\mathrm{d}X_2}{\mathrm{d}\theta}\right)^{-2} \end{aligned} \tag{12.4.12}$$

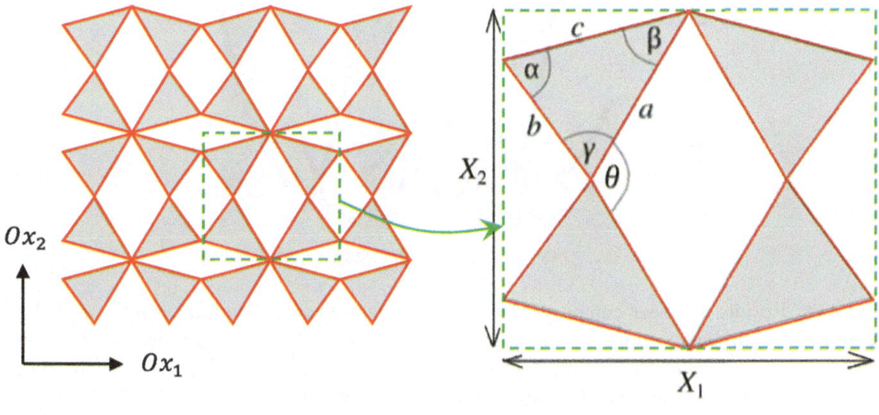

Fig. 12.7 NC study on rigid unit mode by Dudek et al. (2016)

where k_h is the rotational stiffness constant while z is the unit cell dimension measured along the Ox_3 axis. The linear compressibilities in the Ox_1 and Ox_2 directions are thus obtained by substituting Eq. (12.4.12) into Eq. (12.2.19). Figure 12.8 shows the plots of linear compressibilities for three different triangle shapes at $k_h = 12\,\text{kJ}\,\text{mol}^{-1}\,\text{rad}^{-2}$. When the linear compressibilities are added $\alpha_1^{(P)} + \alpha_2^{(P)}$ the results are non-negative, thereby indicating that the area NC is unattainable.

The NC analysis of Type I connected rectangles and Type α connected rhombi, as illustrated in Fig. 11.10.1b, d, has been performed by Attard et al. (2016). The linear compressibilities along the Ox_1 and Ox_2 axes for the Type I connected rectangles are

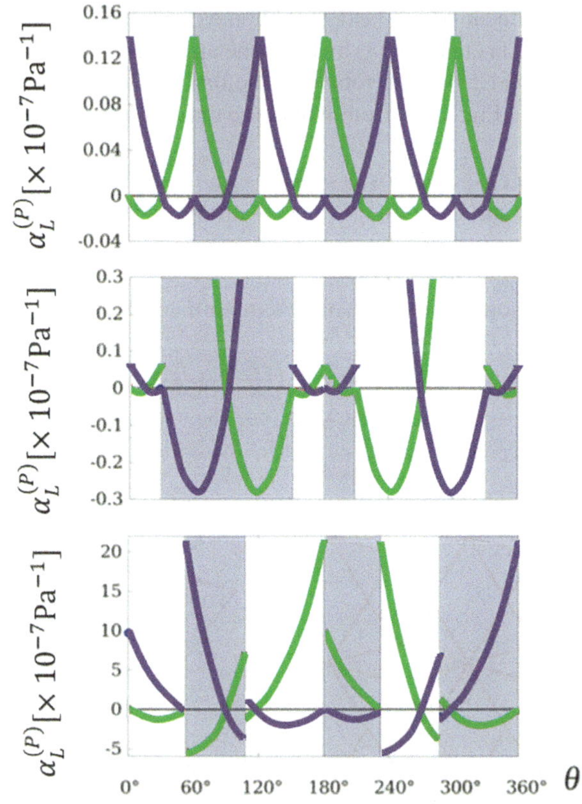

Fig. 12.8 Results of linear compressibility for equilateral triangles with $a = b = c = 1\,\text{nm}$ (top), isosceles triangles with $a = 1\,\text{nm}$, $b = c = 2\,\text{nm}$ (middle) and scalene triangles $a = 6\,\text{nm}$, $b = 3\,\text{nm}$, $c = 4\,\text{nm}$ (bottom) whereby green and purple curves denote $\alpha_1^{(P)}$ and $\alpha_2^{(P)}$, respectively, while dissimilar background shadings distinguish the different forms displayed by the system

$$\begin{aligned}\alpha_1^{(P)} &= -\frac{(a^2+b^2)\cos\theta}{8k_h}\left(\frac{a\tan\frac{\theta}{2}-b}{a+b\tan\frac{\theta}{2}}\right)\\ \alpha_2^{(P)} &= +\frac{(a^2+b^2)\cos\theta}{8k_h}\left(\frac{a-b\tan\frac{\theta}{2}}{a\tan\frac{\theta}{2}+b}\right)\end{aligned} \tag{12.4.13}$$

while the linear compressibilities along the Ox_1 and Ox_2 axes for the Type α connected rhombi are

$$\begin{aligned}\alpha_1^{(P)} &= \frac{a^2\cos\theta}{4k_h}\cot\left(\frac{\phi+\theta}{2}\right)\\ \alpha_2^{(P)} &= \frac{a^2\cos\theta}{4k_h}\tan\left(\frac{\phi-\theta}{2}\right)\end{aligned} \tag{12.4.14}$$

From the on-axes compressibilities described in Eqs. (12.4.13) and (12.4.14), Attard et al. (2016) obtained off-axes compressibilities $\alpha_\zeta^{(P)}$ for both Type I connected rectangles and Type α connected rhombi by transformation of axis. Maps of linear compressibilities $\alpha_\zeta^{(P)}$ for these two types of unit cells are furnished in Fig. 12.9 for demarcating the linear NC regions from the positive compressibility regions, whereby ζ is the off-axis angle. A summary of the criterion for linear NC in Type I connected rectangles and Type α connected rhombi is furnished in Tables 12.1 and 12.2, respectively.

As a final example for this section, we look at a 3D study on the NC properties of rotating squares that was attempted by Grima-Cornish et al. (2020), which also exhibits auxetic behavior (Fig. 12.10). This auxetic-generating "triangular elongation mechanism" is illustrated in Fig. 12.11a through a triangular building block where the base corresponds to the distance between two opposite vertices in the pore of the "rotating squares" structure shown in Fig. 12.11b. As the system is stretched in the Ox_1 direction, the square units located in the $Ox_1 - Ox_2$ plane rotate relative to each other from an angle θ to $\theta + d\theta$ with the consequence that the distance between opposite vertices A and A ($=r$) decreases. This forces the triangular units located in the orthogonal plane to elongate, causing an increase in h and generating a negative Poisson's ratio. Since this model permits deformation in the Ox_3 direction, as exemplified by the Poisson's ratio discussion, it follows that a similar 3D deformation entails in response to changing pressure.

Based on the geometrical descriptions indicated in Fig. 12.10, Grima-Cornish et al. (2020) obtained the Poisson's ratios and Young's moduli expressions

$$\begin{aligned}v_{12} &= \frac{1}{v_{21}} = \begin{cases}-1 & 0<\theta<\pi, \theta \neq \pi/2\\ \text{undefined} & \theta = \pi/2\end{cases}\\ v_{13} &= v_{23} = \frac{1}{v_{31}} = \frac{1}{v_{32}} = -\frac{\sin(\theta)\tan\left(\frac{\pi}{4}+\frac{\theta}{2}\right)}{2\left[\frac{L^2}{l^2}-\cos^2\left(\frac{\theta}{2}\right)\right]}\\ E_i &= \frac{8}{X_3}\left[k_\theta + k_\phi\frac{\sin^2\left(\frac{\theta}{2}\right)}{\frac{L^2}{l^2}-\cos^2\left(\frac{\theta}{2}\right)}\right]\left(\frac{dX_i}{d\theta}\right)^{-2} \quad (i=1,2)\\ E_3 &= \frac{8X_3}{X_1X_2}\left[k_\theta + k_\phi\frac{\sin^2\left(\frac{\theta}{2}\right)}{\frac{L^2}{l^2}-\cos^2\left(\frac{\theta}{2}\right)}\right]\left(\frac{dX_3}{d\theta}\right)^{-2}\end{aligned} \tag{12.4.15}$$

where the unit cell dimensions are

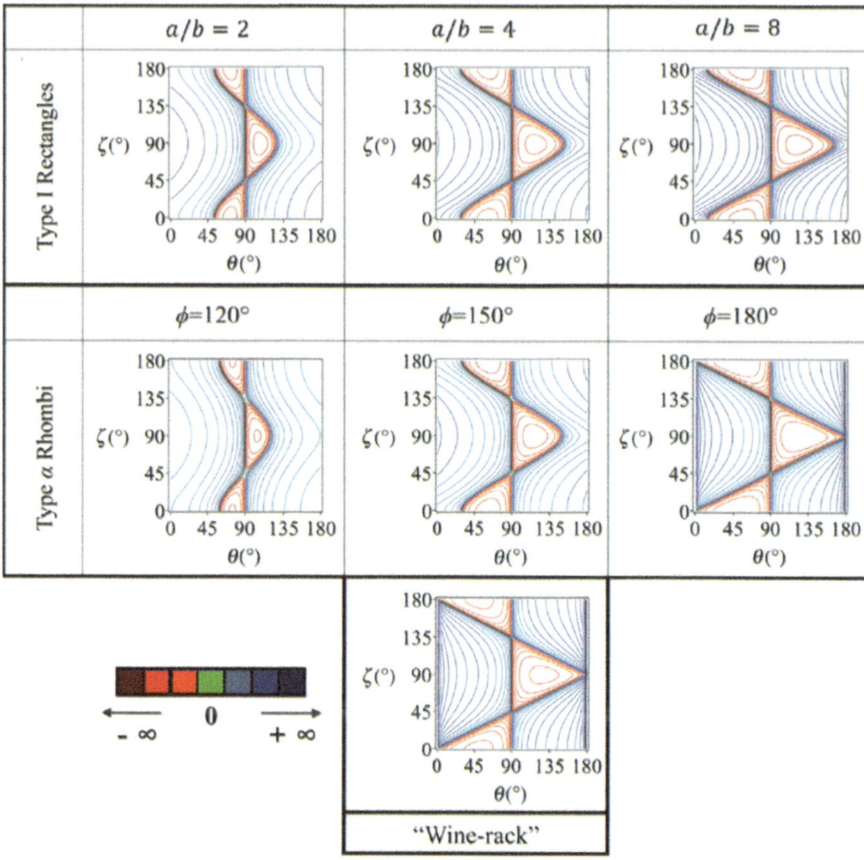

Fig. 12.9 Contour maps of linear compressibility for Type I connected rectangles (top row), Type α connected rhombi (middle row), wine-rack model (bottom). Reprinted with permission from John Wiley and Sons

Table 12.1 On-axes linear NC conditions for Type I connected rectangles

	$a > b$	$a < b$
$\alpha_1^{(P)} < 0$	$2\tan^{-1}\left(\frac{b}{a}\right) < \theta < \frac{\pi}{2}$	$\frac{\pi}{2} < \theta < 2\tan^{-1}\left(\frac{b}{a}\right)$
$\alpha_2^{(P)} < 0$	$\frac{\pi}{2} < \theta < 2\tan^{-1}\left(\frac{a}{b}\right)$	$2\tan^{-1}\left(\frac{a}{b}\right) < \theta < \frac{\pi}{2}$

Table 12.2 On-axes linear NC conditions for Type α connected rhombi

	$\theta < \phi$	$\theta > \phi$
$\alpha_1^{(P)} < 0$	$2\tan^{-1}\left[\cot\left(\frac{\phi}{2}\right)\right] < \theta < \frac{\pi}{2}$	$\frac{\pi}{2} < \theta < 2\tan^{-1}\left[\cot\left(\frac{\phi}{2}\right)\right]$
$\alpha_2^{(P)} < 0$	$\frac{\pi}{2} < \theta < \phi$	$\phi < \theta < \frac{\pi}{2}$

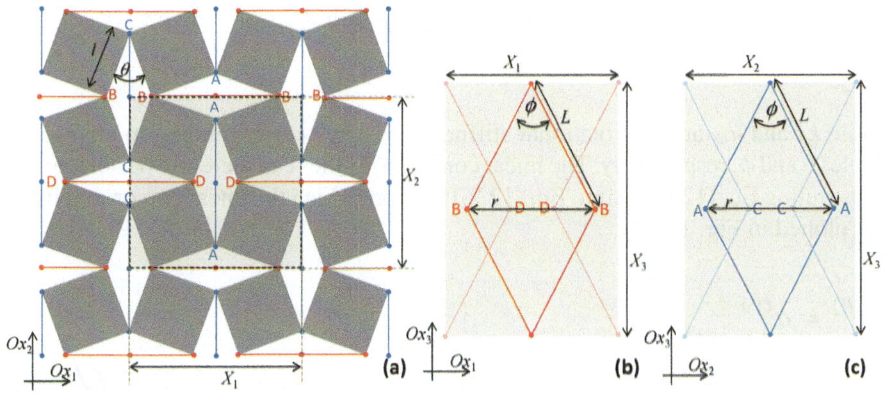

Fig. 12.10 System being modeled by Grima-Cornish et al. (2020), showing the **a** $Ox_1 - Ox_2$ projection with the unit cell highlighted, as well as unit cell projections on **b** $Ox_1 - Ox_3$ plane and **c** $Ox_2 - Ox_3$ plane

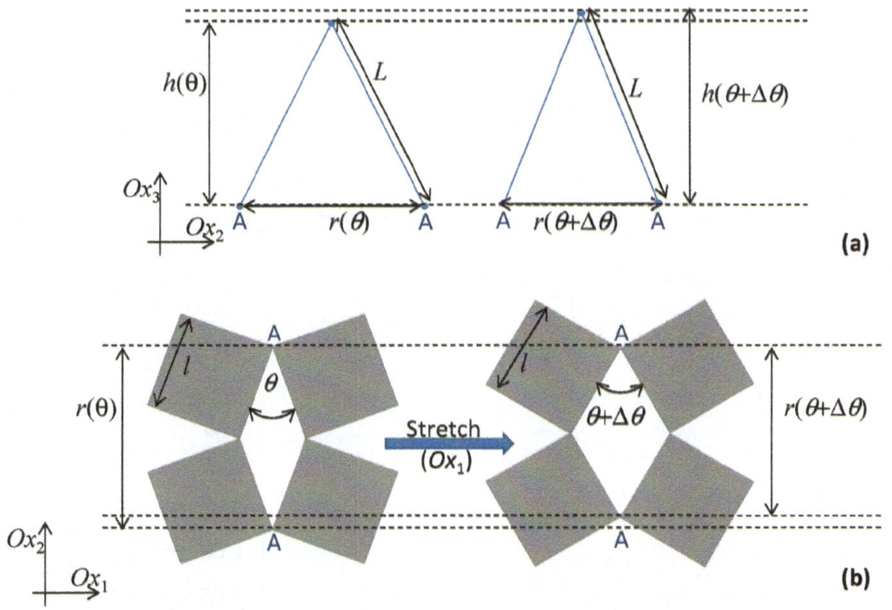

Fig. 12.11 Auxetic-generating "triangular elongation mechanism" proposed by Grima-Cornish et al. (2020) illustrated in **a** through a triangular building block where the base corresponds to the distance between two opposite vertices in the pore of the "rotating squares" structure shown in (**b**)

$$X_1 = X_2 = 2\sqrt{2}l\sin\left(\frac{\pi}{4} + \frac{\theta}{2}\right)$$
$$X_3 = 2l\sqrt{\frac{L^2}{l^2} - \cos^2\left(\frac{\theta}{2}\right)} \tag{12.4.16}$$

while k_θ and k_Φ are the rotational stiffness constants corresponding to changes to angles θ and ϕ, respectively. The linear compressibilities can therefore be obtained by substituting Eq. (12.4.15) into Eq. (12.2.12), from which the linear compressibilities are plotted in Fig. 12.12.

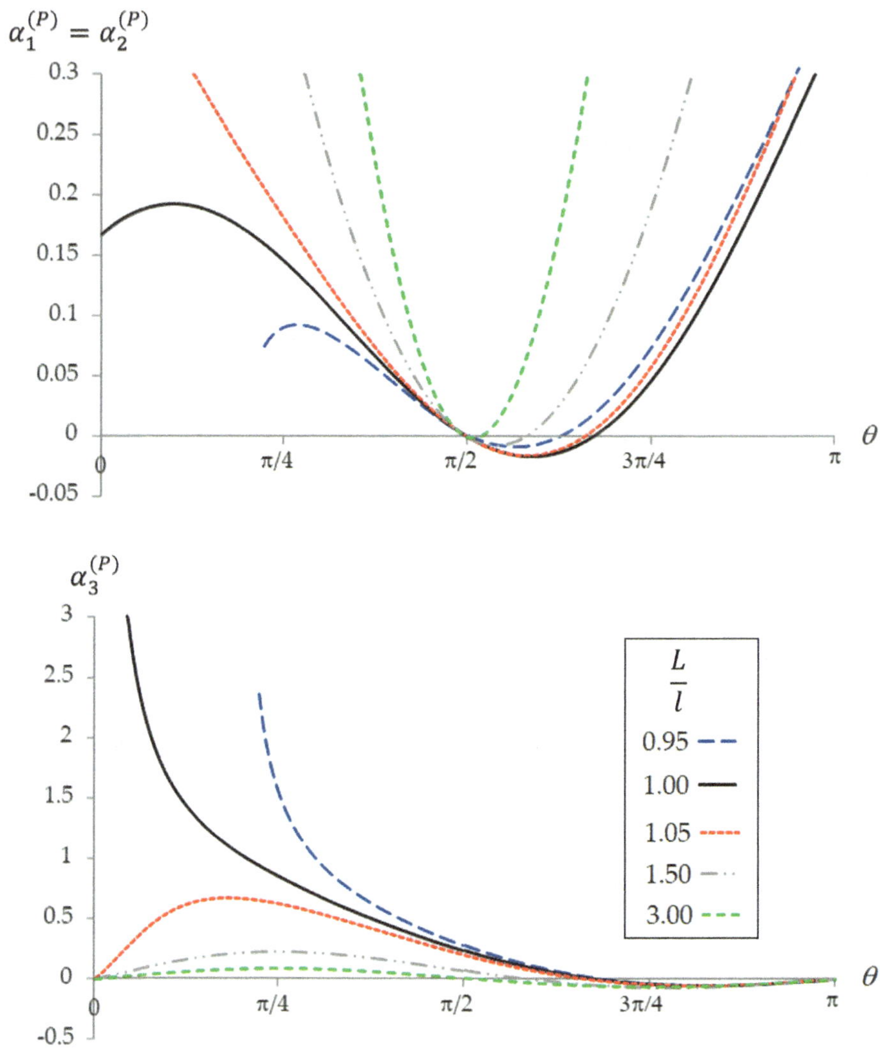

Fig. 12.12 Plots of linear compressibilities in Ox_1 or Ox_2 directions (top) and in the Ox_3 direction (bottom) by Grima-Cornish et al. (2020)

12.5 Example of Negative Compressibilities in Microstructures with Combined Rotational and Stretching Modes of Deformation

A 2D analysis of a hexagonal honeycomb, and its special case of wine-rack model, has been established by Grima et al. (2011) by incorporating both the hinge rotational mode of deformation and the rib stretching mode of deformation. Based on the geometrical descriptions furnished in Fig. 12.13, Grima et al. (2011) gave the Poisson's ratio and Young's moduli as

$$\begin{aligned} v_{12}^h &= \frac{1}{v_{21}^h} = \frac{\cos^2\theta}{\left(\frac{h}{l}+\sin\theta\right)\sin\theta} \\ E_1^h &= k_h \frac{\cos\theta}{b\sin^2\theta\left(\frac{h}{l}+\sin\theta\right)} \\ E_2^h &= k_h \frac{\frac{h}{l}+\sin\theta}{b\cos^3\theta} \end{aligned} \tag{12.5.1}$$

based on hinge rotation mode of deformation, and

$$\begin{aligned} v_{12}^s &= -\frac{\sin\theta}{\frac{h}{l}+\sin\theta} \\ v_{21}^s &= -\frac{\sin\theta\left(\frac{h}{l}+\sin\theta\right)}{2\frac{h}{l}+\sin^2\theta} \\ E_1^s &= k_s \frac{1}{b\cos\theta\left(\frac{h}{l}+\sin\theta\right)} \\ E_2^s &= k_s \frac{\frac{h}{l}+\sin\theta}{b\cos\theta\left(2\frac{h}{l}+\sin^2\theta\right)} \end{aligned} \tag{12.5.2}$$

based on rib stretching mode of deformation, where b is the thickness measured along the Ox_3 direction, while k_h and k_s are the hinge rotational stiffness constant and rib stretching stiffness constant, respectively. The combined Young's moduli E_i^{s+h} and

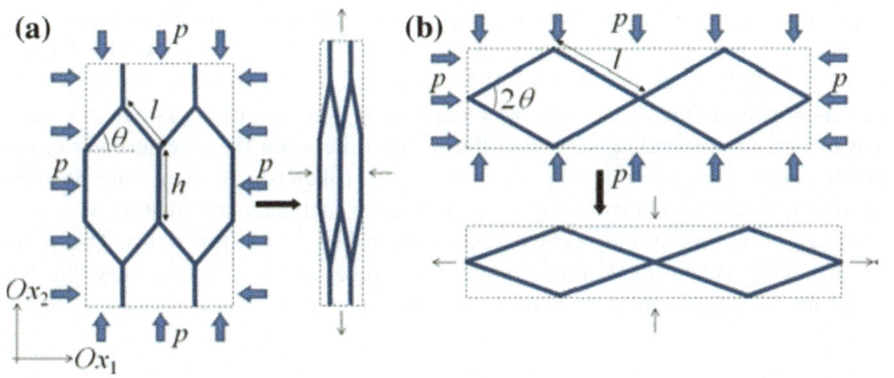

Fig. 12.13 Schematics of hexagonal honeycomb (**a**) and wine-rack model (**b**) undergoing deformation due to increasing pressure, by Grima et al. (2011). Reprinted with permission of Elsevier

Poisson's ratio v_{ij}^{s+h} for $i, j = 1, 2$ under both the rotation and stretching modes of deformation can be obtained from Eq. (12.2.9), which can then be substituted into Eq. (12.2.19) to give the 2D on-axes linear compressibilities

$$\begin{aligned} \alpha_1^{(P)} &= \frac{b}{l}\left(\frac{(h\sin\theta - l\cos 2\theta)\tan\theta}{k_h} + \frac{(2l\sin\theta + h)\cos\theta}{k_s}\right) \\ \alpha_2^{(P)} &= \frac{b\cos\theta}{l\sin\theta + h}\left(\frac{l\cos 2\theta - h\sin\theta}{k_h} + \frac{2l\sin^2\theta + h\sin\theta + 2h}{k_s}\right) \end{aligned} \tag{12.5.3}$$

For the special case of wine-rack model, the substitution of $h = 0$ simplifies Eq. (12.5.3) to

$$\begin{aligned} \alpha_1^{(P)} &= b\left(\frac{-\cos 2\theta\tan\theta}{k_h} + \frac{\sin 2\theta}{k_s}\right) \\ \alpha_2^{(P)} &= \frac{b}{\tan\theta}\left(\frac{\cos 2\theta}{k_h} + \frac{2\sin^2\theta}{k_s}\right) \end{aligned} \tag{12.5.4}$$

Both the on-axes compressibility results, with $k_s = 10k_h$, are shown in Fig. 12.14 for $h = 3l$, $h = l$ and $h = 0$.

12.6 NC System Using Bimaterial Strips in Anti-tetrachiral Arrangement

In addition to the hinge rotation and rib stretching mode, another type of NC system was explored by Gatt and Grima (2008) using bimaterial strips that are arranged in an anti-tetrachiral arrangement, as shown in Fig. 12.15. Both layers are isotropic, while the square nodes are rigid. Suppose material 1 (red) has a greater compressibility than material 2 (blue), then an increase in pressure causes material 1 to contract to a greater extent such that it forms the concave side. By similar argument if there is a decrease in pressure, material 1 expands to a greater extent such that it forms the convex side. Using $a = 1, b = 12, t_1 = t_2 = 0.05$, $E_1 = 2.415 \times 10^9$ Pa, $E_2 = 4.14 \times 10^9$ Pa, and $v_1 = v_2 = 0.35$, Gatt and Grima (2008) simulated the deformed shape of the unit cell, as shown in Fig. 12.16 (top). With decreasing pressure, the square nodes rotate due to bimaterial bending such that the distance between their centers get closer to one another, as shown in Fig. 12.16 (top left). Likewise, an increasing pressure causes the nodes to rotate the opposite direction in such a manner that increases the distances between the node centers, as shown in Fig. 12.16 (top right). This gives NC behavior. When the Young's modulus of material 1 is made to vary, the NC characteristics diminish as its Young's modulus increases.

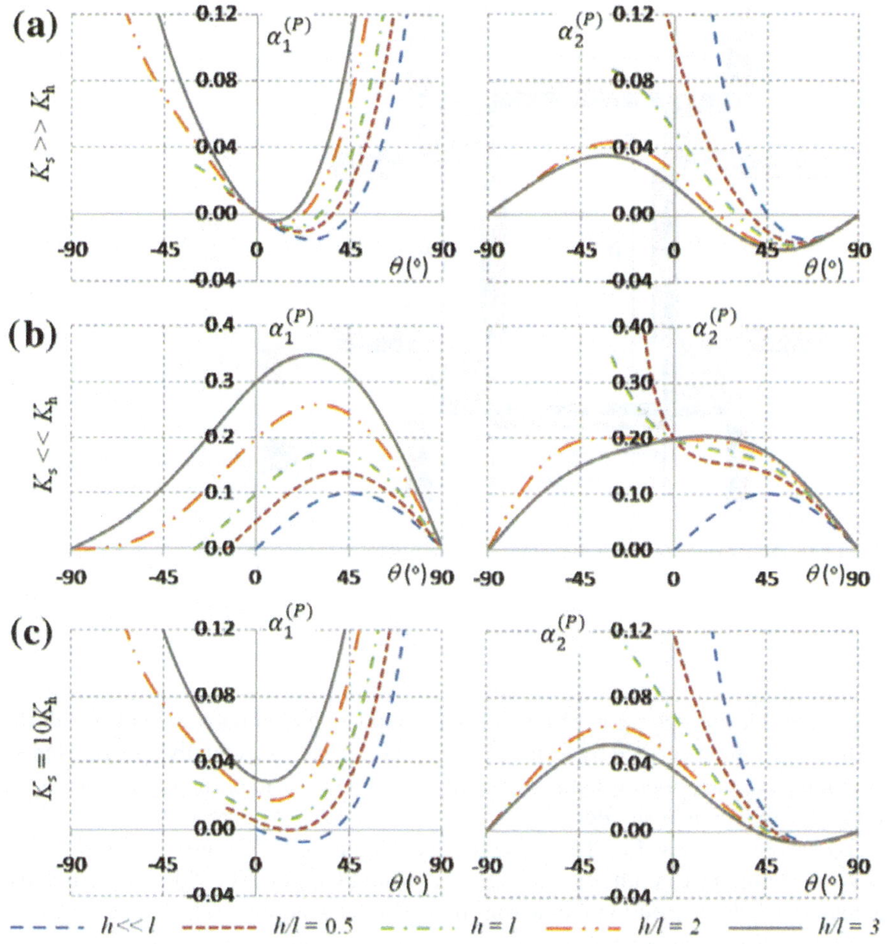

Fig. 12.14 Plots of $\alpha_1^{(P)}$ (left) and $\alpha_2^{(P)}$ (right) versus θ where **a** $k_s \gg k_h$, **b** $k_s \ll k_h$ and **c** $k_s = 10k_h$. Note that systems with $h = 0$ correspond to wine-rack structures. Reprinted with permission of Elsevier

12.7 2D Structures Exhibiting Negative Area Compressibility

This section concerns area compressibility, i.e., the second of Eq. (12.1.1), in which the presence of increasing in-plane pressure reduces the bounded conventional material area but increases the effective unit cell area. Since conventional materials dictate that $\partial A/\partial p$ is a negative value, a positive $\partial A/\partial p$ for the unit cell would therefore give rise to negative area compressibility (NAC) according to the second Eq. (12.1.1), i.e.,

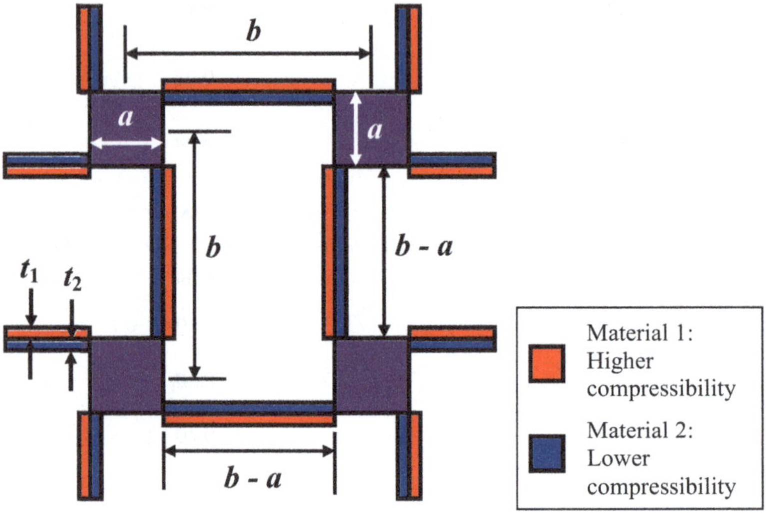

Fig. 12.15 Schematics of bimaterial arranged in anti-tetrachiral arrangement to exhibit planar NC

$$\frac{\partial A}{\partial p} > 0 \Rightarrow \alpha_A^{(P)} < 0 \tag{12.7.1}$$

Recall that for a polygon of n sides, the bounded area is maximized if (i) all the vertices are circumscribed by a circle and (ii) the vertices are equally spaced. Hence, any change to the polygon shape by relative motion of its sides via rotation about the vertices will lead to a smaller area.

By way of example, Fig. 12.17 illustrates two types of NAC consisting of side rods and connecting rods, both of which are of constant lengths. The side rods define the boundary of the conventional material (indicated by shadings). The connecting rods link one conventional material with its adjacent neighbors, with the midpoint of the connecting rods defining the halfway point of one polygon to another, as indicated by the dashed lines. Suppose we begin with octahedron and hexagon, indicated by Fig. 12.17a, b, respectively, as original shapes of the conventional material that are denoted in dark blue. An increasing pressure reduces the material area (shaded area) at constant rod lengths such that the octagonal and hexagonal materials deform toward the shapes of square and triangle, respectively, as displayed in red color. During the course of the side rod motion, the connecting rods get pushed out, thereby increasing the square or rectangular areas bounded by the dashed lines. In other words, the polygonal material area possesses positive compressibility while the linkage mechanism permits the manifestation of NAC. The hinges do not carry moments. These angles are kept in place by the polygonal area such that the change of in-plane pressure changes the polygonal material area, which is facilitated by the change in hinge angles.

Unlike the case of 3D, in which pressure applied on the outer surface of the sample is transmitted to ribs inside the sample, the exposure of both surfaces to pressure in

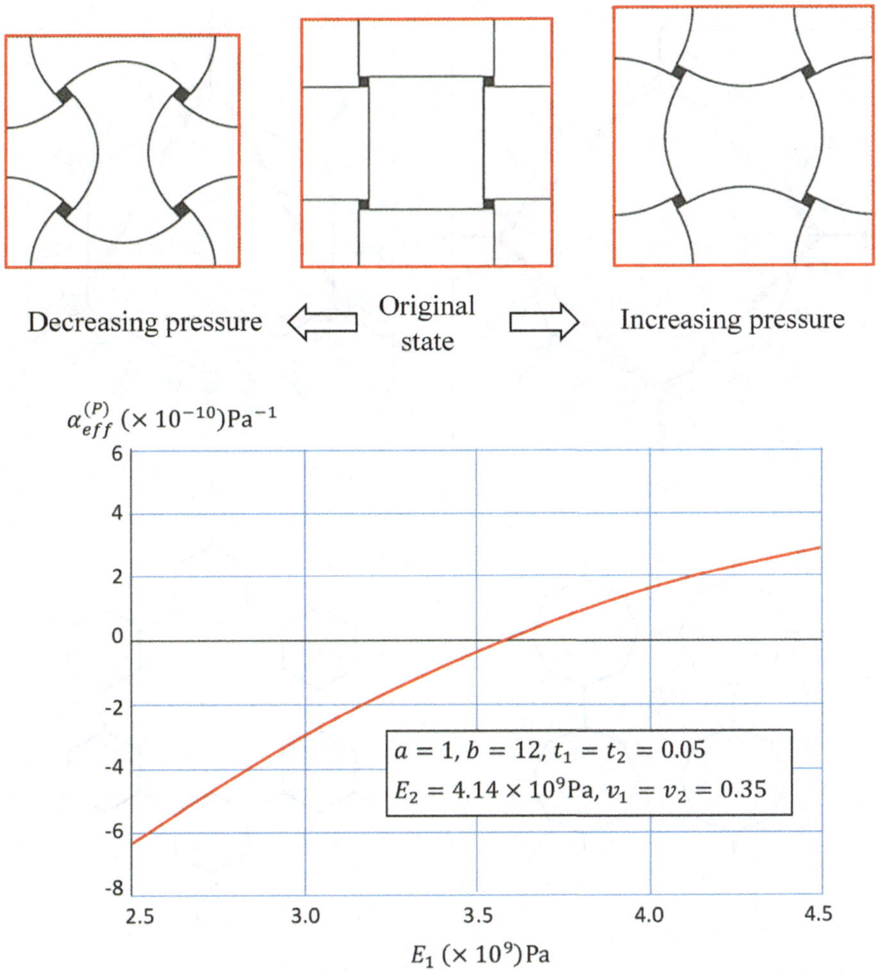

Fig. 12.16 Simulated results for the NC system depicted in Fig. 12.15 with $a = 1$, $b = 12$, $t_1 = t_2 = 0.05$, $E_2 = 4.14 \times 10^9$ Pa and $v_1 = v_2 = 0.35$, showing the deformed unit cell for $E_1 = 2.415 \times 10^9$ Pa (top) and the effective compressibility for variable E_1 (bottom)

2D systems would facilitate entrance of air pressure in the voids. As a result, pressure is applied on the side rods that define the boundary of the material. Inferring from the second of Eq. (12.1.1), we have the compressibilities of the bounded material

$$\alpha_{Am}^{(P)} = -\frac{1}{A_m}\left(\frac{\partial A_m}{\partial p}\right)_T \tag{12.7.2}$$

and that of the effective area

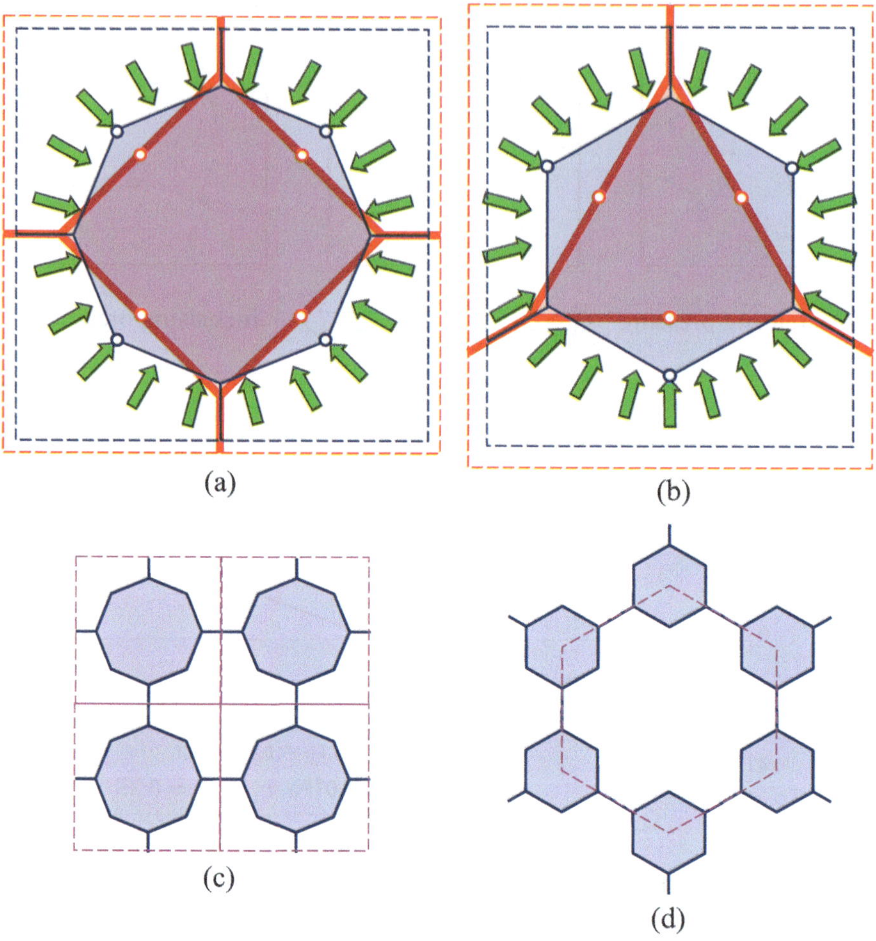

Fig. 12.17 Increased in-plane pressure decreases the material (shaded) area by means of side rod rotations at constant rod lengths, with the connecting rods being pushed out to increase the area enclosed by dashed boundaries: **a** change from octagon material to square material in an NAC of square array, and **b** change from hexagonal material to triangular material in an NAC of hexagonal array. The unit cells are indicated by pink dashed lines in **c** for square array, and **d** for hexagonal array. Green arrow symbolizes in-plane pressure

$$\alpha_{Ae}^{(P)} = -\frac{1}{A_e}\left(\frac{\partial A_e}{\partial p}\right)_T \tag{12.7.3}$$

where A_m and A_e are the material area and the effective area, respectively. The following analyses model the polygonal material area, A_m, and the effective area, A_e, in terms of the characteristics angles θ and ϕ for square and hexagonal arrays, respectively, and rods lengths (l_s, l_c), whereby the side length l_s is defined as the

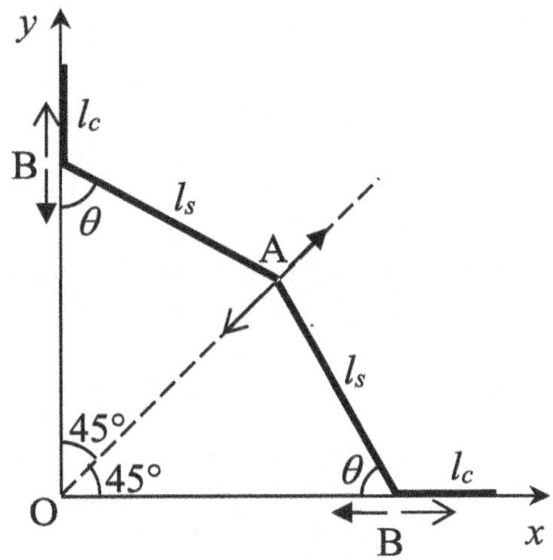

Fig. 12.18 A portion of square array NAC for analysis

length of the polygonal material side while l_c is the half-length of the rod that connects the vertices of two neighboring polygonal materials.

Consider the square array. Let the side rods, which confine the conventional materials, be of length l_s while the connecting rods be of length $2l_c$ between two polygonal materials; hence, the length of each connecting rod associated with each polygon is halved to l_c. A portion of the square array NAC is depicted in Fig. 12.18 for analysis.

By symmetry, hinge *A* moves along the dashed diagonal line shown in Fig. 12.18 while points *B* move horizontally and vertically along the *x* and *y* axes, respectively. The distance of *B* from *O* can be expressed in terms of the *OA* and *AB* distances as

$$\overline{OB} = \frac{1}{\sqrt{2}}\overline{OA} + \overline{AB}\cos\theta \tag{12.7.4}$$

where $\overline{OA}$ is a function of a so-called characteristic angle θ while $\overline{AB} = l_s$. Since

$$\left(\overline{OA}\right)_x = \left(\overline{OA}\right)_y = l_s \sin\theta \tag{12.7.5}$$

substituting

$$\overline{OA} = \sqrt{2}\, l_s \sin\theta \tag{12.7.6}$$

and $\overline{AB} = l_s$ into Eq. (12.7.4) gives

$$\overline{OB} = l_s(\sin\theta + \cos\theta) \tag{12.7.7}$$

With reference to Eqs. (12.7.5) and (12.7.7), the area formed by the triangle *OAB* is

$$A_{OAB} = \frac{1}{2} l_s^2 \sin\theta(\sin\theta + \cos\theta) \tag{12.7.8}$$

Since this triangular area is an eighth of the material polygonal area, we have

$$A_m = 4l_s^2 \sin\theta(\sin\theta + \cos\theta) = 2l_s^2(1 + \sin 2\theta - \cos 2\theta) \tag{12.7.9}$$

Perusal to Fig. 12.18 shows that one half of the side is of the length $\overline{OB} + l_c$. From Eq. (12.7.7), we then have the length of the effective side length

$$L_e = 2l_s(\sin\theta + \cos\theta) + 2l_c \tag{12.7.10}$$

which gives the effective area

$$A_e = 4\left[l_s^2(1 + \sin 2\theta) + 2l_s l_c(\sin\theta + \cos\theta) + l_c^2\right] \tag{12.7.11}$$

Equations (12.7.9) and (12.7.11) will be employed later for obtaining the compressibility of the effective area. For the special case where the conventional material areas are in direct contact with their nearest neighbors, i.e., $l_c = 0$, Eq. (12.7.11) simplifies to

$$A_e = 4l_s^2(1 + \sin 2\theta) \tag{12.7.12}$$

The analysis of the hexagonal array NAC adopts the same set of parameters, except that the choice of characteristic angle, ϕ is made different to prevent confusion and, instead of a quarter, a third of the polygon is used in the following analysis for obvious reason.

Perusal to Fig. 12.19 suggests that

$$\overline{OB} = \frac{1}{2}\overline{OA} + \overline{AB}\cos\phi \tag{12.7.13}$$

where

$$\left(\overline{OA}\right)_x = \sqrt{3}\left(\overline{OA}\right)_y = l_s \sin\phi \tag{12.7.14}$$

Substituting

$$\overline{OA} = \frac{2}{\sqrt{3}} l_s \sin\phi \tag{12.7.15}$$

and $\overline{AB} = l_s$ into Eq. (12.7.13) leads to

$$\overline{OB} = l_s\left(\frac{1}{\sqrt{3}}\sin\phi + \cos\phi\right) \tag{12.7.16}$$

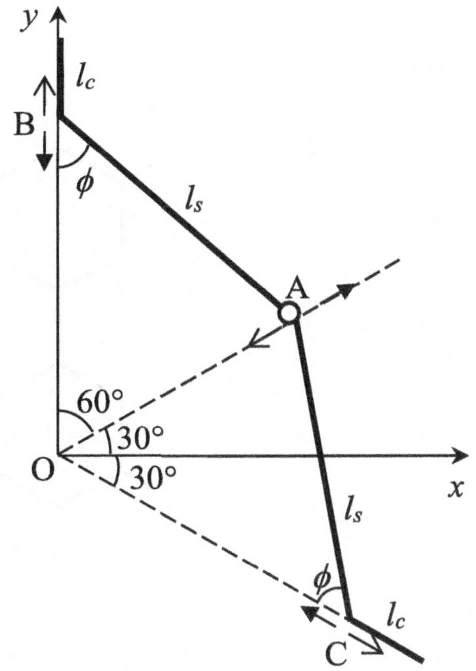

Fig. 12.19 A portion of hexagonal array NAC for analysis

Since the triangular area formed by *OAB* can be obtained using

$$A_{OAB} = \frac{1}{2}\left(\overline{OA}\right)\left(\overline{OB}\right)\sin\frac{\pi}{3} \tag{12.7.17}$$

and recognizing that it forms a sixth of the material polygonal area, we have

$$A_m = 3l_s^2 \sin\phi\left(\frac{1}{\sqrt{3}}\sin\phi + \cos\phi\right) = \frac{\sqrt{3}}{2}l_s^2\left(1 + \sqrt{3}\sin 2\phi - \cos 2\phi\right) \tag{12.7.18}$$

The rectangular boundary that encompasses each hexagonal area and its associated connecting rods is shown in Fig. 12.20, in which the width and height of the rectangles for analysis were determined by symmetry. Although each of these rectangles are not the unit cell, the change in rectangle size that contains the hexagonal material is the same as the change in rectangle size that does not contain the hexagonal material. It follows that the percentage change in the "unit cell" area is the same as the said percentage change in the rectangle area. Specifically, the effective boundary at the midway between two neighboring cells containing the hexagons is determined by symmetry.

With reference to Fig. 12.19, one half of the rectangular boundary width is

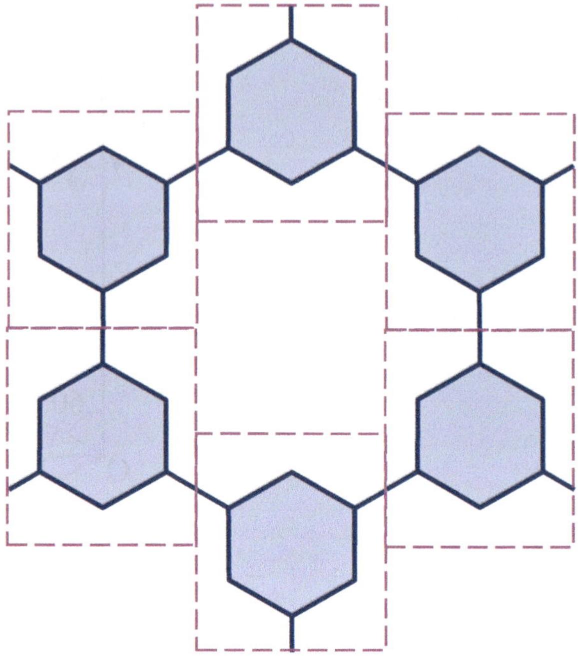

Fig. 12.20 Rectangular enclosures adopted for analysis of hexagonal array NAC

$$\frac{1}{2}(L_e)_x = \left(\overline{AB}\right)_x + \left(\overline{AC}\right)_x + l_c \cos\frac{\pi}{6} \tag{12.7.19}$$

while one half of its height is

$$\frac{1}{2}(L_e)_y = \left(\overline{OB}\right) + l_c \tag{12.7.20}$$

Substituting

$$\begin{aligned} \left(\overline{AB}\right)_x &= l_s \sin\phi \\ \left(\overline{AC}\right)_x &= l_s \cos\left(\phi + \tfrac{\pi}{6}\right) \end{aligned} \tag{12.7.21}$$

and Eq. (12.7.16) into Eqs. (12.7.19) and (12.7.20), respectively, yields

$$(L_e)_x = \frac{\sqrt{3}}{2}(L_e)_y = l_s\left(\sin\phi + \sqrt{3}\cos\phi\right) + \sqrt{3}l_c \tag{12.7.22}$$

so as to give the effective area

$$A_e = \frac{2}{\sqrt{3}}l_s^2\left[2 + \sqrt{3}\sin 2\phi + \cos 2\phi + 2\sqrt{3}\left(\sin\phi + \sqrt{3}\cos\phi\right)\frac{l_c}{l_s} + 3\frac{l_c^2}{l_s^2}\right] \tag{12.7.23}$$

Equations (12.7.18) and (12.7.23) will be used in the subsequent analysis for the compressibility of hexagonal array NAC. For the special case of directly connected material polygons $l_c = 0$, Eq. (12.7.23) reduces to

$$A_e = \frac{2}{\sqrt{3}} l_s^2 \left(2 + \sqrt{3} \sin 2\phi + \cos 2\phi\right) \tag{12.7.24}$$

In developing the compressibility models, recall that the angles θ and ϕ have thus far been used for establishing the effective area for the 8-sided polygonal material in square array and 6-sided polygonal material in hexagonal array, respectively, and are therefore termed the characteristic angle. Suppose we define a generic characteristic angle ψ for a generic polygonal material of n sides and, taking note that the pressure variable is not reflected in the area models of A_m and A_e for both NAC arrays, we express Eqs. (12.7.2) and (12.7.3) as

$$\alpha_{Am}^{(P)} = -\frac{1}{A_m} \frac{\partial A_m}{\partial \psi} \frac{\partial \psi}{\partial p} \tag{12.7.25}$$

and

$$\alpha_{Ae}^{(P)} = -\frac{1}{A_e} \frac{\partial A_e}{\partial \psi} \frac{\partial \psi}{\partial p} \tag{12.7.26}$$

respectively, so as to extract the compressibility ratio

$$\frac{\alpha_{Ae}^{(P)}}{\alpha_{Am}^{(P)}} = \frac{A_m \frac{\partial A_e}{\partial \psi}}{A_e \frac{\partial A_m}{\partial \psi}} \tag{12.7.27}$$

where $\psi = \theta$ and $\psi = \phi$ for square and hexagonal arrays, respectively. Therefore, substituting Eqs. (12.7.9) and (12.7.11) into Eq. (12.7.27) for square array NAC gives

$$\frac{\alpha_{Ae}^{(P)}}{\alpha_{Am}^{(P)}} = \frac{1 + \sin 2\theta - \cos 2\theta}{\cos 2\theta + \sin 2\theta} \frac{\cos 2\theta + (\cos\theta - \sin\theta)\frac{l_c}{l_s}}{1 + \sin 2\theta + 2(\sin\theta + \cos\theta)\frac{l_c}{l_s} + \frac{l_c^2}{l_s^2}} \tag{12.7.28}$$

while the substitution of Eqs. (12.7.18) and (12.7.23) into Eq. (12.7.27) for hexagonal array leads to

$$\frac{\alpha_{Ae}^{(P)}}{\alpha_{Am}^{(P)}} = \frac{1 + \sqrt{3}\sin 2\phi - \cos 2\phi}{\sqrt{3}\cos 2\phi + \sin 2\phi} \times \frac{\sqrt{3}\cos 2\phi - \sin 2\phi + \sqrt{3}\left(\cos\phi - \sqrt{3}\sin\phi\right)\frac{l_c}{l_s}}{2 + \sqrt{3}\sin 2\phi + \cos 2\phi + 2\sqrt{3}\left(\sin\phi + \sqrt{3}\cos\phi\right)\frac{l_c}{l_s} + 3\frac{l_c^2}{l_s^2}} \tag{12.7.29}$$

To express in terms of pressure, we introduce the 2D version of bulk modulus, $K_m^{2D} = -\partial p^{2D}/(\partial A_m/A_m)$, as a material property of the polygonal material such that (Lim 2017)

$$\alpha_{Ae}^{(P)} = \frac{1}{K_m^{2D}} \frac{1+\sin 2\theta - \cos 2\theta}{\cos 2\theta + \sin 2\theta} \frac{\cos 2\theta + (\cos\theta - \sin\theta)\frac{l_c}{l_s}}{1+\sin 2\theta + 2(\sin\theta + \cos\theta)\frac{l_c}{l_s} + \frac{l_c^2}{l_s^2}} \quad (12.7.30)$$

for square array, and

$$\alpha_{Ae}^{(P)} = \frac{1}{K_m^{2D}} \frac{1+\sqrt{3}\sin 2\phi - \cos 2\phi}{\sqrt{3}\cos 2\phi + \sin 2\phi} \times \frac{\sqrt{3}\cos 2\phi - \sin 2\phi + \sqrt{3}\left(\cos\phi - \sqrt{3}\sin\phi\right)\frac{l_c}{l_s}}{2+\sqrt{3}\sin 2\phi + \cos 2\phi + 2\sqrt{3}\left(\sin\phi + \sqrt{3}\cos\phi\right)\frac{l_c}{l_s} + 3\frac{l_c^2}{l_s^2}} \quad (12.7.31)$$

for hexagonal array. Since the polygonal material is taken to be conventional, i.e., $K_m^{2D} > 0$, a negative value for $\alpha_{Ae}^{(P)}$ would indicate negative compressibility.

To pave a way for visualizing the NC properties of the proposed NAC models herein, the effective area is plotted against the material area, as displayed in Fig. 12.21, in terms of dimensionless area. The dimensionless area is defined herein is the ratio of area to the square of the side rod length. Since the application of increased pressure decreases the material area, the NAC range is identified as the region with corresponding increase in the effective area. Therefore, the NAC region, being defined as $A_e/\partial A_m < 0$, is identified in Fig. 12.21 where the slope of A_e/l_s^2 is negative.

To observe the influence of characteristic angle on the occurrence of NAC, a family of dimensionless effective area and the material polygon area curves are plotted against the characteristics angle, as furnished in Fig. 12.22. NAC occurs where the slopes of A_e/l_s^2 and A_m/l_s^2 in Fig. 12.22 possess opposite signs. As evident

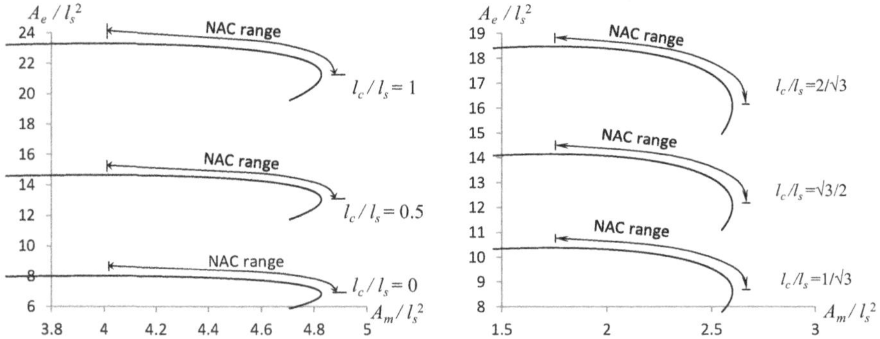

Fig. 12.21 A family of dimensionless effective area plotted against the dimensionless material area at various l_c/l_s ratio for square array (left) and hexagonal array (right)

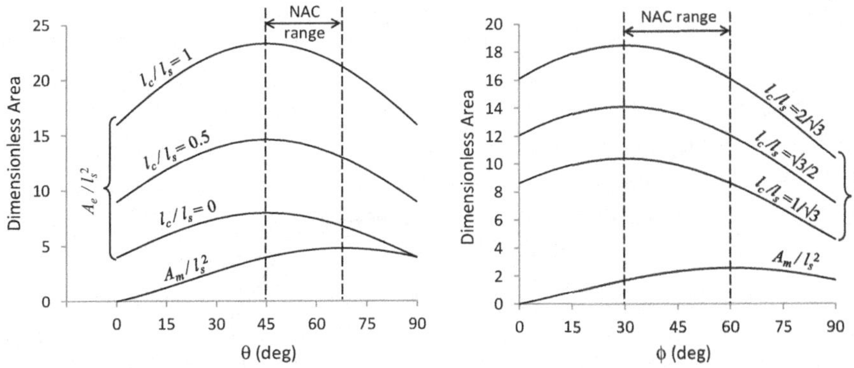

Fig. 12.22 Plots of dimensionless areas versus characteristic angles at various l_c/l_s ratio for square array (left) and hexagonal array (right)

from Fig. 12.22, the NAC takes effect when the characteristic angles are within the range $45° < \theta < 67.5°$ and $30° < \phi < 60°$ for the square and hexagonal arrays, respectively.

The compressibility ratio $\alpha_{Ae}^{(P)}/\alpha_{Am}^{(P)}$ allows the NAC to be inferred; since $\alpha_{Am}^{(P)} > 0$ for conventional material in the polygonal area, a negative value for the compressibility ratio implies $\alpha_{Ae}^{(P)} < 0$. For the sake of retaining the plotted graphs in dimensionless terms, the compressibility ratios are plotted in Fig. 12.23 for both square and hexagonal arrays. Figure 12.23 suggests that regardless of the polygonal material's elastic property, the effective compressibility can be maintained at zero by setting the characteristic angle at the lower NAC limit, i.e.,

$$\lim_{\theta \to 45°} \alpha_{Ae}^{(P)} = \lim_{\phi \to 30°} \alpha_{Ae}^{(P)} = 0 \tag{12.7.32}$$

On the other hand, extreme positive and negative effective compressibilities are achievable at the upper NAC limit, and can be expressed by one-sided limits. Specifically, extreme positive compressibility is attained via a right-handed limit while extreme negative compressibility is obtained through a left-handed limit

$$\begin{aligned} \lim_{\theta \to (67.5°)^+} \alpha_{Ae}^{(P)} &= \lim_{\phi \to (60°)^+} \alpha_{Ae}^{(P)} = +\infty \\ \lim_{\theta \to (67.5°)^-} \alpha_{Ae}^{(P)} &= \lim_{\phi \to (60°)^-} \alpha_{Ae}^{(P)} = -\infty \end{aligned} \tag{12.7.33}$$

Suppose a dimensionless compressibility of the effective area is introduced as a product of the effective compressibility and the 2D "bulk modulus" of the polygonal material $\alpha_{Ae}^{(P)} K_m^{2D}$ and assuming that the 2D "bulk modulus" of the polygonal material is not only a positive value but also a material constant, the value of $\alpha_{Ae}^{(P)} K_m^{2D}$ not only indicates the compressibility magnitude, but also the sign of its compressibility. For the special case where the polygonal material touches its nearest neighbor at the

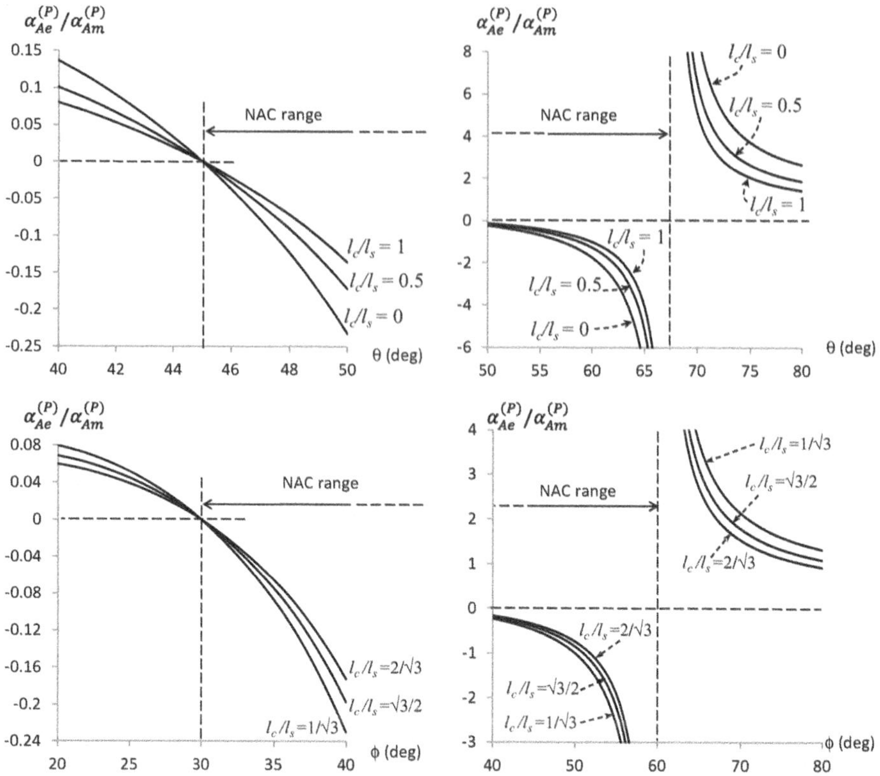

Fig. 12.23 Plots of compressibility ratios for square array (top row) and hexagonal array (bottom row) at small characteristic angle (left column) and large characteristic angle (right column) for various l_c/l_s ratio

vertices, i.e., $l_c = 0$, we have

$$\alpha_{Ae}^{(P)} K_m^{2D} = \frac{1}{1+\tan 2\theta}\left(1 - \frac{\cos 2\theta}{1+\sin 2\theta}\right) \tag{12.7.34}$$

for square array, and

$$\alpha_{Ae}^{(P)} K_m^{2D} = \frac{\sqrt{3}\cos 2\phi - \sin 2\phi}{\sqrt{3}\cos 2\phi + \sin 2\phi}\left(\frac{1+\sqrt{3}\sin 2\phi - \cos 2\phi}{2+\sqrt{3}\sin 2\phi + \cos 2\phi}\right) \tag{12.7.35}$$

for hexagonal array. Graphical representations for the variation of $\alpha_{Ae}^{(P)} K_m^{2D}$ with the characteristic angles at various l_c/l_s ratio are furnished in Fig. 12.24. Regions of $\alpha_{Ae}^{(P)} K_m^{2D} < 0$, indicating NAC, are shown within $45^\circ < \theta < 67.5^\circ$ and $30^\circ < \phi < 60^\circ$ for the square and hexagonal arrays, respectively.

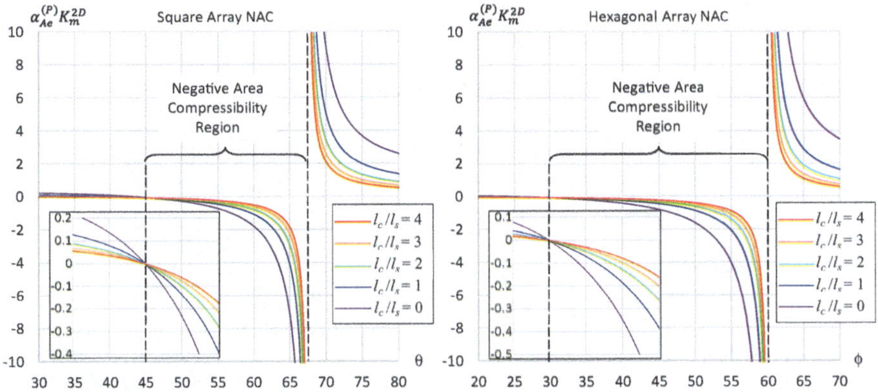

Fig. 12.24 Variation of dimensionless compressibilities with respect to the characteristic angles for different l_c/l_s ratio, showing regions of negative area compressibilities. The insets expand the scales on the vertical axes for clarity while retaining alignment of horizontal axes to those of the main graphs

A pair of negative area compressibility systems has been proposed in this section, each consisting of 2D conventional material in the shape of polygon with connecting rods, in which the characteristic angles determine the negativity of the compressibility. Analyses on both systems of NAC show that negative compressibility is observed when the characteristics angles are of the range $45° < \theta < 67.5°$ and $30° < \phi < 60°$ for the square and hexagonal arrays, respectively (Lim 2017). In addition to indicating the range of NAC and positive area compressibility (PAC) as determined by the characteristic angle, results obtained suggest that (a) the magnitude of effective compressibility is enhanced by the use of conventional material with large compressibility, (b) zero effective compressibility is achieved by the implementation of lower characteristic angle, and (c) infinitely large compressibility is attained by employing the upper characteristic angle. Schematic summaries of both NAC systems are furnished in Figs. 12.25 and 12.26 for the square and hexagonal arrays, respectively, indicating the range of NAC as determined by the characteristic angle.

12.8 Further Readings

Apart from the examples of NC materials and systems discussed in previous sections, the reader is referred to many other NC works in the literature. The author can only mention a few for the sake of brevity. Miller et al. (2015) showed that linear NC in materials originate from the misalignment of polymers/fibers. Lakes and Wojciechowski (2008) gave a detailed discussion on negative compressibility and its relationship with negative Poisson's ratio and stability; they also proposed a

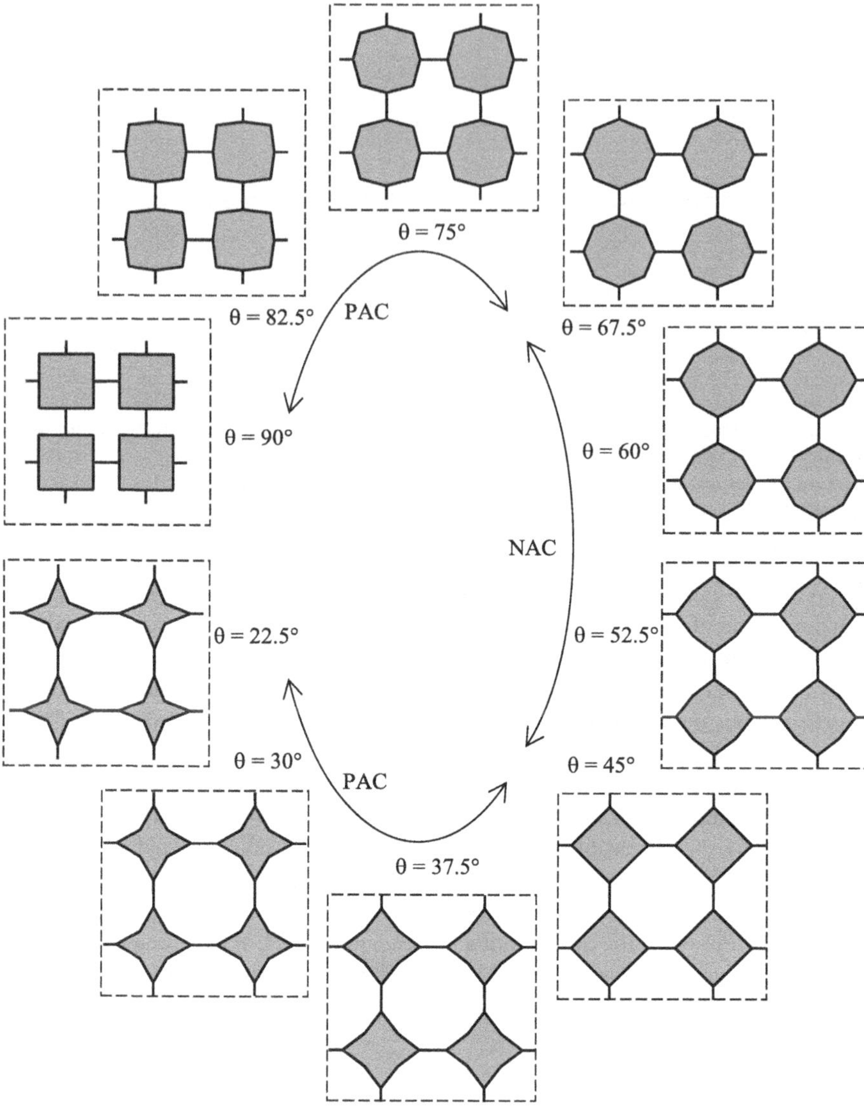

Fig. 12.25 Schematic summary of square array NAC, with dashed squares of equal size superposed to facilitate size comparison, whereby PAC and NAC refer to positive and negative area compressibilities, respectively

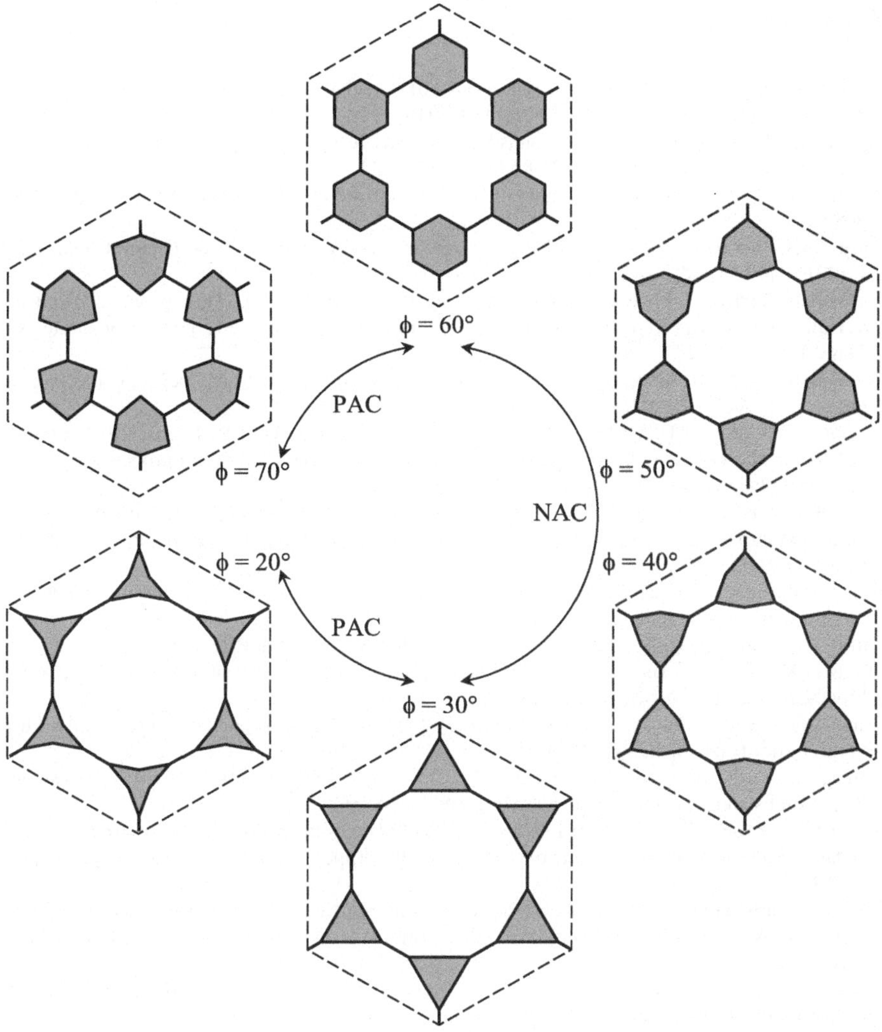

Fig. 12.26 Schematic summary of hexagonal array NAC, with dashed hexagons of equal size superposed to aid size comparison, in which PAC and NAC correspond to positive and negative area compressibilities, respectively

constrained microscopic model which exhibits NC. Imre (2014) showed that the destabilization of metastable equilibrium by Nicolaou and Motter (2012) can be found in nature (water). The occurrence of sign-switching in the coefficient of compressibility has been applied on newly designed composites (Lim 2019, 2020) For review on NC, the reader is referred to the works of Cairns and Goodwin (2015), for their effort in presenting a mechanistic understanding on NC.

References

Attard D, Caruana-Gauci R, Gatt R, Grima JN (2016) Negative linear compressibility from rotating rigid units. Phys Status Solidi B 253(7):1410–1418

Baughman RH, Stafstrom S, Cui C, Dantas SO (1998) Materials with negative compressibilities in one or more dimensions. Science 279(5356):1522–1524

Bridgman PW (1922) The compressibility of metals at high pressures. Proc Natl Acad Sci 8(12):361–365

Cairns AB, Goodwin AL (2015) Negative linear compressibility. Phys Chem Chem Phys 17(32):20449–20465

Cairns AB, Catafesta J, Levelut C, Rouquette J, van der Lee A, Peters L, Thompson AL, Dmitriev V, Haines J, Goodwin AL (2013) Giant negative linear compressibility in zinc dicyanoaurate. Nat Mater 12(3):212–216

Colmenero F (2019) Silver oxalate: mechanical properties and extreme negative mechanical phenomena. Adv Theory Simul 2(6):1900040

Degabriele EP, Attard D, Grima-Cornish JN, Caruana-Gauci R, Gatt R, Evans KE, Grima JN (2019) On the compressibility properties of the wine-rack-like carbon allotropes and related poly(phenylacetylene). Phys Status Solidi B 256:1800572

Dudek KK, Attard D, Caruana-Gauci R, Wojciechowski KW, Grima JN (2016) Unimode metamaterials exhibiting negative linear compressibility and negative thermal expansion. Smart Mater Struct 25(2):025009

Fortes ADS, Suard E, Knight KS (2011) Negative linear compressibility and massive anisotropic thermal expansion in methanol monohydrate. Science 331(6018):742–746

Gatt R, Grima JN (2008) Negative compressibility. Phys Status Solidi RRL 2(5):236–238

Goodwin AL, Keen DA, Tucker MG (2008) Large negative linear compressibility of $Ag_3[Co(CN)_6]$. Proc Natl Acad Sci 105(48):18708–18713

Grima JN, Attard D, Gatt R (2008) Truss-type systems exhibiting negative compressibility. Phys Status Solidi B 245(11):2405–2414

Grima JN, Attard D, Caruana-Gauci R, Gatt R (2011) Negative linear compressibility of hexagonal honeycombs and related systems. Scr Mater 65(7):565–568

Grima JN, Caruana-Gauci R, Attard D, Gatt R (2012) Three-dimensional cellular structures with negative Poisson's ratio and negative compressibility properties. Proceedings of Royal Society A 468(2146):3121–3138

Grima JN, Caruana-Gauci R, Wojciechowski KW, Evans KE (2013) Smart hexagonal truss systems exhibiting negative compressibility through constrained angle stretching. Smart Mater Struct 22(8):084015

Grima-Cornish JN, Grima JN, Attard D (2020) A novel mechanical metamaterial exhibiting auxetic behavior and negative compressibility. Materials 13(1):79

Imre AR (2014) Metamaterials with negative compressibility—a novel concept with a long history. Mater Sci Pol 32(2):126–129

Lakes R, Wojciechowski KW (2008) Negative compressibility, negative Poisson's ratio, and stability. Phys Status Solidi B 245(3):545–551

Lim TC (2017) 2D structures exhibiting negative area compressibility. Phys Status Solidi B 254(12):1600682

Lim TC (2019) Composite metamaterial with sign-switchable coefficients of hygroscopic, thermal and pressure expansions. Adv Compos Hybrid Mater 2(4):657–669

Lim TC (2020) Composite metamaterial square grids with sign-flipping expansion coefficients leading to a type of Islamic design. SN Appl Sci 2(5):918

Ma M, Zhou X, Li J, Li H, He S (2019) The negative compressibility in 3D cellular elongated octahedron model. IOP Conf Ser Ear Env Sci 252(2):022140

Miller W, Evans KE, Marmier A (2015) Negative linear compressibility in common materials. Appl Phys Lett 106(23):231903

Nicolaou ZG, Motter AE (2012) Mechanical metamaterials with negative compressibility transitions. Nat Mater 11(7):608–613
Zhou X, Zhang L, Zhang H, Liu Q, Ren T (2016) 3D cellular models with negative compressibility through the wine-rack-type mechanism. Phys Status Solidi B 253(10):1977–1993
Zhou X, Li J, Ma M, Yang L (2018) The comparison of negative linear compressibility of hexagonal honeycomb under different layouts. IOP Conf Ser Mater Sci Eng 394(3):032076

Chapter 13
Negative Moisture Expansion, Negative Hygrothermal Expansion, and Negative Environmental Expansion

Abstract This chapter introduces negative moisture expansion (NME), which is also known as negative hygroscopic expansion and negative swelling. A comparison is made between NME, negative thermal expansion (NTE), and negative compressibility (NC). Thereafter, the concepts of negative hygrothermal expansion (NHTE) and negative environmental expansion (NEE) are discussed.

Keywords Negative environmental expansion · Negative hygroscopic expansion · Negative hygrothermal expansion · Negative moisture expansion · Negative swelling

13.1 Introduction

Some materials expand upon absorption of moisture and, but the same argument, contract upon dissipation of moisture. A measure of moisture expansion can be expressed in terms of the linear coefficient of moisture expansion (CME) $\alpha^{(C)}$, also known as coefficient of hygroscopic expansion (CHE) and coefficient of swelling, as

$$\alpha^{(C)} = \frac{1}{L}\frac{\mathrm{d}L}{\mathrm{d}C} = \frac{\varepsilon_C}{\mathrm{d}C} \tag{13.1.1}$$

where L and $\mathrm{d}L$ are the considered length and its change in response to a change in moisture concentration $\mathrm{d}C$ in the material. It is obvious that $\varepsilon_C = \mathrm{d}L/L$ is the strain induced by the change in moisture concentration in the material. To obtain $\mathrm{d}C$, we begin with the definition of moisture concentration C in a material

$$C = \frac{m}{M} \times 100 \tag{13.1.2}$$

where m is the mass of moisture in the material and M is the mass of the dry material. It can be seen that under almost all practical circumstances, M is a constant and that the description of C is a percentage. In response to a change in moisture mass $\mathrm{d}m$, the change in moisture concentration in the material is thus

T. Lim, *Mechanics of Metamaterials with Negative Parameters*, Engineering Materials,
https://doi.org/10.1007/978-981-15-6446-8_13

$$dC = \frac{dm}{M} \times 100 \tag{13.1.3}$$

As such, the CME $\alpha^{(C)}$ for moisture concentration change in a solid is analogous with the CTE $\alpha^{(T)}$ for temperature change and compressibility $\alpha^{(P)}$ for pressure change. By the same reasoning, the change in moisture concentration in a material dC forms analogies with the change in temperature dT and the negative change in pressure $(-dP)$. The negative sign is included for the latter because materials tend to expand with increasing temperature and moisture concentration, but contracts with increasing pressure. While this observation seems to show that thermal expansion is closer to moisture expansion than compressibility, i.e., pressure expansion, it will be shown next that moisture expansion is rather unique in comparison to the other two types of expansions. A change of temperature in the environment dT is equal to a change of temperature in the material dT upon thermal equilibrium. In the first place, both the environment and material possess equal initial temperature T_0, also known as the base temperature. When the temperature of the environment changes to T_f, heat transfer takes place between the environment and the material due to a temperature gradient until thermal equilibrium is reached, for which the material's temperature stabilizes at T_f, and remains so until a further change in environmental temperature occurs. In other words, not only is the temperature change the same for the environment and the material, but their temperatures are also common upon thermal equilibrium. Nevertheless, in calculating the CTE, only the temperature change is influential rather than the temperature itself. For composites or systems that consist of different materials, the change in temperature in all different phases is common, and equal to the change in the surrounding temperature upon thermal equilibrium. In the same manner, the calculation of compressibility requires the change in pressure of the environment. This change in pressure is common to all phases that are exposed to the environment. The same cannot be said so for moisture concentration. The environment itself has its own concentration. Some materials, such as metals, have no moisture content, and this remains so when the moisture concentration in the environment changes. Other materials, such as wood and polymers, possess their intrinsic moisture concentration, whereby different materials contain different moisture concentrations, i.e., $C_A \neq C_B \neq C_C \ldots$ for materials A, B, C, etc., respectively, for an initial environmental moisture concentration C_{env}. Upon a change in environmental moisture concentration to C'_{env}, the moisture concentration of materials A, B, C, etc., change with different rate but regardless of different rates of moisture absorption or dissipation, these materials reach their final moisture concentrations at hygroscopic equilibrium $C'_A \neq C'_B \neq C'_C \ldots$, which are not equal in general. Apart from the rates of moisture absorption or dissipation, a more important parameter is the moisture absorptivity, which governs the amount of moisture absorbed upon hygroscopic equilibrium. In other words, the changes in moisture concentrations are different for the various phases, and also different from the environment $dC_{\text{env}} \neq dC_A \neq dC_B \neq dC_C \ldots$. Notice that for the thermal analogy, the temperature change would have been equal when thermal equilibrium is attained $dT_{\text{env}} = dT_A = dT_B = dT_C \ldots$, and so a change in the environmental temperature can be taken as a change in the material temperature for calculation of the CTE. Similarly, all phases are exposed to a common pressure and so an increase of pressure in the environment $\mathrm{d}p_{\mathrm{env}}$ is equally imposed on all phases, i.e., $\mathrm{d}p_{\mathrm{env}} = \mathrm{d}p_A = \mathrm{d}p_B = \mathrm{d}p_C \ldots$.

A negative moisture expansion (NME) material is defined as a material that shrinks upon absorption of moisture ($\varepsilon_C < 0 < \mathrm{d}C$) or swells with release of moisture $\mathrm{d}C < 0 < \varepsilon_C$, i.e., (Lim 2017a)

$$\alpha^{(C)} < 0 \tag{13.1.4}$$

NME is also known as negative swelling, but the latter has a broader meaning. Since the term moisture, hygroscopic or water is absent, swelling includes the expansion (or contraction) of materials with the absorption (or release) of non-aqueous solutions. As such, moisture or hygroscopic expansion is a subset of swelling, and it therefore follows that NME is a subset of negative swelling. In some practical cases, materials and structures are exposed to fluctuating increase and decrease of steam, i.e., a simultaneous increase in temperature and moisture concentration as well as simultaneous decrease in these conditions. Consideration of both thermal and hygroscopic changes and the corresponding material responses forms the hygrothermal analysis. Consequently, the combination of NTE and NME gives rise to negative hygrothermal expansion (NHTE) behavior. The hygrothermal analysis is important because it gives a more complete understanding on material response when subjected to both thermal and moisture fluctuations. Take, for example, a conventional material whereby an increase in temperature causes existing moisture in the material to be released to the environment. Due to a positive thermal expansivity $\alpha^{(T)} > 0$, there is a positive thermal strain component $\varepsilon_T > 0$ in conjunction with the increased temperature $\mathrm{d}T > 0$. Due to a positive moisture expansivity $\alpha^{(C)} > 0$, the release of moisture $\mathrm{d}m < 0$ and hence a reduction in moisture concentration $\mathrm{d}C < 0$ tends to reduce the size of the material; i.e., there is a negative moisture strain component $\varepsilon_C < 0$. If the moisture strain has a greater magnitude than the thermal strain $|\varepsilon_C| > \varepsilon_T$, then the overall strain is negative $\varepsilon_T + \varepsilon_C < 0$. As such if one were to analyze the problem solely from the thermal expansion point of view, one may be misled into believing that the material is NTE. As such combining the thermal and moisture expansions not only provides more accurate description, but more importantly it prevents erroneous interpretations. Going by this line of argument, it is beneficial to incorporate compressibility, or pressure expansion, analysis into hygrothermal expansion study. For this reason, the phrase negative environmental expansion (NEE) is introduced to take into account the interlacing effect of NTE, NC, and NME. The terms "moisture" and "hygroscopic" are used interchangeably in this book.

13.2 Example: A Non-auxetic 2D Linear and Area NHTE System Using Reinforced Kite-Shaped Microstructure

The microstructure that is being considered consists of pin-jointed rods to facilitate free rotations at the joints and the rods do not encounter bending. The rods are aligned along five directions, as numbered in Fig. 13.1a, wherein rod 1 and rod 2 are mirror images to rod 4 and rod 5, respectively; with rod 3 on the axis of symmetry. Rods 1, 2, 4, and 5, which form the shape of a kite, are rigid. Rod 3 is expansible such that it expands with increasing temperature or with the absorption of moisture. Since rods 1, 2, 4, and 5 are rigid, i.e., only rod 3 is elastic, this system is non-auxetic (Lim 2019a). Due to symmetry, only rods 1, 2, and 3 are required for analysis.

With reference to Fig. 13.1b, rods 1, 2, 3 are assigned length of l_1, l_2, l_3 while their inclinations are indicated by the angles $\theta_1, \theta_2, \theta_3$ being measured anticlockwise from the x-axis, such that $\theta_3 = 0° < \theta_1 < 90° < \theta_2 < 180°$. Upon thermal or moisture expansion of rod 3 by $\mathrm{d}l_3$ arising from environmental temperature change of $\mathrm{d}T$ or environmental moisture concentration change of $\mathrm{d}C$, we have its strain

$$\varepsilon_3 = \frac{\mathrm{d}l_3}{l_3} = \alpha_3^{(T)}\mathrm{d}T_3 = \alpha_3^{(C)}\mathrm{d}C_3 \tag{13.2.1}$$

which is equivalent to the strain along the x-axis, whereby $\alpha_3^{(T)}$ and $\alpha_3^{(C)}$ are the CTE and CME, respectively, of rod 3, while $\mathrm{d}T_3$ and $\mathrm{d}C_3$ are the changes in temperature and moisture concentration, respectively, in rod 3. It must be noted herein that the change of temperature in rod 3 is the same as the change in environmental temperature upon thermal equilibrium $\mathrm{d}T_3 = \mathrm{d}T$ but the change in moisture concentration in rod 3 is not the same as that in the environment $\mathrm{d}C_3 \neq \mathrm{d}C$. This is because the moisture concentration in rod 3 is dependent on its coefficient of absorption. Since rods 1 and 2 have been defined as being rigid, their lengths remain constant—i.e.,

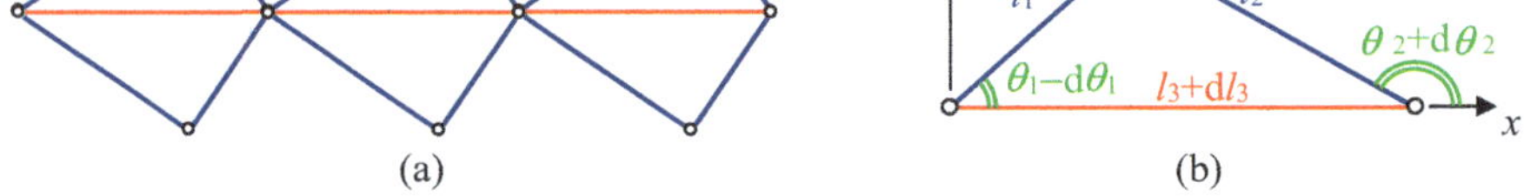

Fig. 13.1 **a** A kite-shaped microstructure consisting of rods aligned in four different directions with the reinforcement rod placed at the axis of symmetry, and **b** a representative half-kite unit before expansion (top) and after expansion (bottom) of rod 3 due to temperature or moisture concentration increase in its surrounding environment

zero strain—therefore there is no requirement for dT_i and dC_i $(i = 1, 2)$. The strain along the x-axis can alternatively be expressed in terms of rod 1 and rod 2. Perusal to Fig. 13.1b gives the original dimension along the x-axis

$$x_0 = l_1 \cos\theta_1 + l_2 \cos(\pi - \theta_2) = l_1 \cos\theta_1 - l_2 \cos\theta_2 \tag{13.2.2}$$

while its change in dimension along the same direction

$$dx = l_1 \cos(\theta_1 - d\theta_1) - l_1 \cos\theta_1 + l_2 \cos[\pi - (\theta_2 + d\theta_2)] - l_2 \cos(\pi - \theta_2) \tag{13.2.3}$$

can be simplified for infinitesimal deformation ($\sin d\theta_i \approx d\theta_i$ and $\cos d\theta_i \approx 1$ for $i = 1, 2$) to give

$$dx \approx l_1 d\theta_1 \sin\theta_1 + l_2 d\theta_2 \sin\theta_2 \tag{13.2.4}$$

Therefore, the strain along the x-axis is

$$\varepsilon_x = \frac{dx}{x_0} = \frac{l_1 d\theta_1 \sin\theta_1 + l_2 d\theta_2 \sin\theta_2}{l_1 \cos\theta_1 - l_2 \cos\theta_2} = \varepsilon_3 \tag{13.2.5}$$

In the same manner, reference to Fig. 13.1b gives the original dimension along the y-axis

$$y_0 = l_1 \sin\theta_1 = l_2 \sin\theta_2 \tag{13.2.6}$$

and its change in dimension along the y-direction

$$dy = l_1 \sin(\theta_1 - d\theta_1) - l_1 \sin\theta_1 = l_2 \sin[\pi - (\theta_2 + d\theta_2)] - l_2 \sin(\pi - \theta_2) \tag{13.2.7}$$

can be reduced for infinitesimal deformation to yield

$$dy \approx -l_1 d\theta_1 \cos\theta_1 = l_2 d\theta_2 \cos\theta_2 \tag{13.2.8}$$

This gives the stain in y-axis

$$\varepsilon_y = \frac{dy}{y_0} = -\frac{d\theta_1}{\tan\theta_1} = \frac{d\theta_2}{\tan\theta_2} \tag{13.2.9}$$

The relationship between the lengths of rod 1 and rod 2 can be established from Eq. (13.2.6) to give

$$l_2 = l_1 \frac{\sin\theta_1}{\sin\theta_2} \tag{13.2.10}$$

while their rotation relationship can be obtained from Eq. (13.2.9) as

$$\mathrm{d}\theta_2 = -\mathrm{d}\theta_1 \frac{\tan\theta_2}{\tan\theta_1} \tag{13.2.11}$$

Substituting Eqs. (13.2.10) and (13.2.11) into Eq. (13.2.5) gives rise to

$$\mathrm{d}\theta_1 = \frac{\cos\theta_1 - \frac{\sin\theta_1}{\tan\theta_2}}{\sin\theta_1 - \cos\theta_1\tan\theta_2}\varepsilon_3 \tag{13.2.12}$$

which permits the strain in the y-direction to be expressed as

$$\frac{\varepsilon_y}{\varepsilon_3} = -\frac{\cos\theta_1 - \frac{\sin\theta_1}{\tan\theta_2}}{\sin\theta_1(\tan\theta_1 - \tan\theta_2)} \tag{13.2.13}$$

Defining the area strain as $\varepsilon_A = \mathrm{d}A/A_0$ where $A_0 = x_0 y_0$ and

$$\mathrm{d}A = A - A_0 = (x_0 + \mathrm{d}x)(y_0 + \mathrm{d}y) - x_0 y_0 = y_0\mathrm{d}x + x_0\mathrm{d}y + \mathrm{d}x\mathrm{d}y \tag{13.2.14}$$

we have $\varepsilon_A = \varepsilon_x + \varepsilon_y + \varepsilon_x\varepsilon_y$ or

$$\varepsilon_A = \varepsilon_x + \varepsilon_y \tag{13.2.15}$$

for infinitesimal deformation. On the basis of $\varepsilon_x = \varepsilon_3$ and Eq. (13.2.13), we have (Lim 2019a)

$$\frac{\varepsilon_A}{\varepsilon_3} = 1 - \frac{\cos\theta_1 - \frac{\sin\theta_1}{\tan\theta_2}}{\sin\theta_1(\tan\theta_1 - \tan\theta_2)} \tag{13.2.16}$$

Since $T_3 = \mathrm{d}T$, the vertical strain ratio $\varepsilon_y/\varepsilon_3$ can be written as $\alpha_y^{(T)}/\alpha_3^{(T)}$ for environmental temperature change, while the same ratio is $\alpha_y^{(C)}\mathrm{d}C/\alpha_3^{(C)}\mathrm{d}C_3$ in response to environmental moisture concentration change because $\mathrm{d}C_3 \neq \mathrm{d}C$, whereby $\alpha_y^{(T)}$ and $\alpha_y^{(C)}$ indicate the effective CTE and CME, respectively, along the y-axis. By similar reasoning, the area strain ratio $\varepsilon_A/\varepsilon_3$ can be expressed as $\alpha_A^{(T)}/\alpha_3^{(T)}$ and $\alpha_A^{(C)}\mathrm{d}C/\alpha_3^{(C)}\mathrm{d}C_3$ where $\alpha_A^{(T)}$ and $\alpha_A^{(C)}$ are the effective areal CTE and CME, respectively, on the x-y plane.

We shall now consider two special cases: (I) the kite is in the form of a "right kite"; i.e., rod 1 and rod 2 are perpendicular to each other, and (II) the kite is in the form of a rhombus; i.e., rod 1 and rod 2 are of equal length. For special case (I), we have $\theta_2 = \theta_1 + 90°$ such that

$$\tan\theta_2 = -\frac{1}{\tan\theta_1} \tag{13.2.17}$$

Substituting Eq. (13.2.17) into Eqs. (13.2.13) and (13.2.16) gives $\varepsilon_y = -\varepsilon_3$ and $\varepsilon_A = 0$, respectively. The latter demarcates the areal NTE or NME region from the region of positive CTE or CME. For special case (II), we have $l_1 = l_2$ or $\theta_1 = \pi - \theta_2$. As a result, Eqs. (13.2.13) and (13.2.16) simplify to

$$\frac{\varepsilon_y}{\varepsilon_3} = -\frac{1}{\tan^2\theta_1} \tag{13.2.18}$$

and

$$\frac{\varepsilon_A}{\varepsilon_3} = 1 - \frac{1}{\tan^2\theta_1} \tag{13.2.19}$$

respectively.

All the rods considered herein are assumed conventional; i.e., they are either rigid or possess positive CTE and CME. That being the case, the effective CTE and CME along the x-axis are always positive. On the other hand, the effective CTE and CME along the y-axis for the special case of $l_1 = l_2$ are always negative for $0° < \theta_1 < 90°$, as implied from Eq. (13.2.18), while the effective areal CTE and CME on the x-y plane can be either positive or negative; reference to Eq. (13.2.19) for $l_1 = l_2$ shows that areal NTE and NME are attained when $\theta_1 < 45°$. Discussions on the effective CTE and CME for other values of rod inclination $\theta_1 \neq 45°$ at $l_1 = l_2$ as well as when $l_1 \neq l_2$ are furnished in the following results. We consider ε_y instead of $\alpha_y^{(T)}$ and $\alpha_y^{(C)}$, as well as ε_A instead of $\alpha_A^{(T)}$ and $\alpha_A^{(C)}$ to remove repetition. Again, to reduce the number of plotted graphs, we consider strain ratios $\varepsilon_y/\varepsilon_3$ and $\varepsilon_A/\varepsilon_3$ instead of ε_y, ε_A and ε_3. As the relationship between areal strain and strain along the y-axis is $\varepsilon_A/\varepsilon_3 = 1 + \varepsilon_y/\varepsilon_3$, it suffices to plot either graphs of $\varepsilon_A/\varepsilon_3$ or $\varepsilon_y/\varepsilon_3$ as one can be easily inferred from the other.

Having shown that $\alpha_A^{(T)} = 0$ and $\alpha_A^{(C)} = 0$ when $\theta_1 = 45°$ for the $l_1 = l_2$ special case in the previous section, the variation of initial inclination angle θ_1 under the same special condition is plotted in Fig. 13.2 for $0° < \theta_1 < 90°$. Reference to Fig. 13.2 shows that $\varepsilon_A/\varepsilon_3$ increases with θ_1 in a diminishing manner. This observation is not surprising since $\varepsilon_A/\varepsilon_3 = -\infty$ as $\theta_1 \to 0°$ and $\varepsilon_A/\varepsilon_3 = 1$ as $\theta_1 \to 90°$ with reference to Eq. (13.2.19). For the generic case where $l_1 \neq l_2$, perusal to Eq. (13.2.16) reveals that zero areal expansion is observed when $\theta_2 = \theta_1 + 90°$. In addition,

$$\begin{gathered} \lim_{\theta_1 \to 0°} \tfrac{\varepsilon_A}{\varepsilon_3} = \lim_{\theta_2 \to 180°} \tfrac{\varepsilon_A}{\varepsilon_3} = -\infty \\ \lim_{\theta_1 \to 90°} \tfrac{\varepsilon_A}{\varepsilon_3} = \lim_{\theta_2 \to 90°} \tfrac{\varepsilon_A}{\varepsilon_3} = 1 \end{gathered} \tag{13.2.20}$$

As such, positive areal CTE and CME are observed when $\theta_2 - \theta_1 < 90°$, but NTE and NME characteristics are manifested when $\theta_2 - \theta_1 > 90°$. A family of of $\varepsilon_y/\varepsilon_3$ curves, as a function of $\theta_1 \in (0°, 90°)$ and $\theta_2 \in (90°, 180°)$, are furnished in the contour plot of Fig. 13.3. Note that the diagonal defined by $\theta_1 + \theta_2 = 180°$ in

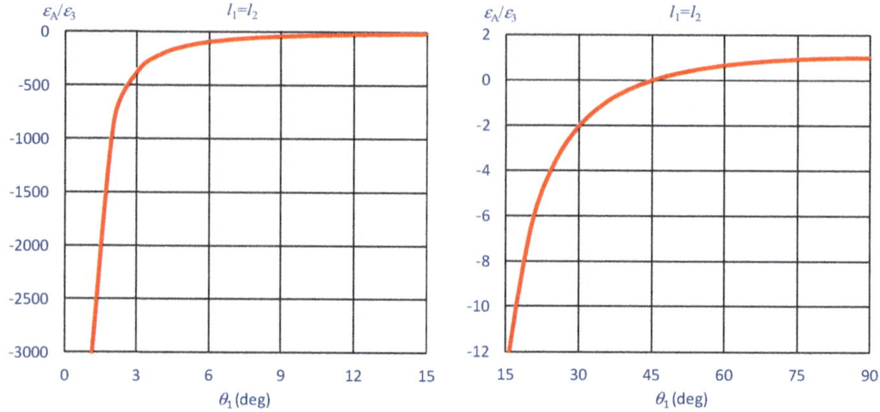

Fig. 13.2 Plot of $\varepsilon_A/\varepsilon_3$ versus θ_1 with $l_1 = l_2$

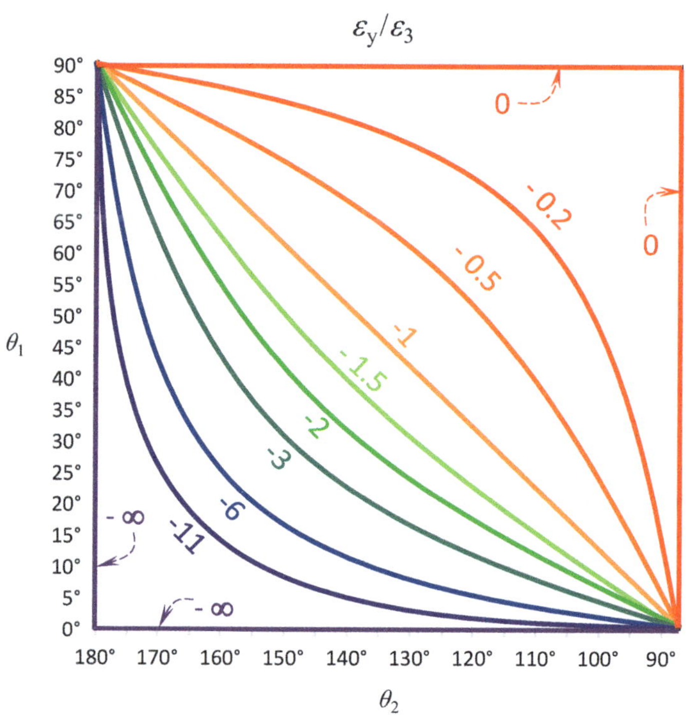

Fig. 13.3 Contour plot of $\varepsilon_y/\varepsilon_3$ versus θ_1 and θ_2

the dimensionless vertical strain of Fig. 13.3 corresponds to the dimensionless areal strain plotted in Fig. 13.2.

13.3 Example: A Volumetric NME System Based on 3D Trusses

The microstructure being considered as shown in Fig. 13.4 (left) consists of two types of rods of equal cross-sectional dimension but possessing different lengths and materials, with its representative volume element (RVE) indicated in Fig. 13.4 (right). The apex rods are of length a and coefficient of moisture expansion (CME) $\alpha_a^{(C)}$, while the base rods are of length b and coefficient of moisture expansion (CME) $\alpha_b^{(T)}$. Such a truss structure is achievable by the use of pin-jointed rods in which the apex and based rods are made of different materials; at microstructural level, this is also achievable with the availability of multi-materials 3D printing technology (Lim 2018).

The change in apex and base rods' initial lengths (a_0, b_0) to their final lengths $\left(a_f, b_f\right)$ as a result of moisture expansion is accompanied by a change of the RVE size from its initial volume V_0 to its final volume V_f, i.e.,

$$V_f = V_0 + \mathrm{d}V \tag{13.3.1}$$

in which the volumetric strain is defined by

$$\varepsilon_V = \frac{\mathrm{d}V}{V_0} \tag{13.3.2}$$

Alternatively, the volumetric strain is related to the volumetric CME, $\alpha_V^{(C)}$ as

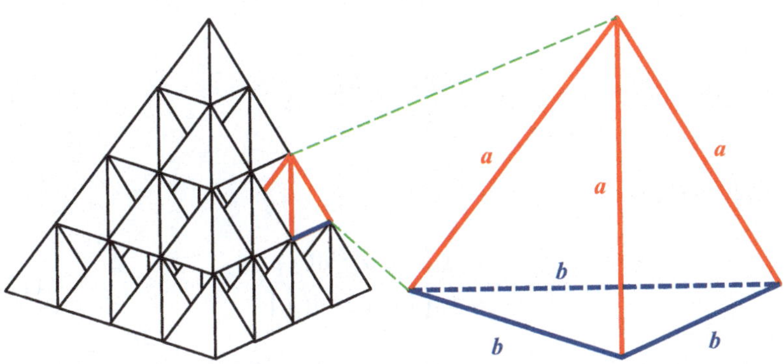

Fig. 13.4 A truss microstructure (left) and its RVE (right), where the apex and base rods are of lengths a and b, respectively. Hidden lines on left figure have been removed for clarity

$$\varepsilon_V = \alpha_V^{(C)} \mathrm{d}C_{\mathrm{eff}} \tag{13.3.3}$$

where the effective change in moisture concentration in the RVE, $\mathrm{d}C_{\mathrm{eff}}$ is related to the changes in the moisture concentration in the apex rod, $\mathrm{d}C_a$ and base rod, $\mathrm{d}C_b$ as

$$\mathrm{d}C_{\mathrm{eff}} = \frac{a_0 \mathrm{d}C_a + b_0 \mathrm{d}C_b}{a_0 + b_0} \tag{13.3.4}$$

if the cross-sectional areas of the rods are equal. It must be borne in mind that as the moisture concentration in the atmosphere changes, the changes to the moisture concentration in the apex and base rods are different due to different moisture absorption coefficients. Suppose both volumetric strains are equated, we have

$$\mathrm{d}V = \alpha_V^{(C)} V_0 \mathrm{d}C_{\mathrm{eff}} \tag{13.3.5}$$

which gives the final RVE volume as

$$V_f = V_0 \left(1 + \alpha_V^{(C)} \mathrm{d}C_{\mathrm{eff}}\right) \tag{13.3.6}$$

Therefore, the volumetric strain due to moisture expansion

$$\varepsilon_V = \frac{V_f}{V_0} - 1 \tag{13.3.7}$$

can be written in terms of rod lengths before and after the moisture expansion if the initial and final volumes of the RVE are expressed as

$$V_0 = \frac{b_0^3}{12} \sqrt{3 \frac{a_0^2}{b_0^2} - 1} \tag{13.3.8}$$

and

$$V_f = \frac{b_f^3}{12} \sqrt{3 \frac{a_f^2}{b_f^2} - 1} \tag{13.3.9}$$

respectively, where

$$\begin{aligned} a_f &= a_0 \left(1 + \alpha_a^{(C)} \mathrm{d}C_a\right) \\ b_f &= b_0 \left(1 + \alpha_b^{(C)} \mathrm{d}C_b\right) \end{aligned} \tag{13.3.10}$$

Substitution of Eqs. (13.3.8)–(13.3.10) into Eq. (13.3.7) leads to the RVE volumetric strain in terms of the rod geometry, properties, and moisture concentration change

$$\varepsilon_V = -1 + \left(1 + \alpha_b^{(C)} \mathrm{d}C_b\right)^2 \sqrt{\frac{3\frac{a_0^2}{b_0^2}\left(1 + \alpha_a^{(C)} \mathrm{d}C_a\right)^2 - \left(1 + \alpha_b^{(C)} \mathrm{d}C_b\right)^2}{3\frac{a_0^2}{b_0^2} - 1}} \tag{13.3.11}$$

If we replace the moisture expansion coefficients with thermal expansion coefficients for both types of rods $\left(\alpha_a^{(C)}, \alpha_b^{(C)}\right) \rightarrow \left(\alpha_a^{(T)}, \alpha_b^{(T)}\right)$ and assume that the change in moisture concentration for both rods types are equal so as to permit $\mathrm{d}C_a = \mathrm{d}C_b = \mathrm{d}T$, then Eq. (13.3.11) converts to its thermal strain equivalence described in Eq. (11.12.12). For the truss microstructure to exhibit overall and incremental NME, the conditions $\varepsilon_V < 0$ and $\mathrm{d}\varepsilon_V < 0$ must be fulfilled. It follows that the overall and incremental CVHE of the RVE can be obtained as

$$\alpha_V^{(C)} = \frac{\varepsilon_V}{\mathrm{d}C_{\mathrm{eff}}} \tag{13.3.12}$$

and

$$\alpha_V^{(C)} = \frac{\partial \varepsilon_V}{\partial (\mathrm{d}C_{\mathrm{eff}})} \tag{13.3.13}$$

respectively. By way of example, we herein consider the special case of rigid apex rods, i.e., apex rod being either non-expansible $\alpha_a^{(C)} = 0$ or non-absorptive $\mathrm{d}C_a = 0$ such that Eqs. (13.3.4) and (13.3.11) simplify to

$$\mathrm{d}C_{\mathrm{eff}} = \frac{b_0}{a_0 + b_0} \mathrm{d}C_b \tag{13.3.14}$$

and

$$\varepsilon_V = -1 + \left(1 + \alpha_b^{(C)} \mathrm{d}C_b\right)^2 \sqrt{\frac{3\frac{a_0^2}{b_0^2} - \left(1 + \alpha_b^{(C)} \mathrm{d}C_b\right)^2}{3\frac{a_0^2}{b_0^2} - 1}} \tag{13.3.15}$$

respectively. In addition, we adopt a dimensionless form of volumetric CME by normalizing it against the CME of the base rods, i.e., $\alpha_V^{(C)}/\alpha_b^{(C)}$. The dimensionless overall and incremental volumetric CME can be obtained as (Lim 2018)

$$\frac{\alpha_V^{(C)}}{\alpha_b^{(C)}} = \frac{1}{\alpha_b^{(C)}} \cdot \frac{\varepsilon_V}{\mathrm{d}C_{\mathrm{eff}}} = \frac{1 + \frac{a_0}{b_0}}{\alpha_b^{(C)} \mathrm{d}C_b} \left\{ 1 + \left(1 + \alpha_b^{(C)} \mathrm{d}C_b\right)^2 \sqrt{\frac{3\frac{a_0^2}{b_0^2} - \left(1 + \alpha_b^{(C)} \mathrm{d}C_b\right)^2}{3\frac{a_0^2}{b_0^2} - 1}} \right\} \tag{13.3.16}$$

and

$$\frac{\alpha_V^{(C)}}{\alpha_b^{(C)}} = \frac{1}{\alpha_b^{(C)}} \cdot \frac{\partial \varepsilon_V}{\partial (\mathrm{d}C_{\mathrm{eff}})} = \frac{1}{\alpha_b^{(C)}} \cdot \frac{\partial \varepsilon_V}{\partial (\mathrm{d}C_b)} \cdot \frac{\partial (\mathrm{d}C_b)}{\partial (\mathrm{d}C_{\mathrm{eff}})}$$

$$= 2\left(1 + \alpha_b^{(C)}\mathrm{d}C_b\right)\left(1 + \frac{a_0}{b_0}\right)\sqrt{\frac{3\frac{a_0^2}{b_0^2} - \left(1 + \alpha_b^{(C)}\mathrm{d}C_b\right)^2}{3\frac{a_0^2}{b_0^2} - 1}}$$

$$- \left(1 + \alpha_b^{(C)}\mathrm{d}C_b\right)^3 \frac{1 + \frac{a_0}{b_0}}{3\frac{a_0^2}{b_0^2} - 1}\sqrt{\frac{3\frac{a_0^2}{b_0^2} - 1}{3\frac{a_0^2}{b_0^2} - \left(1 + \alpha_b^{(C)}\mathrm{d}C_b\right)^2}} \quad (13.3.17)$$

respectively. For infinitesimal change in moisture concentration, both Eq. (13.3.16) and (13.3.17) reduce to

$$\lim_{\mathrm{d}C_b \to 0} \frac{\alpha_V^{(C)}}{\alpha_b^{(C)}} = 3\left(1 + \frac{a_0}{b_0}\right)\frac{2\frac{a_0^2}{b_0^2} - 1}{3\frac{a_0^2}{b_0^2} - 1} \quad (13.3.18)$$

i.e., the dimensionless infinitesimal CVHE is purely in terms of rod length ratio. Note that L'Hopital's rule needs to be applied on Eq. (13.3.16) to obtain Eq. (13.3.18). When the change in the base rod's moisture concentration is infinitesimal, Eq. (13.3.18) reveals that NME is attained under the following geometrical condition

$$\lim_{\mathrm{d}C_{\mathrm{eff}} \to 0} \frac{\alpha_V^{(C)}}{\alpha_b^{(C)}} < 0 \Leftarrow \frac{a_0}{b_0} < \frac{1}{\sqrt{2}} \quad (13.3.19)$$

Figure 13.5 (left) displays plots of volumetric strain versus base rod moisture strain for various initial apex-to-base rod length ratio, from which the slope taken at its origin gives the infinitesimal dimensionless volumetric CME shown in Fig. 13.5 (right) using Eqs. (13.3.15) and (13.3.18), respectively. A family of dimensionless overall and incremental dimensionless CME curves in Fig. 13.6 were plotted against the moisture strain of the base rod for various initial apex-to-base rod length ratios using Eqs. (13.3.16) and (13.3.17), respectively, indicating that the incremental volumetric CME gives a more negative volumetric moisture expansion than the overall volumetric CME. The condition of NME is fully attainable when $3^{-1/2} < a_0/b_0 < 2^{-1/2}$ for infinitesimal strain; the RVE volume being undefined for $a_0/b_0 \le 3^{-1/2}$. For $a_0/b_0 > 2^{-1/2}$, the dimensionless NME is attainable for higher values of $\alpha_b^{(C)}\mathrm{d}C_b$.

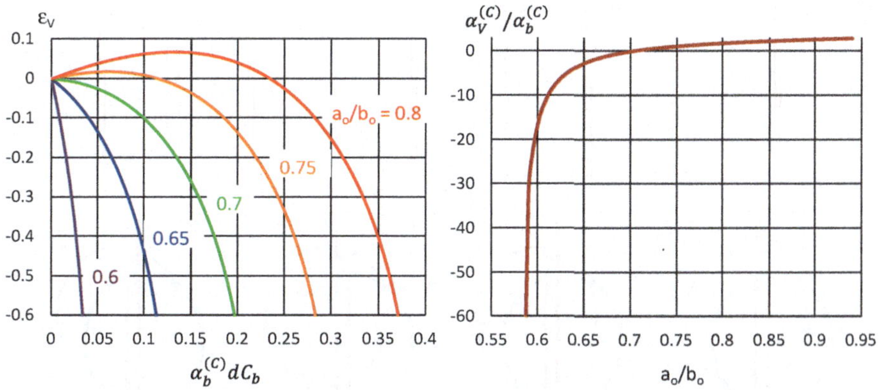

Fig. 13.5 Plots of volumetric strain versus base rod moisture strain for various initial apex-to-base rod length ratio (left) and infinitesimal dimensionless volumetric CME versus the initial apex-to-base rod length ratio

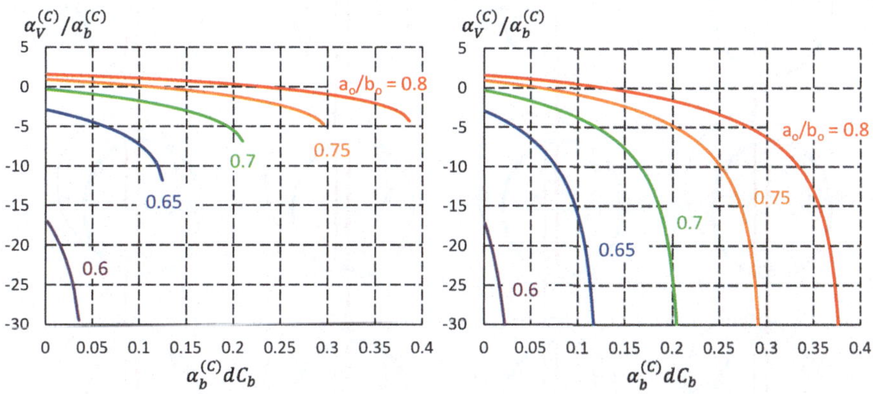

Fig. 13.6 A family of dimensionless overall volumetric CME (left) and incremental volumetric CME (right) curves plotted against the moisture strain of the base rod for various initial apex-to-base rod length ratios

13.4 Example: A 3D NEE Analysis for Interconnected Array of Rings and Sliding Rods System

The present example revisits the ring-rod assembly that was discussed in Sect. 2.5 in terms of its auxeticity and in Sect. 11.11 in terms of NTE (Lim 2017b). For this section, the example explores NEE, i.e., the incorporation of NC and NME in addition to NTE. Furthermore, this example removes two simplifying assumptions that were adopted previously; i.e., the rods were assumed rigid, and the ring's thickness was neglected. These simplifying assumptions are removed in this section (Lim 2019b). With reference to Fig. 13.7a for the ring-rod structure whereby the expansivity of

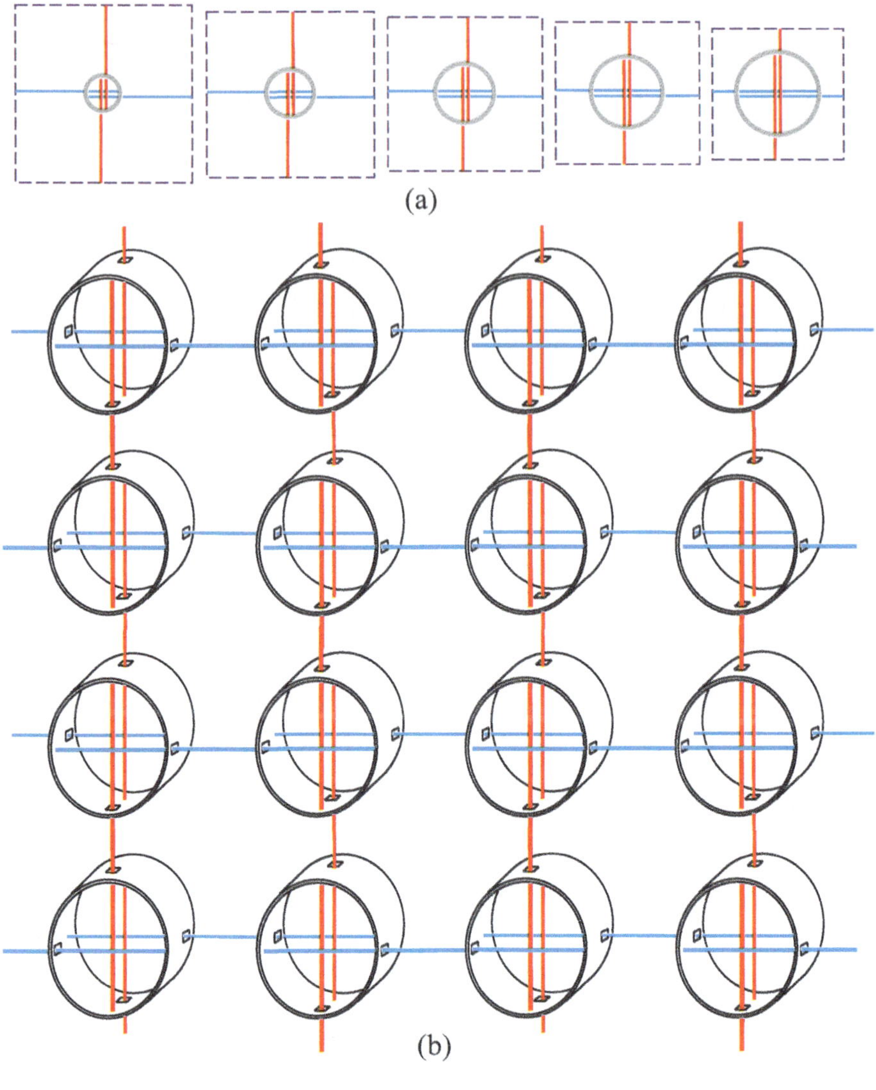

Fig. 13.7 **a** A unit of ring-rod structure showing unit cell contraction due to draw-in of sliding rods arising from ring expansion (from left to right) or unit cell expansion due to push out of rods due ring contraction (from right to left), and **b** arrangement to prevent turning of rings about any axis

the ring is much greater than that of the rods, the expansion of the ring draws the rods into the ring. The movement of the rod's outer end toward the ring results in contraction of the unit cell boundary, and hence, the ring expansion leads to overall unit cell contraction. To prevent rotation of the rings about the rod axes, the rods are arranged in a symmetrical manner shown in Fig. 13.7b.

Figure 13.8 shows a ring of inner radius *r* and outer radius *R*, with rods of half-

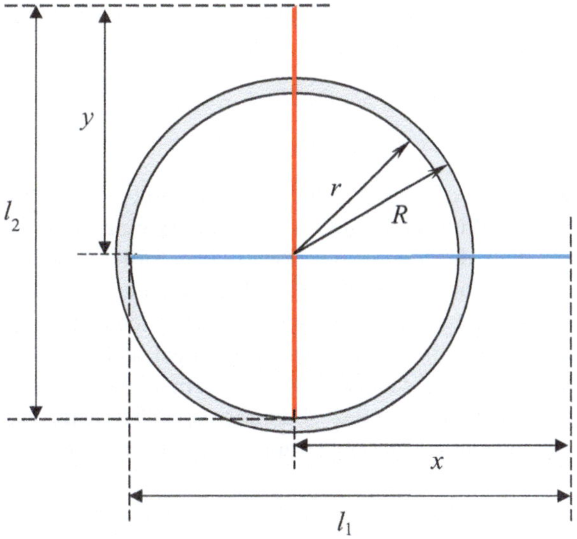

Fig. 13.8 Schematic view of a ring and two rods for analysis. The other two rods are not displayed for clarity

lengths l_1 and l_2—aligned along the x- and y-axes, respectively—and attached to the ring's inner surface but protrudes from the ring through apertures diametrically from the fixed end. As such the horizontal length $x = l_1 - r$ and the vertical length $y = l_2 - r$, as measured from the origin, form a quarter of a unit cell that is sufficient for analysis, with the ring's inner circumference being $2\pi r$.

Arising from a change in the environment–such as increase in temperature or moisture, or a decrease in pressure–the horizontal and vertical rods and the ring circumference expand as $\mathrm{d}x = \mathrm{d}l_1 - \mathrm{d}r$, $\mathrm{d}y = \mathrm{d}l_2 - \mathrm{d}r$ and $2\pi \mathrm{d}r$, respectively, to give the following strains

$$\varepsilon_x = \frac{l_1\varepsilon_1 - r\varepsilon_O}{l_1 - r} \tag{13.4.1}$$

$$\varepsilon_y = \frac{l_2\varepsilon_2 - r\varepsilon_O}{l_2 - r} \tag{13.4.2}$$

where $\varepsilon_1 = \mathrm{d}l_1/l_1$ and $\varepsilon_2 = \mathrm{d}l_2/l_2$ are the linear strains of the horizontal and vertical rods, respectively, while $\varepsilon_O = \mathrm{d}r/r$ is the circumferential strain of the ring. The strains ε_x and ε_y are in terms of r, but not R, because the rods are attached to the ring's inner surface, and it is further assumed that the ring remains circular throughout expansion and contraction due to the absence of constraint. From the original area $A = xy$, we have its increment $\mathrm{d}A = y\mathrm{d}x + x\mathrm{d}y + \mathrm{d}x\mathrm{d}y$ such that the areal strain is obtained as

$$\varepsilon_A = \frac{\mathrm{d}A}{A} = \varepsilon_x + \varepsilon_y + \varepsilon_x\varepsilon_y \approx \varepsilon_x + \varepsilon_y \tag{13.4.3}$$

upon neglecting the second-order term. By similar approach, the volumetric strain can be approximated as

$$\varepsilon_V = \frac{\mathrm{d}V}{V} \approx \varepsilon_x + \varepsilon_y + \varepsilon_z \tag{13.4.4}$$

upon neglecting the second- and third-order terms. Due to the adoption of engineering strain definition as well as the simplifying assumptions set out in Eqs. (13.4.3) and (13.4.4), the models developed henceforth are valid for small strains and are only approximate for large strains. For the latter case, the true strain—also known as the logarithmic strain—definition must be used instead, as well as the retention of higher order terms in the corresponding areal and volumetric strains. In this section, we assign $\varepsilon^{(T)}$, $\varepsilon^{(C)}$ and $\varepsilon^{(P)}$ as the thermal, moisture and pressure strains, respectively, while $\alpha^{(T)}$, $\alpha^{(C)}$ and $\alpha^{(P)}$ denote the CTE, CME, and compressibility, respectively. Let the overall thermal strains along the rod axes be written as

$$\begin{aligned} \varepsilon_x^{(T)} &= \alpha_x^{(T)}\mathrm{d}T \\ \varepsilon_y^{(T)} &= \alpha_y^{(T)}\mathrm{d}T \end{aligned} \tag{13.4.5}$$

where $\alpha_x^{(T)}$ and $\alpha_y^{(T)}$ are the effective CTEs along the x- and y-axes, respectively, with $\mathrm{d}T$ being the change in temperature. Substituting the thermal strains of the ring and rods

$$\begin{aligned} \varepsilon_O^{(T)} &= \alpha_O^{(T)}\mathrm{d}T \\ \varepsilon_1^{(T)} &= \alpha_1^{(T)}\mathrm{d}T \\ \varepsilon_2^{(T)} &= \alpha_2^{(T)}\mathrm{d}T \end{aligned} \tag{13.4.6}$$

into Eqs. (13.4.1) and (13.4.2) allows the effective CTEs to be obtained as

$$\alpha_x^{(T)} = \frac{l_1\alpha_1^{(T)} - r\alpha_O^{(T)}}{l_1 - r} \tag{13.4.7}$$

and

$$\alpha_y^{(T)} = \frac{l_2\alpha_2^{(T)} - r\alpha_O^{(T)}}{l_2 - r} \tag{13.4.8}$$

while the out-of-plane thermal strain

$$\varepsilon_z^{(T)} = \alpha_O^{(T)}\mathrm{d}T \tag{13.4.9}$$

is determined by CTE of the ring material. For the special case where the rod is rigid $\left(\alpha_1^{(T)} = \alpha_2^{(T)} = 0\right)$, Eqs. (13.4.7) and (13.4.8) simplify to

$$\alpha_x^{(T)} = -\alpha_O^{(T)} \frac{r}{l_1 - r} \tag{13.4.10}$$

and

$$\alpha_y^{(T)} = -\alpha_O^{(T)} \frac{r}{l_2 - r} \tag{13.4.11}$$

respectively. Since it is a geometrical requirement that $l_i > R + r > 2r$ for both rods ($i = 1, 2$), i.e., $l_i > r$, the effective CTEs in Eqs. (13.4.10) and (13.4.11) are negative. Introducing the in-plane area thermal strain

$$\varepsilon_A^{(T)} = \alpha_A^{(T)} \mathrm{d}T \tag{13.4.12}$$

where $\alpha_A^{(T)}$ is the area CTE, and substituting it along with Eqs. (13.4.5), (13.4.7), and (13.4.8) into Eq. (13.4.3) gives

$$\alpha_A^{(T)} = \alpha_x^{(T)} + \alpha_y^{(T)} = \frac{l_1\alpha_1^{(T)} - r\alpha_O^{(T)}}{l_1 - r} + \frac{l_2\alpha_2^{(T)} - r\alpha_O^{(T)}}{l_2 - r} \tag{13.4.13}$$

Substituting $\alpha_1^{(T)} = \alpha_2^{(T)} = 0$ into Eq. (13.4.13) for the special case of rigid rods reduces the expression to

$$\alpha_A^{(T)} = -r\alpha_O^{(T)} \left(\frac{1}{l_1 - r} + \frac{1}{l_2 - r} \right) \tag{13.4.14}$$

which indicates area NTE. Introducing the volumetric thermal strain

$$\varepsilon_V^{(T)} = \alpha_V^{(T)} \mathrm{d}T \tag{13.4.15}$$

where $\alpha_V^{(T)}$ is the volumetric CTE, and substituting it together with Eqs. (13.4.5) and (13.4.7)–(13.4.9) into Eq. (13.4.4) leads to

$$\alpha_V^{(T)} = \frac{l_1\alpha_1^{(T)} - r\alpha_O^{(T)}}{l_1 - r} + \frac{l_2\alpha_2^{(T)} - r\alpha_O^{(T)}}{l_2 - r} + \alpha_O^{(T)} \tag{13.4.16}$$

For the special case of rigid rods, Eq. (13.4.16) simplifies to

$$\alpha_V^{(T)} = \alpha_O^{(T)} - r\alpha_O^{(T)} \left(\frac{1}{l_1 - r} + \frac{1}{l_2 - r} \right) \tag{13.4.17}$$

Unlike Eqs. (13.4.10) or (13.4.11) and (13.4.14), which indicate NTE for rigid rods, Eq. (13.4.17) does not necessarily indicate NTE since its first term plays an important role in determining the sign of $\alpha_V^{(T)}$. Nevertheless, a volumetric NTE requirement in the case of rigid rods can be determined by imposing $\alpha_V^{(T)} < 0$

on Eq. (13.4.17) to yield the following condition for meeting the volumetric NTE requirement

$$\frac{1}{r} < \frac{1}{l_1 - r} + \frac{1}{l_2 - r} \tag{13.4.18}$$

In the case of CTE analysis, it is well understood that at steady state the various solid phases are in thermal equilibrium not only with each other but also with the environment. With a change in environmental temperature by $\mathrm{d}T$, the temperature of the solid phases also changes during transient state such that at steady state the solid phases possess the same temperature as that of the environment. This condition does not apply in the case of moisture concentration in solids, which is defined as

$$C = \frac{m}{M} \times 100 \tag{13.4.19}$$

where m is the mass of water in the considered material and M is the mass of the dry material. Due to different absorptivity level, various materials absorb differing amount of moisture from the environment. As such, a change in environmental moisture concentration $\mathrm{d}C$ leads to changes in moisture concentration in the ring material $\mathrm{d}C_\mathrm{O}$, and in the rods ($\mathrm{d}C_1$, $\mathrm{d}C_2$). Therefore, $\mathrm{d}C \neq \mathrm{d}C_\mathrm{O} \neq \mathrm{d}C_1 \neq \mathrm{d}C_2$ in general. In the first instance, before a change in the environmental moisture concentration takes place, the moisture concentration in the environment and in the constituent materials need not necessarily be equal $C \neq C_\mathrm{O} \neq C_1 \neq C_2$. For this reason, the models for effective CTEs developed in the previous section cannot be directly applied by replacing $\alpha^{(T)}$ and $\mathrm{d}T$ with $\alpha^{(C)}$ and $\mathrm{d}C$ respectively, where $\alpha^{(C)}$ is the coefficient of moisture expansion (CME). Note that the moisture concentration in the environment is typically expressed in terms of absolute, relative, and specific humidities; for the sake of consistency, this book adopts the moisture concentration in the environment in the form described by Eq. (13.4.19), where in this case M is the mass of dry air that takes up the same volume as that for the mass of environmental moisture m. In the subsequent analysis, we define the changes to the moisture concentration in the environment, the ring and the rods as

$$\begin{array}{l} \mathrm{d}C = 100\frac{\mathrm{d}m}{M} \\ \mathrm{d}C_\mathrm{O} = 100\frac{\mathrm{d}m_\mathrm{O}}{M_\mathrm{O}} \\ \mathrm{d}C_1 = 100\frac{\mathrm{d}m_1}{M_1} \\ \mathrm{d}C_2 = 100\frac{\mathrm{d}m_2}{M_2} \end{array} \tag{13.4.20}$$

where $\mathrm{d}m$, $\mathrm{d}m_\mathrm{O}$, $\mathrm{d}m_1$ and $\mathrm{d}m_2$ indicate the changes in moisture mass in the environmental cuboid enclosing the ring-rod structure, the ring, the horizontal rod, and the vertical rod, respectively, while M, M_O, M_1, and M_2 refer to the mass of dry air in the cuboid enclosure, as well as the mass of dry ring, horizontal rod, and vertical rod, respectively. The size of the cuboid enclosure can be taken arbitrarily as long as the environmental moisture is uniformly distributed. Introducing the overall moisture strains along the rod axes, arising from a change in environmental moisture concentration, as

$$\begin{aligned} \varepsilon_x^{(C)} &= \alpha_x^{(C)} \mathrm{d}C \\ \varepsilon_y^{(C)} &= \alpha_y^{(C)} \mathrm{d}C \end{aligned} \tag{13.4.21}$$

where $\alpha_x^{(C)}$ and $\alpha_y^{(C)}$ are the effective CMEs along the x- and y-axes, respectively, and substituting the hygroscopic strain the ring and the rods

$$\begin{aligned} \varepsilon_O^{(C)} &= \alpha_O^{(C)} \mathrm{d}C_O \\ \varepsilon_1^{(C)} &= \alpha_1^{(C)} \mathrm{d}C_1 \\ \varepsilon_2^{(C)} &= \alpha_2^{(C)} \mathrm{d}C_2 \end{aligned} \tag{13.4.22}$$

into Eqs. (13.4.1) and (13.4.2) allows the effective CMEs along the x- and y-axes to be obtained as

$$\alpha_x^{(C)} = \frac{l_1 \alpha_1^{(C)} \mathrm{d}C_1 - r\alpha_O^{(C)} \mathrm{d}C_O}{(l_1 - r)\mathrm{d}C} \tag{13.4.23}$$

and

$$\alpha_y^{(C)} = \frac{l_2 \alpha_2^{(C)} \mathrm{d}C_2 - r\alpha_O^{(C)} \mathrm{d}C_O}{(l_2 - r)\mathrm{d}C} \tag{13.4.24}$$

while the out-of-plane moisture strain

$$\varepsilon_z^{(C)} = \alpha_O^{(C)} \mathrm{d}C_O \tag{13.4.25}$$

is determined by CME of the ring material. For the special case where the rods are rigid $\alpha_1^{(C)} = \alpha_2^{(C)} = 0$ and/or non-absorptive $\mathrm{d}C_1 = \mathrm{d}C_2 = 0$, Eqs. (13.4.23) and (13.4.24) abridge to

$$\alpha_x^{(C)} = -\alpha_O^{(C)} \frac{r}{l_1 - r} \frac{\mathrm{d}C_O}{\mathrm{d}C} \tag{13.4.26}$$

and

$$\alpha_y^{(C)} = -\alpha_O^{(C)} \frac{r}{l_2 - r} \frac{\mathrm{d}C_O}{\mathrm{d}C} \tag{13.4.27}$$

respectively. Since it is geometrically necessary that $l_i > r$ for $i = 1, 2$ as previously mentioned, the CMEs in Eqs. (13.4.26) and (13.4.27) are negative. Introducing the in-plane area moisture strain

$$\varepsilon_A^{(C)} = \alpha_A^{(C)} \mathrm{d}C \tag{13.4.28}$$

where $\alpha_A^{(C)}$ is the area CME, and substituting it along with Eqs. (13.4.21), (13.4.23) and (13.4.24) into Eq. (13.4.3) gives

$$\alpha_A^{(C)} = \alpha_x^{(C)} + \alpha_y^{(C)} = \frac{l_1\alpha_1^{(C)}\mathrm{d}C_1 - r\alpha_O^{(C)}\mathrm{d}C_O}{(l_1 - r)\mathrm{d}C} + \frac{l_2\alpha_2^{(C)}\mathrm{d}C_2 - r\alpha_O^{(C)}\mathrm{d}C_O}{(l_2 - r)\mathrm{d}C} \tag{13.4.29}$$

When the rods are rigid $\left(\alpha_1^{(C)} = \alpha_2^{(C)} = 0\right)$ and/or non-absorptive ($\mathrm{d}C_1 = \mathrm{d}C_2 = 0$), Eq. (13.4.29) becomes

$$\alpha_A^{(C)} = -r\alpha_O^{(C)} \frac{\mathrm{d}C_O}{\mathrm{d}C}\left(\frac{1}{l_1 - r} + \frac{1}{l_2 - r}\right) \tag{13.4.30}$$

thereby denoting an in-plane area NME. Introducing the volumetric moisture strain

$$\varepsilon_V^{(C)} = \alpha_V^{(C)} \mathrm{d}C \tag{13.4.31}$$

where $\alpha_V^{(C)}$ is the volumetric CME, and substitution it together with Eqs. (13.4.21) and (13.4.23)–(13.4.25) into Eq. (13.4.4) leads to

$$\alpha_V^{(C)} = \frac{l_1\alpha_1^{(C)}\mathrm{d}C_1 - r\alpha_O^{(C)}\mathrm{d}C_O}{(l_1 - r)\mathrm{d}C} + \frac{l_2\alpha_2^{(C)}\mathrm{d}C_2 - r\alpha_O^{(C)}\mathrm{d}C_O}{(l_2 - r)\mathrm{d}C} + \alpha_O^{(C)} \tag{13.4.32}$$

For the special case of zero rod expansion, due to either rigid or non-absorptive rods, Eq. (13.4.32) can be expressed as

$$\alpha_V^{(C)} = \alpha_O^{(C)} - r\alpha_O^{(C)} \frac{\mathrm{d}C_O}{\mathrm{d}C}\left(\frac{1}{l_1 - r} + \frac{1}{l_2 - r}\right) \tag{13.4.33}$$

As with Eq. (13.4.17), Eq. (13.4.33) does not necessarily indicate volumetric NME. However, volumetric NME in the case of non-expansible and/or non-absorptive rods is attained if the following condition

$$\frac{1}{r}\frac{\mathrm{d}C}{\mathrm{d}C_O} < \frac{1}{l_1 - r} + \frac{1}{l_2 - r} \tag{13.4.34}$$

is met by imposing $\alpha_V^{(C)} < 0$ on Eq. (13.4.33). Instead of the negativity of bulk modulus K, we herein consider the negativity of compressibility $\alpha_V^{(P)}$. Compressibility is the reciprocal of bulk modulus and is expressed as

$$\alpha_V^{(P)} = \frac{1}{K} = -\frac{1}{V}\frac{\mathrm{d}V}{\mathrm{d}P} \tag{13.4.35}$$

where dP indicates the change in environmental pressure, and the negative sign caters for a positive value to $\alpha_V^{(P)}$ when an increasing pressure causes decreasing volume. While the bulk modulus represents the resistance to change in volume, the compressibility corresponds to the compliance to change in volume in response to pressure change and would therefore allow a more direct comparison with $\alpha_V^{(T)}$ and $\alpha_V^{(C)}$. Substituting $\mathrm{d}V/V = \varepsilon_V$ into Eq. (13.4.35) allows it to be expressed as

$$\varepsilon_V^{(P)} = -\alpha_V^{(P)}\mathrm{d}P \tag{13.4.36}$$

for convenient comparison with its counterpart on volumetric thermal and moisture responses as described by Eqs. (13.4.15) and (13.4.31), respectively. For an isotropic material under uniform pressure, the strain is non-directional for every solid phase. Since the volumetric strain is the summation of strains in all three orthogonal directions under the given condition, it follows that $3\varepsilon^{(P)} = -\alpha_V^{(P)}\mathrm{d}P$; i.e., the linear compressibility is

$$\varepsilon^{(P)} = -\frac{1}{3}\alpha_V^{(P)}\mathrm{d}P = -\alpha^{(P)}\mathrm{d}P \tag{13.4.37}$$

for each solid phase. Introducing the overall pressure-induced strains along the rod axes in the same manner

$$\begin{aligned} \varepsilon_x^{(P)} &= -\alpha_x^{(P)}\mathrm{d}P \\ \varepsilon_y^{(P)} &= -\alpha_y^{(P)}\mathrm{d}P \end{aligned} \tag{13.4.38}$$

where $\alpha_x^{(P)}$ and $\alpha_y^{(P)}$ are the effective compressibilities of the ring-rod structure along the x- and y-axes, respectively, and substituting the pressure-induced strain on the ring and rods

$$\begin{aligned} \varepsilon_O^{(P)} &= -\alpha_O^{(P)}\mathrm{d}P \\ \varepsilon_1^{(P)} &= -\alpha_1^{(P)}\mathrm{d}P \\ \varepsilon_2^{(P)} &= -\alpha_2^{(P)}\mathrm{d}P \end{aligned} \tag{13.4.39}$$

into Eqs. (13.4.1) and (13.4.2) allows the effective compressibilities along the x- and y-axes to be calculated as

$$\alpha_x^{(P)} = \frac{l_1\alpha_1^{(P)} - r\alpha_O^{(P)}}{l_1 - r} \tag{13.4.40}$$

and

$$\alpha_y^{(P)} = \frac{l_2\alpha_2^{(P)} - r\alpha_O^{(P)}}{l_2 - r} \tag{13.4.41}$$

while the out-of-plane pressure-induced strain

$$\varepsilon_z^{(P)} = \alpha_O^{(P)} \mathrm{d}P \tag{13.4.42}$$

is determined by compressibility of the ring material. Since the linear effective compressibilities obtained in Eqs. (13.4.40) and (13.4.41) are analogous to the linear effective CTEs furnished in Eqs. (13.4.7) and (13.4.8), respectively, it follows that the effective areal and volumetric compressibilities are

$$\alpha_A^{(P)} = \alpha_x^{(P)} + \alpha_y^{(P)} = \frac{l_1\alpha_1^{(P)} - r\alpha_O^{(P)}}{l_1 - r} + \frac{l_2\alpha_2^{(P)} - r\alpha_O^{(P)}}{l_2 - r} \tag{13.4.43}$$

and

$$\alpha_V^{(P)} = \frac{l_1\alpha_1^{(P)} - r\alpha_O^{(P)}}{l_1 - r} + \frac{l_2\alpha_2^{(P)} - r\alpha_O^{(P)}}{l_2 - r} + \alpha_O^{(P)} \tag{13.4.44}$$

respectively.

It has been shown that the linear CTE in any of its principal direction is negative when the rods are rigid. In general, the CTE negativity of this (micro)structure can be obtained by substituting $\alpha_x^{(T)} < 0$ into Eq. (13.4.7) and $\alpha_y^{(T)} < 0$ into Eq. (13.4.8) to give the following NTE conditions

$$l_1\alpha_1^{(T)} < r\alpha_O^{(T)} \tag{13.4.45}$$

and

$$l_2\alpha_2^{(T)} - r\alpha_O^{(T)} \tag{13.4.46}$$

respectively. To give a visual representation, a contour map to exhibit the dimensionless ratio of the effective CTE $\alpha_x^{(T)}/\alpha_O^{(T)}$ as a function dimensionless rod CTE $\alpha_1^{(T)}/\alpha_O^{(T)}$ and the dimensionless rod length l_1/r is furnished in Fig. 13.9. These dimensionless parameters have been normalized against the corresponding properties of the ring. The choice of range for $\alpha_1^{(T)}/\alpha_O^{(T)}$ pertains to the case of rigid rods $\alpha_1^{(T)}/\alpha_O^{(T)} = 0$, which was earlier shown to be NTE, and rods made from the same material as the ring $\alpha_1^{(T)}/\alpha_O^{(T)} = 1$, which would obviously exhibit a conventional behavior. The boundary between NTE and conventional behavior, i.e., $\alpha_x^{(T)}/\alpha_O^{(T)} = 0$ would therefore be formed between $\alpha_1^{(T)}/\alpha_O^{(T)} = 0$ and $\alpha_1^{(T)}/\alpha_O^{(T)} = 1$. Perusal to Fig. 13.9 also indicates that the negativity of CTE is primarily influenced by the $\alpha_1^{(T)}/\alpha_O^{(T)}$ ratio, and secondarily by the l_1/r ratio.

By similar reasoning, the following NC conditions

$$l_1\alpha_1^{(P)} < r\alpha_O^{(P)} \tag{13.4.47}$$

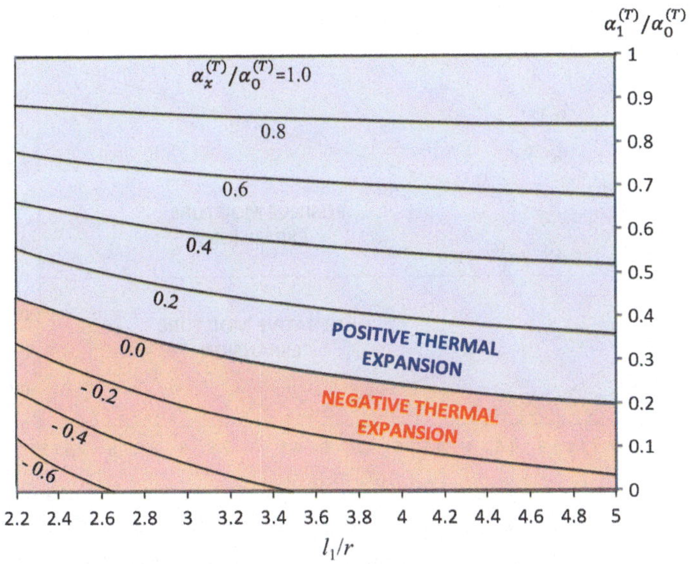

Fig. 13.9 Thermal expansion map in terms of dimensionless effective linear CTE $\alpha_x^{(T)}/\alpha_O^{(T)}$ for varying dimensionless rod CTE $\alpha_1^{(T)}/\alpha_O^{(T)}$ and rod-to-ring dimension ratio l_1/r

and

$$l_2\alpha_2^{(P)} < r\alpha_O^{(P)} \tag{13.4.48}$$

apply along the x- and y-principal axes, respectively. Likewise, the contour plot of Fig. 13.9 is valid for the compressibility analogy $\alpha_x^{(P)}/\alpha_O^{(P)}$ as a function of $\alpha_1^{(P)}/\alpha_O^{(P)}$ and l_1/r, but not for the moisture expansion analogy $\alpha_x^{(C)}/\alpha_O^{(C)}$ unless $\mathrm{d}C = \mathrm{d}C_O = \mathrm{d}C_1 = \mathrm{d}C_2$. It has been shown that the linear CME in any of its principal direction is negative not only when the rods are rigid, but also when the rods are non-absorptive. The generic negativity for the ring-rod structure's CME is obtained by substituting $\alpha_x^{(C)} < 0$ into Eq. (13.4.23) and $\alpha_y^{(C)} < 0$ into Eq. (13.4.24) to give the following NME conditions

$$l_1\alpha_1^{(C)}\mathrm{d}C_1 < r\alpha_O^{(C)}\mathrm{d}C_O \tag{13.4.49}$$

and

$$l_2\alpha_2^{(C)}\mathrm{d}C_2 < r\alpha_O^{(C)}\mathrm{d}C_O \tag{13.4.50}$$

respectively. To give a visual representation on the influence of changes in moisture concentration, a contour map of a dimensionless effective CME $\alpha_x^{(C)}/\alpha_O^{(C)}$ for varying dimensionless change in rod moisture concentration $\mathrm{d}C_1/\mathrm{d}C_O$ and dimensionless

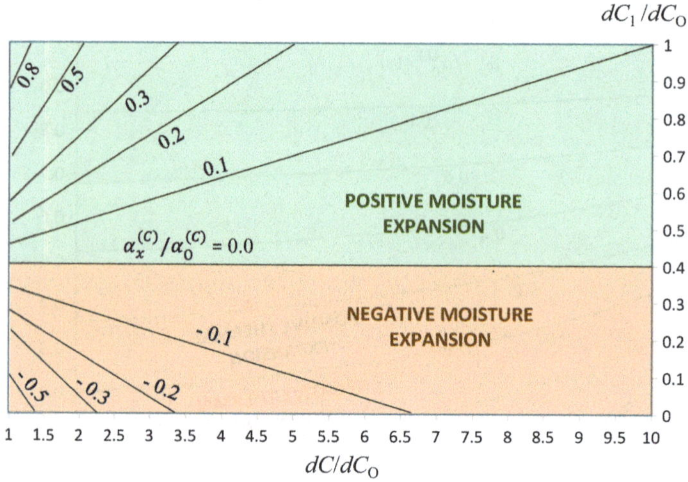

Fig. 13.10 Moisture expansion map in terms of dimensionless CME $\alpha_x^{(C)}/\alpha_\mathrm{O}^{(C)}$ for varying moisture concentrations in rod, ring and environment when the rod and ring possess equal moisture expansivity $\left(\alpha_1^{(C)} = \alpha_\mathrm{O}^{(C)}\right)$ at $l_1/r = 2.5$

change in environmental moisture concentration $\mathrm{d}C/\mathrm{d}C_\mathrm{O}$ is furnished in Fig. 13.10 for equal CMEs of the rods and ring $\left(\alpha_1^{(C)} = \alpha_2^{(C)} = \alpha_\mathrm{O}^{(C)}\right)$ at $l_1/r = 2.5$. It can be observed that the negativity of the dimensionless CME is enhanced by decreasing $\mathrm{d}C_1/\mathrm{d}C_\mathrm{O}$ ratio while the effective dimensionless CME's magnitude increases with decreasing $\mathrm{d}C/\mathrm{d}C_\mathrm{O}$ ratio.

In the category of areal CTE and CME, the conditions for NTE and NME are obtained by substituting $\alpha_A^{(T)} < 0$ into Eq. (13.4.13) to give

$$\frac{\frac{l_1}{r}\frac{\alpha_1^{(T)}}{\alpha_\mathrm{O}^{(T)}} - 1}{\frac{l_1}{r} - 1} + \frac{\frac{l_2}{r}\frac{\alpha_2^{(T)}}{\alpha_\mathrm{O}^{(T)}} - 1}{\frac{l_2}{r} - 1} < 0 \tag{13.4.51}$$

and $\alpha_A^{(C)} < 0$ into Eq. (13.4.29) to yield

$$\frac{\frac{l_1}{r}\frac{\alpha_1^{(C)}}{\alpha_\mathrm{O}^{(C)}}\frac{\mathrm{d}C_1}{\mathrm{d}C_\mathrm{O}} - 1}{\left(\frac{l_1}{r} - 1\right)\frac{\mathrm{d}C}{\mathrm{d}C_\mathrm{O}}} + \frac{\frac{l_2}{r}\frac{\alpha_2^{(C)}}{\alpha_\mathrm{O}^{(C)}}\frac{\mathrm{d}C_2}{\mathrm{d}C_\mathrm{O}} - 1}{\left(\frac{l_2}{r} - 1\right)\frac{\mathrm{d}C}{\mathrm{d}C_\mathrm{O}}} < 0 \tag{13.4.52}$$

For square array ($l_1 = l_1 = l$), Eqs. (13.4.51) and (13.4.52) greatly simplify to

$$\frac{\alpha_1^{(T)} + \alpha_2^{(T)}}{2\alpha_\mathrm{O}^{(T)}} < \frac{r}{l} \tag{13.4.53}$$

and

$$\frac{\alpha_1^{(C)} \mathrm{d}C_1 + \alpha_2^{(C)} \mathrm{d}C_2}{2\alpha_O^{(C)} \mathrm{d}C_O} < \frac{r}{l} \tag{13.4.54}$$

respectively. Performing similarly for the category of volumetric CTE and CME, the conditions for negativity of thermal and moisture expansions are obtained by the substitution of $\alpha_V^{(T)} < 0$ into Eq. (13.4.16) to result in

$$\frac{\frac{l_1}{r}\frac{\alpha_1^{(T)}}{\alpha_O^{(T)}} - 1}{\frac{l_1}{r} - 1} + \frac{\frac{l_2}{r}\frac{\alpha_2^{(T)}}{\alpha_O^{(T)}} - 1}{\frac{l_2}{r} - 1} + 1 < 0 \tag{13.4.55}$$

and the substitution of $\alpha_V^{(C)} < 0$ into Eq. (13.4.32) leads to

$$\frac{\frac{l_1}{r}\frac{\alpha_1^{(C)}}{\alpha_O^{(C)}}\frac{\mathrm{d}C_1}{\mathrm{d}C_O} - 1}{\left(\frac{l_1}{r} - 1\right)\frac{\mathrm{d}C}{\mathrm{d}C_O}} + \frac{\frac{l_2}{r}\frac{\alpha_2^{(C)}}{\alpha_O^{(C)}}\frac{\mathrm{d}C_2}{\mathrm{d}C_O} - 1}{\left(\frac{l_2}{r} - 1\right)\frac{\mathrm{d}C}{\mathrm{d}C_O}} + 1 < 0 \tag{13.4.56}$$

As with the case of areal NTE and NME, the conditions for volumetric NTE and NME in Eqs. (13.4.55) and (13.4.56) greatly reduce to

$$\frac{\alpha_1^{(T)} + \alpha_2^{(T)} + \alpha_O^{(T)}}{3\alpha_O^{(T)}} < \frac{r}{l} \tag{13.4.57}$$

and

$$\frac{\alpha_1^{(C)} \mathrm{d}C_1 + \alpha_2^{(C)} \mathrm{d}C_2 + \alpha_O^{(C)} \mathrm{d}C}{\alpha_O^{(C)}(2\mathrm{d}C_O + \mathrm{d}C)} < \frac{r}{l} \tag{13.4.58}$$

respectively, for square array.

If, instead of 2D array, we have a 3D array of double-rings interconnected by sliding rods in all three orthogonal directions as shown in Fig. 13.11 (detailed schematics are furnished Fig. 11.38d), then inference from Eq. (13.4.13) gives the volumetric CTE

$$\alpha_V^{(T)} = \sum_{i=1}^{3} \frac{l_i \alpha_i^{(T)} - r\alpha_O^{(T)}}{l_i - r} \tag{13.4.59}$$

where l_3 is the half-length of the sliding rod that is orthogonal to the x-y plane and $\alpha_3^{(T)}$ is its CTE, while inferring from Eq. (13.4.29) implies the corresponding volumetric CME

Fig. 13.11 Simplified schematic view of a double-ring with three orthogonal pairs of sliding rods in a 3D array of NEE structure

$$\alpha_V^{(C)} = \sum_{i=1}^{3} \frac{l_i \alpha_i^{(C)} \mathrm{d}C_i - r\alpha_O^{(C)} \mathrm{d}C_O}{(l_i - r)\mathrm{d}C} \tag{13.4.60}$$

where $\alpha_3^{(C)}$ and $\mathrm{d}C_3$ are the CME and change in the moisture concentration, respectively, for the third set of sliding rods. For cube array ($l_1 = l_2 = l_3 = l$), Eqs. (13.4.59) and (13.4.60) greatly shorten to

$$\frac{\alpha_1^{(T)} + \alpha_2^{(T)} + \alpha_3^{(T)}}{3\alpha_O^{(T)}} < \frac{r}{l} \tag{13.4.61}$$

for NTE and

$$\frac{\alpha_1^{(C)} \mathrm{d}C_1 + \alpha_2^{(C)} \mathrm{d}C_2 + \alpha_3^{(C)} \mathrm{d}C_3}{3\alpha_O^{(C)} \mathrm{d}C_O} < \frac{r}{l} \tag{13.4.62}$$

for NME conditions, respectively. The corresponding descriptions for linear, areal, and volumetric NC can be obtained by performing the following substitutions $\left(\alpha^{(T)}, \mathrm{d}T\right) \to \left(\alpha^{(P)}, \mathrm{d}P\right)$ on the CTE and NTE expressions, i.e., the volumetric compressibility

$$\alpha_V^{(P)} = \sum_{i=1}^{3} \frac{l_i \alpha_i^{(P)} - r\alpha_O^{(P)}}{l_i - r} \tag{13.4.63}$$

where $\alpha_3^{(P)}$ is the compressibility of the rods orthogonal to the x-y plane, while the stipulation of NC under the special case of cube array yields

$$\frac{\alpha_1^{(P)} + \alpha_2^{(P)} + \alpha_3^{(P)}}{3\alpha_O^{(P)}} < \frac{r}{l} \tag{13.4.64}$$

Where the various environmental changes occur simultaneously, the principle of superposition can be applied. However, there is no overall environmental coefficient that combines the responses from thermal expansion, moisture expansion, and pressure compressibilities; this can be attested from the different units for the coefficient of thermal expansion, the coefficient of moisture expansion and the compressibility. The principle of superposition can, nevertheless, be implemented for parameters that possess similar units; as such the cumulative responses from variation of environmental temperature, moisture concentration, and pressure can be accounted for in terms of strains. One may therefore write the linear environmental strain as a superposition from the three separate environmental strain components $\varepsilon_i = \varepsilon_i^{(T)} + \varepsilon_i^{(C)} + \varepsilon_i^{(P)}$ for $i = x, y$.

$$\varepsilon_x = \frac{l_1\alpha_1^{(T)} - r\alpha_O^{(T)}}{l_1 - r}\mathrm{d}T + \frac{l_1\alpha_1^{(C)}\mathrm{d}C_1 - r\alpha_O^{(C)}\mathrm{d}C_O}{l_1 - r} - \frac{l_1\alpha_1^{(P)} - r\alpha_O^{(P)}}{l_1 - r}\mathrm{d}P \tag{13.4.65}$$

$$\varepsilon_y = \frac{l_2\alpha_2^{(T)} - r\alpha_O^{(T)}}{l_2 - r}\mathrm{d}T + \frac{l_2\alpha_2^{(C)}\mathrm{d}C_2 - r\alpha_O^{(C)}\mathrm{d}C_O}{l_2 - r} - \frac{l_2\alpha_2^{(P)} - r\alpha_O^{(P)}}{l_2 - r}\mathrm{d}P \tag{13.4.66}$$

based on Eqs. (13.4.7), (13.4.8), (13.4.23), (13.4.24), (13.4.40), and (13.4.41), while the areal environmental strain can again be expressed as a superposition $\varepsilon_A = \varepsilon_A^{(T)} + \varepsilon_A^{(C)} + \varepsilon_A^{(P)}$

$$\begin{aligned}\varepsilon_A = &\left(\frac{l_1\alpha_1^{(T)} - r\alpha_O^{(T)}}{l_1 - r} + \frac{l_2\alpha_2^{(T)} - r\alpha_O^{(T)}}{l_2 - r}\right)\mathrm{d}T \\ &- \left(\frac{l_1\alpha_1^{(P)} - r\alpha_O^{(P)}}{l_1 - r} + \frac{l_2\alpha_2^{(P)} - r\alpha_O^{(P)}}{l_2 - r}\right)\mathrm{d}P \\ &+ \left(\frac{l_1\alpha_1^{(C)}\mathrm{d}C_1 - r\alpha_O^{(C)}\mathrm{d}C_O}{l_1 - r} + \frac{l_2\alpha_2^{(C)}\mathrm{d}C_2 - r\alpha_O^{(C)}\mathrm{d}C_O}{l_2 - r}\right)\end{aligned} \tag{13.4.67}$$

with reference to Eqs. (13.4.13), (13.4.29), and (13.4.43), or by adding Eqs. (13.4.65) and (13.4.66). The volumetric strain in the case of 2D array is

$$\varepsilon_V = \left(\frac{l_1\alpha_1^{(T)} - r\alpha_O^{(T)}}{l_1 - r} + \frac{l_2\alpha_2^{(T)} - r\alpha_O^{(T)}}{l_2 - r} + \alpha_O^{(T)}\right)\mathrm{d}T$$

$$-\left(\frac{l_1\alpha_1^{(P)}-r\alpha_{\mathrm{O}}^{(P)}}{l_1-r}+\frac{l_2\alpha_2^{(P)}-r\alpha_{\mathrm{O}}^{(P)}}{l_2-r}+\alpha_{\mathrm{O}}^{(P)}\right)\mathrm{d}P$$
$$+\left(\frac{l_1\alpha_1^{(C)}\mathrm{d}C_1-r\alpha_{\mathrm{O}}^{(C)}\mathrm{d}C_{\mathrm{O}}}{l_1-r}+\frac{l_2\alpha_2^{(C)}\mathrm{d}C_2-r\alpha_{\mathrm{O}}^{(C)}\mathrm{d}C_{\mathrm{O}}}{l_2-r}+\alpha_{\mathrm{O}}^{(C)}\mathrm{d}C_{\mathrm{O}}\right) \tag{13.4.68}$$

by virtue of Eqs. (13.4.16), (13.4.32), and (13.4.44), while the volumetric strain for 3D array of double-ring and rods can be written as

$$\varepsilon_V=\mathrm{d}T\sum_{i=1}^{3}\frac{l_i\alpha_{\mathrm{i}}^{(T)}-r\alpha_{\mathrm{O}}^{(T)}}{l_i-r}+\sum_{i=1}^{3}\frac{l_i\alpha_{\mathrm{i}}^{(C)}\mathrm{d}C_i-r\alpha_{\mathrm{O}}^{(C)}\mathrm{d}C_{\mathrm{O}}}{l_i-r}$$
$$-\mathrm{d}P\sum_{i=1}^{3}\frac{l_i\alpha_{\mathrm{i}}^{(P)}-r\alpha_{\mathrm{O}}^{(P)}}{l_i-r} \tag{13.4.69}$$

on the basis of Eqs. (13.4.59), (13.4.60), and (13.4.63). For the special case where the rings and double-rings are arranged in square and cube arrays, respectively, while the rods are rigid, Eqs. (13.4.65)–(13.4.67) and (13.4.69) are greatly simplified and can be combined as

$$\varepsilon_x=\varepsilon_y=\frac{1}{2}\varepsilon_A=\frac{1}{3}\varepsilon_V=-\left(\alpha_{\mathrm{O}}^{(T)}\mathrm{d}T+\alpha_{\mathrm{O}}^{(C)}\mathrm{d}C_{\mathrm{O}}-\alpha_{\mathrm{O}}^{(P)}\mathrm{d}P\right)\frac{r}{l-r} \tag{13.4.70}$$

Since it is well known that ring expansion occurs with increasing temperature ($\mathrm{d}T>0$), increasing moisture concentration ($\mathrm{d}C_{\mathrm{O}}>0$) and decreasing pressure ($\mathrm{d}P<0$), therefore the linear, areal, and volumetric strains in Eq. (13.4.70) are negative (Lim 2019b). In general, conditions for the negativity of strains listed in Eqs. (13.4.65)–(13.4.69) can be achieved by imposing $\varepsilon<0$ under the given environmental changes.

In the analysis, the ring-rod structure can be made of three materials in general or two materials if the rods aligned in both directions are to be of the same material. Any combination of materials can be employed to construct the ring-rod structure. However, the negativities are enhanced if the ring material has a much greater expansivity than that of the rod material. As the various environmental effects take place at different rate of change—e.g., the temperature and pressure change more rapidly while equilibration of the moisture content takes a relatively greater time frame—it follows that the thermal- and pressure-induced strain components inflict a quick response to the system, while the hygroscopic-induced strain component responds gradually. The separation of each effective strain expressions into three terms, such as those described in Eqs. (13.4.65)–(13.4.70), permits the overall strain to take into account the non-monotonically varying environmental temperature, pressure, and moisture concentration. Nevertheless, the change of the moisture concentration in solids may result in a change in the material properties of the components. In such a case, the developed models are still valid, bearing in mind that the CTEs, CMEs, and compressibilities of the ring-rod components are no longer constants, but are instead functions of moisture concentration.

13.5 Example: An Auxetic NHTE System Using Reinforced Double Arrowhead Microstructure

This section illustrates a type of truss system in the microstructure that can exhibit negative hygrothermal expansion (NHTE) behavior and, in so doing, avails the solutions for zero hygrothermal expansion at the boundary of conventional and NHTE behavior. The double arrowhead configuration is adopted due to its ease in achieving large strain by controlling the angles. The reinforced double arrowhead structure is shown in Fig. 13.12a where reinforcing rods–indicated as dashed lines–are inserted so as to join the tips of the arrowheads, which consists of the longer inclined rods and the shorter inclined rods (hereinafter known as "long rods" and "short rods", respectively, for brevity). The usual auxetic double arrowhead structure is recovered when the reinforcing rods diminish. With the insertion of the reinforcing rods, the microstructure consists of rods being aligned in five directions. Rod 3, which is the reinforcing rod, forms the axis of symmetry from which rod 1' and rod 2' are mirror images of rod 1 and rod 2, respectively. Due to this symmetry, only half of the arrowhead, as shaded in Fig. 13.12b, is required for analysis. As far as hygrothermal absorption is concerned, there is no directionality within the rod itself as we consider isotropic rods. The microstructure is sufficiently porous to permit uniform temperature change and uniform moisture absorption throughout the lattice. There is, however, directionality in terms of the overall structural response due to the anisotropic behavior arising from the lattice geometry. The schematic view in Fig. 13.12c facilitates analysis. Let the initial lengths of the long rods, short rods, and reinforcing rods be l_1, l_2 and l_3, respectively, whereby the long and short rods are subtended by angles θ_1 and θ_2, respectively, from the reinforcing rods. Arising from a change in the hygroscopic concentration in the environment, dC we write the final change in the hygroscopic concentrations in the long, short, and reinforcing rods as dC_1, dC_2 and dC_3, respectively, at steady state. Unlike the case of thermal expansion whereby the temperatures of the various solid components attain the environmental temperature at thermal equilibrium, various materials of the same shape and size absorb different amounts of moisture mass at moisture diffusion equilibrium. Hence, it is not possible to obtain the change in material dimension directly from the change in environmental hygroscopic concentration.

The different changes to the moisture concentration is in recognition of the fact that the change in material moisture concentration is lower than that in the environment, and that for different materials that change in moisture concentration varies due to different moisture absorptivity. The change in moisture content in a material can be measured by comparing its weight—at equilibrium state—before and after the

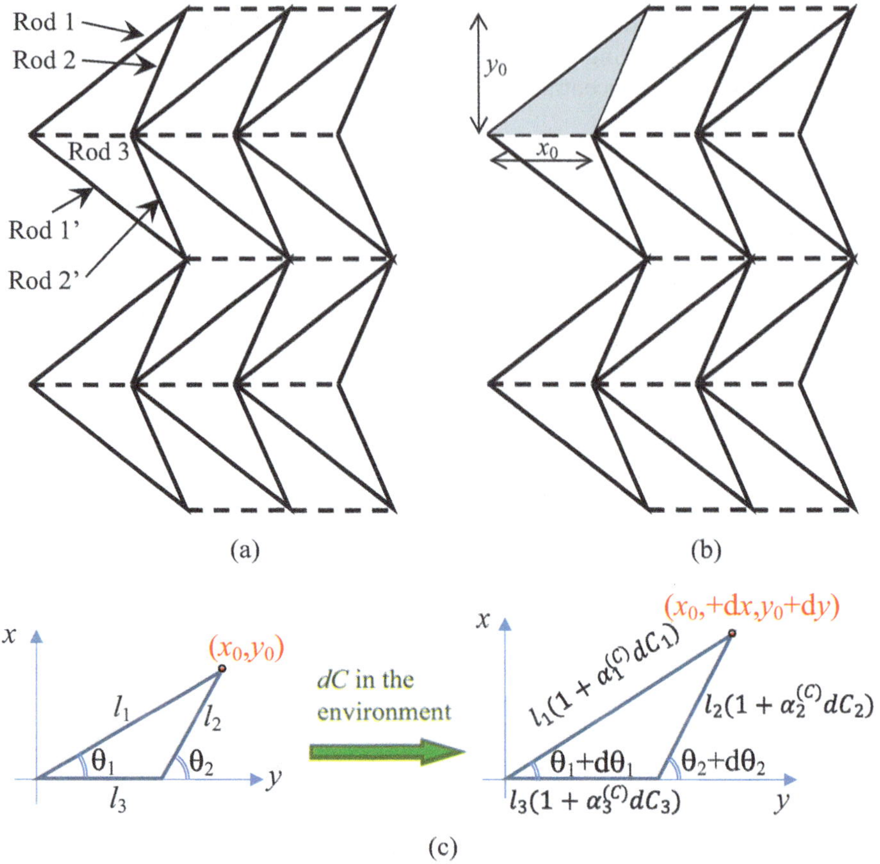

Fig. 13.12 **a** Overview of a double arrowhead structure as continuous lines with insertion of reinforcing rods as dashed lines, showing all 5 directions of the rod alignments, indicated as rod 1, rod 2, rod 3, rod 1', and rod 2', **b** definitions of x_0 and y_0, and **c** changes in the subtending angles arising from changes in rod lengths

change in environmental moisture concentration. The hygroscopic strains in the long, short, and reinforcing rods are therefore $\alpha_1^{(C)}\mathrm{d}C_1, \alpha_2^{(C)}\mathrm{d}C_2$, and $\alpha_3^{(C)}\mathrm{d}C_3$, respectively, where $\alpha_1^{(C)}$, $\alpha_2^{(C)}$, and $\alpha_3^{(C)}$ are the corresponding coefficients of moisture expansion (CME) for the rods. Accordingly, the subtending angles for the long and short rods change by $\mathrm{d}\theta_1$ and $\mathrm{d}\theta_2$, respectively. As a result from the changes in the rod lengths and their inclination angles, the original location of the sharp edge (x_0, y_0) relocates to $(x_0+\mathrm{d}x, y_0+\mathrm{d}y)$, as indicated by Fig. 13.12c. In what follows, the conditions for attaining NME as well as NTE along the y-direction are established; it is assumed that the rods are connected by pin joints to facilitate free rotations at the vertices.

By geometry, the horizontal displacement of the sharp edge is

$$dx = l_1\left(1 + \alpha_1^{(C)}dC_1\right)\cos(\theta_1 + d\theta_1) - l_1\cos\theta_1 \tag{13.5.1}$$

based on the long rod, and

$$dx = l_2\left(1 + \alpha_2^{(C)}dC_2\right)\cos(\theta_2 + d\theta_2) - l_2\cos\theta_2 + l_3\alpha_3^{(C)}dC_3 \tag{13.5.2}$$

with reference to the short and reinforcing rods, while the vertical displacement is

$$dy = l_1\left(1 + \alpha_1^{(C)}dC_1\right)\sin(\theta_1 + d\theta_1) - l_1\sin\theta_1 \tag{13.5.3}$$

on the basis of the long rods, while perusal to the short rod gives

$$dy = l_2\left(1 + \alpha_2^{(C)}dC_2\right)\sin(\theta_2 + d\theta_2) - l_2\sin\theta_2 \tag{13.5.4}$$

The above trigonometric descriptions can be expanded followed by substitution of $\sin d\theta = d\theta$ and $\cos d\theta = 1$ for infinitesimal deformation to give the infinitesimal horizontal displacement of the sharp edge

$$dx = -l_1 d\theta_1 \sin\theta_1\left(1 + \alpha_1^{(C)}dC_1\right) + l_1\alpha_1^{(C)}dC_1\cos\theta_1 \tag{13.5.5}$$

via the long rod, and

$$dx = -l_2 d\theta_2 \sin\theta_2\left(1 + \alpha_2^{(C)}dC_2\right) + l_2\alpha_2^{(C)}dC_2\cos\theta_2 + l_3\alpha_3^{(C)}dC_3 \tag{13.5.6}$$

on the basis of the short and reinforcing rods, while the corresponding infinitesimal vertical displacement

$$dy = l_1 d\theta_1 \cos\theta_1\left(1 + \alpha_1^{(C)}dC_1\right) + l_1\alpha_1^{(C)}dC_1\sin\theta_1 \tag{13.5.7}$$

as regards to the long rod or

$$dy = l_2 d\theta_2 \cos\theta_2\left(1 + \alpha_2^{(C)}dC_2\right) + l_2\alpha_2^{(C)}dC_2\sin\theta_2 \tag{13.5.8}$$

pertaining to the short rod. Recognizing that the original location of the sharp edge (x_0, y_0) can be expressed by the rod lengths and inclination angles as

$$x_0 = l_1\cos\theta_1 = l_2\cos\theta_2 + l_3 \tag{13.5.9}$$

and

$$y_0 = l_1\sin\theta_1 = l_2\sin\theta_2 \tag{13.5.10}$$

one can then obtain the ratio of horizontal displacement $\mathrm{d}x$ to the original horizontal position of the sharp edge as

$$\frac{\mathrm{d}x}{x_0} = -\mathrm{d}\theta_1 \tan\theta_1 \left(1 + \alpha_1^{(C)} \mathrm{d}C_1\right) + \alpha_1^{(C)} \mathrm{d}C_1 \tag{13.5.11}$$

in terms of the long rod or

$$\frac{\mathrm{d}x}{x_0} = -\mathrm{d}\theta_2 \frac{\sin\theta_2}{\cos\theta_2 + \frac{l_3}{l_2}} \left(1 + \alpha_2^{(C)} \mathrm{d}C_2\right) + \frac{\cos\theta_2}{\cos\theta_2 + \frac{l_3}{l_2}} \alpha_2^{(C)} \mathrm{d}C_2 + \frac{\frac{l_3}{l_2}}{\cos\theta_2 + \frac{l_3}{l_2}} \alpha_3^{(C)} \mathrm{d}C_3 \tag{13.5.12}$$

by means of the short and reinforcing rods, while the vertical strain $\varepsilon_y = \mathrm{d}y/y_0$ can be expressed in terms of the long rod as

$$\varepsilon_y = \frac{\mathrm{d}\theta_1}{\tan\theta_1} \left(1 + \alpha_1^{(C)} \mathrm{d}C_1\right) + \alpha_1^{(C)} \mathrm{d}C_1 \tag{13.5.13}$$

or in terms of the short rod as

$$\varepsilon_y = \frac{\mathrm{d}\theta_2}{\tan\theta_2} \left(1 + \alpha_2^{(C)} \mathrm{d}C_2\right) + \alpha_2^{(C)} \mathrm{d}C_2 \tag{13.5.14}$$

It should be cautioned here that the ratio $\mathrm{d}x/x_0$ in Eqs. (13.5.11) and (13.5.12) are not termed the horizontal strain because ε_x has to be defined by the elongation ratio of the reinforcement rod. Nevertheless, the $\mathrm{d}x/x_0$ ratio is important as it allows the angular changes $\mathrm{d}\theta_1$ and $\mathrm{d}\theta_2$ to be simultaneously calculated in conjunction with the vertical strain, as will be shown shortly. Since the $\mathrm{d}x/x_0$ descriptions in Eqs. (13.5.11) and (13.5.12) are the same, they can be equated to give

$$\mathrm{d}\theta_1 \tan\theta_1 \left(1 + \alpha_1^{(C)} \mathrm{d}C_1\right) - \alpha_1^{(C)} \mathrm{d}C_1 = \frac{\mathrm{d}\theta_2 \sin\theta_2 \left(1 + \alpha_2^{(C)} \mathrm{d}C_2\right)}{\cos\theta_2 + \frac{l_3}{l_2}} - \frac{\cos\theta_2 \alpha_2^{(C)} \mathrm{d}C_2}{\cos\theta_2 + \frac{l_3}{l_2}} - \frac{\frac{l_3}{l_2} \alpha_3^{(C)} \mathrm{d}C_3}{\cos\theta_2 + \frac{l_3}{l_2}} \tag{13.5.15}$$

or, on the basis of Eqs. (13.5.9) and (13.5.10), the inclined rod length ratio

$$\frac{l_3}{l_2} = \frac{\sin\theta_2}{\tan\theta_1} - \cos\theta_2 \tag{13.5.16}$$

is substituted into Eq. (13.5.15) to give

$$d\theta_1 \tan\theta_1\left(1+\alpha_1^{(C)}dC_1\right) - d\theta_2 \tan\theta_1\left(1+\alpha_2^{(C)}dC_2\right)$$
$$= \alpha_1^{(C)}dC_1 - \frac{\tan\theta_1}{\tan\theta_2}\alpha_2^{(C)}dC_2 - \left(1-\frac{\tan\theta_1}{\tan\theta_2}\right)\alpha_3^{(C)}dC_3 \tag{13.5.17}$$

Likewise, the commonality of the vertical strains allows Eqs. (13.5.13) and (13.5.14) to be equated, thereby leading to

$$\frac{d\theta_1}{\tan\theta_1}\left(1+\alpha_1^{(C)}dC_1\right) - \frac{d\theta_2}{\tan\theta_2}\left(1+\alpha_2^{(C)}dC_2\right) = -\alpha_1^{(C)}dC_1 + \alpha_2^{(C)}dC_2 \tag{13.5.18}$$

The angular changes of $d\theta_1$ and $d\theta_2$ can then be simultaneously solved from Eqs. (13.5.17) and (13.5.18) to give

$$d\theta_1\left(1+\alpha_1^{(C)}dC_1\right) = -\frac{1+\tan\theta_1\tan\theta_2}{\tan\theta_2-\tan\theta_1}\alpha_1^{(C)}dC_1 + \frac{\tan\theta_2+\frac{1}{\tan\theta_2}}{\frac{\tan\theta_2}{\tan\theta_1}-1}\alpha_2^{(C)}dC_2$$
$$+\frac{\alpha_3^{(C)}dC_3}{\tan\theta_2} \tag{13.5.19}$$

and

$$d\theta_2\left(1+\alpha_2^{(C)}dC_2\right) = -\frac{\tan\theta_1+\frac{1}{\tan\theta_1}}{1-\frac{\tan\theta_1}{\tan\theta_2}}\alpha_1^{(C)}dC_1 + \frac{1+\tan\theta_1\tan\theta_2}{\tan\theta_2-\tan\theta_1}\alpha_2^{(C)}dC_2$$
$$+\frac{\alpha_3^{(C)}dC_3}{\tan\theta_1} \tag{13.5.20}$$

Although Eqs. (13.5.19) and (13.5.20) can be further simplified using trigonometric identities, such as

$$\frac{1+\tan\theta_1\tan\theta_2}{\tan\theta_2-\tan\theta_1} = \frac{1}{\tan(\theta_2-\theta_1)} \tag{13.5.21}$$

the longer form is retained therein for the convenience of calculating the special cases later. The substitution of Eqs. (13.5.19) or (13.5.20) into Eqs. (13.5.13) or (13.5.14), respectively, gives a common solution for the strain in the y-direction

$$\varepsilon_y = -\frac{\tan\theta_1+\frac{1}{\tan\theta_1}}{\tan\theta_2-\tan\theta_1}\alpha_1^{(C)}dC_1 + \frac{\tan\theta_2+\frac{1}{\tan\theta_2}}{\tan\theta_2-\tan\theta_1}\alpha_2^{(C)}dC_2 + \frac{\alpha_3^{(C)}dC_3}{\tan\theta_1\tan\theta_2} \tag{13.5.22}$$

Since $\varepsilon_y = \alpha_{Cy}dC$, we have the CME along the y-direction

$$\alpha_{Cy} = \frac{-\left(\tan\theta_1 + \frac{1}{\tan\theta_1}\right)\alpha_1^{(C)}\mathrm{d}C_1 + \left(\tan\theta_2 + \frac{1}{\tan\theta_2}\right)\alpha_2^{(C)}\mathrm{d}C_2 + \left(\frac{\tan\theta_2 - \tan\theta_1}{\tan\theta_1\tan\theta_2}\right)\alpha_3^{(C)}\mathrm{d}C_3}{(\tan\theta_2 - \tan\theta_1)\mathrm{d}C} \tag{13.5.23}$$

For attaining NME in the y-direction, we impose $\alpha_{Cy} < 0$ to give (Lim 2020)

$$\left(\tan\theta_1 + \frac{1}{\tan\theta_1}\right)\alpha_1^{(C)}\mathrm{d}C_1 - \left(\tan\theta_2 + \frac{1}{\tan\theta_2}\right)\alpha_2^{(C)}\mathrm{d}C_2 > \left(\frac{\tan\theta_2 - \tan\theta_1}{\tan\theta_1\tan\theta_2}\right)\alpha_3^{(C)}\mathrm{d}C_3 \tag{13.5.24}$$

The condition for positive hygroscopic expansion in the y-direction is described when the inequality sign in Eq. (13.5.24) is reversed. By similar reasoning, the condition of zero hygroscopic expansion in the y-direction is characterized when the inequality sign is replaced by the equality sign.

We herein consider three special cases to aid subsequent illustrations:

(a) Special Case I where $\theta_1 = 30°$ and $\theta_2 = 60°$,
(b) Special Case II where $\theta_1 = 30°$ and $\theta_2 = 45°$, and
(c) Special Case III where $\theta_1 = 45°$ and $\theta_2 = 60°$,

as depicted in Fig. 13.13.

Substituting $\theta_1 = 30°$ and $\theta_2 = 60°$ into Eqs. (13.5.23) and (13.5.24) for Special Case I gives the effective CME in the y-direction

$$\alpha_y^{(C)} = \frac{-2\alpha_1^{(C)}\mathrm{d}C_1 + 2\alpha_2^{(C)}\mathrm{d}C_2 + \alpha_3^{(C)}\mathrm{d}C_3}{\mathrm{d}C} \tag{13.5.25}$$

and the condition for attaining NME

$$2\alpha_1^{(C)}\mathrm{d}C_1 - 2\alpha_2^{(C)}\mathrm{d}C_2 > \alpha_3^{(C)}\mathrm{d}C_3 \tag{13.5.26}$$

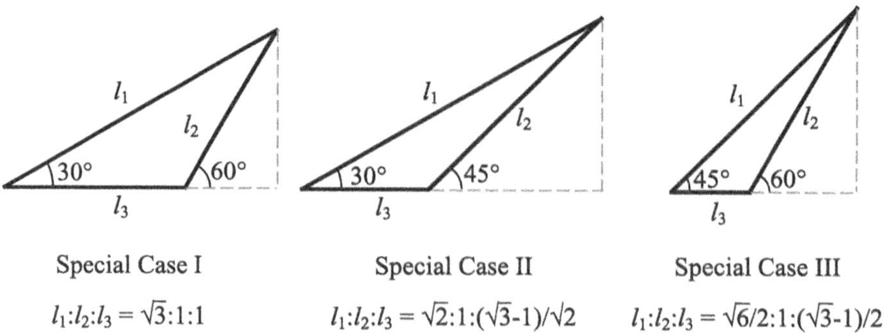

Fig. 13.13 Consideration of three special cases for illustration

respectively. For Special Case II, prescribing $\theta_1 = 30°$ and $\theta_2 = 45°$ into Eqs. (13.5.23) and (13.5.24) yields the CME

$$\alpha_y^{(C)} = \frac{-4\alpha_1^{(C)}\mathrm{d}C_1 + 2\sqrt{3}\alpha_2^{(C)}\mathrm{d}C_2 + \sqrt{3}\left(\sqrt{3}-1\right)\alpha_3^{(C)}\mathrm{d}C_3}{\left(\sqrt{3}-1\right)\mathrm{d}C} \tag{13.5.27}$$

and the corresponding NME requirement

$$4\alpha_1^{(C)}\mathrm{d}C_1 - 2\sqrt{3}\alpha_2^{(C)}\mathrm{d}C_2 > \sqrt{3}\left(\sqrt{3}-1\right)\alpha_3^{(C)}\mathrm{d}C_3 \tag{13.5.28}$$

Specifying $\theta_1 = 45°$ and $\theta_2 = 60°$ for Special Case III leads to the CME

$$\alpha_y^{(C)} = \frac{-2\sqrt{3}\alpha_1^{(C)}\mathrm{d}C_1 + 4\alpha_2^{(C)}\mathrm{d}C_2 + \left(\sqrt{3}-1\right)\alpha_3^{(C)}\mathrm{d}C_3}{\sqrt{3}\left(\sqrt{3}-1\right)\mathrm{d}C} \tag{13.5.29}$$

and condition for NME

$$2\sqrt{3}\alpha_1^{(C)}\mathrm{d}C_1 - 4\alpha_2^{(C)}\mathrm{d}C_2 > \left(\sqrt{3}-1\right)\alpha_3^{(C)}\mathrm{d}C_3. \tag{13.5.30}$$

Recall that the moisture concentration change in the environment is different from those in materials and that the different absorptivity of various materials leads to different changes in moisture concentrations for different materials. Where thermal expansion is concerned, a change in environmental temperature of $\mathrm{d}T$ leads to a common change in solid temperature by the same amount upon attaining thermal equilibrium. The governing equations for moisture expansion can be converted to those of thermal expansion by implementing the following changes

$$\begin{matrix} \alpha_y^{(C)} \to \alpha_y^{(T)} \\ \alpha_1^{(C)} \to \alpha_1^{(T)} \\ \alpha_2^{(C)} \to \alpha_2^{(T)} \\ \alpha_3^{(C)} \to \alpha_3^{(T)} \end{matrix}, \quad \begin{matrix} \mathrm{d}C \to \mathrm{d}T \\ \mathrm{d}C_1 \to \mathrm{d}T \\ \mathrm{d}C_2 \to \mathrm{d}T \\ \mathrm{d}C_3 \to \mathrm{d}T \end{matrix} \tag{13.5.31}$$

so as to yield the effective CTE in the y-direction

$$\alpha_y^{(T)} = \frac{-\left(\tan\theta_1 + \frac{1}{\tan\theta_1}\right)\alpha_1^{(T)} + \left(\tan\theta_2 + \frac{1}{\tan\theta_2}\right)\alpha_2^{(T)} + \left(\frac{\tan\theta_2 - \tan\theta_1}{\tan\theta_1\tan\theta_2}\right)\alpha_3^{(T)}}{(\tan\theta_2 - \tan\theta_1)} \tag{13.5.32}$$

For attaining NTE in the y-direction, we impose $\alpha_{Ty} < 0$ to give (Lim 2020)

$$\left(\tan\theta_1 + \frac{1}{\tan\theta_1}\right)\alpha_1^{(T)} - \left(\tan\theta_2 + \frac{1}{\tan\theta_2}\right)\alpha_2^{(T)} > \left(\frac{\tan\theta_2 - \tan\theta_1}{\tan\theta_1 \tan\theta_2}\right)\alpha_3^{(T)} \tag{13.5.33}$$

Accordingly, the effective CMEs along the y-direction for the three special cases and their corresponding conditions for NHE, as described by Eqs. (13.5.25)–(13.5.30), can be transformed into their thermal analogies by means of Eq. (13.5.31), i.e.,

$$\alpha_y^{(T)} = -2\alpha_1^{(T)} + 2\alpha_2^{(T)} + \alpha_3^{(T)} \tag{13.5.34}$$

for Special Case I with its NTE condition

$$2\alpha_1^{(T)} - 2\alpha_2^{(T)} > \alpha_3^{(T)} \tag{13.5.35}$$

and

$$\alpha_y^{(T)} = \frac{-4\alpha_1^{(T)} + 2\sqrt{3}\alpha_2^{(T)} + \sqrt{3}\left(\sqrt{3}-1\right)\alpha_3^{(T)}}{\sqrt{3}-1} \tag{13.5.36}$$

for Special Case II with its corresponding NTE condition

$$4\alpha_1^{(T)} - 2\sqrt{3}\alpha_2^{(T)} > \sqrt{3}\left(\sqrt{3}-1\right)\alpha_3^{(T)} \tag{13.5.37}$$

and finally

$$\alpha_y^{(T)} = \frac{-2\sqrt{3}\alpha_1^{(T)} + 4\alpha_2^{(T)} + \left(\sqrt{3}-1\right)\alpha_3^{(T)}}{\sqrt{3}\left(\sqrt{3}-1\right)} \tag{13.5.38}$$

for Special Case III with its NTE condition

$$2\sqrt{3}\alpha_1^{(T)} - 4\alpha_2^{(T)} > \left(\sqrt{3}-1\right)\alpha_3^{(T)} \tag{13.5.39}$$

When the short and reinforcing rods are rigid, i.e., $\alpha_2^{(C)} = \alpha_3^{(C)} = 0$ and $\alpha_2^{(T)} = \alpha_3^{(T)} = 0$, the effective CMEs and CTEs along the y-direction are listed in Table 13.1, indicating definite NME and NTE, respectively.

Finally, it can be seen that if the CMEs of the individual rods are nonnegative, then the negativity of $\alpha_y^{(C)}$ can be maximized if Rod 2 and Rod 3 are either non-swelling $\alpha_2^{(C)} = \alpha_3^{(C)} = 0$ or non-absorptive $\mathrm{d}C_2 = \mathrm{d}C_3 = 0$. Under either of these condition, Eq. (13.5.23) reduces to

Table 13.1 Summary of CMEs and CTEs along y-direction when the short and reinforcing rods are rigid, indicating definite NHE and NTE, respectively

θ_1	θ_2	$\alpha_y^{(C)}$	$\alpha_y^{(T)}$
30°	60°	$-2\alpha_1^{(C)} \frac{dC_1}{dC}$	$-2\alpha_1^{(T)}$
30°	45°	$-\frac{4}{\sqrt{3}-1}\alpha_1^{(C)} \frac{dC_1}{dC}$	$-\frac{4}{\sqrt{3}-1}\alpha_1^{(T)}$
45°	60°	$-\frac{2}{\sqrt{3}-1}\alpha_1^{(C)} \frac{dC_1}{dC}$	$-\frac{2}{\sqrt{3}-1}\alpha_1^{(T)}$

$$\frac{\alpha_y^{(C)}}{\alpha_1^{(C)}} = \frac{\tan\theta_1 + (1/\tan\theta_1)}{\tan\theta_1 - \tan\theta_2} \frac{dC_1}{dC}. \tag{13.5.40}$$

Bearing in mind that the moisture concentration in Rod 1 (dC_1) increases and decreases with the rise and drop of moisture concentration in the environment (dC), the ratio dC_1/dC in Eq. (13.5.40) is positive. Since $0° < \theta_1 < \theta_2 < 90°$, the negativity of Eq. (13.5.40) is definite. This negativity applies for its corresponding thermal analogy, wherein

$$\frac{\alpha_y^{(T)}}{\alpha_1^{(T)}} = \frac{\tan\theta_1 + (1/\tan\theta_1)}{\tan\theta_1 - \tan\theta_2} \tag{13.5.41}$$

bearing in mind that $dT_1 \rightarrow dT$ at thermal equilibrium.

The expansion coefficients in the x-direction have so far not being established because it is defined by the change in Rod 3 length. Hence, if one assumes that the rods are conventional, i.e., possessing positive moisture expansion and positive thermal expansion, then the overall moisture and thermal expansions along the x-axis are never negative. Nevertheless, $\alpha_x^{(C)}$ and $\alpha_x^{(T)}$ are furnished hereafter for the sake of completeness. Since the strain in x-direction is defined by Rod 3, one may equate the hygroscopic strain of Rod 3, i.e., $\alpha_3^{(C)} dC_3$, with the effective hygroscopic strain in the x-direction $\alpha_x^{(C)} dC$ to give

$$\alpha_x^{(C)} = \alpha_3^{(C)} \frac{dC_3}{dC} \tag{13.5.42}$$

Equation (13.5.42) suggests that even if the CME of Rod 3 is high, the effective CME in the x-direction is reduced if Rod 3 has low moisture absorptivity, which results in $dC_3 < dC$ at hygroscopic equilibrium. In terms of thermal expansion, perusal to Eq. (13.5.31) gives

$$\alpha_x^{(T)} = \alpha_3^{(T)} \tag{13.5.43}$$

This is because when thermal equilibrium is achieved, the final temperature of all rods (including Rod 3) reaches the temperature of the environment; since the initial temperatures are common, the changes in environmental and Rod 3 temperatures are equal.

Equations (13.5.26), (13.5.28), and (13.5.30) specify the conditions by which the CME along the y-direction is negative for Special Cases I, II, and III, respectively. In order to visualize the occurrence of NHE based on the relative values of $\alpha_1^{(C)}\mathrm{d}C_1$, $\alpha_2^{(C)}\mathrm{d}C_2$, and $\alpha_3^{(C)}\mathrm{d}C_3$, contour plots are furnished in Fig. 13.14 for these three special cases, whereby NME along y-direction is attained when $\alpha_1^{(C)}\mathrm{d}C_1$ occurs above the surfaces, while positive moisture expansion is obtained when $\alpha_1^{(C)}\mathrm{d}C_1$ falls below the surfaces. Needless to say, zero moisture expansion is achieved when $\alpha_1^{(C)}\mathrm{d}C_1$ takes place on the surfaces. Perusal to Fig. 13.14 suggests that, for the considered geometries, $\alpha_2^{(C)}\mathrm{d}C_2$ and $\alpha_3^{(C)}\mathrm{d}C_3$ play the primary and secondary roles, respectively, in influencing the sign of $\alpha_y^{(C)}$. The effect from $\alpha_2^{(C)}\mathrm{d}C_2$ vis-à-vis $\alpha_3^{(C)}\mathrm{d}C_3$ becomes stronger as the length of the reinforcing rod shortens with respect to the other two rods or as the length of the short rod approaches that of the long rod.

With reference to Eq. (13.5.40) for case where Rod 2 and Rod 3 are rigid, four contour plots of a dimensionless $\alpha_y^{(C)}/\alpha_1^{(C)}$ versus θ_1 (vertical axis) and θ_2 (horizontal axis) are shown in Fig. 13.15, whereby $0° < \theta_1 < \theta_2 < 90°$ in complying to the double arrowhead geometry, hence resulting in triangular plots. In addition, we note

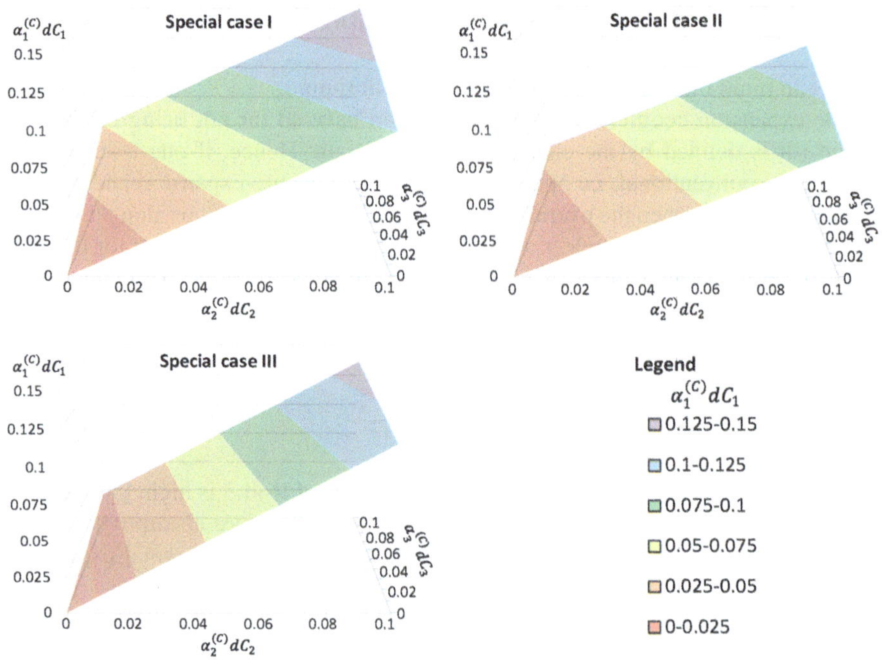

Fig. 13.14 Contour plots as functions of $\alpha_2^{(C)}\mathrm{d}C_2$ and $\alpha_3^{(C)}\mathrm{d}C_3$ for indicating the solutions of zero moisture expansion along y-direction when $\alpha_1^{(C)}\mathrm{d}C_1$ falls on the surface, while positive and negative moisture expansion is obtained when $\alpha_1^{(C)}\mathrm{d}C_1$ is located below and above the surfaces, respectively. These surface plots are also valid for the solutions of zero thermal expansion along y-direction by substituting $\left(\alpha_1^{(C)}\mathrm{d}C_1, \alpha_2^{(C)}\mathrm{d}C_2, \alpha_3^{(C)}\mathrm{d}C_3\right) \to \left(\alpha_1^{(T)}\mathrm{d}T, \alpha_2^{(T)}\mathrm{d}T, \alpha_3^{(T)}\mathrm{d}T\right)$

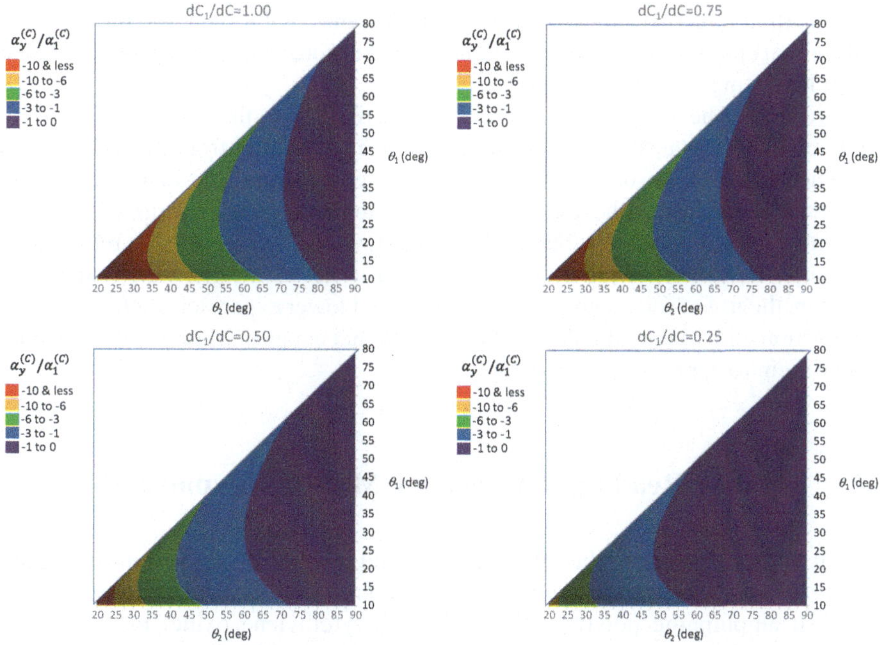

Fig. 13.15 Contour plots of $\alpha_y^{(C)}/\alpha_1^{(C)}$ when Rod 2 and Rod 3 possess zero CMEs $\alpha_2^{(C)} = \alpha_3^{(C)} = 0$ or zero absorptivity $\mathrm{d}C_2 = \mathrm{d}C_3 = 0$ at $\mathrm{d}C_1 = \mathrm{d}C$ (top left), $\mathrm{d}C_1 = 0.75\mathrm{d}C$ (top right), $\mathrm{d}C_1 = 0.5\mathrm{d}C$ (bottom left) and $\mathrm{d}C_1 = 0.25\mathrm{d}C$ (bottom right) for various angles of θ_1 and θ_2

that $0 < \mathrm{d}C_1 < \mathrm{d}C$. The lower limit of $\mathrm{d}C_1$ indicates the possibility of Rod 1 being non-absorptive while the upper bound indicates that the change in moisture concentration in Rod 1 is equal to the change in moisture concentration in the environment. This need not necessarily be so even at equilibrium, and so lower values such as $\mathrm{d}C_1 = 0.75\mathrm{d}C$, $\mathrm{d}C_1 = 0.5\mathrm{d}C$, and $\mathrm{d}C_1 = 0.25\mathrm{d}C$ are included in Fig. 13.15.

With reference to Eqs. (13.5.13) and (13.5.14), the strain along the y-direction due to temperature change can be added by virtue of Eq. (13.5.31) to give

$$\varepsilon_y = \alpha_1^{(C)}\mathrm{d}C_1 + \alpha_1^{(T)}\mathrm{d}T + \left(2 + \alpha_1^{(C)}\mathrm{d}C_1 + \alpha_1^{(T)}\mathrm{d}T\right)\frac{\mathrm{d}\theta_1}{\tan\theta_1} \tag{13.5.44}$$

and

$$\varepsilon_y = \alpha_2^{(C)}\mathrm{d}C_2 + \alpha_2^{(T)}\mathrm{d}T + \left(2 + \alpha_2^{(C)}\mathrm{d}C_2 + \alpha_2^{(T)}\mathrm{d}T\right)\frac{\mathrm{d}\theta_2}{\tan\theta_2} \tag{13.5.45}$$

based on the long and shorts rods, respectively. To design against hygrothermal strains in unconstrained systems or against hygrothermal stresses in constrained systems, one needs to select materials and inclination angles such that $\varepsilon_y = 0$ based on either Eq. (13.5.44) or Eq. (13.5.45). This is a more flexible approach in reducing the hygrothermal effects by allowing nonzero CTEs and CMEs such that the moisture and thermal effects offset each other, rather than the more rigid or restrictive approach of implementing both $\alpha_y^{(T)} = 0$ and $\alpha_y^{(C)} = 0$.

Throughout the analysis, the rods are considered thin so that the high surface-to-volume ratio facilitates rapid moisture transport into and out from the rods as well as to achieve uniform distribution of moisture concentration in comparison with thick rods; in the case of the latter, moisture distribution is non-uniform, with greater concentration at the surface than at the core. However, the use of thin rods has its limitations as it hinders the ability for the rod to act as a rigid unit. Under such situation, the use of rods with greater modulus and lower expansion coefficients—in conjunction with rods of smaller modulus and higher expansion coefficients—would enable them to approximate rigid rods.

13.6 Further Readings on Negative Moisture Expansions

Barrett et al. (2013) described the synthesis and characterization of mechanically tough zero or negative swelling mussel-inspired surgical adhesives based on catechol-modified amphiphilic poly(propylene oxide)-poly(ethylene oxide) block copolymers. The following examples consider the design of NME, or negative swelling, by the use of differential swelling in bimaterial layers that leads to curving. A 2D example was furnished by Zhang et al. (2018) and shown in Fig. 13.16. The work by Liu et al. (2016) employed square array, hexagonal array, and triangular array models with alternating orientation of the bimaterial layers, in which one of the square array models adopted the alternating bimaterial layers that resemble Fig. 11.8, except that the out-of-plane dimension is in the same order as the unit cell. As such the bimaterial layers, which are plates rather than beams, undergo surface curving to form shells of double curvatures instead of beams with single curvature. For this reason, the "2D microstructure" in the original state due to the prismatic model undergoes 3D deformation and would hence be properly classified as being 3D. Another 3D model was proposed by Curatolo (2018) as shown in Fig. 13.17 (top left) which consists of hollow cubes in cubic array and connected by rigid rods at the center of the square outer walls. Differential swelling of the wall materials induces transformation of the flat walls into curved walls toward the center of the hollow cubes such that the centers of the deformed cubes move close to one another (Fig. 13.17 top right). Further swelling of the hollow cube materials is followed by continued surface curving of the cube walls and the resulting convergence of the cubes (Fig. 13.17 bottom left) until the sharp corners of the neighboring cubes come into contact (Fig. 13.17 bottom right).

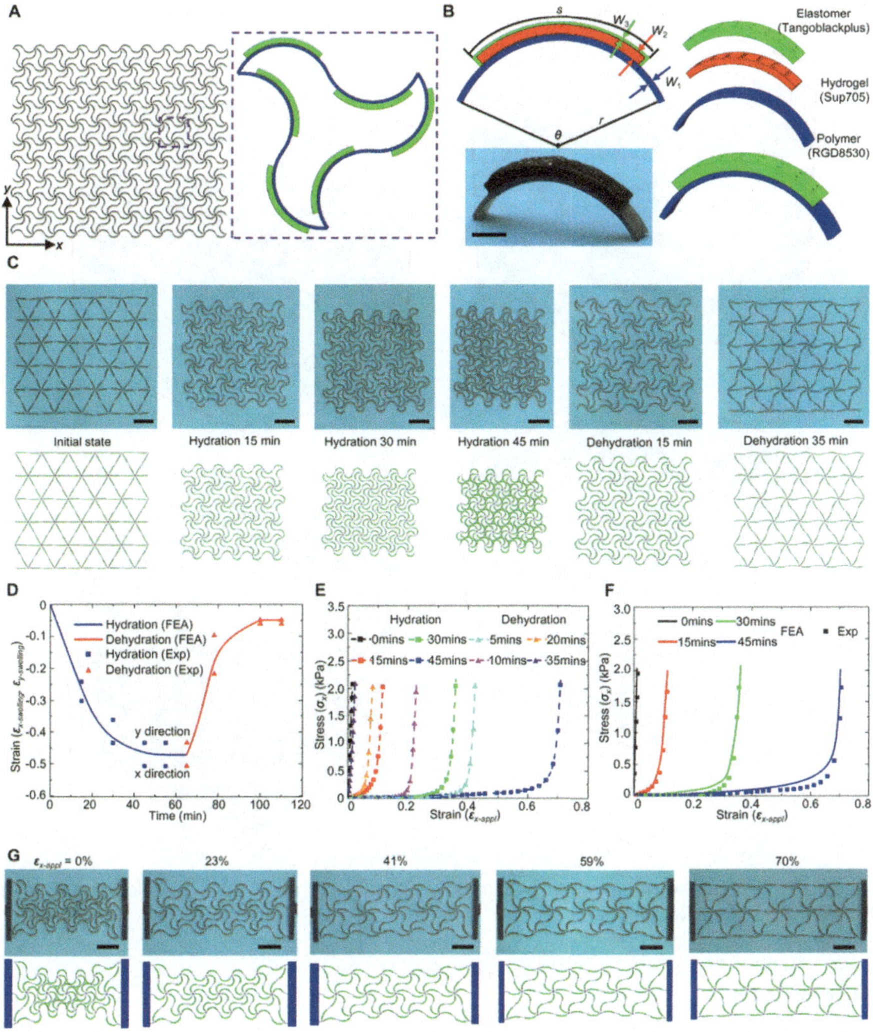

Fig. 13.16 Design concepts of soft mechanical metamaterials with large negative swelling ratios and tunable stress-strain curves. See Zhang et al. (2018) for details.

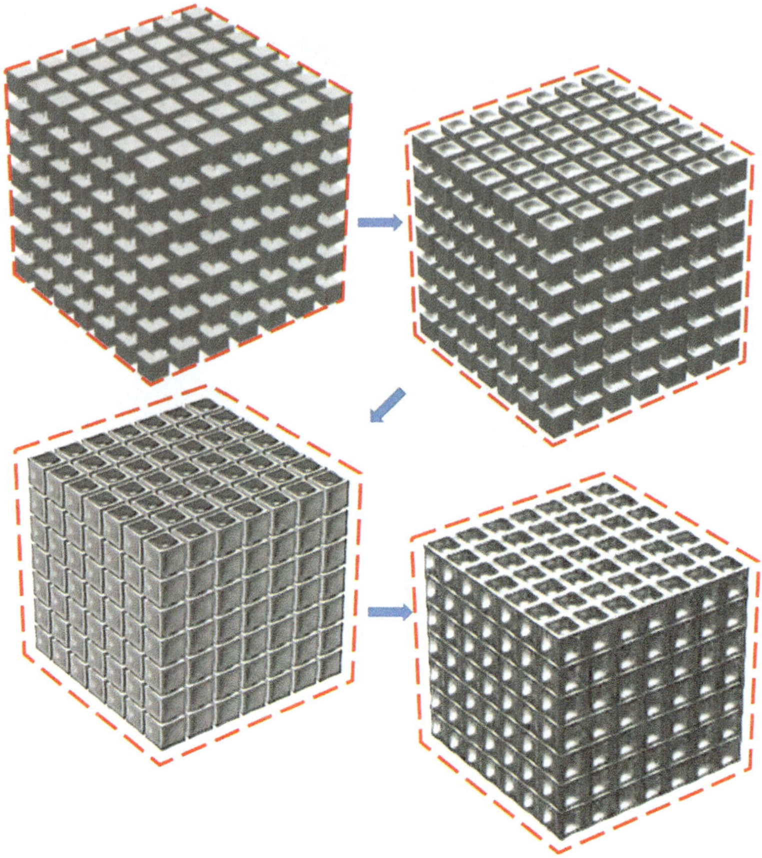

Fig. 13.17 Negative swelling system by Curatolo (2018). Red "boundaries" provide visual aid on the extent of overall contraction

References

Barrett DG, Bushnell GG, Messersmith PB (2013) Mechanically robust, negative-swelling, mussel-inspired tissue adhesives. Adv Healthc Mater 2(5):745–755

Curatolo M (2018) Effective negative swelling of hydrogel-solid composites. Extr Mech Lett 25:46–52

Lim TC (2017a) Negative moisture expansion. In: Abstract book of the 8th international conference on Auxetics and other materials and models with "negative" characteristics, and the 13th international workshop on Auxetics and related systems, Heraklion, p 33

Lim TC (2017b) Auxetic and negative thermal expansion structure based on interconnected array of rings and sliding rods. Phys Status Solidi B 254(12):1600775

Lim TC (2018) A negative hygroscopic expansion material. Mater Sci Forum 928:277–282

Lim TC (2019a) A reinforced kite-shaped microstructure with negative linear and area hygrothermal expansions. Key Eng Mater 803:272–277

Lim TC (2019b) Negative environmental expansion for interconnected array of rings and sliding rods. Phys Status Solidi B 256(1):1800032

Lim TC (2020) Negative hygrothermal expansion of reinforced double arrowhead microstructure. Phys Status Solidi B 257:1800055

Liu J, Gu T, Shan S, Kang SH, Weaver JC, Bertoldi K (2016) Harnessing buckling to design architected materials that exhibit effective negative swelling. Adv Mater 28(31):6619–6624

Zhang H, Gou X, Wu J, Fang D, Zhang Y (2018) Soft mechanical metamaterials with unusual swelling behavior and tunable stress-strain curves. Sci Adv 4(6):eaar8535

Chapter 14
Negative Stiffness

Abstract This chapter reviews a few representative examples of negative stiffness structures (NS) that exhibit NS in one direction, two directions, and all three orthogonal directions based on 2D and 3D microstructures.

Keyword Negative stiffness

14.1 Introduction

Materials and structures typically exhibit positive stiffness. When a load is applied on a material or structure, there is an opposing force. In other words, when one applies a pushing force on a structure it pushes back, or when one pulls a material it pulls back. Negative stiffness implies that when a pushing force is applied, there will come a stage when the material pulls away in the same direction as the pushing force, or vice versa. An understanding of negative stiffness can be easily visualized from the concept of snap-through shown in Fig. 14.1. The application of a force (indicated by arrow) on a pre-buckled beam, as shown in Fig. 14.1 (left), would initially give a conventional or positive stiffness as the buckled beam tends to oppose the applied force. A state of instability is attained (Fig. 14.1 center) wherein the buckled beam snaps through (Fig. 14.1 right). If the arrow is replaced by a spring, then the spring is in the state of compression from Fig. 14.1 (left) to Fig. 14.1 (center), and in a state of tension from Fig. 14.1 (center) to Fig. 14.1 (right).

In 1992, Roderic S. Lakes discovered that the compressive behavior of the tetrakaidecahedron and octahedron cell models were not monotonic, in contrast to the behavior of the foam as a whole. The load-deformation curve displayed a negative slope over a range of strain; outside this range the slope was positive (Lakes 1993). Lakes (2001a, b) and Lakes et al. (2001) discussed the extreme damping in composite materials possessing negative stiffness phase or inclusions, while Wang and Lakes (2001) explored the extreme thermal expansion, piezoelectricity, and other coupled

T. Lim, *Mechanics of Metamaterials with Negative Parameters*, Engineering Materials,
https://doi.org/10.1007/978-981-15-6446-8_14

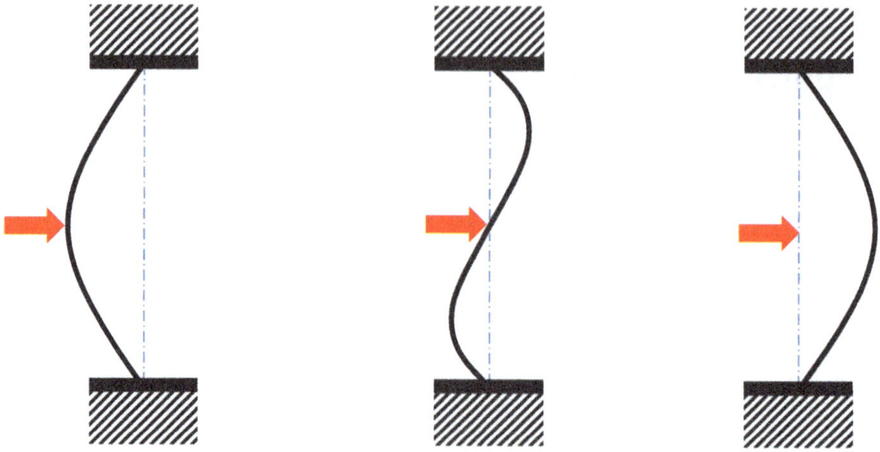

Fig. 14.1 An illustration of negative stiffness using snap-through concept

field properties in composites with a negative stiffness phase. It was also shown that extreme stiffness occurs in composites with negative stiffness phase (Lakes and Drugan 2002) and in systems with negative stiffness elements (Wang and Lakes 2004a). The reader is also referred to viscoelastic studies of negative stiffness systems and components by Wang and Lakes (2004b, 2005) and Wang et al. (2006), as well as to the works of Moore et al. (2006) on negative incremental bulk modulus in foams.

14.2 Example: Negative Stiffness in 2D Microstructures

Although 3D printed, this example by Che et al. (2017) can be categorized as "2D microstructure" due to its prismatic structure and the negative stiffness is 1D as it is manifested in only one dimension—the direction of compressive loading. Figure 14.2 shows the dimensionless deformation for one of the unit cells versus the dimensionless total deformation for two unit cells arranged in series. The dimensionless deformation for the other unit cell is represented by the other curve. An illustration for the progressive compression for a 7-by-5 array of unit cells—i.e., 7 columns and 5 rows—is shown in Fig. 14.3. Che et al. (2017) identified two methods to control the collapse sequence of the microstructure: (a) by varying the thickness of the thin curved beams and (b) by incorporating mode imperfection size that varies for different rows and performed simulation and experimental verification using 5-by-5 array of unit cells. By identifying these two methods of varying the geometrical parameters across, the different rows (while keeping the geometrical parameters constant within the same row) Che et al. (2017) gave reliable approaches for determining and controlling the sequence of snap-through collapse in their multistable metamaterial with negative stiffness.

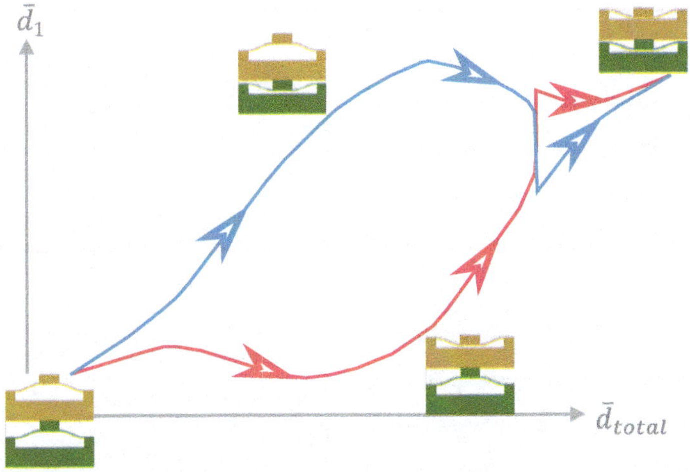

Fig. 14.2 Plots of dimensionless deformation for one unit versus the dimensionless total deformation for both units. For superposition of strain energy contour plot, see Che et al. (2017)

The 2D microstructure by Zhang et al. (2019), illustrated in Fig. 14.4, provides the possibility of 2D negative stiffness. These phase transforming cellular materials (PXCMs) are a class of periodic cellular materials that exhibit solid-state energy absorption and dissipation and are comparable to honeycombs, especially at low plateau stresses. The shape of the elementary beam, shown in Fig. 14.4 (a), is that of the first buckling mode of a straight prismatic beam under axial loading, while the mechanical responses in the form of force-displacement (F-d) and energy-displacement (U-d) responses are displayed in Fig. 14.4b. The various hierarchical levels of the PXCMs are shown in Fig. 14.4c, d, e for levels 0, 1, and 2, respectively. The compressive load-unload cycles performances for the square-type PXCM samples loaded in the 0° or 90° directions (Fig. 14.5 top) and loaded in the 45° or 135° directions (Fig. 14.5 bottom) as well as for the triangular-type PXCM samples loaded in the 0°, 60°, or 120° directions (Fig. 14.6 top) and loaded in the 30°, 90°, or 150° directions (Fig. 14.6 bottom) by Zhang et al. (2019) are described in (a) for their boundary and loading conditions, (b) for the phase transformation sequence by finite element simulation, (c) force-displacement relations comparison between simulated results and experimental verification, and (d) photographs of the original states and final deformed states. Perusal to Fig. 14.5 shows that when the square-type PXCM is loaded in the on-axes directions, the snap-through, or negative stiffness manifestation, occurs in one direction. However, when loaded in the off-axis 45° direction, biaxial snap-through occurs, thereby manifesting negative stiffness in 2D. The numerical and experimental results indicate that these materials exhibit similar solid-state energy dissipation for loads applied along the various axes of reflectional symmetry of the material. Zhang et al. (2019) reported that the specific energy dissipation capacity of the triangular-type is slightly greater and less sensitive to the loading

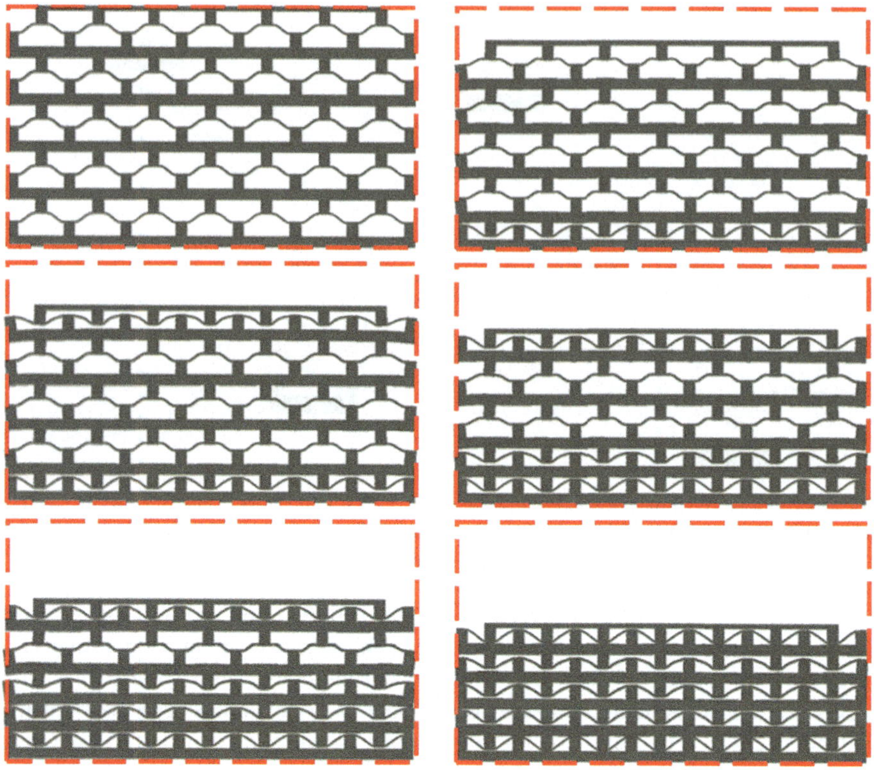

Fig. 14.3 Sequential snap-through with gradual compression for a 7-by-5 array of unit cells. Two methods have been identified by Che et al. (2017) for determining the sequential collapse by rows using 5-by-5 array

direction than the square-type under most of the loading directions. Nevertheless, both types of materials were proven to be very effective in dissipating energy.

14.3 Example: Negative Stiffness in 3D Microstructures

Two examples are shown herein for 3D microstructures that permit snap-through in one direction (Ha et al. 2018) and in all three orthogonal directions (Ha et al. 2019). In the first example, Ha et al. (2018) proposed a unit cell that can be assembled to form a lattice that manifests energy absorption by means of snap-through behavior. See Fig. 14.7 (top). The material used for both the beams and plates was identical. Using materials with Young's modulus $E = 0.717\text{GPa}$, yield strength $\sigma_Y = 31.6\text{MPa}$, Poisson's ratio $v = 0.24$ and density $\rho = 0.962\text{g/cm}^3$, Ha et al. (2018) plotted four graphs–at (a) $r_t/2r_b = 10$, (b) $r_t/2r_b = 15$, (c) $r_t/2r_b = 20$ and (d) $r_t/2r_b = 25$–of downward force versus normalized downward displacement for various inclination

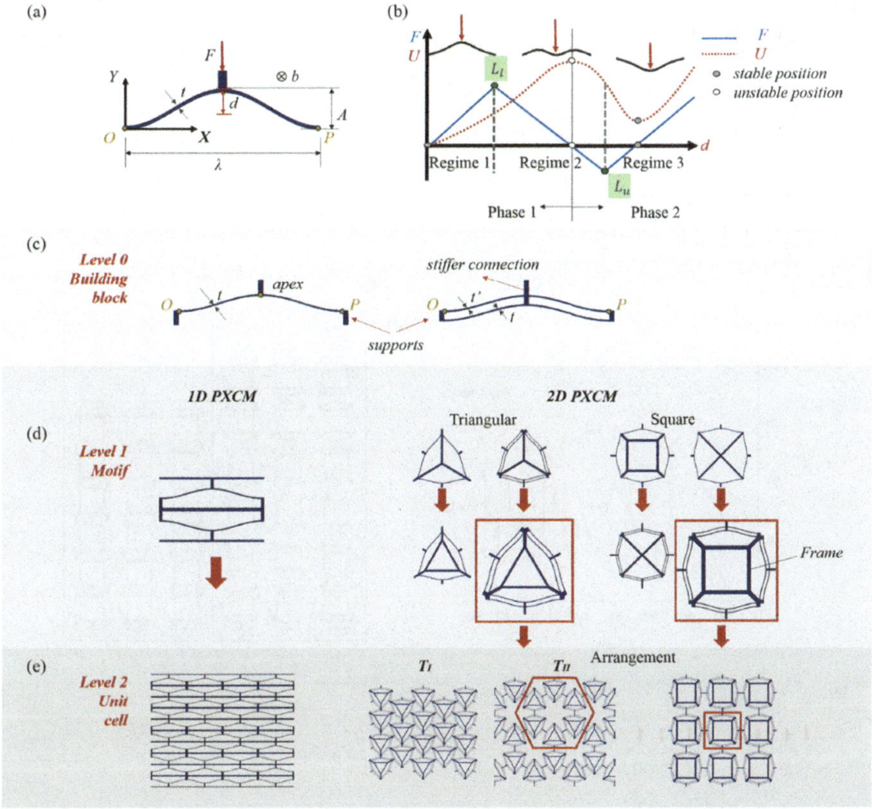

Fig. 14.4 Hierarchical construction of functionally two-dimensional PXCMs by Zhang et al. (2019), displaying **a** geometry of the elementary bent beam, **b** schematic representation of the force-displacement (F-d) and energy-displacement (U-d) responses of a bistable bent beam, and **c–e** Levels 0–2 of the hierarchical structure of the 2D PXCMs studied in their work

angles of the inclined rods with reference to the horizontal plane, where r_t is the diagonal distance from the center of the unit cell to its corner and r_b is the radius of the inclined rods. See Fig. 14.7 (bottom). A newer structure by Ha et al. (2019) exhibits a truly 3D cubic negative stiffness lattice structure that can achieve energy absorption and recover its original configuration under cyclic loading in excess of a strain of approximately 20%. See Fig. 14.8 (top) for a unit cell and Fig.14.8 (bottom) for an assembly. Using the same material properties as Ha et al. (2018) and geometrical details outlined in Ha et al. (2019), the loading and unloading stress–strain cycles were obtained. In the case of loading in the same direction and in the transverse direction for 3D array of unit cells, Ha et al. (2019) show various plots of force-deformation curves for describing the negative stiffness characteristics indicated by multiple snap-throughs Fig. 14.9 (top). Photographs showing the evolution of deformation in the assembly is furnished in Fig. 14.9 (bottom). Their results also

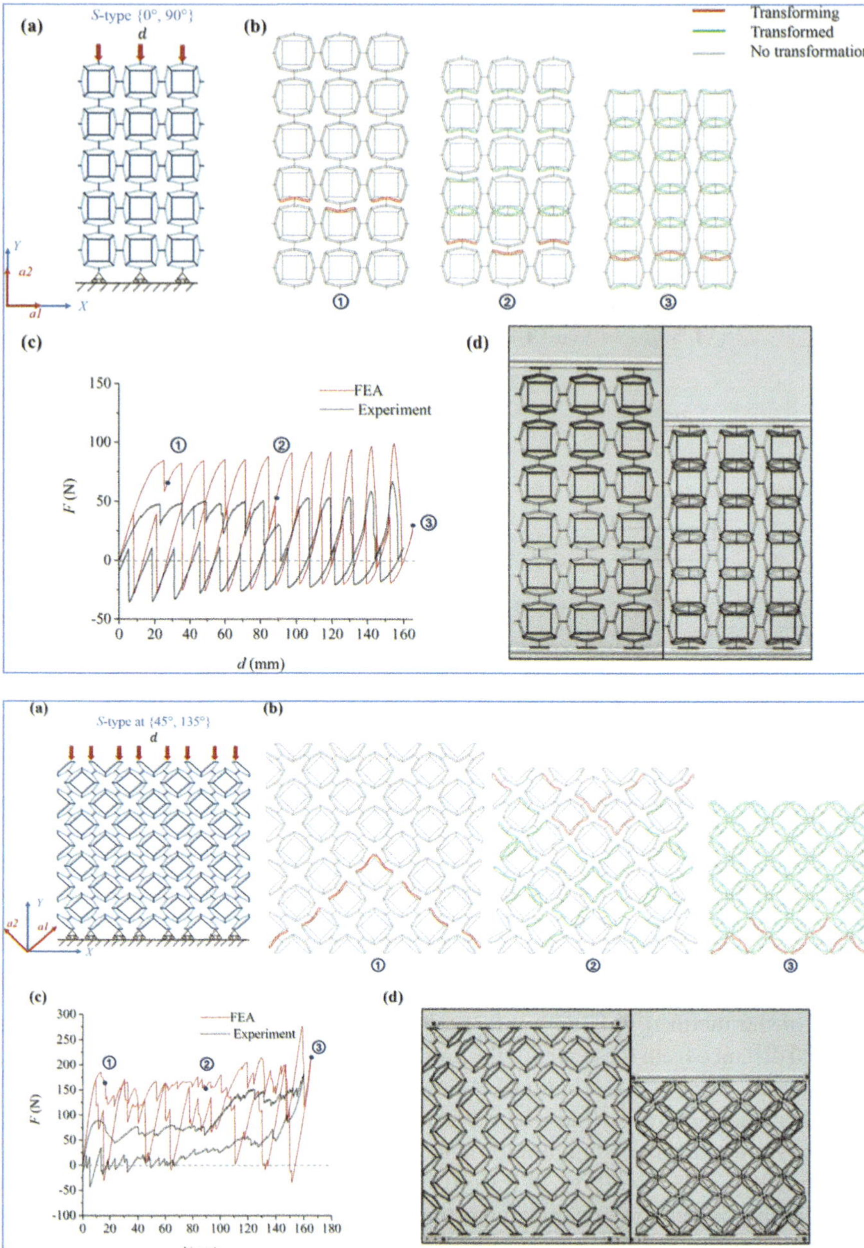

Fig. 14.5 Performance of square-type PXCM sample under compressive load-unload cycles at {0°, 90°} (top half) and {45°, 135°} (bottom half) by Zhang et al. (2019), showing the **a** samples under uniaxial loading condition and supported by rollers at bottom, **b** phase transformation sequence of the three characteristic states from FE simulation, **c** force-displacement relations from FE simulation and experiment, and **d** initial state and final deformed configurations

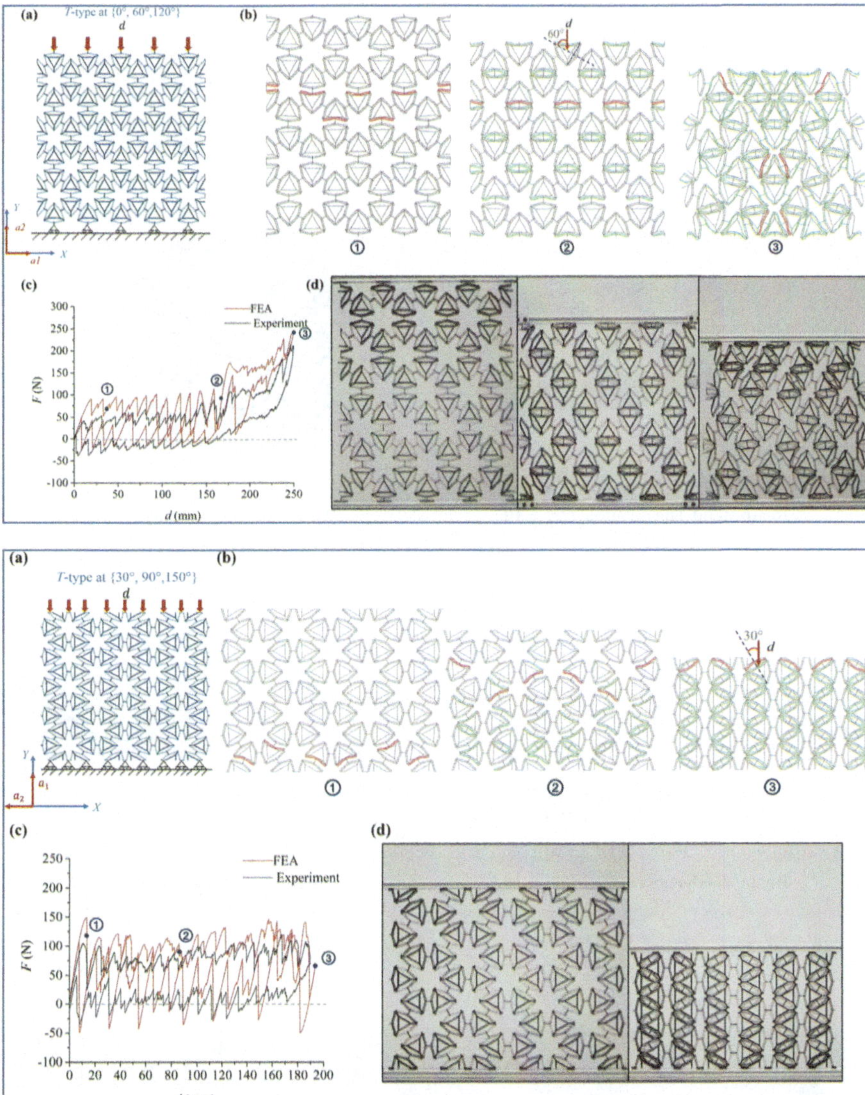

Fig. 14.6 Performance of triangular-type PXCM samples under compressive load-unload cycles at {0°, 60°, 120°} (top half) and {30°, 90°, 150°} (bottom half) by Zhang et al. (2019), showing the **a** samples under uniaxial loading condition and supported by rollers at bottom, **b** phase transformation sequence of the three characteristic states from FE simulation, **c** force-displacement relations from FE simulation and experiment, and **d** initial state and final deformed configurations

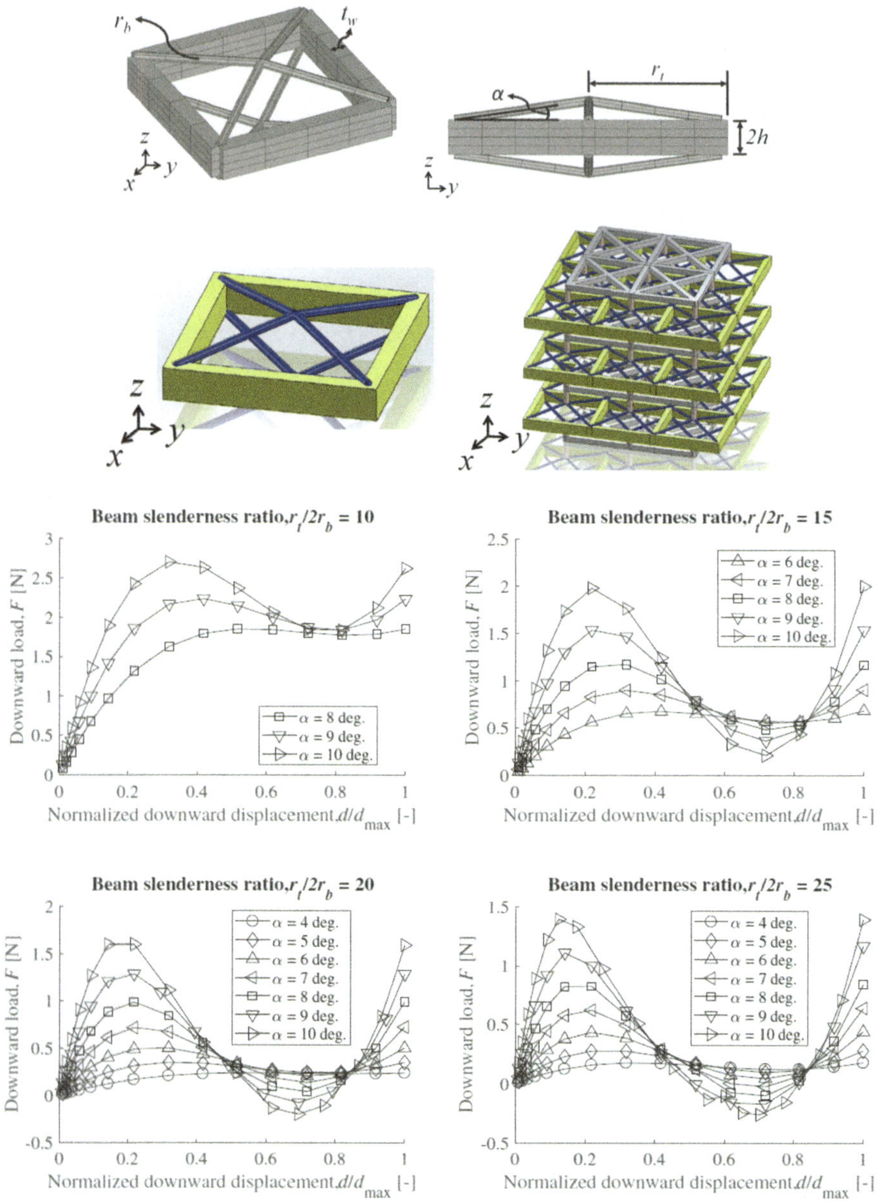

Fig. 14.7 An energy absorption lattice by Ha et al. (2018), showing single units and an assembly (top), with plots of force versus displacements (bottom). Reprinted with permission from Elsevier

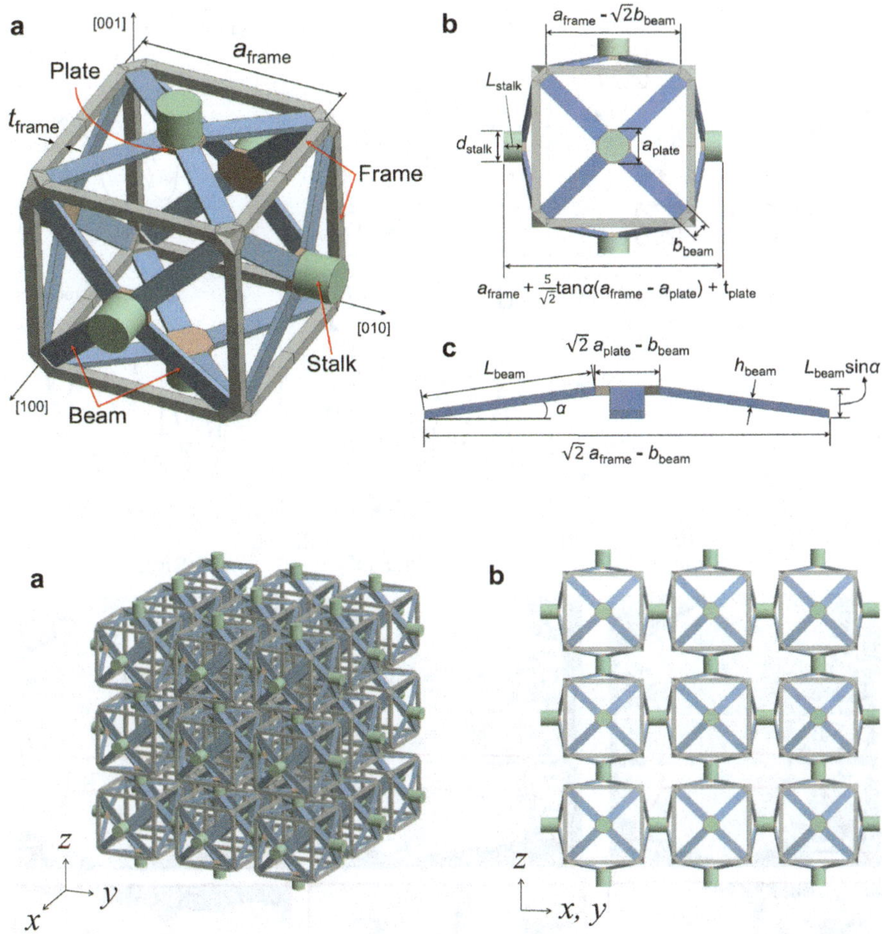

Fig. 14.8 A unit of cubic negative stiffness lattice for energy absorption (top) and a 3 × 3 × 3 array (bottom) by Ha et al. (2019). Reprinted with permission from Elsevier

clearly showed that their designed structure is capable of absorbing mechanical energy effectively with a full recovery of geometry in three principal directions and that the amount of energy absorbed during cyclic loading increases with its size.

14.4 Further Readings on Negative Stiffness Systems

Although similar to the 1D snap-through mechanism by Che et al. (2017) in some aspects, the snap-through mechanism by Rafsanjani is attained by reversing the pre-buckled beams such that snap-through occurs upon application of tensile forces.

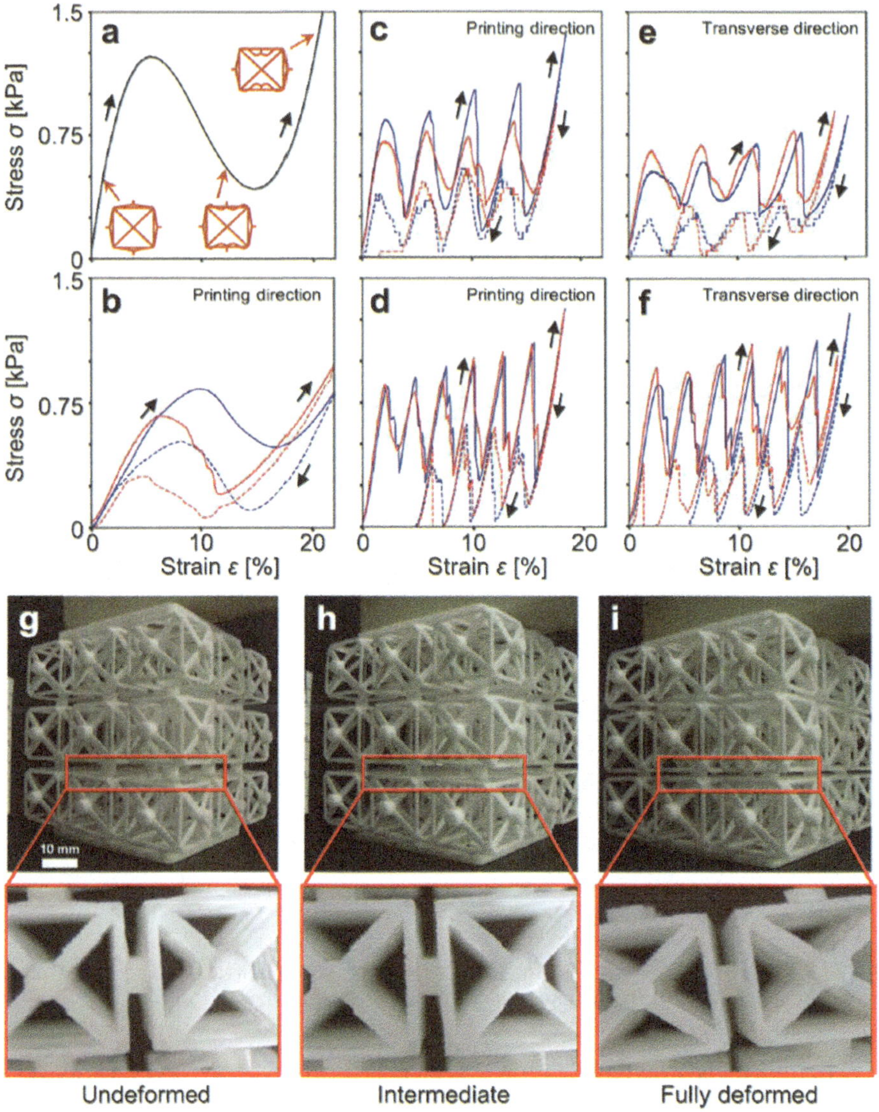

Fig. 14.9 Stress–strain hysteresis loops of negative stiffness lattice structures with a different number of cells per side. **a** FE simulation of a single unit cell; an evolution of deformation configurations at different strains is also shown. **b** Single unit cell in printing direction (experiment). **c–d** 2 × 2 × 2 and 3 × 3 × 3 lattice structures in printing direction, respectively (experiment). **e–f** 2 × 2 × 2 and 3 × 3 × 3 lattice structures in transverse direction, respectively (experiment). Blue and red curves represent responses of samples with the flat and half-sphere end conditions, respectively. All samples were capable of self-recovery with a desired energy dissipation after the removal of loading. **g–i** Photographs displaying an evolution of deformation for a 3 × 3 × 3 negative stiffness lattice structure. Reprinted with permission from Elsevier (Ha et al. 2019)

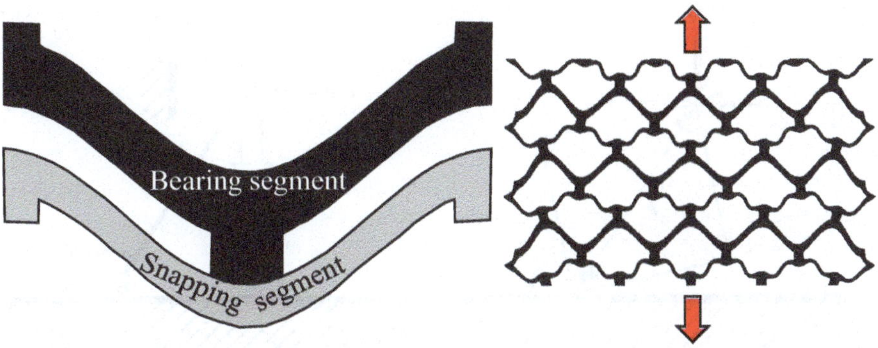

Fig. 14.10 A unit cell of the tensile snap-through mechanism by Rafsanjani et al. (2015) consisting of a pair of bearing and snapping segments (left) and an idealized illustration for the deformed microstructure upon snapping with the application of tensile load (right)

This is attained by the use of a pair of bearing and snapping segments as shown in Fig. 14.10 (left). The bearing segment is relatively rigid such that the application of tensile force brings about the deformed microstructural shape as shown in Fig. 14.10 (right).

A novel concept of negative stiffness by means of rotation of non-spherical particles was introduced by Dyskin and Pasternak (2012) and further discussed by Karachevtseva et al. (2019). See Fig. 14.11 (top left). Taking moment about the contact point on the ground "O", as shown in Fig. 14.11 (top right), we have $Fl\ \sin(180° - \theta) = Nl\ \cos(180° - \theta)$ or

$$F\ \sin\theta = -N\ \cos\theta \tag{14.4.1}$$

from which

$$\mathrm{d}F\ \sin\theta + F\ \cos\theta \mathrm{d}\theta = N\ \sin\theta \mathrm{d}\theta \tag{14.4.2}$$

From the kinematics of the rod rotation shown in Fig. 14.11 (bottom left), we have

$$\mathrm{d}\theta = -\frac{\mathrm{d}u}{l\ \sin\theta} \tag{14.4.3}$$

Substituting Eqs. (14.4.1) and (14.4.3) into Eq. (14.4.2) gives a familiar spring force equation

$$\mathrm{d}F = -k^{(-)}\mathrm{d}u \tag{14.4.4}$$

where

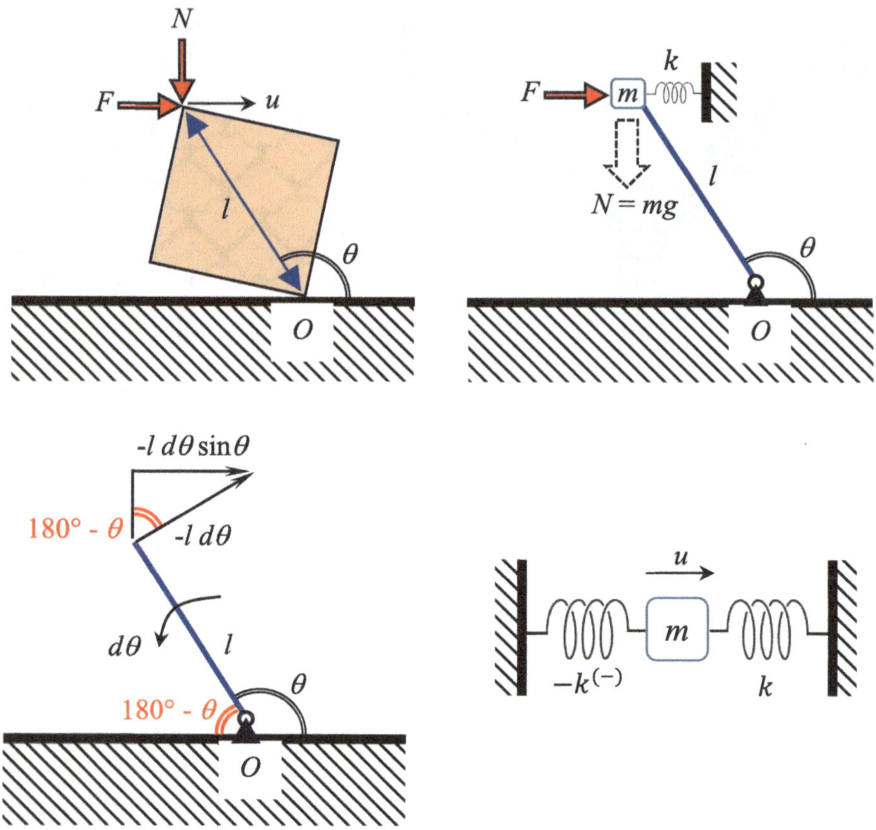

Fig. 14.11 Apparent negative stiffness produced by rotating blocks by Karachevtseva et al. (2019), demonstrating rotation of a single square block (top left), conceptualization as an inverted pendulum from the block corners in contact (top right), kinematic relationship (bottom left), and negative stiffness model of the inverted pendulum (bottom right)

$$k^{(-)} = \frac{N}{l\ \sin^3 \theta} \tag{14.4.5}$$

In studying the effects of negative stiffness produced by rotation of non-spherical particles on frictional sliding, Karachevtseva et al. (2019) demonstrated that the concept of negative stiffness permits formulation of a simple and accurate model of inverted pendulum and rotating non-spherical particles. In addition, they showed that a set of randomly sized rotating non-spherical particles create fluctuations in the friction force, which can form a mechanism of experimentally observed friction force fluctuations.

A highly innovative work on negative (and positive) stiffness metamaterial was reported by Dudek et al. (2018) and shown in Fig. 14.12 (left). They discussed the concept associated with the insertion of magnets into their mechanical metamaterials

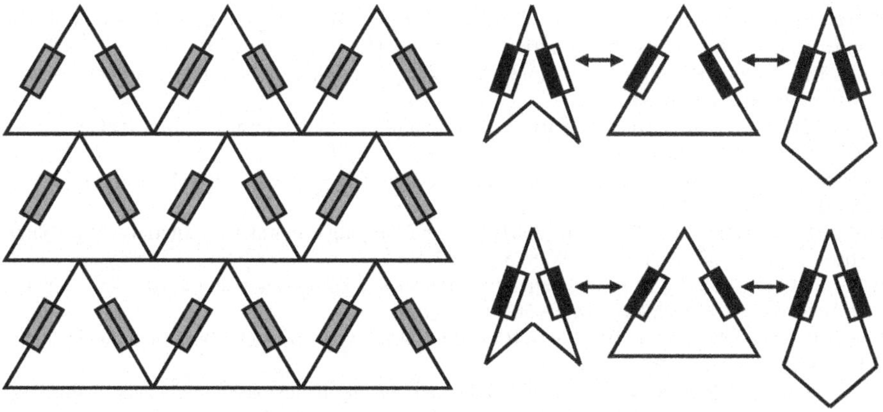

Fig. 14.12 System discovered by Dudek et al. (2018), showing magnetic attachments (left), as well as deformation of a single unit with magnetic attraction (top right) and magnetic repulsion (bottom right)

in order to obtain negative stiffness. Specifically, it was shown that the investigated system may exhibit either positive or negative stiffness, depending solely on the orientation of magnets within the system. See Fig. 14.12 (right). Dudek et al. (2018) pointed out that the system can manifest negative stiffness and auxetic property at the same time. They also discussed the system's stability, wherein the systems consisting of attracting and repelling magnets are unstable and stable, respectively.

In addition to the above, the reader is also referred to the works of Balch and Lakes (2017) on the amelioration of waves and microvibrations by microbuckling in open-celled foam and the works of Wang et al. (2019) on negative stiffness plates. A comprehensive discourse on metamaterials with inclusions having auxetic or negative stiffness has been given by Pasternak and Dyskin (2019).

References

Balch and Lakes RS (2017) Amelioration of waves and microvibrations by micro-buckling in open celled foam. Cellular Polymers 36(1):1–12

Che K, Yuan C, Wu J, Qi HJ, Meaud J (2017) Three-dimensional-printed multistable mechanical metamaterials with a deterministic deformation sequence. J Appl Mech 84(1):011004

Dyskin AV, Pasternak E (2012) Mechanical effect of rotating non-spherical particles on failure in compression. Phil Mag 92(28–30):3451–3473

Dudek KK, Gatt R, Dudek MR, Grima JN (2018) Negative and positive stiffness in auxetic magneto-mechanical metamaterials. Proceedings of Royal Society A 474(2215):20180003

Ha CS, Lakes RS, Plesha ME (2018) Design, fabrication, and analysis of lattice exhibiting absorption via snap through behavior. Mater Des 141:426–437

Ha CS, Lakes RS, Plesha ME (2019) Cubic negative stiffness lattice structure for energy absorption: Numerical and experimental studies. Int J Solids Struct 178–179:127–135

Karachevtseva I, Pasternak E, Dyskin AV (2019) Negative stiffness produced by rotation of non-spherical particles and its effect on frictional sliding. Phys Status Solidi B 256(1):1800003
Lakes RS, Rosakis P, Ruina A (1993) Microbuckling instability in elastomeric cellular solids. J Mater Sci 28(17):4667–4672
Lakes RS (2001a) Extreme damping in compliant composites with a negative-stiffness phase. Philos Mag Lettters 81(2):95–100
Lakes RS (2001b) Extreme damping in composite materials with a negative stiffness phase. Phys Rev Lett 86(13):2897–2900
Lakes RS, Lee T, Bersie A, Wang YC (2001) Extreme damping in composite materials with negative stiffness inclusions. Nature 410(6828):565–567
Lakes RS, Drugan WJ (2002) Dramatically stiffer elastic composite materials due to a negative stiffness phase? J Mech Phys Solids 50(5):979–1009
Moore B, Jaglinski T, Stone DS, Lakes RS (2006) Negative incremental bulk modulus in foams. Philos Mag Lett 86(10):651–659
Pasternak E, Dyskin AV (2019) Architectured materials with inclusions having negative Poisson's ratio or negative stiffness. In: Estrin Y, Brechet Y, Dunlop J, Fratzl P (eds) Architectured materials in nature and engineering. Springer, Switzerland, pp 51–87
Rafsanjani A, Akbarzadeh A, Pasini D (2015) Snapping mechanical metamaterials under tension. Adv Mater 27(39):5931–5935
Wang YC, Lakes RS (2001) Extreme thermal expansion, piezoelectricity, and other coupled field properties in composites with a negative stiffness phase. J Appl Phys 90(12):6458–6465
Wang YC, Lakes RS (2004a) Extreme stiffness systems due to negative stiffness elements. Am J Phys 72(1):40–50
Wang YC, Lakes RS (2004b) Negative stiffness-induced extreme viscoelastic mechanical properties: stability and dynamics. Phil Mag 84(35):3785–3801
Wang YC, Lakes RS (2005) Stability of negative stiffness viscoelastic systems. Q Appl Math 63(1):34–55
Wang YC, Swadener JG, Lakes RS (2006) Two-dimensional viscoelastic discrete triangular system with negative-stiffness components. Philos Mag Lett 86(2):99–112
Wang YC, Lai HW, Shen MW (2019) Effects of cracks on anomalous mechanical behavior and energy dissipation of negative-stiffness plates. Phys Status Solidi B 256(1):1800489
Zhang Y, Restrepo D, Velay-Lizancos M, Mankame ND, Zavattieri PD (2019) Energy dissipation in functionally two-dimensional phase transforming cellular materials. Sci Rep 9:12581

Chapter 15
Sign-Switching of Metamaterial Properties

Abstract Sign-switching refers to materials that can exhibit both positive and negative properties in situ as a result of changing loading direction or opposing changes in environmental conditions, without active control or the need impose changes to the microstructural architecture. Two concepts for sign-switching are briefly introduced, microstructural reshape and microstructural duality, which are elaborated in the remaining chapters of the book.

Keywords Sign-flipping · Sign-reversible · Sign-switching · Sign toggling

15.1 Introduction

We have so far considered the usefulness of architecting the microstructural lattice geometry of metamaterials for producing negative mechanical properties. In some applications, it is beneficial for these properties to be positive under certain conditions but reverse into negative properties under opposing conditions without the need to impose further processing of the materials. For the sake of practical applications, it is useful for the switch in material property signs to take place without active control. For example, the sign of Poisson's ratio can be changed in graphene by hydrogenation (Jiang et al. 2016), by applied strain along the zigzag direction of semi-fluorinated graphene (Qin et al. 2017), and in blue phosphorus oxide (Zeng et al. 2019). The sign of Poisson's ratio in composites has also been shown to be tunable via adjustment of the temperature by Li et al. (2016), and by Jopek and Strek (2018). In addition, the sign-switching of CTE of $YFe(CN)_6$-based Prussian blue analog was demonstrated by Gao et al. (2017) by the introduction of guest ions $\left(K^+\right)$ and molecules (H_2O). It is within the context of smart metamaterials for their overall mechanical properties to alter (drastically or otherwise) by sensing the change in the external stimuli. In order to appreciate the sign-switching properties, as opposed to both the conventional (positive) and negative properties, recall that the simplest conventional responses to stimuli are linear ones, as shown in Fig. 15.1 (top left). The material properties are taken from the slope and, in the case of linear plots, the properties are positive constants, as shown in Fig. 15.1 (top right). Some examples of stimuli and their corresponding responses and properties are listed in Table 15.1. Assuming again the

T. Lim, *Mechanics of Metamaterials with Negative Parameters*, Engineering Materials,
https://doi.org/10.1007/978-981-15-6446-8_15

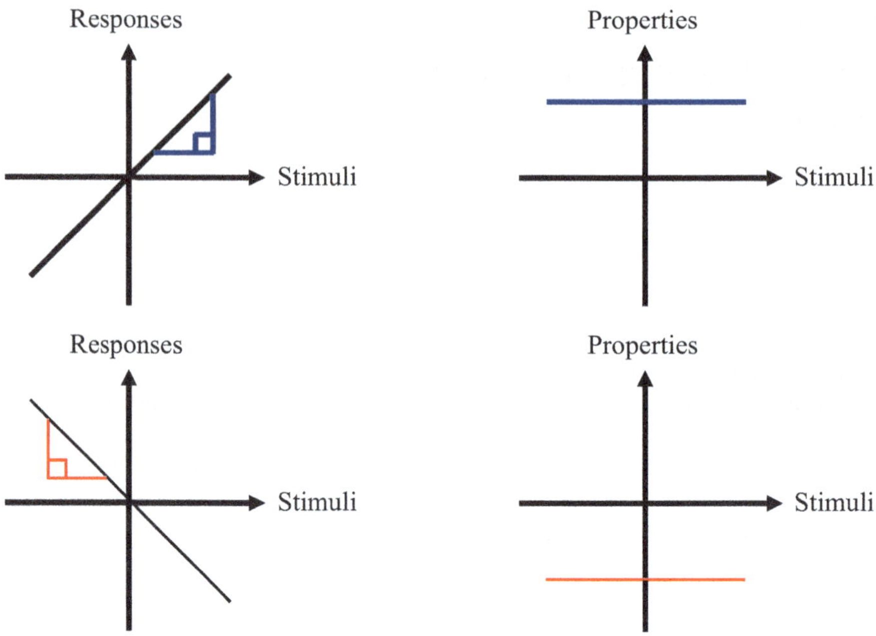

Fig. 15.1 Generic linear responses versus stimuli plots (left) and their corresponding properties versus stimuli plots (right) for conventional or positive systems (top) and negative systems (bottom)

Table 15.1 List of stimuli with corresponding responses and properties

Stimuli	Responses	Properties
$\varepsilon_{\mathrm{Loading}}$	$-\varepsilon_{\mathrm{Tranverse}}$	v
$\mathrm{d}T$	$\varepsilon^{(T)}$	$\alpha^{(T)}$
$-\mathrm{d}P$	$\varepsilon^{(P)}$	$\alpha^{(P)}$
$\mathrm{d}C$	$\varepsilon^{(C)}$	$\alpha^{(C)}$

simplest category of linear responses to stimuli, the corresponding responses and properties for negative materials with reference to stimuli are shown in Fig. 15.1 (bottom). This is not the case of sign-switching metamaterials, which are briefly introduced in Sects. 15.2 and 15.3 for two broad categories.

15.2 Microstructural Reshape

Microstructural reshape refers to the gradual transformation of the unit cell shape such that the imposition of stimuli changes its shape, thereby resulting in altering the overall metamaterial properties (Lim 2019a, 2019b, 2019c, 2019d, 2019e, 2020a, 2020b, 2020c). There are two sub-categories for sign-switching metamaterial by microstructural reshape. In the first sub-category, the original shape of the unit cell is chosen such that it conforms to the characteristic shape that exhibits zero material properties. Upon a positive change in the stimulus, the unit cells reshape to resemble those that exhibit negative properties while a negative change in stimulus brings about unit cell reshape that are similar to those that show positive properties. This results in the plots illustrated in Fig. 15.2 (top). When the response is the opposite, i.e., upon a positive change in the stimulus the unit cells' shape shift to conform to those that manifest positive properties while a negative change in stimulus brings about unit cell alteration to align with those that show negative properties. As a consequence, we have the plots as shown in Fig. 15.2 (bottom). Since the response within each sub-category is similar regardless of whether the change in stimulus is positive or negative (Fig. 15.2, left), therefore, there is a change in the material property sign. This change in material property sign takes place gradually (Fig. 15.2, right).

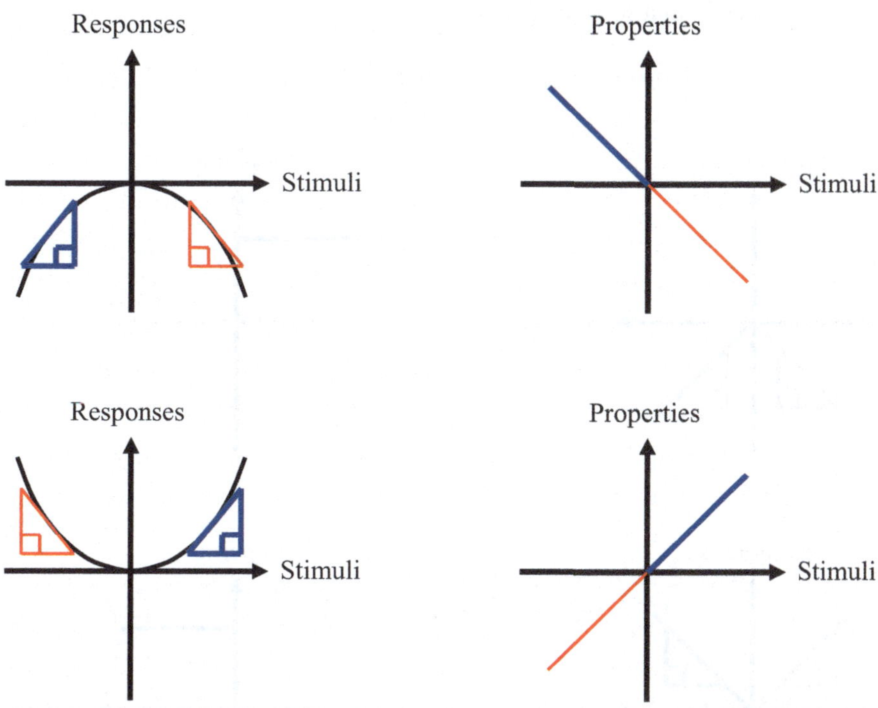

Fig. 15.2 Generic nonlinear responses versus stimuli plots (left) and their corresponding properties versus stimuli plots (right) for metamaterials with gradual change in material properties

15.3 Microstructural Duality

Unlike the sign-switching material properties by microstructural reshape, which occurs gradually, the sign-switching by means of microstructural duality comes about abruptly. This is emphasized in Fig. 15.3 (left), whereby the plots of responses versus stimuli are continuous throughout but non-differentiable at the original state

$$\lim_{\text{Stimulus}\to 0^-}\left(\frac{\partial(\text{Response})}{\partial(\text{Stimulus})}\right) \neq \lim_{\text{Stimulus}\to 0^+}\left(\frac{\partial(\text{Response})}{\partial(\text{Stimulus})}\right) \tag{15.3.1}$$

such that the material property is non-existent when there is no change in stimuli; instead, there exists two sets of material properties (Fig. 15.3, right), one set when the change in stimulus is positive and another set when the change in stimulus is negative. The difference in both sets of material properties are not only in terms of their magnitudes, but also in their signs, i.e.,

$$(\text{Property})_{(\text{Stimulus})<0} \times (\text{Property})_{(\text{Stimulus})>0} < 0 \tag{15.3.2}$$

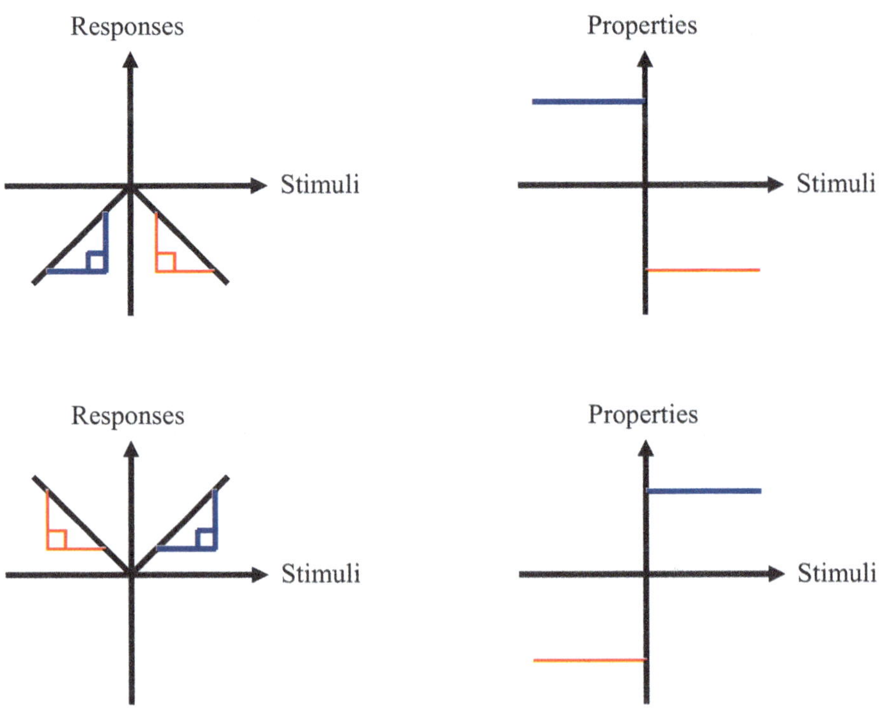

Fig. 15.3 Generic continuous piecewise linear responses versus stimuli plots (left) and their corresponding properties versus stimuli plots (right) for metamaterials with sudden change in material properties

The sudden change in property can be implemented by means of microstructural duality (Lim 2019f, g). But how can a single microstructure exhibit duality? By means of functional and redundant parts! In other words, the microstructure is deliberately overdesigned to cater for redundancies. Under a positive change in stimulus, some parts of the microstructure are functional while other parts are non-functional. When the change in stimuli is negative, the functional parts become redundant while the non-functional parts perform. As a result of such microstructural functional-redundant parts swopping, there is no gradual evolution of the microstructural shape from a negative change to a positive change in the stimulus. Under nonlinear case, there is gradual change in the material property with respect to the change in stimulus but not so at the original state.

References

Gao Q, Chen J, Sun Q, Chang D, Huang Q, Wu H, Sanson A, Milazzo R, Zhu H, Li Q, Liu Z, Deng J, Xing X (2017) Switching between giant positive and negative thermal expansions of a $YFe(CN)_6$-based Prussian blue analogue induced by guest species. Angew Chem 56(31):9023–9028

Jiang JW, Chang T, Guo X (2016) Tunable negative Poisson's ratio in hydrogenated graphene. Nanoscale 8(35):15948–15953

Jopek H, Strek T (2018) Thermoauxetic behavior of composite structures. Materials 11(2):294

Li D, Ma J, Lakes RS (2016) A bi-material structure with Poisson's ratio tunable from positive to negative via temperature control. Mater Lett 181:285–288

Lim TC (2019a) Composite microstructures with Poisson's ratio sign switching upon stress reversal. Compos Struct 209:34–44

Lim TC (2019b) A composite metamaterial with sign switchable elastic and hygrothermal properties induced by stress direction and environmental change reversals. Compos Struct 220:185–193

Lim TC (2019c) A class of shape-shifting composite metamaterial honeycomb structures with thermally-adaptive Poisson's ratio signs. Compos Struct 226:111256

Lim TC (2019d) 2D metamaterial with in-plane positive and negative thermal expansion and thermal shearing based on interconnected alternating biomaterials. Mater Res Express 6(11):115804

Lim TC (2019e) Composite metamaterial with sign-switchable coefficients of hygroscopic, thermal and pressure expansions. Adv Compos Hybrid Mater 2(4):657–669

Lim TC (2019f) Metamaterials with Poisson's ratio sign toggling by means of microstructural duality. SN Appl Sci 1(2):176

Lim TC (2019g) A 2D auxetikos system based on interconnected shurikens. SN Appl Sci 1(11):1383

Lim TC (2020a) Metacomposite with auxetic and in situ sign reversible thermal expansivity upon temperature fluctuation. Compos Commun 19:30–36

Lim TC (2020b) Composite metamaterial square grids with sign-flipping expansion coefficients leading to a type of Islamic design. SN Appl Sci 2(5):918

Lim TC (2020c) Metacomposite structure with sign-changing coefficients of hygrothermal expansions inspired by Islamic motif. Compos Struct 251:112660

Qin R, Zheng J, Zhu W (2017) Sign-tunable Poisson's ratio in semi-fluorinated graphene. Nanoscale 9(1):128–133

Zeng B, Long M, Dong Y, Xiao J, Zhang S, Yi Y, Gao Y (2019) Stress-sign-tunable Poisson's ratio in monolayer blue phosphorus oxide. J Phys Condens Matter 31(29):295702

Chapter 16
Sign-Switching of Poisson's Ratio with Stress Reversal

Abstract This chapter considers two examples of Poisson's ratio sign-switching upon stress reversal. In the first example, the strain that is transverse to the loading direction is persistently negative for two types of microstructures. The microstructures were designed based on microstructural reshape with the direction of applied strain. In addition, these microstructures reveal NTE and ZTE properties under unconstrained and constrained boundaries when specific conditions are met. In the second example, the strain that is transverse to the loading direction is persistently positive for two types of microstructures. The microstructures were designed based on microstructural duality with the direction of applied strain. Specifically, these microstructures employ alternating lock and slide mechanism.

Keywords Poisson's ratio · Sign-flipping · Sign-reversible · Sign-switching · Sign toggling · Stress reversal

16.1 Sign-Switching of Poisson's Ratio with Persistently Negative Transverse Strain

Two types of 2D composite metamaterials—rectangular cells in triangular array composite microstructure (Fig. 16.1, top row) and triangular cells in rectangular array composite microstructure (Fig. 16.1, bottom row)—under the influence of uniaxial loading are shown to possess both conventional and auxetic behavior, depending on the sign of the applied stress (Lim 2019a). For brevity, the rectangular cells in triangular array composite microstructure and the triangular cells in rectangular array composite microstructures are hereinafter termed rectangular microstructure and triangular microstructure, respectively. Each of these composite microstructures consists of short rigid rods and long flexible rods connected by freely rotating hinges, as shown in Fig. 16.1 (left column). The rectangular microstructure is geometrically akin to the centro-symmetric honeycomb with T-shaped joints by Cauchi et al. (2013), except that the said honeycomb is made from a single material, while both the rectangular and triangular microstructures considered herein consist of two constituent materials and are therefore composite structures.

T. Lim, *Mechanics of Metamaterials with Negative Parameters*, Engineering Materials,
https://doi.org/10.1007/978-981-15-6446-8_16

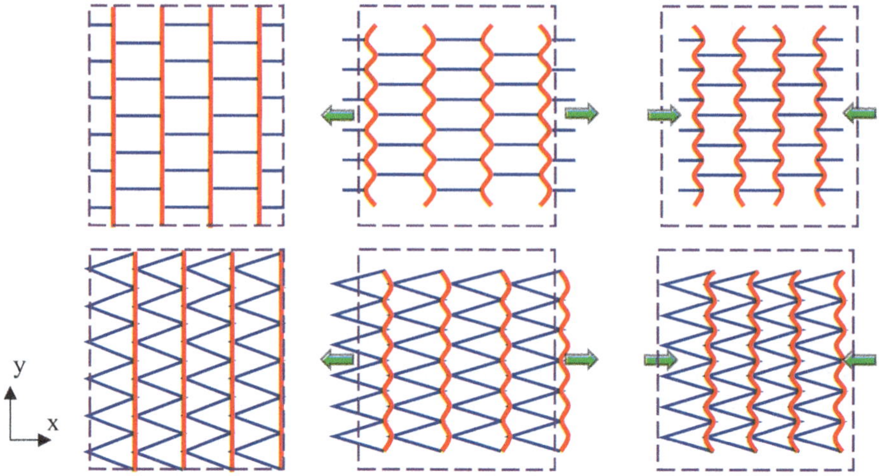

Fig. 16.1 Rectangular microstructure (top row) and triangular microstructure (bottom row) before deformation (left), exhibiting $\upsilon_{xy} > 0$ for $\sigma_x > 0$ (middle), and $\upsilon_{xy} < 0$ for $\sigma_x < 0$ (right). Dashed rectangles indicate dimension of the undeformed structure

Application of horizontal tensile stress $\sigma_x > 0$ (i.e., $\varepsilon_x > 0$) or horizontal compressive stress $\sigma_x < 0$ (i.e., $\varepsilon_x < 0$) on these microstructures causes flexure to the long flexible rods that leads to shorter projected length on the vertical axis, which gives $\varepsilon_y < 0$, thereby resulting in conventional ($\upsilon_{xy} > 0$) and auxetic ($\upsilon_{xy} < 0$) properties, respectively, as indicated in Fig. 16.1 (middle and right columns). In other words, the sign of υ_{xy} switches with the reversal of σ_x direction.

The deformation analysis of the long flexible rods for both microstructures as shown in Fig. 16.1 (left) can be made by considering moderately large deflection of simply supported beam of length L, Young's modulus E, and second moment area I about its neutral axis under a concentrated load P at mid-span, as illustrated in Fig. 16.2, which gives the deflection profile

Fig. 16.2 Moderately large deflection of simply supported beam under a concentrated load at mid-span

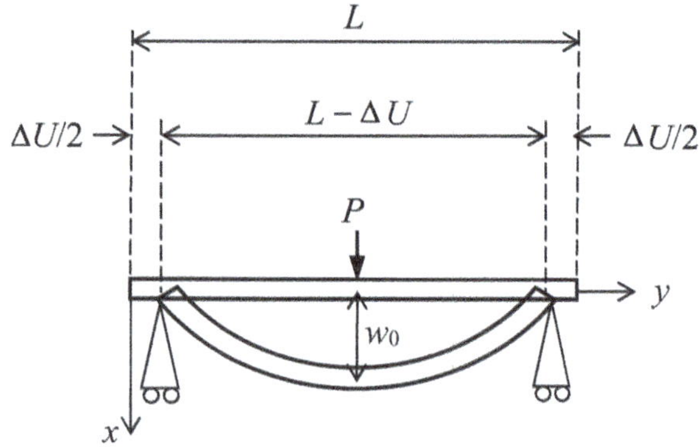

$$w = w_0\left(3\frac{y}{L} - 4\frac{y^3}{L^3}\right); \quad 0 \le \frac{y}{L} \le \frac{1}{2} \tag{16.1.1}$$

where the maximum deflection in the case of central lateral load

$$w_0 = \frac{PL^3}{48EI} \tag{16.1.2}$$

occurs at the mid-span, while the deflection on the other half of the beam is obtained on the basis of mirror image at the beam mid-span. In general, the decrease in projected length ΔU can be obtained from

$$\frac{dU}{dy} = \frac{1}{2}\left(\frac{dw}{dy}\right)^2 \Rightarrow \Delta U = \frac{1}{2}\int_0^L \left(\frac{dw}{dy}\right)^2 dy \tag{16.1.3a}$$

However, the validity range of the deflection description in Eq. (16.1.1) applies for $0 \le y \le L/2$ only. As such,

$$\Delta U = \int_0^{L/2} \left(\frac{dw}{dy}\right)^2 dy = 2.4\frac{w_0^2}{L} \tag{16.1.3b}$$

These details are then implemented for the deformed microstructures of Fig. 16.1. It can be seen that the "wavelength" of the deformed flexible rod in both microstructures is $2L$, hence, resulting in a maximum deflection of $2w_0$. In general, the strains in the y- and x-directions are

$$\varepsilon_y = -\frac{\Delta U}{L} \tag{16.1.4}$$

and

$$\varepsilon_x = \frac{W_1}{W} - 1 \tag{16.1.5}$$

respectively, and the y-direction strain for both microstructures under σ_x loading can be obtained by substituting Eq. (16.1.3b) into Eq. (16.1.4) to yield

$$\varepsilon_y = -2.4\left(\frac{w_0}{L}\right)^2 \tag{16.1.6}$$

The x-direction strain for the rectangular microstructure (Fig. 16.3, top left) is easily obtained by considering only the maximum deflections of the flexible rods since the short rigid rods remain parallel to the x-axis. Hence, the consideration of the maximum deflection of the long flexible rods gives

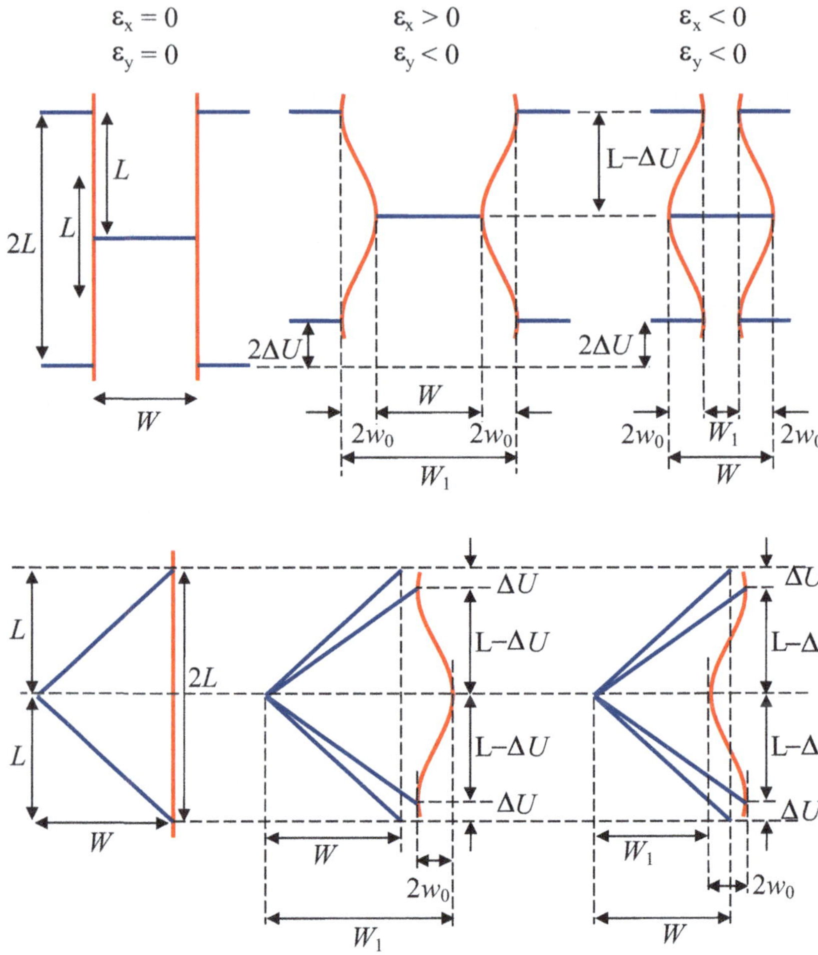

Fig. 16.3 Representative units of rectangular microstructure (top row) and triangular microstructure (bottom row) for analysis, showing the undeformed state (left column), as well as deformed state under $\sigma_x > 0$ stress (middle column) and $\sigma_x < 0$ stress (right column). The middle and right columns will also be revisited later when considering buckling under $\sigma_y < 0$ stress

$$\varepsilon_x = \frac{4w_0}{W} \tag{16.1.7}$$

where the w_0 is positive or negative when deflected outward (Fig. 16.3, top middle) or inward (Fig. 16.3, top right), respectively. These strain expressions give

$$v_{xy} = -\frac{\varepsilon_y}{\varepsilon_x} = \frac{3W}{5L}\left(\frac{w_0}{L}\right) \tag{16.1.8}$$

for the rectangular microstructure. For the triangular microstructure under σ_x loading (Fig. 16.3, bottom left), the strain in x-direction can be established by firstly considering length constancy of the short rigid rods during rotation

$$\sqrt{(W_1 - 2w_0)^2 + (L - \Delta U)^2} = \sqrt{W^2 + L^2} \tag{16.1.9}$$

as indicated by Fig. 16.3 (bottom middle). For compressive stress in x-direction (Fig. 16.3, bottom right), w_0 possesses negative value. Substituting Eq. (16.1.3b) into Eq. (16.1.9) and solving for W_1, we have

$$W_1 = 2w_0 \pm \sqrt{W^2 + 4.8w_0^2 - 5.76\frac{w_0^4}{L^2}} \tag{16.1.10}$$

Selecting the lower solution would give $W_1 < 0$, which is physically inadmissible. Hence, substituting the upper solution of Eq. (16.1.10) into Eq. (16.1.5) gives

$$\varepsilon_x = -1 + 2\frac{w_0}{W} + \sqrt{1 + 4.8\frac{w_0^2}{W^2} - 5.76\frac{w_0^4}{W^2 L^2}} \tag{16.1.11}$$

which, together with Eq. (16.1.6), leads to

$$\upsilon_{xy} = -\frac{2.4\left(\frac{w_0}{L}\right)^2}{1 - 2\frac{w_0}{L}\left(\frac{L}{W}\right) - \sqrt{1 + 4.8\left(\frac{w_0}{L}\right)^2\left(\frac{L}{W}\right)^2 - 5.76\left(\frac{w_0}{L}\right)^4\left(\frac{L}{W}\right)^2}} \tag{16.1.12}$$

where in the case of triangular microstructure, the ratio L/W provides the inclination angle of the short rigid rods, $\tan^{-1}(L/W)$, with reference to the x-axis.

Perusal to the rectangular microstructure Fig. 16.4 (top left) indicates that when tensile stress is applied in the y-direction, the flexible rods elongate while the short rigid rods remain aligned parallel to the x-axis and displace almost only parallel to the y-axis (Fig. 16.4, top right). As such

$$\upsilon_{yx} = -\frac{\varepsilon_x}{\varepsilon_y} \approx 0 \tag{16.1.13}$$

for the rectangular microstructure since $\varepsilon_x \approx 0$. When a similar stress is applied on the triangular microstructure (Fig. 16.4, bottom left), the elongation of the flexible rods induces rotation of the short rigid rods such that the latter become more oriented to the y-axis, as shown in Fig. 16.4 (bottom right). This results in shortening of the entire structure along the x-direction, thereby implying that $\upsilon_{yx} > 0$ since $\varepsilon_x < 0 < \varepsilon_y$.

With reference to Fig. 16.5, the Poisson's ratio for the triangular microstructure under $\sigma_y > 0$ can be made by considering again the length constancy of the rigid rods

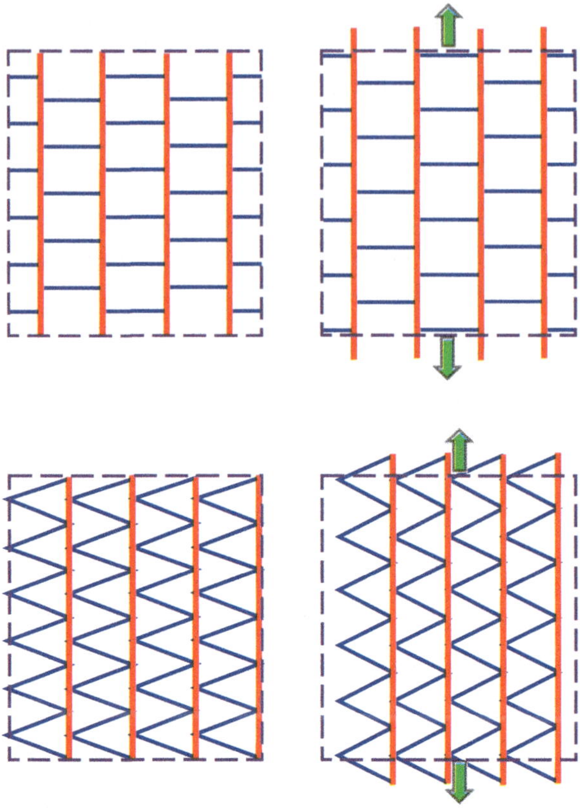

Fig. 16.4 Application of $\sigma_y > 0$ gives $\upsilon_{yx} \approx 0$ and $\upsilon_{yx} > 0$ for the rectangular (top) and triangular (bottom) microstructures, respectively

$$\sqrt{W_1^2 + (L + \Delta L)^2} = \sqrt{W^2 + L^2} \tag{16.1.14}$$

to give

$$W_1 = \sqrt{W^2 - 2L\Delta L - (\Delta L)^2} \tag{16.1.15}$$

where $\Delta L > 0$ is the elongation with respect to each length L of the flexible rod. Needless to say, the strain in y-direction is

$$\varepsilon_y = \frac{\Delta L}{L} > 0 \tag{16.1.16}$$

while substituting Eq. (16.1.15) into Eq. (16.1.5) gives

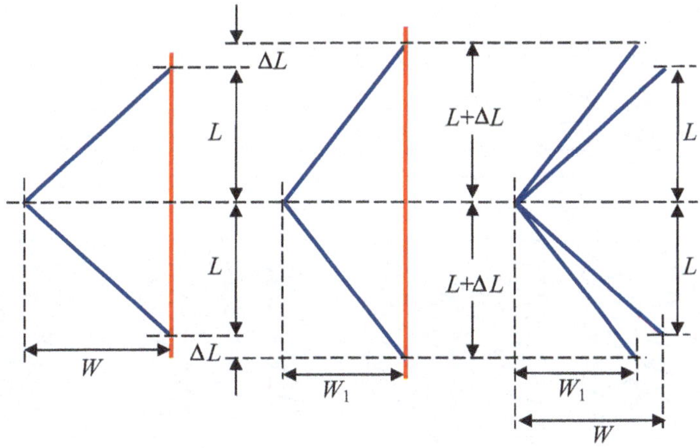

Fig. 16.5 Representative unit of the triangular microstructure (left), under $\sigma_y > 0$ stress (middle), and a comparison for Poisson's ratio analysis (right)

$$\varepsilon_x = -1 + \sqrt{1 - 2\frac{L}{W}\left(\frac{\Delta L}{W}\right) - \left(\frac{\Delta L}{W}\right)^2} \tag{16.1.17}$$

Therefore, we have the Poisson's ratio

$$\upsilon_{yx} = \frac{1 - \sqrt{1 - 2\frac{L}{W}\left(\frac{\Delta L}{W}\right) - \left(\frac{\Delta L}{W}\right)^2}}{\frac{\Delta L}{L}} \tag{16.1.18}$$

Since

$$0 < \sqrt{1 - 2\frac{L}{W}\left(\frac{\Delta L}{W}\right) - \left(\frac{\Delta L}{W}\right)^2} < 1 \tag{16.1.19}$$

it follows that $\upsilon_{yx} > 0$ for the triangular microstructure when $\sigma_y > 0$. In other words, auxeticity is non-existent for both microstructures when tensile stress is applied parallel to the long flexible rods.

While auxeticity is non-existent when $\sigma_y > 0$, auxeticity is indeterminate when $\sigma_y < 0$. The application of compressive stress in y-direction on these microstructures may lead to Poisson's ratio of either sign. In the case of the rectangular microstructure as shown in Fig. 16.6 (top row), the Poisson's ratio can be positive or negative if the immediate neighbor of the buckled flexible rods are anti-phase, while $\upsilon_{yx} \approx 0$ if all the buckled flexible rods are in phase. In the case of the triangular microstructure, the Poisson's ratio υ_{yx} can be of extreme positive or negative, as shown in Fig. 16.6 (bottom far left and far right, respectively) if all the buckled flexible rods are in phase. Poisson's ratio υ_{yx} between these two extremes are observed in the presence

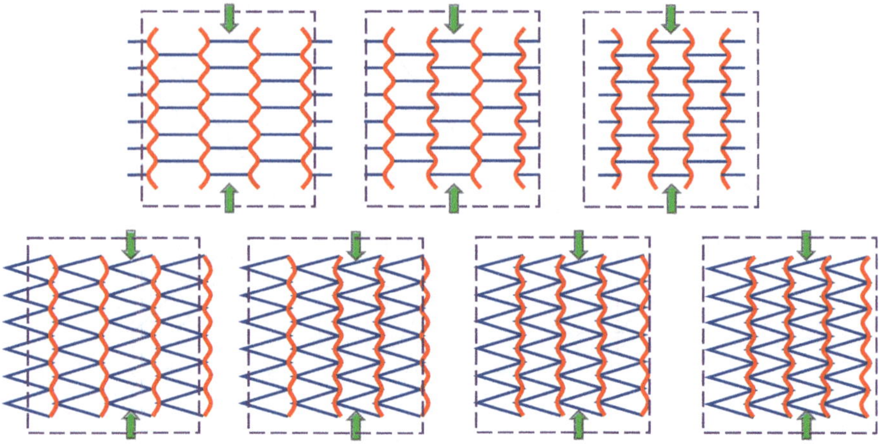

Fig. 16.6 Rectangular microstructures (top row) exhibiting $v_{yx} > 0$ (left), $v_{yx} \approx 0$ (center), and $v_{yx} < 0$ (right), as well as triangular microstructures (bottom row) demonstrating $v_{yx} \gg 0$ (far left), $v_{yx} > 0$ (mid-left), $v_{yx} \approx 0$ (mid-right), and $v_{yx} < 0$ (far right) upon application of $\sigma_y < 0$. Dashed rectangles indicate original dimension

of in phase and anti-phase buckled flexible rods, as illustrated in Fig. 16.6 (bottom, mid-left and mid-right).

To pave a way for evaluating the Poisson's ratio of these microstructures under buckling load, the displacement of the loaded tip in Fig. 16.7 (left) is considered. The available exact coordinates, in terms of the x/l and y/l for the tip of the loaded column, have been given by Timoshenko and Gere (1961), where $l = L/2$. The travel path of the column tip can be expressed by performing curve-fitting on the exact data, in which the coordinate data for buckling on the other side (i.e., $x/l < 0$)

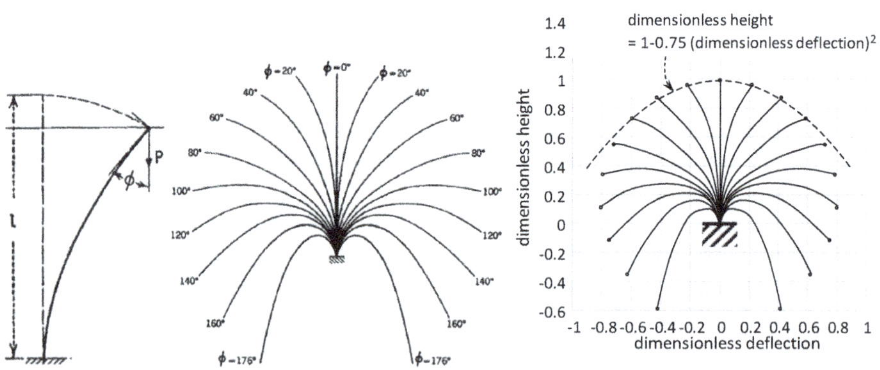

Fig. 16.7 Profile of a column of length l under a buckling load P, with ϕ as the tip angle with reference to the vertical axis (left), profiles for various ϕ (middle), and a semi-empirical equation for the path of travel for the free end of the column (right). The left and middle figures are modified from Timoshenko and Gere (1961) for clarity

is reproduced in order to comply with the boundary condition

$$\left.\frac{\mathrm{d}y}{\mathrm{d}x}\right|_{x=0} = 0 \tag{16.1.20}$$

to yield

$$\frac{y}{l} = 1 - \frac{3}{4}\left(\frac{x}{l}\right)^2 \tag{16.1.21}$$

which is valid for tilt angle of up to 60° with respect to *y*-axis. The decrease in the projected height is then $\Delta u = l - y$, or

$$\frac{\Delta u}{l} = 1 - \frac{y}{l} = \frac{3}{4}\left(\frac{x}{l}\right)^2 \tag{16.1.22}$$

in dimensionless form, where $\Delta u = \Delta U/2$. Therefore, substituting $l = L/2$, $\Delta u = \Delta U/2$ and $x = w_0$, where

$$w_0 = \frac{1}{2k}\int_0^{\phi} \frac{\sin\theta}{\sqrt{\sin^2\left(\frac{\phi}{2}\right) - \sin^2\left(\frac{\theta}{2}\right)}}\mathrm{d}\theta \tag{16.1.23}$$

with $k^2 = P/(EI)$ (Timoshenko and Gere 1961) into Eq. (16.1.22) gives the dimensionless reduction in projected length

$$\frac{\Delta U}{L} = 3\left(\frac{w_0}{L}\right)^2 \tag{16.1.24}$$

With reference to Fig. 16.3 (top row) for compression of the rectangular microstructure in *y*-direction, substitution of Eq. (16.1.24) into Eq. (16.1.4) gives

$$\varepsilon_y = -3\left(\frac{w_0}{L}\right)^2 \tag{16.1.25}$$

which, together with Eq. (16.1.7), gives

$$v_{yx} = -\frac{\varepsilon_x}{\varepsilon_y} = \frac{4}{3}\frac{L}{W}\left(\frac{w_0}{L}\right)^{-1} \tag{16.1.26}$$

whereby positive and negative values of w_0 indicate outward and inward bulging of the buckled flexible rods, respectively. The expression for ε_y in Eq. (16.1.25) also applies to the triangular microstructure under $\sigma_y < 0$ loading, because it similarly describes the profile of the flexible rod under buckling load. With reference to Fig. 16.3 (bottom row) for compression of the triangular microstructure in *y*-direction, substitution of Eq. (16.1.24) into Eq. (16.1.9) gives

$$W_1 = 2w_0 \pm \sqrt{W^2 + 6w_0^2 - 9\frac{w_0^4}{L^2}} \tag{16.1.27}$$

Again, the choice of lower solution would lead to a negative value for W_1. Hence, substituting the upper solution of Eq. (16.1.27) into Eq. (16.1.5) gives

$$\varepsilon_x = -1 + 2\frac{w_0}{W} + \sqrt{1 + 6\left(\frac{w_0}{W}\right)^2 - 9\frac{w_0^4}{W^2L^2}} \tag{16.1.28}$$

which, together with Eq. (16.1.25), leads to

$$v_{yx} = \frac{-1 + 2\frac{w_0}{L}\left(\frac{L}{W}\right) + \sqrt{1 + 6\left(\frac{w_0}{L}\right)^2\left(\frac{L}{W}\right)^2 - 9\left(\frac{w_0}{L}\right)^4\left(\frac{L}{W}\right)^2}}{3\left(\frac{w_0}{L}\right)^2} \tag{16.1.29}$$

Arising from an increase in temperature of $\mathrm{d}T > 0$, the rectangular microstructure elongates in the y-direction by a thermal strain of $\varepsilon_y^{(T)} = \alpha_r^{(T)}\mathrm{d}T$, i.e., a length of L increases to $L\left(1 + \alpha_r^{(T)}\mathrm{d}T\right)$ where $\alpha_r^{(T)}$ is the coefficient of thermal expansion (CTE) of the flexible rods, while the length of the short rigid rods remains unchanged, thereby leading to $\varepsilon_x^{(T)} = 0$. As such the rectangular microstructure is said to possess zero thermal expansion (ZTE) and positive thermal expansion (PTE) along the x- and y-directions, respectively, as indicated in Fig. 16.8 (top left).

If this structure is constrained from expanding in y-direction, then the thermally elongated rod of length $L\left(1 + \alpha_r^{(T)}\mathrm{d}T\right)$ is compressed to its overall original dimension such that buckling takes place, as shown in the remaining parts of Fig. 16.8. Proportionally, the maximum deflection is expressed as $w_0\left(1 + \alpha_r^{(T)}\mathrm{d}T\right)$ for a length of $L\left(1 + \alpha_r^{(T)}\mathrm{d}T\right)$ where, as before, a positive and negative values of $w_0\left(1 + \alpha_r^{(T)}\mathrm{d}T\right)$ correspond to outward and inward deflections, respectively. Therefore, the structure is said to exhibit ZTE along the x-direction if the buckled flexible rods are in phase as shown in Fig. 16.8 (bottom left), due to cancelation effect from the positive and negative values on both sides. However, if the buckled flexible rods are anti-phase, then the structure can exhibit PTE or NTE if the flexible rods buckle outward (Fig. 16.8 top right) or inward (Fig. 16.8 bottom right), respectively.

With reference to Fig. 16.9, it can be seen that the prescription of constraint in y-direction is synonymous to the compression of thermally elongated flexible rods. Substituting

$$L \to L\left(1 + \alpha_r^{(T)}\mathrm{d}T\right) \tag{16.1.30}$$

and

$$w_0 \to w_0\left(1 + \alpha_r^{(T)}\mathrm{d}T\right) \tag{16.1.31}$$

into Eq. (16.1.25) for prescribing zero thermal strain in y-direction gives

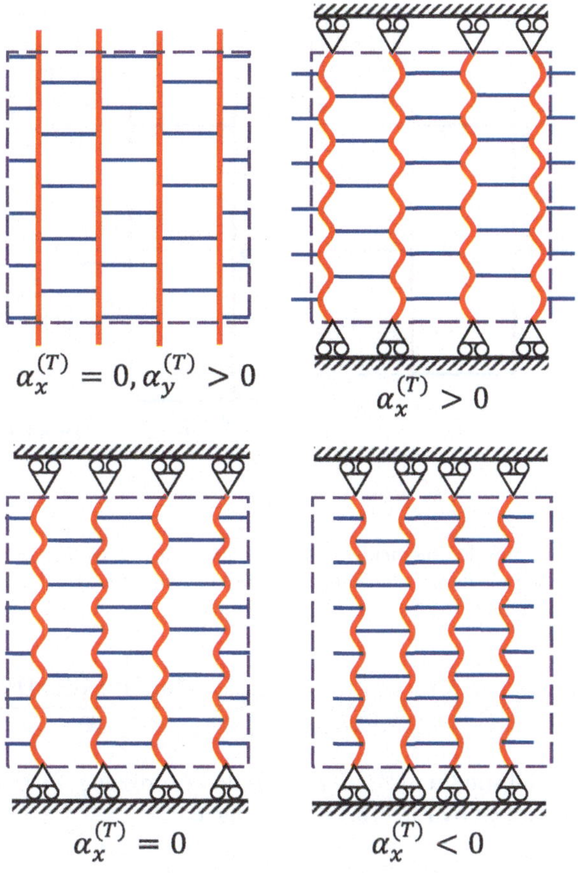

Fig. 16.8 Manifestations of ZTE in x-direction for unconstrained rectangular microstructure (top left), as well as PTE (top right), ZTE (bottom left), and NTE (bottom right) in x-direction with y-direction constraint for $\mathrm{d}T > 0$

$$\varepsilon_y = -3\left(\frac{w_0\left(1+\alpha_r^{(T)}\mathrm{d}T\right)}{L\left(1+\alpha_r^{(T)}\mathrm{d}T\right)}\right)^2 = -3\left(\frac{w_0}{L}\right)^2 \tag{16.1.32}$$

i.e., the same expression of ε_y under pure buckling. Hence, equating the magnitudes of these two expressions for strain in y-direction, i.e., the thermal strain $\varepsilon_y^{(T)} = \alpha_r^{(T)}\mathrm{d}T$ when unconstrained, and the prescribed $\varepsilon_y = -3(w_0/L)^2$ for imposing zero strain in y-direction when constrained, we have

$$w_0 = \pm L\sqrt{\frac{\alpha_r^{(T)}\mathrm{d}T}{3}} \tag{16.1.33}$$

where the upper and lower signs denote outward and inward buckling patterns, respectively. Substituting Eq. (16.1.31) into Eq. (16.1.7) for the strain in x-direction resulting from the applied constraint in y-direction gives

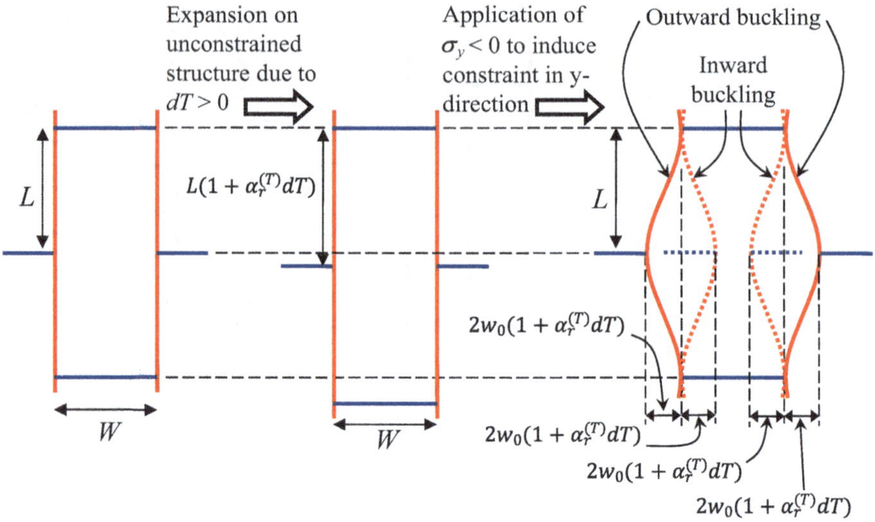

Fig. 16.9 Representative unit of rectangular microstructure in original state (left), for thermal expansion analysis in *y*-direction when unconstrained (middle), and occurrence of buckling when constraint is imposed in *y*-direction (right)

$$\varepsilon_x^{(T)} = \frac{4w_0\left(1+\alpha_r^{(T)}\mathrm{d}T\right)}{W} \tag{16.1.34}$$

Substituting $\varepsilon_x^{(T)} = \alpha_x^{(T)}\mathrm{d}T$ and expressing w_0 in terms of flexible rod CTE, as described by Eq. (16.1.33), we have

$$\alpha_x^{(T)}\mathrm{d}T = \pm\frac{4}{\sqrt{3}}\frac{L}{W}\left(1+\alpha_r^{(T)}\mathrm{d}T\right)\sqrt{\alpha_r^{(T)}\mathrm{d}T} \tag{16.1.35}$$

where $\alpha_x^{(T)}$ in Eq. (16.1.35) denotes the effective CTE in *x*-direction of the rectangular microstructure when constrained in *y*-direction. The term $\alpha_x^{(T)}\mathrm{d}T$ is a dimensionless form of $\alpha_x^{(T)}$.

Consider similarly an increase in temperature of $\mathrm{d}T > 0$, the triangular microstructure also elongates in the *y*-direction by a strain of $\varepsilon_y^{(T)} = \alpha_r^{(T)}\mathrm{d}T$, i.e., a length of L increases to $L\left(1+\alpha_r^{(T)}\mathrm{d}T\right)$, while the short rigid rods rotate such that they become more aligned to the *y*-axis, as illustrated in Fig. 16.10 (top left). With reference to Fig. 16.11 (middle), substituting

$$\Delta L = L\alpha_r^{(T)}\mathrm{d}T \tag{16.1.36}$$

into Eq. (16.1.17) gives

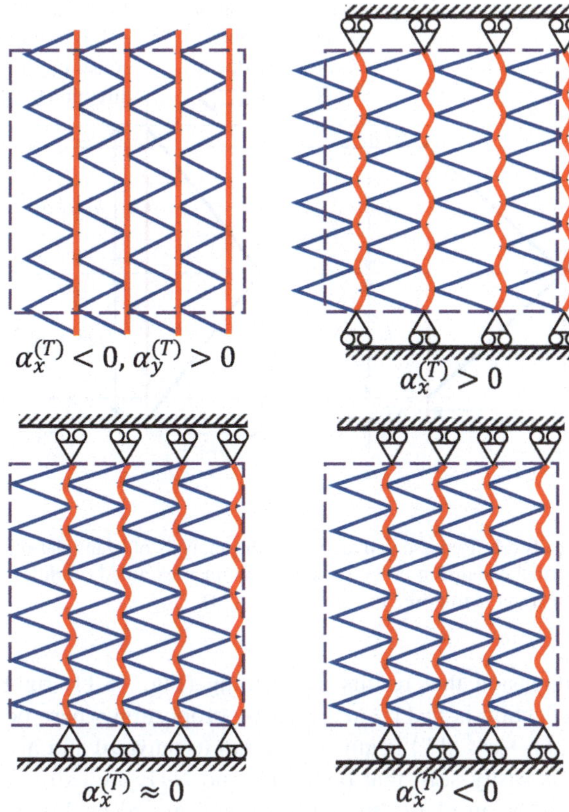

Fig. 16.10 Manifestations of NTE in x-direction for unconstrained triangular microstructure (top left), as well as PTE (top right), ZTE (bottom left), and NTE (bottom right) in x-direction with y-direction constraint for $\mathrm{d}T > 0$

$$\varepsilon_x^{(T)} = -1 + \sqrt{1 - 2\left(\frac{L}{W}\right)^2 \alpha_r^{(T)}\mathrm{d}T - \left(\frac{L}{W}\right)^2 \left(\alpha_r^{(T)}\mathrm{d}T\right)^2} \tag{16.1.37a}$$

or

$$\alpha_x^{(T)} = -\frac{1}{\mathrm{d}T} + \sqrt{\frac{1}{(\mathrm{d}T)^2} - 2\frac{\alpha_r^{(T)}}{\mathrm{d}T}\left(\frac{L}{W}\right)^2 - \left(\alpha_r^{(T)}\right)^2\left(\frac{L}{W}\right)^2} \tag{16.1.37b}$$

Since

$$0 < \sqrt{1 - 2\left(\frac{L}{W}\right)^2 \alpha_r^{(T)}\mathrm{d}T - \left(\frac{L}{W}\right)^2 \left(\alpha_r^{(T)}\mathrm{d}T\right)^2} < 1 \tag{16.1.38}$$

it follows that $\alpha_x^{(T)} < 0$. As such, the triangular cell is said to possess NTE and PTE along the x- and y-directions, respectively, as indicated in Fig. 16.10 (top left). If this structure is constrained from expanding in y-direction, then a thermally elongated flexible rod of length $L\left(1 + \alpha_r^{(T)}\mathrm{d}T\right)$ is compressed to its overall original

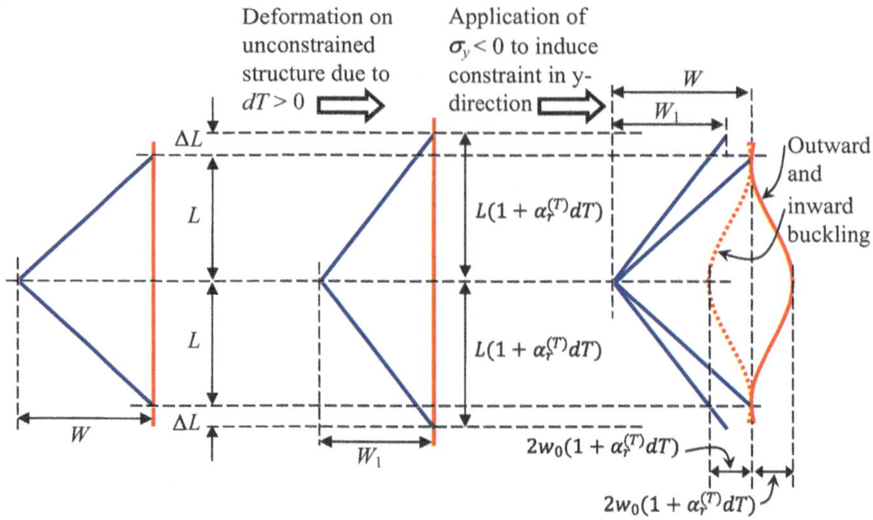

Fig. 16.11 Representative unit for the analysis of triangular microstructure in original state (left), thermal expansion in y-direction when unconstrained (middle), and occurrence of buckling when constraint is imposed in y-direction (right)

dimension that results in buckling (Fig. 16.11, right). As before, the maximum deflection is $w_0\left(1+\alpha_r^{(T)}\mathrm{d}T\right)$ by proportion where positive and negative values of $w_0\left(1+\alpha_r^{(T)}\mathrm{d}T\right)$ again correspond to outward and inward deflections, respectively. It is easily seen that both PTE and NTE are exhibited when the buckled flexible rods are in phase; the outward and inward buckling of the flexible rods give rise to $\varepsilon_x > 0$ (Fig. 16.10, top right) and $\varepsilon_x < 0$ (Fig. 16.10, bottom right), respectively. Equation (16.1.33) for describing the deflection of the buckled flexible rods for prescribing y-direction constraint in rectangular microstructure also applies in the case of triangular microstructure.

Substituting Eq. (16.1.31) into Eq. (16.1.28) gives

$$\varepsilon_x^{(T)} = -1 + 2\frac{w_0\left(1+\alpha_r^{(T)}\mathrm{d}T\right)}{W} + \sqrt{1 + 6\left(\frac{w_0\left(1+\alpha_r^{(T)}\mathrm{d}T\right)}{W}\right)^2 - 9\frac{w_0^4\left(1+\alpha_r^{(T)}\mathrm{d}T\right)^4}{W^2L^2}} \quad (16.1.39)$$

followed by the substitution of $\varepsilon_x^{(T)} = \alpha_x^{(T)}\mathrm{d}T$ and Eq. (16.1.33) into Eq. (16.1.39) leads to

$$\alpha_x^{(T)}\mathrm{d}T = -1 \pm \frac{2}{\sqrt{3}}\frac{L}{W}\left(1+\alpha_r^{(T)}\mathrm{d}T\right)\sqrt{\alpha_r^{(T)}\mathrm{d}T}$$

$$+\sqrt{1+2\left(\frac{L}{W}\right)^2\alpha_r^{(T)}\mathrm{d}T\left(1+\alpha_r^{(T)}\mathrm{d}T\right)^2-\left(\frac{L}{W}\right)^2\left(\alpha_r^{(T)}\mathrm{d}T\right)^2\left(1+\alpha_r^{(T)}\mathrm{d}T\right)^4} \tag{16.1.40}$$

where $\alpha_x^{(T)}$ in Eq. (16.1.40) refers to the effective CTE of the triangular microstructure when constrained in y-direction.

Based on the developed models of Poisson's ratio under various loading conditions and the developed effective CTE under free and constrained conditions, the following plotted results graphically show how these effective mechanical properties vary with the microstructural shape, specifically the L/W ratio. For the triangular microstructure, the ratio $L/W = 1$ indicates a 45° inclination of the short rigid rods to either axis, while the ratios of $L/W = \sqrt{3}$ and $L/W = 1/\sqrt{3}$ denote the short rod inclination of 60° and 30°, respectively, to the x-axis. As such, the Poisson's ratio and effective CTEs of the triangular microstructure are plotted using these L/W ratios. For the sake of consistency, the same set of L/W ratios are adopted for the rectangular microstructure. Note that when $L/W = 1/\sqrt{3}$, the rectangular microstructure is arranged in equilateral triangular array. Plots of ε_y versus ε_x accompany those of v_{xy} or v_{yx} versus w_0/L. The Poisson's ratios are plotted against w_0/L as the latter is a dimensionless measure of the maximum deflection.

A family of ε_y versus ε_x and its corresponding v_{xy} versus w_0/L curves for rectangular microstructures under compressive and tensile stresses in x-direction are plotted in Fig. 16.12 (top) using Eqs. (16.1.6) to (16.1.8), while the same results for triangular microstructures are plotted in Fig. 16.12 (bottom) using Eqs. (16.1.6), (16.1.11), and (16.1.12). The plots of ε_y versus ε_x for rectangular microstructure are symmetrical because the strain in x-direction is independent from the rotation of short rigid rods, and are solely due to the deflection of the long flexible rods, which possess equal magnitudes for both compression and tension. Although the deflection component magnitude of the long flexible rods is the same in the case of triangular microstructure for both compression and tension, both loading directions cause the short rigid rods to be more aligned to the x-axis, thereby giving an additional positive strain component to ε_x. As a result, the magnitude of ε_x under tension is greater than that under compression for equal ε_y, thereby contributing to the non-symmetric distribution of strain in the case of σ_x loading on the triangular microstructure.

Since both microstructures do not exhibit auxeticity for tensile loading in y-direction, consideration is made on compressive loading of these microstructures. A family of ε_y versus ε_x and its corresponding v_{yx} versus w_0/L curves for rectangular microstructures under compressive stress in y-direction are plotted in Fig. 16.13 (top) using Eqs. (16.1.7), (16.1.25), and (16.1.26), while the same results for triangular microstructures are plotted in Fig. 16.13 (bottom) using Eqs. (16.1.25), (16.1.28), and (16.1.29). Again, the observation of symmetry and non-symmetry in the plots of ε_y versus ε_x for the $\sigma_y < 0$ loading in the rectangular and triangular microstructures, respectively, is attributed to the same reason as with the σ_x loading. However, the plots of positive and negative ε_x in the case of rectangular microstructure are based on Fig. 16.6 (top left and right, respectively), i.e., they represent the most positive

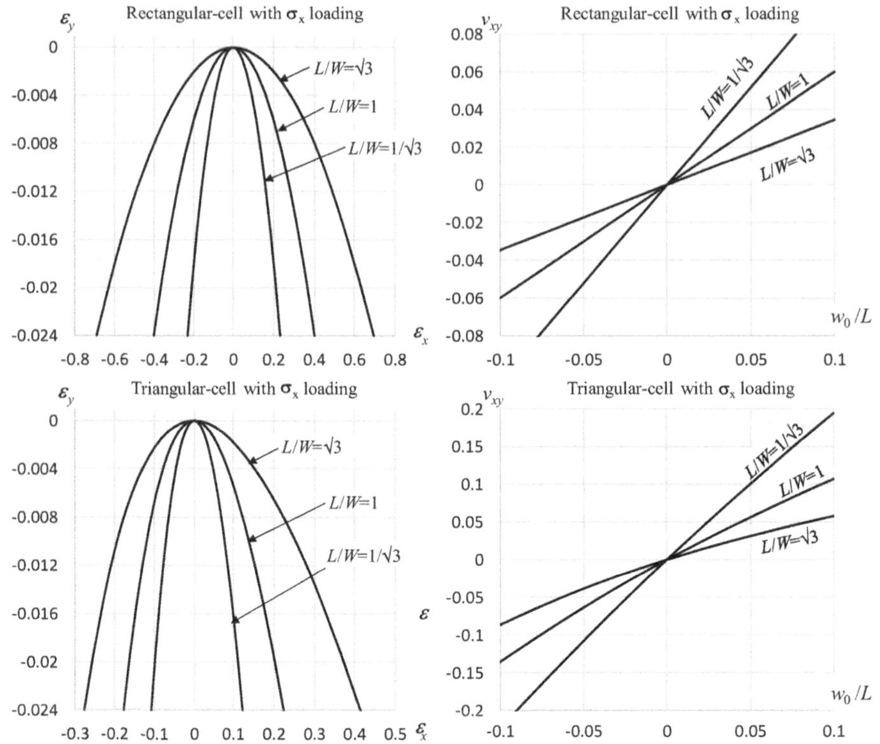

Fig. 16.12 Plots of ε_y versus ε_x (left column) and v_{xy} versus w_0/L (right column) for rectangular microstructure (top row) and triangular microstructure (bottom row) under compressive and tensile stresses in x-direction

and negative values of ε_x for a given ε_y, whereby the neighboring buckled flexible rods are anti-phase. For other combinations of flexible rods buckling patterns, the values of ε_x under a given ε_y fall between these extremes; an example would be in the case where $\varepsilon_x = 0$ for any ε_y if the buckled flexible rods are in phase. Similarly, for the case of $\sigma_y < 0$ loading on triangular microstructure, in which the results of Fig. 16.13 (bottom) are based on the most positive and most negative values of ε_x for a given ε_y in Fig. 16.6 (bottom far left and far right, respectively), whereby the buckled flexible rods are in phase. When some of the buckled flexible rods are not in phase, the ε_x values are in between those extreme positive and negative plotted results.

Plots of dimensionless effective CTE in x-direction, $\alpha_x^{(T)}\mathrm{d}T$ against the dimensionless CTE of the long flexible rods $\alpha_r^{(T)}\mathrm{d}T$ with constraint in y-direction, i.e., $\varepsilon_y = 0$, at increasing temperature $\mathrm{d}T > 0$ for the rectangular and triangular microstructures are furnished in Fig. 16.14 using Eqs. (16.1.35) and (16.1.40), respectively. For every L/W ratio in each microstructure indicated in Fig. 16.14, the two curves at $\alpha_x^{(T)}\mathrm{d}T > 0$ and $\alpha_x^{(T)}\mathrm{d}T < 0$ correspond to the most positive effective CTE, $\alpha_x^{(T)}$

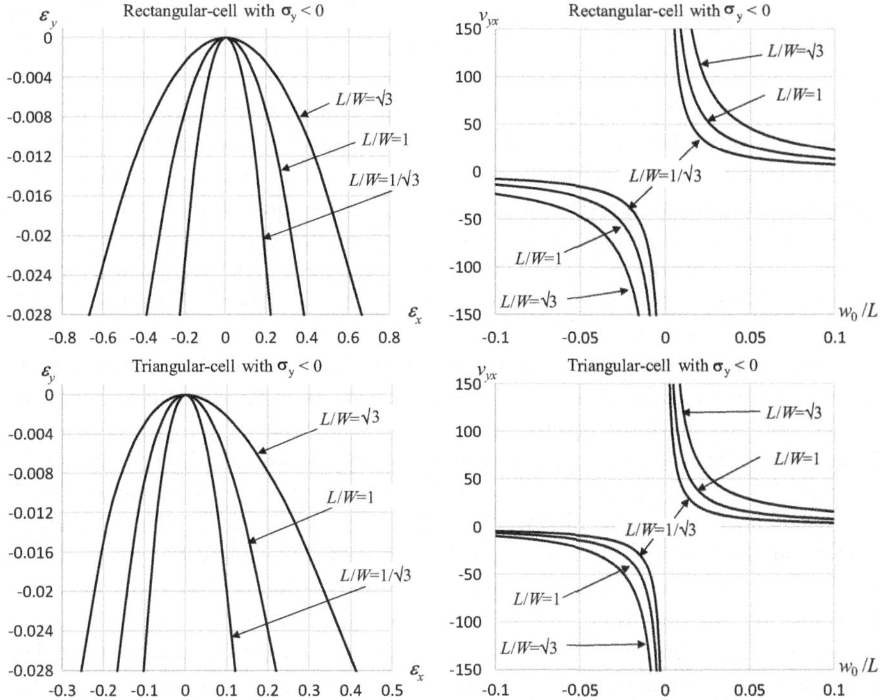

Fig. 16.13 Plots of ε_y versus ε_x (left column) and v_{yx} versus w_0/L (right column) for rectangular microstructure (top row) and triangular microstructure (bottom row) under compressive stress in y-direction

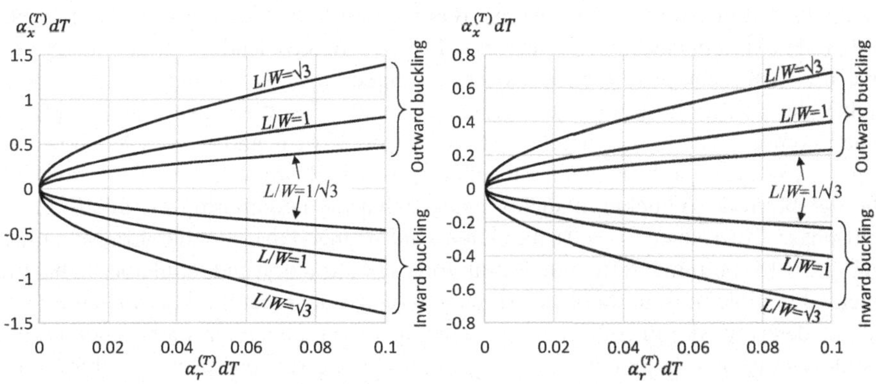

Fig. 16.14 Plots of dimensionless effective CTE in x-direction versus the dimensionless CTE of the long flexible rods, with constraint in y-direction $\left(\varepsilon_y = 0\right)$ at increasing temperature ($\mathrm{d}T > 0$), for the rectangular microstructure (left) and triangular microstructure (right)

due to total outward buckling and the most negative effective CTE, $\alpha_x^{(T)}$ due to total inward buckling of the flexible rods, respectively. The total outward buckling for the rectangular and triangular microstructures are illustrated in Fig. 16.8 (top right) and Fig. 16.10 (top right), respectively, while the total inward buckling for the rectangular and triangular microstructures are indicated in Fig. 16.8 (bottom right) and Fig. 16.10 (bottom right), respectively. When both the outward and inward buckling of the flexible rods are present, the effective $\alpha_x^{(T)}$ falls within the plotted upper and lower bounds. A special case exists for the rectangular cells when the number of outwardly and inwardly buckled flexible rods are equal; under such a situation the strain in x-direction is zero, which gives $\alpha_x^{(T)} = 0$, i.e., zero thermal expansion (ZTE) in x-direction.

In the analysis and discussion of both microstructures under compressive σ_y loading, it has been shown that the Poisson's ratio is indeterminate and that one can only predict the most positive and the most negative v_{yx}, which depends on the number of flexible rods undergoing inward and outward buckling. Likewise, the effective CTE $\alpha_x^{(T)}$ with constraint $\varepsilon_y = 0$, for increasing temperature-induced buckling, is indeterminate as the effective CTE is dependent on the number of inwardly and the number of outwardly buckled flexible rods. One possible way to control the relative number of the inwardly and outwardly buckled flexible rods, and hence determine the exact v_{yx} (under $\sigma_y < 0$) as well as $\alpha_x^{(T)}$ (with $\varepsilon_y = 0$ for $dT > 0$), is by creating the propensity of the flexible rods to buckle in the desired manner. This can be attained by attaching objects with charges to create forces of attraction and repulsion. The forces should not be sufficient to induce flexure, but the presence of these forces would lead the flexible rods to buckle in a predictable manner.

With reference to Fig. 16.15 (top left) for the rectangular microstructure, the locations of the charged objects are indicated as circles. When the charges are arranged as shown in Fig. 16.15 (top right), the repulsion of like charges causes the flexible rods to buckle outward when compressed by σ_y, thereby leading to the most positive v_{yx}. If the charges on each row are arranged as shown in Fig. 16.15 (bottom right), the attraction of opposite charges causes the flexible rods to buckle inward when compressed by σ_y, thereby leading to the most negative v_{yx}. As shown in Fig. 16.15 (right), both arrangements of charges, which are in rectangular array, lead the flexible rods to buckle in such a manner that the neighboring flexible rods are anti-phase. To produce $v_{yx} = 0$, the charges are arranged in triangular array as shown in Fig. 16.15 (bottom left) to induce both forces of attraction and repulsion so that the buckled flexible rods are in phase. Having demonstrated that the Poisson's ratio v_{yx} can be determined *a priori* via charge attachment, a rectangular microstructure that exhibits Poisson's ratio sign-switching upon stress direction reversal—along both the x- and y-directions—can therefore be achieved by ordering the charges as shown in Fig. 16.15 (bottom right).

With reference to Fig. 16.16 (top left) for the triangular microstructure, the locations of the charged objects are indicated as circles. When the charges are arranged as shown in Fig. 16.16 (top right), the repulsion of positive charges causes the flexible rods to buckle outward when compressed by σ_y, thereby leading to the most positive v_{yx}. The same effect is achieved by replacing all the positive charges with

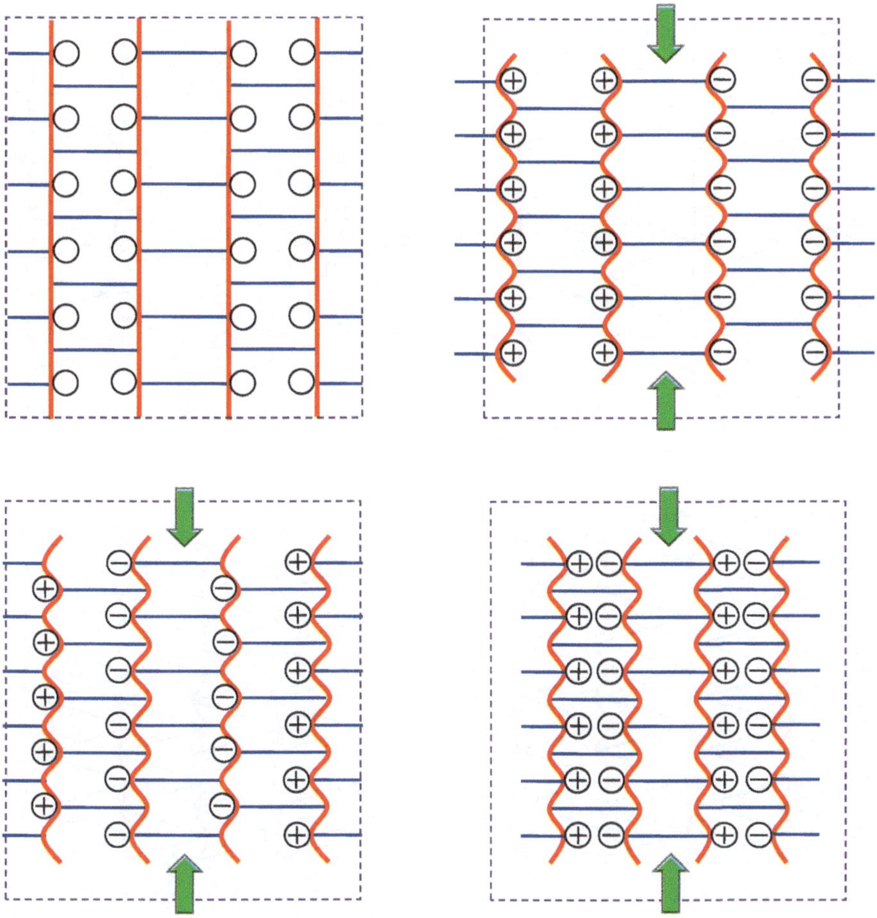

Fig. 16.15 Rectangular microstructure with the location of attached objects with charges denoted by circles (top left) arranged in such a way to produce the most positive (top right) and the most negative (bottom right) Poisson's ratio v_{yx} under compressive σ_y loading. The locations of the charges are different to create zero v_{yx} (bottom left)

negative ones. If the charges on each row are arranged as shown in Fig. 16.16 (bottom right), the attraction of opposite charges causes the flexible rods to buckle inward when compressed by σ_y, thereby leading to the most negative v_{yx}. As shown in Fig. 16.16 (right), both arrangements of charges lead the flexible rods to buckle in such a manner that they are in phase. The buckled flexible rods can exhibit anti-phase pattern from one flexible rod to its immediate neighbor if the charges are arranged in the manner depicted in Fig. 16.16 (bottom left), which gives a lower positive value of v_{yx}. Since the Poisson's ratio v_{yx} can be established by charge attachment prior to compression, a triangular microstructure that exhibits Poisson's ratio sign-switching upon stress direction reversal—along both the x- and y-directions—can hence be

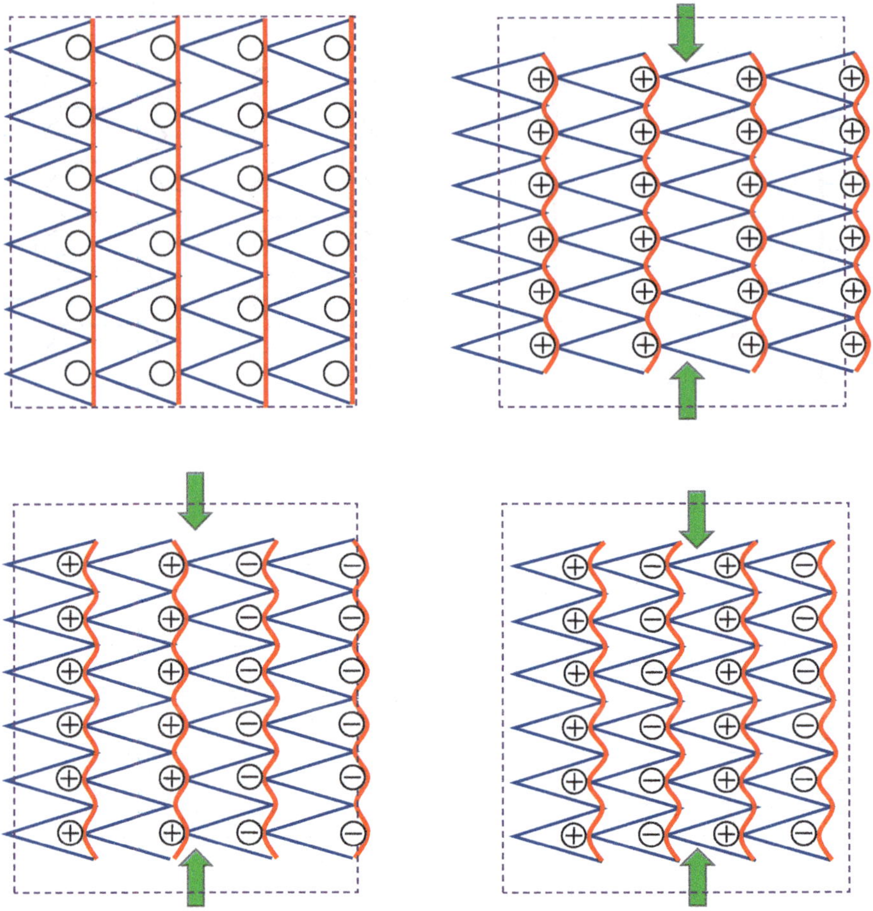

Fig. 16.16 Triangular microstructure with the location of attached objects with charges denoted by circles (top left) arranged in such a way to produce the most positive (top right) and the most negative (bottom right) Poisson's ratio v_{yx} under compressive σ_y loading. An example of intermediate v_{yx} between these two extremes is also shown (bottom left)

attained by arranging the charges as shown in Fig. 16.16 (bottom right). Apart from the use of charged objects, one may also create the likelihood for the long flexible rods to buckle in predictable pattern by mechanical, magnetic, and other approaches. The mechanical approach can be attained by introducing imperfections, but this will alter the overall stiffness of the microstructure. The attachment of magnets may possibly work in similar way as that of charged objects. However, magnets could be geometrically cumbersome as they are dipoles rather than monopoles.

Having shown that the Poisson's ratio v_{yx}, as a consequent of σ_y compression, can be controlled by means of charged attachment for both microstructures, it follows that under $\varepsilon_y = 0$ and $\mathrm{d}T > 0$ conditions, the ambiguity of σ_x for the rectangular and triangular microstructures under the given constraint as shown in Figs. 16.8 and 16.10, respectively, can be predetermined by means of charged attachment. This is because under the condition of increasing temperature, the elongated flexible rods undergo buckling when constrained in the y-direction. A summary of Poisson's ratio and effective CTEs—the latter under free and constrained conditions—is furnished in Table 16.1.

16.2 Sign-Switching of Poisson's Ratio with Persistently Positive Transverse Strain

The basic hybrid rhombic–re-entrant metamaterial microstructure is shown Fig. 16.17a, which consists of a pair of slot and slider to effect simultaneous jam and slide mechanism. Each column consists of a zigzag structure with alternating slots and sliders pointing to opposite directions, as shown in Fig. 16.17b, wherein the slots are joined to the sliders in a neighboring zigzag structure while the sliders are joined to the slots in another neighboring zigzag structure. The zigzag vertices are pin joints to permit free rotations with each rod member being rigid. In the relaxed state, i.e., original state as indicated in Fig. 16.17b, the sliders lightly touch the open ends of the slots while the vertices from neighboring zigzags also lightly touch one another. Upon application of compressive stress in the x-direction, as shown in Fig. 16.17c, the vertices are prevented from motion along the x-direction due to symmetric constraints from neighboring vertices, and therefore negative strain in x-direction is made possible through motion of the sliders along the slots toward the closed ends, and a simultaneous rotation of the inclined rods. In other words, the contact points of the vertices are effectively pin joints while the slot-slider pairs are effectively non-structural. The effective microstructure is hence that of rhombic geometry, which is a special case of hexagonal honeycomb with diminished horizontal ribs. When a tensile stress is applied in the x-direction as indicated in Fig. 16.17d, the interlock between the sliders and the open ends of the slots retains the total dimension of the horizontal parts such that positive strain in x-direction is made possible through rotation of the inclined rods. The slot-slider pairs effectively become rigid horizontal rods and the effective microstructure is therefore that of re-entrant geometry. In other words, the hybrid rhombic–re-entrant metamaterial exhibits microstructural duality, and hence behavioral duality with opposing Poisson's ratio signs for opposing stress directions (Lim 2019b).

A unit of the hybrid kite-arrowhead metamaterial microstructure is illustrated in Fig. 16.18a, each consisting of two slots and two sliders. The inner and outer sliders are placed in the inner and outer slots, respectively, of the neighboring parts, as illustrated in Fig. 16.18b. In this original state, both the sliders are in light contact with the closed end of the slots. The outer slot-slider pairs are locked with the application of compressive stress in the x-direction, while the inner sliders move toward the open ends of their corresponding slots, as denoted in Fig. 16.18c. Essentially, the outer

Table 16.1 Summary of Poisson's ratio and effective CTE of the considered microstructures (Lim 2019a)

Type 2 of loading	Rectangular microstructure	Triangular microstructure	Remarks
σ_x	$v_{xy} = \frac{3}{5}\frac{W w_0}{L^2}$	$v_{xy} = -\frac{2.4\left(\frac{w_0}{L}\right)^2}{1-2\frac{w_0}{W}-\sqrt{1+4.8\left(\frac{w_0}{W}\right)^2-5.76\frac{w_0^4}{L^2W^2}}}$	Conventional when $\sigma_x > 0$ due to $w_0/L > 0$, and auxetic when $\sigma_x < 0$ due to $w_0/L < 0$. For mid-span concentrated load, w_0 is defined by Eq. (16.1.2)
$\sigma_y > 0$	$v_{yx} = 0$	$v_{yx} = \frac{1-\sqrt{1-2\frac{L\Delta L}{W^2}-\left(\frac{\Delta L}{W}\right)^2}}{\frac{\Delta L}{L}}$	Conventional
$\sigma_y < 0$	$v_{yx} = \frac{4}{3}\frac{L^2}{W w_0}$	$v_{yx} = \frac{-1+2\frac{w_0}{W}+\sqrt{1+6\left(\frac{w_0}{W}\right)^2-9\frac{w_0^4}{L^2W^2}}}{3\left(\frac{w_0}{L}\right)^2}$	Auxeticity is indeterminate, unless the propensity to buckle in predictable manner is created. These most positive and most negative v_{yx} are based on $w_0 > 0$ and $w_0 < 0$, respectively. For buckling load, w_0 is defined by Eq. (16.1.23)
$\mathrm{d}T \neq 0$, $\sigma_y = 0$	$\alpha_x^{(T)} = 0, \alpha_y^{(T)} = \alpha_r^{(T)}$	$\alpha_x^{(T)} = -\frac{1}{\mathrm{d}T} + \sqrt{\frac{1}{(\mathrm{d}T)^2} - 2\frac{\alpha_r^{(T)}}{\mathrm{d}T}\left(\frac{L}{W}\right)^2 - \left(\alpha_r^{(T)}\frac{L}{W}\right)^2}, \alpha_y^{(T)} = \alpha_r^{(T)}$	Conventional

(continued)

Table 16.1 (continued)

Type 2 of loading	Rectangular microstructure	Triangular microstructure	Remarks
$\mathrm{d}T > 0$, $\varepsilon_y = 0$	$\alpha_x^{(T)} = \frac{\pm 4L}{\sqrt{3}W}\sqrt{\frac{\alpha_r^{(T)}}{\mathrm{d}T}}\left(1 + \alpha_r^{(T)}\mathrm{d}T\right)$	$\alpha_x^{(T)} = -\frac{1}{\mathrm{d}T} \pm \frac{2}{\sqrt{3}}\frac{L}{W}\left(1 + \alpha_r^{(T)}\mathrm{d}T\right)\sqrt{\frac{\alpha_r^{(T)}}{\mathrm{d}T}} + \sqrt{\frac{1}{(\mathrm{d}T)^2} + 2\left(\frac{L}{W}\right)^2\frac{\alpha_r^{(T)}}{\mathrm{d}T}\left(1 + \alpha_r^{(T)}\mathrm{d}T\right)^2 - \left(\alpha_r^{(T)}\frac{L}{W}\right)^2\left(1 + \alpha_r^{(T)}\mathrm{d}T\right)^4}$	NTE in x-direction is indeterminate, unless the likelihood to buckle in predictable manner is made

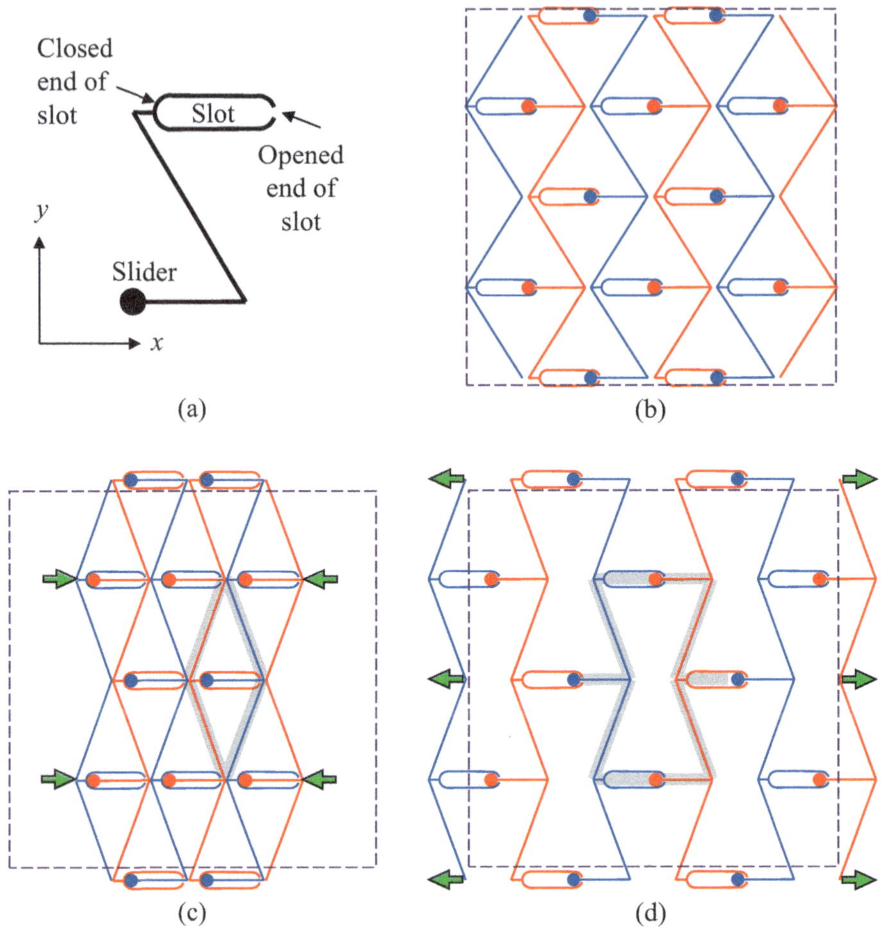

Fig. 16.17 Hybrid rhombic–re-entrant metamaterial microstructure, showing **a** a slot and a slider in a symmetrical half, **b** original microstructure, **c** conversion to rhombic microstructure under compressive load, and **d** conversion to re-entrant microstructure under tensile load. The effective parts in the middle units of **c** and **d** are indicated in gray. A dashed rectangle that encompasses the microstructure **b** is transposed on **c** and **d** to aid visual comparison

slot-slider pairs form rigid rods while the inner slot-slider pairs are non-structural so as to effectively form a network of kite microstructural geometry. When tensile load is applied in the x-direction as shown in Fig. 16.18d, the inner slot-slider pairs are being locked, while the outer sliders move toward the open ends of the outer slots. The inner slot-slider pairs become rigid rods while the outer slot-slider pairs are non-load bearing. This results in a modified double arrowhead geometry. For this reason, it can be said that the hybrid kite-arrowhead metamaterial demonstrates microstructural duality, and therefore behavioral duality with opposite Poisson's ratio signs being manifested under opposing stress directions (Lim 2019b).

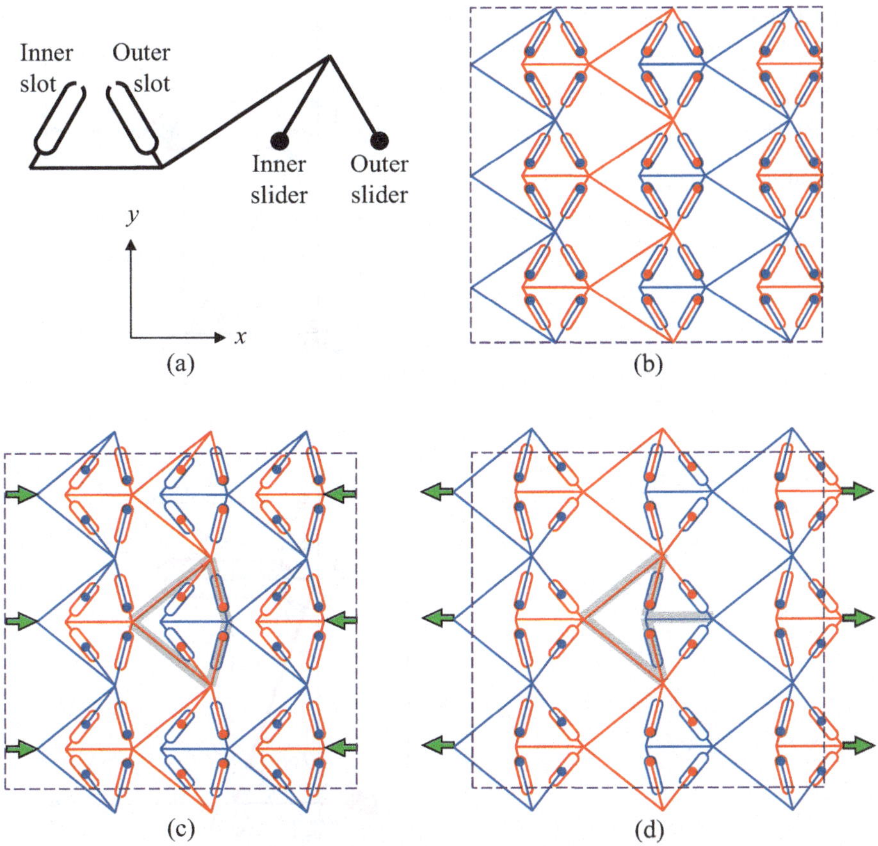

Fig. 16.18 Hybrid kite-arrowhead metamaterial microstructure showing **a** a pair of slots and a pair of sliders in a symmetrical half, **b** original microstructure, **c** conversion to kite microstructure under compressive load, and **d** conversion to double arrowhead-like microstructure under tensile load. The effective parts in the middle units of **c** and **d** are indicated in gray. A dashed rectangle that encompasses the original microstructure **b** is transposed on **c** and **d** to aid visual comparison

In what follows, the representative units for both metamaterials' microstructures are isolated for the purpose of analysis, whereby the origin is denoted by "O" while the vertices "A" and "B" are freely rotating pin-jointed vertices as indicated in Figs. 16.19 and 16.20. For the hybrid rhombic–re-entrant microstructure, "O" is located at the mid-span of a horizontal slot-slider mechanism, whereas "C" is located midway in the nearest horizontal slot-slider; both are not freely rotating pin joints for this microstructure. For the hybrid kite-arrowhead microstructure, "O" is a freely rotating joint while "C" is also a freely rotating joint, but forms the origin of the neighboring representative unit. The original states of both microstructures are shown in Figs. 16.19a and 16.20a. The geometries as illustrated in Figs. 16.19b and 16.20b refer to the case where compressive stress is applied along the x-direction, while those

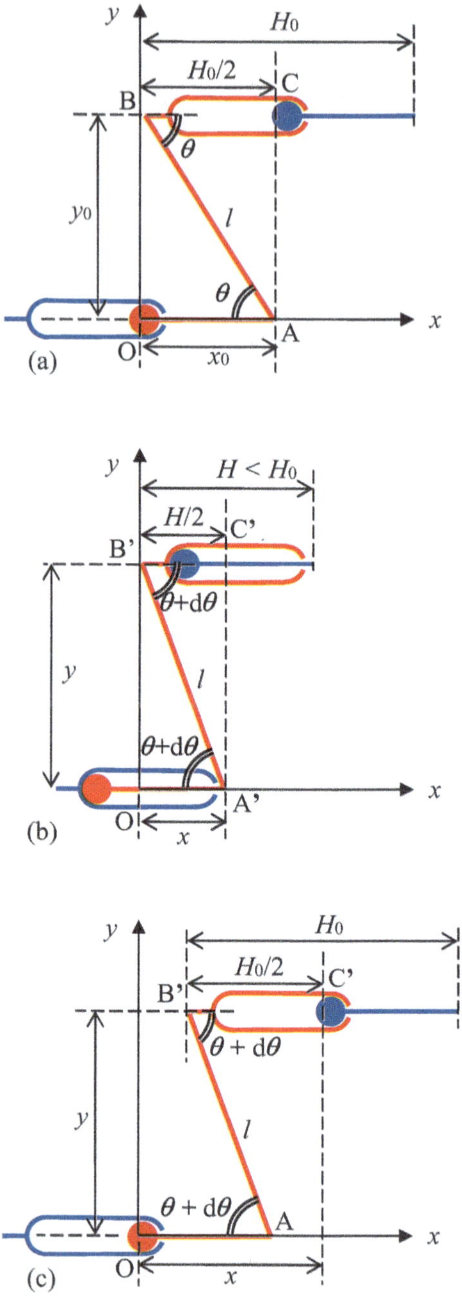

Fig. 16.19 Representative unit of the hybrid rhombic–re-entrant metamaterial for analysis: **a** original state, **b** compressed, and **c** stretched along x-direction

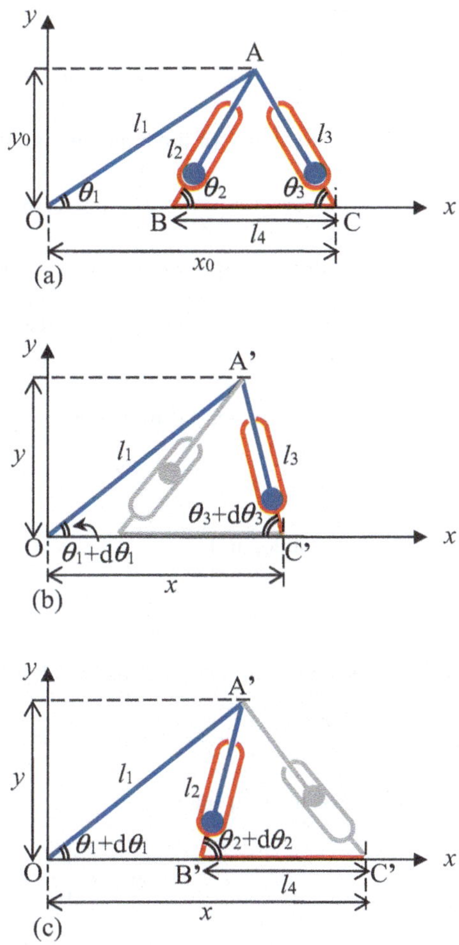

Fig. 16.20 Representative unit of the hybrid kite-arrowhead metamaterial for analysis: **a** original state, **b** compressed, and **c** stretched along x-direction. Redundant parts are indicated in gray

in Figs. 16.19c and 16.20c correspond to the case where tensile stress is applied in the y-direction.

The analysis of compression in x-direction for the hybrid rhombic–re-entrant metamaterial can be established by comparing Fig. 16.19b against Fig. 16.19a, in which B is prevented from moving to the left due to a symmetrically opposing vertex; overall contraction in x-direction is therefore made possible with the sliding of vertices A and C in the slots toward O and B, respectively, to their new locations A' and C', with the vertex B shifting upward to B'. This causes the inclined rod of length l to rotate clockwise by an amount $d\theta$ such that the projections y_0 elongates to y while x_0 shortens to x.

From the projected dimension along the y-axis

$$y_0 = l \sin\theta \tag{16.2.1}$$

in the original state, and

$$y = l\ \sin(\theta + \mathrm{d}\theta) \approx l(\sin\theta + \mathrm{d}\theta\ \cos\theta) \tag{16.2.2}$$

in the deformed state, we have the change in projected dimension along y-direction

$$\mathrm{d}y = y - y_0 = l\mathrm{d}\theta\ \cos\theta \tag{16.2.3}$$

so as to give the corresponding strain component

$$\varepsilon_y = \frac{\mathrm{d}y}{y_0} = \frac{\mathrm{d}\theta}{\tan\theta} \tag{16.2.4}$$

Based on the projected dimension along the x-axis

$$x_0 = l\ \cos\theta \tag{16.2.5}$$

in the original state, and

$$x = l\ \cos(\theta + \mathrm{d}\theta) \approx l(\cos\theta - \mathrm{d}\theta\ \sin\theta) \tag{16.2.6}$$

in the deformed state, one obtains the change of projected dimension in x-direction

$$\mathrm{d}x = x - x_0 = -l\mathrm{d}\theta\ \sin\theta \tag{16.2.7}$$

to yield its strain component

$$\varepsilon_x = \frac{\mathrm{d}x}{x_0} = -\mathrm{d}\theta\ \tan\theta \tag{16.2.8}$$

This gives the Poisson's ratio of the hybrid rhombic–re-entrant microstructure for $\sigma_x < 0$

$$v_{xy} = -\frac{\varepsilon_y}{\varepsilon_x} = \frac{1}{\tan^2\theta} \tag{16.2.9}$$

which indicates conventional behavior for $\sigma_x < 0$.

The analysis of tension in x-direction for the hybrid rhombic–re-entrant metamaterial can be obtained by comparing Fig. 16.19c vis-à-vis Fig. 16.19a. Since the sliders are locked on the right side of the slots, extension in x-direction is made possible by clockwise rotation of the inclined rod by an angle $\mathrm{d}\theta$ such that BC undergoes curvilinear motion to B'C', i.e., B'C' remains horizontal and retains the same length as that for BC. Considering the original state, the projected length on y-axis is given by Eq. (16.2.1), and for a clockwise rotation of $\mathrm{d}\theta$ the projected length at the same axis is the same as that in Eq. (16.2.2), thereby leading to the incremental change of dimension in y-direction and its strain as furnished in Eqs. (16.2.3) and (16.2.4), respectively. Likewise, the projected dimension along the x-axis in the original state is that given in Eq. (16.2.5). Upon rotation of the inclined rod by $\mathrm{d}\theta$, the projected dimension along the x-direction increases, and can be obtained from Fig. 16.19c as

$$x = 2x_0 - l\ \cos(\theta + \mathrm{d}\theta) \approx l(\cos\theta + \mathrm{d}\theta\ \sin\theta) \quad (16.2.10)$$

so as to give

$$\mathrm{d}x = x - x_0 = l\mathrm{d}\theta\ \sin\theta \quad (16.2.11)$$

and its corresponding strain

$$\varepsilon_x = \frac{\mathrm{d}x}{x_0} = \mathrm{d}\theta\ \tan\theta \quad (16.2.12)$$

thereby leading to

$$v_{xy} = -\frac{\varepsilon_y}{\varepsilon_x} = -\frac{1}{\tan^2\theta} \quad (16.2.13)$$

which indicates auxetic behavior for $\sigma_x > 0$.

The analysis of compression in x-direction for the hybrid kite-arrowhead metamaterial can be established by contrasting Fig. 16.20b with reference to Fig. 16.20a, whereby the length l_3 remains constant as A and C displace to A' and C', respectively. During this time, OA rotates anticlockwise to OA' by an angle $\mathrm{d}\theta_1$ while AC rotates clockwise to A'C' by $\mathrm{d}\theta_3$. Therefore, the compression analysis considers the movcment of linkage OAC while the linkange ABC is redundant.

Perusal to Fig. 16.20a for the projection on y-axis

$$y_0 = l_1 \sin\theta_1 = l_3 \sin\theta_3 \quad (16.2.14)$$

at the original state, and consideration of $y = l_1 \sin(\theta_1 + \mathrm{d}\theta_1) = l_3 \sin(\theta_3 + \mathrm{d}\theta_3)$ from Fig. 16.20b for similar projection gives

$$y = l_1(\sin\theta_1 + \mathrm{d}\theta_1 \cos\theta_1) = l_3(\sin\theta_3 + \mathrm{d}\theta_3 \cos\theta_3) \quad (16.2.15)$$

based on infinitesimal deformation. This gives the incremental change in dimension measured along the y-axis

$$\mathrm{d}y = y - y_0 = l_1\mathrm{d}\theta_1 \cos\theta_1 = l_3\mathrm{d}\theta_3 \cos\theta_3 \quad (16.2.16)$$

and its corresponding strain

$$\varepsilon_y = \frac{\mathrm{d}y}{y_0} = \frac{\mathrm{d}\theta_1}{\tan\theta_1} = \frac{\mathrm{d}\theta_3}{\tan\theta_3} \tag{16.2.17}$$

To pave a way for obtaining the strain in x-direction, it is useful at this stage to write

$$\frac{l_3}{l_1} = \frac{\sin\theta_1}{\sin\theta_3} \tag{16.2.18}$$

based on Eq. (16.2.14), and

$$\sin\theta_1 + \mathrm{d}\theta_1\cos\theta_1 = \frac{l_3}{l_1}(\sin\theta_3 + \mathrm{d}\theta_3\cos\theta_3) \tag{16.2.19}$$

from Eq. (16.2.15). Substituting Eq. (16.2.18) into Eq. (16.2.19) gives

$$\frac{\mathrm{d}\theta_3}{\mathrm{d}\theta_1} = \frac{\tan\theta_3}{\tan\theta_1} \tag{16.2.20}$$

Reference to Fig. 16.20a gives the projection along the x-axis through OAC as $x_0 = l_1\cos\theta_1 + l_3\cos\theta_3$ or, using Eq. (16.2.18),

$$x_0 = l_1\left(\cos\theta_1 + \frac{\sin\theta_1}{\tan\theta_3}\right) \tag{16.2.21}$$

for the original state, while perusal to Fig. 16.20b yields a similar projection through OA'C' as $x = l_1\cos(\theta_1 + \mathrm{d}\theta_1) + l_3\cos(\theta_3 + \mathrm{d}\theta_3)$, or

$$x = l_1(\cos\theta_1 - \mathrm{d}\theta_1\sin\theta_1) + l_3(\cos\theta_3 - \mathrm{d}\theta_3\sin\theta_3) \tag{16.2.22}$$

on the basis of infinitesimal rod rotation. Substituting Eqs. (16.2.18) and (16.2.20) into Eq. (16.2.22) eliminates l_3 and $\mathrm{d}\theta_3$ to yield

$$x = l_1\left(\cos\theta_1 - \mathrm{d}\theta_1\sin\theta_1 + \frac{\sin\theta_1}{\tan\theta_3} - \cos\theta_1\tan\theta_3\mathrm{d}\theta_1\right) \tag{16.2.23}$$

Therefore, we have the incremental change in dimension along the x-direction

$$\mathrm{d}x = x - x_0 = -l_1\mathrm{d}\theta_1(\sin\theta_1 + \cos\theta_1\tan\theta_3) \tag{16.2.24}$$

and strain in the same direction

$$\varepsilon_x = \frac{\mathrm{d}x}{x_0} = -\mathrm{d}\theta_1\tan\theta_3 \tag{16.2.25}$$

This gives the Poisson's ratio

$$\upsilon_{xy} = -\frac{\varepsilon_y}{\varepsilon_x} = \frac{1}{\tan\theta_1 \tan\theta_3} \tag{16.2.26}$$

which denotes conventional behavior for $\sigma_x < 0$. For the special case, where $l_1 = l_3$ (or $\theta_1 = \theta_3$), Eq. (16.2.26) simplifies to

$$\upsilon_{xy} = \frac{1}{\tan^2\theta_1} \tag{16.2.27}$$

which is analogous to Eq. (16.2.9) for the compression of the hybrid rhombic–re-entrant microstructure in the x-direction.

The analysis of tensile load application in x-direction on the hybrid kite-arrowhead metamaterial is carried out by comparing Fig. 16.20c with respect to Fig. 16.20a, whereby the length l_2 remains constant as A, B, and C displace to A', B', and C', respectively. During this time, OA and AB rotate anticlockwise to OA' and A'B' by $\mathrm{d}\theta_1$ and $\mathrm{d}\theta_2$, respectively, while BC moves to B'C' by translation along the x-axis. In other words, the tensile analysis accounts for the motion of linkage OABC, with the link AC being redundant. Perusal to Fig. 16.20a for the projection on y-axis at the original state

$$y_0 = l_1 \sin\theta_1 = l_2 \sin\theta_2 \tag{16.2.28}$$

and consideration of $y = l_1 \sin(\theta_1 + \mathrm{d}\theta_1) = l_2 \sin(\theta_2 + \mathrm{d}\theta_2)$ from Fig. 16.20c for similar projection gives rise to

$$y = l_1(\sin\theta_1 + \mathrm{d}\theta_1 \cos\theta_1) = l_2(\sin\theta_2 + \mathrm{d}\theta_2 \cos\theta_2) \tag{16.2.29}$$

on the assumption of infinitesimal deformation. This gives the incremental change in dimension measured along the y-axis

$$\mathrm{d}y = y - y_0 = l_1 \mathrm{d}\theta_1 \cos\theta_1 = l_2 \mathrm{d}\theta_2 \cos\theta_2 \tag{16.2.30}$$

and its strain

$$\varepsilon_y = \frac{\mathrm{d}y}{y_0} = \frac{\mathrm{d}\theta_1}{\tan\theta_1} = \frac{\mathrm{d}\theta_2}{\tan\theta_2} \tag{16.2.31}$$

As before, it is beneficial to express

$$\frac{l_2}{l_1} = \frac{\sin\theta_1}{\sin\theta_2} \tag{16.2.32}$$

based on Eq. (16.2.28) and

$$\sin\theta_1 + \mathrm{d}\theta_1 \cos\theta_1 = \frac{l_2}{l_1}(\sin\theta_2 + \mathrm{d}\theta_2 \cos\theta_2) \tag{16.2.33}$$

from Eq. (16.2.29). Substituting Eq. (16.2.32) into Eq. (16.2.33) gives

$$\frac{d\theta_2}{d\theta_1} = \frac{\tan\theta_2}{\tan\theta_1} \tag{16.2.34}$$

In addition,

$$l_4 = l_2\cos\theta_2 + l_3\cos\theta_3 \tag{16.2.35}$$

Reference to Fig. 16.20a gives the projection along the x-axis through OABC as $x_0 = l_1\cos\theta_1 - l_2\cos\theta_2 + l_4$ or, using Eq. (16.2.32),

$$x_0 = l_1\left(\cos\theta_1 - \frac{\sin\theta_1}{\tan\theta_2}\right) + l_4 \tag{16.2.36}$$

for the original state, while perusal to Fig. 16.20c yields a similar projection through OA'B'C' as $x = l_1\cos(\theta_1 + d\theta_1) - l_2\cos(\theta_2 + d\theta_2) + l_4$, or

$$x = l_1(\cos\theta_1 - d\theta_1\sin\theta_1) - l_2(\cos\theta_2 - d\theta_2\sin\theta_2) + l_4 \tag{16.2.37}$$

based on infinitesimal rod rotation. Substituting Eqs. (16.2.32) and (16.2.34) into Eq. (16.2.37) eliminates l_2 and $d\theta_2$ to yield

$$x = l_1\left(\cos\theta_1 - d\theta_1\sin\theta_1 - \frac{\sin\theta_1}{\tan\theta_2} + \cos\theta_1\tan\theta_2 d\theta_1\right) + l_4 \tag{16.2.38}$$

Therefore, we have the incremental change in dimension along the x-direction

$$dx = x - x_0 = l_1 d\theta_1(-\sin\theta_1 + \cos\theta_1\tan\theta_2) \tag{16.2.39}$$

and strain in the corresponding direction

$$\varepsilon_x = \frac{dx}{x_0} = \frac{d\theta_1\tan\theta_2\left(\frac{1}{\tan\theta_1} - \frac{1}{\tan\theta_2}\right)}{\frac{1}{\tan\theta_1} - \frac{1}{\tan\theta_2} + \frac{l_4}{l_1\sin\theta_1}} \tag{16.2.40}$$

Substituting Eq. (16.2.35) into Eq. (16.2.40) gives

$$\varepsilon_x = d\theta_1\tan\theta_3\frac{\tan\theta_2 - \tan\theta_1}{\tan\theta_3 + \tan\theta_1} \tag{16.2.41}$$

thereby leading to the Poisson's ratio

$$v_{xy} = -\frac{\varepsilon_y}{\varepsilon_x} = \frac{1}{\tan\theta_1\tan\theta_3}\left(\frac{\tan\theta_1 + \tan\theta_3}{\tan\theta_1 - \tan\theta_2}\right) \tag{16.2.42}$$

which indicates auxetic behavior for $\sigma_x > 0$ because $\tan\theta_1 < \tan\theta_2$. Although AC is redundant in the case of tensile σ_x, the angle θ_3 comes into play arising from the need to include the length l_4 for evaluating x, i.e., OC', as shown in Fig. 16.20c. Equation (16.2.42) is comparable to Eq. (16.2.26) except for the terms in the parenthesis.

Perusal to Eqs. (16.2.9) and (16.2.13) for the Poisson's ratio υ_{xy} of the hybrid rhombic–re-entrant metamaterial, as a result from σ_x loading, indicates that the magnitudes of υ_{xy} is gigantic if the inclined rod is highly aligned to the x-axis, i.e.,

$$\lim_{\theta\to 0^\circ} \upsilon_{xy} = \begin{cases} +\infty \Leftrightarrow \sigma_x < 0 \\ -\infty \Leftrightarrow \sigma_x > 0 \end{cases} \tag{16.2.43}$$

but diminishes when the inclined rod becomes oriented toward the y-axis

$$\lim_{\theta\to 90^\circ} \upsilon_{xy} = 0 \tag{16.2.44}$$

The variation of Poisson's ratio for other values of θ is plotted in Fig. 16.21 to aid visual observation.

With reference to Eqs. (16.2.26) and (16.2.42) for the Poisson's ratio υ_{xy} of the hybrid kite-arrowhead metamaterial arising from σ_x loading, we observe that

$$\lim_{\theta_1\to 0^\circ} \upsilon_{xy} = \begin{cases} +\infty \Leftrightarrow \sigma_x < 0 \\ -\infty \Leftrightarrow \sigma_x > 0 \end{cases} \tag{16.2.45}$$

and

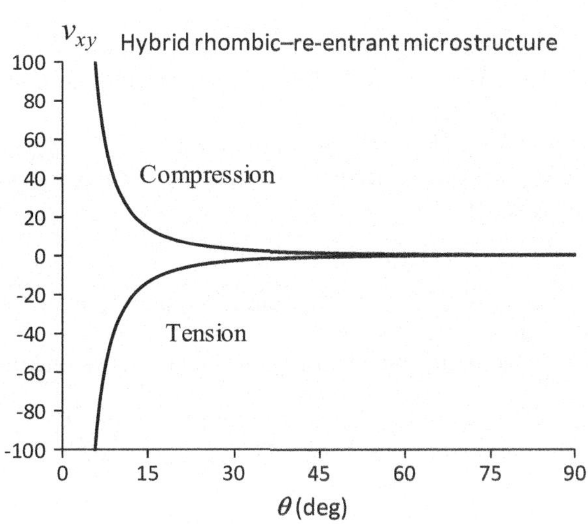

Fig. 16.21 Switching of Poisson's ratio υ_{xy} sign for the hybrid rhombic–re-entrant metamaterial due to σ_x direction reversal

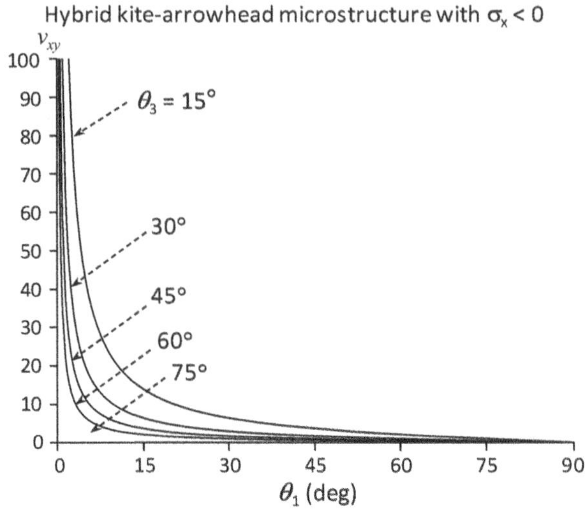

Fig. 16.22 Positive Poisson's ratio υ_{xy} for the hybrid kite-arrowhead metamaterial under compressive σ_x

$$\lim_{\theta_1 \to 90^\circ} \upsilon_{xy} = 0 \Leftrightarrow \sigma_x < 0 \tag{16.2.46}$$

which resemble the hybrid rhombic–re-entrant metamaterial characteristic. Unlike the hybrid rhombic–re-entrant metamaterial, perusal to Eq. (16.2.42) for the hybrid kite-arrowhead metamaterial under tensile load further indicates that

$$\lim_{\theta_1 \to \theta_2} \upsilon_{xy} = -\infty \Leftrightarrow \sigma_x > 0 \tag{16.2.47}$$

In addition to these extremes, the Poisson's ratio for other values of θ_1 are furnished in Figs. 16.22 and 16.23 to facilitate visual observation under compressive and tensile σ_x, respectively.

For Special Case I where the joints ABC in Fig. 16.20a form the corners of an isosceles triangle such that $l_3 = l_2$ (or $\theta_3 = \theta_2$) as the original condition, Eq. (16.2.42) reduces to

$$\upsilon_{xy} = \frac{1}{\tan\theta_1 - \tan\theta_2}\left(\frac{1}{\tan\theta_1} + \frac{1}{\tan\theta_2}\right) \tag{16.2.48}$$

To facilitate comparison with Eq. (16.2.48), one may express Eq. (16.2.26) as

$$\upsilon_{xy} = \frac{1}{\tan\theta_1 + \tan\theta_3}\left(\frac{1}{\tan\theta_1} + \frac{1}{\tan\theta_3}\right) \tag{16.2.49}$$

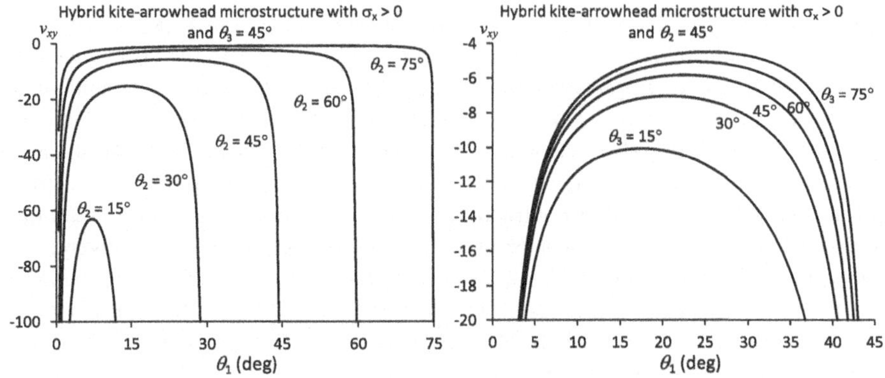

Fig. 16.23 Negative Poisson's ratio υ_{xy} for the hybrid kite-arrowhead metamaterial under tensile σ_x, with variation of θ_2 at fixed θ_3 (left) and variation of θ_3 at fixed θ_2 (right)

For Special Case II, where the angle BAC in Fig. 16.20a forms a right angle, we have $\theta_2 + \theta_3 = 90°$ such that applying the relation $\tan\theta_2 \tan\theta_3 = 1$ on Eq. (16.2.42) gives $\upsilon_{xy} = -\tan(\theta_1 + \theta_3)/\tan\theta_1$ or

$$\upsilon_{xy} = \frac{1}{\tan\theta_1}\frac{1}{\tan(\theta_1 - \theta_2)} \quad (16.2.50)$$

which is comparable to Eq. (16.2.26). The Poisson's ratio plots of these two special cases are shown in Fig. 16.24.

A note on application would be useful at this juncture. For the purposes of wrapping a flat sheet onto a curved surface, a positive, a zero or a negative Poisson's ratio

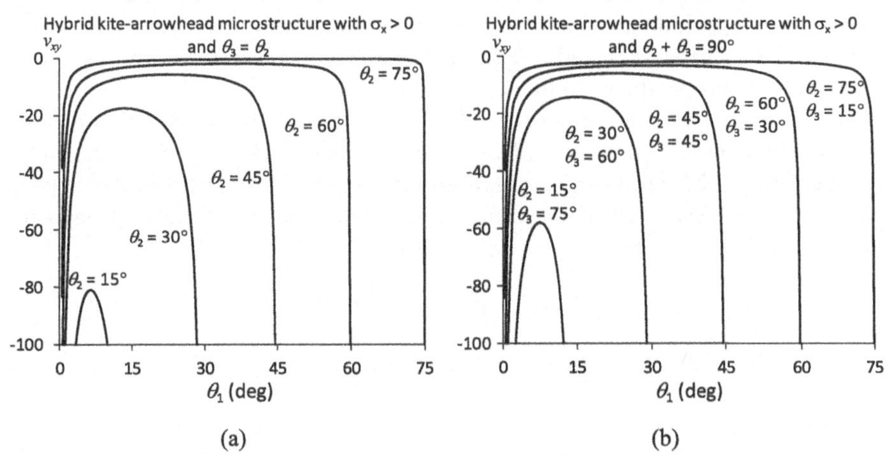

Fig. 16.24 Negative Poisson's ratio υ_{xy} for the hybrid kite-arrowhead metamaterial under tensile σ_x, under **a** Special Case I ($\theta_3 = \theta_2$) and **b** Special Case II ($\theta_2 + \theta_3 = 90°$)

material is advised if the surface takes the form of an anti-clastic shape, a cylindrical shape or a synclastic shape in order to reduce bending stress in the sheet material. This line of reasoning implies that the choice of Poisson's ratio sign is dependent on the application. Likewise, the choice of auxetic fiber is useful to resist fiber pull-out from the matrix material due to the self-locking mechanism in the form of radial expansion during axial pulling; however, auxetic fibers are easily pushed out from matrix material due to radial contraction. In fact, it is the conventional fibers that resist push-out due to radial expansion, although it is also known that conventional fibers are easily pulled out due to the resulting radial contraction. If the fiber behaves as auxetic material during fiber pull-out ($\sigma_z > 0 \Rightarrow \upsilon_{zr} < 0$) but becomes conventional material during fiber push-out ($\sigma_z < 0 \Rightarrow \upsilon_{zr} > 0$), then such a fiber resists both pull-out and push-out as a result of its duality.

16.3 Conclusions

It can be concluded from Sect. 16.1 that the Poisson's ratio υ_{xy} sign-switching—triggered by the reversal of σ_x direction—is attained by designing microstructures that stand in between two microstructures that give conventional and auxetic properties, such that σ_x loading in opposite directions change the original microstructure to those associated with Poisson's ratio of opposite signs (Lim 2019a). While both microstructures are non-auxetic under σ_y tension, the Poisson's ratio υ_{yx} is indeterminate in the case of σ_y compression and can range from positive to negative. This ambiguity can be removed by incorporation of charged attachments so as to create the propensity of the flexible rods in buckle in predictable manner, thereby clearing the way for the Poisson's ratio υ_{yx} to be determinable. By alternating the charge signs along x-direction for both microstructures, the Poisson's ratio υ_{yx} also exhibits sign-switching upon reversal of σ_y direction. In addition to auxeticity, these microstructures reveal NTE and ZTE properties under unconstrained and constrained boundaries when specific conditions are met.

Unlike Sect. 16.1, two metamaterial microstructures—the hybrid rhombic–re-entrant metamaterial and the hybrid kite-arrowhead metamaterial—have been introduced in Sect. 16.2 to produce sign toggling of Poisson's ratio upon reversal of applied stress direction. Specifically, these metamaterials exhibit the following opposing Poisson's ratio properties

$$\sigma_x < 0 \Rightarrow \upsilon_{xy} > 0$$

$$\sigma_x > 0 \Rightarrow \upsilon_{xy} < 0$$

or collectively as $\upsilon_{xy}\sigma_x < 0$ (Lim 2019b). While the approach of attaining Poisson's ratio sign-switching upon stress reversal in Sect. 16.1 can be achieved by catering for lateral contraction upon axial loading to give $\upsilon_{xy}\sigma_x > 0$, the opposing sense of

Poisson's ratio toggling elucidated in Sect. 16.2 is made possible by means of simultaneous jam and slide mechanism in both metamaterials. The capability of materials to switch their Poisson's ratio signs by mere stress direction reversal permits them to act in ways that can only be partially fulfilled by auxetic and conventional materials. While this chapter discussed the change in Poisson's ratio signs with the reversal of applied stress direction, it will be shown in the next chapter by way of examples how metamaterias can be designed such that their expansion coefficients toggle between positive and negative signs upon reversal of (hygro)thermal changes such that the (hygro)thermal strains in the direction(s) of interest are either persistently negative (Lim 2019c) or persistently positive (Lim 2019d).

References

Cauchi R, Attard D, Grima JN (2013) On the mechanical properties of centro-symmetric honeycombs with T-shaped joints. Phys Status Solidi B 250(10):2002–2011

Lim TC (2019a) Composite microstructures with Poisson's ratio sign switching upon stress reversal. Compos Struct 209:34–44

Lim TC (2019b) Metamaterials with Poisson's ratio sign toggling by means of microstructural duality. SN Appl Sci 1(2):176

Lim TC (2019c) A composite metamaterial with sign switchable elastic and hygrothermal properties induced by stress direction and environmental change reversals. Composite Structures 220:185–193

Lim TC (2019d) A 2D auxetikos system based on interconnected shurikens. SN Applied Sciences 1(11):1383

Timoshenko SP, Gere JM (1961) Theory of elastic stability. McGraw-Hill, New York

Chapter 17
Sign-Switching of Environmental Expansion Coefficients with Environmental Change Reversals

Abstract This chapter considers two examples of expansion coefficient sign-switching upon environmental change reversal. For the first example, the thermal and moisture strains in one direction are persistently negative. In the first instance, it is shown that these microstructures exhibit Poisson's ratio sign-toggling with stress direction reversal in the same manner as that discussed in Sect. 16.1. In the second example, the in-plane thermal strain is shown to be persistently positive, based on the principle of microstructural duality.

Keywords Moisture expansion · Poisson's ratio · Thermal expansion · Sign-flipping · Sign-reversible · Sign-switching · Sign-toggling · Stress reversal

17.1 Sign-Switching of Expansion Coefficients with Persistently Negative Hygrothermal Strain in One Direction

With reference to the microstructure shown in Fig. 17.1a, the application of tensile load and compressive load in the horizontal direction, i.e., parallel to the x-axis, gives contraction in the vertical direction, i.e., parallel to the y-axis, as shown in Fig. 17.1b, c. In other words, this microstructure exhibits conventional and auxetic properties depending on the sign of applied load σ_x. Specifically, $v_{xy} > 0$ for $\sigma_x > 0$ but $v_{xy} < 0$ when $\sigma_x < 0$. The quantitative description can be obtained using l_1 and l_2 as the lengths of the inclined rod (or rod 1) and vertical rod (or rod 2), respectively, such that rod 1 is inclined at an angle of θ_1 from the horizontal axis, while rod 2 is perpendicular to this axis, as shown in Fig. 17.1d. The application of loads along the x-axis brings about rotation of rod 1 and rod 2 by an amount $\mathrm{d}\theta_1$ and $\mathrm{d}\theta_2$. It can be seen from Fig. 17.1e, f that both tensile and compressive loads along the x-direction would rotate rod 1 in the same direction, with decrease to the θ_1 angle by an amount $\mathrm{d}\theta_1$, hence the final angle being $\theta_1 - \mathrm{d}\theta_1$. On the other hand, the application of tensile and compressive load would rotate rod 2 by an amount $\mathrm{d}\theta_2$ in opposing directions, thereby leading to the final angles of $\theta_2 + \mathrm{d}\theta_2$ and $\theta_2 - \mathrm{d}\theta_2$, respectively. Since $\theta_2 = 90°$, it is sufficient to indicate only $\mathrm{d}\theta_2$ in Fig. 17.1e, f. The following

T. Lim, *Mechanics of Metamaterials with Negative Parameters*, Engineering Materials,
https://doi.org/10.1007/978-981-15-6446-8_17

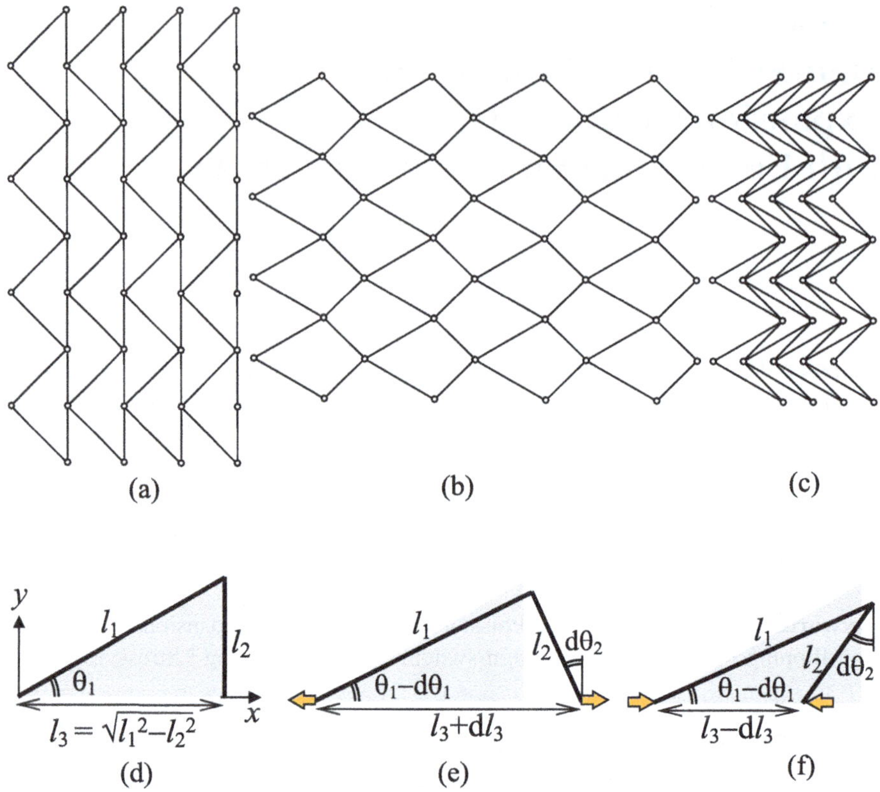

Fig. 17.1 **a** A microstructure that gives dual properties under direct axial load along horizontal direction, i.e., **b** conventional property under tensile load, and **c** auxetic property under compressive load; **d** a representative unit for analysis, **e** representative unit under tensile load, and **f** representative unit under compressive load. The gray shade in (**d**), which is enclosed by rod 1, rod 2 and the x-axis, is replicated in (**e**) and (**f**) to facilitate comparison

analysis in this section assumes that rod 1 and rod 2 are rigid and permitted to rotate at the pin joints (Lim 2019a).

The strain along the x-direction can be arrived at by considering the change in the horizontal length l_3 by the amount dl_3, which is also dx. With reference to Fig. 17.1d–f, the horizontal dimension of

$$x_0 = l_1 \cos \theta_1 \tag{17.1.1}$$

encounters a change in dimension by an amount

$$dx = l_1 \cos(\theta_1 - d\theta_1) \pm l_2 \sin d\theta_2 - l_1 \cos \theta_1 \tag{17.1.2}$$

where the upper and lower signs correspond to the application of horizontal tensile and compressive loads, respectively. Dividing Eq. (17.1.3) by Eq. (17.1.1) gives the horizontal strain

$$\varepsilon_x = \cos d\theta_1 - 1 + \tan\theta_1(\sin d\theta_1 \pm \sin d\theta_2) \tag{17.1.3}$$

Perusal to Fig. 17.1d–f again shows that the horizontal dimension

$$y_0 = l_1 \sin\theta_1 = l_2 \tag{17.1.4}$$

experiences a change in dimension of

$$dy = l_1 \sin(\theta_1 - d\theta_1) - l_1 \sin\theta_1 = l_2(\cos d\theta_2 - 1) \tag{17.1.5}$$

Dividing Eq. (17.1.5) with Eq. (17.1.4) gives the vertical strain

$$\varepsilon_y = \cos d\theta_1 - \frac{\sin d\theta_1}{\tan\theta_1} - 1 \tag{17.1.6}$$

based on rod 1, or

$$\varepsilon_y = \cos d\theta_2 - 1 \tag{17.1.7}$$

on the basis of rod 2. Since both descriptions of vertical strains are common, they are equated to express the rotation of rod 2 in terms of rod 1 angle and its rotation

$$d\theta_2 = \cos^{-1}\left(\cos d\theta_1 - \frac{\sin d\theta_1}{\tan\theta_1}\right) \tag{17.1.8}$$

such that substitution of Eq. (17.1.8) into Eq. (17.1.3) gives

$$\varepsilon_x = \cos d\theta_1 - 1 + \tan\theta_1\left\{\sin d\theta_1 \pm \sin\left[\cos^{-1}\left(\cos d\theta_1 - \frac{\sin d\theta_1}{\tan\theta_1}\right)\right]\right\} \tag{17.1.9}$$

where, as before, the upper and lower signs refer to the application of tensile and compressive loads, respectively. For the special case of $\theta_1 = 45°$, the strain descriptions reduce to

$$\varepsilon_x = \cos d\theta_1 - 1 + \sin d\theta_1 \pm \sin\left[\cos^{-1}(\cos d\theta_1 - \sin d\theta_1)\right] \tag{17.1.10}$$

and

$$\varepsilon_y = \cos d\theta_1 - \sin d\theta_1 - 1 \tag{17.1.11}$$

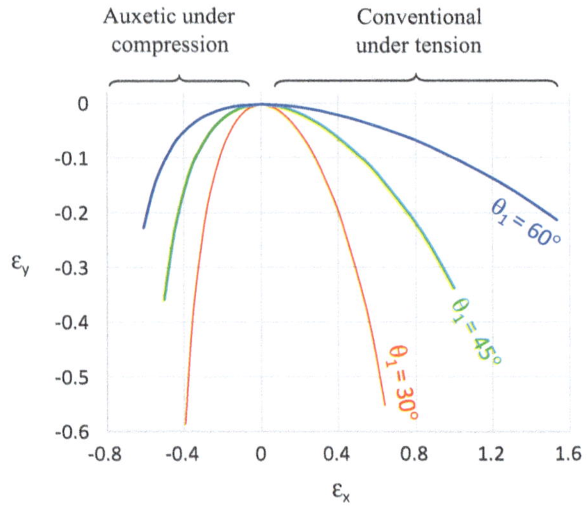

Fig. 17.2 Curves of ε_y versus ε_x for $\theta_1 = 30°,\ 45°,\ 60°$ manifesting auxetic and conventional behavior initiated by the prescription of $\varepsilon_x < 0$ and $\varepsilon_x > 0$, respectively

The values of $\mathrm{d}\theta_1$ and $\mathrm{d}\theta_2$ are positive due to the way they are denoted in Fig. 17.1e, f.

Plots of vertical strain versus horizontal strain are furnished in Fig. 17.2, whereby auxeticity is demonstrated during compression while conventional behavior is shown during tension. A discrimination between tensile and compressive loads is not made through the sign of $\Delta\theta_1$ since this angular change is defined as being positive for both tensile and compressive loads. The discrimination between tensile and compressive loads is effected by the choice of upper and lower signs in Eq. (17.1.9). Reference to Fig. 17.2 also suggests that the condition of zero Poisson's ratio (ZPR) is achieved when the deformation is infinitesimal.

The deformation pattern shown in Fig. 17.1 can be achieved by means of thermal or moisture expansions if the microstructure is modified through the insertion of a reinforcing rod aligned parallel to the x-axis, as shown in Fig. 17.3a. Let this reinforced rod be called rod 3, the length l_3 now represents the original length of rod 3 and $\mathrm{d}l_3$ denotes the change in the length of this rod. Suppose rod 1 and rod 2 are again rigid, then a change in temperature $\mathrm{d}T$ or moisture concentration $\mathrm{d}C$ in the surrounding environment leads to a change in temperature $\mathrm{d}T$ or moisture concentration $\mathrm{d}C_3$ in rod 3 at steady state. Let the coefficient of thermal expansion (CTE) and the coefficient of moisture expansion (CME) of rod 3 be $\alpha_3^{(T)}$ and $\alpha_3^{(C)}$, respectively, then

$$\mathrm{d}l_3 = l_3 \alpha_3^{(T)} \mathrm{d}T \tag{17.1.12}$$

due to thermal expansion, or

$$\mathrm{d}l_3 = l_3 \alpha_3^{(C)} \mathrm{d}C_3 \tag{17.1.13}$$

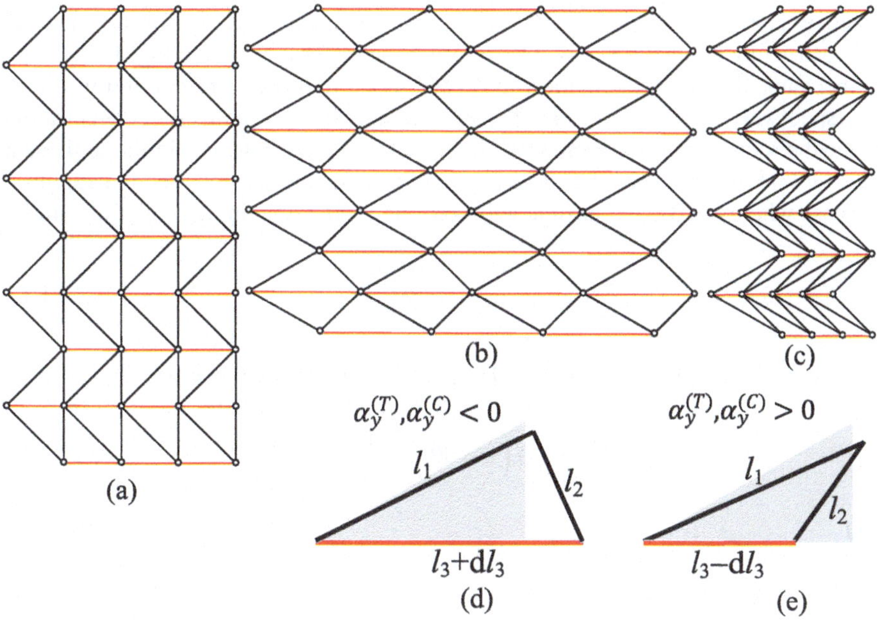

Fig. 17.3 **a** Modified microstructure from Fig. 17.1a by incorporation of horizontally aligned rods so as to produce overall vertical contraction with **b** expansion of rod 3 and **c** contraction of rod 3; the representative units of (**b**) and (**c**) are displayed in (**d**) and (**e**), respectively, with the shaded region representing the area enclosed by rod 1, rod 2, and rod 3 before hygrothermal strain

due to moisture expansion. Since rod 3 is aligned along the x-axis, it follows that its strain is representative of the horizontal strain for the microstructure. Writing in terms of strain

$$\varepsilon_x = \varepsilon_3 = \frac{\mathrm{d}l_3}{l_3} \tag{17.1.14}$$

we have the overall CTE and CME of the microstructure along the horizontal direction

$$\alpha_x^{(T)} = \frac{\varepsilon_x^{(T)}}{\mathrm{d}T} = \alpha_3^{(T)} \tag{17.1.15}$$

and

$$\alpha_x^{(C)} = \frac{\varepsilon_x^{(C)}}{\mathrm{d}C}, \quad \alpha_3^{(C)} = \frac{\varepsilon_3^{(C)}}{\mathrm{d}C_3} \tag{17.1.16}$$

respectively.

It is obvious that the effective CTE and the CME of this modified microstructure along the x-axis have to be positive if one assumes that the CTE or CME of rod 3 to be positive. However, a negative CTE or negative CME can be found for this modified microstructure along the y-axis. The deformation patterns shown in Fig. 17.1b, c due to mechanical loading along the x-axis correspond to those furnished in Fig. 17.3b, c, respectively, arising from the change in temperature or moisture concentration. Since the rotations of rod 1 and rod 2 in Figs. 17.3d and 17.4e are similar to those in Fig. 17.1e, f, respectively, the strain expressions from the previous section applies to the case where rod 3 is expansible with rod 1 and rod 2 remaining rigid. Substituting

$$\begin{aligned} \varepsilon_y^{(T)} &= \alpha_y^{(T)} \mathrm{d}T \\ \varepsilon_y^{(C)} &= \alpha_y^{(C)} \mathrm{d}C \end{aligned} \tag{17.1.17}$$

into Eq. (17.1.6) gives the overall CTE and CME along the y-axis as

$$\alpha_y^{(T)} = \frac{1}{\mathrm{d}T}\left(\cos \mathrm{d}\theta_1 - \frac{\sin \mathrm{d}\theta_1}{\tan \theta_1} - 1\right) \tag{17.1.18}$$

and

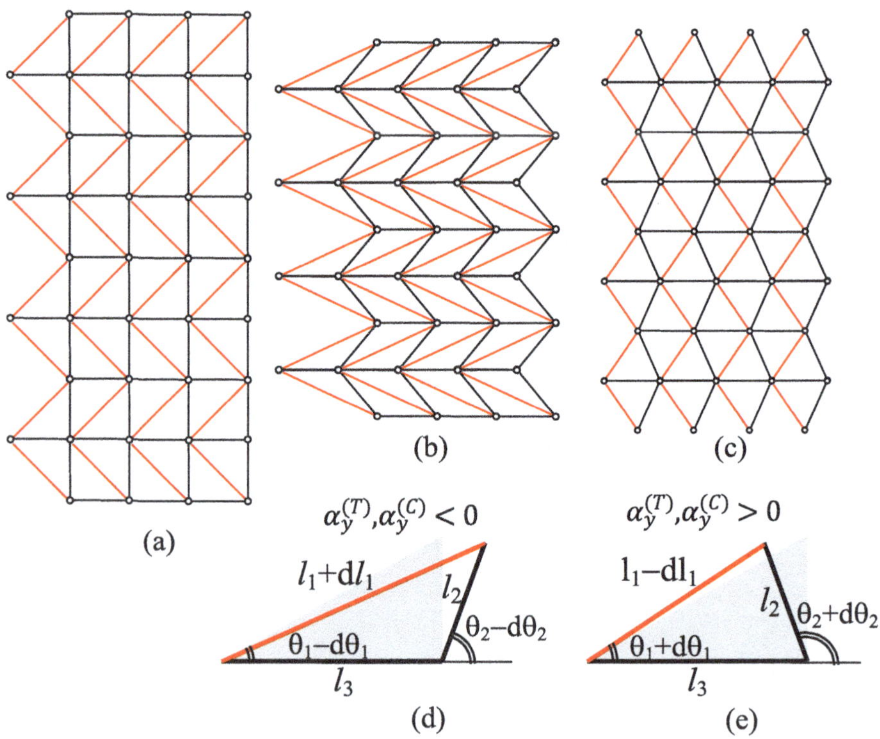

Fig. 17.4 **a** The same modified microstructure can be made to exhibit overall vertical contraction due to **b** expansion of rod 1 and **c** contraction of rod 1; the representative units for (**b**) and (**c**) are furnished in (**d**) and (**e**), respectively

$$\alpha_y^{(C)} = \frac{1}{\mathrm{d}C}\left(\cos \mathrm{d}\theta_1 - \frac{\sin \mathrm{d}\theta_1}{\tan \theta_1} - 1\right) \tag{17.1.19}$$

respectively. For this reason, Fig. 17.2 applies for the modified microstructure of Fig. 17.3a, whereby the vertical and horizontal axes in Fig. 17.2 can be replaced by $\alpha_y^{(T)}$ and $\alpha_x^{(T)}$, respectively, for thermal fluctuation. In the case of moisture concentration changes, the vertical and horizontal axes of Fig. 17.2 can be replaced by $\alpha_y^{(C)}$ and $\alpha_x^{(C)}$, respectively, for the modified microstructure of Fig. 17.3a.

Reference to Fig. 17.4 indicates that negative expansion can also be observed along the y-direction without strain along the x-axis, i.e., it exhibits zero thermal expansion (ZTE) and zero moisture expansion (ZME) in the x-direction. The negative expansion along the y-direction is made possible by the expansion or contraction of rod 1 with rod 2 and rod 3 being rigid. Starting from Fig. 17.4a, the expansion and contraction of the inclined rods lead to the deformed microstructure indicated by Fig. 17.4b, c, respectively. Since the angles $\theta_1 - \mathrm{d}\theta_1$ and $\theta_2 - \mathrm{d}\theta_2$ in Fig. 17.4d arising from the expanded length $l_1 + \mathrm{d}l_1$ of rod 1 conversely correspond to angles of $\theta_1 + \mathrm{d}\theta_1$ and $\theta_2 + \mathrm{d}\theta_2$ in Fig. 17.4e arising from the contracted length $l_1 - \mathrm{d}l_1$ of rod 1, it suffices to consider only one of them in analysis. By the length constancy of rod 3 as shown in Fig. 17.4d,

$$l_3 = (l_1 + \mathrm{d}l_1)\cos(\theta_1 - \mathrm{d}\theta_1) - l_2 \sin \mathrm{d}\theta_2 = l_1 \cos \theta_1 \tag{17.1.20}$$

Dividing Eq. (17.1.20) by Eq. (17.1.1) and recognizing that $\mathrm{d}l_1/l_1 = \varepsilon_1$ and $l_2/l_1 = \sin \theta_1$, the angular change of rod 2 inclination is obtained as

$$\mathrm{d}\theta_2 = \sin^{-1}\left[-\frac{1}{\tan \theta_1} + (1 + \varepsilon_1)\left(\sin \mathrm{d}\theta_1 + \frac{\cos \mathrm{d}\theta_1}{\tan \theta_1}\right)\right] \tag{17.1.21}$$

The change in vertical dimension

$$\mathrm{d}y = (l_1 + \mathrm{d}l_1)\sin(\theta_1 - \mathrm{d}\theta_1) - l_1 \sin \theta_1 = l_2(\cos \mathrm{d}\theta_2 - 1) \tag{17.1.22}$$

leads to the vertical strain by dividing it with Eq. (17.1.4), i.e.,

$$\varepsilon_y = (1 + \varepsilon_1)\left(\cos \mathrm{d}\theta_1 - \frac{\sin \mathrm{d}\theta_1}{\tan \theta_1}\right) - 1 \tag{17.1.23}$$

based on rod 1, and

$$\varepsilon_y = \cos \mathrm{d}\theta_2 - 1 \tag{17.1.24a}$$

on the basis of rod 2. The latter can be expressed in terms of θ_1 by virtue of Eq. (17.1.21), i.e.,

$$\varepsilon_y = \cos\left\{\sin^{-1}\left[-\frac{1}{\tan\theta_1} + (1+\varepsilon_1)\left(\sin \mathrm{d}\theta_1 + \frac{\cos \mathrm{d}\theta_1}{\tan\theta_1}\right)\right]\right\} - 1 \tag{17.1.24b}$$

It is obvious that the strain of rod 1 is related to its angular rotations of rod 1 and rod 2. To establish a relationship of ε_1 with $\mathrm{d}\theta_1$ and $\mathrm{d}\theta_2$, we apply the sine rule on Fig. 17.4d, i.e.,

$$\frac{l_1 + \mathrm{d}l_1}{\sin(90^\circ + \mathrm{d}\theta_2)} = \frac{l_2}{\sin(\theta_1 - \mathrm{d}\theta_1)} \tag{17.1.25}$$

to give

$$\varepsilon_1 = \frac{\cos \mathrm{d}\theta_2}{\cos \mathrm{d}\theta_1 - \frac{\sin \mathrm{d}\theta_1}{\tan\theta_1}} - 1 \tag{17.1.26}$$

Calculations for the vertical strain can be made by providing a starting solution for $\mathrm{d}\theta_2$ in Eq. (17.1.21) with $\varepsilon_1 = 0$ in the first instance. The obtained $\mathrm{d}\theta_2$ is then substituted into Eq. (17.1.26) to obtain a better estimate of ε_1, which is then used for extracting a more accurate $\mathrm{d}\theta_2$ which, in turn, facilitates a refined ε_1. This iteration process continues until no appreciable change in either $\mathrm{d}\theta_2$ or ε_1 is observed. The final value of ε_1 is then substituted into Eq. (17.1.24b) to obtain ε_y. For the special case of $\theta_1 = 45^\circ$, the general expressions are greatly simplified to

$$\mathrm{d}\theta_2 = \sin^{-1}[-1 + (1+\varepsilon_1)(\sin \mathrm{d}\theta_1 + \cos \mathrm{d}\theta_1)] \tag{17.1.27}$$

$$\varepsilon_1 = \frac{\cos \mathrm{d}\theta_2}{\cos \mathrm{d}\theta_1 - \sin \mathrm{d}\theta_1} - 1 \tag{17.1.28}$$

$$\varepsilon_y = \cos\{\sin^{-1}[-1 + (1+\varepsilon_1)(\sin \mathrm{d}\theta_1 + \cos \mathrm{d}\theta_1)]\} - 1 \tag{17.1.29}$$

Figure 17.5 (left) shows the iterative process in the case of $\theta_1 = 45^\circ$, i.e., a starting value of $\mathrm{d}\theta_2$ (with $\varepsilon_1 = 0$) is put forth in the first instance using Eq. (17.1.27), whose result is then used in Eq. (17.1.28) for calculating the starting or first value of ε_1 in Fig. 17.5 (right). As shown in Fig. 17.5, the results quickly converge by the first iteration; however, computation is continued until the fifth iteration for the sake of accuracy in subsequent calculation of the vertical strain described in Eq. (17.1.29). The profiles of ε_y versus $\mathrm{d}\theta_1$ at $\theta_1 = 30^\circ$, 45°, 60° are plotted in Fig. 17.6. On the basis of Eq. (17.1.17), the vertical strain as induced by changes to the thermal and moisture concentration can be expressed in terms of CTE and CME as

$$\alpha_y^{(T)} = \frac{1}{\mathrm{d}T}\left\{\cos\left\{\sin^{-1}\left[-\frac{1}{\tan\theta_1} + (1+\varepsilon_1)\left(\sin \mathrm{d}\theta_1 + \frac{\cos \mathrm{d}\theta_1}{\tan\theta_1}\right)\right]\right\} - 1\right\} \tag{17.1.30}$$

and

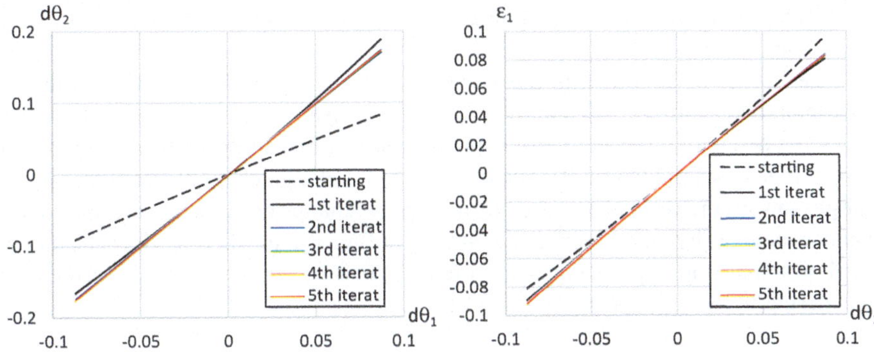

Fig. 17.5 Iterative plots of $\mathrm{d}\theta_2$ (left) and ε_1 (right) with reference to $\mathrm{d}\theta_1$, demonstrating rapid convergence

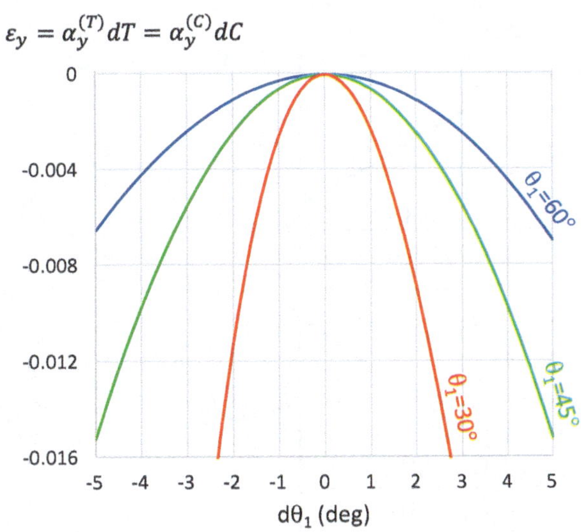

Fig. 17.6 Curves of ε_y against $\mathrm{d}\theta_1$ for $\theta_1 = 30°, 45°, 60°$ based on the microstructure of Fig. 17.4, which caters to expansion and contraction of rod 1, with rod 2 and rod 3 being rigid

$$\alpha_y^{(C)} = \frac{1}{\mathrm{d}C}\left\{\cos\left\{\sin^{-1}\left[-\frac{1}{\tan\theta_1} + (1+\varepsilon_1)\left(\sin\mathrm{d}\theta_1 + \frac{\cos\mathrm{d}\theta_1}{\tan\theta_1}\right)\right]\right\} - 1\right\} \tag{17.1.31}$$

respectively. One can then see that the condition of $\varepsilon_1 > 0$ (which implies $\mathrm{d}T > 0$ and/or $\mathrm{d}C > 0$) suggests that $\alpha_y^{(T)} < 0$ and/or $\alpha_y^{(C)} < 0$, respectively, due to $\varepsilon_y < 0$; on the other hand, the condition of $\varepsilon_1 < 0$ (which implies $\mathrm{d}T < 0$ and/or $\mathrm{d}C < 0$) suggests that $\alpha_y^{(T)} > 0$ and/or $\alpha_y^{(C)} > 0$, respectively, due to $\varepsilon_y < 0$. More importantly, perusal to Fig. 17.6 for infinitesimal $\mathrm{d}T$ and/or $\mathrm{d}C$ suggests that $\alpha_y^{(T)} \approx 0$ and/or $\alpha_y^{(C)} \approx 0$, which is an important characteristic to arrest the occurrence of stain in structures that are exposed to unpredictable environmental fluctuation of small magnitude.

It has earlier been shown in the Poisson's ratio analysis that the application of horizontal loads with $\sigma_x > 0$ and $\sigma_x < 0$ lead to conventional ($\upsilon_{xy} > 0$) and auxetic ($\upsilon_{xy} < 0$) behaviors, respectively. Suppose one were to apply vertical load, it can be seen that when the load is tensile $\sigma_y > 0$, rod 2 elongates by $\mathrm{d}l_2$ such that rod 1 realigns toward the vertical direction by an angular change of $\mathrm{d}\theta_1$ while the horizontal distance changes by an amount $\mathrm{d}l_3$, as shown in Fig. 17.7a, whereby $\mathrm{d}l_3$ is negative. The Poisson's ratio υ_{yx} arising from a tensile load $\sigma_y > 0$ can be established by considering the length constancy of rod 1, i.e.,

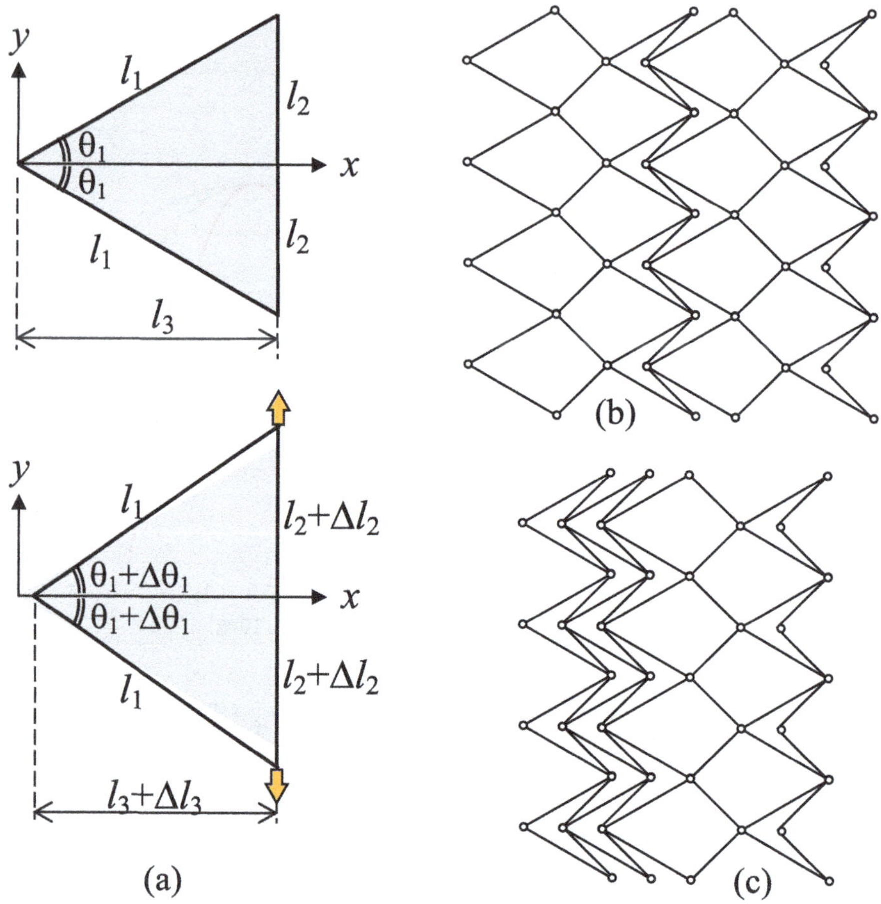

Fig. 17.7 Application of load along the *y*-direction on the microstructure given in Fig. 17.1a, showing **a** before (top) and during (bottom) tensile load; examples of microstructural changes with compressive vertical load are shown in (**b**) and (**c**), as well as in Fig. 17.1b, c

Table 17.1 Summary of microstructure properties given in Fig. 17.1

	x-direction loading	y-direction loading
Positive normal stress	$\sigma_x > 0 \Rightarrow \upsilon_{xy} > 0$ (non-auxetic)	$\sigma_y > 0 \Rightarrow \upsilon_{yx} > 0$ (non-auxetic)
Negative normal stress	$\sigma_x < 0 \Rightarrow \upsilon_{xy} < 0$ (auxetic)	$\sigma_y < 0 \Rightarrow \upsilon_{yx} = \pm$ (indeterminate)

$$\sqrt{l_2^2 + l_3^2} = \sqrt{(l_2 + \mathrm{d}l_2)^2 + (l_3 + \mathrm{d}l_3)^2} \tag{17.1.32}$$

Expanding the RHS terms and neglecting the highest order gives

$$l_2(\mathrm{d}l_2) + l_3(\mathrm{d}l_3) = 0 \tag{17.1.33}$$

Dividing by l_2 and recognizing that $l_2/l_3 = \tan\theta_1$, $\mathrm{d}l_2/l_2 = \varepsilon_y$ and $\mathrm{d}l_3/l_3 = \varepsilon_x$, we have

$$\upsilon_{yx} = \tan^2\theta_1 \tag{17.1.34}$$

which is conventional. Note that this ε_x is negative because $\mathrm{d}l_3 < 0$ for $\sigma_y > 0$.

On the other hand, the application of compressive load $\sigma_y < 0$ causes buckling collapse, which can take on a number of possible Poisson's ratio. Two possible extremes for which buckling collapse take place are furnished in Fig. 17.1b, c, which are conventional ($\upsilon_{yx} > 0$) and auxetic ($\upsilon_{yx} < 0$), respectively. Under these two extremes, the Poisson's ratio can be written as $\upsilon_{yx} = -\varepsilon_x/\varepsilon_y$, where ε_y and ε_x are defined by Eqs. (17.1.6) and (17.1.9), respectively, such that the upper and lower signs for Eq. (17.1.9) correspond to the deformation shown in Fig. 17.1b, c, respectively. In addition to these two extremes, the buckling collapse can also take on any hybrid deformation pattern that combines the two extremes; two such examples are shown in Fig. 17.7b, c. Table 17.1 summarizes the mixed auxetic-conventional characteristics of the proposed microstructure arising from the application of tensile and compressive loads along the x-direction as well as in the y-direction.

So far, in the analysis of hygrothermal expansion, it has been shown that the sole extension of rod 3 or rod 1 leads to contraction in the y-direction. This is not so in the case of sole extension of rod 2, as illustrated in Fig. 17.8. From its original state of Fig. 17.8a with rod 1 and rod 3 being rigid, expansion, and contraction of rod 2 leads to the microstructural geometries of Fig. 17.8b, c, respectively. An increase in temperature and/or moisture concentration causes an expansion of rod 2, as shown in Fig. 17.8d, such that both rod 1 and rod 2 rotate anti-clockwise. At the same time, the increase in the vertical dimension implies positive CTE and/or CME. By similar reasoning, a decrease in temperature and/or moisture concentration brings about contraction of rod 2, as denoted in Fig. 17.8e, so that both rod 1 and rod 2 rotate clockwise with a decrease in the vertical dimension. Again, this indicates positive CTE and/or CME.

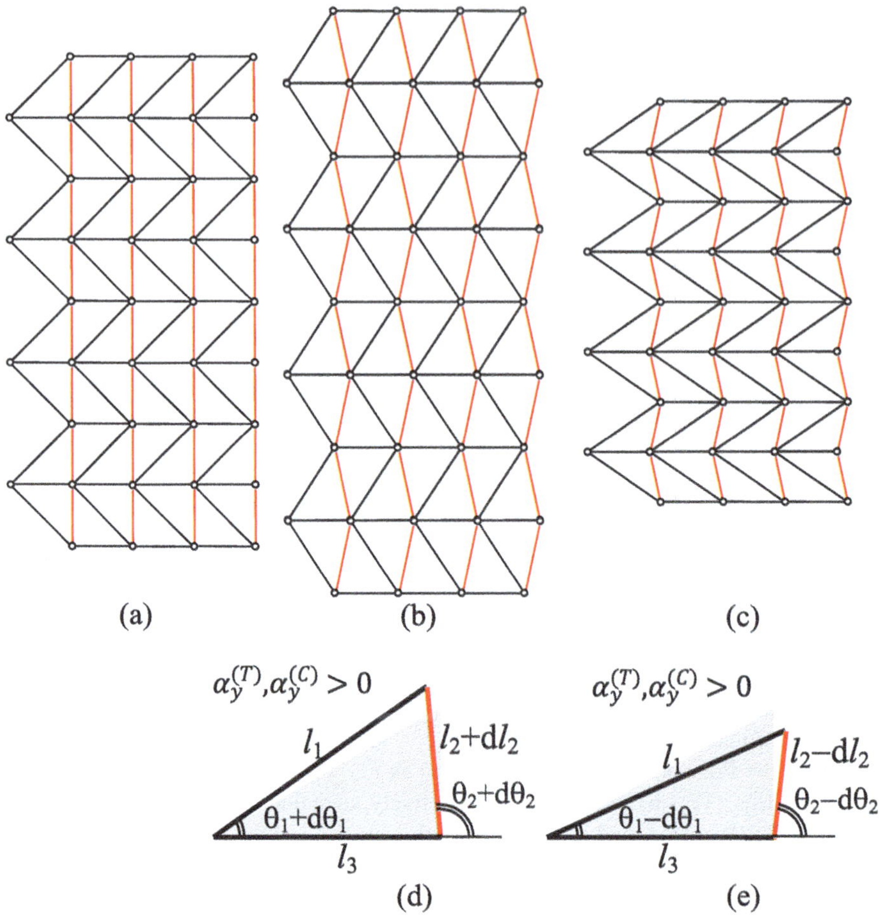

Fig. 17.8 **a** Modified microstructure that exhibits **b** overall vertical expansion with expansion of rod 2 and **c** overall vertical contraction with contraction of rod 2; the representative units for (**b**) and (**c**) are given in (**d**) and (**e**), respectively

An analytical quantification can be made as follows. Upon expansion of rod 2, the subtending angles of rod 1 and rod 2 increase to $\theta_1 + \mathrm{d}\theta_1$ and $\theta_2 + \mathrm{d}\theta_2$, respectively, while the contraction of rod 2 decreases the subtending angles of rod 1 and rod 2 to $\theta_1 - \mathrm{d}\theta_1$ and $\theta_2 - \mathrm{d}\theta_2$ respectively. The choice of nomenclature in Fig. 17.8d, e indicates that $\mathrm{d}\theta_1$ and $\mathrm{d}\theta_2$ take on positive values. The change in dimension along the y-direction is

$$\mathrm{d}y = l_1 \sin(\theta_1 \pm \mathrm{d}\theta_1) - l_1 \sin\theta_1 = (l_2 \pm \mathrm{d}l_2)\cos\mathrm{d}\theta_2 - l_2 \tag{17.1.35}$$

where the upper and lower signs correspond to the expansion and contraction of rod 2, respectively. Dividing Eq. (17.1.35) with Eq. (17.1.4) gives the vertical strain

$$\varepsilon_y = \cos \mathrm{d}\theta_1 \pm \frac{\sin \mathrm{d}\theta_1}{\tan \theta_1} - 1 \tag{17.1.36}$$

Substituting $\cos \mathrm{d}\theta_1 \approx 1$ and $\sin \mathrm{d}\theta_1 \approx \mathrm{d}\theta_1$ into Eq. (17.1.36) for small angular change yields

$$\varepsilon_y = \pm \frac{\mathrm{d}\theta_1}{\tan \theta_1} \tag{17.1.37}$$

which gives $\varepsilon_y > 0$ with the expansion of rod 2 due to increases in temperature and/or moisture concentration, and $\varepsilon_y < 0$ with the contraction of rod 2 due to decreases in temperature and/or moisture concentration. Either way, we have $\alpha_y^{(T)} > 0$ and/or $\alpha_y^{(C)} > 0$, i.e., neither NTE nor NME is observed when rod 2 expands or contracts in the presence of rigid rod 1 and rod 3. Table 17.2 summarizes the mixed hygrothermal properties of the modified microstructure with emphasis on the sign reversal of their mechanical property with the sign reversal of $\mathrm{d}T$ or $\mathrm{d}C$ (Lim 2019a).

17.2 Sign-Switching of CTE with Persistently Positive Thermal Strain

From the Greek work αὔξησις (auxesis), a noun meaning "increase," comes the word αὐξητικός (auxetikos), which means "that which tends to increase"; the latter leads to the term "auxetics," which has been coined for referring to materials and structures that exhibit negative Poisson's ratio. Section 17.1 has shown that the CTE sign can be switched between positive and negative values *in situ*, in response to thermal fluctuation, specifically the effective CTE in y-direction is positive upon cooling, but the effective CTE sign in the same direction flips to negative upon heating, thereby always giving negative thermal strain regardless of whether the temperature increases or decreases, which is not auxetikos. A graphical description that shows how the present auxetikos system relates with positive thermal expansion (PTE) and negative thermal expansion (NTE) systems is illustrated in Fig. 17.9. Note that the thermal strain for auxetikos system is always positive to indicate it having the tendency to increase in dimension. This translates into two distinct slopes of the plot of thermal strain versus temperature, and the corresponding abrupt change in CTE at the original state (Lim 2019b).

This example shows the development of a 2D auxetikos material system that tends to increase in dimension under both heating and cooling effect, i.e., $\alpha_{\mathrm{eff}}^{(T)} > 0$ when $\mathrm{d}T > 0$, but the effective CTE sign changes to $\alpha_{\mathrm{eff}}^{(T)} < 0$ when $\mathrm{d}T < 0$. The implementation of alternating CTE sign with thermal fluctuation is attained herein by incorporating a set of rotating rods that are redundant during heating but functional during cooling. These are indicated as inclined linkages that are parallel to the shuriken sides in the original state illustrated in Fig. 17.10 (top), while each shuriken (together with its rotating rods) is connected to its four closest neighbors

Table 17.2 Summary on expansion negativity of modified microstructure properties

Modified microstructure	$\mathrm{d}T,\ \mathrm{d}C > 0$ (x-direction)	$\mathrm{d}T, \mathrm{d}C < 0$ (x-direction)	$\mathrm{d}T, \mathrm{d}C > 0$ (y-direction)	$\mathrm{d}T, \mathrm{d}C < 0$ (y-direction)
Figure 17.3 (rod 1 and rod 2 rigid)	$\varepsilon_x > 0, \therefore \alpha_x^{(T)}, \alpha_x^{(C)} > 0$ Conventional	$\varepsilon_x < 0, \therefore \alpha_x^{(T)}, \alpha_x^{(C)} > 0$ Conventional	$\varepsilon_y < 0, \therefore \alpha_y^{(T)}, \alpha_y^{(C)} < 0$ NTE, NME	$\varepsilon_y < 0, \therefore \alpha_y^{(T)}, \alpha_y^{(C)} > 0$ Conventional
Figure 17.4 (rod 2 and rod 3 rigid)	$\varepsilon_x = 0, \therefore \alpha_x^{(T)}, \alpha_x^{(C)} = 0$ ZTE, ZME	$\varepsilon_x < 0, \therefore \alpha_x^{(T)}, \alpha_x^{(C)} = 0$ ZTE, ZME	$\varepsilon_y < 0, \therefore \alpha_y^{(T)}, \alpha_y^{(C)} < 0$ NTE, NME	$\varepsilon_y < 0, \therefore \alpha_y^{(T)}, \alpha_y^{(C)} > 0$ Conventional
Figure 17.8 (rod 1 and rod 3 rigid)	$\varepsilon_x = 0, \therefore \alpha_x^{(T)}, \alpha_x^{(C)} = 0$ ZTE, ZME	$\varepsilon_x = 0, \therefore \alpha_x^{(T)}, \alpha_x^{(C)} = 0$ ZTE, ZME	$\varepsilon_y > 0, \therefore \alpha_y^{(T)}, \alpha_y^{(C)} > 0$ Conventional	$\varepsilon_y < 0, \therefore \alpha_y^{(T)}, \alpha_y^{(C)} > 0$ Conventional

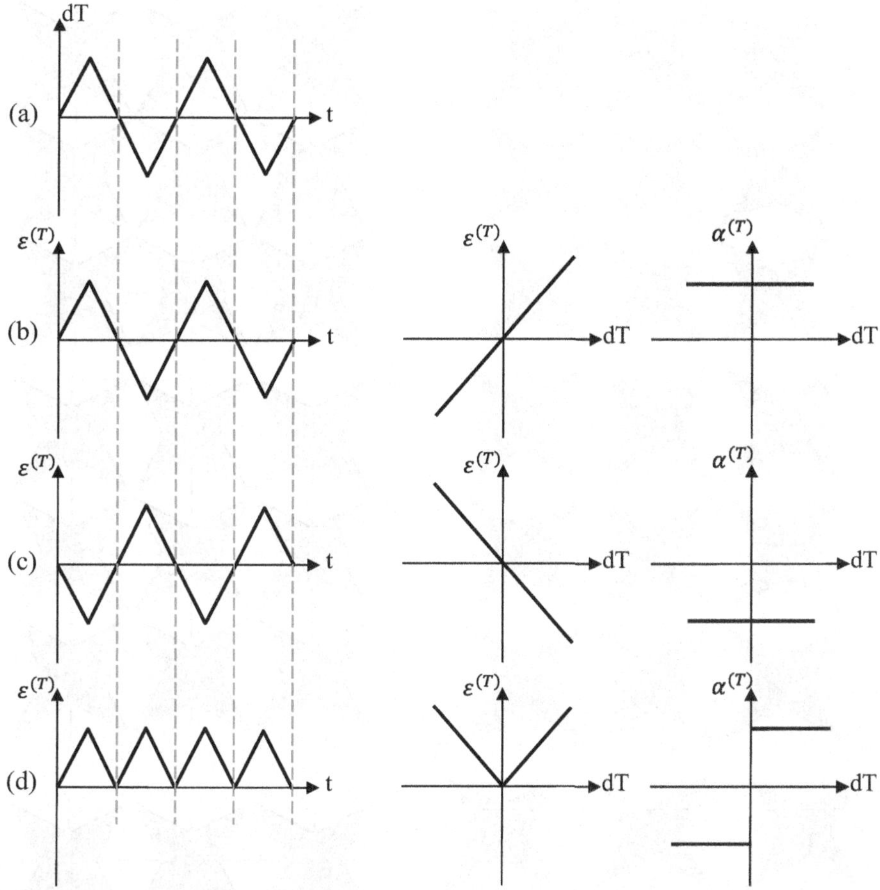

Fig. 17.9 For a given **a** temperature variation with time, the responses for thermal strain with reference to time (left column), thermal strain with temperature (middle column), and CTE with temperature (right column) are illustrated for **b** PTE, **c** NTE and **d** auxetikos systems

by the horizontally and vertically aligned connecting rods. Upon cooling, all rods contract such that the hinges at the corners move away from the shuriken to a greater extent than the shortening of the connecting rods (Fig. 17.10, left), thereby creating an overall distancing between the shurikens. This translates to effective NTE behavior during cooling. Upon heating, only the connecting rods elongate (Fig. 17.10, right); the rotating rods encounter build-up of compressive thermal stress as the hinges are locked at the shuriken corners. For this reason, the analysis consists of two parts for catering to two different overall expansion mechanisms.

We consider a network of interconnected shuriken whereby every shuriken possesses four axes of symmetry, i.e., two axes of symmetry with each axis passing through the opposing sharp edges, and two axes of symmetry with each axis of symmetry passing through the opposing corners. Figure 17.11 (top) shows a unit cell

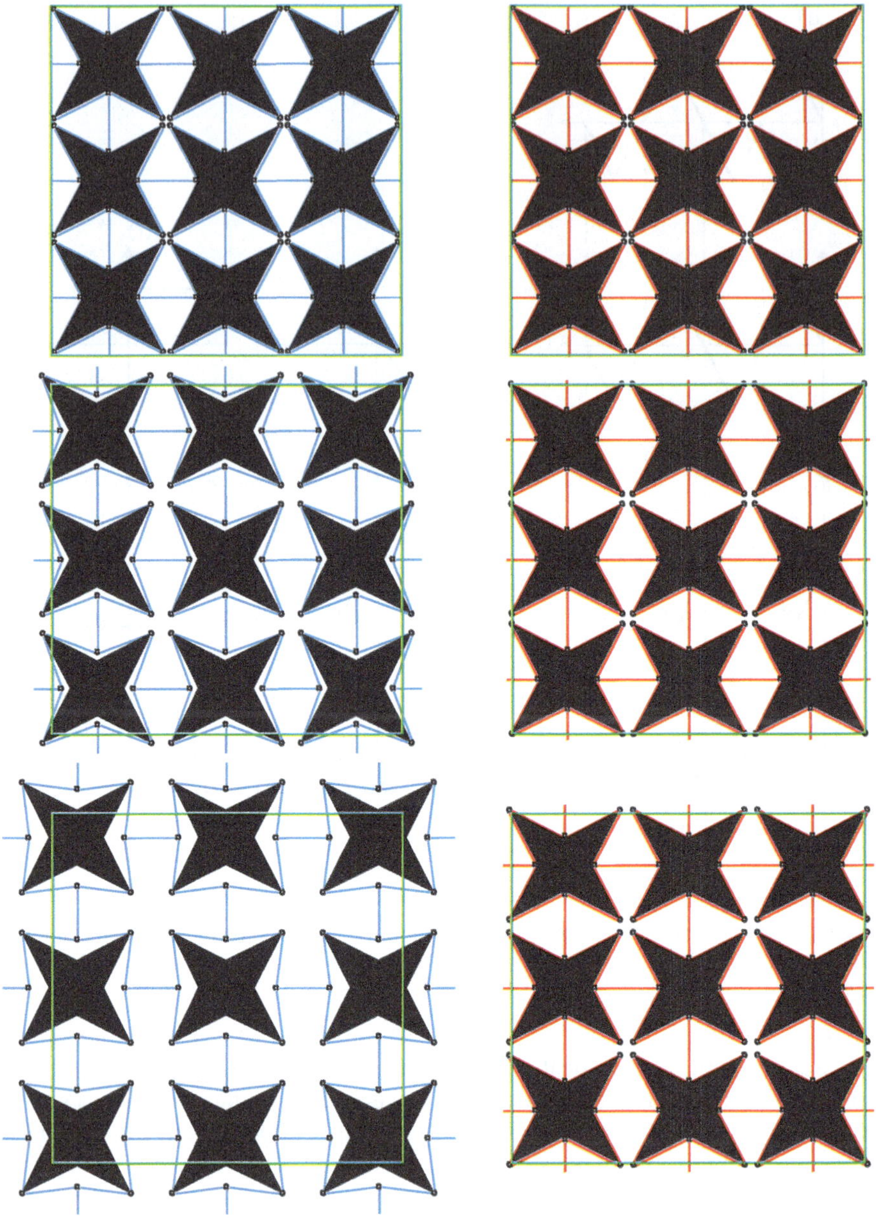

Fig. 17.10 Schematics of 3-by-3 interconnected shuriken network upon cooling (left) and heating (right), with rod strain magnitudes of 0% (top), 5% (middle) and 10% (bottom). Green squares indicate original size of 3-by-3 unit cells for comparison

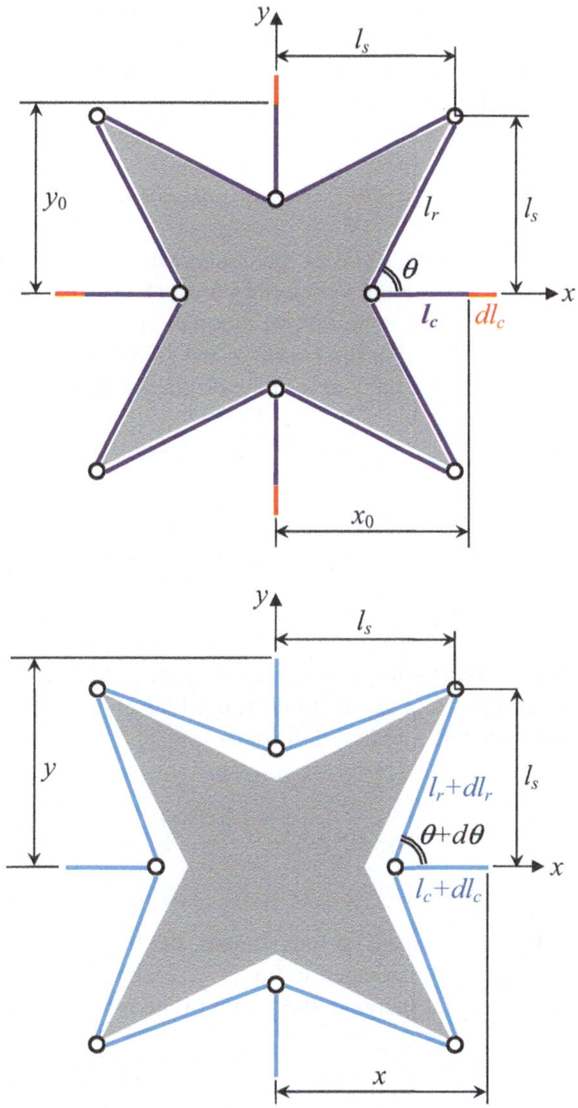

Fig. 17.11 A unit of the shuriken network in original state indicated by purple rods and extension of connecting rods by dl_c indicated in red upon heating (top), and contraction of all rods upon cooling indicated in blue, where dl_r and dl_c take on negative values (bottom)

of the interconnected shuriken network, where the rotating rods and connecting rods in the original state are indicated by purple color, forming an angle of θ between them. The rotating rod is of length l_r while the entire connecting rod is of length $2l_c$ for connecting two shurikens. The half-length of each connecting rod l_c is therefore assigned to each connecting rod within the boundary of the unit cell. The horizontal and vertical distances between neighboring hinges on the sharp edges of each shuriken are $2l_s$, so that its half-length

$$l_s = l_r \sin\theta \tag{17.2.1}$$

can be expressed by the rotating rod length and its angle formed with the connecting rod. Upon an increase in temperature of $\mathrm{d}T$, the half-length of each connecting rod increases by

$$\mathrm{d}l_c = l_c \alpha_c^{(T)} \mathrm{d}T \tag{17.2.2}$$

where $\alpha_c^{(T)}$ is the CTE of the connecting rod. However, there is no increase in the length of the rotating rods, as they are locked in place by the corners, thereby resulting in a build-up of compressive thermal stress. Suppose the origin of the coordinate system is placed at the center of the shuriken, the half-lengths of the unit cells as measured along the x-axis in the original state is

$$x_0 = l_s - l_r \cos\theta + l_c \tag{17.2.3}$$

while the same half-length upon heating is

$$x = l_s - l_r \cos\theta + l_c\left(1 + \alpha_c^{(T)} \mathrm{d}T\right) \tag{17.2.4}$$

upon considering Eq. (17.2.2). If, in addition to the four axes of symmetry imposed on the shuriken, we let $y_0 = x_0$ to give square array, then the in-plane strain is equibiaxial. Substituting Eq. (17.2.1) into Eqs. (17.2.3) and (17.2.4) gives the thermal strain in x-direction $\varepsilon_x^{(T)} = (x - x_0)/x_0$ as

$$\varepsilon_x^{(T)} = \alpha_c^{(T)} \mathrm{d}T \frac{\frac{l_c}{l_r}}{\sin\theta - \cos\theta + \frac{l_c}{l_r}} \tag{17.2.5}$$

upon heating.

Upon a decrease in temperature, the half-length of the connecting rod changes by the amount indicated in Eq. (17.2.2), but this value is negative because $\mathrm{d}T < 0$. Likewise, the change in the length of the rotating rod

$$\mathrm{d}l_r = l_r \alpha_r^{(T)} \mathrm{d}T \tag{17.2.6}$$

where $\alpha_r^{(T)}$ is the CTE of the rotating rod takes on a negative value for the same reason. Due to the contraction of the rotating rod, its angle formed with the connecting rod increases to $\theta + \mathrm{d}\theta$ as shown in Fig. 17.11 (bottom). For these changes in rod lengths and angles, we have the updated half-length of the unit cell in x-direction

$$x = l_s - l_r\left(1 + \alpha_r^{(T)} \mathrm{d}T\right)\cos(\theta + \mathrm{d}\theta) + l_c\left(1 + \alpha_c^{(T)} \mathrm{d}T\right) \tag{17.2.7}$$

Substituting $\cos \mathrm{d}\theta = 1$ and $\sin \mathrm{d}\theta = \mathrm{d}\theta$ as $\mathrm{d}\theta \to 0$ for infinitesimal deformation, we have

$$x = l_s - l_r\left(1 + \alpha_r^{(T)}\mathrm{d}T\right)[\cos\theta - \mathrm{d}\theta\sin\theta] + l_c\left(1 + \alpha_c^{(T)}\mathrm{d}T\right) \tag{17.2.8}$$

The angular change can be eliminated by equating the vertical components of the rotating rod before and after cooling

$$l_r\sin\theta = l_r\left(1 + \alpha_r^{(T)}\mathrm{d}T\right)\sin(\theta + \mathrm{d}\theta) \tag{17.2.9}$$

to give

$$\frac{1}{1 + \alpha_r^{(T)}\mathrm{d}T} = \cos\mathrm{d}\theta + \frac{\sin\mathrm{d}\theta}{\tan\theta} \tag{17.2.10}$$

Considering infinitesimal deformation again, we have

$$\mathrm{d}\theta = \left(\frac{1}{1 + \alpha_r^{(T)}\mathrm{d}T} - 1\right)\tan\theta \tag{17.2.11}$$

which upon substitution into Eq. (17.2.8) gives rise to

$$x = l_s - l_r\left(1 + \alpha_r^{(T)}\mathrm{d}T\right)\left[\cos\theta - \sin\theta\tan\theta\left(\frac{1}{1 + \alpha_r^{(T)}\mathrm{d}T} - 1\right)\right] + l_c\left(1 + \alpha_c^{(T)}\mathrm{d}T\right) \tag{17.2.12}$$

Using Eq. (17.2.1) and recalling the original dimension described by Eq. (17.2.3), one obtains the thermal strain

$$\varepsilon_x^{(T)} = -\alpha_r^{(T)}\mathrm{d}T\frac{\sin\theta\tan\theta + \cos\theta - \frac{l_c}{l_r}\frac{\alpha_c^{(T)}}{\alpha_r^{(T)}}}{\sin\theta - \cos\theta + \frac{l_c}{l_r}} \tag{17.2.13}$$

upon cooling.

Based on the definition of CTE, one can obtain the effective CTE of the interconnected shuriken network from Eqs. (17.2.5) and (17.2.13) for heating and cooling, respectively. When normalized against the CTE of the rotating rods, we have the dimensionless effective CTE

$$\frac{\alpha_x^{(T)}}{\alpha_r^{(T)}} = \frac{\frac{l_c}{l_r}\frac{\alpha_c^{(T)}}{\alpha_r^{(T)}}}{\sin\theta - \cos\theta + \frac{l_c}{l_r}};\quad \mathrm{d}T > 0 \tag{17.2.14}$$

for heating, and

$$\frac{\alpha_x^{(T)}}{\alpha_r^{(T)}} = -\frac{\sin\theta\tan\theta + \cos\theta - \frac{l_c}{l_r}\frac{\alpha_c^{(T)}}{\alpha_r^{(T)}}}{\sin\theta - \cos\theta + \frac{l_c}{l_r}};\quad \mathrm{d}T < 0 \tag{17.2.15}$$

for cooling.

Effects from the various geometrical and mechanical properties of the interconnected shuriken network are investigated for rotating rod CTE of $\alpha_{\mathrm{r}}^{(T)} = 20 \times 10^{-6}\,\mathrm{K}^{-1}$, connecting-to-rotating rod length ratio of $l_{\mathrm{c}}/l_{\mathrm{r}} = 1$ and the angle of $\theta = 60°$ between these two rods. The effective thermal strains in x-direction $\varepsilon_x^{(T)}$ are plotted in Fig. 17.12 (left column) while the corresponding effective CTEs in the same direction are furnished in Fig. 17.12 (right column) with reference to the change in temperature $\mathrm{d}T$. Here, effects of the connecting rod CTE $\alpha_{\mathrm{c}}^{(T)}$ are shown in Fig. 17.12 (top row) using rod CTE ratio of $\alpha_{\mathrm{c}}^{(T)}/\alpha_{\mathrm{r}}^{(T)} = 0,\ 0.25,\ 0.5,\ 0.75,\ 1$. Influence from the connecting-to-rotating rod subtending angle is furnished in Fig. 17.12 (middle row) for $\theta = 50°,\ 55°,\ 60°,\ 65°,\ 70°$ while the effect from these rods' length ratio are plotted in Fig. 17.12 (bottom row) for $l_{\mathrm{c}}/l_{\mathrm{r}} = 0.5,\ 0.75,\ 1,\ 1.25,\ 1.5$.

The continuous effect of $\alpha_{\mathrm{c}}^{(T)}/\alpha_{\mathrm{r}}^{(T)}$, θ and $l_{\mathrm{c}}/l_{\mathrm{r}}$ on the dimensionless effective CTE is displayed in Fig. 17.13. It shows that the negativity of the effective CTE upon cooling can be intensified (or the positivity of the effective CTE upon heating can be reduced) by decreasing $\alpha_{\mathrm{c}}^{(T)}/\alpha_{\mathrm{r}}^{(T)}$ or $l_{\mathrm{c}}/l_{\mathrm{r}}$, or by increasing θ. It is of interest to note that while the material system fulfills the auxetikos criterion—as evident from the increasing thermal strain regardless of whether $\mathrm{d}T$ is positive or negative, and hence this system exhibits PTE under heating but reverses to NTE under cooling—the negativity of the CTE is of a greater extent in comparison to its positive counterpart. The existence of two curves, one each for heating and cooling, suggests that the effective CTE for the investigated material system herein is undefined when there is no change in temperature, for the CTE and its sign come into play only when one specifies the condition of temperature change—whether increasing or decreasing.

17.3 Conclusions

From Sect. 17.1, it can be concluded that a microstructure consisting of pin-jointed rods that exhibits conventional behavior upon tensile loading has been shown to manifest auxetic behavior upon reversal of loading direction, similar to the behavior of both metamaterials discussed in Sect. 6.1. By a slight modification to this microstructure through incorporation of an additional set of reinforcing rods, unique hygrothermal properties can be observed. A decrease in temperature or moisture concentration leads to dimensional contraction, or positive hygrothermal coefficients, as one would expect in conventional solids; under certain special cases, however, an increase in temperature or moisture concentration also brings about dimensional contraction, or negative hygrothermal coefficients. These reversals of properties as activated by the reversal of mechanical and hygrothermal loads is being made possible through the design of the microstructure that stands at the boundary of auxetic and conventional behavior (Lim 2019a). At the point where the reversal of property sign takes place, the hygrothermomechanical properties are zero for infinitesimal deformation. This

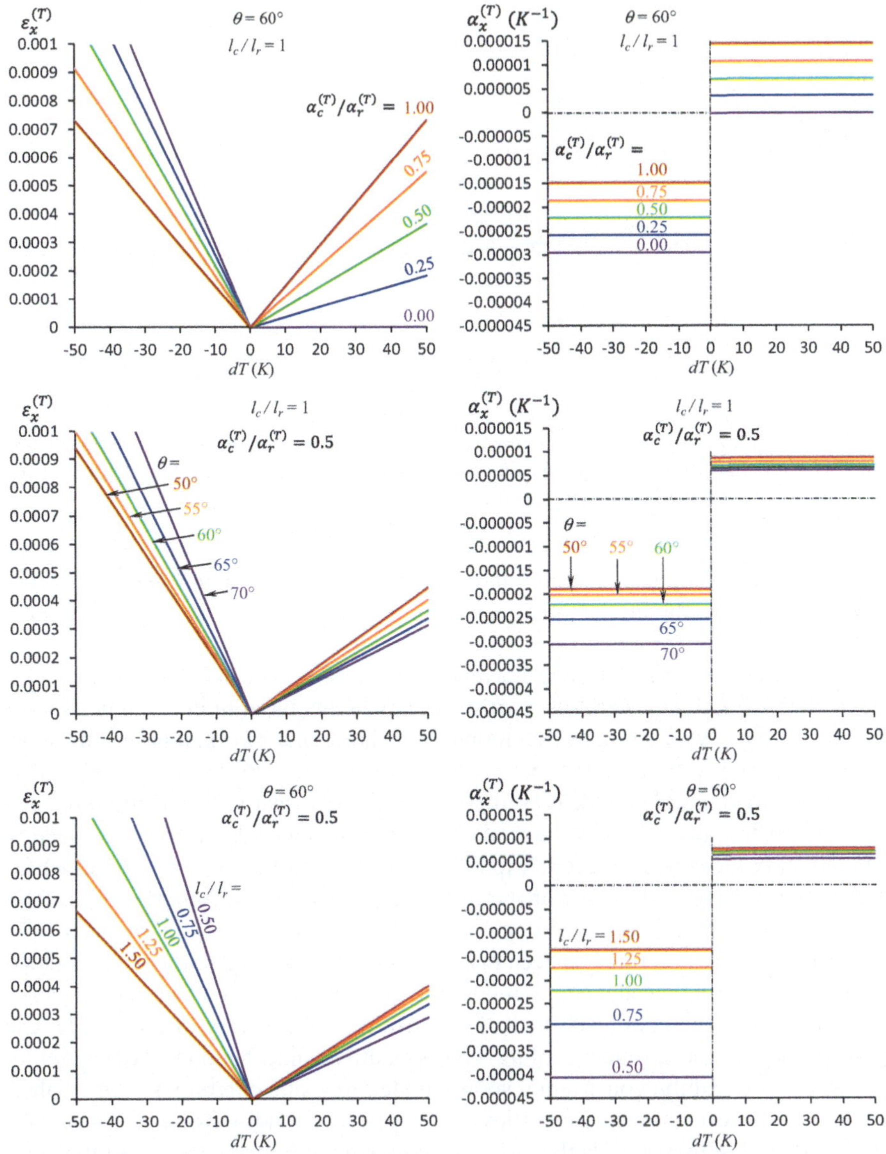

Fig. 17.12 Plots of thermal strain (left) and effective CTE (right) against thermal change for various connecting-to-rotating rod CTE ratio (top), rotating-to-connecting rod angles (middle), and connecting-to-rotating rod length ratio (bottom)

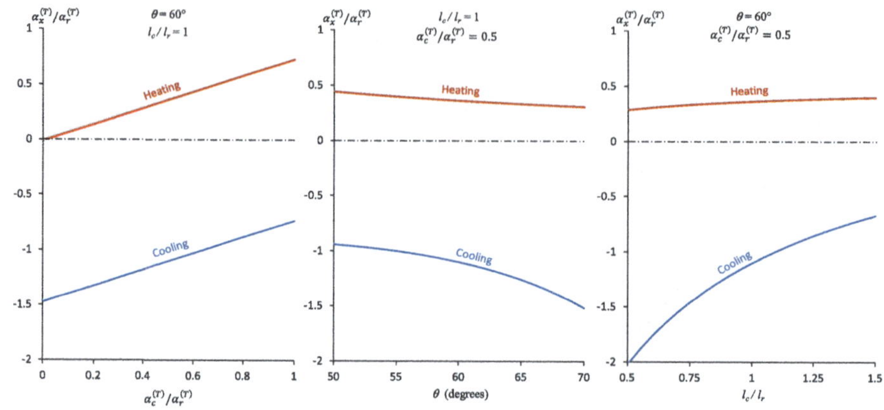

Fig. 17.13 Variation of dimensionless effective CTE with reference to the connecting-to-rotating rod CTE ratio (left), rotating-to-connecting rod angles (center), and connecting-to-rotating rod length ratio (right) under heating and cooling

is useful for reducing lateral strains and hygrothermal strains in unconstrained structures, as well as reducing lateral stresses and hygrothermal stresses in constrained ones.

An auxetikos material, i.e., one that tends to increase, has been defined in Sect. 17.2 as a material system that exhibits positive strain whether the stimulus is of positive or negative value. By adopting the interconnected shuriken network, it has been shown that the material system expands based on the magnitude of the thermal change. This has been made possible by the presence of two sets of effective microstructure. One set of microstructure, whereby the rotating rod is redundant, takes effect upon heating wherein only the connecting rods expand. The other set of microstructure, in which the rotating and connecting rods are permitted to contract simultaneously, takes effect upon cooling. The capability of demonstrating two effective microstructures from just a single microstructure has been made possible through the implementation of a jamming or locking mechanism to render the rotating rods redundant under heating but functional under cooling (Lim 2019b). Results reveal that the effective CTE can be made more negative (under cooling) or less positive (under heating). The capability for materials to exhibit consistent response under opposing conditions—due to their ability to manifest opposing properties under opposing conditions—paves a way for engineers to design materials that are able to change their behavior to suit the environment. So far we have considered the sign-switching of Poisson's ratio with stress reversal in Chapter 16 (Lim 2019c, d) while this chapter deals with sign-switching in the coefficient of (hygro)thermal expansions with (hygro)thermal fluctuation. In the next chapter we shall explore metamaterials that, in addition to changes to CTEs in response to temperature variation, can exhibit sign-switching of Poisson's ratio not by stress direction reversal but by thermal undulation (Lim 2019e, f).

References

Lim TC (2019a) A composite metamaterial with sign switchable elastic and hygrothermal properties induced by stress direction and environmental change reversals. Compos Struct 220:185–193
Lim TC (2019b) A 2D auxetikos system based on interconnected shurikens. SN Appl Sci 1(11):1383
Lim TC (2019c) Composite microstructures with Poisson's ratio sign switching upon stress reversal. Compos Struct 209:34–44
Lim TC (2019d) Metamaterials with Poisson's ratio sign toggling by means of microstructural duality. SN Appl Sci 1(2):176
Lim TC (2019e) A class of shape-shifting composite metamaterial honeycomb structures with thermally-adaptive Poisson's ratio signs. Compos Struct 226:111256
Lim TC (2019f) 2D metamaterial with in-plane positive and negative thermal expansion and thermal shearing based on interconnected alternating bimaterials. Mater Res Express 6(11):115804

Chapter 18
Sign-Switching of Poisson's Ratio with Temperature Change Reversals

Abstract This chapter considers two examples of Poisson's ratio sign-switching upon temperature change reversal using bimaterial strips with alternating orientation. In the first example, alternating signs of temperature change switches the microstructure between hexagonal-like cells and re-entrant-like cells. The former and latter are known for exhibiting positive and negative Poisson's ratios, respectively. In addition to the CTE analysis, Poisson's ratio analysis is included. In the second example, the cell walls are also made from alternating bimaterial strips, and are arranged in rectangular array with interconnecting rigid rods that are joined at the centers of the bimaterial strips. Fluctuating temperature flips the microstructure shape between octagon-like and star-like cells. Although only the CTE analysis is furnished for the second example, it is known that the interconnected star array manifests auxetic property. In spite of only thermal analysis, the effective CTE models developed can be converted to effective compressibility and effective CME models.

Keywords Poisson's ratio · Thermal expansion · Sign-flipping · Sign-reversible · Sign-switching · Sign toggling

18.1 Thermally Activated Toggle Between Hexagonal and Re-entrant Microstructural Reshape

This section explores the use of alternating bimaterial strips to form rectangular cells in triangular array, so as to control the signs of both the CTE and Poisson's ratio based on the sign of temperature change. Specifically, this example primarily aims to design a class of metamaterial that can perform in the following ways:

- $v_{xy} > 0$ for $\mathrm{d}T > 0$ but $v_{xy} < 0$ for $\mathrm{d}T < 0$ (O-Type)
- $v_{xy} < 0$ for $\mathrm{d}T > 0$ but $v_{xy} > 0$ for $\mathrm{d}T < 0$ (X-Type)

for the first two types of this metamaterial class. For completeness' sake, another two types (CC-Type and SS-Type) are discussed thereafter. The use of alternating bimaterial strips was established by Ha et al. (2015) for developing controllable thermal expansion of large magnitude in chiral arrangement. With reference to Fig. 18.1, the rigid crossbeams are aligned horizontally and indicated in black while

T. Lim, *Mechanics of Metamaterials with Negative Parameters*, Engineering Materials,
https://doi.org/10.1007/978-981-15-6446-8_18

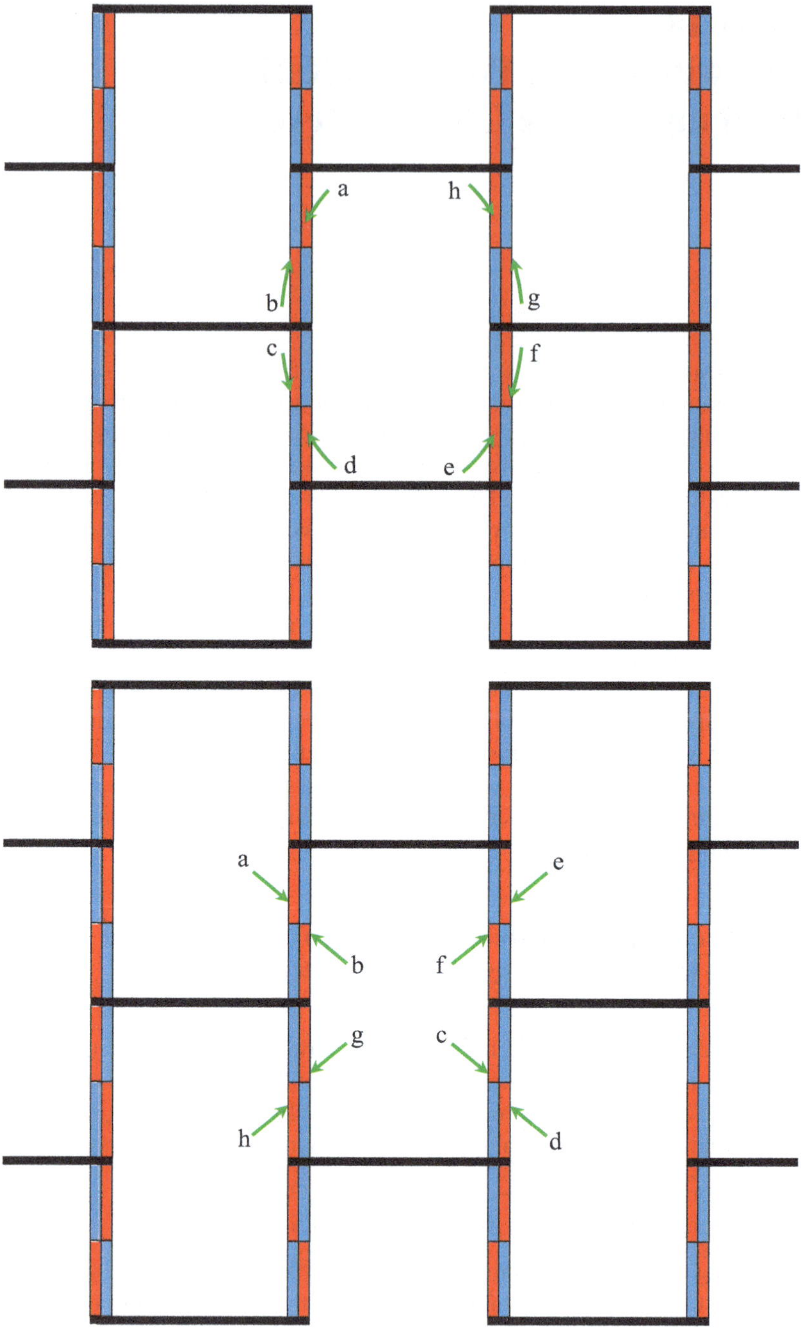

Fig. 18.1 Rectangular-shaped cells arranged in triangular array using rigid crossbeams (black color) and bimaterial strips made from metals of higher CTE (red color) and lower CTE (blue color). The "more active" material, which has a higher CTE as indicated by red color, can be arranged in a loop to form the O-Type (top) and in a criss-cross to form the X-Type (bottom)

the bimaterials are vertically oriented and indicated in red and blue for higher and lower CTEs, respectively. While straight bimaterials encounter transformation into a C-shaped curve upon change in temperature, the alternating arrangement of materials in each bimaterial would transform it into an S-shaped curve. Perusal to Fig. 18.1 shows two types of such metamaterial with five unit cells each. Unlike the model by Li et al. (2016) whereby the ends of the crossbeams are joined to the surfaces of the bimaterial, in the current model, the ends of the bimaterials are joined to the surfaces of the crossbeams instead. When the material strip with higher CTE is arranged in a loop, we call this the O-Type as indicated in Fig. 18.1 (top). If the material with higher CTE is arranged in a criss-cross manner furnished in Fig. 18.1 (bottom), we call this the X-Type.

The negativity of thermal expansion and Poisson's ratio can be qualitatively visualized for the O-Type in Fig. 18.2 (top row) and for the X-Type in Fig. 18.2 (bottom row), wherein the original states before temperature change is placed at Fig. 18.2 (left column). Under the influence of decreasing temperature, the materials of higher and lower CTE form the concave and convex parts, respectively, but reverse for increasing temperature $\mathrm{d}T > 0$. For this reason, the O-Type and X-Type metamaterials deform into hourglass- and barrel-shaped cells, respectively, with $\mathrm{d}T < 0$ as illustrated in Fig. 18.2 (middle column), but the shapes reverse for $\mathrm{d}T > 0$, as indicated in Fig. 18.2 (right column). It is easily seen that the dimensional decrease in y-direction for $\mathrm{d}T \neq 0$ for Fig. 18.2 (middle and right columns) implies PTE along the y-direction for both metamaterials when $\mathrm{d}T < 0$, but reverse to NTE when $\mathrm{d}T > 0$. The contraction (due to $\mathrm{d}T < 0$) and expansion (resulting from $\mathrm{d}T > 0$) along the x-direction for the O-Type metamaterial suggest that it exhibits PTE in the x-direction; on the other hand, the expansion (as a result of $\mathrm{d}T < 0$) and contraction (arising from $\mathrm{d}T > 0$) along the x-direction for the X-Type metamaterial suggest that it exhibits NTE in the x-direction. In regard to the Poisson's ratio, it is worth noticing that the hourglass and barrel shapes approximate the re-entrant and hexagonal shapes, respectively, thereby indicating that the hourglass- and barrel-shaped microstructures manifest negative and positive Poisson's ratios, respectively. A summary of the qualitative descriptions of CTE and, more importantly, the Poisson's ratio sign-switching by means of change in temperature sign is furnished in Fig. 18.3 (Lim 2019a). Having discussed in a qualitative manner, the following analysis develops the quantitative descriptions for the CTE and Poisson's ratio under temperature change.

Figure 18.4 (top) illustrates a straight bimaterial cantilever of length l_h and total thickness $h = h_1 + h_2$ transforming into a bent cantilever with radius of curvature r upon a temperature change of $\mathrm{d}T$; the curvature $1/r$ has been given by Timoshenko (1925) as

$$\frac{1}{r} = \frac{\left(\alpha_1^{(T)} - \alpha_2^{(T)}\right)\mathrm{d}T}{\frac{h}{2} + \frac{2}{h}(E_1 I_1 + E_2 I_2)\left(\frac{1}{E_1 h_1} + \frac{1}{E_2 h_2}\right)} \tag{18.1.1}$$

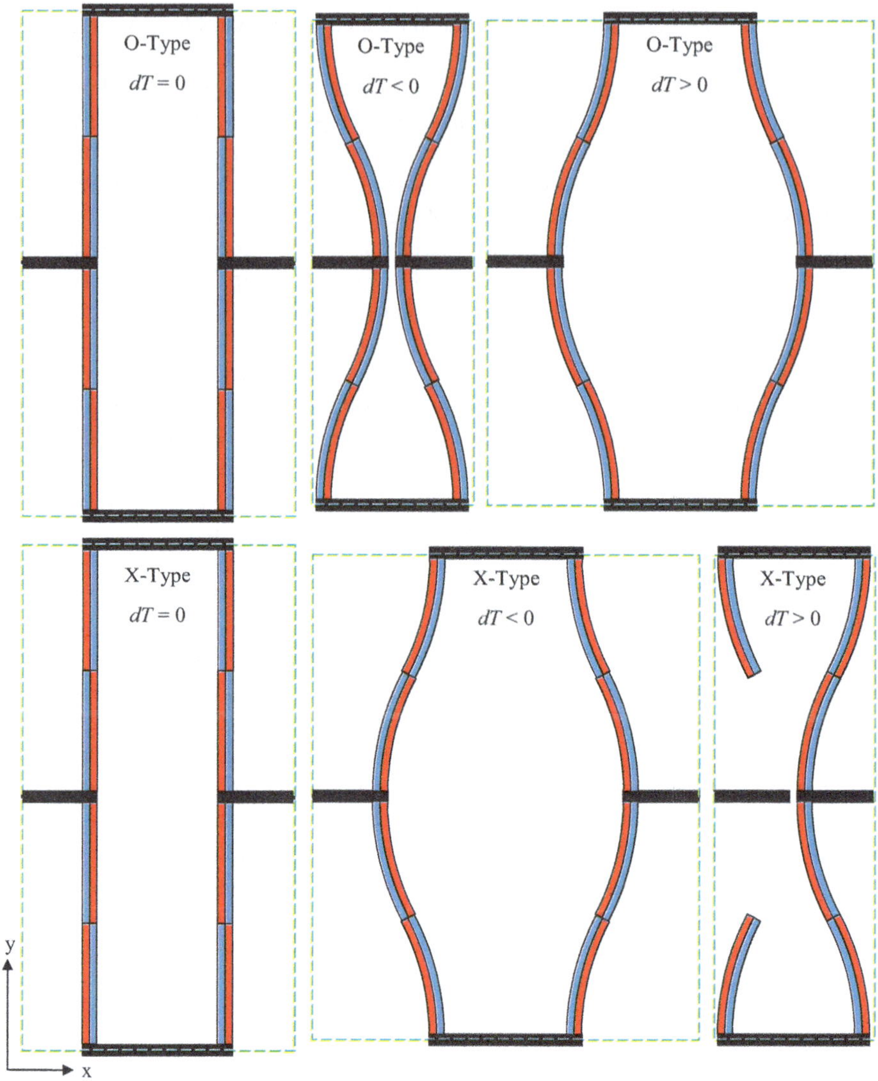

Fig. 18.2 A unit of O-Type (top row) and X-Type (bottom row) metamaterials before temperature change (left column), as well as after temperature decrease (middle column) and increase (right column)

where $\alpha_n^{(T)}$, E_n, I_n. and h_n are the CTEs, Young's moduli, second moment areas, and thicknesses, respectively, for layers $n = 1, 2$ of the bimaterial strip with

$$I_n = \frac{h_n^3}{12} \tag{18.1.2}$$

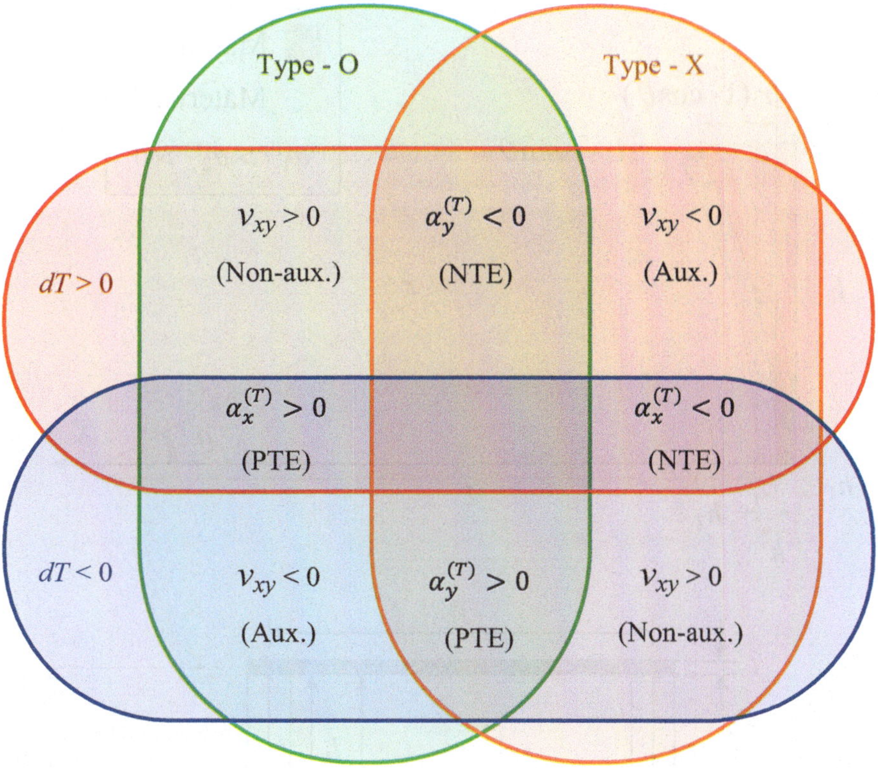

Fig. 18.3 A summary of the qualitative descriptions of CTEs and Poisson's ratio for the O-Type and X-Type metamaterials

It must be pointed out that this Timoshenko bimaterial strip model applies to Euler–Bernoulli beams only, which confines its validity to slender beams.

If both layers of the bimaterial share equal Young's moduli and equal thicknesses, then Eq. (18.1.1) greatly simplifies to

$$\frac{1}{r} = \frac{3}{2}\frac{\alpha_1^{(T)} - \alpha_2^{(T)}}{h}\mathrm{d}T \tag{18.1.3a}$$

Suppose only the bimaterial strips are of equal thicknesses so as to permit different Young's moduli ratio, Eq. (18.1.1) becomes

$$\frac{1}{r} = \frac{24\left(\alpha_1^{(T)} - \alpha_2^{(T)}\right)\mathrm{d}T}{h\left(14 + \frac{E_1}{E_2} + \frac{E_2}{E_1}\right)} \tag{18.1.3b}$$

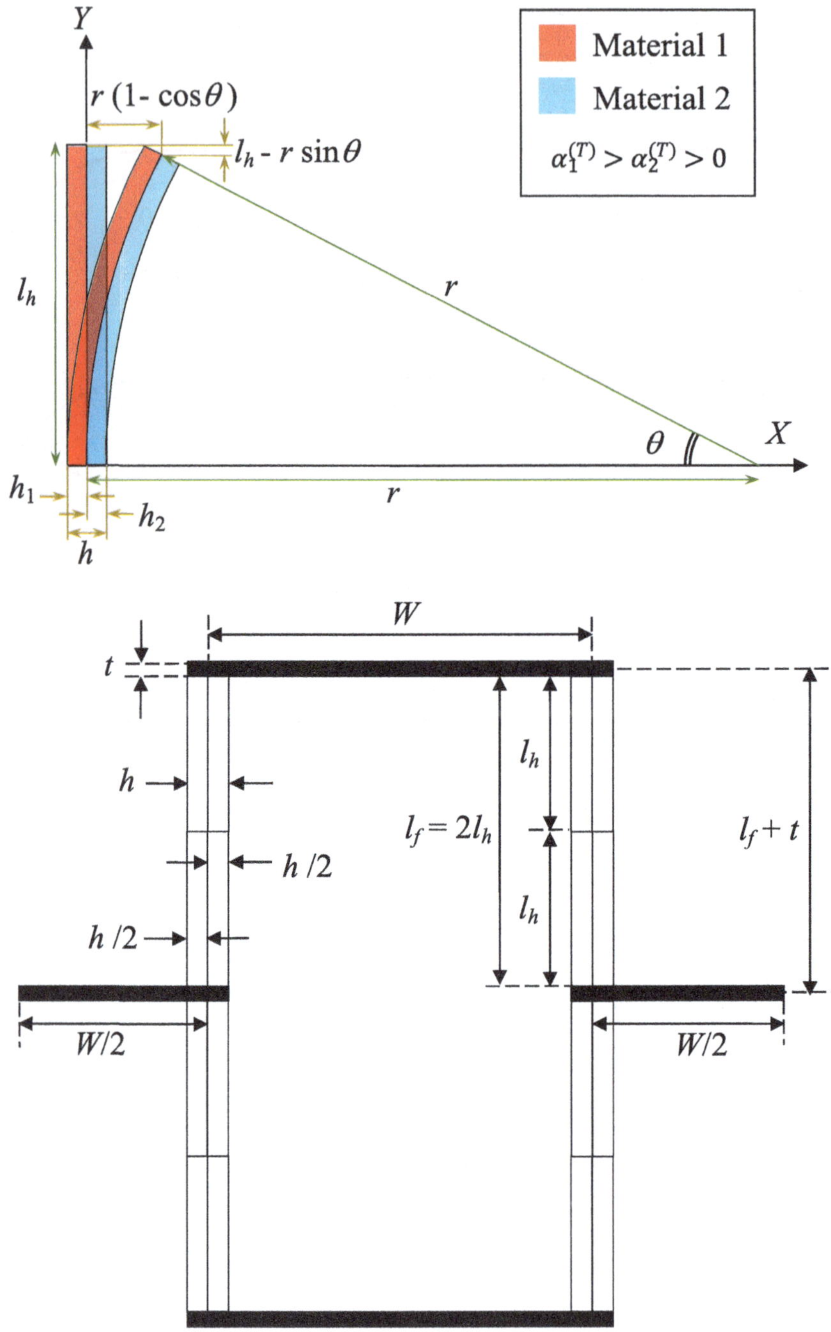

Fig. 18.4 Thermal deflection of a cantilever bimaterial strip of half-length l_h (top), and a unit cell with geometrical parameters (bottom)

Arising from the curving of the bimaterial, a sectional angle θ is formed at the center of the curvature, as illustrated in Fig. 18.4 (top). This gives a decrease in the projected height of the cantilever bimaterial as $l_h - r\,\sin\theta$ while the horizontal deflection of the cantilever tip is $r(1-\cos\theta)$. The subtending angle relation $\theta = l_h/r$ assumes that the arc length after thermal change remains the same as the original length, and is therefore not exact, i.e., this relation is true only at small angles. Nevertheless, this assumption is valid insofar as the present scope of analysis is concerned. See Appendix A.1. In the following analysis, we consider the rectangular unit cell represented in Fig. 18.4 (bottom) where $h_1 = h_2 = h/2$, while the width of the rectangle is W such that the overall width of the unit cell is $2W$. The height of the cantilever beam shown in Fig. 18.4 (top) is the half-length, l_h, such that the full-length of the deformable rib, l_f, is $2l_h$. Taking into account the thickness t of the rigid crossbeams, we have the height of the unit cell being $2(l_f + t)$. By virtue of symmetry, only one quarter of the unit cell needs to be considered for analysis. From this quarter unit cell, we have its original width measured along the x-axis

$$x_0 = W \tag{18.1.4}$$

and its original height measured along the y-axis

$$y_0 = l_f + t \tag{18.1.5}$$

Writing the magnitudes of changes in the width

$$|\mathrm{d}x| = 2r\left[1 - \cos\left(\frac{l_h}{r}\right)\right] \tag{18.1.6}$$

and height

$$|\mathrm{d}y| = 2\left[l_h - r\,\sin\left(\frac{l_h}{r}\right)\right] \tag{18.1.7}$$

we have the thermal strains

$$\varepsilon_x^{(T)} = \pm\frac{|\mathrm{d}x|}{x_0} = \pm\frac{2r}{W}\left[1 - \cos\left(\frac{l_h}{r}\right)\right] \tag{18.1.8}$$

in the x-direction, and

$$\varepsilon_y^{(T)} = -\frac{|\mathrm{d}y|}{y_0} = -\frac{2}{2l_h + t}\left[l_h - r\,\sin\left(\frac{l_h}{r}\right)\right] \tag{18.1.9}$$

in the y-direction. The upper and lower signs in Eq. (18.1.8) correspond to the O-Type and X-Type, respectively, for $\mathrm{d}T > 0$, in complying to width expansion in Fig. 18.2 (top right) and width contraction in Fig. 18.2 (bottom right), while the negative sign in Eq. (18.1.9) indicates height reduction for both types of metamaterials. A separate analysis for $\mathrm{d}T < 0$ is unnecessary as this can be effected by implementing negative values to $\mathrm{d}T$. From the definition of the CTE, we obtain

$$\alpha_x^{(T)} = \frac{\varepsilon_x^{(T)}}{\mathrm{d}T} = \pm \frac{2r}{W\mathrm{d}T}\left[1 - \cos\left(\frac{l_h}{r}\right)\right] \tag{18.1.10}$$

and

$$\alpha_y^{(T)} = \frac{\varepsilon_y^{(T)}}{\mathrm{d}T} = -\frac{2}{(2l_h + t)\mathrm{d}T}\left[l_h - r\sin\left(\frac{l_h}{r}\right)\right] \tag{18.1.11}$$

For Special Case I where $E_1/E_2 = h_1/h_2 = 1$, substitution of Eq. (18.1.3a) into Eq. (18.1.10) gives

$$\alpha_x^{(T)} = \pm \frac{l_f}{W} \frac{1 - \cos\left(\frac{3l_h}{2h}\left(\alpha_1^{(T)} - \alpha_2^{(T)}\right)\mathrm{d}T\right)}{\frac{3l_h}{2h}\left(\alpha_1^{(T)} - \alpha_2^{(T)}\right)(\mathrm{d}T)^2} \tag{18.1.12}$$

while the substitution of Eq. (18.1.3a) into Eq. (18.1.11) leads to

$$\alpha_y^{(T)} = -\frac{1}{\left(1 + \frac{1}{2}\frac{t}{l_h}\right)\mathrm{d}T}\left[1 - \frac{\sin\left(\frac{3l_h}{2h}\left(\alpha_1^{(T)} - \alpha_2^{(T)}\right)\mathrm{d}T\right)}{\frac{3l_h}{2h}\left(\alpha_1^{(T)} - \alpha_2^{(T)}\right)\mathrm{d}T}\right] \tag{18.1.13}$$

For Special Case II where only $h_1/h_2 = 1$, substitution of Eq. (18.1.3b) into Eq. (18.1.10) gives

$$\alpha_x^{(T)} = \pm \frac{l_f}{W} \frac{1 - \cos\left(\frac{3l_h}{2h}\frac{16\left(\alpha_1^{(T)} - \alpha_2^{(T)}\right)\mathrm{d}T}{14 + \frac{E_1}{E_2} + \frac{E_2}{E_1}}\right)}{\frac{3l_h}{2h}\frac{16\left(\alpha_1^{(T)} - \alpha_2^{(T)}\right)(\mathrm{d}T)^2}{14 + \frac{E_1}{E_2} + \frac{E_2}{E_1}}} \tag{18.1.14}$$

while the substitution of Eq. (18.1.3b) into Eq. (18.1.11) leads to

$$\alpha_y^{(T)} = -\frac{1}{\left(1 + \frac{1}{2}\frac{t}{l_h}\right)\mathrm{d}T}\left[1 - \frac{\sin\left(\frac{3l_h}{2h}\frac{16\left(\alpha_1^{(T)} - \alpha_2^{(T)}\right)\mathrm{d}T}{14 + \frac{E_1}{E_2} + \frac{E_2}{E_1}}\right)}{\frac{3l_h}{2h}\frac{16\left(\alpha_1^{(T)} - \alpha_2^{(T)}\right)\mathrm{d}T}{14 + \frac{E_1}{E_2} + \frac{E_2}{E_1}}}\right] \tag{18.1.15}$$

The ambiguity of the radius of curvature, r, in Eqs. (18.1.10) and (18.1.11) is removed by substitution from Eqs. (18.1.3a) and (18.1.3b). To find the Poisson's ratio after the temperature-activated change in microstructure, we restore the thermally deformed shapes back to their original shapes. As a result, the thermally deformed shapes illustrated in Fig. 18.2 (right column) are the updated shapes, in which mechanical load is applied along the x-axis to return the shape to those indicated in Fig. 18.2 (left column). For this reason, we have the updated width

$$x_0' = W \pm |\mathrm{d}x| \tag{18.1.16}$$

where the upper and lower signs provide updates to the width of Fig. 18.2 (top right) and Fig. 18.2 (bottom right), respectively, while the updated height

$$y_0' = (l_f + t) - |\mathrm{d}y| \tag{18.1.17}$$

is common for both O-Type and X-Type. Therefore, the prescribed mechanical strains

$$\varepsilon_x = \frac{\mp|\mathrm{d}x|}{x_0'} = \frac{\mp|\mathrm{d}x|}{W \pm |\mathrm{d}x|} \tag{18.1.18}$$

and

$$\varepsilon_y = \frac{|\mathrm{d}y|}{y_0'} \approx \frac{|\mathrm{d}y|}{l_f - |\mathrm{d}y|} \tag{18.1.19}$$

give rise to the Poisson's ratio

$$v_{xy} = -\frac{\varepsilon_y}{\varepsilon_x} = -\frac{|\mathrm{d}y|}{2l_h - |\mathrm{d}y|}\frac{W \pm |\mathrm{d}x|}{\mp|\mathrm{d}x|} \tag{18.1.20}$$

Substituting Eqs. (18.1.6) and (18.1.7) into Eq. (18.1.20), we have the Poisson's ratio in terms of the dimensionless curvature l_h/r and array aspect ratio of W/l_f (Lim 2019a)

$$v_{xy} = \pm\left[\frac{\frac{l_h}{r}}{\sin\left(\frac{l_h}{r}\right)} - 1\right]\left[\frac{\frac{W}{l_f}\frac{l_h}{r}}{1 - \cos\left(\frac{l_h}{r}\right)} \pm 1\right] \tag{18.1.21}$$

Substituting Eq. (18.1.3a) into Eq. (18.1.21) for Special Case I gives

$$v_{xy} = \pm\left[\frac{\frac{3l_h}{2h}\left(\alpha_1^{(T)} - \alpha_2^{(T)}\right)\mathrm{d}T}{\sin\left(\frac{3l_h}{2h}\left(\alpha_1^{(T)} - \alpha_2^{(T)}\right)\mathrm{d}T\right)} - 1\right]\left[\frac{\frac{W}{l_f}\frac{3l_h}{2h}\left(\alpha_1^{(T)} - \alpha_2^{(T)}\right)\mathrm{d}T}{1 - \cos\left(\frac{3l_h}{2h}\left(\alpha_1^{(T)} - \alpha_2^{(T)}\right)\mathrm{d}T\right)} \pm 1\right] \tag{18.1.22}$$

while using Eq. (18.1.3b) for Special Case II leads to

$$v_{xy} = \pm\left[\frac{\frac{3l_h}{2h}\frac{16\left(\alpha_1^{(T)}-\alpha_2^{(T)}\right)dT}{14+\frac{E_1}{E_2}+\frac{E_2}{E_1}}}{\sin\left(\frac{3l_h}{2h}\frac{16\left(\alpha_1^{(T)}-\alpha_2^{(T)}\right)dT}{14+\frac{E_1}{E_2}+\frac{E_2}{E_1}}\right)} - 1\right]\left[\frac{\frac{W}{l_f}\left(\frac{3l_h}{2h}\frac{16\left(\alpha_1^{(T)}-\alpha_2^{(T)}\right)dT}{14+\frac{E_1}{E_2}+\frac{E_2}{E_1}}\right)}{1-\cos\left(\frac{3l_h}{2h}\frac{16\left(\alpha_1^{(T)}-\alpha_2^{(T)}\right)dT}{14+\frac{E_1}{E_2}+\frac{E_2}{E_1}}\right)} \pm 1\right] \tag{18.1.23}$$

In order to select realistic material properties of the bimaterials, consideration is made on metallic materials due to their ease of joining in comparison to ceramics and their larger range of working temperature in comparison to polymers. In the case of metals, the choice of $\alpha_1^{(T)} = 25 \times 10^{-6}\,\mathrm{K}^{-1}$ would put it among a group of metals of high CTEs such as lead $\left(29 \times 10^{-6}\,\mathrm{K}^{-1}\right)$, magnesium $\left(26 \times 10^{-6}\,\mathrm{K}^{-1}\right)$, and brass $\left(19 \times 10^{-6}\,\mathrm{K}^{-1}\right)$, while $\alpha_2^{(T)}$ is selected to be of values 0, 1/3 and 2/3 of $\alpha_1^{(T)}$ for Special Cases I and II. In the following illustrations using Special Case I, the array aspect ratio of $W/l_f = \sqrt{3}$ is adopted to represent equilateral triangular array, with a typical bimaterial aspect ratio order of $l_h/h = 100$.

Figure 18.5 (left) shows curves of exact $\alpha_x^{(T)}$ versus dT for the abovementioned $\alpha_2^{(T)}/\alpha_1^{(T)}$ ratios using Eq. (18.1.12) with $\alpha_1^{(T)} = 25 \times 10^{-6}\,\mathrm{K}^{-1}$, $W/l_f = \sqrt{3}$ and $l_h/h = 100$. The plotted curves not only confirm the manifestation of PTE and NTE in the x-direction for the O-Type and X-Type metamaterials, respectively, but also seem to suggest that the CTEs are constants that are independent from the change in temperature. This observation can be validated by performing Taylor series expansion on the cosine function

$$\cos\left(\frac{l_h}{r}\right) = 1 - \frac{1}{2!}\left(\frac{l_h}{r}\right)^2 + \cdots \tag{18.1.24}$$

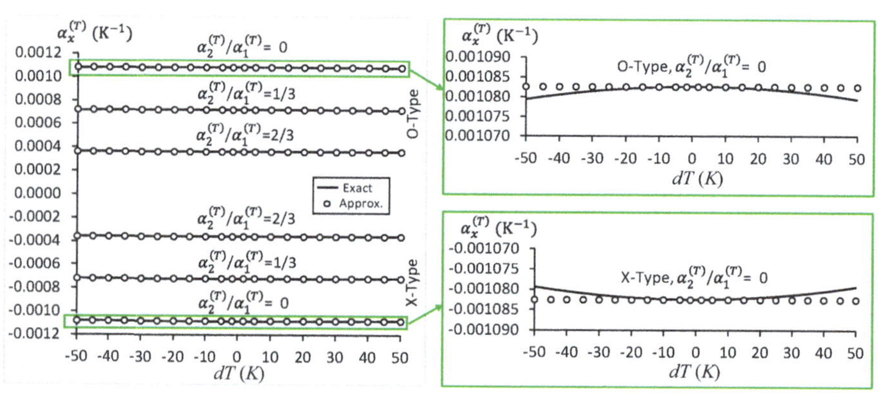

Fig. 18.5 Plots of CTE in x-direction for the O-Type and X-Type metamaterials at $E_1 = E_2$, $h_1 = h_2$, $\alpha_1^{(T)} = 25 \times 10^{-6}\,\mathrm{K}^{-1}$, $W/l_f = \sqrt{3}$ and $l_h/h = 100$, showing insignificant effect of dT (left), and that the magnitude of the approximated $\alpha_x^{(T)}$ slightly overestimates the actual value (right)

and retaining only the first two terms for substitution into Eq. (18.1.12) in Special Case I, whereby $E_1 = E_2$ and $h_1 = h_2$, to give

$$\alpha_x^{(T)} \approx \pm\frac{3}{4}\frac{l_f}{W}\frac{l_h}{h}\left(\alpha_1^{(T)} - \alpha_2^{(T)}\right) \tag{18.1.25}$$

i.e., the approximated $\alpha_x^{(T)}$, plotted as discrete points in Fig. 18.5, is dependent on the array aspect ratio W/l_f, aspect ratio of the bimaterial l_h/h, and the difference of the CTEs of the bimaterial strips $\left(\alpha_1^{(T)} - \alpha_2^{(T)}\right)$, but independent from the change in temperature dT. Nevertheless, the magnitude of the exact CTE in x-direction decreases with the magnitude of temperature change, as evident in Fig. 18.5 (right), thereby indicating that the magnitude of the approximated $\alpha_x^{(T)}$ slightly overestimates the actual magnitude.

For the y-direction, curves of exact $\alpha_y^{(T)}$ are plotted in Fig. 18.6 (left) using Eq. (18.1.13) based on $t/l_h = 1/100$ and other conditions similar for calculating $\alpha_x^{(T)}$ in Fig. 18.5. Both the O-Type and X-Type metamaterials exhibit NTE and PTE in the y-direction for increasing and decreasing temperatures, respectively, in an almost linear manner. This observation is not surprising, as the substitution of the following Taylor series expansion for the sine function

$$\sin\left(\frac{l_h}{r}\right) = \left(\frac{l_h}{r}\right) - \frac{1}{3!}\left(\frac{l_h}{r}\right)^3 + \cdots \tag{18.1.26}$$

into Eq. (18.1.13) for Special Case I where at $E_1 = E_2$ and $h_1 = h_2$, and further assuming that $t/l_h \to 0$ for simplicity, leads to

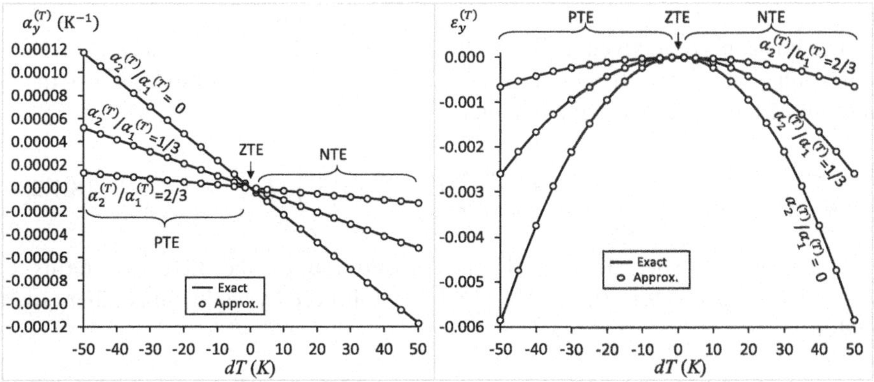

Fig. 18.6 Plots of CTE in y-direction for the O-Type and X-Type metamaterials at $E_1 = E_2$, $h_1 = h_2$, $\alpha_1^{(T)} = 25 \times 10^{-6}\,\text{K}^{-1}$, $W/l_f = \sqrt{3}$, $l_h/h = 100$ and $t/l_h = 1/100$ (left) and the thermal strains in the same direction (right)

$$\alpha_y^{(T)} \approx -\frac{3}{8}\left(\frac{l_h}{r}\right)^2 \left(\alpha_1^{(T)} - \alpha_2^{(T)}\right)^2 \mathrm{d}T \tag{18.1.27}$$

i.e., the approximated $\alpha_y^{(T)}$, plotted in Fig. 18.6 (left) as discrete points, varies linearly with $\mathrm{d}T$. In addition to showing that the sign of $\alpha_y^{(T)}$ opposes the sign of $\mathrm{d}T$, this distribution provides an almost exact and exactly quadratic variations of exact and approximate thermal strains, respectively, in the y-direction, as furnished in Fig. 18.6 (right). Suppose only the first term of Eq. (18.1.26) is retained for substitution into Eq. (18.1.13), we have $\alpha_y^{(T)} = 0$, which is indicated in Fig. 18.6 as zero thermal expansion (ZTE) for infinitesimal thermal deformation.

Considering again Special Case I where $E_1 = E_2$ and $h_1 = h_2$, and substituting $\alpha_1^{(T)} = 25 \times 10^{-6}\,\mathrm{K}^{-1}$ into Eq. (18.1.22), families of exact Poisson's ratio v_{xy} versus change in temperature $\mathrm{d}T$ curves for various array aspect ratio W/l_f, bimaterial aspect ratio l_h/h, and CTE ratio $\alpha_2^{(T)}/\alpha_1^{(T)}$ are plotted in Figs. 18.7 and 18.8 for the O-Type and X-Type metamaterials, respectively. Plotted results for the O-Type metamaterial indicate that the slope $\mathrm{d}v_{xy}/\mathrm{d}T > 0$ such that it increases with the dimensionless cell width W/l_f and the dimensionless height of the bimaterial l_h/h, but decreases with the CTE ratio $\alpha_2^{(T)}/\alpha_1^{(T)}$. The trend reverses for the X-Type metamaterial, i.e., the slope $\mathrm{d}v_{xy}/\mathrm{d}T > 0$ whereby $\left|\mathrm{d}v_{xy}/\mathrm{d}T\right|$ increases with W/l_f and l_h/h, but decreases with $\alpha_2^{(T)}/\alpha_1^{(T)}$. Perusal to Figs. 18.7 and 18.8 suggests that the v_{xy} versus $\mathrm{d}T$ curves are approximately linear. This can be shown for Special Case I by substituting the first two terms of Eqs. (18.1.24) and (18.1.26) into Eq. (18.1.22) and thereafter neglecting higher orders to yield

$$v_{xy} = \pm \frac{W\left(\alpha_1^{(T)} - \alpha_2^{(T)}\right)\mathrm{d}T}{4h} \tag{18.1.28a}$$

The Poisson's ratio given in Eq. (18.1.28a), while sufficiently abridged, makes no indication on the array aspect ratio W/l_f and bimaterial aspect ratio l_h/h; to express in terms of these geometrical parameters, we write

$$v_{xy} = \pm \frac{1}{2}\frac{W}{l_f}\frac{l_h}{h}\left(\alpha_1^{(T)} - \alpha_2^{(T)}\right)\mathrm{d}T \tag{18.1.28b}$$

Likewise, for Special Case II, the substitution of the first two terms of Eqs. (18.1.24) and (18.1.26) into Eq. (18.1.23), followed by neglecting higher orders leads to

$$v_{xy} = \pm 4 \frac{\frac{W}{h}\left(\alpha_1^{(T)} - \alpha_2^{(T)}\right)\mathrm{d}T}{14 + \frac{E_1}{E_2} + \frac{E_2}{E_1}} \tag{18.1.29a}$$

or, for more geometrical insights,

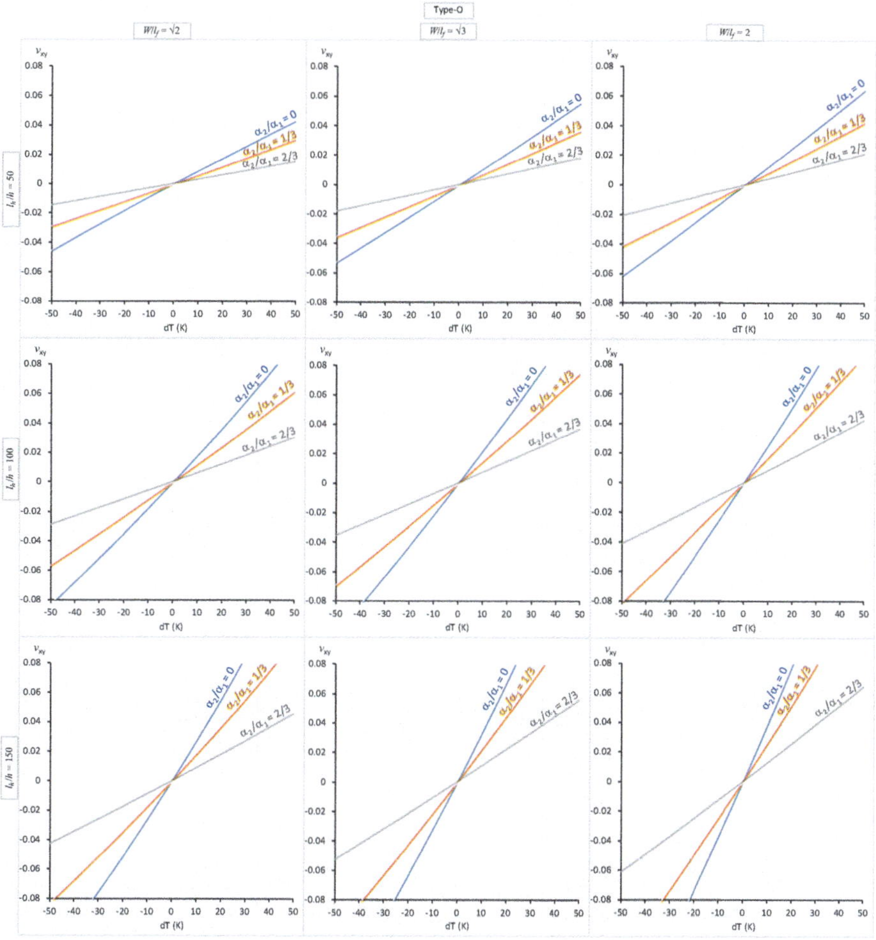

Fig. 18.7 Families of exact υ_{xy} versus dT curves for the O-Type metamaterial at $E_1 = E_2, h_1 = h_2$ and $\alpha_1^{(T)} = 25 \times 10^{-6}\,\text{K}^{-1}$

$$\upsilon_{xy} = \pm 8 \frac{W}{l_f} \frac{l_h}{h} \frac{\left(\alpha_1^{(T)} - \alpha_2^{(T)}\right) \text{d}T}{14 + \frac{E_1}{E_2} + \frac{E_2}{E_1}} \tag{18.1.29b}$$

The validity of the approximate, or linearized, Poisson's ratio model is attested in Fig. 18.9 for both the O-Type and X-Type metamaterials with $E_1 = E_2, h_1 = h_2$, $\alpha_1^{(T)} = 25 \times 10^{-6}\,\text{K}^{-1}$, $W/l_f = \sqrt{3}$ and $l_h/h = 100$.

To visualize the pure effect of E_1/E_2, Special Case II with $\alpha_2^{(T)}/\alpha_1^{(T)} = 0$ is employed using the same set of geometrical parameters. Reference to the numerator of either Eqs. (18.1.3a) or (18.1.3b) indicates that for Special Case I, the difference between the CTEs of the bimaterial strip layers $\left(\alpha_1^{(T)} - \alpha_2^{(T)}\right)$ plays a significant role

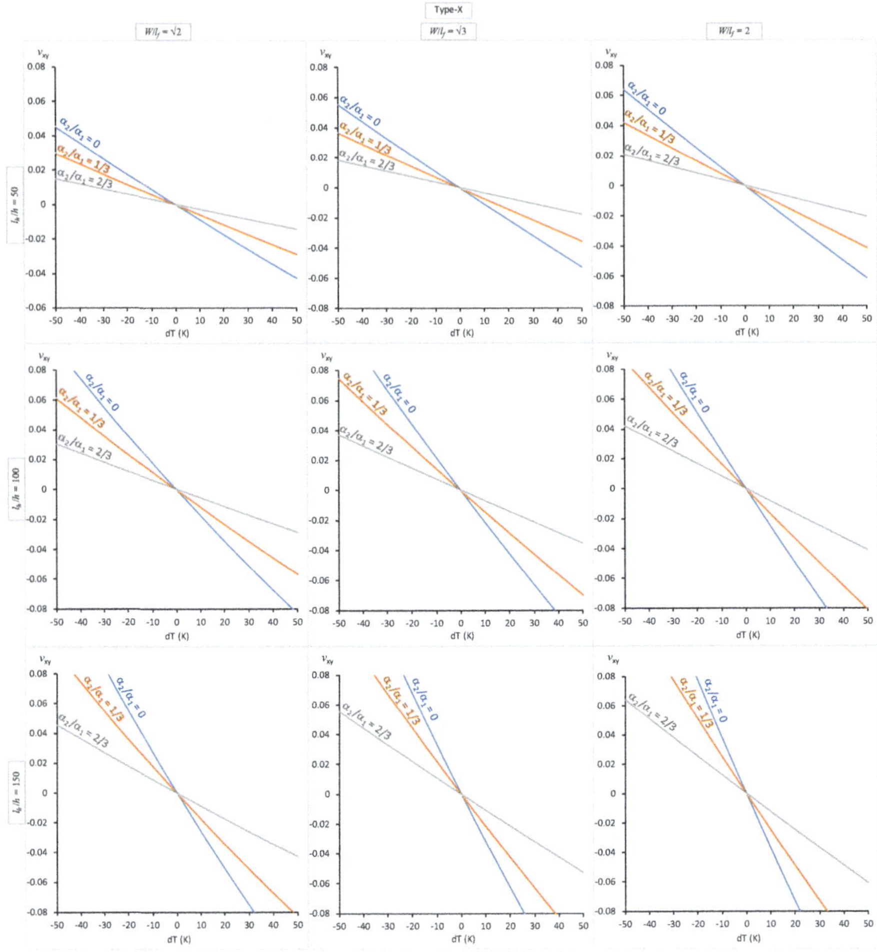

Fig. 18.8 Families of exact υ_{xy} versus dT curves for the X-Type metamaterial at $E_1 = E_2, h_1 = h_2$ and $\alpha_1^{(T)} = 25 \times 10^{-6}\,\text{K}^{-1}$

in influencing the magnitudes of the effective CTEs and Poisson's ratio. On the other hand, perusal to the denominator of Eq. (18.1.3b) for Special Case II reveals that the Young's modulus ratio of the bimaterial layers E_1/E_2 also plays an important role in controlling the effective CTEs and Poisson's ratio. Specifically, the Young's moduli ratio influences these effective properties in the form of $E_1/E_2 + E_2/E_1$. For this reason, the various effective property plots at any particular E_1/E_2 value is similar to those of E_2/E_1 with the same value.

The effective CTE in the x-direction is plotted in Fig. 18.10 (top left) for various E_1/E_2 ratio, instead of $\alpha_2^{(T)}/\alpha_1^{(T)}$ ratio in Fig. 18.5, to show the influence of the Young's moduli ratio using Eq. (18.1.14) for at $h_1 = h_2$, $\alpha_1^{(T)} = 25 \times 10^{-6}\,\text{K}^{-1}$,

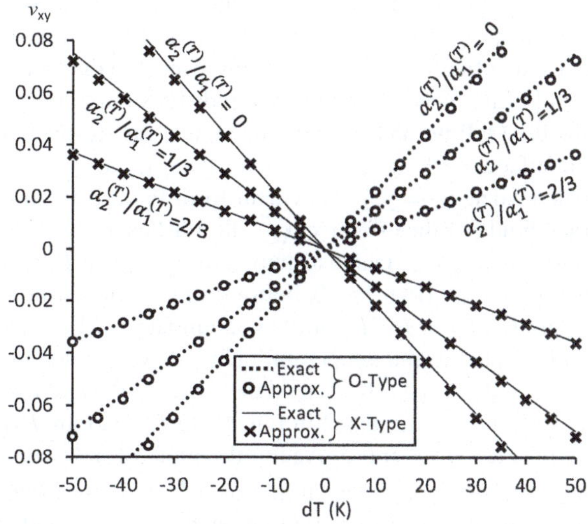

Fig. 18.9 Comparison between the exact (curves) and linearized (discrete points) Poisson's ratio for both O-Type and X-Type metamaterials with $E_1 = E_2$, $h_1 = h_2$, $\alpha_1^{(T)} = 25 \times 10^{-6}\,\mathrm{K}^{-1}$, $W/l_f = \sqrt{3}$ and $l_h/h = 100$

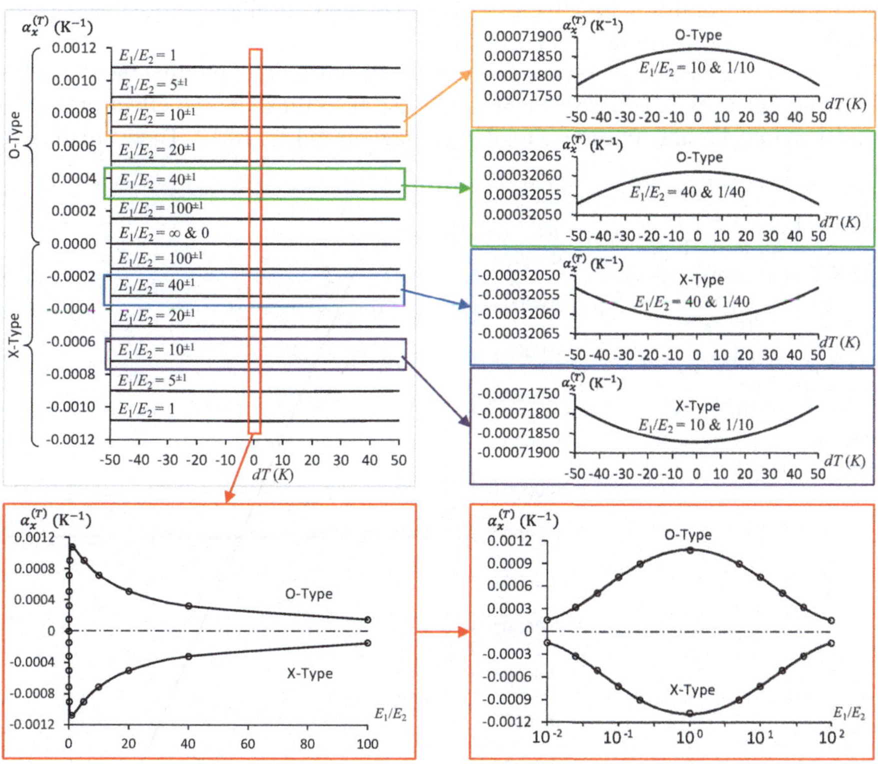

Fig. 18.10 Plots of CTE in x-direction for the O-Type and X-Type metamaterials at $h_1 = h_2$, $\alpha_1^{(T)} = 25 \times 10^{-6}\,\mathrm{K}^{-1}$, $\alpha_2^{(T)}/\alpha_1^{(T)} = 0$, $W/l_f = \sqrt{3}$ and $l_h/h = 100$ for variation with dT (top), and variation with E_1/E_2 (bottom)

$\alpha_2^{(T)}/\alpha_1^{(T)} = 0$, $W/l_f = \sqrt{3}$ and $l_h/h = 100$. Similar to the various $\alpha_2^{(T)}/\alpha_1^{(T)}$ curves in Fig. 18.5, the various E_1/E_2 curves in Fig. 18.10 (top left) indicate almost constant effective $\alpha_x^{(T)}$ values with respect to the change in temperature, and that the trend of its insignificant change indicates maximum and minimum $\alpha_x^{(T)}$ at $\mathrm{d}T = 0$ for the O-Type and X-Type metamaterials, respectively (Fig. 18.10, top right). If the effective $\alpha_x^{(T)}$ is plotted against E_1/E_2 at $\mathrm{d}T = 0$, the optimal condition takes place when $E_1 = E_2$, as evident from Fig. 18.10 (bottom left). In other words, the magnitude of the effective $\alpha_x^{(T)}$ diminishes as the E_1/E_2 ratio deviates from 1. The observed symmetrical distribution of $\alpha_x^{(T)}$ about $E_1/E_2 = 1$ when the horizontal axis is logarithmic (Fig. 18.10, bottom right) is not surprising due to the presence of the terms $E_1/E_2 + E_2/E_1$ at the denominator of Eq. (18.1.3b). Plots of effective $\alpha_y^{(T)}$ with respect to the temperature change were made using Eq. (18.1.15) at $h_1 = h_2$, $\alpha_1^{(T)} = 25 \times 10^{-6}\,\mathrm{K}^{-1}$, $\alpha_2^{(T)}/\alpha_1^{(T)} = 0$, $W/l_f = \sqrt{3}$, $l_h/h = 100$ and $t/l_h = 1/100$, and are furnished in Fig. 18.11 for various E_1/E_2 ratio, unlike Fig. 18.6 for various $\alpha_2^{(T)}/\alpha_1^{(T)}$ ratio. Again, the plots of $\alpha_y^{(T)}$ versus $\mathrm{d}T$ is almost linear with a negative slope. The magnitude of $\alpha_y^{(T)}/\mathrm{d}T$ slope begins with zero at $E_1/E_2 = 0$, which increases to its maximum at $E_1/E_2 = 1$, and thereafter decreases to zero as $E_1/E_2 \to \infty$. For this reason, there exists loci of theoretically inadmissible regions, which are indicated as gray in Fig. 18.11.

Adopting again the same set of geometrical parameters for plotting the Poisson's ratio laid out in Eq. (18.1.23) for Special Case II to illustrate the effect of E_1/E_2 ratio, one may broadly observe a total of six regions in Fig. 18.12. There are the two

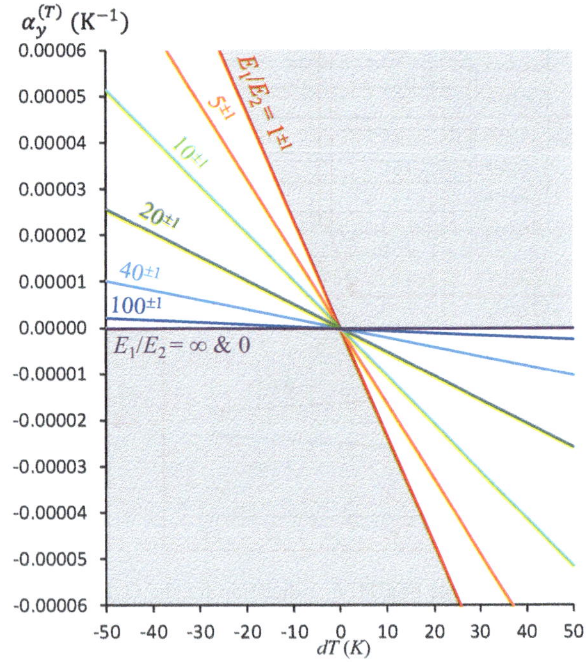

Fig. 18.11 Plots of CTE in y-direction for the O-Type and X-Type metamaterials at $h_1 = h_2$, $\alpha_1^{(T)} = 25 \times 10^{-6}\,\mathrm{K}^{-1}$, $\alpha_2^{(T)}/\alpha_1^{(T)} = 0$, $W/l_f = \sqrt{3}$, $l_h/h = 100$ and $t/l_h = 1/100$; the shaded parts indicate theoretically inadmissible region

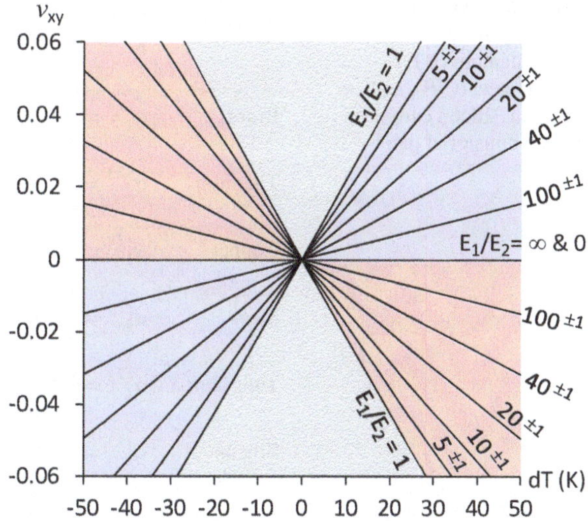

Fig. 18.12 Poisson's ratio for metamaterials of O-Type (blue regions) and X-Type (pink regions) at $h_1 = h_2$, $\alpha_1^{(T)} = 25 \times 10^{-6}\,\mathrm{K}^{-1}$, $\alpha_2^{(T)}/\alpha_1^{(T)} = 0$, $W/l_f = \sqrt{3}$ and $l_h/h = 100$, with the theoretically inadmissible region indicated in gray

regions for the O-Type metamaterial indicated in blue, two regions for the X-Type metamaterial indicated in pink, and two theoretically inadmissible regions shown in gray. The magnitudes of the $v_{xy}/\mathrm{d}T$ slopes decrease as the E_1/E_2 deviates from 1 such that the Poisson's ratio for both the O-Type and X-Type metamaterials merge as $E_1/E_2 \to 0$ and as $E_1/E_2 \to \infty$. On the other hand, the magnitudes of the $v_{xy}/\mathrm{d}T$ slopes increase as $E_1/E_2 \to 1$ but do not merge, such that the resulting gap defines the theoretically impossible region.

Insofar as the previous illustrations have been made on the basis of Special Cases I and II, no actual material properties has been employed. The reason for doing so is to establish fundamental plots that illustrate to the fullest extent the variation of effective CTEs and Poisson's ratios arising from the change in geometrical parameters and from the change in material properties $\alpha_2^{(T)}/\alpha_1^{(T)}$ at $E_1/E_2 = 1$ and E_1/E_2 at $\alpha_2^{(T)}/\alpha_1^{(T)} = 0$, as well as their accompanying theoretically inadmissible regions. This having been established, we now turn our attention to the seemingly mundane but equally important task of estimating what effective CTEs and Poisson's ratios that can be practically achieved using pairs of common materials. Three pairs of bimaterial strips are considered herein, namely the brass–titanium, the copper–steel, and the tungsten–silicon carbide bimaterials. The CTE and Young's modulus data for these materials are summarized in Table 18.1. For illustration purpose, we select the mid-point values in cases where the material properties are given as a range in avaiable data.

Curves of effective CTEs in the x-direction for the O-Type and X-Type metamaterials based on brass–titanium, copper–steel, and tungsten–silicon carbide bimaterials at $h_1 = h_2$, $W/l_f = \sqrt{3}$ and $l_h/h = 100$ are separately plotted in Fig. 18.13a for $-50\,\mathrm{K} \le \mathrm{d}T \le 50\,\mathrm{K}$, and are consolidated in Fig. 18.13b showing that the effective

Table 18.1 Material properties for the brass–titanium (BT), copper–steel (C-S), and tungsten–silicon carbide (T-SC) bimaterial pairs

Material	CTE ($\times 10^{-6}$ K^{-1})	Young's modulus (GPa)	Remarks
Brass	$\alpha_1^{(T)} = 19$	$E_1 = 112.5$	Material 1 in B-T bimaterial
Titanium	$\alpha_2^{(T)} = 8.6$	$E_2 = 110.3$	Material 2 in B-T bimaterial
Copper	$\alpha_1^{(T)} = 17$	$E_1 = 117$	Material 1 in C-S bimaterial
Steel	$\alpha_2^{(T)} = 12$	$E_2 = 200$	Material 2 in C-S bimaterial
Tungsten	$\alpha_1^{(T)} = 4.5$	$E_1 = 405$	Material 1 in T-SC bimaterial
Silicon carbide	$\alpha_2^{(T)} = 2.77$	$E_2 = 450$	Material 2 in T-SC bimaterial

$\alpha_x^{(T)}$ using the brass–titanium, copper–steel, and the tungsten–silicon carbide bimaterials are $\pm 450 \times 10^{-6}\ K^{-1}$, $\pm 213 \times 10^{-6}\ K^{-1}$ and $\pm 74.9 \times 10^{-6}\ K^{-1}$, respectively. These effective CTE values are at least an order greater than those of the individual material CTEs. While the choice of the O-Type and X-Type arrangements determine the sign of the effective $\alpha_x^{(T)}$, their magnitudes are enhanced by the choice of greater CTE difference $\left(\alpha_1^{(T)} - \alpha_2^{(T)}\right)$ and its E_1/E_2 ratio (or its reciprocal) being closer to 1. Although the selection of O-Type and X-Type arrangements does not influence the sign of the effective $\alpha_y^{(T)}$, which is negative, the choice of higher CTE difference $\left(\alpha_1^{(T)} - \alpha_2^{(T)}\right)$ and its E_1/E_2 ratio (or its reciprocal) being closer to 1 leads to greater the magnitude of the $\alpha_y^{(T)}/dT$ slope, as indicated in Fig. 18.14.

Plots of Poisson's ratio for metamaterials of O-Type and X-Type arrangements using brass–titanium, copper–steel, and tungsten–silicon carbide bimaterial strips with $h_1 = h_2$, $W/l_f = \sqrt{3}$ and $l_h/h = 100$ are given in Fig. 18.15. The exact v_{xy} values have been plotted using Eq. (18.1.23) and are shown as continuous curves, while the approximated v_{xy} values have been plotted using Eq. (18.1.29b) and are denoted as discrete data points. The validity of the simplified approximated Poisson's ratio is attested by its correlation with the exact results, thereby providing a convenient Poisson's ratio design equation.

Finally, it is of interest to explore the remaining types of metamaterials that can be formed by applying different arrangements to the alternating bimaterials. A group of five unit cells each are furnished in Fig. 18.16 (top) for the CC-Type metamaterial and in Fig. 18.16 (bottom) for the SS-Type metamaterial. As shown in the unit cells located in the middle, the CC-Type is so named because the material with higher CTE forms a double C within each rectangular cell, while the SS-Type is named as such due to the material with higher CTE forming a double S in each rectangular cell. The deformation patterns arising from the decrease and increase in temperature are

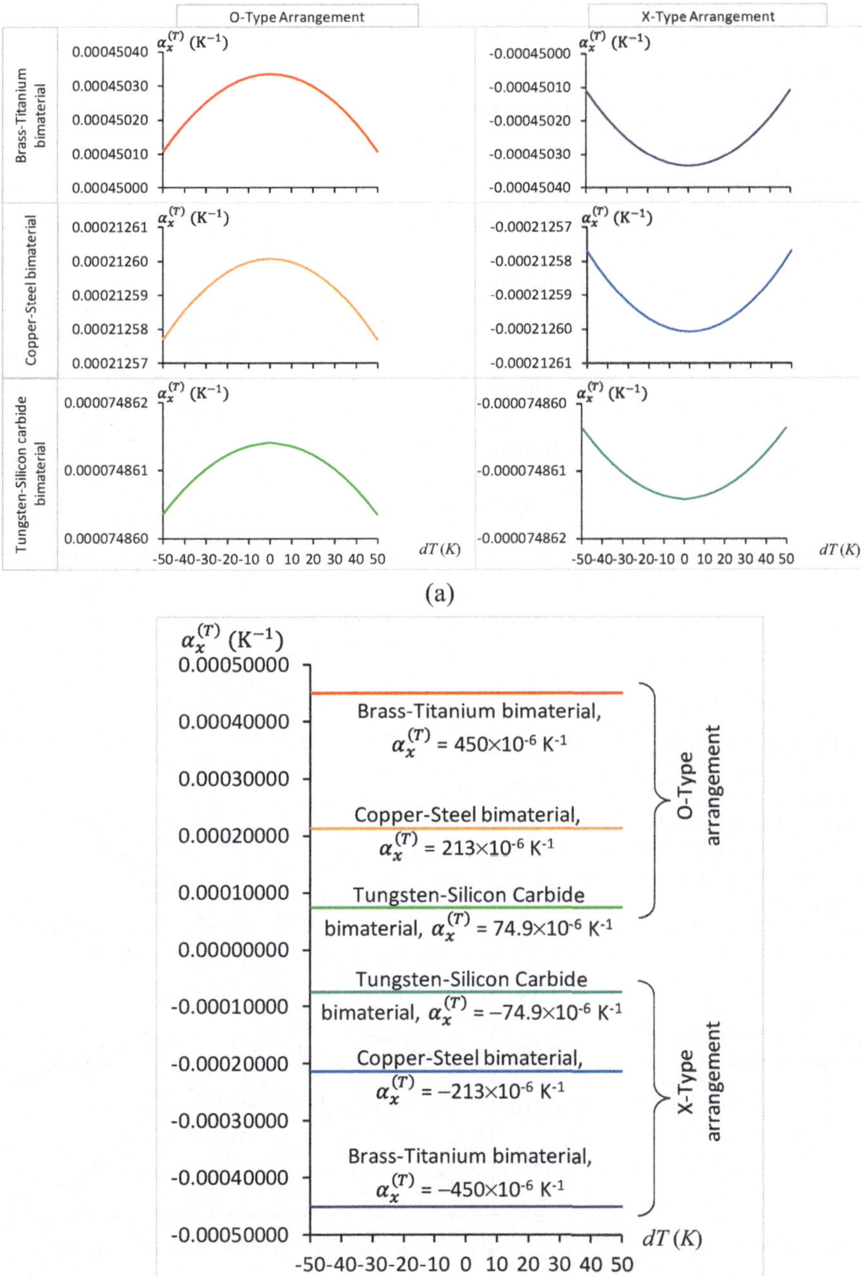

(a)

(b)

Fig. 18.13 Plots of CTE in x-direction for the: **a** O-Type arrangement (left column) and X-Type arrangement (right column) metamaterials based on brass–titanium (top row), copper–steel (middle row), and tungsten–silicon carbide (bottom row) bimaterials at $h_1 = h_2$, $W/l_f = \sqrt{3}$ and $l_h/h = 100$ and **b** consolidated plots showing an almost constant effective $\alpha_x^{(T)}$ with temperature variation

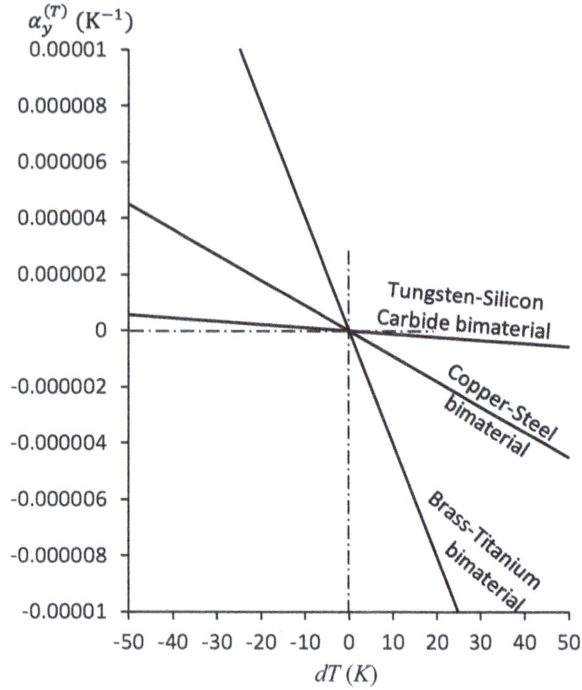

Fig. 18.14 Plots of CTE in y-direction for the O-Type and X-Type metamaterials based on brass–titanium, copper–steel, and tungsten–silicon carbide bimaterials at $h_1 = h_2$, $W/l_f = \sqrt{3}$, $l_h/h = 100$ and $t/l_h = 1/100$

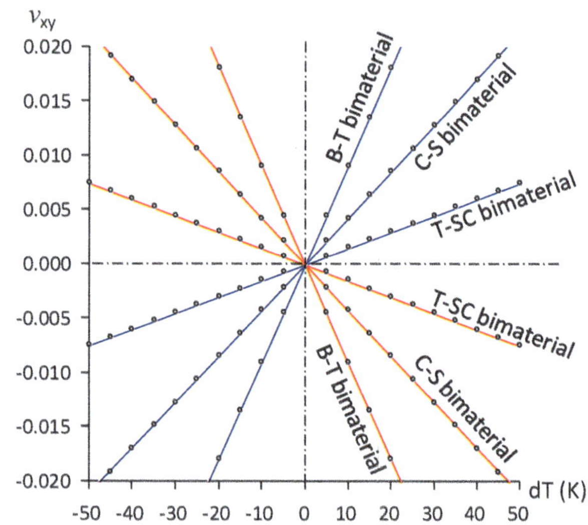

Fig. 18.15 Poisson's ratio for metamaterials of O-Type (blue curves) and X-Type (red curves) arrangements using brass–titanium (BT), copper–steel (CS), and tungsten–silicon carbide (T-SC) bimaterial strips with $h_1 = h_2$, $W/l_f = \sqrt{3}$ and $l_h/h = 100$. Simplified approximate values are indicated as circles

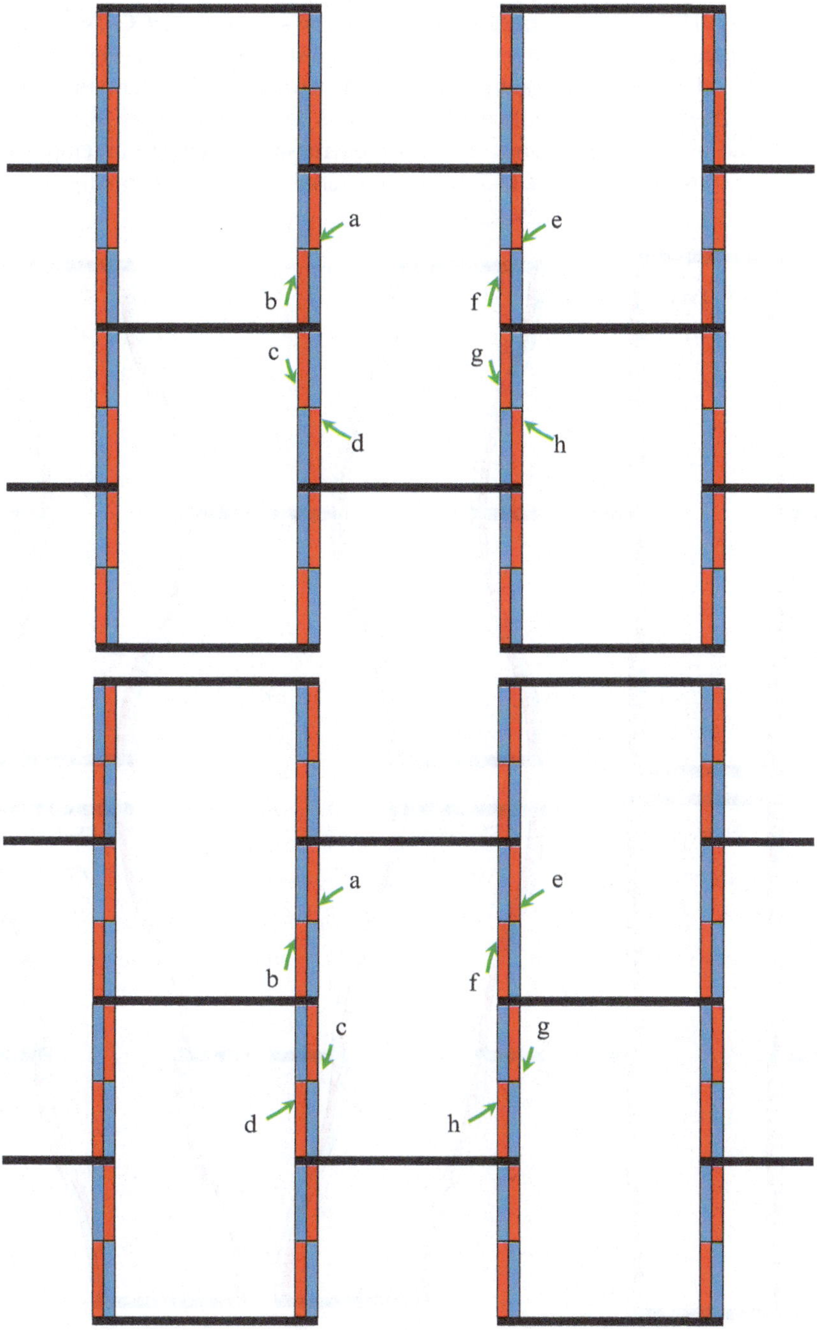

Fig. 18.16 Five unit cells each for the CC-Type (top) and the SS-Type (bottom), whereby the material with higher CTE is shown to form double C and double S, respectively, in each unit cell

shown in Fig. 18.17, wherein the top and bottom rows refer to the CC-Type and SS-Type metamaterials, respectively. From the original state indicated in Fig. 18.17 (left column), the change in temperature brings about curvature to the alternating bimaterial strips, whereby the thermally deformed shapes are shown in Fig. 18.17 (middle column) and in Fig. 18.17 (right column) for decrease and increase in temperature, respectively. Specifically, the CC-Type metamaterial undergoes thermally activated

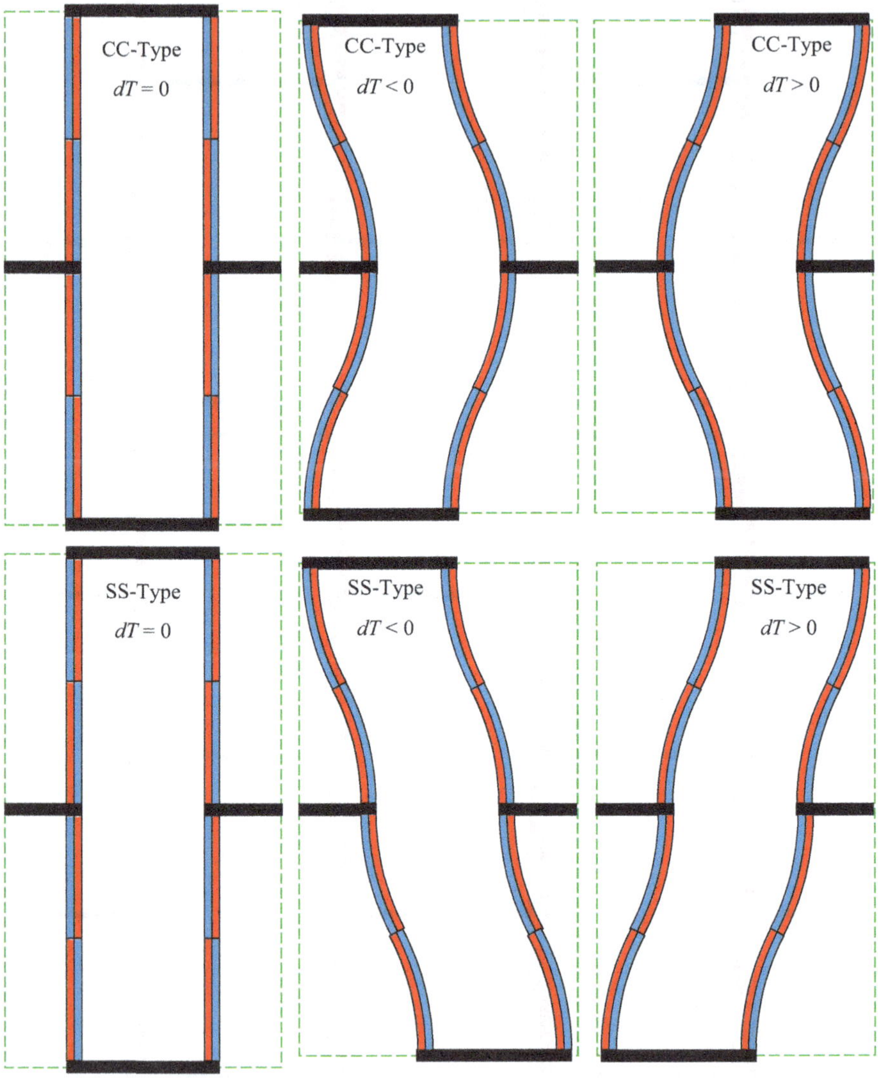

Fig. 18.17 A unit of CC-Type (top row) and SS-Type (bottom row) metamaterials before temperature change (left column), as well as after temperature decrease (middle column) and increase (right column)

deformation to form parallel curves, while the SS-Type metamaterial undergoes curving with overall shearing; the latter resembling thermal shearing (Lim 2005). Similar to the O-Type and X-Type metamaterials, the CC-Type and SS-Type metamaterials exhibit negative thermal strain $\varepsilon_y^{(T)} < 0$ in the y-direction for any change in temperature, regardless of whether the temperature decreases or increases. Therefore, the CC-Type and SS-Type metamaterials exhibit PTE $\alpha_y^{(T)} > 0$ and NTE $\alpha_y^{(T)} < 0$ in the y-direction corresponding to decreasing and increasing temperature, respectively, just like the O-Type and X-Type metamaterials. The CTE in the y-direction is therefore generally written as Eq. (18.1.11) and is described by Eqs. (18.1.13) and (18.1.15) for Special Cases I and II, respectively. Unlike the O-Type and X-Type metamaterials, the CC-Type and SS-Type metamaterials show zero thermal strain $\varepsilon_x^{(T)} = 0$, and therefore ZTE $\alpha_x^{(T)} = 0$, in the x-direction. The Poisson's ratio of the CC-Type and SS-Type metamaterials can be evaluated by returning the thermally deformed shapes shown in Fig. 18.17 (middle and right columns), by application of $\sigma_y > 0$, to their original conditions furnished in Fig. 18.17 (left column). Since the thermally induced strain in the x-direction is zero, it follows that $v_{yx} = 0$. In fact, the thermally deformed CC-Type metamaterial is analogous to the type honeycomb that combines hexagonal and re-entrant walls within each unit cell by Grima et al. (2010) and Attard and Grima (2011), which has been proven to exhibit zero Poisson's ratio.

18.2 Thermally Activated Toggle Between Connected-Octagonal and Connected-Star Microstructural Reshape

Two possible arrangements for rectangular cells in rectangular array are furnished in Fig. 18.18 whereby the bimaterial layers of higher and lower coefficients of thermal expansion (CTE) are colored in red and blue, respectively, while the rigid rods are indicated in black. Based on these opposing arrangements of bimaterial strips, the unit cells denoted by Fig. 18.18 (top) exhibit positive thermal expansion (PTE) while Fig. 18.18 (bottom) manifests negative thermal expansion (NTE); for this reason, they are herein termed the P-Type and N-Type metamaterials, respectively. Under the influence of increasing temperature, the P-Type and N-Type deform into the shapes shown in Fig. 18.19 (top) and (bottom), respectively (Lim 2019b). These shapes swap places for decreasing temperature. It can be seen that the deformed shape illustrated in Fig. 18.19 (bottom) resembles the interconnected stars network, which has been proven to exhibit auxetic behavior. By the same reasoning, an interconnected star network is obtained from the P-Type metamaterial upon cooling. In other words,

$$
\begin{aligned}
dT > 0 &\Rightarrow v^{\text{P-Type}} > 0 \\
dT < 0 &\Rightarrow v^{\text{P-Type}} < 0 \\
dT > 0 &\Rightarrow v^{\text{N-Type}} < 0 \\
dT < 0 &\Rightarrow v^{\text{N-Type}} > 0
\end{aligned}
\tag{18.2.1}
$$

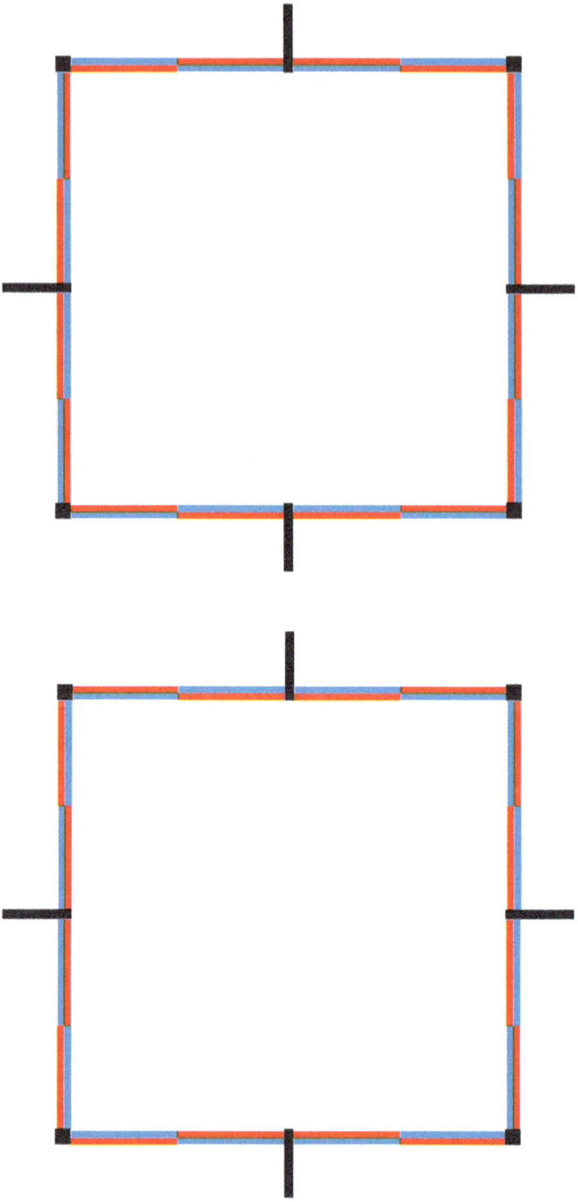

Fig. 18.18 Two possible arrangements of alternating bimaterials in the currently considered class of metamaterials with biaxial PTE (top) and NTE (bottom)

or

$$\upsilon^{\text{P-Type}} \mathrm{d}T > 0$$
$$\upsilon^{\text{N-Type}} \mathrm{d}T < 0 \qquad (18.2.2)$$

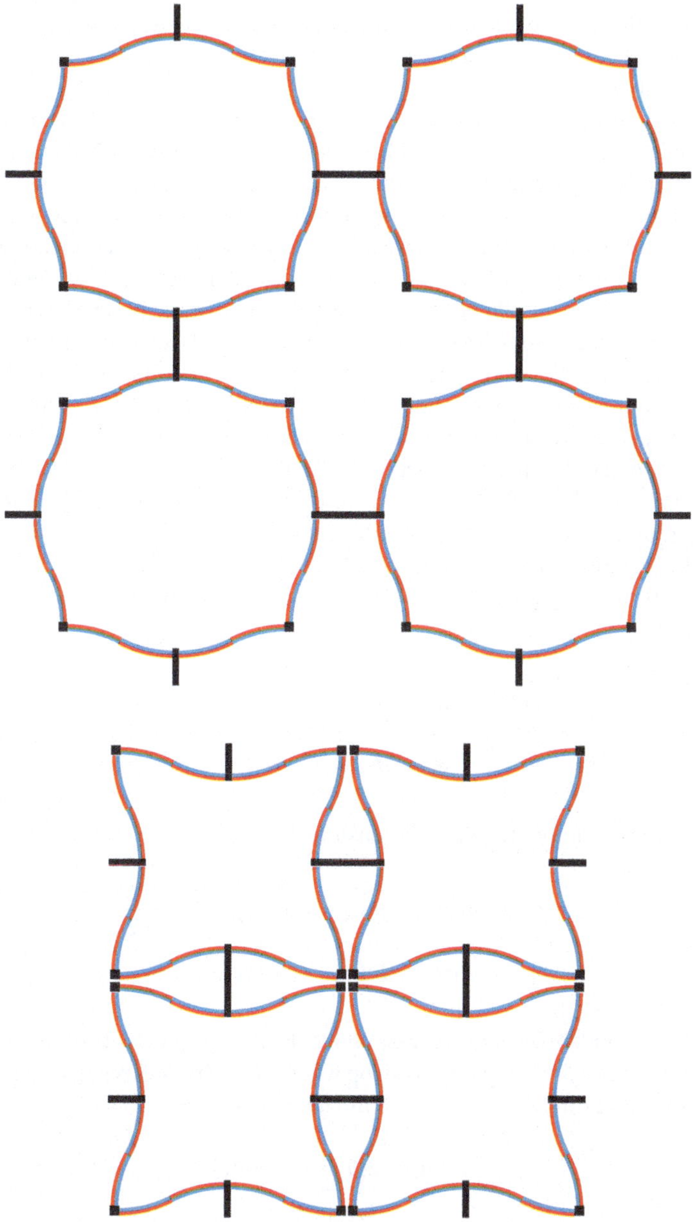

Fig. 18.19 Illustrations of 2-by-2 unit cells of P-Type (top) and N-Type (bottom) metamaterials upon increased temperature

Therefore, the P-Type and N-Type metamaterials in this section are analogous to the O-Type and X-Type metamaterials, respectively, in Sect. 18.1. For brevity, the following analyses explore only the effective CTEs.

Again, we consider an alternating bimaterial strip where l_h indicates its half-length; hence l_{hx} and l_{hy} furnished in Fig. 18.20 (top) refer to the half-lengths of the bimaterials aligned parallel to the x- and y-axes, respectively. Likewise, l_{cx} and l_{cy} denote the half-lengths of the connecting rods that connect each rectangular cell with its nearest neighbor along the x- and y-axes, respectively; these connecting rods are rigid and are of thickness t. Square nodes, of sides h, are also rigid and are placed at the corners of the rectangular cells for joining the ends of the bimaterial rods that are at right angles to each other. With reference to Fig. 18.20 (top), the bimaterial is also of thickness h, where h_1 and h_2 are the thicknesses of both layers in the bimaterial strips. Upon a change in temperature, each alternating bimaterial strip transforms into an S-shaped curve such that its center is an inflexion point and each half of the alternating bimaterial strip curves according to Eq. (18.1.1). See Fig. 18.20 (middle) for the N-Type and Fig. 18.20 (bottom) for the P-Type metamaterials in the case of increasing temperature, with the subtending angles being $\theta_x = l_{hx}/r$ and $\theta_y = l_{hy}/r$; these are the rotations at the mid-points of the alternating bimaterials.

Starting from the original dimensions of the quarter unit cell measured along the x- and y-axes

$$\begin{aligned} x_0 &= 2l_{hx} + l_{cx} + \frac{t}{2} \\ y_0 &= 2l_{hy} + l_{cy} + \frac{t}{2} \end{aligned} \tag{18.2.3}$$

a change in temperature alters the dimensions to

$$\begin{aligned} x &= 2r\,\sin\theta_x \pm 2r\left(1 - \cos\theta_y\right) + l_{cx} + \frac{t}{2} \\ y &= 2r\,\sin\theta_y \pm 2r(1 - \cos\theta_x) + l_{cy} + \frac{t}{2} \end{aligned} \tag{18.2.4}$$

where the upper and lower signs correspond to the P-Type and N-Type metamaterials, respectively, when $\mathrm{d}T > 0$ (the signs swop when $\mathrm{d}T < 0$). As a result of the thermally induced change in dimension, we have the thermal strains

$$\begin{aligned} \varepsilon_x^{(T)} &= \frac{r\,\sin\theta_x \pm r\left(1 - \cos\theta_y\right) - l_{hx}}{l_{hx} + \frac{l_{cx}}{2} + \frac{t}{4}} \\ \varepsilon_y^{(T)} &= \frac{r\,\sin\theta_y \pm r(1 - \cos\theta_x) - l_{hy}}{l_{hy} + \frac{l_{cy}}{2} + \frac{t}{4}} \end{aligned} \tag{18.2.5}$$

and their corresponding CTEs (Lim 2019b)

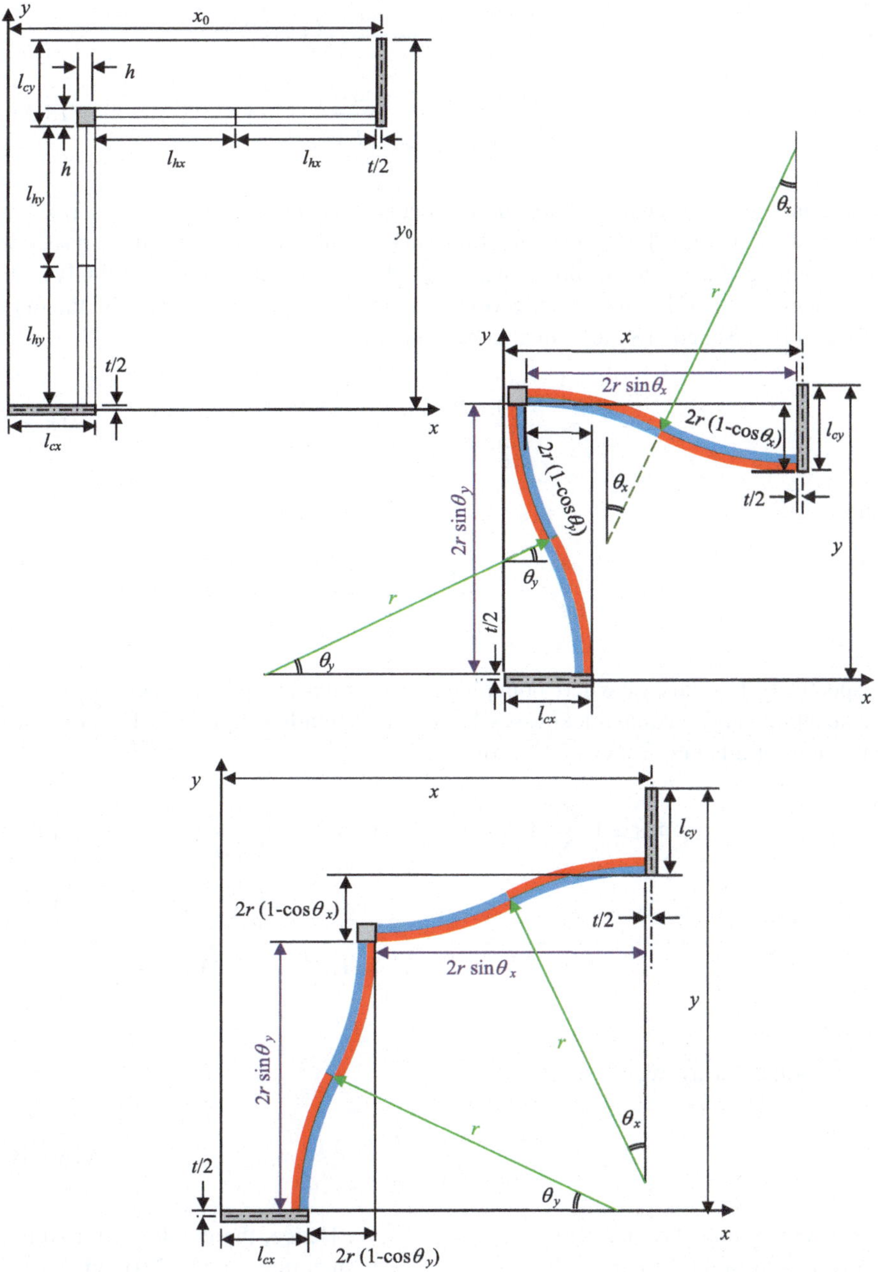

Fig. 18.20 A quarter of unit cell at original state (top), as well as after temperature increase in the N-Type (middle) and P-Type (bottom) metamaterials for analysis

$$\alpha_x^{(T)} = \frac{r\sin\theta_x \pm r\left(1-\cos\theta_y\right) - l_{hx}}{\left(l_{hx} + \frac{l_{cx}}{2} + \frac{t}{4}\right)\mathrm{d}T}$$
$$\alpha_y^{(T)} = \frac{r\sin\theta_y \pm r(1-\cos\theta_x) - l_{hy}}{\left(l_{hy} + \frac{l_{cy}}{2} + \frac{t}{4}\right)\mathrm{d}T} \tag{18.2.6}$$

where the radius of curvature for the thermally bent bimaterials, r, is described in Eq. (18.1.1). As the CTE in the y-direction can be obtained from that in the x-direction by swopping the x- and y-subscripts, it is sufficient for the following analysis to consider only the CTE in the x-direction. Since $\theta_x = l_{hx}/r$ and $\theta_y = l_{hy}/r$, the firsts of Eqs. (18.2.5) and (18.2.6) can also be expressed as

$$\varepsilon_x^{(T)} = \frac{\pm 1 \mp \cos\frac{l_{hy}}{r} + \sin\frac{l_{hx}}{r} - \frac{l_{hx}}{r}}{\frac{l_{hx}}{r} + \frac{1}{2}\frac{l_{cx}}{l_{hx}}\frac{l_{hx}}{r} + \frac{1}{4}\frac{t}{l_{hx}}\frac{l_{hx}}{r}} \tag{18.2.7}$$

and

$$\alpha_x^{(T)} = \frac{\varepsilon_x^{(T)}}{\mathrm{d}T} = \frac{\pm 1 \mp \cos\frac{l_{hy}}{r} + \sin\frac{l_{hx}}{r} - \frac{l_{hx}}{r}}{\left(1 + \frac{1}{2}\frac{l_{cx}}{l_{hx}} + \frac{1}{4}\frac{t}{l_{hx}}\right)\frac{l_{hx}}{r}\mathrm{d}T} \tag{18.2.8}$$

respectively. For the case where both phases of the alternating bimaterial strips are of equal moduli and of equal thicknesses, Eq. (18.1.1) greatly simplifies to Eq. (18.1.3a). Substituting this into Eq. (18.2.8) gives

$$\alpha_x^{(T)} = \frac{\varepsilon_x^{(T)}}{\mathrm{d}T} = \frac{\pm 1 \mp \cos\left(\frac{3l_{hy}}{2h}\left(\alpha_1^{(T)} - \alpha_2^{(T)}\right)\mathrm{d}T\right) + \sin\left(\frac{3l_{hx}}{2h}\left(\alpha_1^{(T)} - \alpha_2^{(T)}\right)\mathrm{d}T\right) - \left(\frac{3l_{hx}}{2h}\left(\alpha_1^{(T)} - \alpha_2^{(T)}\right)\mathrm{d}T\right)}{\left(1 + \frac{1}{2}\frac{l_{cx}}{l_{hx}} + \frac{1}{4}\frac{t}{l_{hx}}\right)\left(\frac{3l_{hx}}{2h}\left(\alpha_1^{(T)} - \alpha_2^{(T)}\right)\mathrm{d}T\right)\mathrm{d}T} \tag{18.2.9}$$

For square array, we have

$$\frac{2l_{hy} + l_{cy} + \frac{t}{2}}{2l_{hx} + l_{cx} + \frac{t}{2}} = 1 \tag{18.2.10}$$

while square cells are denoted as $l_{hx} = l_{hy} = l_h$. Hence, the condition for square array, in addition to square cells, is $l_{cx} = l_{cy} = l_c$ such that Eq. (18.2.9) reduces to

$$\alpha_x^{(T)} = \alpha_y^{(T)} = \frac{\pm 1 \mp \cos\left(\frac{3l_h}{2h}\left(\alpha_1^{(T)} - \alpha_2^{(T)}\right)\mathrm{d}T\right) + \sin\left(\frac{3l_h}{2h}\left(\alpha_1^{(T)} - \alpha_2^{(T)}\right)\mathrm{d}T\right) - \left(\frac{3l_h}{2h}\left(\alpha_1^{(T)} - \alpha_2^{(T)}\right)\mathrm{d}T\right)}{\left(1 + \frac{1}{2}\frac{l_c}{l_h} + \frac{1}{4}\frac{t}{l_h}\right)\left(\frac{3l_h}{2h}\left(\alpha_1^{(T)} - \alpha_2^{(T)}\right)\mathrm{d}T\right)\mathrm{d}T} \tag{18.2.11}$$

To visualize the effective in-plane CTEs of the currently considered metamaterial, results were plotted to show the effect of rod thickness, bimaterial CTE ratio, bimaterial aspect ratio, cell aspect ratio, and array aspect ratio on the in-plane thermal strain and in-plane CTE with and without variation in temperature. The effect of cell aspect ratio was investigated at square array while the effect of array aspect ratio was investigated for square cells. The effect of rod thickness, bimaterial CTE ratio, and bimaterial aspect ratio was generated for square cells in square array. We consider the case of equal bimaterial moduli $E_1 = E_2$ and thicknesses $h_1 = h_2 = h/2$ so that Eq. (18.2.9) applies for rectangular cells in rectangular array; in the case of square cells in square array, Eq. (18.2.11) holds. In the subsequent plotted results, we adopt $\alpha_1^{(T)} = 25 \times 10^{-6}\ \mathrm{K}^{-1}$ as this value is typical to materials with high CTEs, while the range of temperature change is selected as $-50\ \mathrm{K} \le \mathrm{d}T \le 50\ \mathrm{K}$ because a range of 100 K is sufficiently realistic.

The effects of bimaterial and geometrical properties are evaluated as follows. Based on Eq. (18.2.11), the use of dimensionless terms such as l_h/h, $\alpha_2^{(T)}/\alpha_1^{(T)}$, l_c/l_h and t/l_h are useful to reduce the number of required plots. Figure 18.21 (left) shows the effect of the relative rod thickness, t/l_h, on the in-plane CTE with the change in temperature at fixed $l_h/h = 60$, $\alpha_2^{(T)}/\alpha_1^{(T)} = 1/2$ and $l_c/l_h = 2$. The almost constant CTEs with respect to the change in temperature are obviously due to the almost linear

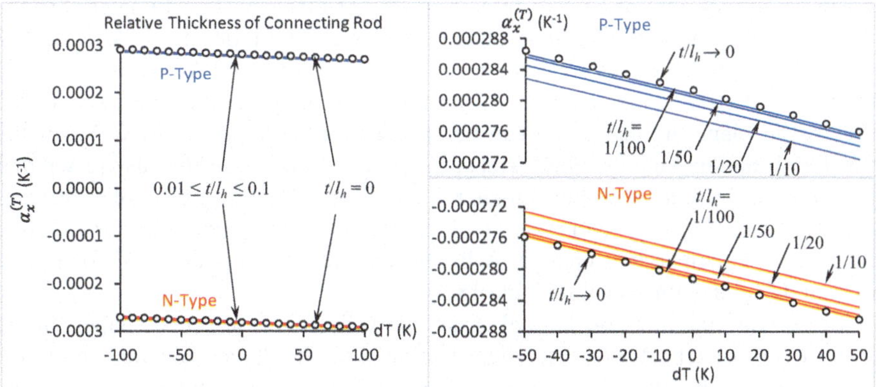

Fig. 18.21 Effect of connecting rod thickness on the effective in-plane CTE (left) with zoomed-in view for the P-Type (top right) and N-Type (bottom right) metamaterials with change in temperature for square cells in square array at $l_h/h = 60$, $\alpha_2^{(T)}/\alpha_1^{(T)} = 1/2$ and $l_c/l_h = 2$

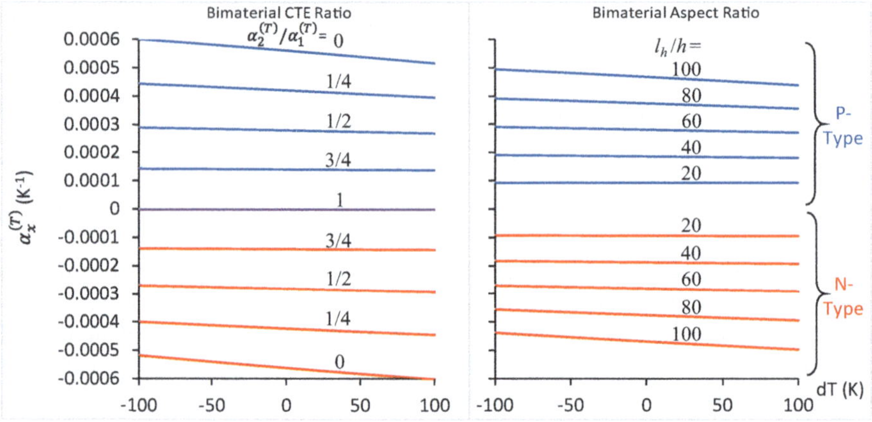

Fig. 18.22 Effect of bimaterial CTE ratio $\alpha_2^{(T)}/\alpha_1^{(T)}$ at $l_h/h = 60$ (left) and bimaterial aspect ratio l_h/h at $\alpha_2^{(T)}/\alpha_1^{(T)} = 1/2$ (right) on the effective in-plane CTE for the P-Type (blue) and N-Type (red) metamaterials with change in temperature for square cells in square array at $l_c/l_h = 2$ and $t/l_h = 1/100$

thermal strain with reference to the temperature change. More importantly, the plots reveal that the assumption of $t = 0$ is valid. In order to discriminate the effects from the relative rod thickness, zoomed-in views of the effective CTEs, $\alpha_x^{(T)}$ are shown in Fig. 18.21 (top right) for P-Type and in Fig. 18.21 (bottom right) for N-Type. The plots of CTE magnitude for $t = 0$ theoretically demarcates a locus by which any effective CTE beyond the boundary is practically impossible.

Adopting Eq. (18.2.11) again, the influence from the bimaterial's thermal and geometrical properties is plotted in Fig. 18.22. Specifically, the plots of effective CTE magnitudes show the influence of bimaterial CTE ratio $\alpha_2^{(T)}/\alpha_1^{(T)}$ at a fixed bimaterial aspect ratio of $l_h/h = 60$ in Fig. 18.22 (left) as well as bimaterial shape in terms of its aspect ratio l_h/h at a fixed bimaterial CTE ratio of $\alpha_2^{(T)}/\alpha_1^{(T)} = 1/2$ in Fig. 18.22 (right) for relative connecting rod length and thickness of $l_c/l_h = 2$ and $t/l_h = 1/100$, respectively. In addition to exhibiting that the magnitude of the effective in-plane thermal strain can be enhanced by contrasting the CTEs of the bimaterials and making the bimaterial strip slenderer, plotted results also reveal that the effect of cooling is slightly greater for P-Type while the effect of heating is slightly greater for N-Type in regard to their effective CTE magnitudes for the same magnitude of temperature change. A list of the effective in-plane CTE for square array at $dT = 0$ is furnished in Table 18.2.

To generate the plots of thermal strain and CTE for various cell shapes, in the form of cell aspect ratio l_{hy}/l_{hx}, there is a need to simultaneously vary the ratios l_{cy}/l_{cx} and l_{cx}/l_{hx} in order to maintain a fixed array. To maintain square array, Eq. (18.2.10) is used for obtaining appropriate values of l_{cy}/l_{cx} and l_{cx}/l_{hx}. For generating results of thermal strain and CTE with various array aspect ratio, there is a requirement to concurrently alter other geometrical ratios. In the case of retaining

Table 18.2 In-plane CTE for square cells in square array at $\alpha_1^{(T)} = 25 \times 10^{-6}\,\mathrm{K}^{-1}$ and $\mathrm{d}T = 0\,\mathrm{K}$

t/l_h	$\alpha_2^{(T)}/\alpha_1^{(T)}$	l_h/h	$\pm\alpha_x^{(T)}(\times 10^{-6}\,\mathrm{K}^{-1})$
0.00	1/2	60	281.26
0.01	1/2	60	280.91
0.02	1/2	60	280.56
0.05	1/2	60	279.51
0.10	1/2	60	277.78
0.01	0	60	561.79
0.01	1/4	60	421.36
0.01	1/2	60	280.91
0.01	3/4	60	140.45
0.01	1	60	000.00
0.01	1/2	100	468.17
0.01	1/2	80	374.53
0.01	1/2	60	280.91
0.01	1/2	40	187.25
0.01	1/2	20	93.59

the cell shape as squares $l_{hx} = l_{hy} = l_h$, the change in aspect ratio was implemented by assigning numerical values to $(2l_h + l_{cy} + t/2)/(2l_h + l_{cx} + t/2)$ followed by extracting ratios of l_{cy}/l_{cx} and l_{cx}/l_{hx}. The geometries for prescribing change in one type of aspect ratio while maintaining the other type of aspect ratio fixed are summarized in Table 18.3. Substituting these geometrical parameters into Eq. (18.2.9), plots of effective thermal strains and CTEs, as measured in the x-direction, are furnished in Fig. 18.23. Results also show that the CTE magnitudes in the x-direction can be

Table 18.3 In-plane CTE for rectangular cells in rectangular array at $t/l_{hx} = 0.01$, $\alpha_2^{(T)}/\alpha_1^{(T)} = 1/2$, $l_{hx}/h = 60$, $\alpha_1^{(T)} = 25 \times 10^{-6}\,\mathrm{K}^{-1}$ and $\mathrm{d}T = 0\,\mathrm{K}$

l_{hy}/l_{hx}	l_{cy}/l_{cx}	l_{cx}/l_{hx}	$\frac{2l_{hy}+l_{cy}}{2l_{hx}+l_{cx}}$	$\frac{2l_{hy}+l_{cy}+t/2}{2l_{hx}+l_{cx}+t/2}$	$\pm\alpha_x^{(T)}(\times 10^{-6}\,\mathrm{K}^{-1})$
25/9 = 2.77778	5/21	14/3	1	1	1301.09
25/16 = 1.5625	5/8	3	1	1	548.78
1	1	2	1	1	280.91
16/25 = 0.64	8/5	6/5	1	1	143.80
9/25 = 0.36	21/5	2/5	1	1	60.65
1	35/3	2/5	25/9 = 2.77778	2.77408	467.79
1	5/2	6/5	25/16 = 1.5625	1.56162	351.02
1	1	2	1	1	280.91
1	2/5	3	16/25 = 0.64	0.64036	224.78
1	3/35	14/3	9/25 = 0.36	0.36048	168.63

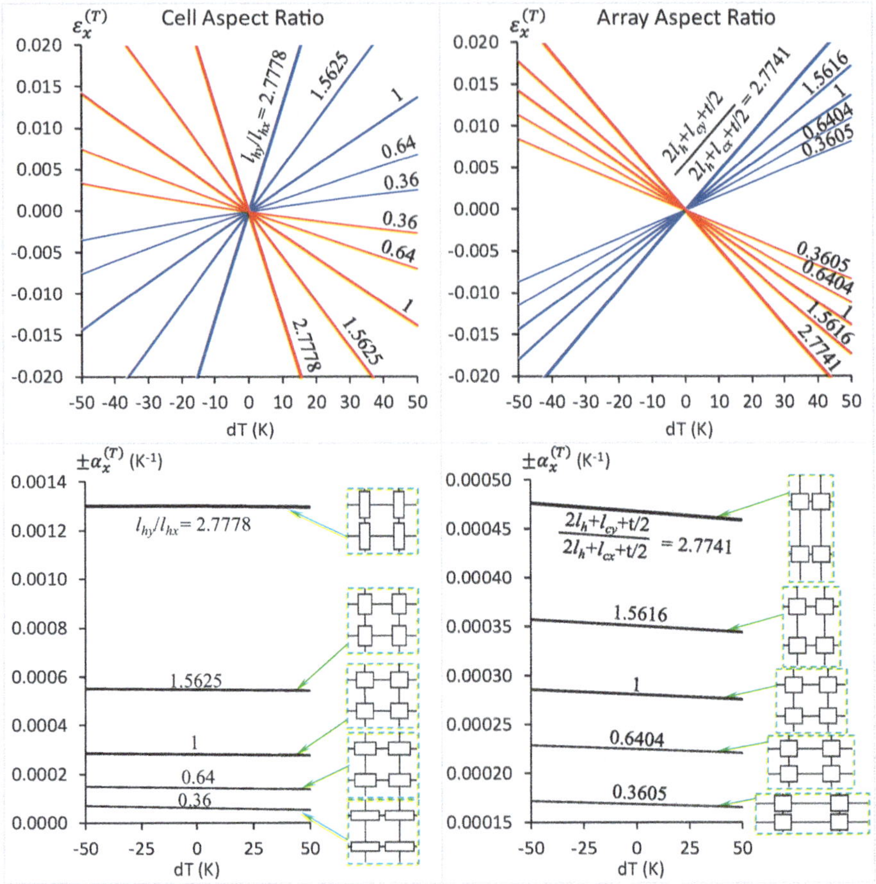

Fig. 18.23 Effect of cell aspect ratio at square array (left) and array aspect ratio with square cells (right) on the in-plane thermal strain (top) and in-plane CTE magnitude (bottom) for bimaterial properties of $\alpha_2^{(T)}/\alpha_1^{(T)} = 1/2$, $l_{hx}/h = 60$ and $t/l_{hx} = 1/100$

heightened by increasing l_{hy} and decreasing l_{cx}, as indicated in Fig. 18.23 (bottom left) and Fig. 18.23 (bottom right), respectively. The former observation is attributed to the fact that the transverse deflection of the bimaterials has a greater magnitude than the axial displacement of its ends, while the latter observation is attributed simply to the denominator effect.

So far the generated effective thermal strains and CTEs were plotted with respect to the change in temperature for a few values of bimaterial strip slenderness, contrasting CTEs of the bimaterial layers and the connecting rod thickness. Figure 18.24 shows the continuous variation of the effective CTEs with respect to the bimaterial strip slenderness, contrasting CTEs of the bimaterial layers and the connecting rod thickness in terms of their dimensionless ratios l_h/h, $\alpha_2^{(T)}/\alpha_1^{(T)}$ and t/l_h, respectively.

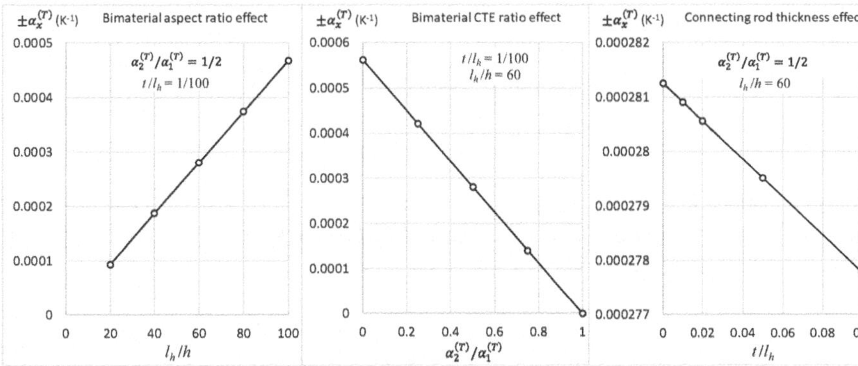

Fig. 18.24 Effect of bimaterial aspect ratio (left), bimaterial CTE ratio (middle), and connecting rod thickness (right) on the CTE magnitudes for square cells in square array

The almost linear variation of the CTE with l_h/h and $\alpha_2^{(T)}/\alpha_1^{(T)}$ can be made obvious by substituting

$$\begin{aligned} \sin\left(\frac{3l_h}{2h}\left(\alpha_1^{(T)}-\alpha_2^{(T)}\right)\mathrm{d}T\right) &\approx \frac{3l_h}{2h}\left(\alpha_1^{(T)}-\alpha_2^{(T)}\right)\mathrm{d}T \\ \cos\left(\frac{3l_h}{2h}\left(\alpha_1^{(T)}-\alpha_2^{(T)}\right)\mathrm{d}T\right) &\approx 1-\frac{1}{2!}\left(\frac{3l_h}{2h}\left(\alpha_1^{(T)}-\alpha_2^{(T)}\right)\mathrm{d}T\right)^2 \end{aligned} \tag{18.2.12}$$

into Eq. (18.2.11) to give

$$\alpha_x^{(T)}=\alpha_y^{(T)}\approx\pm\frac{3\frac{l_h}{h}\alpha_1^{(T)}\left(1-\frac{\alpha_2^{(T)}}{\alpha_1^{(T)}}\right)}{4+2\frac{l_c}{l_h}+\frac{t}{l_h}} \tag{18.2.13}$$

Applying Maclaurin expansion on Eq. (18.2.13) gives

$$\begin{aligned} \alpha_x^{(T)}=\alpha_y^{(T)}\approx\pm&\frac{3\frac{l_h}{h}\alpha_1^{(T)}\left(1-\frac{\alpha_2^{(T)}}{\alpha_1^{(T)}}\right)}{4+2\frac{l_c}{l_h}} \\ &\left[1-\frac{\frac{t}{l_h}}{4+2\frac{l_c}{l_h}}+\cdots\right];\quad\left|\frac{\frac{t}{l_h}}{4+2\frac{l_c}{l_h}}\right|<1 \end{aligned} \tag{18.2.14}$$

thereby accounting for the approximately linear trend of CTE with variation in t/l_h. Substituting $l_h/h=60$, $\alpha_1^{(T)}=25\times10^{-6}\,\mathrm{K}^{-1}$, $l_c/l_h=2$ and $t/l_h=0$ into either Eqs. (18.2.13) or (18.2.14) gives $\alpha_x^{(T)}=\alpha_y^{(T)}\approx\pm281.25\times10^{-6}\,\mathrm{K}^{-1}$, which is a good estimation to the effective CTE plots of Fig. 18.21 (right) as $\mathrm{d}T\to0$.

Since the choice of cell aspect ratio l_{hy}/l_{hx} at square array contains reciprocal values, one may infer the thermal strains and CTEs along the y-direction based on those along the x-direction

$$\begin{Bmatrix} \varepsilon_y^{(T)} \\ \alpha_y^{(T)} \end{Bmatrix}_{l_{hx}/l_{hy}} = \begin{Bmatrix} \varepsilon_x^{(T)} \\ \alpha_x^{(T)} \end{Bmatrix}_{l_{hy}/l_{hx}} \tag{18.2.15}$$

for square array. By similar reasoning on the reciprocals of array aspect ratio with square cells, the thermal strains and CTEs parallel to the y-axis can be obtained from those parallel to the x-axis

$$\begin{Bmatrix} \varepsilon_y^{(T)} \\ \alpha_y^{(T)} \end{Bmatrix}_{\frac{2l_h+l_{cx}+t/2}{2l_h+l_{cy}+t/2}} = \begin{Bmatrix} \varepsilon_x^{(T)} \\ \alpha_x^{(T)} \end{Bmatrix}_{\frac{2l_h+l_{cy}+t/2}{2l_h+l_{cx}+t/2}} \tag{18.2.16}$$

for square cells. Figure 18.25 furnishes the calculated $\alpha_x^{(T)}$ alongside the inferred $\alpha_y^{(T)}$ by means of the abovementioned correspondence of indices. Specifically, the CTE plots for variation in cell shape at square array and for variation in cell array with square cells are furnished in Fig. 18.25 (top left) and Fig. 18.25 (bottom left), respectively. Since $\alpha_x^{(T)}$ and $\alpha_y^{(T)}$ are swopped when the axes are rotated by 90° with respect to the microstructural geometry, it follows that the curves of $\alpha_x^{(T)}$ and $\alpha_y^{(T)}$ become mirror images of each other about the axis of unity aspect ratio, and hence preserving symmetry, when the aspect ratios are plotted in any logarithmic scale. Figure 18.25 (right) shows examples of CTE curves plotted against both aspect ratios in natural logarithmic scale.

We shall now consider a different arrangement of bimaterial strips—the S-Type metamaterial as furnished in Fig. 18.26 (top)—such that it is neither biaxial PTE nor biaxial NTE upon temperature change, but instead, it exhibits thermal shearing as shown in Fig. 18.26 (bottom), thereby manifesting both PTE and NTE in the diagonal directions. The sheared shapes reverse when temperature decreases. Finally, we consider yet another assembly—not by rearranging the bimaterial strips within each unit cell—but by alternating the P-Type and N-Type cells in 2D array, to give a P&N-Type. An example is illustrated in Fig. 18.27 using a 2-by-2 unit cells whereby the P-Type takes the top left and bottom right positions while the N-Type takes the top right and bottom left positions; the P-Type and N-Type cells expand and contract, respectively, upon heating. Upon cooling, the P-Type and N-Type cells shrink and expand, respectively.

It is of interest to note that the P-Type and N-Type metamaterials deform into shapes that approximate interconnected stars upon decreasing and increasing temperatures, respectively; the interconnected star networks have been shown to exhibit auxetic behavior. On the other hand, the S-Type metamaterial exhibits thermal shearing, thereby giving positive and negative thermal expansion behaviors along its diagonals. Finally, the combination of P-Type and N-Type cells in alternating pattern, i.e., the P&N-Type, gives rise to NTE property for increasing temperature, but PTE with decreasing temperature. Although the expansion in P-Type cells appears to be offset by contraction of N-Type cells, there is an overall decrease in

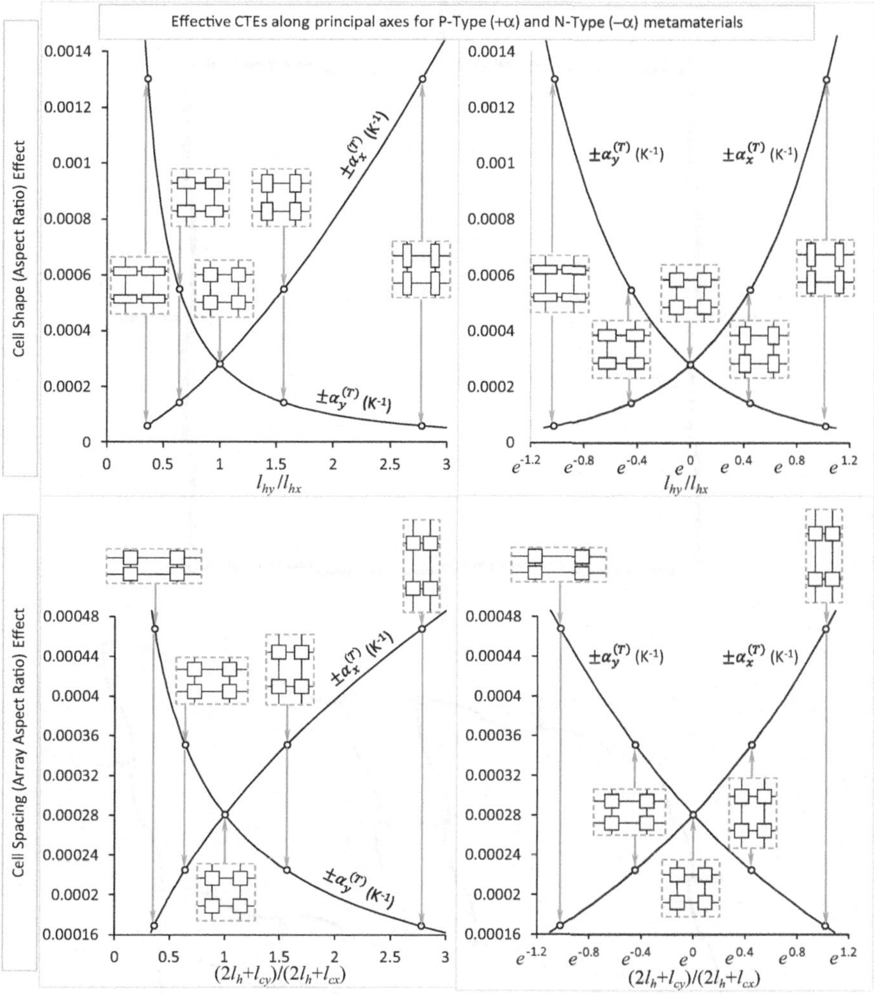

Fig. 18.25 Effects of cell aspect ratio at square array (top row) and array aspect ratio for square cells (bottom row) on the CTE magnitudes with $t/l_{hx} = 1/100$ and $\alpha_2^{(T)}/\alpha_1^{(T)} = 1/2$, plotted against the linear scale (left column) and natural logarithmic scale (right column)

size regardless of whether temperature increases or decreases due to the curving of all cell walls. This is due to the curving of the bimaterial strips that shorten the distance between the cell corners. Therefore, the P&N-Type metamaterial manifests sign-switching of CTE with persistently negative thermal strain regardless of whether temperature increases or decreases. Suppose the rigid cross bars of the P&N-Type metamaterial diminish such that the bimaterial walls of neighboring cells merge,

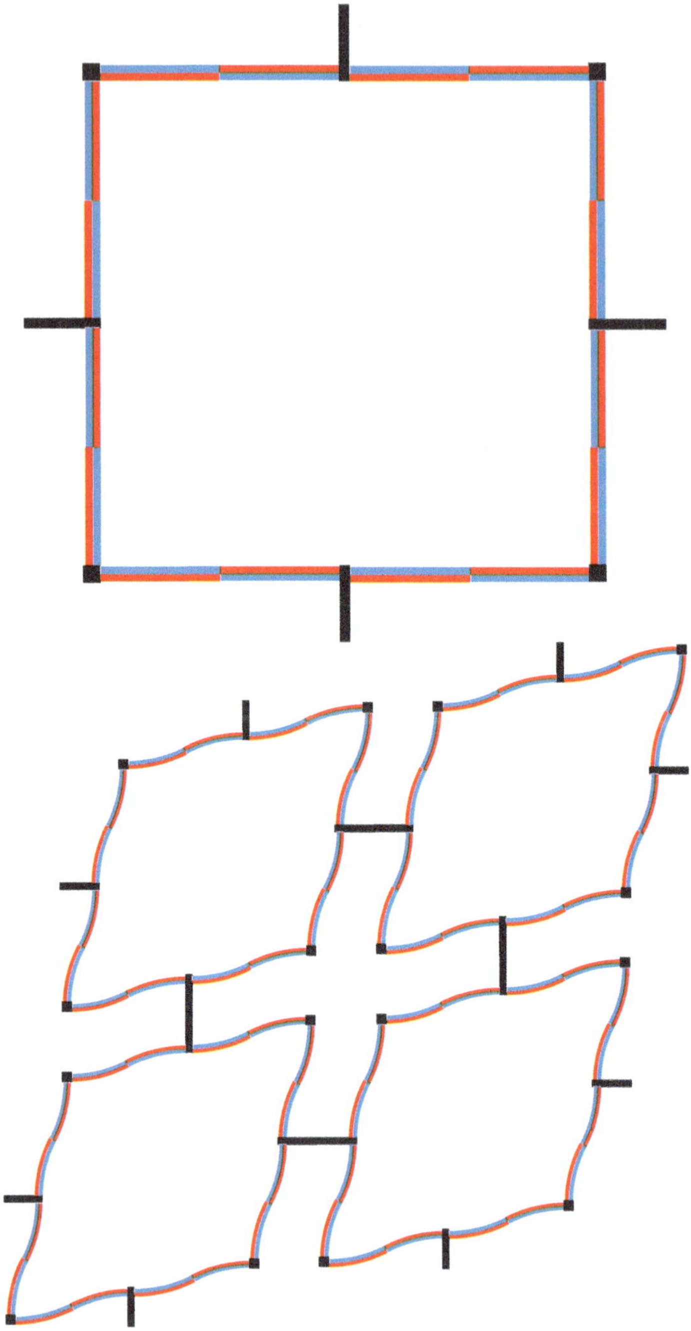

Fig. 18.26 A single unit of S-Type metamaterial showing the bimaterial arrangement of higher (red) and lower (blue) CTE layers (top), and an illustration of 2-by-2 units undergoing thermal shearing upon increased temperature (bottom)

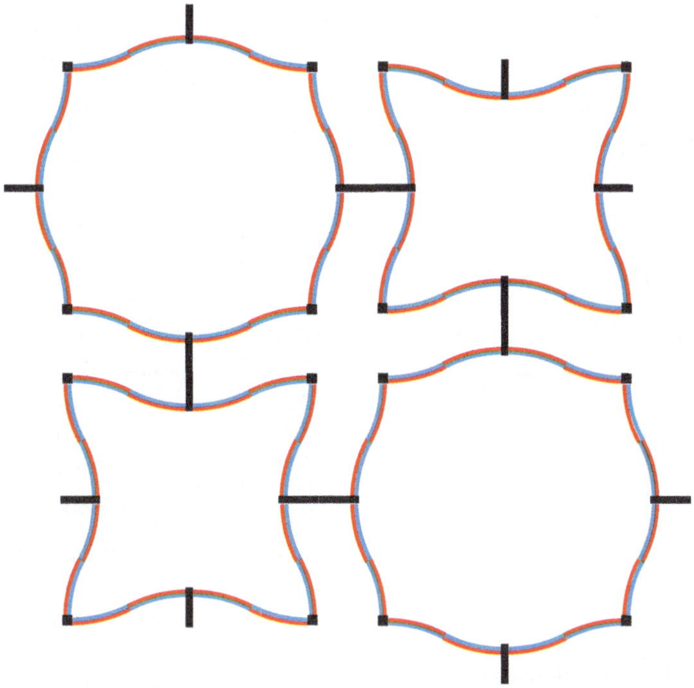

Fig. 18.27 Illustration of NTE in a 2-by-2 unit cells of a P&N-Type metamaterial upon increased temperature

then the thermally-deformed metamaterial resembles the type of environmentally-deformed metamaterial discussed in Sect. 20.2 (Lim 2020a) and approximates the hygrothermally-deformed metamaterial analyzed in Sect. 20.3 (Lim 2020b).

Regardless of the type of bimaterial arrangements—be they P-Type, N-Type, S-Type or P&N-Type—the Poisson's ratio of the structures *at original state*, when *infinitesimally* stretched along either axes, are near zero due to the retention of the rectangular cell shape. For large stretching, however, the bimaterials that are perpendicular to the force direction will curve outwards such that the projected length becomes shorter. As a consequence of bending, the accompanying rotations at the cell corners induce bending at the ends of the bimaterial strips that are parallel to the force direction, thereby causing them to bend inwards. This brings about inward bending of the bimaterials that are parallel to the force, which in turn leads to inward displacements of the connecting rods that are at right angles to the loading direction. In other words, the microstructure at original state exhibits positive Poisson's ratio when it undergoes large stretching along either axes. See Fig. 18.28.

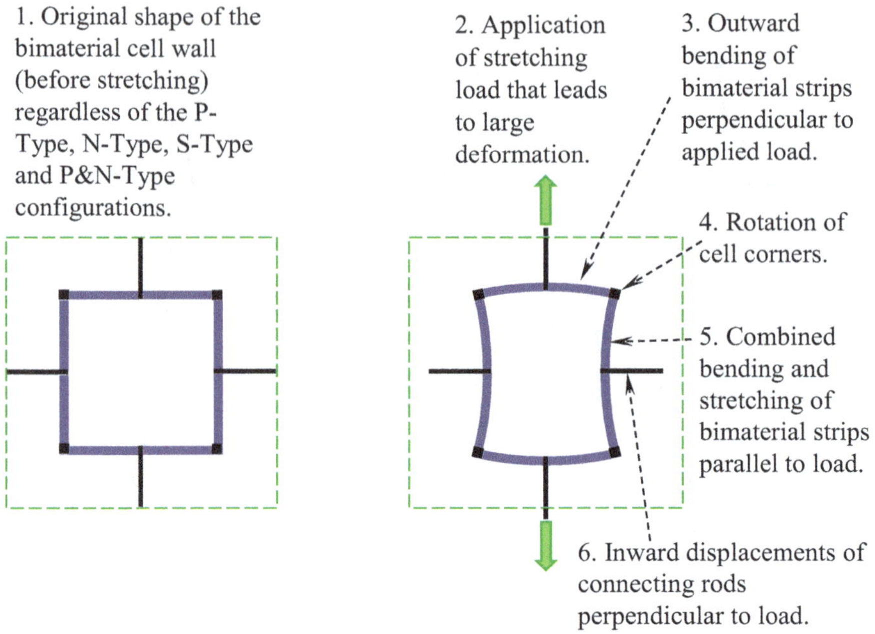

Fig. 18.28 Qualitative illustration of positive Poisson's ratio arising large stretching along one of the axes. The dashed green square aids comparison

18.3 Conclusions

Recall from Sect. 18.1 that by judicious arrangements of the alternating bimaterial strips, the metamaterials can be made to exhibit (a) fixed positive or negative values of some properties regardless of the sign of dT and, more interestingly, (b) alternating positive or negative values of other properties depending of the sign of dT. Specifically, the originally rectangular cells of the O-Type and X-Type metamaterials alternate between barrel- and hourglass-shaped cells for alternating dT signs, thereby alternating their Poisson's ratio signs in situ as a consequence. For completeness' sake, the CC-Type and SS-Type metamaterials were also discussed; these exhibit zero thermal expansion (ZTE) in the x-direction and zero Poisson's ratio (ZPR) for loading in the y-direction, and therefore are good candidates for application where dimensional stability is required. A summarized overview with emphasis on the Poisson's ratio and CTE sign-switchability by means of thermal-toggling for the proposed class of metamaterial is furnished in Table 18.4. In this analysis, the crossbeams have been assumed rigid. Since the ends of the crossbeams are partially built-in, i.e., clamped at the top and bottom with the remaining two sides being free surfaces, the use of an exact deflection solution with strong boundary conditions (Lim 2021) for the clamped portion of the crossbeams is recommended for future analysis that involves bending of the crossbeams.

Table 18.4 Sign-switchability of CTE and Poisson's ratio by means of temperature change

Metamaterial type	$T \downarrow$	$T \uparrow$	Remarks on thermal fluctuation effect
O-Type	$\alpha_x^{(T)} = +$	$\alpha_x^{(T)} = +$	PTE (non-switchable)
	$\alpha_y^{(T)} = +$	$\alpha_y^{(T)} = -$	*Sign switchable*
	$v_{xy} = -$	$v_{xy} = +$	*Sign switchable*
X-Type	$\alpha_x^{(T)} = -$	$\alpha_x^{(T)} = -$	NTE (non-switchable)
	$\alpha_y^{(T)} = +$	$\alpha_y^{(T)} = -$	*Sign switchable*
	$v_{xy} = +$	$v_{xy} = -$	*Sign switchable*
CC- and SS-Types	$\alpha_x^{(T)} = 0$	$\alpha_x^{(T)} = 0$	ZTE (non-switchable)
	$\alpha_y^{(T)} = +$	$\alpha_y^{(T)} = -$	*Sign switchable*
	$v_{yx} = 0$	$v_{yx} = 0$	ZPR (non-switchable)

The various thermal expansion and Poisson's ratio signs for the class of metamaterials discussed in Sect. 18.2 are summarized in Table 18.5. It can thus be concluded that due to the consistent CTE signs for the P-Type and N-Type metamaterials and the *in situ* sign-flipping of the Poisson's ratio due to temperature fluctuation, this class of metamaterial offers to the composite material engineer not only the choice for assigning materials with fixed positive or negative values of CTEs but also the choice of enabling the Poisson's ratio sign to toggle in adaptation to environmental

Table 18.5 Summary of thermally deformed shapes and resulting properties of the P-Type, N-Type, S-Type, and combined P and N-Type metamaterials based on originally square cells

Metamaterial (initial cell shape)	Decreasing temperature	Increasing temperature
P-Type	• Contracts into interconnected stars network • Positive thermal expansion (PTE) • Auxetic	• Expands into interconnected octagons network • Positive thermal expansion (PTE) • Non-auxetic
N-Type	• Expands into interconnected octagons network • Negative thermal expansion (NTE) • Non-auxetic	• Contracts into interconnected stars network • Negative thermal expansion (NTE) • Auxetic
S-Type	• Negative shearing • NTE and PTE in 45° and –45° directions, respectively	• Positive shearing • PTE and NTE in 45° and –45° directions, respectively
P&N-Type	• P-Type and N-Type cells contract and expand, respectively. Overall contraction • PTE	• P-Type and N-Type cells expand and contract, respectively. Overall contraction • NTE

changes. While the metamaterials in this chapter includes discussion on sign-toggling of Poisson's ratio in response to thermal changes, the next chapter explores the sign-toggling of thermal and other environmental expansion coefficients in metamaterials that imply persistently negative Poisson's ratio (Lim 2019c, 2020c).

References

Attard D, Grima JN (2011) Modelling of hexagonal honeycombs exhibiting zero Poisson's ratio. Phys Status Solidi B 248(1):52–59

Grima JN, Oliveri L, Attard D, Ellul B, Gatt R, Cicala G, Recca G (2010) Hexagonal honeycombs with zero Poisson's ratios and enhanced stiffness. Adv Eng Mater 12(9):855–862

Ha CS, Hestekin E, Li J, Plesha ME, Lakes RS (2015) Controllable thermal expansion of large magnitude in chiral negative Poisson's ratio lattices. Phys Status Solidi B 252(7):1431–1434

Li D, Ma J, Dong L, Lakes RS (2016) A bi-material structure with Poisson's ratio tunable from positive to negative via temperature control. Mater Lett 181:285–288

Lim TC (2005) Anisotropic and negative thermal expansion behavior in a cellular microstructure. J Mater Sci 40(12):3275–3277

Lim TC (2019a) A class of shape-shifting composite metamaterial honeycomb structures with thermally-adaptive Poisson's ratio signs. Compos Struct 226:111256

Lim TC (2019b) 2D metamaterial with in-plane positive and negative thermal expansion and thermal shearing based on interconnected alternating bimaterials. Mater Res Express 6(11):115804

Lim TC (2019c) Composite metamaterial with sign-switchable coefficients of hygroscopic, thermal and pressure expansions. Adv Compos Hybrid Mater 2(4):657–669

Lim TC (2020a) Composite metamaterial square grids with sign-flipping expansion coefficients leading to a type of Islamic design. SN Appl Sci 2(5):918

Lim TC (2020b) Metacomposite structure with sign-changing coefficients of hygrothermal expansions inspired by Islamic motif. Compos Struct 251:112660

Lim TC (2020c) Metacomposite with auxetic and in situ sign reversible thermal expansivity upon temperature fluctuation. Compos Commun 19:30–36

Lim TC (2021) An exact deflection solution to a type of cantilever with partially built-in end using strong boundary conditions. Int J Mech Eng Educ (in press). https://doi.org/10.1177/0306419019851363

Timoshenko SP (1925) Analysis of bi-metal thermostats. J Opt Soc Am 11(3):233–255

Chapter 19
Sign-Switching of Expansion Coefficients with Auxetic Behavior

Abstract This chapter considers two examples of microstructures with sign-switching coefficients of expansion by microstructural reshape. However, the general shape of the environmentally deformed microstructure bears certain similarities regardless of whether the environmental (thermal, pressure, moisture concentration) changes are positive or negative. The first example is based on pin-jointed truss system that exhibits in-plane isotropy, while the second example is based on hybrid bimaterial strip and pin-jointed truss system that gives in-plane anisotropy. In both examples, the in-plane environmental strain is negative regardless of increasing or decreasing environmental parameters. Some auxetic aspects are observed for both microstructures. In the first example, the driven or secondary cells are analogous to the rotating squares while the driving or primary cells are analogous to the empty spaces between the squares. In the second example, the environmentally deformed microstructure bear resemblance with the rotating squares model, anti-tetrachiral model, and the instability-induced auxeticity of square grids.

Keywords Compressibility · Poisson's ratio · Moisture expansion · Thermal expansion · Sign-flipping · Sign-reversible · Sign-switching · Sign-toggling

19.1 A Metamaterial with Sign-Switching Expansion Coefficient, In-Plane Isotropy, and Apparent Rotating Squares

This section demonstrates a lightweight material such that by making judicious modifications to its 2D truss microstructural system with pin-jointed rods, sign-switching occurs in its CTE, CME, and compressibility by means of microstructural reshape, i.e.,

- $\alpha_x^{(T)} = \alpha_y^{(T)} < 0$ for $\mathrm{d}T > 0$ but $\alpha_x^{(T)} = \alpha_y^{(T)} > 0$ for $\mathrm{d}T < 0$
- $\alpha_x^{(C)} = \alpha_y^{(C)} < 0$ for $\mathrm{d}C > 0$ but $\alpha_x^{(C)} = \alpha_y^{(C)} > 0$ for $\mathrm{d}C < 0$
- $\alpha_x^{(P)} = \alpha_y^{(P)} > 0$ for $\mathrm{d}P > 0$ but $\alpha_x^{(P)} = \alpha_y^{(P)} < 0$ for $\mathrm{d}P < 0$

A pin-jointed 2D truss microstructure that can achieve such sign-switchable expansion coefficient is furnished in Fig. 19.1 (center), which consists of side rods

T. Lim, *Mechanics of Metamaterials with Negative Parameters*, Engineering Materials,
https://doi.org/10.1007/978-981-15-6446-8_19

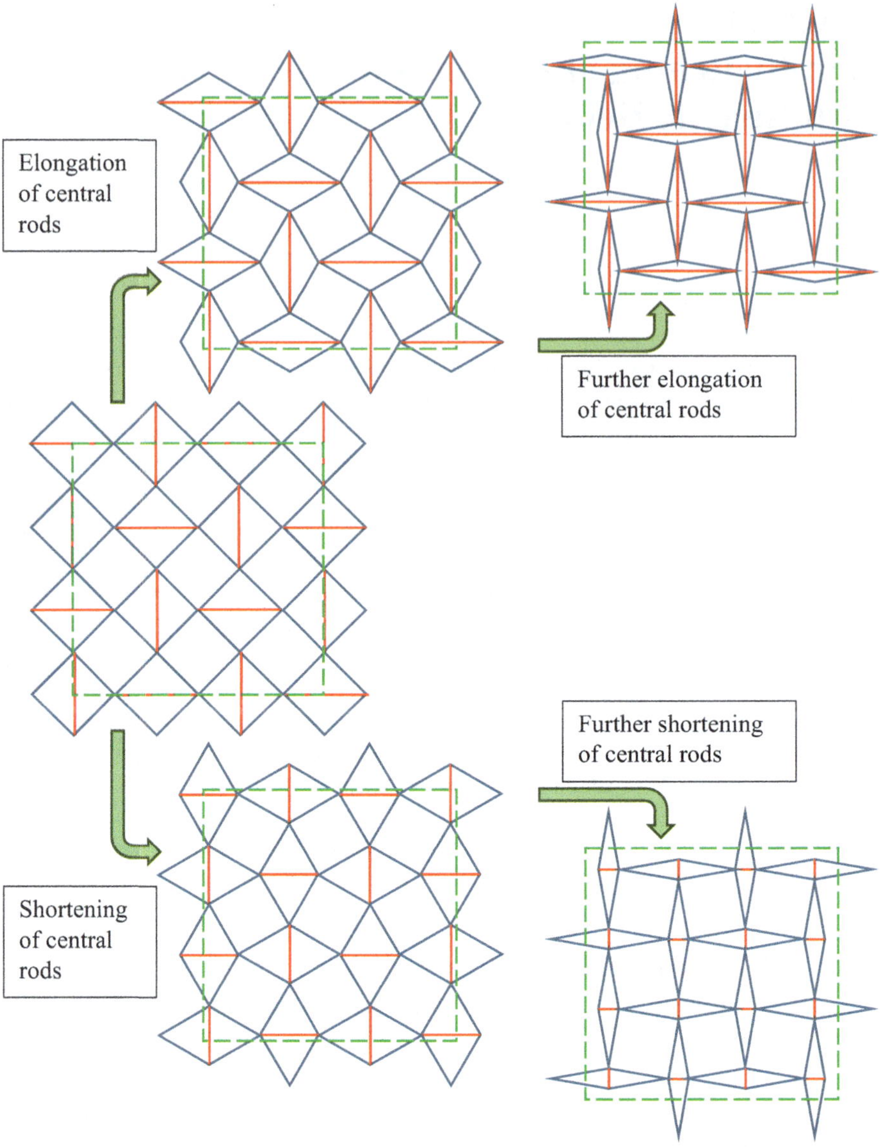

Fig. 19.1 Microstructural evolution in terms of elongation and shortening of central rods as a result of environmental changes, with fixed side rod length, leading to overall contraction of the microstructure. Dashed squares aid visual comparison of final size to the original size

(indicated in blue) and central rods (denoted as red) that are arranged into primary and secondary square cells. Each primary cell consists of 4 side rods that form a square with a central rod attached diagonally to the square, while the secondary cell has no central rod. Suppose the side rods are rigid while the central rods are free to elongate or shorten, then under the influence of environmental fluctuations the primary squares transform into rhombic cells while the secondary cells retain their size and shape. Specifically, elongation and shortening of the central rods cause them to form the major and minor axes, respectively, of the rhombic cells, as indicated in Fig. 19.1 (top and bottom). Either way, the area contraction of the primary cells reduces the overall area of the 2D truss system, thereby leading to negative in-plane isotropic strain. The assumption of rigid side rods is a special case rather than a generic one. The negativity of the area strain diminishes if the side rods are expandable. Under the more generic condition non-rigid side rods, the effective in-plane strain, and hence the effective expansion coefficient(s), are ascertained by analysis. The following analysis considers moisture expansion in the first instance before simplifying to thermal expansion and compressibility considerations. The reason for this is that the change in the rods' moisture concentration is not equal to the change in the environmental moisture concentration, and that the differing moisture absorptivity leads to different moisture concentrations for dissimilar materials, whereas the temperature of dissimilar rod materials tends to the environmental temperature at equilibrium while the pressure experienced by the dissimilar rods materials are inherently that of the environmental pressure.

We begin the analysis with the assertion that both the central and side rods possess positive moisture expansion, i.e., these rods elongate and shorten with the absorption and dissipation of moisture, respectively. Let the original half-length of the central rod l_{c0} be l_0 and the original length of the side rod l_{s0} be $\sqrt{2}l_0$, as shown in Fig. 19.2. Recall that the moisture concentration in solids is defined as $C = 100$ m/M where $m =$ moisture mass in the solid while $M =$ mass of the dry solid. Hence, a change in moisture concentration in the solid is $\mathrm{d}C = 100\ \mathrm{d}m/M$. Extending the definitions of C and $\mathrm{d}C$ for the moisture concentration and its change in the environment, it follows that m/M refers to the ratio of moisture mass to the dry air mass for the same control volume while $\mathrm{d}m/M$ indicates the corresponding change in moisture mass to dry air mass ratio.

Arising from a change of moisture concentration $\mathrm{d}C_{\mathrm{env}}$ in the environment, which results in changes of moisture concentration in the central rods $\mathrm{d}C_{\mathrm{c}}$ and in the side rods $\mathrm{d}C_{\mathrm{s}}$, we have the incremental moisture strains in the central rods

$$\mathrm{d}\varepsilon_{\mathrm{c}}^{(C)} = \frac{\mathrm{d}l_{\mathrm{c}}}{l_{\mathrm{c}}} = \alpha_{\mathrm{c}}^{(C)}\mathrm{d}C_{\mathrm{c}} \tag{19.1.1}$$

and in the side rods

$$\mathrm{d}\varepsilon_{\mathrm{s}}^{(C)} = \frac{\mathrm{d}l_{s}}{l_{\mathrm{s}}} = \alpha_{\mathrm{s}}^{(C)}\mathrm{d}C_{\mathrm{s}} \tag{19.1.2}$$

Fig. 19.2 Isolation of two units of primary cells for analyzing the distance between their centroids before (top) and after (bottom) the central and side rods experience moisture strains

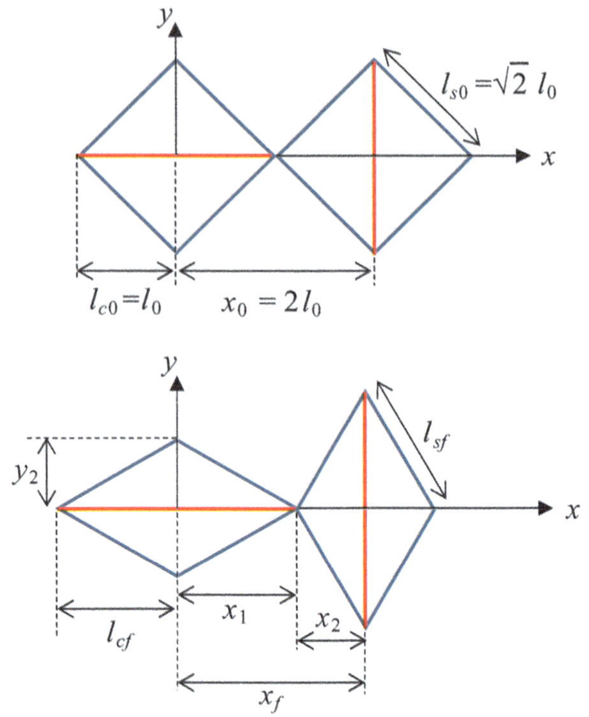

where dC_c and dC_s are the incremental changes of C_c and C_s, respectively, while $\alpha_c^{(C)}$ and $\alpha_s^{(C)}$ denote the coefficients of moisture expansion for the central and side rods, respectively. Taking integral to obtain finite strains for considerable changes of moisture concentrations in the rods gives the following moisture strains

$$\varepsilon_c^{(C)} = \ln \frac{dl_{cf}}{l_{c0}} = \alpha_c^{(C)} \Delta C_c \tag{19.1.3}$$

and

$$\varepsilon_s^{(C)} = \ln \frac{dl_{sf}}{l_{s0}} = \alpha_s^{(C)} \Delta C_s \tag{19.1.4}$$

where l_{cf} = final half-length of the central rod, while l_{sf} = final length of the side rod. The final lengths are thus

$$l_{cf} = l_{c0} e^{\alpha_c^{(C)} \Delta C_c} = x_1 \tag{19.1.5}$$

and

$$l_{\mathrm{sf}} = l_{\mathrm{s0}} \mathrm{e}^{\alpha_{\mathrm{s}}^{(C)} \Delta C_{\mathrm{s}}} \tag{19.1.6}$$

By symmetry consideration $y_2 = x_2$,

$$x_2 = \sqrt{l_{\mathrm{s0}}^2 \mathrm{e}^{2\alpha_{\mathrm{s}}^{(C)} \Delta C_{\mathrm{s}}} - l_{\mathrm{c0}}^2 \mathrm{e}^{2\alpha_{\mathrm{c}}^{(C)} \Delta C_{\mathrm{c}}}} \tag{19.1.7}$$

so that

$$x_{\mathrm{f}} = l_{\mathrm{c0}} \mathrm{e}^{\alpha_{\mathrm{c}}^{(C)} \Delta C_{\mathrm{c}}} + \sqrt{l_{\mathrm{s0}}^2 \mathrm{e}^{2\alpha_{\mathrm{s}}^{(C)} \Delta C_{\mathrm{s}}} - l_{\mathrm{c0}}^2 \mathrm{e}^{2\alpha_{\mathrm{c}}^{(C)} \Delta C_{\mathrm{c}}}} \tag{19.1.8}$$

We note that the in-plane strain is isotropic $\varepsilon_x = \varepsilon_y$, and hence it is sufficient to employ only one direction to represent the effective strain

$$\varepsilon_{\mathrm{eff}}^{(C)} = \ln \frac{x_f}{x_0} = \ln \frac{l_{\mathrm{c0}} \mathrm{e}^{\alpha_{\mathrm{c}}^{(C)} \Delta C_{\mathrm{c}}} + \sqrt{l_{\mathrm{s0}}^2 \mathrm{e}^{2\alpha_{\mathrm{s}}^{(C)} \Delta C_{\mathrm{s}}} - l_{\mathrm{c0}}^2 \mathrm{e}^{2\alpha_{\mathrm{c}}^{(C)} \Delta C_{\mathrm{c}}}}}{2 l_0} \tag{19.1.9}$$

Substitution of $l_{\mathrm{c0}} = l_0$ and $l_{\mathrm{s0}} = \sqrt{2} l_0$ leads to

$$\varepsilon_{\mathrm{eff}}^{(C)} = \ln \left\{ \frac{1}{2} e^{\alpha_{\mathrm{c}}^{(C)} \Delta C_{\mathrm{c}}} + \frac{1}{2} \sqrt{2 \mathrm{e}^{2\alpha_{\mathrm{s}}^{(C)} \Delta C_{\mathrm{s}}} - e^{2\alpha_{\mathrm{c}}^{(C)} \Delta C_{\mathrm{c}}}} \right\} \tag{19.1.10}$$

While Eq. (19.1.10) expresses the effective moisture strain in terms of moisture strains of the central rods $\alpha_{\mathrm{c}}^{(C)} \Delta C_{\mathrm{c}}$ and of the side rods $\alpha_{\mathrm{s}}^{(C)} \Delta C_{\mathrm{s}}$ it is desirable to also express it as an effective coefficient of moisture expansion $\alpha_{\mathrm{eff}}^{(C)}$ by the relation $\varepsilon_{\mathrm{eff}}^{(C)} = \alpha_{\mathrm{eff}}^{(C)} \Delta C_{\mathrm{env}}$

$$\alpha_{\mathrm{eff}}^{(C)} = \frac{1}{\Delta C_{\mathrm{env}}} \ln \left\{ \frac{1}{2} e^{\alpha_{\mathrm{c}}^{(C)} \Delta C_{\mathrm{c}}} + \frac{1}{2} \sqrt{2 e^{2\alpha_{\mathrm{s}}^{(C)} \Delta C_{\mathrm{s}}} - e^{2\alpha_{\mathrm{c}}^{(C)} \Delta C_{\mathrm{c}}}} \right\} \tag{19.1.11}$$

To preserve dimensionless form, it is useful—and physically meaningful—to take Eq. (19.1.11) as a ratio of the central rod's coefficient of moisture expansion. This permits the non-dimensionalized effective coefficient of moisture expansion (Lim 2019a)

$$\frac{\alpha_{\mathrm{eff}}^{(C)}}{\alpha_{\mathrm{c}}^{(C)}} = \frac{1}{\alpha_{\mathrm{c}}^{(C)} \Delta C_{\mathrm{c}}} \left(\frac{\Delta C_{\mathrm{c}}}{\Delta C_{\mathrm{env}}} \right) \ln \left\{ \frac{1}{2} e^{\alpha_{\mathrm{c}}^{(C)} \Delta C_{\mathrm{c}}} + \frac{1}{2} \sqrt{2 e^{2\alpha_{\mathrm{s}}^{(C)} \Delta C_{\mathrm{s}}} - e^{2\alpha_{\mathrm{c}}^{(C)} \Delta C_{\mathrm{c}}}} \right\} \tag{19.1.12}$$

to be expressed in dimensionless groups, specifically, the ratio of moisture concentration change in the central rods to that in the environment $\Delta C_{\mathrm{c}} / \Delta C_{\mathrm{env}}$, as well as the moisture strains in the central rods $\alpha_{\mathrm{c}}^{(C)} \Delta C_{\mathrm{c}}$ and in the side rods $\alpha_{\mathrm{s}}^{(C)} \Delta C_{\mathrm{s}}$. Although it has been earlier stated that the rods' coefficients of moisture expansions $\alpha_{\mathrm{c}}^{(C)}$ and $\alpha_{\mathrm{s}}^{(C)}$ are non-negative, the corresponding strains can be both positive and negative as a consequent of increasing and decreasing moisture concentrations.

It is beneficial at this juncture to consider some special cases for the purpose of better clarity on the influence from each parameter. For Special Case I where the side rods are rigid either due to it being non-expandable $\left(\alpha_{\mathrm{s}}^{(C)} = 0\right)$ and/or non-absorptive $(\Delta C_{\mathrm{s}} = 0)$, Eqs. (19.1.10) and (19.1.12) reduce to

$$\varepsilon_{\mathrm{eff}}^{(C)} = \ln\left\{\frac{1}{2}e^{\alpha_{\mathrm{c}}^{(C)}\Delta C_{\mathrm{c}}} + \frac{1}{2}\sqrt{2 - \mathrm{e}^{2\alpha_{\mathrm{c}}^{(C)}\Delta C_{\mathrm{c}}}}\right\} \tag{19.1.13}$$

and

$$\frac{\alpha_{\mathrm{eff}}^{(C)}}{\alpha_{\mathrm{c}}^{(C)}} = \frac{1}{\alpha_{\mathrm{c}}^{(C)}\Delta C_{\mathrm{c}}}\left(\frac{\Delta C_{\mathrm{c}}}{\Delta C_{\mathrm{env}}}\right)\ln\left\{\frac{1}{2}\mathrm{e}^{\alpha_{\mathrm{c}}^{(C)}\Delta C_{\mathrm{c}}} + \frac{1}{2}\sqrt{2 - \mathrm{e}^{2\alpha_{\mathrm{c}}^{(C)}\Delta C_{\mathrm{c}}}}\right\} \tag{19.1.14}$$

respectively. For Special Case II where both the central and side rods encounter common change in moisture concentration $\Delta C_{\mathrm{c}} = \Delta C_{\mathrm{s}}$, Eqs. (19.1.10) and (19.1.12) simplify to

$$\varepsilon_{\mathrm{eff}}^{(C)} = \ln\left\{\frac{1}{2}e^{\alpha_{\mathrm{c}}^{(C)}\Delta C_{\mathrm{c}}} + \frac{1}{2}\sqrt{2\mathrm{e}^{2\alpha_{\mathrm{c}}^{(C)}\Delta C_{\mathrm{c}}\left(\alpha_{\mathrm{s}}^{(C)}/\alpha_{\mathrm{c}}^{(C)}\right)} - \mathrm{e}^{2\alpha_{\mathrm{c}}^{(C)}\Delta C_{\mathrm{c}}}}\right\} \tag{19.1.15}$$

and

$$\frac{\alpha_{\mathrm{eff}}^{(C)}}{\alpha_{\mathrm{c}}^{(C)}} = \frac{1}{\alpha_{\mathrm{c}}^{(C)}\Delta C_{\mathrm{c}}}\left(\frac{\Delta C_{\mathrm{c}}}{\Delta C_{\mathrm{env}}}\right)\ln\left\{\frac{1}{2}\mathrm{e}^{\alpha_{\mathrm{c}}^{(C)}\Delta C_{\mathrm{c}}} + \frac{1}{2}\sqrt{2\mathrm{e}^{2\alpha_{\mathrm{c}}^{(C)}\Delta C_{\mathrm{c}}\left(\alpha_{\mathrm{s}}^{(C)}/\alpha_{\mathrm{c}}^{(C)}\right)} - \mathrm{e}^{2\alpha_{\mathrm{c}}^{(C)}\Delta C_{\mathrm{c}}}}\right\} \tag{19.1.16}$$

respectively, so as to facilitate observation on the effects from central rod's moisture strain $\alpha_{c}^{(C)}\Delta C_{c}$ and the side-to-central rod CME ratio $\alpha_{\mathrm{s}}^{(C)}/\alpha_{\mathrm{c}}^{(C)}$. These special cases are useful for visualizing the effects from the various dimensionless groups.

The developed effective moisture strain and dimensionless coefficient of moisture expansion laid out in Eqs. (19.1.10) and (19.1.12), respectively, can be converted for use under the influence of changing temperature as well as pressure. It should be noted that the moisture concentration in a solid is not equal to that in the environment; for multiconstituent truss microstructure considered herein, the moisture concentrations and their changes in the central and side rods are, in general, not equal. More importantly, the changes in moisture concentration in the environment and in the rods of different materials are distinct in general. This is not so in the case when considering thermal strain, whereby temperature change in the environment is the same as that in the solids at thermal equilibrium. As such, performing the following substitutions on Eqs. (19.1.10) and (19.1.12)

$$\begin{Bmatrix} \alpha_{\mathrm{eff}}^{(C)} \\ \alpha_c^{(C)} \\ \alpha_s^{(C)} \\ \Delta C_{\mathrm{env}} \\ \Delta C_{\mathrm{c}} \\ \Delta C_{\mathrm{s}} \end{Bmatrix} \mapsto \begin{Bmatrix} \alpha_{\mathrm{eff}}^{(T)} \\ \alpha_{\mathrm{c}}^{(T)} \\ \alpha_{\mathrm{s}}^{(T)} \\ \Delta T \\ \Delta T \\ \Delta T \end{Bmatrix} \tag{19.1.17}$$

where $\alpha_{\mathrm{eff}}^{(T)}$, $\alpha_{\mathrm{c}}^{(T)}$, and $\alpha_{\mathrm{s}}^{(T)}$ are the coefficients of thermal expansion for the microstructure, the central rods, and the side rods, respectively, and recognizing that ΔT is the common temperature change in the environment and in all rods, we have the effective thermal strain on the microstructure

$$\varepsilon_{\mathrm{eff}}^{(T)} = \ln\left\{\frac{1}{2}e^{\alpha_{\mathrm{c}}^{(T)}\Delta T} + \frac{1}{2}\sqrt{2e^{2\alpha_{\mathrm{s}}^{(T)}\Delta T} - e^{2\alpha_{\mathrm{c}}^{(T)}\Delta T}}\right\} \tag{19.1.18}$$

and the corresponding dimensionless effective coefficient of thermal expansion

$$\frac{\alpha_{\mathrm{eff}}^{(T)}}{\alpha_c^{(T)}} = \frac{1}{\alpha_{\mathrm{c}}^{(T)}\Delta T}\ln\left\{\frac{1}{2}e^{\alpha_{\mathrm{c}}^{(T)}\Delta T} + \frac{1}{2}\sqrt{2e^{2\alpha_{\mathrm{s}}^{(T)}\Delta T} - e^{2\alpha_{\mathrm{c}}^{(T)}\Delta T}}\right\} \tag{19.1.19}$$

respectively.

A similar analogy can be made for the case of compressibility. Recall that the pressure-induced volumetric strain is $\varepsilon_{\mathrm{V}}^{(P)} = -\alpha_{\mathrm{V}}^{(P)}\Delta P$ where $\alpha_{\mathrm{V}}^{(P)}$ is the compressibility while ΔP, the environmental pressure change, is the common pressure change experienced by both the central and side rods. Unlike responses to moisture and temperature, increasing pressure $\Delta P > 0$ induces volumetric contraction $\varepsilon_{\mathrm{V}}^{(P)} < 0$, and so a negative sign is required to preserve positive $\alpha_{\mathrm{V}}^{(P)}$ in conventional sense. Since the volumetric strain is three times the linear strain $\varepsilon_{\mathrm{V}}^{(P)} = 3\varepsilon_{\mathrm{L}}^{(P)}$ for isotropic solids under pressure change, we have the pressure-induced strain $\varepsilon_{\mathrm{L}}^{(P)} = -\alpha_{\mathrm{V}}^{(P)}\Delta P/3$. However, the considered metamaterial is 2D and so $\alpha_{\mathrm{V}}^{(P)}$ is undefined for the present analysis. As such, a linear compressibility $\alpha_{\mathrm{L}}^{(P)}$ is introduced such that the pressure-induced strain is $\varepsilon_{\mathrm{L}}^{(P)} = -\alpha_{\mathrm{L}}^{(P)}\Delta P$, so as to permit the following substitutions

$$\begin{Bmatrix} \alpha_{\mathrm{eff}}^{(C)} \\ \alpha_{\mathrm{c}}^{(C)} \\ \alpha_{\mathrm{s}}^{(C)} \\ \Delta C_{\mathrm{env}} \\ \Delta C_{\mathrm{c}} \\ \Delta C_{\mathrm{s}} \end{Bmatrix} \to \begin{Bmatrix} \alpha_{\mathrm{eff}}^{(P)} \\ \alpha_{\mathrm{c}}^{(P)} \\ \alpha_{\mathrm{s}}^{(P)} \\ -\Delta P \\ -\Delta P \\ -\Delta P \end{Bmatrix} \tag{19.1.20}$$

on Eqs. (19.1.10) and (19.1.12) to yield the pressure-induced effective strain on the microstructure

$$\varepsilon_{\text{eff}}^{(P)} = \ln\left\{\frac{1}{2}\mathrm{e}^{-\alpha_c^{(P)}\Delta P} + \frac{1}{2}\sqrt{2\mathrm{e}^{-2\alpha_s^{(P)}\Delta P} - \mathrm{e}^{-2\alpha_c^{(P)}\Delta P}}\right\} \tag{19.1.21}$$

and the corresponding dimensionless effective compressibility

$$\frac{\alpha_{\text{eff}}^{(P)}}{\alpha_c^{(P)}} = \frac{1}{-\alpha_c^{(P)}\Delta P}\ln\left\{\frac{1}{2}\mathrm{e}^{-\alpha_c^{(P)}\Delta P} + \frac{1}{2}\sqrt{2\mathrm{e}^{-2\alpha_s^{(P)}\Delta P} - \mathrm{e}^{-2\alpha_c^{(P)}\Delta P}}\right\} \tag{19.1.22}$$

respectively.

Since the changes in the environment's moisture concentration, temperature, and pressure take place non-monotonically, the overall effect from these environmental fluctuations can be accounted for by taking superposition of the effective strains arising from each environmental change $\varepsilon_{\text{eff}} = \varepsilon_{\text{eff}}^{(C)} + \varepsilon_{\text{eff}}^{(T)} + \varepsilon_{\text{eff}}^{(P)}$, i.e.,

$$\begin{aligned}\varepsilon_{\text{eff}} = \ln\Bigg\{&\left(\frac{\mathrm{e}^{\alpha_c^{(C)}\Delta C_c}}{2} + \sqrt{\frac{\mathrm{e}^{2\alpha_s^{(C)}\Delta C_s}}{2} - \frac{\mathrm{e}^{2\alpha_c^{(C)}\Delta C_c}}{4}}\right)\\ &\left(\frac{\mathrm{e}^{\alpha_c^{(T)}\Delta T}}{2} + \sqrt{\frac{\mathrm{e}^{2\alpha_s^{(T)}\Delta T}}{2} - \frac{\mathrm{e}^{2\alpha_c^{(T)}\Delta T}}{4}}\right)\left(\frac{\mathrm{e}^{-\alpha_c^{(P)}\Delta P}}{2} + \sqrt{\frac{\mathrm{e}^{-2\alpha_s^{(P)}\Delta P}}{2} - \frac{\mathrm{e}^{-2\alpha_c^{(P)}\Delta P}}{4}}\right)\Bigg\}\end{aligned} \tag{19.1.23}$$

which is simplified to

$$\begin{aligned}\varepsilon_{\text{eff}} = \ln\Bigg\{&\left(\frac{\mathrm{e}^{\alpha_c^{(C)}\Delta C_c}}{2} + \sqrt{\frac{1}{2} - \frac{\mathrm{e}^{2\alpha_c^{(C)}\Delta C_c}}{4}}\right)\left(\frac{e^{\alpha_c^{(T)}\Delta T}}{2} + \sqrt{\frac{1}{2} - \frac{\mathrm{e}^{2\alpha_c^{(T)}\Delta T}}{4}}\right)\\ &\left(\frac{\mathrm{e}^{-\alpha_c^{(P)}\Delta P}}{2} + \sqrt{\frac{1}{2} - \frac{\mathrm{e}^{-2\alpha_c^{(P)}\Delta P}}{4}}\right)\Bigg\}\end{aligned} \tag{19.1.24}$$

for rigid and/or non-absorptive side rods.

So far the analysis has adopted straight lines to represent the distance between the hinges. In a practical sense, the sizes of the rods and their locations along the thickness direction must be defined. The rods are laid in such a manner that the central rods are placed at the middle layer; all sides' rods that are aligned in one direction are placed above the central rods while those oriented in another direction are positioned below the central rods, as shown in Fig. 19.3 (top). The sizes of the rods can be as small as that which can be achieved by the latest technology, and hence we herein discuss the length ratios as it would be more meaningful in understanding the limitations of deformation. As with beams or rods, the hinge-to-hinge lengths of l_{s0} for the side rods or $\sqrt{2}l_0$ can be made about an order higher than the width of the side rod w_0, as shown in Fig. 19.3 on the basis of Special Case I. Since the microstructure decreases in size regardless of whether the central

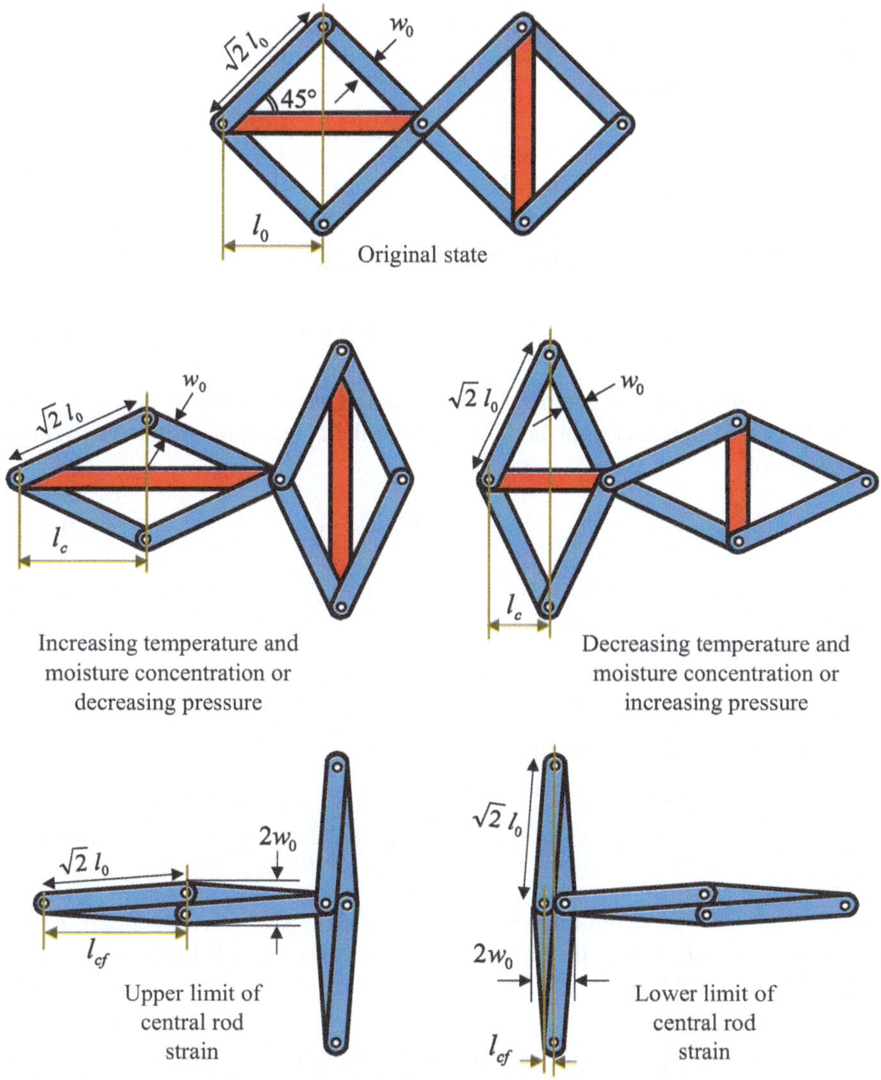

Fig. 19.3 Strains limits illustrated for Special Case I considering two units of the primary cells

rod expands and contracts, there is only one theoretical limit, which is the lower bound. The "upper bound" is formed by the original state. However, there exist two limits for the central rods. The theoretical upper limit and the theoretical lower limit or the central rods are attained under the influence of environmental change when the opposing side rods come into contact, as shown in Fig. 19.3 (bottom left) and Fig. 19.3 (bottom right), respectively. Taking the width of the side rods into account for limiting the deformation mechanism, the theoretical upper limit of the central rod strain is obtained from $l_0^2\left(1+\varepsilon_c^{\mathrm{upp}}\right)^2 = 2l_0^2 - w_0^2/4$ while the theoretical lower limit is obtained from $l_0\left(1+\varepsilon_c^{\mathrm{low}}\right) = w_0/2$ to give the theoretical bounds of the central rod strain for rigid side rods

$$\frac{1}{2}\frac{w_0}{l_0} < 1 + \varepsilon_c < \sqrt{2 - \frac{1}{4}\left(\frac{w_0}{l_0}\right)^2} \tag{19.1.25}$$

If $w_0 \ll l_0$, the range is expressed in terms of constants

$$-1 < \varepsilon_c < \left(-1 + \sqrt{2}\right) \approx 0.4142 \tag{19.1.26}$$

At either limits of the central rods, there exist two rhombi of equal size and shape, but each aligned at right angles to one another. Hence, the original horizontal dimension of

$$L_0 = 4l_0 \tag{19.1.27}$$

changes to a summation of the major and minor diameters of the fully closed rhombi

$$L_f = 2\sqrt{2l_0^2 - \frac{1}{4}w_0^2} + w_0 \tag{19.1.28}$$

to give the theoretical limiting strain

$$\varepsilon_{\mathrm{lim}} = \frac{1}{2}\sqrt{2 - \frac{1}{4}\left(\frac{w_0}{l_0}\right)^2} + \frac{1}{4}\frac{w_0}{l_0} - 1 \tag{19.1.29}$$

thereby establishing the limiting strain as a function of side rod aspect ratio w_0/l_0. Neglecting the side rod width for $w_0 \ll l_0$ simplifies Eq. (19.1.29) to a negative constant

$$\varepsilon_{\mathrm{lim}} \approx \frac{1}{\sqrt{2}} - 1 = -0.2929 \tag{19.1.30}$$

Plotted results can be made by considering the variation of the developed effective moisture strain as well as the coefficient of moisture expansion—as representations of the generic environmental strain and expansion coefficient, respectively. To clearly observe the effects from the moisture strain of the central rods, the effective moisture strain is plotted in Fig. 19.4 (left) under Special Case I, i.e., using Eq. (19.1.13), wherein the side rods' lengths are preserved either due to the their intrinsically non-expansion arising from $\alpha_s^{(C)} = 0$, or due to the absence of moisture concentration change $\Delta C_s = 0$ as a consequence of non-absorbance of environmental moisture and non-dissipation of the side rods' moisture content. The combined influence from the moisture strain of the central rods and the ratio of change in the moisture

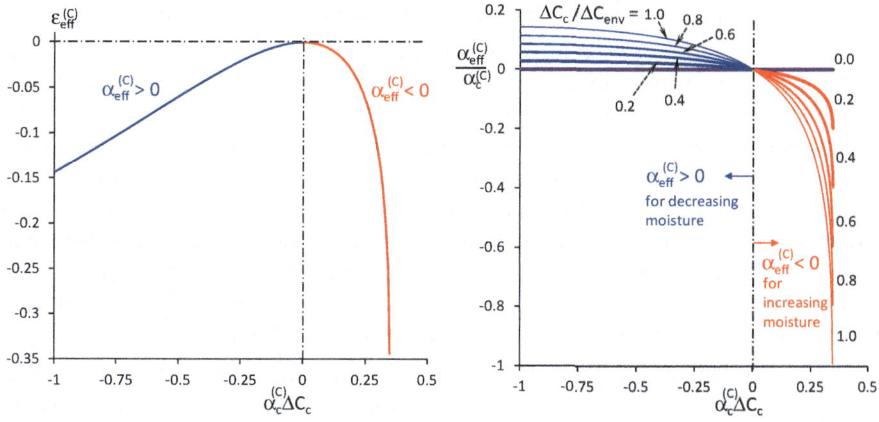

Fig. 19.4 Plots of effective moisture strain (left) and dimensionless coefficients of moisture expansion (right) as a function of central rod moisture strain for various $\Delta C_c/\Delta C_{env}$ ratio at fixed side rod length

concentration of the central rod to that of the surrounding environment $\Delta C_c/\Delta C_{env}$ on the dimensionless effective coefficient of moisture expansion $\alpha_{eff}^{(C)}/\alpha_c^{(C)}$ is plotted in Fig. 19.4 (right) under the same special case using Eq. (19.1.14).

Based on the substitutions described by Eqs. (19.1.17) and (19.1.20), Fig. 19.4 can be converted to its analogy for representing thermal and pressure fluctuations, as shown in Fig. 19.5. Specifically, the sole curve in Fig. 19.4 (left) applies for the effective thermal strain versus $\alpha_c^{(T)}\Delta T$ and the effective pressure strain versus $\alpha_c^{(P)}\Delta P$, while only the curve indicated as $\Delta C_c/\Delta C_{env} = 1$ in the plots of dimensionless

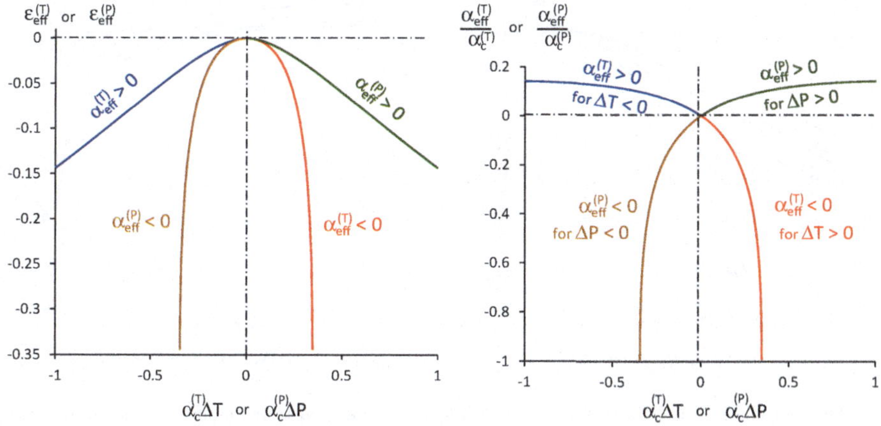

Fig. 19.5 Plots of effective thermal or pressure strains (left) and dimensionless coefficient of thermal expansion or dimensionless compressibility (right) versus central rods' thermal or pressure strains at constant side rod length

coefficient of moisture Fig. 19.4 (right) can be extended to plots of $\alpha_{\mathrm{eff}}^{(T)}/\alpha_{\mathrm{c}}^{(T)}$ and $\alpha_{\mathrm{eff}}^{(P)}/\alpha_{\mathrm{c}}^{(P)}$ versus $\alpha_{\mathrm{c}}^{(T)}\Delta T$ and $\alpha_{\mathrm{c}}^{(P)}\Delta P$, respectively, since the environmental and rod temperature changes are common at equilibrium, and that the pressure change subjected to the rods is none other than the change in environmental pressure.

Observation on the interlacing effects from the central rods' moisture strain and the ratio of the side rod to the central rod coefficients of moisture expansion $\alpha_{\mathrm{s}}^{(C)}/\alpha_{\mathrm{c}}^{(C)}$ can be made on the basis of Special Case II, i.e., using Eq. (19.1.15) wherein it is assumed that the change in moisture concentration in the central rods is that same as that in the side rods $\Delta C_{\mathrm{c}} = \Delta C_{\mathrm{s}}$. This is shown in Fig. 19.6 (left) whereby the quadrants $\alpha_{\mathrm{c}}^{(C)}\Delta C_{\mathrm{c}} < 0 \cap \varepsilon_{\mathrm{eff}}^{(C)} < 0$ and $\alpha_{\mathrm{c}}^{(C)}\Delta C_{\mathrm{c}} > 0 \cap \varepsilon_{\mathrm{eff}}^{(C)} > 0$ indicate conventional, or positive moisture expansion, regions, while the quadrant $\alpha_{\mathrm{c}}^{(C)}\Delta C_{\mathrm{c}} > 0 \cap \varepsilon_{\mathrm{eff}}^{(C)} < 0$ denotes a region of negative moisture expansion. For completeness' sake, the region $\alpha_{c}^{(C)}\Delta C_{c} < 0 \cap \varepsilon_{eff}^{(C)} > 0$ is identified as being negative moisture expansion although no curves pass through that quadrant. Within the same special case, plots of dimensionless coefficient of moisture expansion against the central rods' moisture strain for various $\alpha_{\mathrm{s}}^{(C)}/\alpha_{\mathrm{c}}^{(C)}$ ratio is shown in Fig. 19.6 (right) using Eq. (19.1.16) with an additional simplification of $\Delta C_{\mathrm{c}}/\Delta C_{\mathrm{env}} = 1$ since this ratio has been varied in Fig. 19.4 (right).

Again, by the substitutions of Eqs. (19.1.17) and (19.1.20), the various $\alpha_{\mathrm{s}}^{(C)}/\alpha_{\mathrm{c}}^{(C)}$ ratios for moisture fluctuation are valid for $\alpha_{\mathrm{s}}^{(T)}/\alpha_{\mathrm{c}}^{(T)}$ and $\alpha_{\mathrm{s}}^{(P)}/\alpha_{\mathrm{c}}^{(P)}$ ratios under environmental thermal and pressure changes, respectively. Since the plots of $\varepsilon_{\mathrm{eff}}^{(P)}$ and $\alpha_{\mathrm{eff}}^{(P)}/\alpha_{\mathrm{c}}^{(P)}$ versus $\alpha_{\mathrm{c}}^{(P)}\Delta P$ are symmetrical to those of thermal $\varepsilon_{\mathrm{eff}}^{(T)}$ and $\alpha_{\mathrm{eff}}^{(T)}/\alpha_{\mathrm{c}}^{(T)}$ versus $\alpha_{\mathrm{c}}^{(T)}\Delta T$, respectively, about $\alpha_{\mathrm{c}}^{(P)}\Delta P = \alpha_{\mathrm{c}}^{(T)}\Delta T = 0$ in Fig. 19.5, the plots in Fig. 19.6 for various $\alpha_{s}^{(T)}/\alpha_{c}^{(T)}$ and $\alpha_{\mathrm{s}}^{(P)}/\alpha_{\mathrm{c}}^{(P)}$ ratios can be coincided by using

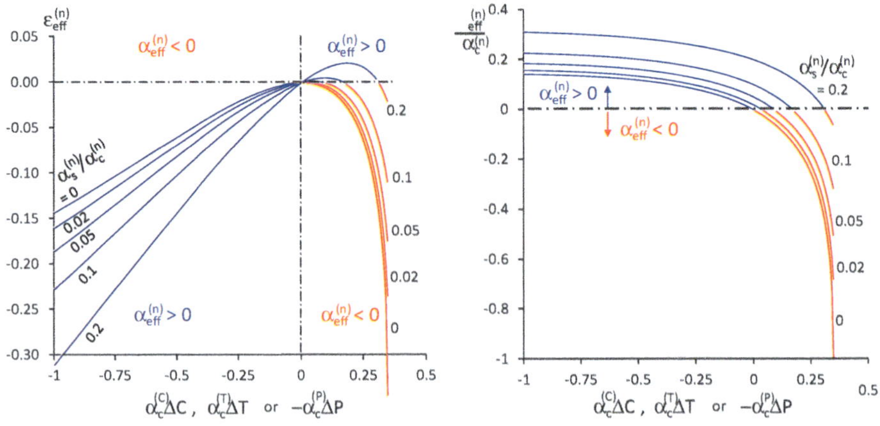

Fig. 19.6 Plots of effective environmental strains (left) and their dimensionless expansion coefficients for various $\alpha_{s}^{(n)}/\alpha_{\mathrm{c}}^{(n)}$ ratios (right) as a function of central rods' environmental strains, where n = moisture concentration (C), temperature (T), or pressure (P). For $n = C$, the condition of $\Delta C_{\mathrm{c}} = \Delta C_{\mathrm{env}}$ is imposed to correspond with $n = T, P$

$-\alpha_{\mathrm{c}}^{(P)}\Delta P$ instead of $\alpha_{\mathrm{c}}^{(P)}\Delta P$ as the abscissa in addition to $\alpha_{\mathrm{c}}^{(T)}\Delta T$. Perusal to Fig. 19.6 also shows that

$$\left.\frac{\partial \varepsilon_{\mathrm{eff}}^{(n)}}{\partial \varepsilon_{\mathrm{c}}^{(n)}}\right|_{\varepsilon_{\mathrm{c}}^{(n)}=0} = 0 \Leftarrow \alpha_{\mathrm{s}}^{(n)} = 0 \tag{19.1.31}$$

for $n = C, T, P$, thereby suggesting that the effective environmental strain is negligible, and hence the environmental stress under constrained system is small, for moderate change in environmental condition if the side rods are rigid.

Having shown that the sign-switchable effective coefficient of moisture expansion can be converted into effective coefficient of thermal expansion and effective compressibility, the considered microstructure also exhibits negative Poisson's ratio. Recall that the microstructure consists of primary and secondary cells. In the absence of environmental changes, the presence of central rods preserves the size and shape of the primary cells while the secondary cells are subjected to changes in shape and area due to the absence of central rods. Therefore, in the absence of environmental changes the primary cells behave as rigid squares whose rotation gives rise to auxetic, or negative Poisson's ratio, behavior with isotropic in-plane Poisson's ratio of -1 (Grima and Evans 2000). In the presence of environmental changes, the primary cells become rhombi so as to give Type α rotating rhombi (Grima et al. 2015). In the absence of mechanical loading, the change in environmental condition alters the primary cells into rhombi while the secondary cells rotate but retain their square shapes, such that the secondary and primary cells are analogous to the rigid rotating squares and their adjacent empty spaces, respectively.

19.2 A Metamaterial with Sign-Switching Expansion Coefficient, In-Plane Anisotropy, and Analogous to Various Auxetic Systems

This section deals with a metamaterial that exhibits positive thermal expansion (PTE) upon cooling but NTE upon heating, as with the previous section. However, regardless of whether the temperature increases or decreases, the negativity of Poisson's ratio v_{xy} increases with the magnitude of the temperature change. In particular, the currently proposed metamaterial exhibits the following: (a) $\alpha_x^{(T)} = \alpha_y^{(T)} = 0$ and $v_{xy} \approx -\frac{1}{3}$ under buckling load for $\mathrm{d}T \approx 0$, (b) $\alpha_x^{(T)}, \alpha_y^{(T)} < 0$ for $\mathrm{d}T > 0$ but $\alpha_x^{(T)}, \alpha_y^{(T)} > 0$ for $\mathrm{d}T < 0$, and (c) $v_{xy} < 0$ for $\mathrm{d}T \neq 0$.

Although discussed in terms of thermal change and CTE, the developed model herein can be extended to pressure change and compressibility, as well as moisture concentration change and CME. It suffices to discuss in terms of thermal change and CTE, as conversion to the other two environmental changes can be implemented as discussed in the previous section.

The considered metamaterial displays biaxial CTE sign-switching with in-plane anisotropy. Such a metamaterial is illustrated in Fig. 19.7 (top) in its original state for 3 by 4 unit cells, consisting of rigid rods (shown in black) oriented parallel to the x-axis while the bimaterials are aligned in the y-direction. The rigid rods are connected by freely rotating pin joints while the ends of each bimaterials are built-in, or fixed joints, into the mid-span of the rigid rods. Suppose the phases with the higher and lower CTEs in Fig. 19.7 are indicated in red and blue, respectively, then a decrease and increase in temperature result in the microstructural shapes shown in Fig. 19.7 (bottom left) and Fig. 19.7 (bottom right), respectively. Therefore, the original rectangle ABCD, indicated by the yellow line in Fig. 19.7 (top), shrinks to A′B′C′D′, denoted by dashed green lines in Fig. 19.7 (bottom). This A′B′C′D′ rectangle is also reproduced in Fig. 19.7 (top) for convenient comparison with the original ABCD before thermal fluctuation.

Since the metamaterial reduces in size with thermal fluctuation, it exhibits PTE during cooling but reverses to NTE upon heating, i.e., sign-switching of its CTE with thermal fluctuation (Lim 2020). In addition, perusal to Fig. 19.7 (bottom) suggests that the application of tensile stress in the x-direction re-orientates the zigzag rigid rods back to their horizontal alignment. The resulting bending moment on the bimaterial strips causes them to be straightened, thereby producing positive strain in the y-direction, which leads to negative Poisson's ratio. The subsequent analysis consists of CTE analysis upon temperature change, followed by Poisson's ratio analysis arising from the thermally deformed microstructure.

Perusal to Fig. 19.8 (top left) shows a unit cell of the metamaterial in its original, or rectangular, shape made from rigid rods of effective length W measured from hinge-to-hinge, and bimaterial strips of full length l_f and total thickness h each, with its ends built into the mid-span of the rigid rods. Since the hinge-to-hinge distance is W before deformation, the width of the unit cell, as measured from the mid-plane of a bimaterial to the mid-plane of its nearest bimaterial, is

$$x_0 = W \tag{19.2.1}$$

As the ends of each bimaterial are built into the mid-span of the rigid rods of thickness t, the effective height of each unit cell is

$$y_0 = 2l_h + t \tag{19.2.2}$$

where $l_h = l_f/2$ is the half-length of the bimaterial, as indicated in Fig. 19.8 (bottom left)

Arising from a change in temperature $\mathrm{d}T$, a straight bimaterial cantilever of length l_h and total thickness $h = h_1 + h_2$ undergo thermal deformation to form a curve of curvature $1/r$. Due to the recurrence of l_h/r in this example, it is expedient to adopt a dimensionless form of the curvature

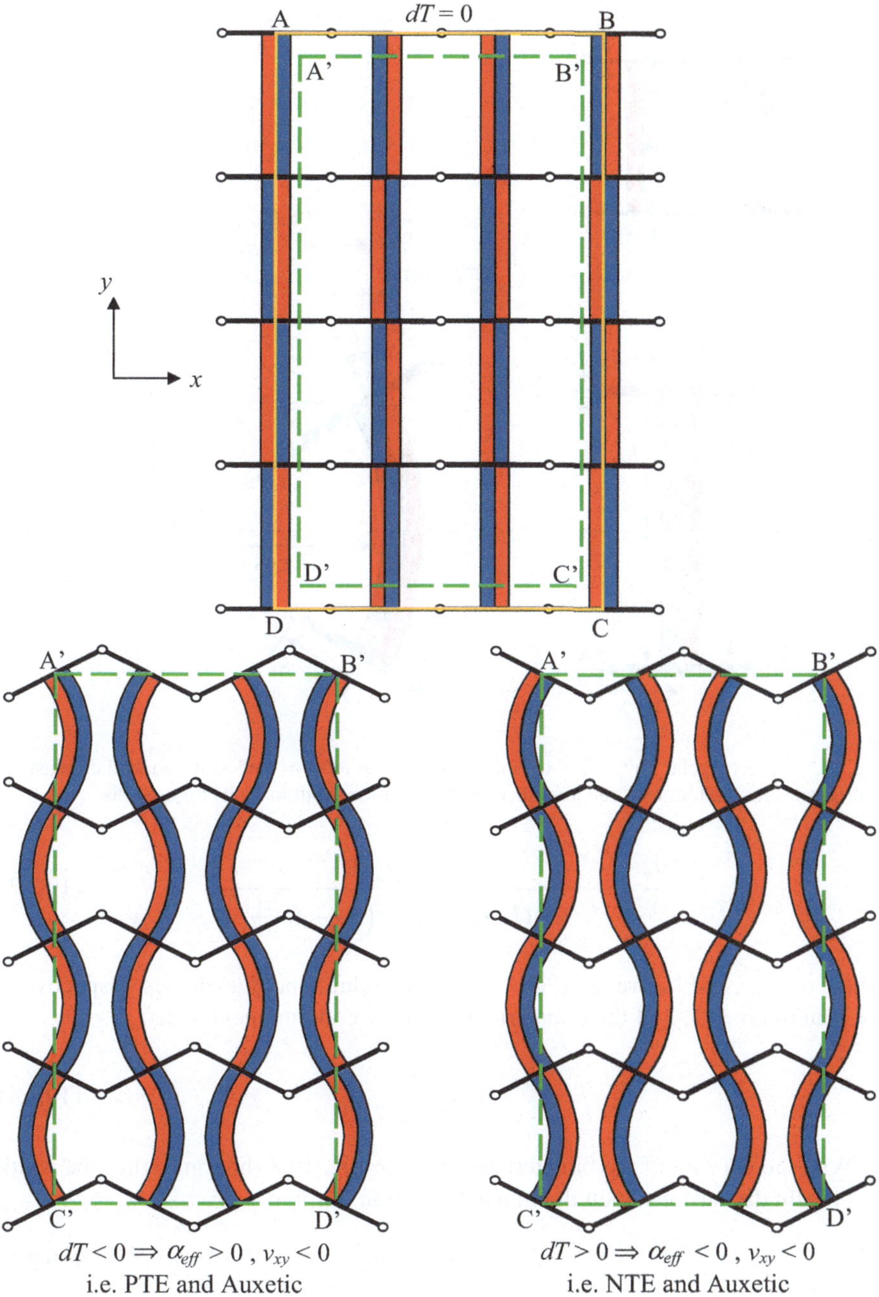

Fig. 19.7 Considered metamaterial in its original state (top) and after temperature decrease (bottom left) and increase (bottom right)

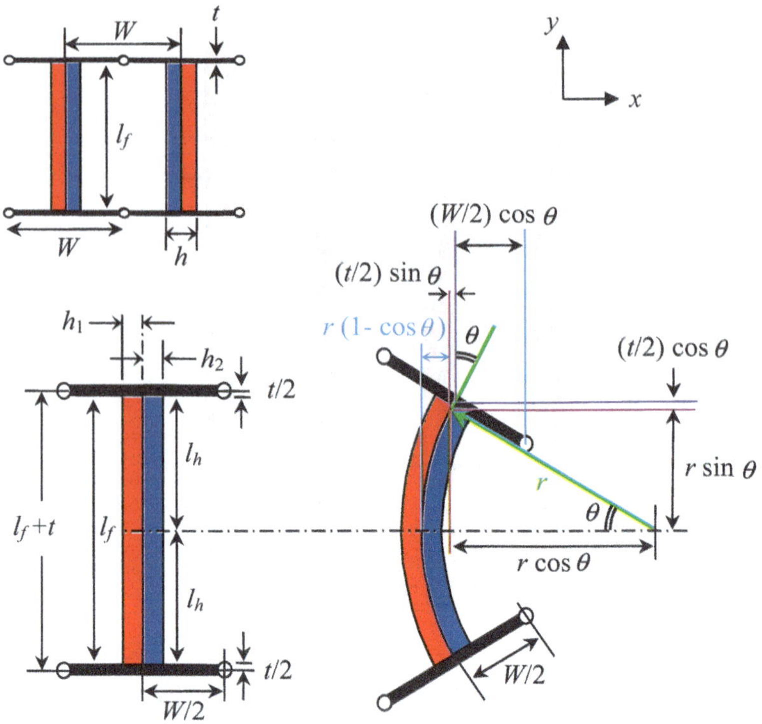

Fig. 19.8 A rectangular unit cell of the metamaterial (top left), as well as its detailed dimensions at original state (bottom left) and after thermal deformation (bottom right) for analysis

$$\frac{l_h}{r} = \frac{l_h(\alpha_1 - \alpha_2)\mathrm{d}T}{\frac{h}{2} + \frac{2}{h}(E_1 I_1 + E_2 I_2)\left(\frac{1}{E_1 h_1} + \frac{1}{E_2 h_2}\right)} \tag{19.2.3}$$

where α_n, E_n, and h_n are the CTEs, Young's moduli, and thicknesses, respectively, for material $n = 1, 2$ of the bimaterial with the second moment areas

$$I_n = \frac{h_n^3}{12} \tag{19.2.4}$$

With the curving of the bimaterials shown in Fig. 19.8 (bottom right), the width and height dimensions upon thermal deformation become

$$x = W \cos\theta \tag{19.2.5}$$

and

$$y = 2r \sin\theta + t \cos\theta \tag{19.2.6}$$

respectively, where the half-angle is defined as

$$\theta = l_h/r \tag{19.2.7}$$

The changes in dimensions, defined as $(\mathrm{d}x, \mathrm{d}y) = (x - x_0, y - y_0)$, are therefore

$$\mathrm{d}x = -W\left[1 - \cos\left(\frac{l_h}{r}\right)\right] \tag{19.2.8}$$

and

$$\mathrm{d}y = -2\left[l_h - r\sin\left(\frac{l_h}{r}\right)\right] - t\left[1 - \cos\left(\frac{l_h}{r}\right)\right] \tag{19.2.9}$$

The thermal strains along the axes, $\left(\varepsilon_x^{(T)}, \varepsilon_y^{(T)}\right) = (\mathrm{d}x/x_0, \mathrm{d}y/y_0)$, can be expressed as

$$\varepsilon_x^{(T)} = -1 + \cos\left(\frac{l_h}{r}\right) \tag{19.2.10}$$

and

$$\varepsilon_y^{(T)} = -1 + \frac{2\sin\left(\frac{l_h}{r}\right) + \frac{t}{r}\cos\left(\frac{l_h}{r}\right)}{2\frac{l_h}{r} + \frac{t}{r}} \tag{19.2.11}$$

which gives their corresponding effective CTEs

$$\alpha_x^{(T)} = -\frac{1}{\mathrm{d}T} + \frac{1}{\mathrm{d}T}\cos\left(\frac{l_h}{r}\right) \tag{19.2.12}$$

and

$$\alpha_y^{(T)} = -\frac{1}{\mathrm{d}T} + \frac{2\sin\left(\frac{l_h}{r}\right) + \frac{t}{r}\cos\left(\frac{l_h}{r}\right)}{\left(2\frac{l_h}{r} + \frac{t}{r}\right)\mathrm{d}T} \tag{19.2.13}$$

respectively.

Upon thermal deformation, Poisson's ratio can be obtained by the application of $\sigma_x > 0$, such that the zigzag conformation of the rigid rods is returned to their original horizontal alignment. As a consequence, the thermally curved bimaterials are straightened. With thermal deformation but before the application of σ_x, we have the updated "original" dimensions $\left(x_0', y_0'\right) = (x, y)$; on prescription of σ_x, the dimensions of the mechanically deformed unit cell are thus returned to $\left(x', y'\right) = (x_0, y_0)$ to give the change in dimension $\left(\mathrm{d}x', \mathrm{d}y'\right) = -(\mathrm{d}x, \mathrm{d}y)$. Therefore,

$$\mathrm{d}x' = W\left[1 - \cos\left(\frac{l_h}{r}\right)\right] \tag{19.2.14}$$

and

$$\mathrm{d}y' = 2\left[l_h - r\sin\left(\frac{l_h}{r}\right)\right] + t\left[1 - \cos\left(\frac{l_h}{r}\right)\right] \tag{19.2.15}$$

which gives the mechanical strains $(\varepsilon_x, \varepsilon_y) = (\mathrm{d}x'/x_0', \mathrm{d}y'/y_0')$ or

$$\varepsilon_x = -1 + \frac{1}{\cos\left(\frac{l_h}{r}\right)} \tag{19.2.16}$$

and

$$\varepsilon_y = -1 + \frac{2\frac{l_h}{r} + \frac{t}{r}}{2\sin\left(\frac{l_h}{r}\right) + \frac{t}{r}\cos\left(\frac{l_h}{r}\right)} \tag{19.2.17}$$

respectively. The resulting Poisson's ratio, $v_{xy} = -\varepsilon_y/\varepsilon_x$ is therefore

$$v_{xy} = -\frac{2\left[\frac{l_h}{r} - \sin\left(\frac{l_h}{r}\right)\right] + \frac{t}{r}\left[1 - \cos\left(\frac{l_h}{r}\right)\right]}{\left[2\tan\left(\frac{l_h}{r}\right) + \frac{t}{r}\right]\left[1 - \cos\left(\frac{l_h}{r}\right)\right]} \tag{19.2.18}$$

Simplifications can be found under two special cases. For Special Case I where the thickness of the rigid rod is negligible $t \ll r$, Eqs. (19.2.11), (19.2.13), and (19.2.18) reduce to

$$\varepsilon_y^{(T)} = -1 + \frac{\sin\left(\frac{l_h}{r}\right)}{\frac{l_h}{r}} \tag{19.2.19}$$

$$\alpha_y^{(T)} = -\frac{1}{\mathrm{d}T} + \frac{\sin\left(\frac{l_h}{r}\right)}{\mathrm{d}T\left(\frac{l_h}{r}\right)} \tag{19.2.20}$$

$$v_{xy} = -\frac{\left(\frac{l_h}{r}\right) - \sin\left(\frac{l_h}{r}\right)}{\tan\left(\frac{l_h}{r}\right)\left[1 - \cos\left(\frac{l_h}{r}\right)\right]} \tag{19.2.21}$$

Special Case II is obtained when both layers of the bimaterial share equal Young's moduli and equal thicknesses ($E_1/E_2 = h_1/h_2 = 1$), such that Eq. (19.2.3) greatly simplifies to

$$\frac{l_h}{r} = \frac{3}{2}\frac{l_h}{h}(\alpha_1 - \alpha_2)\mathrm{d}T \tag{19.2.22}$$

For the purpose of illustration, we adopt Special Case II in the subsequent results section.

Hence, substituting Eq. (19.2.22) into Eqs. (19.2.10)–(19.2.13) and (19.2.18) gives

$$\varepsilon_x^{(T)} = -1 + \cos\left(\frac{3}{2}\frac{l_h}{h}(\alpha_1 - \alpha_2)\mathrm{d}T\right) \tag{19.2.23}$$

$$\varepsilon_y^{(T)} = -1 + \frac{2\sin\left(\frac{3}{2}\frac{l_h}{h}(\alpha_1 - \alpha_2)\mathrm{d}T\right) + \frac{t}{r}\cos\left(\frac{3}{2}\frac{l_h}{h}(\alpha_1 - \alpha_2)\mathrm{d}T\right)}{2\left(\frac{3}{2}\frac{l_h}{h}(\alpha_1 - \alpha_2)\mathrm{d}T\right) + \frac{t}{r}} \tag{19.2.24}$$

$$\alpha_x^{(T)} = -\frac{1}{\mathrm{d}T} + \frac{1}{\mathrm{d}T}\cos\left(\frac{3}{2}\frac{l_h}{h}(\alpha_1 - \alpha_2)\mathrm{d}T\right) \tag{19.2.25}$$

$$\alpha_y^{(T)} = -\frac{1}{\mathrm{d}T} + \frac{2\sin\left(\frac{3}{2}\frac{l_h}{h}(\alpha_1 - \alpha_2)\mathrm{d}T\right) + \frac{t}{r}\cos\left(\frac{3}{2}\frac{l_h}{h}(\alpha_1 - \alpha_2)\mathrm{d}T\right)}{\left(3\frac{l_h}{h}(\alpha_1 - \alpha_2)\mathrm{d}T + \frac{t}{r}\right)\mathrm{d}T} \tag{19.2.26}$$

$$\upsilon_{xy} = -\frac{2\left[\left(\frac{3}{2}\frac{l_h}{h}(\alpha_1 - \alpha_2)\mathrm{d}T\right) - \sin\left(\frac{3}{2}\frac{l_h}{h}(\alpha_1 - \alpha_2)\mathrm{d}T\right)\right] + \frac{t}{r}\left[1 - \cos\left(\frac{3}{2}\frac{l_h}{h}(\alpha_1 - \alpha_2)\mathrm{d}T\right)\right]}{\left[2\tan\left(\frac{3}{2}\frac{l_h}{h}(\alpha_1 - \alpha_2)\mathrm{d}T\right) + \frac{t}{r}\right]\left[1 - \cos\left(\frac{3}{2}\frac{l_h}{h}(\alpha_1 - \alpha_2)\mathrm{d}T\right)\right]} \tag{19.2.27}$$

where

$$\frac{t}{r} = \frac{t}{l_h}\left(\frac{3}{2}\frac{l_h}{h}(\alpha_1 - \alpha_2)\mathrm{d}T\right) \tag{19.2.28}$$

To combine both Special Cases I and II, we let $t = 0$ in Eqs. (19.2.24), (19.2.26), and (19.2.27) to yield

$$\varepsilon_y^{(T)} = -1 + \frac{\sin\left(\frac{3}{2}\frac{l_h}{h}(\alpha_1 - \alpha_2)\mathrm{d}T\right)}{\left(\frac{3}{2}\frac{l_h}{h}(\alpha_1 - \alpha_2)\mathrm{d}T\right)} \tag{19.2.29}$$

$$\alpha_y^{(T)} = -\frac{1}{\mathrm{d}T} + \frac{\sin\left(\frac{3}{2}\frac{l_h}{h}(\alpha_1 - \alpha_2)\mathrm{d}T\right)}{\mathrm{d}T\left(\frac{3}{2}\frac{l_h}{h}(\alpha_1 - \alpha_2)\mathrm{d}T\right)} \tag{19.2.30}$$

$$\upsilon_{xy} = -\frac{\left(\frac{3}{2}\frac{l_h}{h}(\alpha_1 - \alpha_2)\mathrm{d}T\right) - \sin\left(\frac{3}{2}\frac{l_h}{h}(\alpha_1 - \alpha_2)\mathrm{d}T\right)}{\tan\left(\frac{3}{2}\frac{l_h}{h}(\alpha_1 - \alpha_2)\mathrm{d}T\right)\left[1 - \cos\left(\frac{3}{2}\frac{l_h}{h}(\alpha_1 - \alpha_2)\mathrm{d}T\right)\right]} \tag{19.2.31}$$

To evaluate the CTEs and Poisson's ratio of the currently proposed metamaterial, we use $l_h/h = 100$, which is a common aspect ratio for bimaterials, as well as $\alpha_1^{(T)} = 25 \times 10^{-6}\ \mathrm{K}^{-1}$ for the bimaterial phase of higher CTE. This value is typical of metallic materials of higher CTEs, such as lead $\left(\alpha_{\mathrm{Pb}}^{(T)} = 29 \times 10^{-6}\ \mathrm{K}^{-1}\right)$, magnesium $\left(\alpha_{\mathrm{Mg}}^{(T)} = 26 \times 10^{-6}\ \mathrm{K}^{-1}\right)$, and brass $\left(\alpha_{\mathrm{Brass}}^{(T)} = 19 \times 10^{-6}\ \mathrm{K}^{-1}\right)$. A total

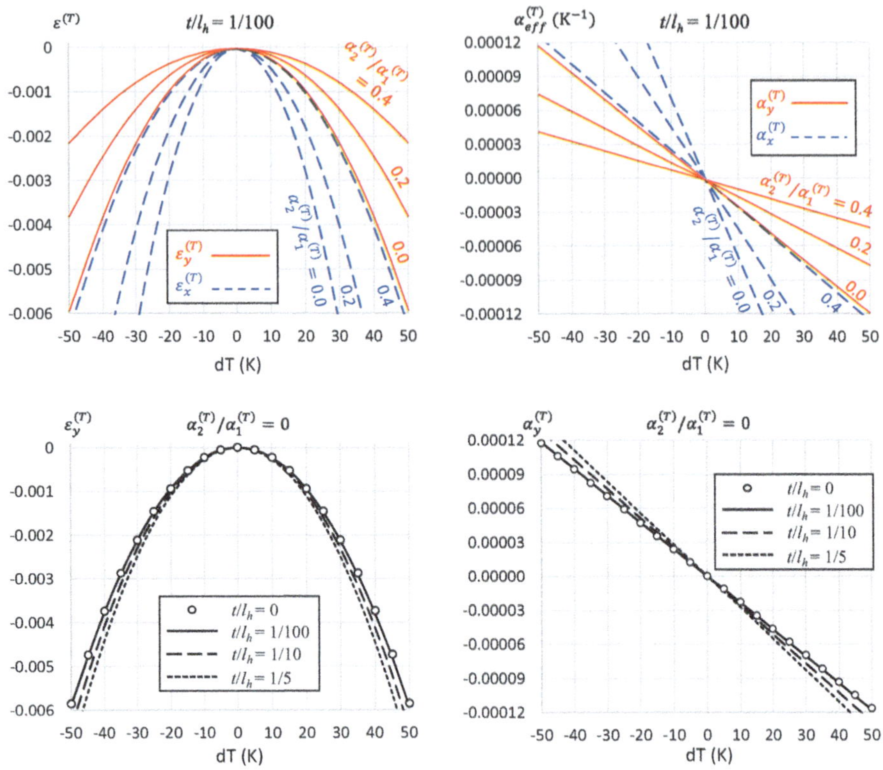

Fig. 19.9 Plots of thermal strains (left) and effective CTEs (right) versus temperature change for various bimaterial CTE phase ratio $\alpha_2^{(T)}/\alpha_1^{(T)}$ (top) and rigid rod thickness ratio t/l_h (bottom) at $\alpha_1^{(T)} = 25 \times 10^{-6} K^{-1}$

temperature difference of 100 K was considered; specifically, the temperature difference range of $-50\text{ K} \leq dT \leq 50\text{ K}$ was computed. Figure 19.9 (top) shows the plots of thermal strains and effective CTEs along the x- and y-axes, using Eqs. (19.2.23)–(19.2.26), for various $\alpha_2^{(T)}/\alpha_1^{(T)}$ ratios at $t/l_h = 1/100$. Regardless of whether $dT < 0$ or $dT > 0$, the thermal strains are negative ($\varepsilon_x^{(T)}, \varepsilon_y^{(T)} < 0$) as plotted in Fig. 19.9 (top left), thereby indicating that the metamaterial exhibits positive thermal expansion (PTE) upon cooling and negative thermal expansion (NTE) upon heating, as shown in Fig. 19.9 (top right). In other words, the sign of in-plane CTE is always opposite to the sign of dT. To evaluate the effect of the rigid rod thickness on the thermal properties, and hence the validity of $t = 0$ simplifying assumption in Eqs. (19.2.29) and (19.2.30), Fig. 19.9 (bottom) shows the plots of $\varepsilon_y^{(T)}$ and $\alpha_y^{(T)}$ using Eqs. (19.2.24) and (19.2.26) for various t/l_h ratios at $\alpha_2^{(T)}/\alpha_1^{(T)} = 0$. When $t = 0$, Eqs. (19.2.24) and (19.2.26) reduce to Eqs. (19.2.29) and (19.2.30), respectively. Plotted curves indicate that the thickness of the rigid rod gives more negative $\varepsilon_y^{(T)}$

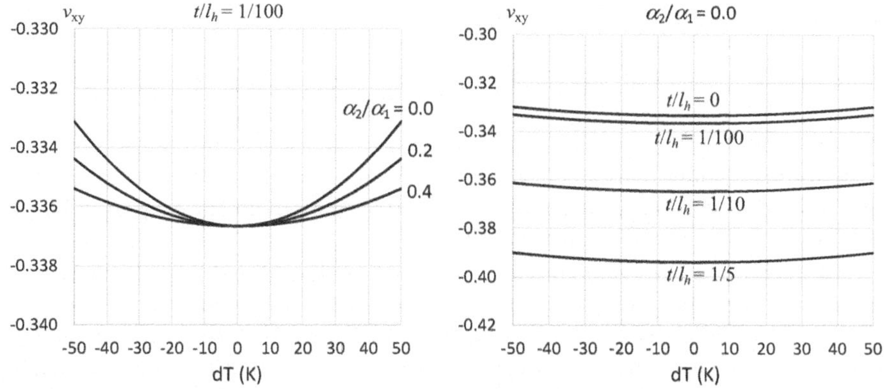

Fig. 19.10 Plots of Poisson's ratio v_{xy} versus temperature change for various bimaterial CTE ratio $\alpha_2^{(T)}/\alpha_1^{(T)}$ (left) and rigid rod thickness ratio t/l_h (right) at $\alpha_1^{(T)} = 25 \times 10^{-6}\text{K}^{-1}$

and greater magnitude of $\alpha_y^{(T)}$, implying that the simplifying assumption gives underestimated magnitudes of $\varepsilon_y^{(T)}$ and $\alpha_y^{(T)}$. Nevertheless, perusal to Fig. 19.9 (bottom) supports the validity of the $t = 0$ assumption when $t/l_h \leq 1/100$.

It is also worthy to mention that although the CTE relation with reference to dT is nonlinear, the plotted results in Fig. 19.9 (right) show an almost linear function. This is because the curvatures are negligible. In the case of $\alpha_2^{(T)}/\alpha_1^{(T)} = 0$ for example, the curvatures of $\alpha_x^{(T)}$ and $\alpha_y^{(T)}$ are of the orders 10^{-9} and 10^{-10}, respectively, for the given range of temperature change. These curvatures are insignificant in comparison to the CTE values themselves. The orders of the curvatures are even lower for greater values of $\alpha_2^{(T)}/\alpha_1^{(T)}$.

Figure 19.10 (left) furnishes the curves of v_{xy}, using Eq. (19.2.27), for various $\alpha_2^{(T)}/\alpha_1^{(T)}$ ratios at $t/l_h = 1/100$. Regardless of whether d$T < 0$ or d$T > 0$, this metamaterial exhibits decreasing auxeticity with the magnitude of thermal change. For assessing the influence of the rigid rod thickness on the elastic properties, and hence the validity of $t = 0$ assumption in Eq. (19.2.31), Fig. 19.10 (right) shows the results of Poisson's ratio using Eq. (19.2.27) for various t/l_h ratios at $\alpha_2^{(T)}/\alpha_1^{(T)} = 0$. At $t = 0$, Eq. (19.2.27) abridges to Eq. (19.2.31). Calculated results reveal that the negativity of v_{xy} is increased by the thickness of the rigid rod. Due to the low curvature of the v_{xy} versus dT plots in comparison with the sensitivity toward the t/l_h ratio, the assumption of $t = 0$ is not valid for v_{xy} except when $t/l_h \approx 0$.

When the rigid rod's thickness is insignificant, the plotted thermal strain ratio $\varepsilon_y^{(T)}/\varepsilon_x^{(T)}$ and CTE ratio $\alpha_y^{(T)}/\alpha_x^{(T)}$ are observed to be approximately 1/3. This can be proven by performing series expansion on the trigonometric terms—listed in Appendix A.2—on Eqs. (19.2.23), (19.2.25), (19.2.29), and (19.2.30), followed by neglecting higher-order terms to give

$$\begin{Bmatrix} \varepsilon_x^{(T)} \\ \varepsilon_y^{(T)} \end{Bmatrix} \approx -\frac{1}{6}\left(\frac{3}{2}\frac{l_h}{h}(\alpha_1 - \alpha_2)\right)^2 (\mathrm{d}T)^2 \begin{Bmatrix} 3 \\ 1 \end{Bmatrix} \tag{19.2.32}$$

and

$$\begin{Bmatrix} \alpha_x^{(T)} \\ \alpha_y^{(T)} \end{Bmatrix} \approx -\frac{1}{6}\left(\frac{3}{2}\frac{l_h}{h}(\alpha_1 - \alpha_2)\right)^2 \mathrm{d}T \begin{Bmatrix} 3 \\ 1 \end{Bmatrix} \tag{19.2.33}$$

Performing similarly for Eq. (19.2.31) for an approximated value of the Poisson's ratio leads to a constant value of $v_{xy} = -1/3$. This approximation agrees with the plots of v_{xy} in Fig. 19.10 when $t/l_h \leq 1/100$.

The establishment of analogies across auxetic models from different geometrical groups has been made by identifying their microstructural counterparts followed by comparison of their deformation mechanisms (Lim 2017). Using this approach, analogies can be established between the rotating square model (Grima and Evans 2000) and the discussed metamaterial, as furnished in Fig. 19.11a (top), whereby two opposing corners of a rotating square coincide with the pin joints of the rigid rods, while the remaining two corners of the rotating square correspond with the mid-spans of the bimaterial strips. Alderson et al. (2010) investigated a number of chiral and anti-chiral structures, one of them being the anti-tetrachiral honeycomb. Upon in-plane compression, the anti-tetrachiral honeycomb is comparable to the currently proposed metamaterial due to thermal deformation, as shown in Fig. 19.11a (bottom). In particular, the regions in which the bimaterial strips are built into the rigid

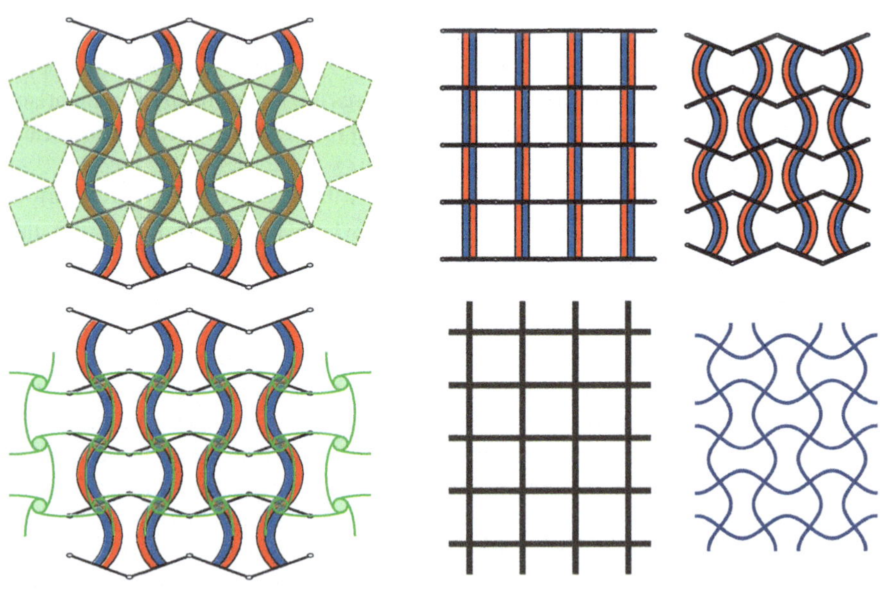

(a) (b)

Fig. 19.11 Analogies between the current metamaterial with **a** the rotating square model by Grima and Evans (2000) (top) and the deformed anti-tetrachiral model by Alderson et al. (2010) (bottom), and **b** the analogy with the buckling study by Haghpanah et al. (2014)

rods for the metamaterial coincide with the cylindrical nodes of the anti-tetrachiral structure. Another analogy can be observed to exist between one of the buckling modes of a square grid by Haghpanah et al. (2014). Prescription of heat on the original states of the current metamaterial in Fig. 19.11b (top left) and application of buckling load on the square grid in Fig. 19.11b (bottom left) lead to the contorted microstructures shown in Fig. 19.11b (right). Specifically, the imposition of heat and buckling load convert the square cavities into cavities that resemble the shapes of "8" and "∞" in alternating sequence. However, the application of tensile load in the on-axes directions when the metamaterial is in its original state, i.e. rectangular grid, will obviously not show sign of auxeticity.

19.3 Conclusions

The metamaterial considered in Sect. 19.1 can be designed in such a manner that it always contracts regardless of how the environmental condition changes, if the following conditions

$$\begin{aligned} \alpha_{\mathrm{s}}^{(C)}\Delta C_s &< 0.5\ln\left[2 - \mathrm{e}^{\alpha_{\mathrm{c}}^{(C)}\Delta C_c}\left(2 - \mathrm{e}^{\alpha_{\mathrm{c}}^{(C)}\Delta C_c}\right)\right] \\ \alpha_{\mathrm{s}}^{(T)}\Delta T &< 0.5\ln\left[2 - \mathrm{e}^{\alpha_{\mathrm{c}}^{(T)}\Delta T}\left(2 - \mathrm{e}^{\alpha_{\mathrm{c}}^{(T)}\Delta T}\right)\right] \\ -\alpha_{\mathrm{s}}^{(P)}\Delta P &< 0.5\ln\left[2 - \mathrm{e}^{-\alpha_{\mathrm{c}}^{(P)}\Delta P}\left(2 - \mathrm{e}^{-\alpha_{\mathrm{c}}^{(P)}\Delta P}\right)\right] \end{aligned} \tag{19.3.1}$$

are met (Lim 2019a). Where the advantages of negative materials over conventional ones under specific conditions become disadvantageous under opposing conditions, the availability of sign-switching of expansion coefficients permits materials to perform in an advantageous manner regardless of how the environmental condition changes.

Recall that the analysis and design of auxetic plates have been investigated for load-bearing applications (Gorodtsov et al. 2018), wave transmission applications (Lim 2019b, c, d), and medical applications (Mehmood et al. 2015). However, not all situations require fixed auxetic properties. Hence, the capability of adjusting auxeticity by temperature change, as demonstrated in Fig. 19.10 for the metamaterial discussed in Sect. 19.2, permits the 2D structure's behavior to be optimized for the abovementioned applications. As for load-bearing applications, it has previously been shown that the maximum bending moment—and hence the corresponding maximum bending stress—in a uniformly loaded simply supported circular plate is minimized when $v = -1/3$ due to its optimum distribution of the bending moment (Lim 2013). Likewise, the bending moment in one of the directions for a uniformly loaded simply supported equilateral triangular plate is minimized when $v = -1/3$ (Lim 2016). Hence, the use of the discussed metamaterial in Sect. 19.2, which exhibits this value of Poisson's ratio, is useful for lowering the maximum bending stress via more uniform spreading of the bending moment throughout the entire plate. It can therefore be concluded that arising from the thermally induced curving of the bimaterial strips and resulting rotation of the rigid rods, the ever-decreasing in-plane area with the magnitude of temperature change produces sign-switching of the in-plane CTE (Lim 2020). However, regardless of whether the temperature increases or decreases, the metamaterial is consistently auxetic, i.e., its Poisson's ratio is not sign-switchable.

References

Alderson A, Alderson KL, Attard D, Evans KE, Gatt R, Grima JN, Miller W, Ravirala N, Smith CW, Zeid K (2010) Elastic constants of 3-, 4- and 6-connected chiral and anti-chiral honeycombs subject to uniaxial in-plane loading. Compos Sci Technol 70(7):1042–1048

Gorodtsov VA, Lisovenko DS, Lim TC (2018) Three-layered plate exhibiting auxeticity based on stretching and bending modes. Compos Struct 194:643–651

Grima JN, Evans KE (2000) Auxetic behavior from rotating squares. J Mater Sci Lett 19:1563–1565

Grima JN, Bajada M, Attard D, Dudek KK, Gatt R (2015) Maximizing negative thermal expansion via rigid unit modes: a geometry-based approach. Proc R Soc A 471(2179):20150188

Haghpanah B, Papadopoulos J, Mousanezhad D, Nayeb-Hashemi H, Vaziri A (2014) Buckling of regular, chiral and hierarchical honeycombs under a general macroscopic stress state. Proc R Soc A 470(2167):20130856

Lim TC (2013) Circular auxetic plates. J Mech 29(1):121–133

Lim TC (2016) Bending stresses in triangular auxetic plates. J Eng Mater Technol 138(1):014501

Lim TC (2017) Analogies across auxetic models based on deformation mechanism. Phys Status Solidi RRL 11(6):1600440

Lim TC (2019a) Composite metamaterial with sign-switchable coefficients of hygroscopic, thermal and pressure expansions. Adv Compos Hybrid Mater 2(4):657–669

Lim TC (2019b) Longitudinal wave speed in auxetic plates with elastic constraint in width direction. Arch Appl Mech 89(4):659–668

Lim TC (2019c) Longitudinal wave speed in cylindrical auxetic rods with elastic constraint in radial direction. Eur J Mech A/Solids 75:443–449

Lim TC (2019d) Longitudinal wave speed in rectangular slabs with mixed restraints in lateral directions. Proc Inst Civil Eng Eng Comput Mech 172(4):153–161

Lim TC (2020) Metacomposite with auxetic and in situ sign reversible thermal expansivity upon temperature fluctuation. Compos Commun 19:30–36

Mehmood S, Ali MN, Ansari U, Mir M, Khan MA (2015) Auxetic polymeric bone plate as internal fixator for long bone fractures: design, fabrication and structural analysis. Technol Health Care 23(6):819–833

Chapter 20
Metamaterials and Islamic Geometric Patterns

Abstract This chapter introduces the science and art of metamaterials, particularly in the design of negative materials. Proceeding from auxetic metamaterials inspired from Islamic motifs, the rest of the chapter deals with sign-switching of expansion coefficients in metamaterials whose microstructural geometries do not readily exhibit Islamic geometric patterns in their original state, but bloom into an Islamic motif as a consequence of environmental changes. Both sign-switching metamaterials deform to form 2D arrays of 8-pointed stars; one of them employs bimaterial strips while the other adopts linkage mechanism.

Keywords Auxetic microstructures · Hygrothermal properties · Compressibility · Islamic design · Sign-flipping · Sign-switching · Sign-toggling

20.1 Introduction: Auxetic Metamaterials with Islamic Motifs

Before proceeding with the scope of this chapter, it is beneficial to reflect that there are occasions when progress in science is made based upon inspiration from nature. Apart from inspiration from biology, one may also derive innovation from the arts and social sciences. Of late, new designs of metamaterial microstructures have been inspired by ancient art forms with religious significance. For example, Rafsanjani and Pasini (2016a) discovered that auxetic microstructures can be developed based on certain ancient Islamic motifs. Figure 20.1 shows brick decorations based in the Kharraqan twin tomb towers in Western Iran, which inspired the building blocks and the corresponding unit cells in the undeformed and deformed states of bistable auxetics with square and triangular rotating units for circular and parallel cut motifs by Rafsanjani and Pasini (2016b). The experimental and finite element plots of force and Poisson's ratio versus displacement for the square and triangular bistable auxetics under tensile and compressive loads are shown in Fig. 20.2.

A family of stress, Poisson's ratio, and strain energy density curves for both the square and triangular unit cells plotted against strain is furnished by Rafsanjani and Pasini (2016b) in Fig. 20.3. The unit cells are subjected to uniaxial extension until they are fully expanded. Upon load removal, the unit cells return to stable states

T. Lim, *Mechanics of Metamaterials with Negative Parameters*, Engineering Materials,
https://doi.org/10.1007/978-981-15-6446-8_20

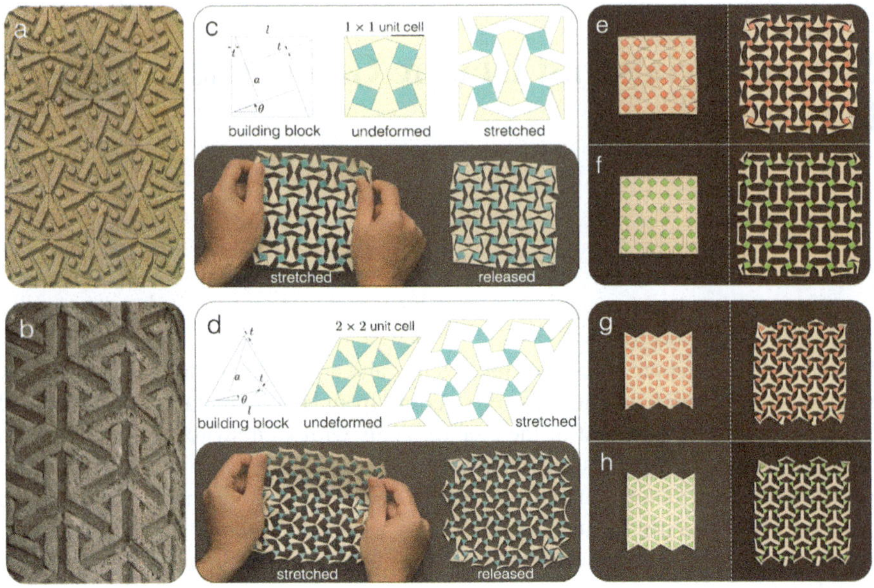

Fig. 20.1 Bistable auxetics by Rafsanjani and Pasini (2016b): Brick decorations based on **a** square and **b** triangular motifs in the Kharraqan twin tomb towers in Western Iran. The building block and the corresponding unit cell in its undeformed and deformed state comprises **c** square and **d** triangular rotating units with tilted cut motifs, which can maintain the stable deformation pattern after the load release. **e**–**h** Undeformed and deformed states of bistable auxetics with square and triangular rotating units for circular and parallel cut motifs. Reprinted with permission from Elsevier

(whose unit cells are shown in the insets) of local minimum energy (markers) where Poisson's ratio equals to -1. In the next two sections, inspiration is drawn from another ancient Islamic geometric pattern, in the form of 2D array of 8-pointed stars, to design a metamaterial that exhibits sign-toggling of thermal, moisture, and compressibility properties.

20.2 Bimaterial-Based Sign-Switching Metamaterial with Islamic Motifs

In this section and in the next section, we draw inspiration from an ancient Islamic geometric pattern, in the form of 2D array of 8-pointed stars, to design a metamaterial that exhibits sign-toggling of environmental properties. The 8-pointed stars and lozenges containing squares is one of the earliest geometrical forms in Islamic art. These date from the year 836 in the Great Mosque of Kairouan, Tunisia. The 8-pointed stars, alongside the 6-pointed stars, are also found in the Ibn Tulun Mosque, Cairo, built in year 879, as well as in the Towers of Kharraqan at Qazvin, Persia (built in 1067) and the Al-Juyushi Mosque, Egypt (1085), and thereafter spread throughout

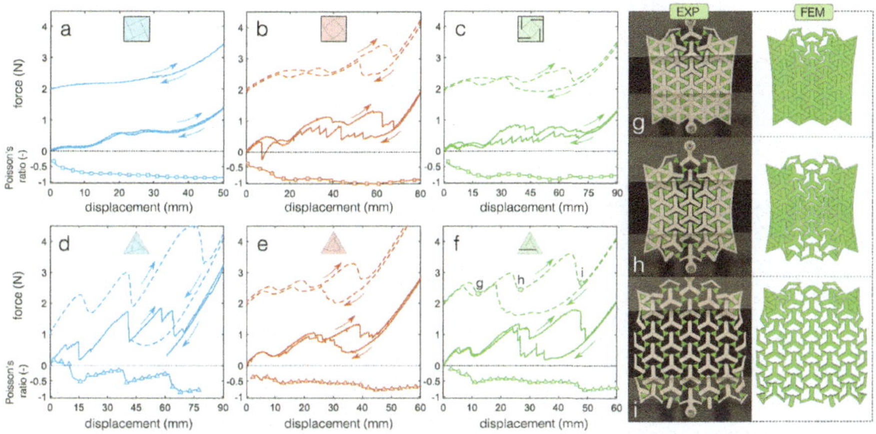

Fig. 20.2 Mechanical response of bistable auxetics to uniaxial loading by Rafsanjani and Pasini (2016b). **a–c** Experimental (solid line) and FE (dashed lines, shifted vertically for readability) force-displacement curves during loading (tension) and unloading (compression) and experimentally characterized Poisson's ratio (symbols) of square bistable auxetics ($a/l = 1/2, t/l = 0.05$) respectively for tilted (blue), circular (red) and parallel (green) square cut motifs. **d–e** Similar results as above for triangular bistable auxetics. Note that the numerical curves are shifted vertically for better readability. **g–i** Comparison of experimental and FE deformed shape of a triangular design specimen with parallel cut motifs at selected displacements shown in diagram (**f**). Reprinted with permission from Elsevier

the entire Islamic world (Abdullahi and Embi 2013). A sample of later 8-pointed stars in architecture and interior designs are shown in Fig. 20.4 (top row), as well as in tiles from thirteenth-century Iran, especially from Kashgar (Fig. 20.4, middle row). The array of interconnected 8-pointed stars can be designed as overlapping ribbons (e.g., Fig. 20.4 bottom left).

In this section, a class of square grids made from bimaterial strips that transforms into arrays of 8-pointed stars is introduced, of which a simplified design is illustrated in Fig. 20.4 (bottom right). The motivation of choosing this design, therefore, is not only confined to the potential for creating metamaterials with physical properties that switch between positive and negative values by environmental changes and without any active control, but also to pay homage to an ancient art that is associated with a great religion. This geometry is also chosen because the art forms are hidden at the reference environmental condition and are only manifested in full splendor when the environmental condition changes, thereby creating a dynamic microlattice structure with esthetic feature. As illustrated in Fig. 20.5, this example explores a 2D metamaterial in which the increase in temperature $dT > 0$ and/or moisture concentration $dC > 0$ as well as decrease in pressure $dP < 0$ lead to material contraction, thereby exhibiting negative expansivity. However, the decrease in temperature $dT < 0$ and/or moisture concentration $dC < 0$ as well as increase in pressure $dP > 0$ also leads to material contraction, thereby exhibiting positive expansivity instead (Lim 2020a).

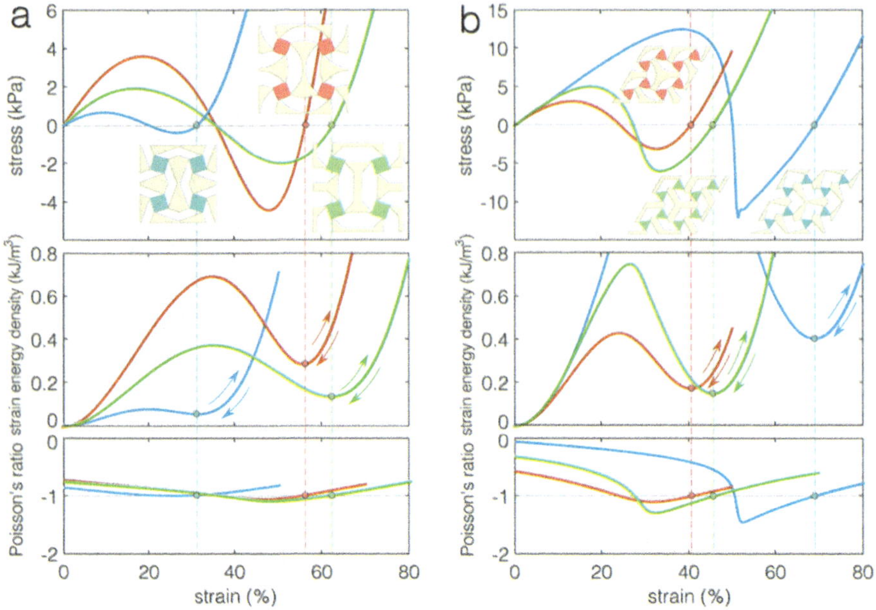

Fig. 20.3 Mechanical response of unit cells obtained by FE simulations by Rafsanjani and Pasini (2016b). Stress, strain energy density and Poisson's versus strain respectively for **a** square and **b** triangular unit cells ($a/l = 1/2, t/l = 0.025$) with tilted (blue), circular (red) and parallel (green) cut motifs calculated by FE simulations under periodic boundary conditions. Reprinted with permission from Elsevier

Two configurations of bimaterial strips within the square grids are identified and illustrated in Figs. 20.6 and 20.7 to demonstrate the manner in which the bending of the straight bimaterial strips in response to changing environmental condition can transform the square grids into arrays that approximate 8-pointed stars found in Islamic art. Specifically, we term the configurations furnished in Figs. 20.6 (top) and 20.7 (top) as the Type A and Type B, respectively. The bimaterial strips in both types consist of two materials, with material 1 (indicated by yellow) possessing a higher expansion coefficient than material 2 (indicated by green), such that different extent of expansion leads to materials 1 and 2 being convex and concave, respectively, as shown in Figs. 20.6 (bottom right) and 20.7 (bottom right). Conversely, under an opposing change in environmental condition, the differential contraction leads to materials 1 and 2 being concave and convex, respectively, as furnished in Figs. 20.6 (bottom left) and 20.7 (bottom left). In both Type A and Type B metamaterials, one end of each bimaterial strip is built into rigid squares (shown in red) and each square is attached to four bimaterial strips. Unlike Type B, the configuration in Type A permits free rotation at the other end of each bimaterial strip, which is attached to a pillow block bearing (indicated in blue) such that the curved outer surface of each pillow block bearing is in contact to another pillow block bearing via connectors (indicated in black) passing through the holes of the pillow block bearings. Unlike

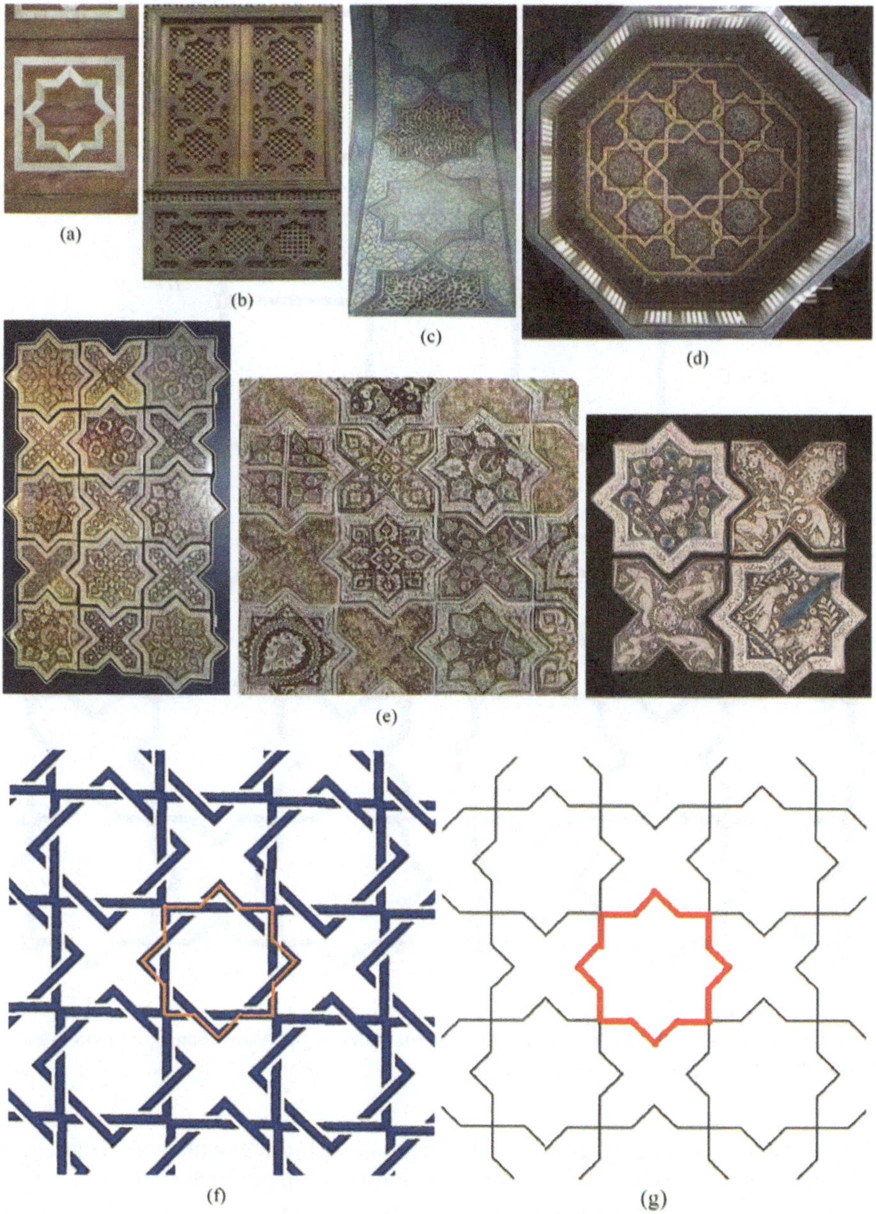

Fig. 20.4 Examples of the Islamic 8-pointed connected star: **a** a single star in sixteenth century Humayun's tomb, India, **b** horizontally and vertically aligned stars in an interior part of the fourteenth century Al-Attarine Madrasa, Egypt, **c** tiled arch at the Madrasa Balasar, Iran, **d** ceiling of the Sultan Ashraf Qaytbay Mosque, Egypt, **e** stars arranged diagonally in tiles from thirteenth-century Iran, **f** an overlapping ribbon type, and **g** a simplified schematic

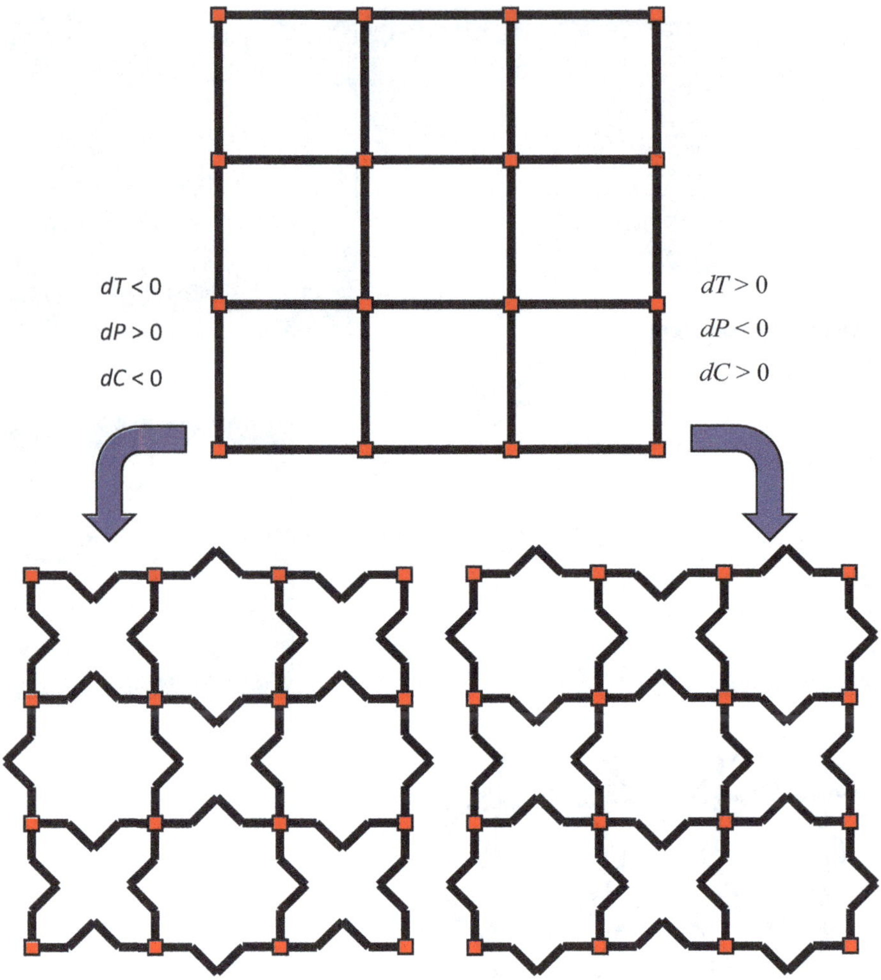

Fig. 20.5 Transformation of a square grid (top) to an 8-pointed star array under a change of environmental condition (bottom left) and its conjugate form under an opposing change of environmental condition (bottom right)

Type A, each bimaterial strip in Type B consists of three segments with length ratio 1:2:1 whereby materials 1 and 2 alternate from one segment to the next.

The bimaterials are permitted to curve based on three types of environmental changes—temperature, pressure, and moisture concentration—through contrasting (1) coefficient of thermal expansion (CTE), (2) compressibility, or coefficient of pressure expansion (CPE), and (3) coefficient of moisture expansion (CME). The volumetric CTE for any matter—be it in the form of solid, liquid or gas—is a measure of the volumetric change in response to temperature change at constant pressure and is expressed as

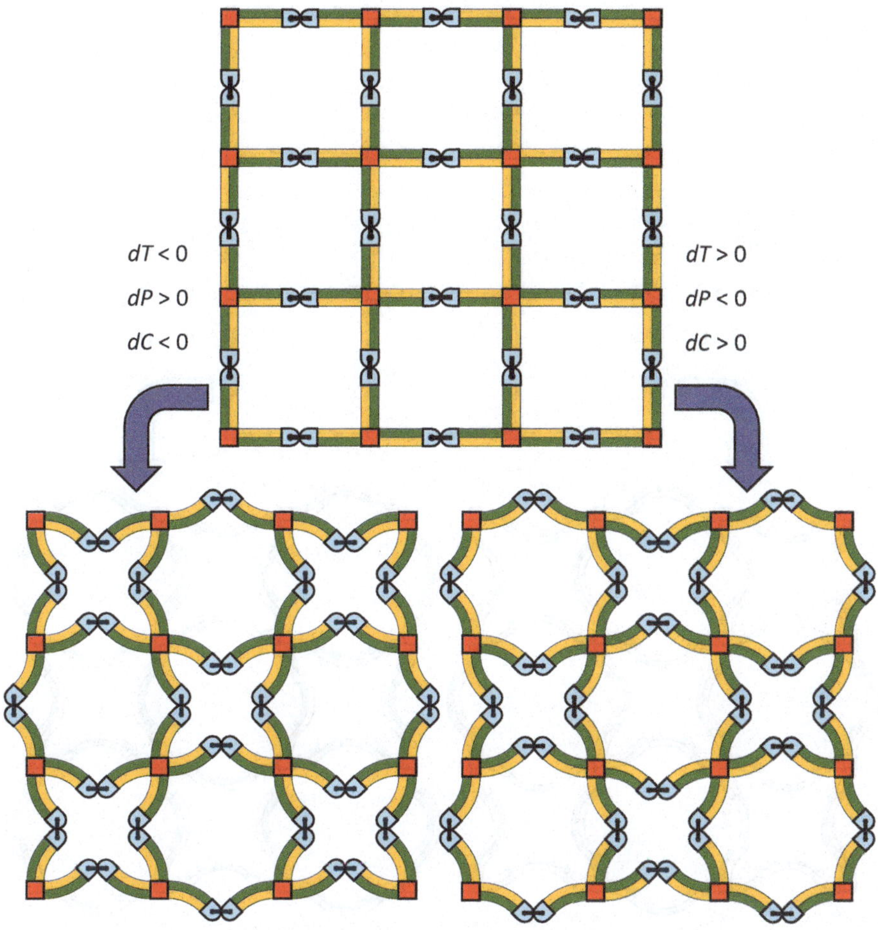

Fig. 20.6 Type A square grid in its original state (top) and its deformation approximating the 8-pointed star under opposing conditions that lead to opposing curvatures in the bimaterial strips (bottom), where the material 1 (yellow) has greater expansion coefficients than material 2 (green)

$$\alpha_{\mathrm{V}}^{(T)} = \frac{1}{V}\left(\frac{\partial V}{\partial T}\right)_{\mathrm{P}} \tag{20.2.1}$$

Due to anisotropy, which is common in solids, it is more meaningful to write separate CTEs in each orthogonal direction. Hence, we have the linear CTE

$$\alpha^{(T)} = \frac{1}{L}\frac{dL}{dT} = \frac{\varepsilon^{(T)}}{dT} \tag{20.2.2}$$

where $\varepsilon^{(T)}$ is thermal strain. In thermofluid mechanics, the coefficient of compressibility, or more conveniently the "compressibility," quantifies the volumetric change

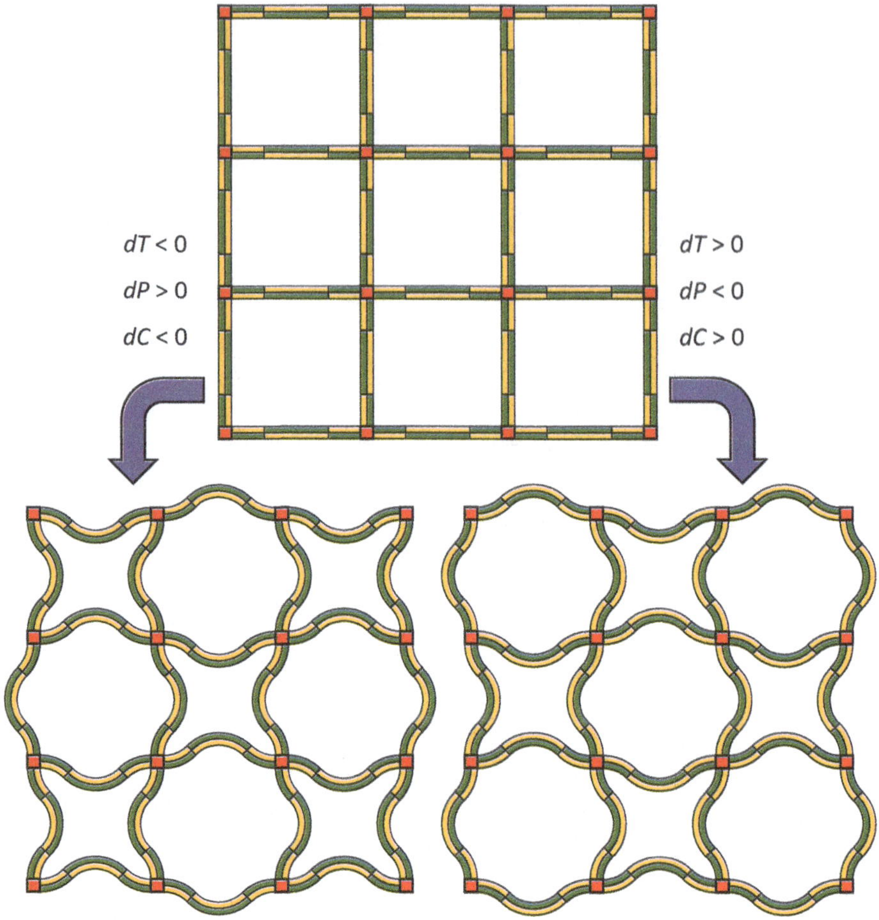

Fig. 20.7 Type B square grid in its original state (top) and its deformation approximating the 8-pointed star under opposing conditions that lead to opposing curvatures in the bimaterial strips (bottom), where the material 1 (yellow) has greater expansion coefficients than material 2 (green)

in response to pressure change. When these changes take place at constant temperature, we have the isothermal compressibility

$$\alpha_{V}^{(P)} = -\frac{1}{V}\left(\frac{\partial V}{\partial P}\right)_{T} \tag{20.2.3}$$

The negative sign indicates a decrease in size with an increase in pressure. For the purpose of consistency with the linear CTE, we herein introduce its linear version as the CPE

$$\alpha^{(P)} = -\frac{1}{L}\frac{dL}{dP} = \frac{\varepsilon^{(P)}}{(-dP)} \tag{20.2.4}$$

where $\varepsilon^{(P)}$ is the strain due to pressure change. To obtain an analogous coefficient due to moisture absorption into or moisture dissipation from a solid, we recall the definition of moisture concentration in solids

$$C = \frac{m}{M} \times 100 \tag{20.2.5}$$

where m is the mass of moisture in the solid and M is the mass of the dry material; hence, the change in moisture concentration is

$$dC = \frac{dm}{M} \times 100 \tag{20.2.6}$$

where dm is the change in moisture mass in the solid. This also applies for the environmental moisture concentration wherein dm is the change in moisture mass per unit volume of the environment while M refers to the mass of dry air in the same volume. Therefore the CME due to the change in moisture concentration in the solid is

$$\alpha^{(C)} = \frac{1}{L}\frac{dL}{dC} = \frac{\varepsilon^{(C)}}{dC} \tag{20.2.7}$$

where $\varepsilon^{(C)}$ is the strain arising from the change in moisture concentration. Due to different absorptivity level, various materials absorb differing amount of moisture from the environment. As such, a change in environmental moisture concentration dC leads to changes in moisture concentration in materials 1 and 2 (dC_1, dC_2) in a two-phase composite wherein $dC \neq dC_1 \neq dC_2$ at moisture transfer equilibrium. This is unlike temperature change where $dT = dT_1 = dT_2$ is attained at thermal equilibrium (Lim 2019a). Analogies can be formed with the various coefficients of expansion and their corresponding strains and environmental changes, as listed in Table 20.1.

Recall that for a straight bimaterial strip with materials 1 and 2 possessing CTEs of $\alpha_1^{(T)}$ and $\alpha_2^{(T)}$, respectively, the radius of curvature $r^{(T)}$ that is formed under a temperature change of dT is

Table 20.1 Analogies between specific coefficients of environmental expansions and related parameters

	Thermal	Pressure	Moisture concentration
Environmental change	dT	$-dP$	dC
Strain response	$\varepsilon^{(P)}$	$\varepsilon^{(P)}$	$\varepsilon^{(C)}$
Expansion coefficient	$\alpha^{(T)}$	$\alpha^{(P)}$	$\alpha^{(C)}$

$$\frac{1}{r^{(T)}} = \frac{\left(\alpha_1^{(T)} - \alpha_2^{(T)}\right) dT}{\frac{h}{2} + \frac{2}{h}(E_1 I_1 + E_2 I_2)\left(\frac{1}{E_1 h_1} + \frac{1}{E_2 h_2}\right)} \tag{20.2.8}$$

where

$$h = h_1 + h_2 \tag{20.2.9}$$

and

$$I_1 = \frac{h_1^3}{12}, \quad I_2 = \frac{h_2^3}{12} \tag{20.2.10}$$

with h_1 and h_2 being the thicknesses of materials 1 and 2, respectively. By similar reasoning on the differential expansions of the bonded strips due to pressure change dP, the resulting curvature $r^{(P)}$ is

$$\frac{1}{r^{(P)}} = \frac{\left(\alpha_1^{(P)} - \alpha_2^{(P)}\right)(-dP)}{\frac{h}{2} + \frac{2}{h}(E_1 I_1 + E_2 I_2)\left(\frac{1}{E_1 h_1} + \frac{1}{E_2 h_2}\right)} \tag{20.2.11}$$

where $\alpha_1^{(P)}$ and $\alpha_2^{(P)}$ are the CPEs of materials 1 and 2, respectively, of the bimaterial strip. As mentioned earlier, the negative sign in Eq. (20.2.4) implies the decrease in dimension with increase in pressure; this results in the negative sign in Eq. (20.2.11), which indicates that the bimaterial strip curves in the opposite direction with increasing pressure when compared to the case of increasing temperature. While the temperature change in bimaterial strip is equal to that in the environment at thermal equilibrium $dT_1 = dT_2 = dT$, and that the pressure change experienced by the bimaterial strips are common to the pressure change in the environment $dP_1 = dP_2 = dP$, the same cannot be said so for the case of moisture change. Due to different levels of moisture concentration change between the environment and materials even at hygroscopic equilibrium, as well as the different extent of moisture retention in each material, the changes in moisture concentration at hygroscopic equilibrium for the environment and in both materials in the bimaterial strip are different. As a result of the different coefficients of moisture expansion $\left(\alpha_1^{(C)}, \alpha_2^{(C)}\right)$ and different extent of moisture concentration change at hygroscopic equilibrium (dC_1, dC_2) in materials 1 and 2, we write the resulting bimaterial curvature $r^{(C)}$ due to a change in the environmental moisture concentration dC as

$$\frac{1}{r^{(C)}} = \frac{\alpha_1^{(C)} dC_1 - \alpha_2^{(C)} dC_2}{\frac{h}{2} + \frac{2}{h}(E_1 I_1 + E_2 I_2)\left(\frac{1}{E_1 h_1} + \frac{1}{E_2 h_2}\right)} \tag{20.2.12a}$$

or

$$\frac{1}{r^{(C)}} = \frac{\left[\alpha_1^{(C)}\left(\frac{dC_1}{dC}\right) - \alpha_2^{(C)}\left(\frac{dC_2}{dC}\right)\right]dC}{\frac{h}{2} + \frac{2}{h}(E_1 I_1 + E_2 I_2)\left(\frac{1}{E_1 h_1} + \frac{1}{E_2 h_2}\right)} \tag{20.2.12b}$$

where (dC_1/dC) and (dC_2/dC) in Eq. (20.2.12b) quantify the ratio of moisture concentration changes in materials 1 and 2 vis-à-vis the moisture concentration change in the surrounding environment. Suppose $dC_1/dC = dC_2/dC = 1$, Eq. (20.2.12) reduces to the form similar to Eq. (20.2.8). In order to focus on the effects of the individual material's expansion coefficient, we consider the bimaterial strips to possess equal thicknesses $(h_1 = h_2 = h/2)$ and equal Young's moduli $(E_1 = E_2)$, such that Eqs. (20.2.8), (20.2.11), and (20.2.12) simplify to

$$\begin{aligned} \frac{l}{r^{(T)}} &= \frac{3}{2}\frac{l}{h}\left(\alpha_1^{(T)} - \alpha_2^{(T)}\right)dT \\ \frac{l}{r^{(P)}} &= \frac{3}{2}\frac{l}{h}\left(\alpha_1^{(P)} - \alpha_2^{(P)}\right)(-dP) \\ \frac{l}{r^{(C)}} &= \frac{3}{2}\frac{l}{h}\left(\alpha_1^{(C)}dC_1 - \alpha_2^{(C)}dC_2\right) \end{aligned} \tag{20.2.13}$$

In modeling any effective coefficient of expansion, consideration is made to the strain induced by the environmental change. As such, reference points are to be identified for tracking relative displacements. These reference points are the center of the rigid square and the center of its nearest neighbor, as furnished in Fig. 20.8.

Due to the different radii of curvatures encountered by the bimaterial strip imposed by the different types of environmental changes, the following analyses begin with modeling for the effective CME, followed by reduction to the effective CTE and effective CPE for both grids. Arising from the large deflection and the corresponding finite relative displacement between the rigid squares, the usual definition of strain for infinitesimal deformation is written as an increment strain

$$d\varepsilon = \frac{dL}{L} \tag{20.2.14}$$

so as to pave a way for describing the total strain as

$$\varepsilon = \int d\varepsilon = \ln\frac{L_\mathrm{f}}{L_\mathrm{i}} \tag{20.2.15}$$

where L_i and L_f are the initial and final distances between reference points, respectively. This gives the moisture strain

$$\varepsilon_\mathrm{C} = \ln\frac{2\left(r^{(C)}\sin\theta + f\cos\theta + R\right) + h}{2(l + f + R) + h} \tag{20.2.16}$$

for Type A (Fig. 20.8, top), and

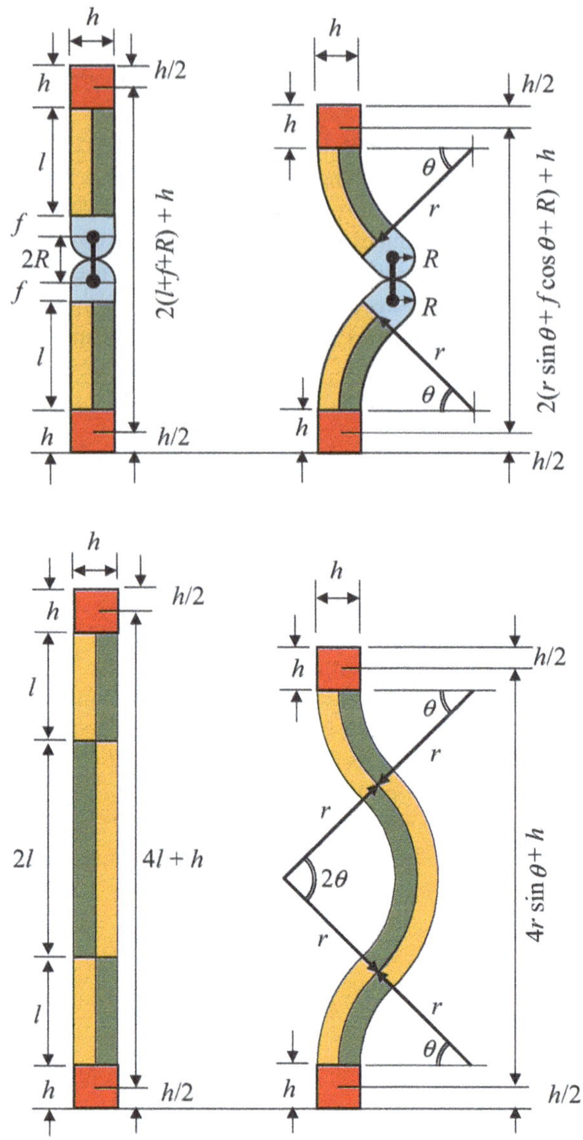

Fig. 20.8 Elements between two rigid blocks for the Type A grid (top) and Type B grid (bottom) in their original state (left) and deformed state (right). Reference points for relative displacements are the centers of rigid squares indicated in red. The thicknesses have been exaggerated for clarity

$$\varepsilon_C = \ln \frac{4r^{(C)} \sin \theta + h}{4l + h} \tag{20.2.17}$$

for Type B (Fig. 20.8, bottom). The length of a bimaterial strip is typically 2 or 3 orders higher than its thickness, which means that $r \sim l \gg h \approx f \approx R$ such that both Eqs. (20.2.16) and (20.2.17) abridge to

$$\varepsilon_C = \ln\left\{\frac{r^{(C)}\sin\theta}{l}\right\} \tag{20.2.18}$$

Substituting $\theta = l/r$ and Eq. (20.2.13) into Eq. (20.2.18) gives

$$\varepsilon_C = \ln\left\{\frac{\sin\left[\frac{3}{2}\frac{l}{h}\left(\alpha_1^{(C)}dC_1 - \alpha_2^{(C)}dC_2\right)\right]}{\left[\frac{3}{2}\frac{l}{h}\left(\alpha_1^{(C)}dC_1 - \alpha_2^{(C)}dC_2\right)\right]}\right\} \tag{20.2.19}$$

With reference to Eq. (20.2.7), we have the effective CME

$$\alpha^{(C)} = \frac{1}{dC}\ln\left\{\frac{\sin\left[\frac{3}{2}\frac{l}{h}\left(\alpha_1^{(C)}dC_1 - \alpha_2^{(C)}dC_2\right)\right]}{\left[\frac{3}{2}\frac{l}{h}\left(\alpha_1^{(C)}dC_1 - \alpha_2^{(C)}dC_2\right)\right]}\right\} \tag{20.2.20}$$

By comparing Eq. (20.2.7) against Eqs. (20.2.2) and (20.2.4)—or by comparing Eq. (20.2.12a) against Eqs. (20.2.8) and (20.2.11)—one may infer the effective CTE and the effective compressibility as

$$\alpha^{(T)} = \frac{1}{dT}\ln\left\{\frac{\sin\left[\frac{3}{2}\frac{l}{h}\left(\alpha_1^{(T)} - \alpha_2^{(T)}\right)dT\right]}{\left[\frac{3}{2}\frac{l}{h}\left(\alpha_1^{(T)} - \alpha_2^{(T)}\right)dT\right]}\right\} \tag{20.2.21}$$

and

$$\alpha^{(P)} = \frac{1}{(-dP)}\ln\left\{\frac{\sin\left[\frac{3}{2}\frac{l}{h}\left(\alpha_1^{(P)} - \alpha_2^{(P)}\right)(-dP)\right]}{\left[\frac{3}{2}\frac{l}{h}\left(\alpha_1^{(P)} - \alpha_2^{(P)}\right)(-dP)\right]}\right\} \tag{20.2.22a}$$

respectively, whereby $(-dP)$ indicates that for a conventional or positive value of compressibility, an increase in pressure results in decrease in size. As the sine function is odd, i.e., $\sin(-z) = -\sin(z)$, Eq. (20.2.22a) can also be written as

$$\alpha^{(P)} = -\frac{1}{dP}\ln\left\{\frac{\sin\left[\frac{3}{2}\frac{l}{h}\left(\alpha_1^{(P)} - \alpha_2^{(P)}\right)dP\right]}{\left[\frac{3}{2}\frac{l}{h}\left(\alpha_1^{(P)} - \alpha_2^{(P)}\right)dP\right]}\right\} \tag{20.2.22b}$$

Since $|\sin z| < |z|$ except when $z = 0$, it follows that $\ln\{\cdots\} < 0$ in Eqs. (20.2.20)–(20.2.22). Therefore the signs of $\alpha^{(C)}$ and $\alpha^{(T)}$ are always opposite to the signs of dC and dT, while the sign of $\alpha^{(P)}$ is always the same as that for dP, in consistency with the overall deformation conceptualized in Figs. 20.6 and 20.7. Having obtained models of expansion coefficients which switch signs for opposing change of environmental condition such that the Type A and Type B metamaterials always exhibit in-plane isotropic contraction based on the magnitude of the environmental change, an attempt in now made to specify the condition by which the deformed grid fits into the 8-pointed star. If the bimaterial deflection is insufficient, the pointed corners of the star along the diagonals are further from the center than those aligned along the vertical and horizontal axes. Beyond a certain value of bimaterial deflection, the diagonal pointed corners are nearer to the center than those lying on the axes. To fit the deformed metamaterial grid onto the 8-pointed star, the distance of the diagonal pointed corners from the center must be equal to those on the axes. This is shown from the superposition of an 8-pointed star (Fig. 20.9, top) onto each unit of the metamaterial (Fig. 20.9, middle).

Geometrical parameters a and b in Fig. 20.9 (bottom) facilitate the determination of conditions for fitting the deformed metamaterial onto the 8-pointed star. The parameter a is half of the distance between the centers of two nearest rigid squares (indicated in red). Therefore, with reference to the vertical distance between the centers of the rigid squares in Fig. 20.8 (right),

$$a = r \sin\theta + f \cos\theta + R + \frac{h}{2} \tag{20.2.23}$$

for Type A, and

$$a = 2r \sin\theta + \frac{h}{2} \tag{20.2.24}$$

for Type B. The parameter b in Fig. 20.9 (bottom left) corresponds to the horizontal distance component between the center of a rigid square and the outermost surface of the pillow block bearing in Fig. 20.8 (top right), i.e.,

$$b = r(1 - \cos\theta) + f \sin\theta + R \tag{20.2.25}$$

for Type A, while the same parameter in Fig. 20.9 (bottom right) refers to the horizontal distance component between the center of a rigid square and the maximum deflection at the mid-point of the bimaterial strip in Fig. 20.8 (bottom right), i.e.,

$$b = 2r(1 - \cos\theta) \tag{20.2.26}$$

for Type B. These give the distances from the center of each deformed grid to the grid rib intersecting the vertical and horizontal axes as

$$a + b = r(1 + \sin\theta - \cos\theta) + f(\sin\theta + \cos\theta) + 2R + \frac{h}{2} \tag{20.2.27}$$

and

$$a + b = 2r(1 + \sin\theta - \cos\theta) + \frac{h}{2} \tag{20.2.28}$$

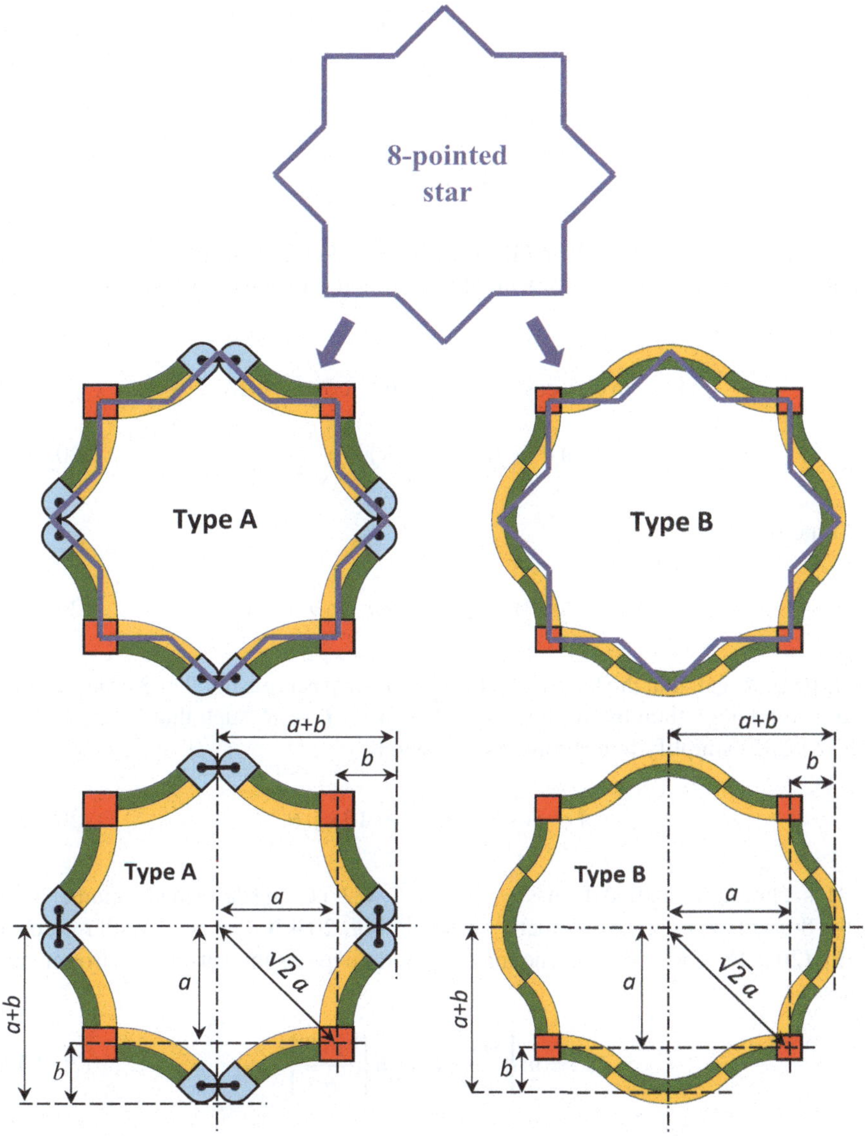

Fig. 20.9 Superposition of an 8-pointed star (top) onto a unit of the deformed metamaterials (middle), with parameters assigned for analysis (bottom)

for Type A and Type B, respectively. Perusal to Fig. 20.9 (bottom) implies that the diagonal distances between the center of the deformed grid and the center of each rigid square are the hypotenuse, i.e.,

$$\sqrt{2}a = \sqrt{2}(r\sin\theta + f\cos\theta + R) + \frac{h}{\sqrt{2}} \tag{20.2.29}$$

and

$$\sqrt{2}a = 2\sqrt{2}r\sin\theta + \frac{h}{\sqrt{2}} \tag{20.2.30}$$

for metamaterials of Types A and B, respectively. Hence to implement equidistance for all the eight points of the star from the deformed grid center, we let $a + b = \sqrt{2}a$ such that

$$\begin{aligned} &r(1+\sin\theta-\cos\theta)+f(\sin\theta+\cos\theta)+2R+\frac{h}{2} \\ &= \sqrt{2}(r\sin\theta+f\cos\theta+R)+\frac{h}{\sqrt{2}} \end{aligned} \tag{20.2.31}$$

for Type A, and

$$2r(1+\sin\theta-\cos\theta)+\frac{h}{2} = 2\sqrt{2}r\sin\theta+\frac{h}{\sqrt{2}} \tag{20.2.32}$$

for Type B. Recall that the length of a bimaterial strip is typically 2 or 3 orders higher than its thickness, thereby implying $r \sim l \gg h \sim f \sim R$, such that Eqs. (20.2.31) and (20.2.32) simplify to a common expression

$$1-\cos\theta = \left(\sqrt{2}-1\right)\sin\theta \tag{20.2.33}$$

This relation is satisfied if $\sin\theta = \cos\theta = 1/\sqrt{2}$, i.e., the deformed metamaterials fit the 8-pointed star array geometry shown in Fig. 20.4 if $\theta = \pi/4$. With reference to Eq. (20.2.18) and $l/r = \theta$, one may elegantly express the various environmental strains as

$$\varepsilon = \ln\left\{\frac{\sin\theta}{l/r}\right\} = \ln\left\{\frac{\sin\theta}{\theta}\right\} \tag{20.2.34}$$

such that substitution of $\theta = \pi/4$ suggests that the effective strain $\varepsilon = \ln\left\{2\sqrt{2}/\pi\right\} = -0.10501$ must be achieved to attain the desired 8-pointed star array. Furthermore, matching Eq. (20.2.34) against Eqs. (20.2.20) to (20.2.22) implies

$$\theta = \frac{3}{2}\frac{l}{h}\left(\alpha_1^{(C)}dC_1 - \alpha_2^{(C)}dC_2\right) = \frac{3}{2}\frac{l}{h}\left(\alpha_1^{(T)} - \alpha_2^{(T)}\right)dT = \frac{3}{2}\frac{l}{h}\left(\alpha_1^{(P)} - \alpha_2^{(P)}\right)dP \tag{20.2.35}$$

which reduces to

$$\alpha_1^{(C)}dC_1 - \alpha_2^{(C)}dC_2 = \left(\alpha_1^{(T)} - \alpha_2^{(T)}\right)dT = \left(\alpha_1^{(P)} - \alpha_2^{(P)}\right)dP = \frac{\pi}{6}\left(\frac{h}{l}\right) \tag{20.2.36}$$

when $\theta = \pi/4$. In other words, if properties of materials 1 and 2 of the bimaterial strip are known, then the array of 8-pointed stars can be achieved from the originally square grids of Type A and Type B metamaterial by controlling the environmental change.

In order to ascertain that the assumption $f = R = h = 0$ is valid, we return to Eqs. (20.2.16) and (20.2.17) to divide all terms at the numerator and denominator with $r^{(C)}$ to obtain

$$\varepsilon_C = \ln \frac{2\left(\sin\theta + \frac{f}{l}\left(\frac{l}{r^{(C)}}\right)\cos\theta + \frac{R}{l}\left(\frac{l}{r^{(C)}}\right)\right) + \frac{h}{l}\left(\frac{l}{r^{(C)}}\right)}{2\left(\frac{l}{r^{(C)}} + \frac{f}{l}\left(\frac{l}{r^{(C)}}\right) + \frac{R}{l}\left(\frac{l}{r^{(C)}}\right)\right) + \frac{h}{l}\left(\frac{l}{r^{(C)}}\right)} \tag{20.2.37}$$

for Type A, and

$$\varepsilon_C = \ln \frac{4\sin\theta + \frac{h}{l}\left(\frac{l}{r^{(C)}}\right)}{4\frac{l}{r^{(C)}} + \frac{h}{l}\left(\frac{l}{r^{(C)}}\right)} \tag{20.2.38}$$

for Type B, such that the substitution of $l/r^{(C)} = \theta = \pi/4$ into Eqs. (20.2.37) and (20.2.38) leads to

$$\varepsilon_C = \ln \frac{\frac{2\sqrt{2}}{\pi} + \frac{1}{\sqrt{2}}\frac{f}{l} + \frac{R}{l} + \frac{1}{2}\frac{h}{l}}{1 + \frac{f}{l} + \frac{R}{l} + \frac{1}{2}\frac{h}{l}} \tag{20.2.39}$$

and

$$\varepsilon_C = \ln \frac{\frac{2\sqrt{2}}{\pi} + \frac{1}{4}\frac{h}{l}}{1 + \frac{1}{4}\frac{h}{l}} \tag{20.2.40}$$

respectively. Since the typical bimaterial strip aspect ratio is such that l is about 2 or 3 orders higher than h, and perusal to Fig. 20.8 indicates that $R = h/2$, we substitute $h/l = 2R/l = 0.01$ into Eq. (20.2.39) to observe the effect of f/l for Type A moisture strain when the 8-pointed star array is attained. Figure 20.10 (left) shows that, under the considered typical parameter values, the percentage error hovers around 1%. Since Type B is independent from f and R, evaluation is made on the effect of h/l ratio on the moisture strain. Figure 20.10 (right) shows that the percentage error for the typical h/l ratio falls within 0.2644%. Even if the bimaterial aspect ratio is increased to the unrealistically high value of $h/l = 0.1$, the percentage error for assuming $h = 0$ is only 2.637%. These observations validate the assumption of $f = R = h = 0$ for the typical bimaterial aspect ratio.

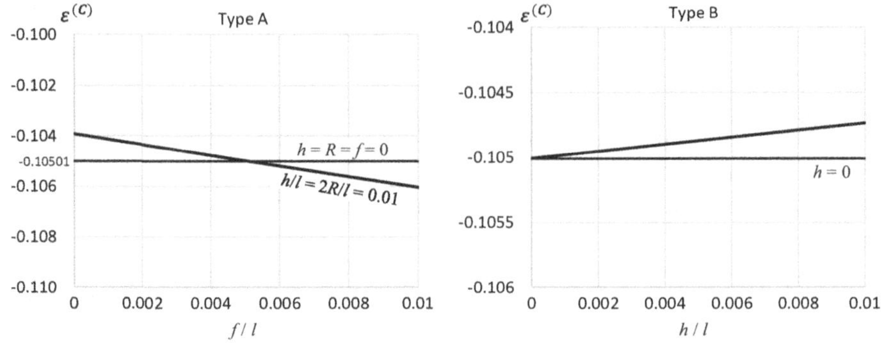

Fig. 20.10 Plots of moisture strain versus for f/l at $h/l = 2R/l = 0.01$ for Type A (left) and versus h/l for Type B (right) for comparison against the simplifying assumptions of $f = R = h = 0$

Having demonstrated the applicability of the developed models when the typical bimaterial aspect ratio is adopted, we shall now turn our attention to the various facets of bimaterial curving in order to attain the array of 8-pointed stars under varying moisture concentration, temperature, and pressure. Typically, the coefficient of moisture expansion in polymers ranges between $\alpha_1^{(C)} = 2 \times 10^{-3}$ and $\alpha_1^{(C)} = 5 \times 10^{-3}$. For the case of moisture expansion, we consider the abovementioned polymers as material 1 while material 2 is of the same polymeric material but with waterproof coating so that $dC_2 = 0$. The choice of same material ensures $E_1 = E_2$ so that the simplification of Eq. (20.2.12) to the third of Eq. (20.2.13) applies. Although $\alpha_1^{(C)} = \alpha_2^{(C)}$, expansion of material 2 is inhibited due to waterproofing so that $\alpha_2^{(C)} dC_2 = 0$. Adopting these values for Eq. (20.2.36) gives

$$dC_1 = -\frac{1}{\alpha_1^{(C)}} \frac{\pi}{6} \left(\frac{h}{l} \right) \tag{20.2.41}$$

This is plotted in Fig. 20.11 for $2 \times 10^{-3} \leq \alpha_1^{(C)} \leq 5 \times 10^{-3}$ within $0 \leq h/l \leq 0.01$.

For the case of thermal expansion, we consider the brass–titanium (B-T), copper–steel (C-S) and tungsten–silicon carbide (T-SC) bimaterial pairs. The motivation for pairing of these materials can be seen from their almost equal Young's modulus $E_1 \approx E_2$ but with sufficient contrast of their individual CTEs so that, instead of using Eq. (20.2.8), one is justified to use the first of Eq. (20.2.13), which is the simplified form. Their thermomechanical properties are listed in Table 20.2 and, using these properties, the required temperature change in order for both metamaterials Type A and B to attain the 8-pointed star array by using pairs of tungsten—silicon carbide, copper—steel and brass—titanium bimaterial strips are plotted in Fig. 20.12 using Eq. (20.2.36).

For the case of compressibility, we consider the magnesium–silicon (M-S), zinc–niobium (Z-N), and manganese–nickel (M-N) bimaterial pairs. The motivation for

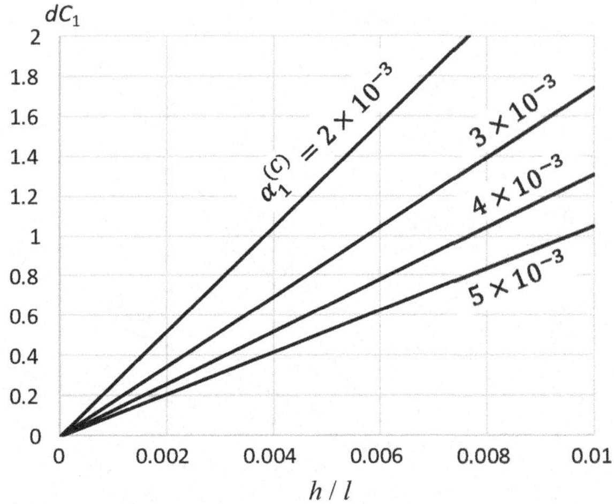

Fig. 20.11 Plots of required moisture absorbed into, or dissipated from, material 1 made from polymers while material 2 is of same material with waterproof coating ($dC_2 = 0$) in order for both metamaterials Type A and Type B to attain the 8-pointed star array

Table 20.2 Material properties for the brass–titanium (B-T), copper–steel (C-S), and tungsten–silicon carbide (T-SC) bimaterial pairs

Material pairs	CTE	Young's modulus
1 = Brass 2 = Titanium	$\alpha_1^{(T)} = 19 \times 10^{-6}\text{K}^{-1}$ $\alpha_2^{(T)} = 8.6 \times 10^{-6}\text{K}^{-1}$	$E_1 = 112.5\,\text{GPa}$ $E_2 = 110.3\,\text{GPa}$
1 = Copper 2 = Steel	$\alpha_1^{(T)} = 17 \times 10^{-6}\text{K}^{-1}$ $\alpha_2^{(T)} = 12 \times 10^{-6}\text{K}^{-1}$	$E_1 = 117\,\text{GPa}$ $E_2 = 200\,\text{GPa}$
1 = Tungsten 2 = Silicon carbide	$\alpha_1^{(T)} = 4.50 \times 10^{-6}\text{K}^{-1}$ $\alpha_2^{(T)} = 2.77 \times 10^{-6}\text{K}^{-1}$	$E_1 = 405\,\text{GPa}$ $E_2 = 450\,\text{GPa}$

selecting these pairs of materials can be seen from their almost equal Young's moduli $E_1 \approx E_2$ but with ample disparity of their individual compressibilities so that, instead of employing Eq. (20.2.11), one is justified to adopt the reduced form given in the second of Eq. (20.2.13). Due to the lack of CPE data, these were converted from their bulk modulus data as follows. From the definition of bulk modulus,

$$K = -V\frac{dP}{dV} \tag{20.2.42}$$

which is a reciprocal of Eq. (20.2.3) whereby $V = L^3$ and $dV = 3L^2dL$, substituting $dV/V = 3dL/L = 3\varepsilon_P$ into Eq. (20.2.42) gives

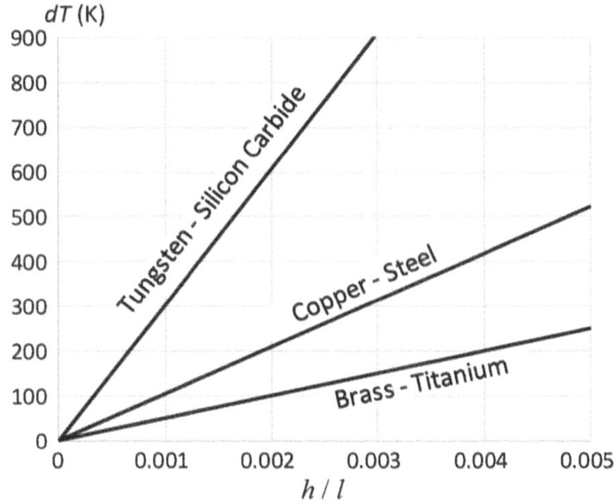

Fig. 20.12 Plots of required temperature change in order for both metamaterials Type A and B to attain the 8-pointed star array by using pairs of tungsten–silicon carbide, copper–steel and brass–titanium bimaterial strips of various aspect ratio

$$K = \frac{-dP}{3\varepsilon_P} \tag{20.2.43}$$

which, upon comparison with Eq. (20.2.4), gives

$$\alpha^{(P)} = \frac{1}{3K} \tag{20.2.44}$$

Their properties are listed in Table 20.3. The required pressure change in order for both metamaterials Type A and Type B to attain the 8-pointed star array by using pairs of magnesium–silicon, zinc–niobium, and manganese–nickel strips is plotted in Fig. 20.13 using Eq. (20.2.36).

An explanation on Figs. 20.12 and 20.13 can now be made in the light of metamaterials concept. As metamaterials are known to demonstrate properties from their artificially designed microarchitecture, the choice of bimaterial strip arrangements

Table 20.3 Material properties for the magnesium–silicon (M-S), zinc–niobium (Z-N), and manganese–nickel (M-N) bimaterial pairs

Material pairs	Young's modulus	Bulk modulus	Compressibility
1 = Magnesium 2 = Silicon	$E_1 = 45\,\text{GPa}$ $E_2 = 47\,\text{GPa}$	$K_1 = 45\,\text{GPa}$ $K_2 = 100\,\text{GPa}$	$\alpha_1^{(P)} = 7.407 \times 10^{-3}\text{GPa}^{-1}$ $\alpha_2^{(P)} = 3.333 \times 10^{-3}\text{GPa}^{-1}$
1 = Zinc 2 = Niobium	$E_1 = 108\,\text{GPa}$ $E_2 = 105\,\text{GPa}$	$K_1 = 70\,\text{GPa}$ $K_2 = 170\,\text{GPa}$	$\alpha_1^{(P)} = 4.762 \times 10^{-3}\text{GPa}^{-1}$ $\alpha_2^{(P)} = 1.961 \times 10^{-3}\text{GPa}^{-1}$
1 = Manganese 2 = Nickel	$E_1 = 198\,\text{GPa}$ $E_2 = 200\,\text{GPa}$	$K_1 = 120\,\text{GPa}$ $K_2 = 180\,\text{GPa}$	$\alpha_1^{(P)} = 2.778 \times 10^{-3}\text{GPa}^{-1}$ $\alpha_2^{(P)} = 1.852 \times 10^{-3}\text{GPa}^{-1}$

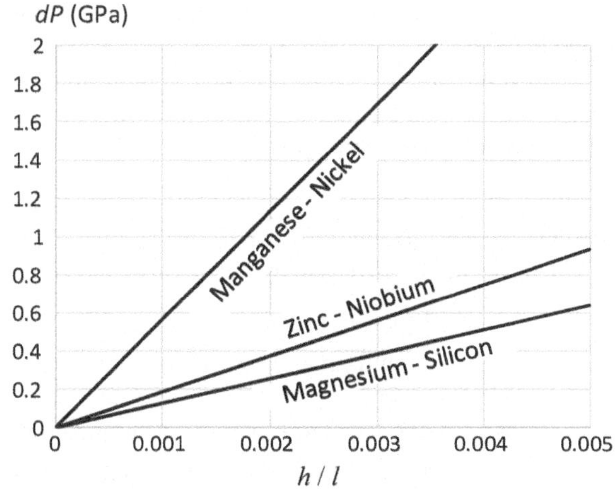

Fig. 20.13 Plots of required pressure change in order for both metamaterials Type A and Type B to attain the 8-pointed star array by using pairs of magnesium–silicon, zinc–niobium, and manganese–nickel bimaterial strips of various aspect ratio

shown in Figs. 20.6 (top) and 20.7 (top) not only permits bending of the straight strips into curves that approximate the Islamic design, but the exact 8-pointed star arrays—displayed Figs. 20.4 and 20.9—are attained when the base properties of the bimaterial strips, environmental changes, and bimaterial aspect ratio h/l fulfill the condition set out in Eq. (20.2.36). To obtain the 2D array of 8-pointed stars featured in Fig. 20.9 in response to temperature change, one will need to refer to a relevant plot in Fig. 20.12 pertaining to one of the bimaterial pairs, for example. If the root mean square of the environment's temperature fluctuation is known, then the required bimaterial aspect ratio h/l can be read from the horizontal axis of Fig. 20.12. Likewise, to obtain the 8-pointed star arrays of Fig. 20.9 due to pressure variation, one can peruse to a related plot in Fig. 20.13 corresponding to one of the bimaterial pairs, for instance. Suppose the root mean square of the environment's pressure undulation is known, then the h/l ratio of the bimaterial strip can be taken from the abscissa of Fig. 20.13. In other words, for a given environmental variation and bimaterial pairs with known properties, the discussed Islamic motif can be achieved by choosing the slenderness of the bimaterial strip. This geometrical control of the microstructures is associated with metamaterials design concept.

Finally, a summary can be made on the responses of both the Type A and Type B units for each variable, and these are categorized into three groups:

1. the geometrical parameters, such as the Type A and Type B arrangements, the bimaterial thickness h and a segment of its length l, and in the case of Type A the geometrical parameters of the connecting parts f and R
2. the material parameters, such as the linear coefficients of moisture expansion $\alpha_1^{(C)}$ and $\alpha_2^{(C)}$, linear coefficients of thermal expansion $\alpha_1^{(T)}$ and $\alpha_2^{(T)}$, and linear compressibilities $\alpha_1^{(P)}$ and $\alpha_2^{(P)}$ of the bimaterial phases 1 and 2; and
3. the environmental parameters, such as the changes of moisture concentration dC_1 and dC_2, temperature dT, and pressure dP on bimaterial phases 1 and 2.

The radius of curvature r formed is not included as it is a function of the parameters listed in (1)–(3).

With reference to Fig. 20.14 for the morphing of the Type A and Type B cell shapes in response to the parameters, the extent of deformation is facilitated by the following groups of parameters:

1. increase in the bimaterial aspect ratio for each segment l/h (purely geometrical parameters)
2. increase in the differences between expansion coefficients $\left|\alpha_1^{(T)} - \alpha_2^{(T)}\right|$ and $\left|\alpha_1^{(P)} - \alpha_2^{(P)}\right|$ (purely material parameters)
3. increase the environmental variations $|dT|$ and $|dP|$ (purely environmental parameters) and
4. increase in the moisture strain difference between the bimaterial phases $\left|\alpha_1^{(C)} dC_1 - \alpha_2^{(C)} dC_2\right|$ (combined material and environmental parameters).

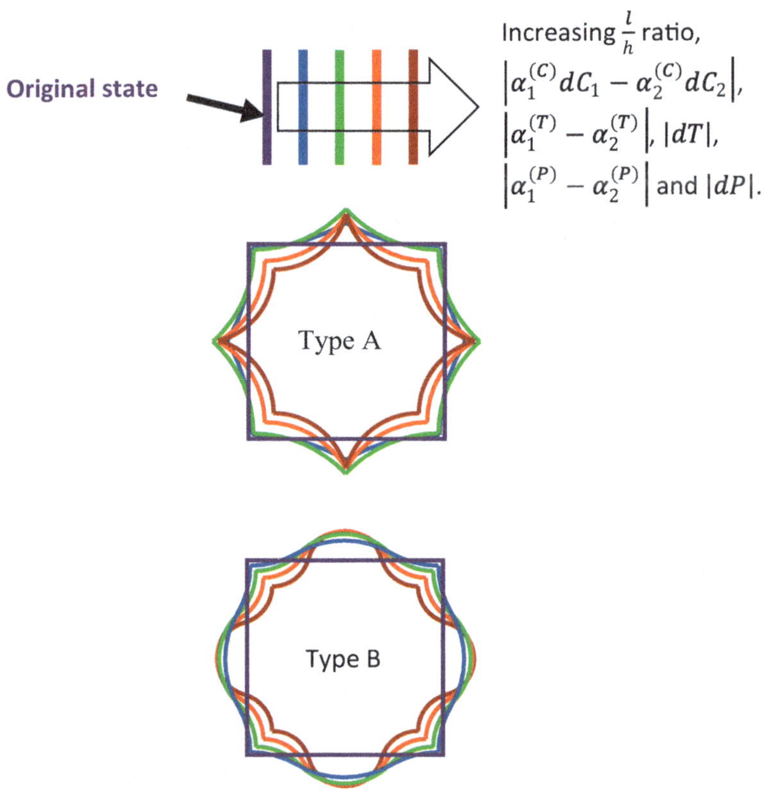

Fig. 20.14 Responses of Type A and Type B units with variation of geometrical, material and environmental parameters

The parameters are described in their magnitudes in order to include negative values. With reference to the diagonal array of 8-pointed stars with alternating cross-shaped cells–such as those shown in Figs. 20.5 (bottom), 20.6 (bottom) or Fig. 20.7 (bottom)—it is the adjacent neighbors that form the 8-pointed stars when the signs are negative. Effects from the geometrical parameters of the connecting parts f and R are negligible for the practical range of h/l with $l \gg h \sim f \sim R$, and as a consequence the effects of selecting either Type A or Type B is also insignificant.

20.3 Linkage-Based Sign-Switching Metamaterial with Islamic Motifs

Unlike the previous section whereby the Islamic motifs are achieved by the use of bimaterial strips, which inherently give curved lines, the metamaterial discussed in this section adopts the linkage mechanism. Consequently, the metamaterial–from its original state to the final state–exhibits straight lines, and is therefore closer to the original art form.

A possible unit cell in its original state is shown in Fig. 20.15 (extreme left), which is made from two types of materials: (1) a cross rod, indicated in blue, that is made from materials with higher CTE $\alpha_{\mathrm{r}}^{(T)}$ and higher CME $\alpha_{\mathrm{r}}^{(C)}$, as well as (2) frame linkage, indicated in black (including the red blocks), that is made from materials with negligible CTE $\alpha_{\mathrm{f}}^{(T)}$ and negligible CME $\alpha_{\mathrm{f}}^{(C)}$. The condition set in (2) implies that each segment of the frame linkage is assumed rigid and therefore can neither expand nor contract, but the frame linkage as a whole can deform via rotation at the hinges. Upon heating and/or moistening the cross rods extend such that the frame linkage deforms into an 8-pointed star illustrated in Fig. 20.15 (top most). Upon cooling and/or drying from its original state, the rods contract such that the frame linkage does not form an 8-pointed star (Fig. 20.15, bottom most). However, if the unit cells are arranged in alternate pattern such that the four nearest neighbors of each unit cell are empty squares, then the deformed shape permits the empty neighboring spaces to form the 8-pointed stars. Figure 20.16 (top) shows the 2D array of metamaterials whereby the unit cells are arranged in alternating pattern by connecting at the rigid block corners in diagonal, then the 8-pointed-star Islamic motifs are obtained regardless of whether the rods expand or contract (Lim 2020c). Under such an arrangement, the effective strain is always negative if the frame linkage possesses zero expansion coefficients. This is because regardless of whether the cross rods lengthen or shorten, the corner blocks always move toward the center of the unit cell (Fig. 20.16, bottom). For this reason, the metamaterial exhibits sign-toggling of hygrothermal expansivity. Specifically, if $\alpha_{\mathrm{f}}^{(T)} = \alpha_{\mathrm{f}}^{(C)} = 0$, then

- $\alpha_{\mathrm{eff}}^{(T)} < 0$ for $dT > 0$ but reverses to $\alpha_{\mathrm{eff}}^{(T)} > 0$ for $dT < 0$
- $\alpha_{\mathrm{eff}}^{(C)} < 0$ for $dC > 0$ but reverses to $\alpha_{\mathrm{eff}}^{(C)} > 0$ for $dC < 0$
- $\alpha_{\mathrm{eff}}^{(T)} \propto -dT$ and $\alpha_{\mathrm{eff}}^{(C)} \propto -dC$

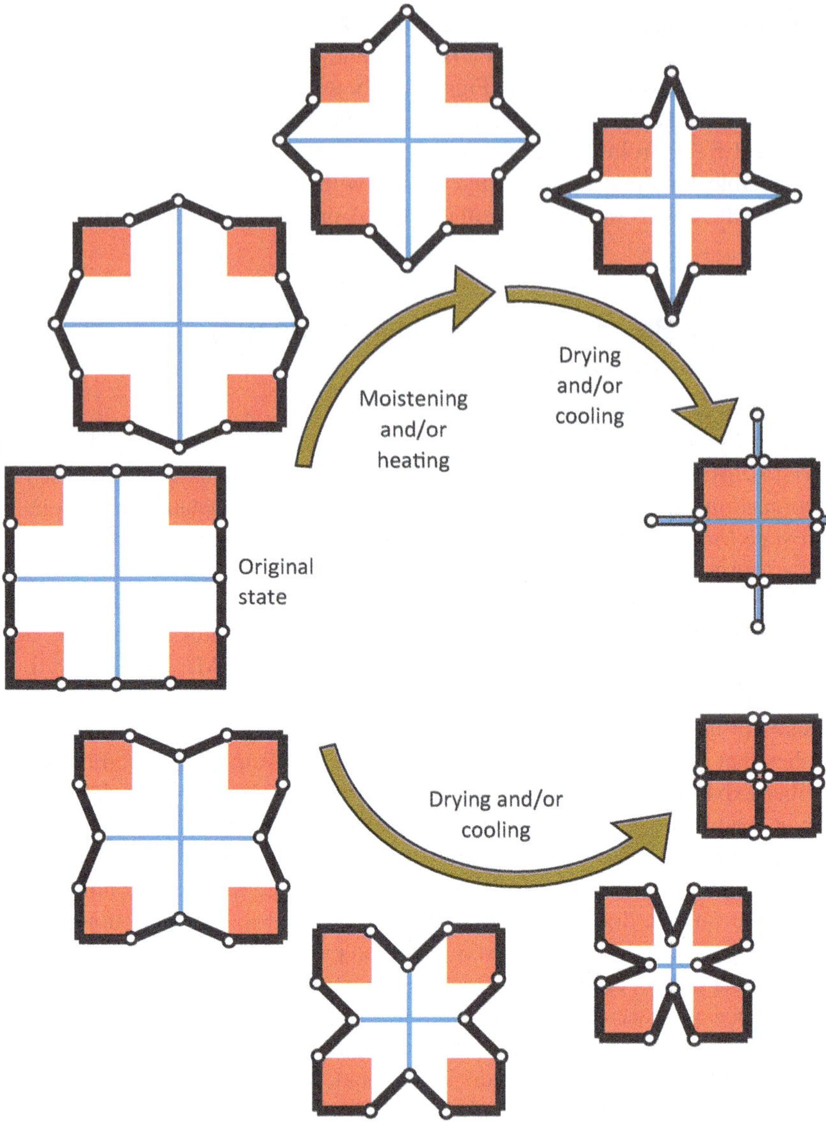

Fig. 20.15 Deformation of the square grid due to increasing and decreasing moisture concentration and temperature. Pin joints are indicated as circles

In addition, it will later be shown that if $\alpha_{\mathrm{f}}^{(T)} = \alpha_{\mathrm{f}}^{(C)} = 0$, then the 2D array of 8-pointed stars can be achieved if the rod strain is either $\varepsilon_{\mathrm{r}} = -1/2 + 1/\sqrt{2}$ or $\varepsilon_{\mathrm{r}} = -1/2$, which are indicated in Fig. 20.16 (bottom right) and (bottom left), respectively. Under both conditions, the effective strains are common, with the value $\varepsilon_{\mathrm{eff}} = -1/2 + 1/\left(2\sqrt{2}\right)$.

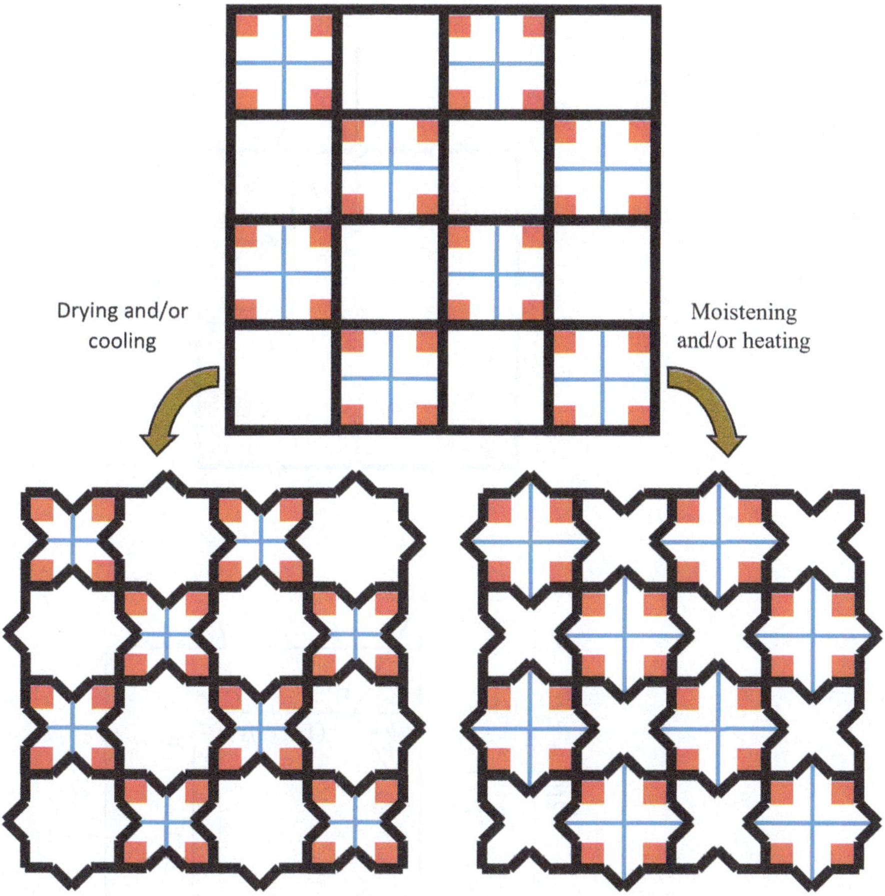

Fig. 20.16 Overall contraction of the square grid (top) into two possible 2D arrays of 8-pointed stars (bottom) with environmental changes. Pin joints are not shown for clarity

To pave a way for the analysis, we consider a unit cell as shown in Fig. 20.17 (top). The x and y axes are placed on this grid such that the origin is located at the center of the cell and therefore at the junction of the centrally joined rods, with the axes being parallel to the sides of the frame linkage. Due to symmetry, only 1/8 of this unit cell is adopted for analysis. We assign points A, B, and C as shown in Fig. 20.17 (top), whereby each rigid square block is of side l while the frame side is $4l$, thereby giving a quarter of the frame side as $2l$. During an increase in temperature and/or moisture absorption of the rod, it lengthens such that point A moves to point A'. Due to symmetry, point C moves to point C' along the diagonal and point B simultaneously moves to point B' in parallel. The overall contraction, as specified by the motion of C to C' during an increase in temperature and/or moisture absorption, indicates negative thermal and/or moisture expansion coefficient(s). During a decrease in temperature

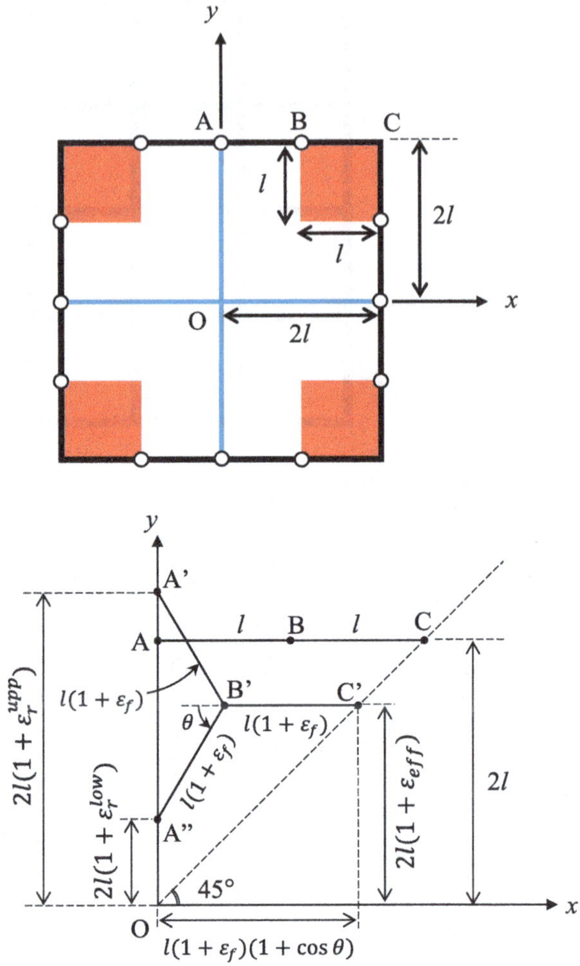

Fig. 20.17 A unit cell showing the locations O, A, B and C in one quadrant for tracking (top) and a detailed schematic showing the displaced locations with hygrothermal strains for analysis (bottom)

and/or moisture dissipation of the rod, it shortens such that point A moves to point A''. By virtue of symmetry, BC undergoes translational motion to $B'C'$. This overall contraction during a decrease in temperature and/or moisture dissipation denotes positive thermal and/or moisture expansion coefficient(s).

The rod and frame linkage strains (ε_r, ε_f) during hygrothermal variation—i.e., temperature change dT as well as moisture concentration changes in the rod dC_r and frame dC_f—consist of two components: the thermal strain and the moisture strain

$$\begin{aligned} \varepsilon_r &= \alpha_r^{(T)} dT + \alpha_r^{(C)} dC_r \\ \varepsilon_f &= \alpha_f^{(T)} dT + \alpha_f^{(C)} dC_f \end{aligned} \tag{20.3.1}$$

where $\alpha_{\mathrm{r}}^{(T)}$ and $\alpha_{\mathrm{r}}^{(C)}$ are the coefficient of thermal expansion (CTE) and coefficient of moisture expansion (CME), respectively, of the rod, while $\alpha_{\mathrm{f}}^{(T)}$ and $\alpha_{\mathrm{f}}^{(C)}$ are the CTE and CME for the frame, respectively. The moisture concentration change in the rod dC_{r} and frame linkage dC_{f} can be obtained based on the general definition of moisture concentration in a solid

$$C_i = \frac{m_i}{M_i} \times 100; \quad (i = r, f) \tag{20.3.2}$$

where m_i is the moisture mass and M_i is the mass of the dry solid corresponding to the cross rods ($i = r$) and frame ($i = f$). Hence, a change in moisture mass dm_i in each solid phase gives the corresponding change in moisture concentration in the solid phase as

$$dC_i = \frac{dm_i}{M_i} \times 100; \quad (i = r, f). \tag{20.3.3}$$

For brevity, the subsequent analysis adopts the rod and frame strains ($\varepsilon_{\mathrm{r}}, \varepsilon_{\mathrm{f}}$)—without referring to the thermal and moisture strains in each part. In addition, the analysis also considers the effective strain $\varepsilon_{\mathrm{eff}}$ as defined from the motion of point *C*. As such, it is possible to express the effective strain as a function of rod and frame strains

$$\varepsilon_{\mathrm{eff}} = f(\varepsilon_{\mathrm{r}}, \varepsilon_{\mathrm{f}}) \tag{20.3.4}$$

or, inversely, express the rod strain in terms of the frame and effective strains

$$\varepsilon_{\mathrm{r}} = f^{-1}(\varepsilon_{\mathrm{f}}, \varepsilon_{\mathrm{eff}}). \tag{20.3.5}$$

Perusal to Fig. 20.17 (bottom) shows that for each prescribed ε_{r} there is only one solution for $\varepsilon_{\mathrm{eff}}$, but for every prescribed $\varepsilon_{\mathrm{eff}}$ there are two solutions for ε_{r}. To contrast the two solutions, we employ the upper solution $\varepsilon_{\mathrm{r}} = \varepsilon_{\mathrm{r}}^{\mathrm{upp}}$ corresponding to heating $dT > 0$ and/or moistening $dC > 0$ as well as the lower solution $\varepsilon_{\mathrm{r}} = \varepsilon_{\mathrm{r}}^{\mathrm{low}}$ for cooling $dT < 0$ and/or drying $dC < 0$ with reference to the original state. This permits the elongated rod length to be expressed as

$$\overline{OA'} = 2l\left(1 + \varepsilon_{\mathrm{r}}^{\mathrm{upp}}\right) \tag{20.3.6}$$

during heating and/or moistening, as well as the shortened rod length

$$\overline{OA''} = 2l\left(1 + \varepsilon_{\mathrm{r}}^{\mathrm{low}}\right) \tag{20.3.7}$$

under cooling and/or drying, whereby

$$\varepsilon_{\mathrm{r}}^{\mathrm{low}} < 0 < \varepsilon_{\mathrm{r}}^{\mathrm{upp}} \tag{20.3.8}$$

For both cases, the rigid squares move toward the origin such that the resolved distance of $\overline{OC'}$, as projected onto the y-axis is

$$\left(\overline{OC'}\right)_{y\text{ - axis}} = 2l(1+\varepsilon_{\text{eff}}) \tag{20.3.9}$$

When projected against the x-axis, we have

$$\left(\overline{OC'}\right)_{x\text{ - axis}} = l(1+\varepsilon_{\text{f}})(1+\cos\theta) \tag{20.3.10}$$

Since C' rests on the $y = x$ line, i.e., $\left(\overline{OC'}\right)_{x\text{ - axis}} = \left(\overline{OC'}\right)_{y\text{ - axis}}$, we obtain

$$\cos\theta = 2\frac{1+\varepsilon_{\text{eff}}}{1+\varepsilon_{\text{f}}} - 1 \tag{20.3.11}$$

The projected length $\overline{OC'}$ on the y-axis can also be made through the $OA'B'$ and $OA''B'$ linkages as

$$\begin{aligned} \left(\overline{OC'}\right)^{OA'B'}_{y\text{ - axis}} &= 2l\left(1+\varepsilon_{\text{r}}^{\text{upp}}\right) + l(1+\varepsilon_{\text{f}})\sin(-\theta) \\ \left(\overline{OC'}\right)^{OA''B'}_{y\text{ - axis}} &= 2l\left(1+\varepsilon_{\text{r}}^{\text{low}}\right) + l(1+\varepsilon_{\text{f}})\sin(+\theta) \end{aligned} \tag{20.3.12}$$

Equation (20.3.12) can be compared with Eq. (20.3.9) to yield

$$\begin{aligned} \varepsilon_{\text{r}}^{\text{upp}} &= \varepsilon_{\text{eff}} + \frac{1}{2}(1+\varepsilon_{\text{f}})\sin\theta \\ \varepsilon_{\text{r}}^{\text{low}} &= \varepsilon_{\text{eff}} - \frac{1}{2}(1+\varepsilon_{\text{f}})\sin\theta \end{aligned} \tag{20.3.13}$$

The angular rotation of link $\overline{AB}$ to $\overline{A'B}$ via $-\theta$ or to $\overline{A''B}$ through $+\theta$ can be written in terms of ε_{eff} by substituting Eq. (20.3.11) into Eq. (20.3.13) to give

$$\begin{aligned} \varepsilon_{\text{r}}^{\text{upp}} &= \varepsilon_{\text{eff}} + \frac{1}{2}(1+\varepsilon_{\text{f}})\sin\left[\cos^{-1}\left(2\frac{1+\varepsilon_{\text{eff}}}{1+\varepsilon_{\text{f}}} - 1\right)\right] \\ \varepsilon_{\text{r}}^{\text{low}} &= \varepsilon_{\text{eff}} - \frac{1}{2}(1+\varepsilon_{\text{f}})\sin\left[\cos^{-1}\left(2\frac{1+\varepsilon_{\text{eff}}}{1+\varepsilon_{\text{f}}} - 1\right)\right] \end{aligned} \tag{20.3.14}$$

The upper and lower strains can alternatively be expressed without the trigonometric functions. Based on Eq. (20.3.11), we deduce from a right-angled triangle that

$$\sin\theta = \frac{2}{1+\varepsilon_{\text{f}}}\sqrt{(1+\varepsilon_{\text{eff}})(\varepsilon_{\text{f}}-\varepsilon_{\text{eff}})} \tag{20.3.15}$$

in order to recast Eq. (20.3.12) as

$$\begin{aligned}\left(\overline{OC'}\right)_{y\text{ - axis}}^{OA'B'} &= 2l\left(1+\varepsilon_{\mathrm{r}}^{\mathrm{upp}}\right) - 2l\sqrt{(1+\varepsilon_{\mathrm{eff}})(\varepsilon_{\mathrm{f}}-\varepsilon_{\mathrm{eff}})}\\ \left(\overline{OC'}\right)_{y\text{ - axis}}^{OA''B'} &= 2l\left(1+\varepsilon_{\mathrm{r}}^{\mathrm{low}}\right) + 2l\sqrt{(1+\varepsilon_{\mathrm{eff}})(\varepsilon_{\mathrm{f}}-\varepsilon_{\mathrm{eff}})}\end{aligned} \tag{20.3.16}$$

When Eq. (20.3.16) is matched against Eq. (20.3.9), we obtain

$$\begin{aligned}\varepsilon_{\mathrm{r}}^{\mathrm{upp}} &= \varepsilon_{\mathrm{eff}} + \sqrt{(1+\varepsilon_{\mathrm{eff}})(\varepsilon_{\mathrm{f}}-\varepsilon_{\mathrm{eff}})}\\ \varepsilon_{\mathrm{r}}^{\mathrm{low}} &= \varepsilon_{\mathrm{eff}} - \sqrt{(1+\varepsilon_{\mathrm{eff}})(\varepsilon_{\mathrm{f}}-\varepsilon_{\mathrm{eff}})}\end{aligned} \tag{20.3.17}$$

Having obtained the rod strain in terms of $\varepsilon_{\mathrm{eff}}$, it can also be shown that $\varepsilon_{\mathrm{eff}}$ can be expressed as a function of ε_{r}. From Eq. (20.3.17), we write

$$\begin{aligned}\varepsilon_{\mathrm{r}}^{\mathrm{upp}} - \varepsilon_{\mathrm{eff}} &= +\sqrt{(1+\varepsilon_{\mathrm{eff}})(\varepsilon_{\mathrm{f}}-\varepsilon_{\mathrm{eff}})}\\ \varepsilon_{\mathrm{r}}^{\mathrm{low}} - \varepsilon_{\mathrm{eff}} &= -\sqrt{(1+\varepsilon_{\mathrm{eff}})(\varepsilon_{\mathrm{f}}-\varepsilon_{\mathrm{eff}})}\end{aligned} \tag{20.3.18}$$

so that squaring both sides give a common solution

$$\left.\begin{aligned}\left(\varepsilon_{\mathrm{r}}^{\mathrm{upp}}\right)^2 - 2\varepsilon_{\mathrm{r}}^{\mathrm{upp}}\varepsilon_{\mathrm{eff}} + (\varepsilon_{\mathrm{eff}})^2\\ \left(\varepsilon_{\mathrm{r}}^{\mathrm{low}}\right)^2 - 2\varepsilon_{\mathrm{r}}^{\mathrm{low}}\varepsilon_{\mathrm{eff}} + (\varepsilon_{\mathrm{eff}})^2\end{aligned}\right\} = (1+\varepsilon_{\mathrm{eff}})(\varepsilon_{\mathrm{f}}-\varepsilon_{\mathrm{eff}}) \tag{20.3.19}$$

thereby implying $\varepsilon_{\mathrm{r}}^{\mathrm{upp}} = \varepsilon_{\mathrm{r}}^{\mathrm{low}}$. As such, we employ a common ε_{r} in Eq. (20.3.19) to solve for $\varepsilon_{\mathrm{eff}}$

$$\varepsilon_{\mathrm{eff}} = \frac{2\varepsilon_{\mathrm{r}} + \varepsilon_{\mathrm{f}} - 1}{4} \pm \frac{1}{4}\sqrt{1 - 4\varepsilon_{\mathrm{r}}(1+\varepsilon_{\mathrm{r}}) + \varepsilon_{\mathrm{f}}(4\varepsilon_{\mathrm{r}} + \varepsilon_{\mathrm{f}} + 6)} \tag{20.3.20}$$

Recognizing that the $\pm$ sign in Eq. (20.3.20) is a consequence of squaring Eq. (20.3.18), and therefore bears no physical meaning, one of the signs is to be discarded on physically inadmissible basis. As a test, consider the original state before any deformation takes place wherein $\varepsilon_{\mathrm{r}} = \varepsilon_{\mathrm{f}} = \varepsilon_{\mathrm{eff}} = 0$. Substituting $\varepsilon_{\mathrm{r}} = \varepsilon_{\mathrm{f}} = 0$ into Eq. (20.3.20) gives two solutions, the upper solution $\varepsilon_{\mathrm{eff}} = 0$ and the lower solution $\varepsilon_{\mathrm{eff}} = -1/2$. Since the latter is not applicable at the original state, the lower solution of Eq. (20.3.20) is therefore discarded to give

$$\varepsilon_{\mathrm{eff}} = \frac{2\varepsilon_{\mathrm{r}} + \varepsilon_{\mathrm{f}} - 1}{4} + \frac{1}{4}\sqrt{1 - 4\varepsilon_{\mathrm{r}}(1+\varepsilon_{\mathrm{r}}) + \varepsilon_{\mathrm{f}}(4\varepsilon_{\mathrm{r}} + \varepsilon_{\mathrm{f}} + 6)} \tag{20.3.21}$$

We shall now consider three special cases: Special Case I where the frame consists of rigid links such that only rotation at the pin joints is permitted, i.e., the links neither expand nor contract, followed by Special Case II where the Islamic motif is achieved, and finally Special Case III where both the rod and frame linkage are made from the same material. Substituting $\varepsilon_{\mathrm{f}} = 0$ into Eqs. (20.3.15), (20.3.17) and (20.3.21) for Special Case I greatly simplify them to

$$\sin\theta = 2\sqrt{-\varepsilon_{eff}\left(1+\varepsilon_{eff}\right)} \tag{20.3.22}$$

$$\begin{aligned} \varepsilon_{\mathrm{r}}^{\mathrm{upp}} &= \varepsilon_{\mathrm{eff}} + \sqrt{-\varepsilon_{\mathrm{eff}}(1+\varepsilon_{\mathrm{eff}})} \\ \varepsilon_{\mathrm{r}}^{\mathrm{low}} &= \varepsilon_{\mathrm{eff}} - \sqrt{-\varepsilon_{\mathrm{eff}}(1+\varepsilon_{\mathrm{eff}})} \end{aligned} \tag{20.3.23}$$

$$\varepsilon_{\mathrm{eff}} = \frac{\varepsilon_{\mathrm{r}}}{2} - \frac{1}{4} + \frac{1}{4}\sqrt{1-4\varepsilon_{\mathrm{r}}(1+\varepsilon_{\mathrm{r}})} \tag{20.3.24}$$

In order to achieve the desired Islamic motif, it is required for the link *AB* to rotate by 45°. With reference to Fig. 20.17 (bottom), selecting anti-clockwise and clockwise rotations for link *AB* would give the patterns shown in Fig. 20.16 (bottom left) and (bottom right), respectively. Equating Eqs. (20.3.10) with (20.3.12) and substituting $\theta = \pi/4$ for Special Case II, therefore, leads to

$$\begin{aligned} \varepsilon_{\mathrm{f}} &= \tfrac{1-\sqrt{2}+2\varepsilon_{\mathrm{r}}}{1+\sqrt{2}}, \quad \varepsilon_r > 0 \\ \varepsilon_{\mathrm{f}} &= 1+2\varepsilon_{\mathrm{r}}, \quad \varepsilon_r < 0 \end{aligned}. \tag{20.3.25}$$

Finally, we consider Special Case III whereby both the rod and frame are made from the same material, i.e., $\alpha_{\mathrm{r}}^{(T)} = \alpha_{\mathrm{f}}^{(T)}$ and $\alpha_{\mathrm{r}}^{(C)} = \alpha_{\mathrm{f}}^{(C)}$, such that both encounter the same hygrothermal strains. Substituting $\varepsilon_{\mathrm{f}} = \varepsilon_{\mathrm{r}}$ into Eq. (20.3.21) gives $\varepsilon_{\mathrm{eff}} = \varepsilon_{\mathrm{r}}$.

In plotting graphs for visually relating the rod and effective strains, limit is placed on what is physically attainable. This is performed in the first instance for Special Case I whereby $\varepsilon_{\mathrm{f}} = 0$ before proceeding with the more generic cases. With reference to Fig. 20.17 (bottom), the displacement of the rigid square such that its inner corner touches the origin implies that B' touches the *y*-axis at a distance *l* from the origin, while A'' also touches the origin. This gives the minimum $\varepsilon_{\mathrm{eff}}$ as $-1/2$, while the minimum $\varepsilon_{\mathrm{r}}^{\mathrm{low}}$ is -1. The maximum $\varepsilon_{\mathrm{eff}}$ is obviously 0 because the rigid squares are at the outermost location in the original state. The maximum $\varepsilon_{\mathrm{r}}^{\mathrm{upp}}$ can be obtained from the first of Eq. (20.3.23) by imposing $d\varepsilon_{\mathrm{r}}^{\mathrm{upp}}/d\varepsilon_{\mathrm{eff}} = 0$ to give

$$1 + 2\varepsilon_{\mathrm{eff}} = 2\sqrt{-\varepsilon_{\mathrm{eff}} - (\varepsilon_{\mathrm{eff}})^2} \tag{20.3.26}$$

which can be solved to give two solutions $\varepsilon_{\mathrm{eff}} = \left(-2 \pm \sqrt{2}\right)/4$. Two solutions are obtained due to the squaring of Eq. (20.3.26) to solve for $\varepsilon_{\mathrm{eff}}$, as such one of the solutions is to be discarded for being non-physical. Since the lower solution $\varepsilon_{\mathrm{eff}} = \left(-2 - \sqrt{2}\right)/4$ falls outside the physically attainable range

$$-\frac{1}{2} < \varepsilon_{\mathrm{eff}} \leq 0 \tag{20.3.27}$$

we select the upper solution by substituting

$$\varepsilon_{\text{eff}} = \frac{-2+\sqrt{2}}{4} \tag{20.3.28}$$

into the first of Eq. (20.3.23) to give the maximum solution of $\varepsilon_{\text{r}}^{\text{upp}}$ as

$$\varepsilon_{\text{r}}^{\text{upp}} = \frac{-1+\sqrt{2}}{2} \tag{20.3.29}$$

The corresponding value for $\varepsilon_{\text{r}}^{\text{low}}$ is obtained by substituting Eq. (20.3.28) into the second of Eq. (20.3.23) to yield

$$\varepsilon_{\text{r}}^{\text{low}} = -\frac{1}{2} \tag{20.3.30}$$

bearing in mind that this value of $\varepsilon_{\text{r}}^{\text{low}}$ is not the lowest value, but one that corresponds with Eq. (20.3.29). Hence in addition to Eq. (20.3.27), we have another range

$$-1 < \varepsilon_{\text{r}} \leq \frac{\sqrt{2}-1}{2} \approx 0.207 \tag{20.3.31}$$

which determines the range of the plotted graphs in Fig. 20.18 (left column). Figure 20.18 (top left) shows the plots of $\varepsilon_{\text{r}}^{\text{upp}}$ and $\varepsilon_{\text{r}}^{\text{low}}$ versus ε_{eff} using Eq. (20.3.23), while Fig. 20.18 (bottom left) furnishes the curve of ε_{eff} with respect to ε_{r} using Eq. (20.3.24). To facilitate comparison, the curve in Fig. 20.18 (bottom left) is prescribed two colors to correspond with Fig. 20.18 (top left). Zoomed-in views are shown in Fig. 20.18 (right column) to indicate the locations of maximum $\varepsilon_{\text{r}}^{\text{upp}}$ and its corresponding ε_{eff} and $\varepsilon_{\text{r}}^{\text{low}}$ values. To facilitate visual description of the deformed metamaterial, a unit of the driving cell is shown at various locations in Fig. 20.18 (right column).

Perusal to Fig. 20.18 also shows that the maximum point in Fig. 20.18 (top) is the end point of Fig. 20.18 (bottom). The absence of the region defined by

$$\begin{aligned} -\tfrac{1}{2} \leq \varepsilon_{\text{eff}} < \tfrac{-2+\sqrt{2}}{4} \\ 0 \leq \varepsilon_{\text{r}} < \tfrac{-1+\sqrt{2}}{2} \end{aligned} \tag{20.3.32}$$

can be elucidated as follows. With reference to Fig. 20.19 (top) as the rod elongates due to increasing temperature and/or moisture concentration point A goes up along the y-axis as indicated by the white arrow until it reaches its maximum point where $\theta = 45°$ while link BC moves diagonally as shown by the white arrow until reaching $B'C'$. This path is represented by the $\varepsilon_{\text{r}}^{\text{upp}}$ curve in Fig. 20.18 (top left) from (A, B, C) to the maximum point (A', B', C'). At this stage, further increase in temperature and/or moisture concentration can no longer push point A' further as it would require point C' to detach from the 45° dash line in preserving symmetry and material continuity about the 45° dashed line. Now if the temperature and/or moisture concentration were to decrease, there are two possible pathways. Although point A' reverses to its

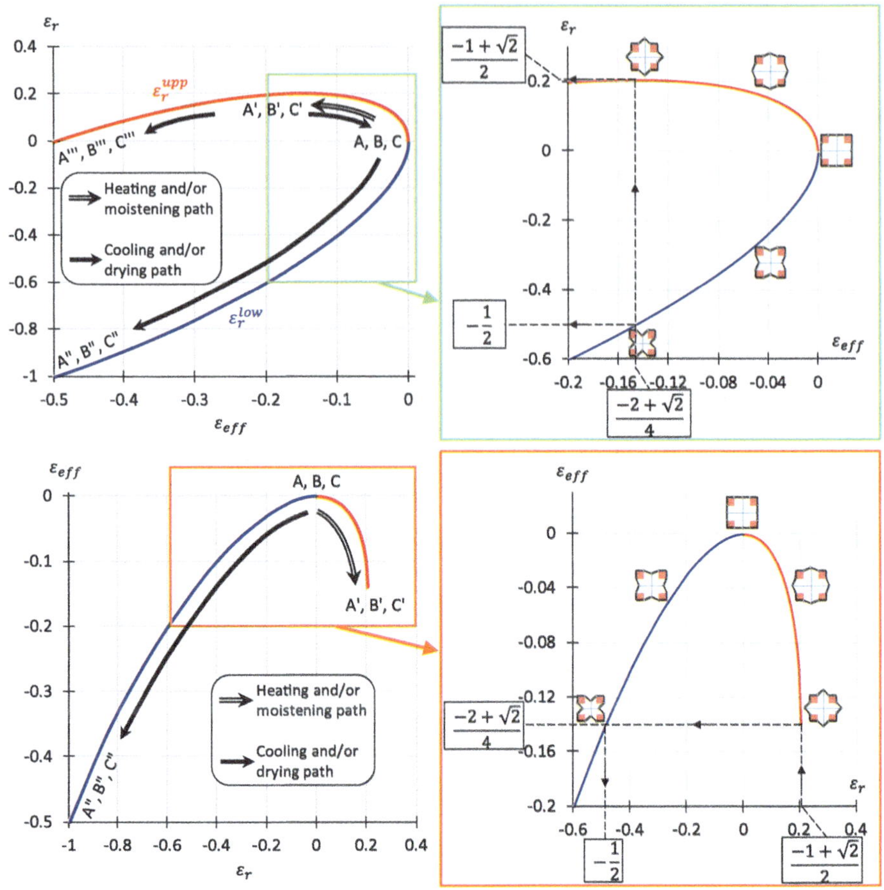

Fig. 20.18 Plots of ε_r^{upp} and ε_r^{low} versus ε_{eff} (top) and ε_{eff} with respect to ε_r (bottom)

original position A, the link $B'C'$ can either return to its original position BC along the same path or continue until reaching $B'''C'''$, as shown in Fig. 20.19 (top). These two pathways are indicated as (A', B', C') to (A, B, C) as well as (A', B', C') to (A, A''', C'''), again along the ε_r^{upp} curve, in Fig. 20.18 (top left). On the other hand, the cooling and/or drying process from its original state is more straightforward and is illustrated in Fig. 20.19 (bottom) where A moves to A'' along the y-axis while BC moves by translational motion to $B''C''$. This pathway is reflected by the mapping of (A, B, C) to (A'', B'', C'') in Fig. 20.18 (top left) along the ε_r^{low} curve.

The appearance of two possible pathways along the ε_r^{upp} curve comes about due to the prescription of ε_{eff} to calculate ε_r. However, since the metamaterial deformation is driven by the change in the rod length, a more physically meaningful approach would be the calculation of ε_{eff} for a prescribed ε_r. The latter has been done for Special Case I wherein the frame links are assumed rigid such that it can only deform by

Fig. 20.19 Visualization of the linkage motion during heating and/moistening from original state until attainment of maximum expansion denoted by white arrows followed by the two possible pathways for subsequent cooling and/or drying indicated by black arrows (top), and linkage motion during cooling and/or drying from original state (bottom)

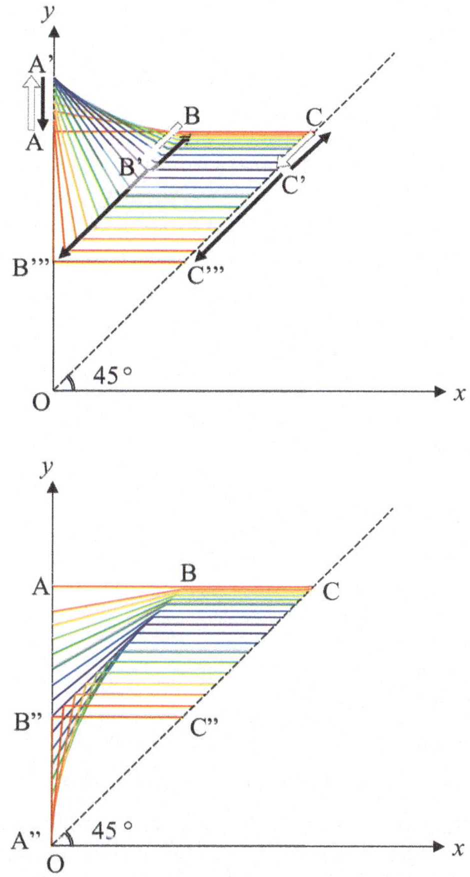

rotation about the pin joints while each segment of the linkage can neither expand nor contract. On the other hand if the rod and frame are made of the same material as specified in Special Case III, then $\varepsilon_{\mathrm{eff}} = \varepsilon_{\mathrm{r}} = \varepsilon_{\mathrm{f}}$. What if $0 < \varepsilon_{\mathrm{f}} < \varepsilon_{\mathrm{r}}$? This is plotted in Fig. 20.20 based on Eq. (20.3.21) for various ε_{f} to ε_{r} ratio, whereby the occurrence of incremental negative hygrothermal expansion is defined by $d\varepsilon_{\mathrm{eff}}/d\varepsilon_{\mathrm{r}} < 0$ and are indicated by red curves. The graphs are plotted using the same range $(\Delta\varepsilon_{\mathrm{eff}}, \Delta\varepsilon_{\mathrm{r}}) = (0.16, 0.4)$ in order to clearly show that the negative hygrothermal expansion region is wider for lower $\varepsilon_{\mathrm{f}}/\varepsilon_{\mathrm{r}}$ ratio.

The four separate plots from Fig. 20.20 can be combined in order to show the locations at which the negative hygrothermal expansion regions occur in the $\varepsilon_{\mathrm{eff}} - \varepsilon_{\mathrm{r}}$ plane. Figure 20.21 (left) shows that as ε_{f} increases, the location of negative hygrothermal expansion occurs further from the origin. The analytical solution for ε_{r} at which the negativity of hygrothermal expansion begins can be obtained as follows. Since the rod strain which demarcates the region of positive hygrothermal expansion from the region of incremental negative hygrothermal expansion is defined

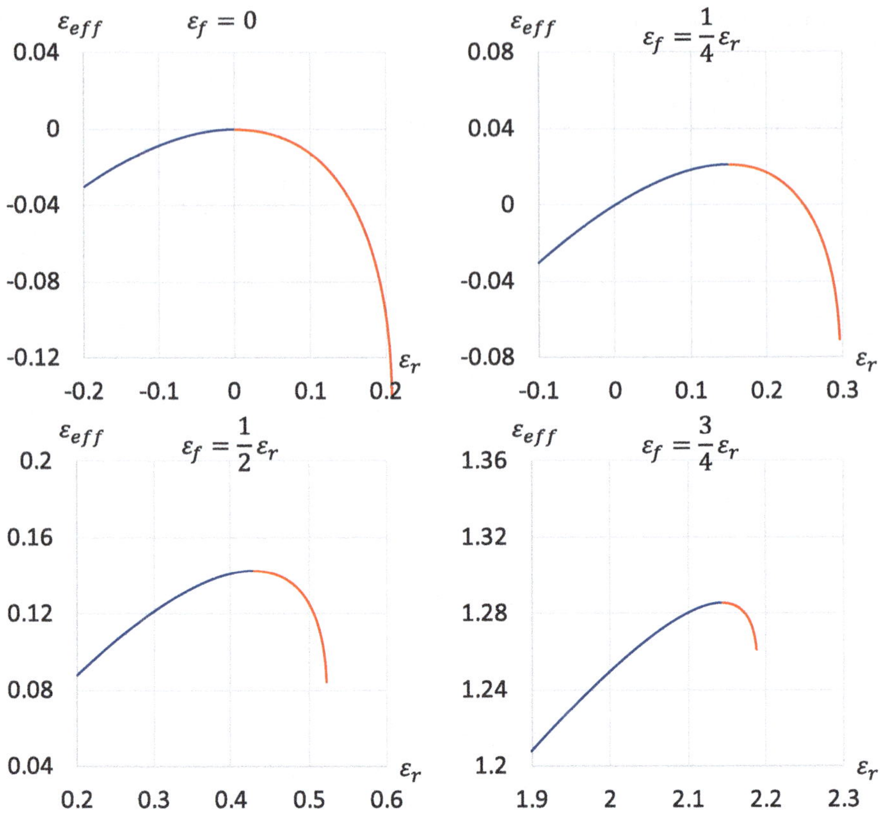

Fig. 20.20 Effective strain $\varepsilon_{\mathrm{eff}}$ curves plotted against the rod strain ε_{r} for various frame strain ε_{f} where the maximum points occur. The negative hygrothermal expansion part of each curve is indicated in red

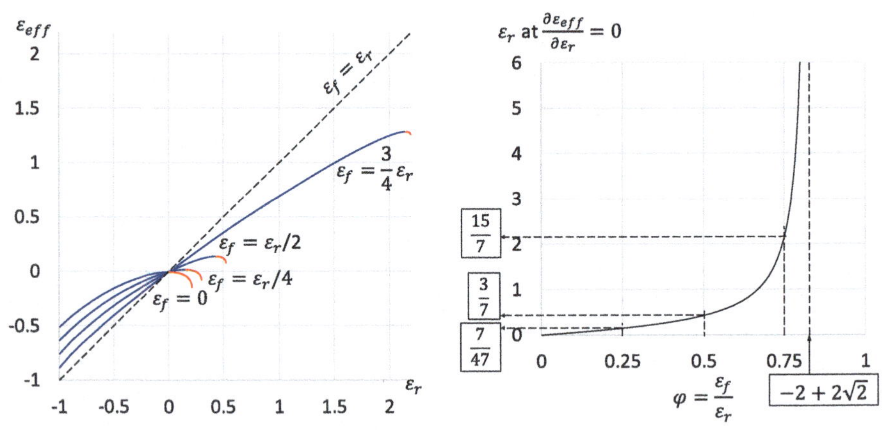

Fig. 20.21 Locations of negative hygrothermal expansion in the $\varepsilon_{\mathrm{eff}} - \varepsilon_{\mathrm{r}}$ plane for various ε_{f} (left) and the limit of negative hygrothermal expansion occurrence (right)

by $d\varepsilon_{\mathrm{eff}}/d\varepsilon_{\mathrm{r}} = 0$, imposition of this first derivative value on Eq. (20.3.21) yields

$$\frac{2 - 3\varphi + \left(4 - 4\varphi - \varphi^2\right)\varepsilon_{\mathrm{r}}}{2 + \varphi} = \sqrt{1 + (6\varphi - 4)\varepsilon_{\mathrm{r}} + \left(\varphi^2 + 4\varphi - 4\right)\varepsilon_{\mathrm{r}}^2} \quad (20.3.33)$$

where $\varphi = \varepsilon_{\mathrm{f}}/\varepsilon_{\mathrm{r}}$. Squaring both sides of Eq. (20.3.33) permits ε_{r} to be solved as

$$\varepsilon_{\mathrm{r}} = \frac{3\varphi - 2 \pm (1 - \varphi)(2 + \varphi)}{4 - 4\varphi - \varphi^2} \quad (20.3.34)$$

Since the lower solution $\varepsilon_r = -1$ is non-physical, it is discarded in favor of the upper solution

$$\varepsilon_{\mathrm{r}} = \frac{2\varphi - \varphi^2}{4 - 4\varphi - \varphi^2} \quad (20.3.35)$$

In addition, the limit at which the negativity of hygrothermal expansion occurs can be obtained from the denominator of Eq. (20.3.35). Let $4 - 4\varphi - \varphi^2 = 0$, we obtain $\varphi = -2 \pm 2\sqrt{2}$. Since the lower solution $\varphi = -2 - 2\sqrt{2}$ is a negative value, which is inconsistent with the requirement that $\varepsilon_{\mathrm{f}}/\varepsilon_{\mathrm{r}} \geq 0$, we choose the upper solution $\varepsilon_{\mathrm{f}}/\varepsilon_{\mathrm{r}} = -2 + 2\sqrt{2}$. In other words, the negativity of hygrothermal expansion in the currently investigated metamaterial exists within the range

$$0 \leq \varepsilon_{\mathrm{f}} < \left(-2 + 2\sqrt{2}\right)\varepsilon_{\mathrm{r}} \quad (20.3.36)$$

A list of ε_{r} strains which separate the incremental negative hygrothermal expansion regions from the conventional ones are furnished in Table 20.4, while a full continuous description of ε_{r} that forms the boundary between positive and negative hygrothermal expansion region is plotted in Fig. 20.21 (right).

As mentioned, the desired Islamic pattern can be achieved through the metamaterial deformation shown in Fig. 20.16 (bottom) such that the frame's link strain that is required for each rod strain as per Special Case II is expressed in Eq. (20.3.25). Here, four sub-special cases can be identified: (a) $\varepsilon_{\mathrm{f}} = 0$ with (i) rod expansion, and

Table 20.4 List of rod and frame strain conditions that determine the positivity or negativity of hygrothermal expansion

Frame strain to rod strain ratio	Condition for incremental positive hygrothermal expansion	Condition for incremental negative hygrothermal expansion
$\varepsilon_{\mathrm{f}} = 0$	$\varepsilon_{\mathrm{r}} < 0$	$\varepsilon_{\mathrm{r}} > 0$
$\varepsilon_{\mathrm{f}} = \frac{1}{4}\varepsilon_r$	$\varepsilon_{\mathrm{r}} < \frac{7}{47}$	$\varepsilon_{\mathrm{r}} > \frac{7}{47}$
$\varepsilon_{\mathrm{f}} = \frac{1}{2}\varepsilon_{\mathrm{r}}$	$\varepsilon_{\mathrm{r}} < \frac{3}{7}$	$\varepsilon_{\mathrm{r}} > \frac{3}{7}$
$\varepsilon_{\mathrm{f}} = \frac{3}{4}\varepsilon_{\mathrm{r}}$	$\varepsilon_{\mathrm{r}} < \frac{15}{7}$	$\varepsilon_{\mathrm{r}} > \frac{15}{7}$

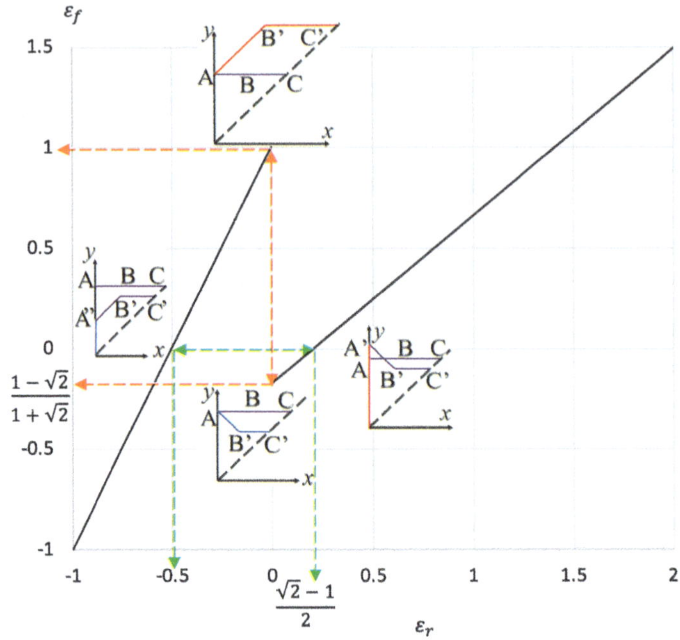

Fig. 20.22 Plots of frame linkage strain versus rod strain for maintaining the 2D array of Islamic 8-pointed star pattern, with the original lengths in purple while the elongated and shortened ones are in red and blue, respectively

(ii) rod contraction, as well as (b) $\varepsilon_r = 0$ with (i) frame linkage expansion, and (ii) frame linkage contraction. The sub-special cases (a) (i) and (ii) are subsets of Special Case I wherein substitution of $\varepsilon_f = 0$ into Eq. (20.3.25) gives $\varepsilon_r = -1/2 + 1/\sqrt{2}$ and $\varepsilon_r = -1/2$, while sub-special cases (b) (i) and (ii) are also obtained from Eq. (20.3.25), but with substitution of $\varepsilon_r = 0$ to give $\varepsilon_f = \left(1-\sqrt{2}\right)/\left(1+\sqrt{2}\right)$ and $\varepsilon_f = 1$. Plots of Eq. (20.3.25) are furnished in Fig. 20.22 whereby the two lines pertain to the opposing rotational directions of link AB, which incorporate the four sub-special cases with their corresponding schematics.

Finally, we consider the possibility of attaining the 2D array of 8-pointed stars each for varying temperature and moisture concentration. Since the units for CTE and CME are different, in which the unit for CTE is K^{-1} while CME is dimensionless, it is not possible to establish a single coeficient of hygrothermal expansion; only the hygrothermal strain can be defined via the summation of thermal and moisture strains. As such the effective CTE and CME must be evaluated separately. In the absence of moisture concentration change, the substititon of Eq. (20.3.1) with $dC_r = dC_f = 0$ and $\varepsilon_{\text{eff}} = \alpha_{\text{eff}}^{(T)} dT$ into Eq. (20.3.21) gives

$$\alpha_{\text{eff}}^{(T)} = \frac{2\alpha_{\text{r}}^{(T)} dT + \alpha_{\text{f}}^{(T)} dT - 1}{4dT} + \frac{1}{4dT}\sqrt{1 - 4\alpha_{\text{r}}^{(T)} dT\left(1 + \alpha_{\text{r}}^{(T)} dT\right) + \alpha_{\text{f}}^{(T)} dT\left(4\alpha_{\text{r}}^{(T)} dT + \alpha_{\text{f}}^{(T)} dT + 6\right)} \tag{20.3.37}$$

while in the absence of temperature change the substitution of Eq. (20.3.1) with $dT = 0$ and $\varepsilon_{\text{eff}} = \alpha_{\text{eff}}^{(C)} dC_{\text{env}}$ into Eq. (20.3.21) leads to

$$\alpha_{\text{eff}}^{(C)} = \frac{2\alpha_{\text{r}}^{(C)} dC_{\text{r}} + \alpha_{\text{f}}^{(C)} dC_{\text{f}} - 1}{4dC_{\text{env}}} + \frac{1}{4dC_{\text{env}}}\sqrt{1 - 4\alpha_{\text{r}}^{(C)} dC_{\text{r}}\left(1 + \alpha_{\text{r}}^{(C)} dC_{\text{r}}\right) + \alpha_{\text{f}}^{(C)} dC_{\text{f}}\left(4\alpha_{\text{r}}^{(C)} dC_{\text{r}} + \alpha_{\text{f}}^{(C)} dC_{\text{f}} + 6\right)} \tag{20.3.38}$$

If the frame linkage material is made from metals or alloys, there is no absorption of moisture. Therefore, substituting $dC_{\text{f}} = 0$ into Eq. (20.3.38) for mettalic frame linkages reduces it to

$$\alpha_{\text{eff}}^{(C)} = \frac{2\alpha_{\text{r}}^{(C)} dC_{\text{r}} - 1}{4dC_{\text{env}}} + \frac{1}{4dC_{\text{env}}}\sqrt{1 - 4\alpha_{\text{r}}^{(C)} dC_{\text{r}}\left(1 + \alpha_{\text{r}}^{(C)} dC_{\text{r}}\right)} \tag{20.3.39}$$

Using Eqs. (20.3.37) and (20.3.39), plots of the effective CTE and CME for the metamaterial are shown in Fig. 20.23 with the regions of positive and negative hygrothermal expansions indicated in blue and red, respectively. Specifically, the effective CTE has been plotted in Fig. 20.23 (top) for frame linkages made from silicon $\alpha_{\text{f}}^{(T)} = 2.56 \times 10^{-6}\,\text{K}^{-1}$, Invar $\alpha_{\text{f}}^{(T)} = 1.2 \times 10^{-6}\,\text{K}^{-1}$, fused quartz $\alpha_{\text{f}}^{(T)} = 0.59 \times 10^{-6}\,\text{K}^{-1}$, Sitall $\alpha_{\text{f}}^{(T)} = (0 \pm 0.15) \times 10^{-6}\,\text{K}^{-1}$ and Zerodur $\alpha_{\text{f}}^{(T)} = (0 \pm 0.007) \times 10^{-6}\,\text{K}^{-1}$, with the rod being made of polypropylene $\alpha_{\text{r}}^{(T)} = 150 \times 10^{-6}\,\text{K}^{-1}$. The CMEs of polymers range from $\alpha_{\text{r}}^{(C)} = 2 \times 10^{-3}$ to $\alpha_{\text{r}}^{(C)} = 5 \times 10^{-3}$, but mostly at around the former. Plots of the effective CMEs are furnished in Fig. 20.23 (bottom left) for rods of various CMEs at fixed moisture absorption, and in Fig. 20.23 (bottom right) for rods of various moisture absorption at fixed CME.

It can be seen that when the CTE of the frame linkage material is zero or almost zero, such as when the frame linkage is made from Sitall and especially Zerodur shown in Fig. 20.23 (top right), sign-toggling of the effective CTE takes place when there is a switch between cooling and heating

$$\alpha_{\text{r}}^{(T)} \gg \alpha_{\text{f}}^{(T)} \approx 0 \Rightarrow \begin{cases} dT > 0 \Rightarrow \alpha_{\text{eff}}^{(T)} < 0 \\ dT < 0 \Rightarrow \alpha_{\text{eff}}^{(T)} > 0 \end{cases} \tag{20.3.40}$$

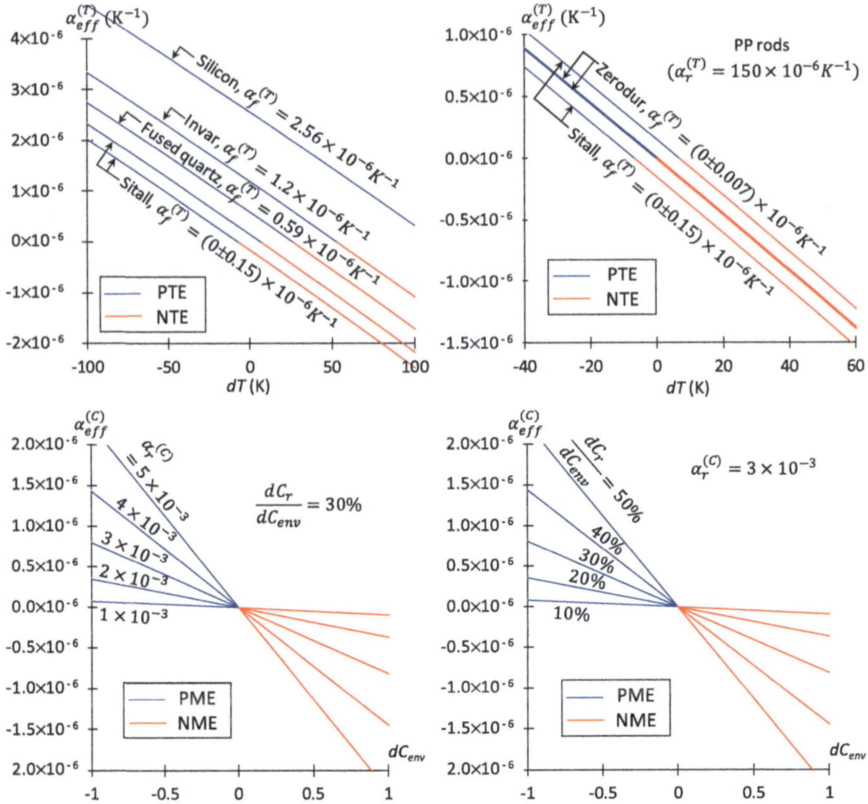

Fig. 20.23 Plots of effective CTE (top) and effective CMEs (bottom) for various combinations of materials and, in the case of CMEs, various moisture absorption

i.e., the sign-flipping of the effective CTE takes place at the original temperature ($dT = 0$) such that the metamaterial always contracts with temperature fluctuation. When the frame linkage is made from other materials, the sign-switching of the effective CTE occurs at higher temperature. With reference to Eq. (20.3.39), the sign-switching effective CTE condition described in Eq. (20.3.40) applies for sign-switching effective CME, i.e.,

$$\alpha_{\mathrm{r}}^{(C)} \gg \alpha_{\mathrm{f}}^{(C)} \approx 0 \Rightarrow \begin{cases} dC_{\mathrm{env}} > 0 \Rightarrow \alpha_{\mathrm{eff}}^{(C)} < 0 \\ dC_{\mathrm{env}} < 0 \Rightarrow \alpha_{\mathrm{eff}}^{(C)} > 0 \end{cases} \tag{20.3.41}$$

In addition, the magnitude of the effective CTE is accentuated with the use of rods with higher CME, enhanced moisture absortivity and greater change in environmental moisture concentration, as implied from Fig. 20.23 (bottom).

20.4 Conclusions

Neither positive nor negative expansion coefficients are fully advantageous across a wide range of structural applications that are exposed to environmental fluctuations. In some cases, it is advantageous for materials to possess duality in material properties so as to take advantage of negative behavior under a change in environmental condition as well as conventional behavior when the condition reverses. These have been made achievable by the use of bimaterial strips (Sect. 20.2) and linkages (Sect. 20.3) that are arranged in a certain manner.

Unlike other works on sign-switching materials properties, which are only interesting from technological viewpoint (Lim 2019b, c, d, e, f, g, h, 2020b), the sign-switchable 2D metamaterials in this chapter are pleasingly esthetic (Lim 2020a, c). Originally in the form of square grids, these metamaterials deform into arrays of 8-pointed stars, which has a unique place in the long and glorious history of Islamic civilization.

References

Abdullahi Y, Embi MRB (2013) Evolution of Islamic geometric patterns. Front Architect Res 2(2):243–251

Lim TC (2019a) Negative environmental expansion for interconnected array of rings and sliding rods. Phys Status Solidi B 256(1):1800032

Lim TC (2019b) Composite microstructures with Poisson's ratio sign switching upon stress reversal. Compos Struct 209:34–44

Lim TC (2019c) A composite metamaterial with sign switchable elastic and hygrothermal properties induced by stress direction and environmental change reversals. Compos Struct 220:185–193

Lim TC (2019d) A class of shape-shifting composite metamaterial honeycomb structures with thermally-adaptive Poisson's ratio signs. Compos Struct 226:111256

Lim TC (2019e) 2D metamaterial with in-plane positive and negative thermal expansion and thermal shearing based on interconnected alternating bimaterials. Mater Res Express 6(11):115804

Lim TC (2019f) Composite metamaterial with sign-switchable coefficients of hygroscopic, thermal and pressure expansions. Adv Compos Hybrid Mater 2(4):657–669

Lim TC (2019g) Metamaterials with Poisson's ratio sign toggling by means of microstructural duality. SN Appl Sci 1(2):176

Lim TC (2019h) A 2D auxetikos system based on interconnected shurikens. SN Appl Sci 1(11):1383

Lim TC (2020a) Composite metamaterial square grids with sign-flipping expansion coefficients leading to a type of Islamic design. SN Appl Sci 2(5):918

Lim TC (2020b) Metacomposite with auxetic and in situ sign reversible thermal expansivity upon temperature fluctuation. Compos Commun 19:30–36

Lim TC (2020c) Metacomposite structure with sign-changing coefficients of hygrothermal expansions inspired by Islamic motif. Compos Struct 251:112660

Rafsanjani A, Pasini D (2016a) Multistable compliant auxetic metamaterials inspired by geometric patterns in Islamic arts. Bull Am Phys Soc 61(2):abstract K40.00008

Rafsanjani A, Pasini D (2016b) Bistable auxetic mechanical metamaterials inspired by ancient geometric motifs. Extreme Mech Lett 9(2):291–216

Appendix

A.1 *Arc Length Analysis of Bimaterial Strip*

In addition to bending of the bimaterial strip due to the differential elongation of the individual layers when unbonded, the arc length of the bent bimaterial strip is longer and shorter than its original length with increased and decreased temperature, respectively. This is because the arc length of the bent bimaterial is bounded by the thermally elongated or shortened lengths of the individual layers when they are unbonded, as shown in Fig. A.1. Upon a temperature change of $\mathrm{d}T$, the original lengths l_h change to $l_1 = l_h\left(1 + \alpha_1^{(T)}\mathrm{d}T\right)$ and $l_2 = l_h\left(1 + \alpha_2^{(T)}\mathrm{d}T\right)$ for layer 1 and layer 2, respectively, in the absence of bonding, thereby generating a length difference of $l_h\left(\alpha_1^{(T)} - \alpha_2^{(T)}\right)\mathrm{d}T$. In the presence of perfect bonding between both layers, this gap is bridged by prescribing internal forces of opposing directions. Arising from the moment imbalance due to the prescribed axial load through the cross-sectional centroids of each layer, equilibrium of moment is maintained by prescribing internal moments to both layers. The result of this approach gives rise the curvature described by Timoshenko (1925). This treatment, however, does not

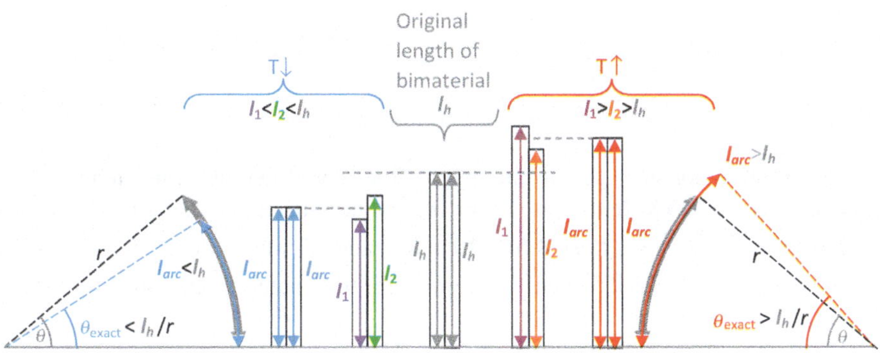

Fig. A.1 Schematics of the change in arc length

T. Lim, *Mechanics of Metamaterials with Negative Parameters*, Engineering Materials,
https://doi.org/10.1007/978-981-15-6446-8

take into account the change in the length of the biomaterial strip. The arc length l_{arc} is typically taken to be equal to that of the original length.

Unlike the analysis of bending whereby the equilibrium of forces and moments are applied, in the arc length analysis only the equilibrium of forces is required. As before, the absence of external force requires the axial forces acting on layer 1 and layer 2 to be of opposing direction but equal in magnitude $\sigma_1 h_1 = \sigma_2 h_2$ with $\sigma_1 = E_1\varepsilon_1$ and $\sigma_2 = E_2\varepsilon_2$ where $\varepsilon_1 = \mathrm{d}l_1/l_1$ and $\varepsilon_2 = \mathrm{d}l_2/l_2$. For equal layer thicknesses $h_1 = h_2$, the force equilibrium becomes

$$\frac{E_1 \mathrm{d}l_1}{l_h\left(1+\alpha_1^{(T)}\mathrm{d}T\right)} = \frac{E_2 \mathrm{d}l_2}{l_h\left(1+\alpha_2^{(T)}\mathrm{d}T\right)} \tag{A1.1}$$

The length gap of $l_h\left(\alpha_1^{(T)} - \alpha_2^{(T)}\right)\mathrm{d}T$ is bridged via the shortening of layer 1 and lengthening of layer 2 by $\mathrm{d}l_1$ and $\mathrm{d}l_2$, respectively, i.e.,

$$\mathrm{d}l_1 + \mathrm{d}l_2 = l_h\left(\alpha_1^{(T)} - \alpha_2^{(T)}\right)\mathrm{d}T \tag{A1.2}$$

Equations (A1.1) and (A1.2) are solved simultaneously to give

$$\mathrm{d}l_1 = \frac{l_h\left(\alpha_1^{(T)} - \alpha_2^{(T)}\right)\mathrm{d}T}{1+\frac{E_1}{E_2}\frac{1+\alpha_2^{(T)}\mathrm{d}T}{1+\alpha_1^{(T)}\mathrm{d}T}} \tag{A1.3a}$$

$$\mathrm{d}l_2 = \frac{l_h\left(\alpha_1^{(T)} - \alpha_2^{(T)}\right)\mathrm{d}T}{1+\frac{E_2}{E_1}\frac{1+\alpha_1^{(T)}\mathrm{d}T}{1+\alpha_2^{(T)}\mathrm{d}T}} \tag{A1.3b}$$

from which one can calculate the arch length as either

$$l_{\text{arc}} = l_h\left(1+\alpha_1^{(T)}\mathrm{d}T\right) - \mathrm{d}l_1 \tag{A1.4a}$$

or

$$l_{\text{arc}} = l_h\left(1+\alpha_2^{(T)}\mathrm{d}T\right) + \mathrm{d}l_2. \tag{A1.4b}$$

We can now consider actual bimaterials considered herein with a temperature change of $\mathrm{d}T = 100\,\mathrm{K}$. In the case of the brass–titanium (BT) bimaterial, the substitution of $\left(\alpha_1^{(T)}, \alpha_2^{(T)}\right) = (19.0, 8.6) \times 10^{-6}\,\mathrm{K}^{-1}$ and $(E_1, E_2) = (112.5, 110.3)$ GPa gives $l_{\text{arc}}/l_h = 1.001385$; for the copper–steel bimaterial, the substitution of $\left(\alpha_1^{(T)}, \alpha_2^{(T)}\right) = (17, 12) \times 10^{-6}\,\mathrm{K}^{-1}$ and $(E_1, E_2) = (117, 200)$ GPa gives $l_{\text{arc}}/l_h = 1.001384$; and for the tungsten–silicon carbide (T-SC) bimaterial, the substitution of $\left(\alpha_1^{(T)}, \alpha_2^{(T)}\right) = \left(4.50\left(\alpha_1^{(T)}, \alpha_2^{(T)}\right) = (4.50, 2.77) \times 10^{-6}\,\text{K}^{-1}, 2.77\right) \times 10^{-6}\,\text{K}^{-1}$ and $(E_1, E_2) = (405, 450)$ GPa gives $l_{\text{arc}}/l_h = 1.000359$. Since $\theta_{\text{exact}} = l_{\text{arc}}/r$ and $\theta = l_h/r$, it follows that the percentage error from the use of θ instead of θ_{exact} is the same as the adoption of l_h in place of l_{arc}, which is defined as

$$\%\,\text{error} = \frac{l_h - l_{\text{arc}}}{l_{\text{arc}}} \times 100\% \tag{A1.5}$$

i.e., the percentage errors are only −0.13829%, −0.13826% and −0.03588% for the B-T, C-S and T-SC bimaterials, respectively. For the case where $E_1 = E_2$ in addition to $h_1 = h_2$, Eq. (A1.3) becomes

$$\left.\begin{aligned} dl_1 &= \frac{l_h\left(\alpha_1^{(T)} - \alpha_2^{(T)}\right)dT}{1 + \frac{1+\alpha_2^{(T)}dT}{1+\alpha_1^{(T)}dT}} \\ dl_2 &= \frac{l_h\left(\alpha_1^{(T)} - \alpha_2^{(T)}\right)dT}{1 + \frac{1+\alpha_1^{(T)}dT}{1+\alpha_2^{(T)}dT}} \end{aligned}\right\} \approx \frac{l_h\left(\alpha_1^{(T)} - \alpha_2^{(T)}\right)dT}{2} \tag{A1.6}$$

which, upon substitution into Eq. (A1.4), leads to

$$l_{\text{arc}} = l_h\left(1 + \frac{\alpha_1^{(T)} + \alpha_2^{(T)}}{2}dT\right) \tag{A1.7}$$

Equation (A1.7) can also be obtained directly. Since $E_1 = E_2$ and $h_1 = h_2$, it follows that the arc length must be a simple average of $l_1 = l_h\left(1 + \alpha_1^{(T)}dT\right)$ and $l_2 = l_h\left(1 + \alpha_2^{(T)}dT\right)$ to give $l_{\text{arc}} = (l_1 + l_2)/2$, which is none other than Eq. (A1.7). In the case of B-T bimaterial, whereby $E_1 \approx E_2$, the use of Eq. (A1.7) gives $l_{\text{arc}}/l_h = 1.001380$, leading to its percentage error of merely −0.13781%. This arc length–and hence subtending angle–analysis shows that while the assumption of $\theta = l_h/r$ is not exact, it is sufficiently accurate for practical applications.

A.2 Some Series Expansion of Trigonometric Functions

The following series are useful for obtaining simplifications from trigonometric functions.

$$\sin z = \sum_{n=0}^{\infty} \frac{(-1)^n z^{2n+1}}{(2n+1)!} = z - \frac{z^3}{3!} + \frac{z^5}{5!} - \cdots \tag{A2.1}$$

$$\cos z = \sum_{n=0}^{\infty} \frac{(-1)^n z^{2n}}{(2n)!} = 1 - \frac{z^2}{2!} + \frac{z^4}{4!} - \cdots \tag{A2.2}$$

$$\begin{aligned} \tan z &= \sum_{n=1}^{\infty} \frac{(-1)^{n-1} 2^{2n} (2^{2n} - 1) B_{2n} z^{2n-1}}{(2n)!} = \sum_{n=1}^{\infty} \frac{(-4)^n (1 - 4^n) B_{2n} z^{2n-1}}{(2n)!} \\ &= z + \frac{z^3}{3} + \frac{2z^5}{15} + \frac{17z^7}{315} + \cdots \quad \text{for } |z| < \frac{\pi}{2} \end{aligned} \tag{A2.3}$$

where B_k is the kth Bernoulli number.

Reference

Timoshenko SP (1925) Analysis of bi-metal thermostats. J Opt Soc Am 11:233–255

The manufacturer's authorised representative in the EU is Springer Nature Customer Service Centre GmbH, Europaplatz 3, 69115 Heidelberg, Germany. If you have any concerns regarding our products, please contact ProductSafety@springernature.com

Printed and bound by CPI Group (UK) Ltd, Croydon, CR0 4YY
30/09/2026
02225333-0017